中国水利学会
第五届青年科技论坛论文集

中国水利学会青年科技工作委员会　编

黄河水利出版社
·郑州·

图书在版编目(CIP)数据

中国水利学会第五届青年科技论坛论文集/中国水利学会青年科技工作委员会编.—郑州:黄河水利出版社,2012.4

ISBN 978-7-5509-0236-7

Ⅰ.中… Ⅱ.中… Ⅲ.水利工程-中国-文集
Ⅳ.TV-53

中国版本图书馆 CIP 数据核字(2012)第 149173 号

组稿编辑:路夷坦　　电话:0371-66026749　　邮箱:hhsllyt@126.com

出 版 社:黄河水利出版社
地址:河南省郑州市金水路 11 号　　邮政编码:450003
发行单位:黄河水利出版社
发行电话:0371-66026940　　传真:0371-66022620
E-mail:hhslcbs@126.com
承印单位:黄河水利委员会印刷厂
开本:787 mm×1 092 mm　1/16
印张:38.5
字数:890 千字　　印数:1—1 000
版次:2012 年 4 月第 1 版　　印次:2012 年 4 月第 1 次印刷

定价:100.00 元

中国水利学会第五届青年科技论坛

顾　问

敬正书　朱尔明　高安泽　陈厚群　马洪琪

组织委员会

主　任　李赞堂

副主任　贾金生　陈建康　李　嘉　曾向辉

委　员　(以姓氏笔画为序)

石　泉　刘　毅　刘晓波　关铁生　阮本清

杨继富　汪在芹　姜乃迁　唐洪武　殷淑华

夏世法　谢红强　戴济群

《中国水利学会第五届青年科技论坛论文集》

编辑委员会

前 言

为促进广大水利青年科技工作者学术与创新思想的交流，培养和造就青年科技人才，并努力为水利青年科技人才提供创新思想碰撞与才华展现的舞台，中国水利学会于2003年设立了中国水利学会青年科技论坛。前四届论坛分别于2003年11月、2005年11月、2007年10月和2008年10月在深圳、西安、成都和北京召开，取得了圆满成功。

为适应我国水利事业快速发展的需要，进一步激发广大水利青年科技工作者的创新思维和动力，应广大水利青年科技工作者的要求，中国水利学会第五届青年科技论坛于2012年4月22~24日在成都召开。本届论坛的主题是“水利改革发展与科技创新”，主要围绕水利改革与科技创新可持续发展，以及水利水电工程关键技术等方面的科技进展和成果进行交流、探讨。论坛得到了水利青年科技工作者的积极响应，共收到论文150多篇，经评审，共有95篇论文被收入论文集，分为7个部分：水资源与节水型社会（14篇）；水环境与水生态（10篇）；防洪减灾与水文学（7篇）；农业节水与农村供水（10篇）；水力学与泥沙（22篇）；岩土工程与结构材料（19篇）；管理、政策及其他（13篇）。这些论文集中反映了近年来全国水利青年科技工作的新成果、新观点、新思路、新举措。

水利部和中国水利学会对第五届青年科技论坛高度重视，给予了极大鼓励。承办单位和协办单位对论坛给予了大量的人力、物力、财力等方面的支持。中国水利学会所属专业委员会和省级水利学会为本届论坛推荐代表并组织投稿，做了许多工作。为此，向所有关心和支持论坛的领导、专家、论文作者表示深深的谢意，对论坛秘书处工作人员及论文评审人员的辛勤劳动表示感谢。

由于时间仓促，书中可能出现疏漏或差错，敬请读者批评指正。

编委会

2012年4月

目 录

一、水资源与节水型社会

二、水环境与水生态

三、防洪减灾与水文学

四、农业节水与农村供水

五、水力学与泥沙

六、岩土工程与结构材料

七、管理、政策及其他

一、水资源与节水型社会

浅析岗、黄水库水价测算遇到的几个技术问题处理

李书锋[1] 崔志刚[1] 王云丽[2]

(1. 河北省黄壁庄水库管理局 石家庄 050224;
2. 石家庄水文水资源勘测局 石家庄 050224)

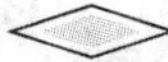

【摘要】 水价的制定和水费的征收政策是水资源开发利用的一项重要经济政策,制定合理的水价和水费征收政策,对于实施水资源的优化配置具有重要意义,本文从岗、黄水库水价测算遇到的几个技术问题处理入手,分析其采用方法的政策依据来源。

【关键词】 水价测算 人员经费 系数确定 公益 供水生产

水是"生命之源、生产之要、生态之基",2011 年中央一号文件对水有了更深层次的定义,同时对水利工作也提出了更高层次的要求。水价的制定和水费的征收政策是水资源开发利用的一项重要经济政策,制定合理的水价和水费征收政策,对于实施水资源的优化配置具有重要意义,同时为更好贯彻落实 2011 年中央一号文件精神,实现水利健康可持续发展的目标打好基础。

1 工程概况

岗南水库、黄壁庄水库(简称岗、黄水库)位于海河流域子牙河水系两大支流之一滹沱河干流上,相距 28 km,同为河北省太行山山区的两座大型梯级水库,总控制流域面积 23 400 km^2,下游黄壁庄水库距离河北省省会石家庄市约 30 km,两库均建于 1958 年,是两座以防洪、供水为主的大(1)型水利枢纽工程。两库呈阶梯式联合防洪、联合供水。总库容 29.14亿 m^3(岗南水库 17.04 亿 m^3,黄壁庄水库 12.1 亿 m^3),其中防洪总库容 16.14 亿 m^3(岗南水库 9.17 亿 m^3,黄壁庄水库 6.97 亿 m^3),兴利总库容 11.57 亿 m^3(岗南水库 7.8 亿 m^3,黄壁庄水库 3.77 亿 m^3),死库容 4.1 亿 m^3(岗南水库 3.41 亿 m^3,黄壁庄水库 0.69 亿 m^3)。水库主要保证石津灌区农业灌溉用水、石家庄城市生活用水和西柏坡电厂工业用水等。

2 联合测算水价基础

为了实现水利健康可持续发展的目标,河北省水利厅价费与管理处将岗、黄水库供水价格测算工作列入 2011 年本处重点工作任务之一,结合工程供水特点,一改过去单库核算方式,变为两库联合测算。分析联合测算水价基础:一方面岗南水库和黄壁庄水库都坐落在滹沱河上,相距只有 28 km,呈阶梯状联合防洪、联合供水,且取水口基本建在下游黄壁庄水

库，上游岗南水库只有少部分农业灌溉供水和城市生活供水；另一方面岗南水库和黄壁庄水库都隶属于河北省水利厅，由水利厅统一下达调水命令。

3 水价测算政策依据

(1)《水利工程供水价格管理办法》(国家发展和改革委员会、水利部第4号)。

(2)《河北省水利工程供水价格管理办法》(省政府183号令)。

(3)《河北省水利工程供水价格管理办法实施细则》(冀价工〔1998〕122号)。

(4)《水利工程供水定价成本监审办法(试行)》(发改价格〔2006〕310号)。

(5)《水利工程供水价格核算规范(试行)》(水财经〔2007〕470号)。

4 测算遇到的几个技术问题处理

4.1 资产的公益部分和生产部分之间的分配

根据《水利工程供水价格核算规范(试行)》第十四条：具有多种功能的综合利用水利工程的共用资产和共同费用，应在各种不同功能之间进行分摊。分摊顺序是：首先在公益服务和生产经营之间进行分摊，再扣除其他生产经营应分摊的部分，得出供水应分摊的资产和生产成本、费用。

《水利工程供水价格核算规范(试行)》第十八条：不同供水对象的共用资产和共同费用，可以采用供水保证率法或其他科学合理的方法进行分摊。岗、黄水库各类供水的保证率分别为：工业供水97%，城市生活供水95%，农业灌溉供水50%。近5年供水实际天数测算农业供水平均60 d，工业供水平均365 d，城市生活供水平均365 d。

4.2 供水量计量依据

《水利工程供水价格核算规范(试行)》第二十一条：水利工程供水一般按产权分界点作为供水和水费结算(收费)计量点；实际水费结算(收费)点和产权分界点不一致的，也可以按照水费结算(收费)点作为供水计量点，但应合理界定不同产权单位的供水生产成本。实际农业灌溉水价测算一般到斗口，而水量计量点一般在水库出口计量，考虑渠系利用系数按50%计算，所以农业灌溉水量以水库出口水量的50%测算。其他供水以工程产权分界点计量数据据实结算。水量核算依据农业灌溉用水取近5年引水量的平均数，城市生活和工业供水一般取近3年引水量的平均数，但是如果近年年际引水变化较大，可以取近5年引水量的平均数。岗、黄水库近5年农业平均供水量为18 725.6万m^3，近5年城市生活平均供水量为7 540.4万m^3，近5年工业平均供水量为2 392.6万m^3。

4.3 人员经费和运行管理费全部作为生产经营的原因

一方面，岗南水库和黄壁庄水库虽然都兼顾防洪与供水，但是在一年中水库绝大部分时间是为供水服务的；另一方面，岗南水库和黄壁庄水库都没有公益收入，每年财政只给十余万元正常防汛费，公益支出得不到补偿。

4.4 人员经费支出核算

4.4.1 人员数量确定

根据《水利工程供水价格核算规范(试行)》第十一条：供水经营者的人员数量应符合国家规定的定员标准，实际人员数量超过定员标准上限的，按定员标准上限核算；实际人员数量小于定员标准下限的，按定员标准的下限核算。

4.4.2 人员支出费用的合理确认

根据《水利工程供水价格核算规范(试行)》第五条:供水生产成本是指正常供水生产过程中发生的职工薪酬、直接材料、其他直接支出、制造费用以及水资源费等。职工薪酬是指水利工程供水运行和生产经营部门职工获得的各种形式的报酬以及其他相关支出,包括职工工资(工资、奖金、津贴、补贴等各种货币报酬)、工会经费、职工教育经费、住房公积金、医疗保险费、养老保险费、失业保险费、工伤保险费、生育保险费等社会基本保险费。

人均工资(含奖金、津贴和补贴)原则上据实核算,但最高不得超过当年统计部门公布的独立核算工业企业(国有经济)平均工资水平的1.2倍;实际工资低于劳动工资管理部门批准的工资标准的,按照批准的工资标准核算。人员工资总额按照核定的人员数量和人均工资核算。

工会经费、职工教育经费、住房公积金以及社会保险费的计提基数按照核定的相应工资标准确定,工会经费、职工教育经费的计提比例按照国家统一规定的比例计提,社会保险费和住房公积金的计提比例按当地政府规定的比例确定。

4.5 供水兴利分摊系数确定

综合利用水库供水兴利分摊系数确定方法一般有库容比例法、工作量比例法、过水量比例法,其中库容比例法主要适用于具有防洪公益服务和生产经营功能的水库工程。所以,岗、黄水库按照库容比例法计算供水兴利分摊系数。

供水兴利分摊系数=(兴利库容+死库容)÷(兴利库容+死库容+防洪库容)

=(11.57+4.1)÷(11.57+4.1+16.14)

=15.67÷31.81

=49.26%

4.6 各类供水分摊折旧费和大修费比例的确定

按供水保证率法,计算各类供水分摊折旧费和大修费比例:

(1)农业分摊比例=(农业用水量×保证率)÷(农业用水量×保证率+工业用水量×保证率+城市生活用水量×保证率)。

(2)工业分摊比例=(工业用水量×保证率)÷(农业用水量×保证率+工业用水量×保证率+城市生活用水量×保证率)。

(3)城市生活分摊比例=(城市生活用水量×保证率)÷(农业用水量×保证率+工业用水量×保证率+城市生活用水量×保证率)。

4.7 各类供水利润和税金的确定原则

《水利工程供水价格核算规范(试行)》第十九条:农业供水价格按补偿供水生产成本、费用的原则核算,不计利润和税金。

《水利工程供水价格核算规范(试行)》第二十条:非农业供水价格在补偿供水生产成本、费用和依法计税的基础上,按供水净资产计提利润。净资产利润率按国内商业银行长期贷款年利率加2~3个百分点核算。国内商业银行长期贷款利率一般按五年贷款期的利率确定。

5 看法与建议

5.1 两点看法

(1)水利工程供水既是与国民经济发展和人民生活关系重大的商品,又是资源稀缺和自然垄断经营的商品,按照《中华人民共和国价格法》规定的定价原则,应由政府定价,按照水价测算(或成本核算)、民主协商、政府定价的程序进行,其中水价测算(或成本核算)成果是民主协商的基础,水价测算和民主协商的结果是政府定价的科学依据。作为水利工程业务主管部门和经营管理单位来说,首要任务是做好调查研究和水价测算工作,为各有关部门、用水户进行民主协商,为政府最后制定水价提供充分可靠的基础资料,为国家合理制定水利工程的供水价格提供科学依据。

(2)水价测算涉及多方面法规政策利用,需要我们在今后的水价测算工作中进一步改进完善。水价形成机制和管理还存在城市供水成本约束机制不完善、农业水价和水费实收率偏低、用水计量设施不健全、末级渠系水价秩序混乱等问题。水价改革的敏感性决定了它必须充分考虑经济发展现状和社会承受能力,运用经济杠杆促进全社会节约用水,已经成为破解石家庄市缺水难题的重要途径。

5.2 建议

建议有关部门组织全国力量(包括管理、科研、规划设计部门)大力开展水价理论和调查研究与分析测算工作。重点要加强以下三个方面的工作:

(1)理论研究。包括水利工程供水特点及其商品属性,制定水价应遵循的原则和计算方法。

(2)选择一批国内有代表性的供水工程(包括不同地区、不同类型的已建和拟建的供水工程),按相同的条件(包括价格水平、筹资方案、成本费用计算项目和计算参数、投资利润率等)进行供水成本和供水价格的测算分析,为国家制定合理的水价改革方案及其实施步骤提供扎实的基础资料和科学依据。

(3)按照国家产业经济政策和水利产业政策的要求,建立完善水利投入机制和水价格体系,把水利供水价格和收费管理工作真真纳入国家物价管理体系,建立和完善一系列价格和收费管理制度,尽快制定和颁布《水利工程供水价格管理办法》,使水利管理工作达到与社会主义市场经济体制要求相适应,促进水利事业的可持续发展。

【作者简介】 李书锋,1973 年 12 月生, 2003 年 6 月毕业于天津大学土木系,本科学历,学士学位,现供职于河北省黄壁庄水库管理局水情调度处,高级工程师,长期从事水库调度及水资源保护等方面的工作。

淮河流域管理数字化系统框架与方法研究

彭顺风　李凤生　黄　云

（淮河水利委员会水文局（信息中心）　蚌埠　233001）

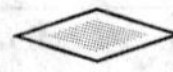

【摘要】 从流域综合管理的需求出发，提出了淮河流域管理数字化及其系统的概念，制定了自下而上由多源数据融合体系、数据管理服务平台、水利专业模型系统、流域管理决策支持系统组成的系统框架，研究了框架中的关键方法，初步建成了淮河流域管理数字化系统。系统在典型地区的成功应用表明，可为流域管理业务提供科学的量化的参考依据，发挥了决策支持的作用，具有广阔的应用前景。

【关键词】 淮河　流域管理数字化　多源数据融合　流域水文模型

1　引言

随着社会经济的发展，流域洪涝、水资源短缺、水体污染、水土流失等涉水问题愈加突出。而这些问题又是相互联系、相互影响的。如果仅对某一地区、某一问题采取措施，可能影响其他问题的解决。例如，在淮北地区平原河道建闸蓄水，解决了局部的水资源短缺问题，却降低了水体的自净能力；对于行洪河道裁弯取直和堤防退建，加快了洪水的下泄速度，却增加了下游的防洪压力。因此，需要从全面、系统的、流域整体的角度出发，综合考虑各种问题之间的有机联系，实现对流域的综合管理，才能更好地解决问题。而基于流域基础空间数据、时间序列数据和水利专业数学模型的数字化系统，是实现流域综合管理的有效手段。因此，近几年来关于流域管理数字化的研究进展迅速、成果丰富、应用效果显著。然而，由于没有形成公认的概念、技术标准、技术框架、边界范围等约束条件，导致了研究成果不能很好地共享。为了保证淮河流域数字化建设成果能在各种流域管理业务之间高度共享，本文研究了流域管理数字化及其系统的概念、系统框架和关键方法，为系统建设提供技术支持。

2　概念与系统框架

顾名思义，流域管理数字化是指基于计算机网络、数据库、水利专业方法和模型等技术，量化流域防洪抗旱、水资源保护与开发利用、水土保持等涉水业务管理的对象和管理行为，实现流域综合管理目标的管理活动和方法。而利用淮河流域数据资源（包括基础地理、遥感、专题电子地图等空间数据资源，以及防洪工程、历史水文、实时水情等水利数据资源）和计算机软、硬件资源，建设具有决策支持功能的计算机系统，即为淮河流域管理数字化系统。为了便于系统建设过程中多专业分工合作、系统集成应用与成果共享，本文提出了四层结构的系统框架（见图1），自下向上分别为多源数据融合体系、数据管理服务平台、水利专业模型体系和流域管理决策支持系统。

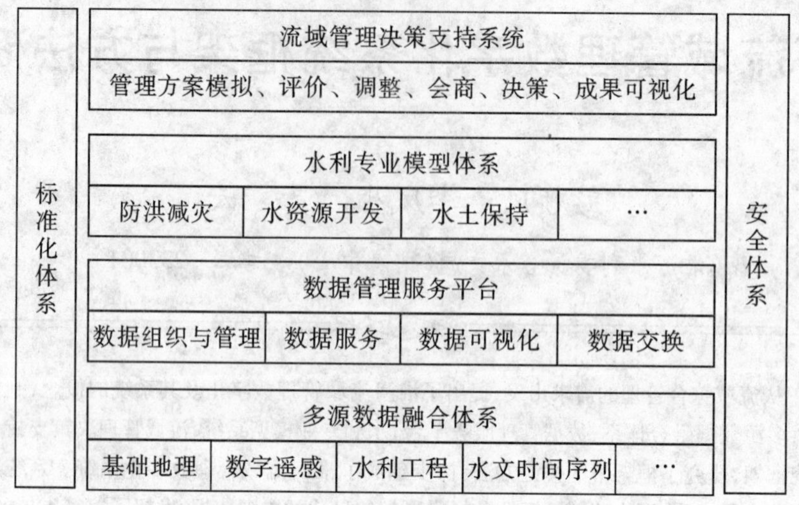

图1 淮河流域管理数字化系统框架

框架最底层的多源数据融合体系是系统的数据基础。所谓多源数据融合体系,就是在国家和行业标准规范的约束下,对流域空间数据标准化加工整理,实现空间数据与关系型水利数据的集成,使之能协同使用。数据管理服务平台是数据、用户、各种应用联系的纽带,是一个综合的数据管理和可视化系统。该系统实现多源、异质、异构、异地数据的组织管理和查询分析应用,为用户提供全面的服务接口,同时实现数据的可视化。水利专业模型体系是整个系统的核心,用于定时定点定量地描述流域的水循环过程,模拟各种管理方法的效果,从而为决策者提供定量的参考依据。最顶层的流域管理决策支持系统用于完成水利专业模型计算结果的评价和调整。决策者在专家经验和专家系统的支持下,对模型的结果进行会商、判断,最后作出决策,从而发挥整个系统的决策支持作用。四层框架在标准化体系和安全体系的保障和约束下分别研究和建设,以保证整个系统协同工作、互为支撑,从而实现数据资源共享、软件组件重用、设施设备共用。

3 关键方法

3.1 多源数据融合方法

淮河流域管理数字化系统面对的数据是纷繁复杂的,如存储于多种存储介质的地形图、遥感影像图、水文气象观测资料、水利工程特征数据和运行管理过程记录数据等。这些数据的生产和管理单位、空间参考系、存储介质、数据格式、存储位置千差万别,这就需要通过多源数据融合方法,实现对这些数据的标准化整理。多源数据融合方法主要包括以下4个方面内容:

(1)标准体系制定与选用方法。目前有大量的关于基础数据的国家和行业标准及规范。为了保证整理后数据的一致性、准确性和协同性,必须制定一个标准体系,明确各种标准的执行顺序。根据科学性、实用性和可扩充性原则,分别以现行标准层级、专业门类、实施时间为标准进行分类。根据标准的层级,把标准划分为国家标准、行业标准、单位规定、惯例;根据专业门类划分为水利、GIS、测绘、遥感四个专业门类;按照标准实施的时间划分为新标准和老标准,新标准优先执行。在多个标准同时可用的情况下,首先选择高层级的标准执行;对于处于同一层级的多个标准,为了保证空间数据与目前已有的关系型水利数据的融合,优先执行水利行业标准。为了发挥数据的空间分析能力,GIS 行业标准执行顺序优先于测绘和遥感方面的标准。这就保证了不同人员、不同时期整理数据的一致性、协同性,也使得数据加工整理工作容易分工合作和集成。

(2)数据指标体系制定方法。数据整理是一个长期的、多方合作的过程。为了保证数据内容一致,便于集成,需要研究和规定数据的指标体系,即规定每一类数据所包含指标的详细列表,规定每个指标的空间要素类型、属性项数据类型、与其他空间要素的拓扑关系、数据长度、填写规则等。这样,就可以保证不同人员、不同时期整理的多源数据能够集成在一起,相互补充,协同使用。

(3)图形要素整理方法。需要根据淮河流域管理数字化系统面对的海量大比例尺原始地形图的实际情况,制定详细的数据整理规则和流程,以保证多源数据整理成果的一致性。主要规则包括:数据的空间参考系采用 1980 西安坐标系,高程基准采用 1985 国家高程基准;地图投影采用高斯-克里格 3°分带投影;数据分幅编号依据《国家基本比例尺地形图分幅和编号》(GB/T 13989—92);图形要素分类与编码依据《基础地理信息要素分类与代码》(GB/T 13923—2006);成果数据中不保留注记数据,该数据转存到图形要素属性中;成果数据中,不再保留可符号化表达的数据(如示坡线等),该数据在 GIS 中通过专门符号来表达;删除图廓整饰内容,其信息保存到元数据库中,例如原始数据图幅编号、生产日期、生产单位、坐标系统、投影系统等内容;点、线、面等图形要素整理成果符合 GIS、测绘方面的标准要求;数据根据淮河水利委员会制定的《淮河流域数字地形图图层分层暂行规定》进行分类分层。

(4)关系型水利数据与空间数据的关联方法。通过水利工程代码与图形要素代码相统一的方法,实现二者关联。对于没有图形要素相对应的水利数据,参考电子地图、高精度遥感影像等相关资料,创建相应图形要素,并赋给相应水利工程的代码,建立关联关系。

3.2 数据管理服务平台建设方法

淮河流域管理数字化系统数据种类多、异构系统多、应用需求多、元数据不全、数据物理位置分散。为了保证数据的安全存储、容易管理、方便展示、方便网络传输、方便多用户并发访问,需要建设数据管理服务平台,以满足应用需求。这里提出了基于通用软件的平台建设方法(见图 2),即选用当前主流的通用数据库管理软件、GIS 软件、中间件为开发平台,以当前主流开发技术进行软件开发。整个平台自下而上分为数据层、中间层和应用层三部分,数据层主要考虑流域综合管理业务公共数据的存储、管理、更新和服务。因为淮河流域管理数字化系统需要访问多种数据库系统,例如 Oracle、SQL Server、Sybase 等,这些数据库的数据结构、组织模型多种多样,而且处在发展变化过程中。为了集成这些数据,该平台设置了中间层,提供了以 ADO. NET 作为数据库管理应用程序的开发软件,实现应用层对数据层的透明访问,保证了平台的通用性、可伸缩性和稳定性。应用层主要基于 ArcGIS Engine 平台,开发系统配置、数据更新、数据抽取、元数据管理等数据操纵功能,实现数据综合管理;基于 ArcGIS Server 和 Skyline TerraGate Suite 开发平台分别实现数据在网络上进行二维、三维发布和服务。

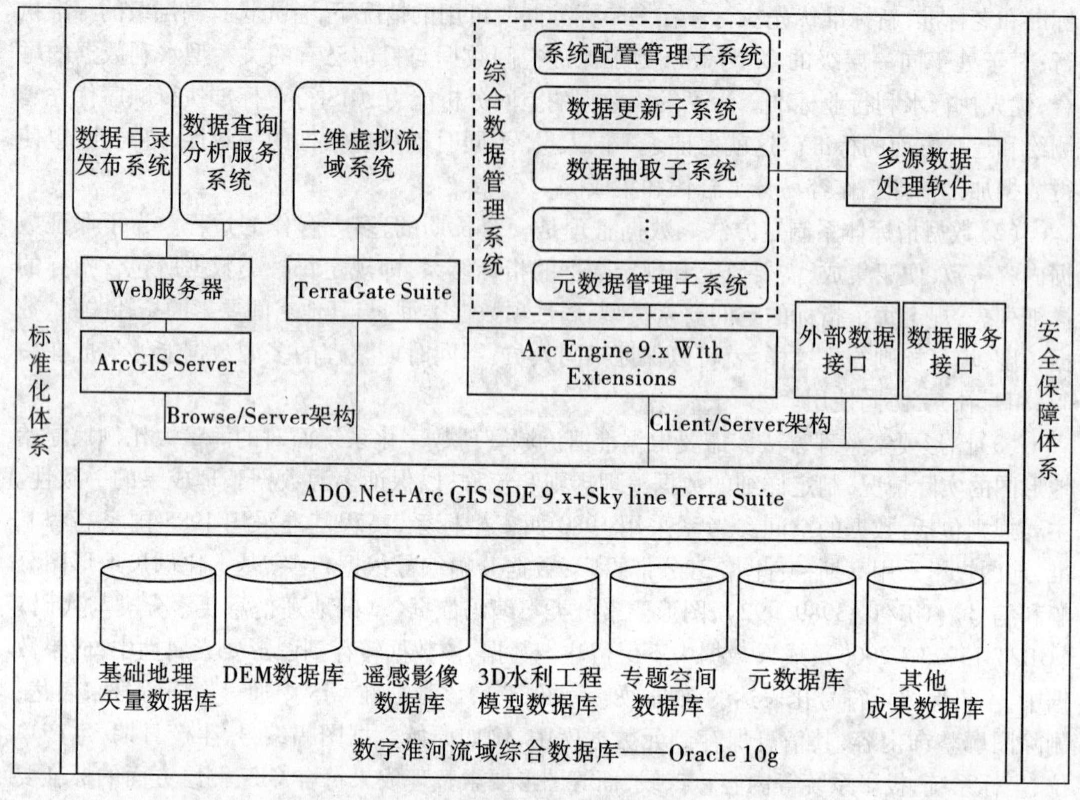

图 2 数据管理服务平台技术框架

3.3 水利专业模型建设方法

为流域综合管理决策提供支持是流域管理数字化系统的主要目标。决策支持系统的核心是水利专业模型的建设。例如在防洪减灾方面,需要建设的洪涝灾情评估模型、流域洪水预报模型、河道与行蓄洪区洪水演进模型等。目前,淮河流域管理业务还是以概念性模型和方法为主要手段,如流域洪水预报使用的降雨径流相关图、单元河道马斯京根法、新安江模型等。这些模型方法计算机化后,在实际工作中发挥了不可替代的作用。然而,这些模型方法的局限性是很明显的。首先,这些模型方法都是基于历史洪水观测过程分析出来的结果,其中间过程是不清楚的,一旦下垫面条件改变了,原来的成果就失去了再使用的基础。其次,经验性的模型方法应用效果依赖于人的经验,而经验很难传承。再次,集总式的计算结果不能满足局部水利工程调度运用需求。因此,需要基于流域水循环的机理,根据下垫面条件和水文气象条件,建设数字化的流域模型体系。建设流域水文模型,首先要从数据管理服务平台提取研究流域的地形、土壤类型、植被类型、土地利用等空间数据,据此数据先把研究流域划分为一系列子流域;再根据产流规律基本相同原则,把各子流域划分为许多水文响应单元,对于每个单元应用水动力学方程,依据上下游关系把这些单元依次连接集成为更大的区域,直至整个计算区域,再在正确的位置嵌入水利工程。这样就构建了数字化分布式水文模型。上述建模过程,可以通过通用流域水文数值模型软件与规范化的时空数据库相结合的方法快速完成。同样,对于河道和行蓄洪区洪水演进模型、污染物运移模型、土壤侵蚀模型、灾情评估模型等,都可采用类似的方法构建。利用该方法,在淮河流域管理数字化系统中,已经构建了史灌河子流域水文模型、中游地区河道与行蓄洪区洪水演进模型以及洪涝灾情评估模型，在实际的流域天然水资源过程模拟、洪水调度预案效果模拟、实时洪涝灾情评估工作中发挥了技术支撑作用。

4 结论

通过对淮河流域管理数字化概念、系统框架和其中关键方法的研究，明确了系统的服务目标、主要功能、技术框架、边界范围和技术路线，清晰地划分了系统的建设模块，使得系统建设可以分步建设、循序渐进，边建设边应用。淮河流域数字化系统据此实现了来源众多、不同标准体系的基础地理数据、数字遥感数据、关系型水利数据的融合，使得数据之间能互相补充、协调使用；初步建成了数据管理服务平台，实现了对异地、异构、异系统数据库系统的综合查询分析和统一服务，具备了空间数据管理维护更新能力；典型地区流域水文模型、水动力学模型、实时灾情评估模型等专业数值模型的构建和应用，为流域综合管理决策提供了支撑作用。然而，流域综合管理需求是复杂的，且随着流域社会经济发展而发展，流域管理需要的基础数据也是海量的，且处在加速增长的过程中，因此流域管理数字化系统的研究和建设将是一个长期的过程，需要流域管理各业务单位和部门协同工作，才能保证数字化系统可持续发展并发挥其作用。

参考文献

[1] 张行南，丁贤荣，张晓祥. 数字流域的内涵和框架探讨[J]. 河海大学学报：自然科学版，2009，37(5)：495-498.
[2] 李佼，吴健平. 基于 Skyline 的三维空间数据网络发布[J]. 测绘科学，2010，35(2)：183-185.
[3] 张行南，彭顺风. 平原区河段洪水演进模拟系统应用于研究[J]. 水利学报，2010，41(7)：803-809.
[4] 彭顺风，李凤生，黄云. 基于 RADARSAT-1 影像的洪涝评估方法[J]. 水文，2008，28(2)：34-37.
[5] 彭顺风，李凤生. 基于数字流域的地表水资源模拟[J]. 水资源保护，2009，25(2)：10-14.

【作者简介】 彭顺风，1969 年 5 月生，博士研究生，2010 年 6 月毕业于河海大学，获工学博士学位，淮河水利委员会水文局（信息中心）信息化处副处长，教授级高级工程师。

An Approach for Estimating Sustainable Yield of Karst Water in Data Sparse Regions

Dan Yin　Fengjun Shi　Li Zhang　Shujie Zhao　Dan Liu　Tao Chen

(Water Conservancy and Hydropower Science Research Institute of Liaoning Province Shenyang　110003)

【Abstract】 This study presents an approach for transferring the qualitative analysis of groundwater sustainability for development to quantitative evaluation by an analogy of two similar regions. A concept of groundwater exploitation sustainability (GES), which is an evaluation index based on water supply capability, eco-geo-environment maintaining capability and the harmony between water and society, is put forward. The Fuzzy Analytic Hierarchy Process (FAHP) method is applied to calculate the GES for the Xiangshan and Dianchang karst groundwater sources in the Huaibei city, Anhui Province, China. The GES of the Xiangshan karst system was calculated to be 0.53 and represents medium exploitation sustainability, while that for the Dianchang is relatively high with a value of 0.70. These two karst systems are separated units but have similar hydrogeological conditions. The Dianchang area had limited groundwater observation data, while the Xiangshan area had long series of observation data, which enabled the computation of the sustainable yield. The sustainable yield of the Xiangshan karst area was used to calibrate the GES, and develop a linear equation between the GES and sustainable yield, which was used to calculate the sustainable yield of the Diangchang karst area as 40.4 million cubic meters.

【Keywords】 sustainable yield　karst water　GES　FAHP　data sparse

1　Introduction

Groundwater is an important part of water resources, and it has a close relationship with ecological and geological environment. How much water can be withdrawn from an aquifer safely and sustainably is an important issue in groundwater science. Sustainable yield of groundwater is a relatively new concept, which evolved from the former concept of safe yield. Lee (1915) first defined the safe yield of groundwater as the quantity of water that can be pumped regularly and permanently without dangerous depletion of the storage reserve (Alley, 2004). The concept expanded over time to include economic feasibility (Meinzer, 1923), degradation of water quality (Conkling, 1946), and the contravention of existing water rights (Banks, 1953). The water demand increased gradually with economic developments. Overexploitation of groundwater during the last decades has caused serious groundwater level declines and eco-geo-environmental problems such as ecological damage, land subsidence, sea water intrusion, vegetation degradation and so on (Holländer et al. ,2009; Waele,2009; León and Parise,2009). Frans (2005) showed that the eco-geo-environmental problems occur even when the pumping rate is lower than safe yield; hence recent research is focused on sustainablc yield rather than safe yield. The sustainable yield of a basin should be a compromised pumping rate, which can be sustained by groundwater recharge and will not cause any unacceptable environmental, economic, or social consequences (Zhou,2009). Sustainable development depends on whether the cone of depression and the reduction of the natural discharge can be accepted by social and environmental considerations. The principal difference between safe yield and sustainable yield is that sustainable yield emphasize on the healthy development of the whole socio-environmental system related to groundwater.

Many hydrogeologists have researchedextensively into the methodology of sustainable yield in recent years (Lutz et al. ,2009). However, its evaluation is still difficult especially in the regions with limited observation data. Heterogeneity and complicated hydrogeological conditions are the characteristics of karst groundwater systems; thus stochastic methods are used, mostly, in areas with long series of observation data for their evaluation (Kurtulus and Razack,2010; Gúrfias-Soliz et al,2010). In some karst groundwater source fields, both the hydrogeological parameters and observation data are often not available; this makes sustainable yield assessment more difficult. This paper proposes a concept of groundwater exploitation sustainability (GES) to illustrate the capacity of an aquifer to deliver water in a sustainable way. The study develops an approach of analogy of the GES for two similar areas to estimate the sustainable yield of the area with limited data and seeks to provide a suitable working methodology in groundwater system. This research will significantly contribute to sustainable groundwater management and determination of how groundwater exploitation impacts on the environment and social economy, especially for regions with limited data.

The study focuses on two karst groundwater sources, which are located in Huaibei, China. The GES of the two areas are evaluated by combining the Analytic Hierarchy Process (AHP) with Fuzzy Comprehensive Evaluation method. A linear equation of GES and sustainable yield was then established for the western part of the area called Xiangshan for which the sustainable yield was calculated using its long series hydrological data. Subsequently, the sustainable yield of the eastern area was calculated with the established equation. This research will, scientifically, guide groundwater management in Huaibei.

2 Study area: karst groundwater systems in Huaibei city

Huaibei islocated in the north of the Anhui Province in China within longitudes 116°23′ E ~ 117°02′ E and latitudes 33°16′ N ~ 34°14′ N. Huaibei has a warm temperate monsoon climate with mean temperature of 14.6 ℃ and an average annual evaporation of 842.2 mm. There are two karst groundwater systems in Huaibei, which are located at Xiangshan and Dianchang, and these have areas of about 278 km^2 and 272 km^2 respectively (Fig. 1). Zhahe, Nantuo and Xiaohe are the main rivers in the area with varying seasonal flows. The peak stream flow in the area typically occurs in August while low flows (and sometimes, no flow) occur in the spring months. The shallow unconfined aquifer in the study area is mainly composed of alluvial soil and sand. The original vegetation cover over the aquifer has been replaced with crops and plants. The two karst aquifers in the area occur mainly in the Ordovician and Cambrian carbonate limestone formation and are highly heterogeneous. Folds, faults and fissures developed in the area control the existence and flow of the groundwater.

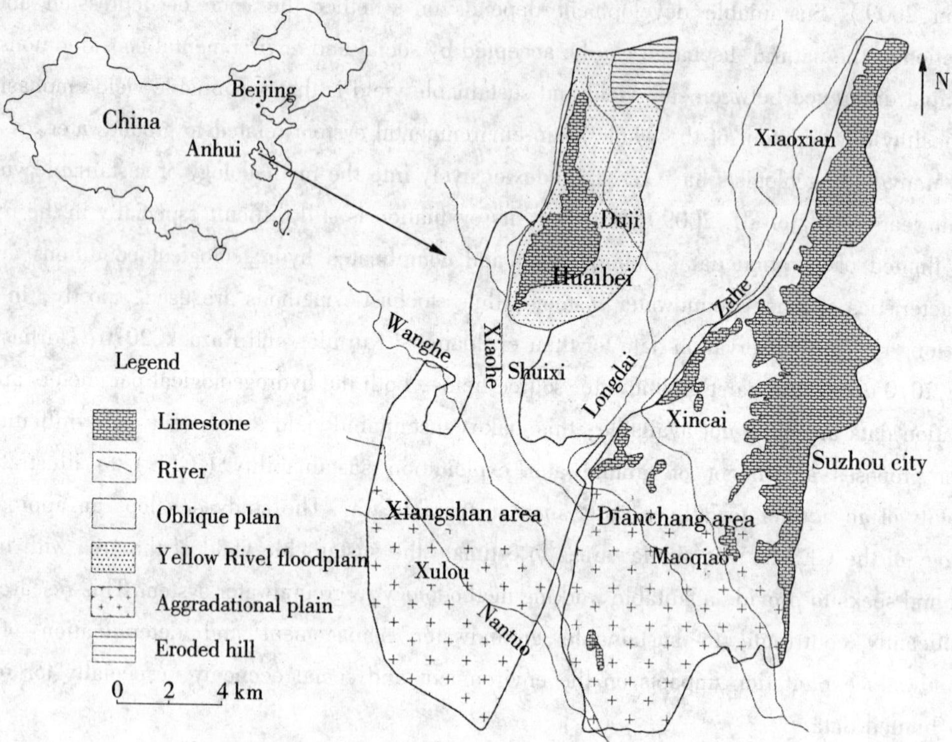

Fig. 1 Location of the two karst groundwater systems in Huaibei, Anhui province, China

The average annual precipitationof the area is 867.0 mm with almost 50 percent of the precipitation in the months June to September. Precipitation is one of the most important components of recharge to the karst aquifer in Huaibei. Fig. 2 shows the relationship between the rainfall and groundwater levels of the Xiangshan area. In the high altitudes of the karst area, rainfall infiltrates into karst aquifer directly through the exposed limestone bedrock. This is observed in the groundwater levels of the area fluctuating closely in relation to the rainfall pattern as shown in Fig. 2. Some parts of the aquifer are covered by the quaternary sediments, which are mainly composed of sand and silt, with the thickness of 2 to 200 meters. The water in this overlying porous unconfined aquifer, also, infiltrates into the karst aquifer as another source of recharge. Groundwater mainly flows from the north to the south of the area. Fig. 3 shows the geologic section of the west Xiangshan karst system.

Exploitation is the main groundwater discharge in both the Xiangshan and Dianchang karst aquifers, which serve as important resource for domestic and industrial water supply in Huaibei. The exploitation of Xiangshan karst groundwater began in the early 1970s, and the pumping rate

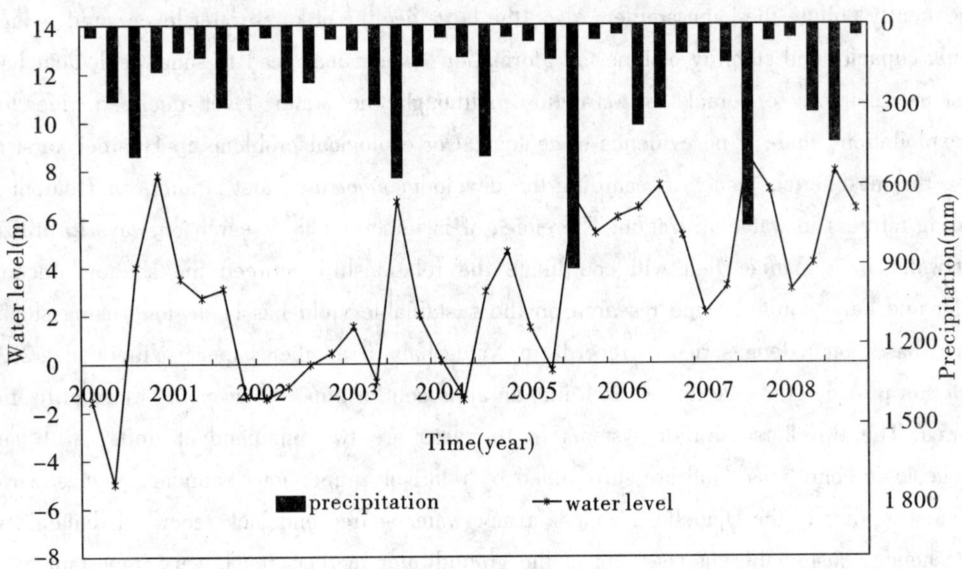

Fig. 2 Temporal variations of karst groundwater level and precipitation in Xiangshan

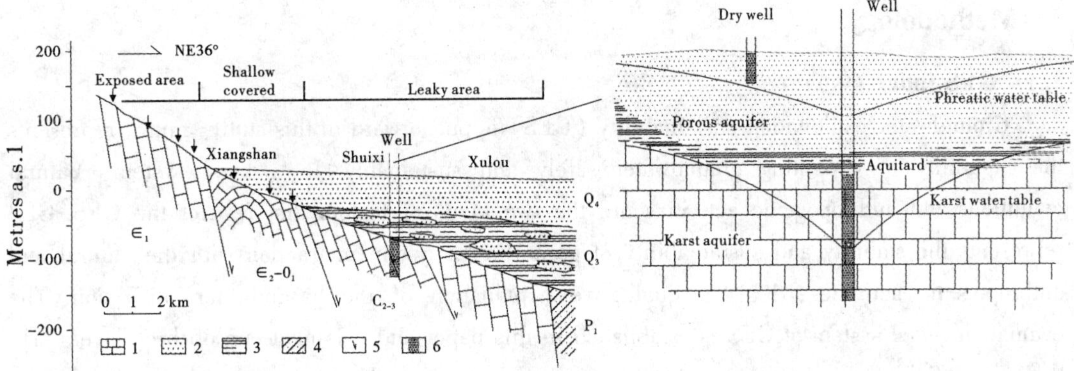

1: limestone; 2: sand; 3: clay; 4: clastic rock; 5: fault; 6: well screen; Q_4: Holocene; Q_3: upper Pleistocene; P_1: lower Permian; C_{2-3}: middle and later Carboniferous; O_1: early Ordovician; $\in_{1,2}$: early and middle Cambrian (Yin et al., 2011)

Fig. 3 Geologic section of Xiangshan karst system and the cone of depression (not drawn to scale) of the two aquifers

increased with economic developments. Overexploitation of the karst aquifer has lead to increased groundwater level decline and the cone of depressions; hence, a large amount of the shallow unconfined water has been leaking into the karst aquifer. The depression of the karst aquifer was about 200 km^2 in 2008, which caused an even larger depression of 300 km^2 in the shallow unconfined aquifer. Agricultural wells in this area are usually shallow and scattered, and most of the pumping wells near the centre of the depression have dried up due to the decline in the shallow unconfined water level. As a result, water utilization conflicts between urban and rural dwellers as well the industrial and agriculture users have become more serious recently. Another potential risk of the groundwater overexploitation is the possible contamination of the shallow aquifer with pesticide and fertilizer from the agricultural activities through infiltration, which could subsequently pollute the karst aquifer. Also, the large decline of karst water level could reduce the bearing capacity and stability of limestone formation, which may lead to some geological hazards like karst collapse or cracks in structure. Although the water level declined due to the overexploitation, there is no evidence of geological or ecological problems in Huaibei karst area. Thus, the most serious concern regarding the development of the karst aquifers in Huaibei is on ensuring fairness in water utilization. Therefore, it is important to research on ways to utilize the karst water in a manner that will coordinate the relationship between the shallow unconfined aquifer and karst aquifer. Some research on the sustainable yield has been done using stochastic method based on a long series of records of Xiangshan in another paper (Yin et al., 2011), which can provide guidance for research into an approach for a more reasonable water utilization in the area. The two karst aquifer systems in the area are two independent units, with similar hydrogeologic conditions, and are surrounded by relatively impervious boundary of clastic rocks. The karst aquifer in the Dianchang area is a new water source and lack record of dynamic water level, hence, sustainable management of the groundwater in this area is very important to avoid overexploitation.

3 Methodology

3.1 GES index system

Groundwater exploitation sustainability (GES) is put forward in this study, and it represents the capability of exploiting groundwater safely and sustainably. In a karst system, natural conditions and anthropogenic activities are the two kinds of factors that restrict the GES. GES represents the stability and sustainability of groundwater system interaction with the natural and anthropogenic activities; it is a qualitative explanation of the groundwater utilization. The evaluation index system of GES is established in this paper and it is made up of three layers viz. the target, field and index layers. The evaluation index system should not only take account of the characteristics of groundwater resource together with the sink and source of the aquifer, but also the eco-geo-environmental problems related to groundwater and harmonious development of social economy. The establishment of the evaluation system needs to consider various factors synthetically, and analyze their influences on the target carefully. The indexes vary with respect to different research objectives (Tang and Zhang, 2001). Some rules like scientific and operability should be considered in the selection of factors (Huang et al., 2006). This paper generalizes the factors in the field layer into three aspects based on the idea of groundwater sustainable utilization.

3.1.1 Water supply capability (WSC)

First of all, the groundwater system must have the capacity to supply water as a resource. Sustainable yield require the groundwater system to have sufficient recharge so that the system can reach a new state of balance in time. The term sustainable is questionable in a groundwater system with virtually no recharge from any source either under natural conditions or following development, and thus it would essentially be a mining venture. The WSC as one of the field aspects mainly takes account of the characteristics of water such as the quality and quantity of groundwater and the renewability of the water resource itself. Three indexes are selected to describe WSC; these are groundwater resource modulus, groundwater quality grade, and area ratio of rainfall recharge. Groundwater resource modulus ($10^4 m^3/km^2$) is an index that represents the water yield property, groundwater quality grade represents the water quality that can impact on water utilization, and the area ratio of rainfall recharge (%) describes the renewability of groundwater system.

3.1.2 Eco-geo-environment maintaining capability (EMC)

As part of the environment system, groundwater has a close relationship with the ecological and geological environments (Parise et al., 2009). Normally, groundwater pumping should not cause excessive depletion of surface water and excessive reduction of discharge to springs, rivers and wetlands. Also, the cone of depression induced by pumping should not cause the intrusion of undesirable quality of water, land subsidence, geo-fractures, karst collapse, and damage to groundwater dependent terrestrial ecosystems (Zhou, 2009). The groundwater system is expected to sustain the eco-geo-environmental health at the time of exploitation. In fact, it may not be possible to completely address all the factors of the environment in many situations. Therefore, some single, correct factors representing sustainable development must be abandoned. Overexploitation of the karst groundwater caused the water level decline and increased the cone of depression in Huaibei, but there is no information to show that the water pollution, drying of rivers or karst collapse is related to overexploitation of the karst water. Three indexes are selected in EMC based on the present situation of the two karst systems in Huaibei. They are ①Depth of groundwater table (m), which indicates that the groundwater system should sustain the environment safely with reasonable water level; ②the ratio of cone of depression (%), which indicates that a large area of the cone of depression sustained for long time will cause geological damages like land subsidence; ③the overexploitation degree.

3.1.3 Harmony between water and society (HWS)

Economic constrains require maximizing groundwater exploitation to satisfy water demand for industrial and agriculture use, but maximum economic benefits would not be obtained by groundwater overdraft. The cone of depression should not go too deep if wells are not to be dried up. Social development should satisfy the demand of potable water and groundwater resource should be distributed equitably for all to have a fair share. People should have equal water right whether they are in the upstream or downstream and inhabiting in rural communities or urban areas. HWS aspect embodies the relationship between social economy and groundwater utilization because it is important to maintain a harmonious development of groundwater and economy. Some of the indexes selected to represent the HWS of Huaibei karst system are ①GDP per cubic meters of water ($Yuan/m^3$), which indicates the average Gross Domestic Product per cubic meters of water generated and shows the efficiency of water use. ②Scarcity ratio of water (%), which is the proportion of water demand that cannot be satisfied by water supply. ③Fairness of water use (%), which indicates the proportionate number of people with the problem of water shortage and reflects the equality of water right.

The evaluation index system of GES is flexibleand must take account of the hydrogeologic condition, the concerns and needs of the inhabitants, and their potential impacts on groundwater quality, the environment and the social economy as much as possible. The index will change in a different study area for the different situations. Based on the above analysis of factors impacting on GES, the evaluation index system of GES in Huaibei was set up. The system includes nine indexes in three layers; namely the target layer, field layer and index layer (Fig. 4).

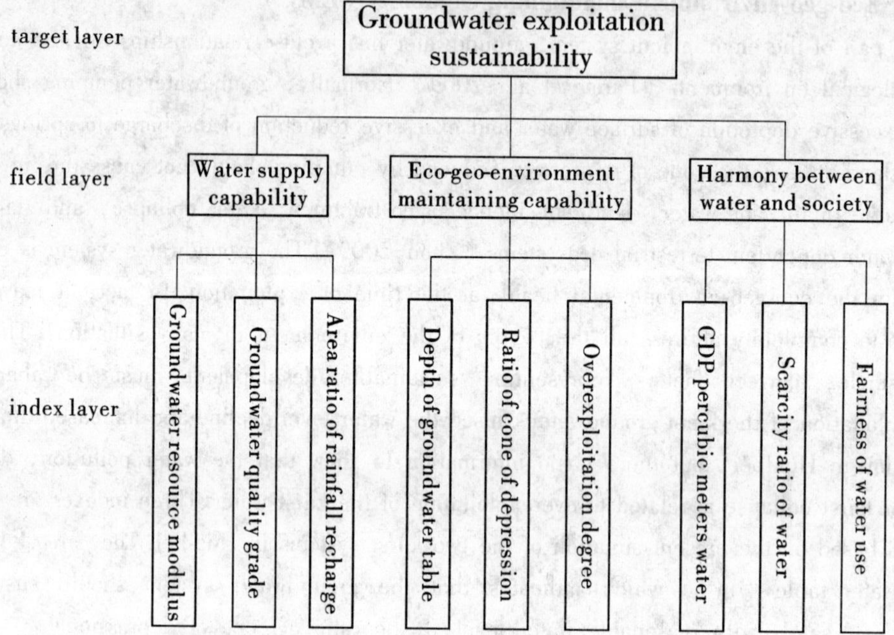

Fig. 4　Structure diagram of GES evaluation index system

3.2　Evaluation of GES

Analytic Hierarchy Process (AHP) combined with fuzzy comprehensive evaluation method was used in this study to evaluate the GES. The GES has many factors with each factor making a different contribution. Thus the Fuzzy comprehensive evaluation method, based on the Fuzzy set theory and can cope with vaguely defined classes or categories (Mohammadi et al., 2009), is used for the evaluation.

3.2.1　Establishment of the set of factors (Liu et al., 2007)

The set of factorsare composed of each index of predetermined assessment objects. Hence, according to the assessment index system in the previous section, the assessment index set is given as: ①The first layer of index set: $U = \{U_1, U_2, U_3\}$, and ②The second layer of index set: $U_i = \{U_{i1}, \cdots, U_{in}\}$ (U_{in} is the nth factor in U_i).

3.2.2　Determination of thecomment set

Considering theeffect degree of evaluation factors have on the GES, the indexes mentioned above are classified into a three level comment set, which is given as $V = \{v_1, v_2, v_3\}$ with a given value of $V = \{0.2, 0.5, 0.8\}$. Table 1 shows the classification value of all the factors. The grade of v_1 indicates a bad situation of groundwater supply potential and environmental sustainability where water cannot sustain the economy and social development; hence management and control of water resources should be emphasized in this state. The grade of v_3 represents a good situation where the groundwater system can sustain the environment, economy and social development with large exploitation potential. The grade of v_2 is the situation between v_1 and v_3. The critical value of each index is a kind of reference state of groundwater sustainability and it should, first, be defined for the evaluation of GES. The groundwater quality grade was assigned using the Quality standard for ground water of the National Standard of China (1994). This standard classifies groundwater quality into five grades: grades 1 and 2 represent the natural low background and normal chemical composition fit for any purpose of water utilization; grade 3 is based on the health of human body and is fit for drinking, industrial and agricultural utilization; grade 4 is fit for industrial and agricultural water utilization; grade 5 is not fit for drinking. The evaluation of water resource for sustainable development (Liu et al., 2009), evaluation of carrying capacity of water resources (Li and Jin, 2009), evaluation of groundwater overexploitation (Dong et al., 2008), and a study on the water environment (Li et al., 2007) defined the critical value of index for their research based on the national and regional standard of China. Thus the national standard of groundwater and some of these water resources for sustainable management experiences were consulted in determining the critical range of index for this paper.

Table 1 Comprehensive evaluation of the value classification

Evaluation factor set U		Judgment set V		
		v_1	v_2	v_3
Groundwater resource modulus ($10^4 m^3/km^2$)	Average groundwater resources per unit area	<15	15–20	>20
Groundwater quality grade	The grade of water quality evaluated by the water quality standard	4	3	2
Area ratio of rainfall recharge (%)	It is the ratio of exposed limestone bedrock to the total karstic area, which represents the renewable capacity of the karst system	<20	20–30	>30
Depth of groundwater table (m)	The vertical distance of groundwater level to the surface	>20	20–10	<10
Ratio of cone of depression (%)	The proportion of depression area of the total karst system area	>40	20–40	<20
Overexploitation degree	The ratio of groundwater withdrawal to the allowable withdrawal	>1.2	0.8–1.2	<0.8
GDP per cubic meters water (Yuan/m^3)	Average Gross Domestic Product per cubic meters of water that can be created in Huaibei	<10	10–20	>20
Scarcity ratio of water (%)	The proportion of water demand which cannot be satisfied by water supply	>15	10–15	<10
Fairness of water use (%)	The proportion number of people with the problem of water shortage	<60	60–80	>80
Scores		0.2	0.5	0.8

3.2.3 Determination of the weight of each factor

The weight isan important degree of one index in the whole index system. This system is a hierarchical structural system; thus the determination of the weight of each factor can be done by the analytic hierarchy process (AHP). The AHP mechanism was proposed by Saaty (1980), and it is a multi-criteria decision making technique based on a 9-point scale. Its versatility in dealing with qualitative factors, multiple objectives, and decision makers has resulted in an impressive array of applications in areas such as energy planning, conflict resolution, banking, etc (Zha, 2006). The AHP is possible to decide weights by comparing the importance of two criteria subjectively. Majumdar (2004), Zha (2006) and Chakraborty et al. (2006) illustrated a more detailed process of the AHP and its applications.

Comparing each two factors, one can obtain a compared matrix. Then from a solution of the eigenvalue of this matrix, the weight set can be acquired. By using the AHP, the weight set of the first subsystem A = $\{A_1, A_2, A_3\}$ can be found with the weight corresponding to each subsystem factor as $A_i = \{a_1, \ldots, a_n\}$, where i is from 1 to 3 and n is the number of the factors corresponding to each subsystem.

Table 2 is the relative importance weight of the indexes, which shows the contribution of each index to the evaluation. The weight of water supply capability is the biggest in the field layer because groundwater is the most important component in any groundwater sustainability development. Groundwater resource modulus is the most important factor in the 9, which also indicates that the abundant degree of groundwater is the predominant factor. However, the other factors are indispensable in achieving the sustainable development.

Table 2 Hierarchy and relative importance weight for field and index factors

Field aspects and weights(A_i)		Index factors and weights(a_i)		Weights order
Water supply capability	0.49	Groundwater resource modulus	0.27	1
		Groundwater quality grade	0.13	3
		Area ratio of rainfall recharge	0.09	4
Eco-geo-environment maintaining capability	0.31	Depth of groundwater table	0.08	6
		Ratio of cone of depression	0.08	7
		Overexploitation degree	0.16	2
Harmony between water and society	0.20	GDP per cubic meters water	0.03	9
		Scarcity ratio of water	0.09	5
		Fairness of water use	0.08	8

3.2.4 The fuzzy judgment matrix R

Given a single factor judgment evaluation on factor U_i and determining its grade of membership (which is expressed as r_{ij}) to judgment grade v_i, the result of a set of single judgment factor $r_i = (r_{i1}, r_{i2}, r_{i3})$ of the ith factor of U_i will be obtained. Then sets of total judgment factors (with amount of m) will create a judgment matrix $\boldsymbol{R}$ (Gong,2009) given by

$$\boldsymbol{R}=\begin{bmatrix}R_1\\R_2\\\vdots\\R_S\end{bmatrix}=\begin{bmatrix}r_{11},r_{12},\cdots,r_{1m}\\r_{21},r_{22},\cdots,r_{2m}\\\vdots\qquad\qquad\vdots\\r_{S1},r_{S2},\cdots,r_{Sm}\end{bmatrix}\tag{1}$$

Therefore, the comprehensive evaluation can be a fuzzy turn around as $B = A\times R$, where A is a fuzzy subset on U and can be expressed as $A=\{a_1, a_2,\cdots,a_n\}$,$0\leqslant a_i\leqslant 1$;$A=\{a_1, a_2,\cdots,a_n\}$ denotes the weight coefficient of all factors on the significance of comprehensive evaluation and $B=\{b_1, b_2,..., b_n\}$ where b_j is the grade of membership in matrix B. The judgment matrix R can, therefore, be derived from the membership function stated.

The GES comprehensive index can be expressed by

$$C=\sum_{i=1}^{n}v_i b_i\tag{2}$$

Where C is the fuzzy synthetic graded value of GES and ranges from 0 to 1, v_i is the given value of each grade, b_i is the membership of each grade, and n is the number of the comment grade.

4 Results and discussion

The current index values of the two karst system in Huaibei, which were used for the evaluation of groundwater exploitation sustainability (GES), are listed in Table 3.

Table 3 Values of evaluation indexes of two karst system in Huaibei

Evaluation index U	Xiangshan	Dianchang
Groundwater resource modulus($10^4 m^3/km^2$)	20.58	24.65
Groundwater quality grade	3	1
Area ratio of rainfall recharge	15.50	52.70
Depth of groundwater table	16.42	7.69
Ratio of cone of depression	80	25
Overexploitation degree	1.10	0.90
GDP per cubic meters water(Yuan/m^3)	105	105
Scarcity ratio of water	8.20	3.00
Fairness of water use	65	82

The calculated comprehensive index of GES of the two karst areas in Huaibei can be seen in Table 4.

Thesynthetic graded value of Xiangshan karst area is 0.53, which signify a medium grade of groundwater exploitation sustainability. The WSC and HWS of this area are 0.56 and 0.60, respectively, and they are all more than the medium level. The EMC is 0.44, which is relatively lower in this area and is the most vulnerable aspect of the GES. The poor eco-geo-environment sustainability in Xiangshan area is mainly because of groundwater overexploitation. Two layers of the cone of depression were formed in this area. The percentage of the cone of depression in Xiangshan karst system was 80% in the year 2008, which seriously exceed the worse critical level of this index. Hence this index affects the evaluation of GES seriously in this area. The decline of the water table has increased the depth to the surface; the depth at Xiangshan is 16.42 m from which 10 m has been the decline during the past decade. This may threaten the structures constructed above or under the ground. There are also some other potential risks like the groundwater pollution and karst collapse. Fairness of water use is 65%, which is close to the worse critical value 60% and reduced the GES. Therefore, it is not suitable to abstract great quantity of groundwater in the Xiangshan area, especially if the focus is on the sustainable development of groundwater.

Table 4 Grade of membership and fuzzy index of the two karst areas

	Aspects	v_1	v_2	v_3	Synthetic value
Xiangshan	WSC	0.14	0.54	0.33	0.56
	EMC	0.19	0.81	0.00	0.44
	HWS	0.10	0.46	0.45	0.60
	GES	0.15	0.61	0.25	0.53
Dianchang	WSC	0.00	0.18	0.82	0.75
	EMC	0.00	0.65	0.35	0.61
	HWS	0.00	0.26	0.74	0.72
	GES	0.00	0.33	0.67	0.70

Remarks: WSC: Water supply capability; EMC: Eco-geo-environment maintaining capability; HWS: Harmony between water and society.

Thesynthetic graded value of Dianchang karst area is 0.70 with the WSC, EMC and HWS greater than 0.6, which shows relatively good groundwater exploitation sustainability. The quality and quantity are both in good conditions, and the karst system also has good water renewability. This is mainly because in almost 50% of the area, the limestone formation is exposed for direct recharge by precipitation. The Dianchang karst area is a relatively new water resource and there are not many environmental problems in the area. Thus, the karst groundwater can be a major water source in the area.

The analyses above are the evaluation of the groundwater exploitation sustainability, and a qualitative characterization of the groundwater resource of the study area, which showed strong or weak sustainability capabilities. The concepts of both groundwater exploitation sustainability and groundwater sustainable yield can be used as a guide for groundwater management. The difference between them is that the former concept is qualitative while the latter is quantitative. In reality, during water resource management procedure, it is important to transfer the qualitative to quantitative to know how much water can be withdrawn from an aquifer safely without damage to the environment and society. Therefore, practitioners care more about the sustainable yield. The natural recharge of the two karst areas in Huaibei can be calculated by the water budget of their respective areas, which are mainly composed of precipitation infiltration, leakage from shallow aquifer and lateral recharge. The recharge components are listed in Table 5. The most important recharge component of Xiangshan is the leakage from shallow porous aquifer while precipitation and lateral recharge are the most important in Dianchang. The sustainable yield of Xiangshan area was calculated by Artificial Neural Network based on a long duration series of records to be 30.05 MCM (Yin et al., 2011).

Table 5 Natural groundwater recharge of the two karst systems in Huaibei (1×10^6 m^3)

	Precipitation infiltration	Lateral recharge	Leakage from shallow aquifer	Total natural recharge
Xiangshan	11.64	9.67	36.90	58.21
Dianchang	26.68	24.00	8.82	59.50

It is assumed that there is some linear relationship between GES and sustainable yield. Thus taking the Xiangshan karst area as an example with a GES of 0.53 and sustainable yield of 30.05 MCM, a simple linear equation can be obtained as follows

$$S=\theta TG \tag{3}$$

Where S is the sustainable yield, T is the total natural recharge, G is groundwater exploitation sustainability, and θ is the adjust coefficient.

θ can be determined by the parameter of Xiangshan area, which is calculated to be 0.97. Hence, equation (3) becomes

$$S=0.97TG \tag{4}$$

It is difficult to assess the sustainable yield of Dianchang karst area due to lack of observation data, but the GES can be calculated using the explained FAHP method. The Xiangshan and Dianchang karst area are in the same city and they are similar in hydrogeological conditions, groundwater exploitation patterns and socio-economic development. Therefore, the analogy method was used to evaluate the sustainable yield of Dianchang karst area based on the GES analysis. The linear equation (4) was used to determine the sustainable yield of Dianchang karst area by substituting the GES and total groundwater amount into the equation to obtain the sustainable yield as 40.4MCM (see Table 6). The exploitation percentage in Table 6 is the ratio of sustainable yield of the total natural recharge. It is 51.6% of the total groundwater recharge that can be pumped in Xiangshan and 67.9% in Dianchang area. These results coincide with the current situation of the karst groundwater in Huaibei.

Table 6 Results of sustainable yield of Huaibei

	Natural recharge ($1\times10^6 m^3$)	GES	Sustainable yield ($1\times10^6 m^3$)	Exploitation percentage
Xiangshan	58.21	0.53	30.05	51.6%
Dianchang	59.50	0.70	40.40	67.9%

5 Conclusion

Estimating groundwater sustainable yield in some karst areas are challenging due to limited data to support the common modeling methods. An approach to estimating the sustainable yield of karstic groundwater has been devised in this study using two karst aquifers in Huaibei as an example to demonstrate the method. Fuzzy combined AHP method was used to evaluate the groundwater exploitation sustainability of both the Xiangshan and Dianchang karst area. Water supply capacity, eco-geo-environment maintaining capacity, and the harmony between water and society were chosen as three key aspects of the evaluation index system. The evaluation result of Xiangshan and Dianchang are 0. 53 and 0. 70 respectively, which shows their sustainability to groundwater exploitation. Xiangshan area is in a relatively medium level situation, mainly due to the long periods of groundwater overexploitation. Therefore, there exist some potential eco-geo-environment risks in that area; hence, it is important to emphasize on groundwater protection there. The GES of Dianchang, however, shows a relatively good water resource supply capacity.

A linear equation was deduced using the GES and the sustainable yield of Xiangshan, which was applied in estimating the sustainable yield of the karst aquifer in the Dianchang area. The equation involved transferring the qualitative data into a quantitative one and is very useful for karst groundwater management. This research serves as a typical example for determining aquifer sustainable yield in regions with limited observation data, which are similar to other regions in their vicinity with available data.

References

[1] Alley W M, Leake S A. The journey from safe yield to sustainability[J]. Ground Water, 2004,42(1):12-16.

[2] Chakraborty S, Banik D. Design of a material handling equipment selection model using analytic hierarchy process[J]. Int. J. Adv. Manuf. Technol,2006,28:1237-1245.

[3] Gong L, Jin CL. Fuzzy comprehensive evaluation for carrying capacity of regional water resources[J]. Water Resour Manage, 2009, 23:2505-2513.

[4] Kalf FRP, Woolley D. Applicability and methodology of determining sustainable yield in groundwater systems[J]. Hydrolgeol. J. ,2005,13:295-312.

[5] Kurtulus B, Razack M. Modeling daily discharge responses of a large karstic aquifer using soft computing methods: Artificial neural network and neuro-fuzzy[J]. J. Hydrol. ,2010,381:101-111.

【作者简介】 殷丹,女,1982 年 7 月生,2011 年 6 月毕业于河海大学,获工学博士学位。现工作于辽宁省水利水电科学研究院,工程师。

指纹识别技术在流域水资源演变归因研究中的应用*

丁相毅　贾仰文　周怀东　贾金生

（中国水利水电科学研究院　北京　100038）

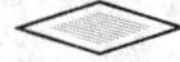

【摘要】 以海河流域为例，将气象气候学中的指纹识别技术与水文水资源学中的二元水循环理论相结合，基于流域二元水循环模拟技术以及气候模式与水文模型耦合技术，对海河流域过去40年的水资源演变进行归因分析，定量区分气候系统的自然变异、温室气体排放导致的气候变暖以及区域高强度人类活动等因素在流域水资源演变过程中的贡献。结果表明，气候系统的自然变异和区域人类活动是导致海河流域过去40年地表水资源量变化的两个因素，并且区域人类活动是主要因素，其贡献约为60%。本文为变化环境下流域水资源演变的归因研究提供了一种新的方法。

【关键词】 指纹识别　二元水循环　水资源演变　归因分析　海河流域

1　引言

由于水资源开发利用和水利工程建设等人类活动的不断增强，流域水循环过程已经从原来的“自然”模式占主导逐渐转变为“自然-人工”二元耦合模式。近年来，以全球变暖为主要特征的气候变化对水循环系统的影响日益凸显，加剧了水循环系统的复杂性，致使对变化环境下的水循环系统进行模拟和预测的难度也在不断加大。由于在气候变化对水循环影响机理方面认识的不足和相关技术的不成熟，现有模型和方法尚不能科学辨识水资源演变过程中气候变化、人类活动以及自然因素的作用，给未来变化环境下的水资源预测和水资源综合管理增加了难度和不确定性。

2008年，美国加州大学Barnett等人在《科学》杂志上发表的相关研究成果表明，气候变化对水循环过程的影响日益加剧，在某些流域其贡献已经超过了自然因素以及人类活动因素。该研究只考虑了气候变化一个因素的影响，无法定量区分取用水、下垫面改变等其他人类活动因素对水资源演变的贡献。国际上相关的归因研究大多是在气象气候学领域中基于指纹识别技术开展的，主要在全球尺度对近地表温度、海洋热容量、地表气压、降水等变量的变化进行归因分析。国内有关该问题的研究则没有一套系统成熟的方法，目前的分项调查法和水文模型法主要以统计、还原和修正等作为基本手段，已经不能满足现代二元驱动力作

* 基金项目：国家重点基础研究发展规划资助项目（2006CB403404），国家水体污染控制与治理科技重大专项（2009ZX07104-001），国家自然科学基金项目（51109223），中国博士后科学基金项目（0135012011）。

用下流域水循环演变中的人类活动效应研究。因此,如何定量区分气候变化、取用水和下垫面改变等人类活动因素及自然因素在水资源演变中的贡献,为水资源综合管理和气候变化应对提供实践指导,已成为现代水文水资源学研究的关键科学问题之一。

作为中国的政治和文化中心,近几十年来,海河流域的气候和环境条件发生了巨大变化。许多观测事实和研究表明,在过去的近 50 年间,海河流域水资源量呈显著减少趋势,并且实测径流量的减少幅度在中国主要江河中位居首位。本文试图对海河流域过去 40 年的水资源演变进行归因分析,定量区分气候变暖、人类活动以及自然因素在流域水资源演变过程中的贡献,为流域水资源综合管理和气候变化应对提供实践指导。

2 研究方法和数据来源

将气象气候学中的指纹识别技术与水文水资源学中的二元水循环理论相结合,利用二元水循环模型分析自然变异、气候变暖以及区域人类活动等情景下的水资源演变状况,利用指纹识别技术计算不同情景下水资源演变的定量评价指标——信号强度,通过将不同情景下水资源演变的信号强度与实际水资源演变的信号强度进行对比,定量区分不同因素在水资源演变过程中的贡献。

2.1 流域二元水循环模型

流域二元水循环模型由分布式水文模型 WEP-L 和水资源配置模型 ROWAS 耦合而成,本文中用来获得不同情景下的水资源演变状况。

WEP-L 模型通过模拟水分在地表、土壤、地下、河道以及人工水循环系统中的运动过程来模拟历史或不同情景下的海河流域水循环状况,重点分析自然水循环过程。ROWAS 模型以对水资源系统的概化为基础,将自然水循环和人工侧支水循环结合为一个整体进行研究,通过配置模拟计算,从时间、空间和用户三个层面上模拟水源到用户的分配,并且在不同层次的分配中考虑各种因素的影响。

2.2 全球气候模式

本文考虑的影响流域水资源演变的因素包括气候系统的自然变异、温室气体排放导致的气候变暖以及包括取用水和下垫面改变在内的区域人类活动等因素,其中自然变异和气候变暖两个情景下的降水和温度数据由全球气候模式提供。

本文选用的气候模式为 PCM(Parallel Climate Model)和其他气候模式(如 CGCMA3、MPI-ECHAM5 等)相比,该模型在海洋和海冰模拟方面具有更高的分辨率,并且模拟的物理过程也更加符合实际,能较好地模拟实际气候情景以及气候的自然变异情况。本文分别选用 PCM 的两个强迫试验来反映自然变异和气候变暖情景下的降水和温度情况。

2.3 统计降尺度模型

全球气候模式和水文模型之间存在着空间尺度不匹配的问题,解决这一问题的方法一般包括动力降尺度和统计降尺度两大类。动力降尺度方法具有较强的物理机制,但花费较高且耗时较长;统计降尺度方法虽然基于统计关系,但简单灵活、计算快捷。相关研究表明:在某些季节和某些区域,动力降尺度和统计降尺度的具体方法各有优劣,总体效果差不多。因此,为方便起见,本研究选用统计降尺度方法来进行全球气候模式输出数据的降尺度处理,具体选用统计降尺度模型 SDSM(Statistical Down Scaling Model)。

SDSM 模型基于多元回归和随机天气发生器相耦合的原理,在建立大尺度气候因子与局地变量之间的统计关系的基础上来模拟局地变化信息或获得未来气候变化情景，是目前国际上应用较为广泛的一个统计降尺度模型。

2.4 指纹识别技术

指纹识别技术是气象气候学中一种对变量的变化进行归因分析的技术方法，采用指纹和信号强度作为变量变化的定量评价指标。某个变量变化的指纹就是对该变量的一系列观测值或模拟值进行经验正交函数（EOF）分解后的第一分量。将该变量的实测或者模拟系列投影到该“指纹”方向，采用最小二乘法计算得出的拟合直线的斜率就称为信号强度，其正负反映变量的增加或减少，其大小反映变量变化程度的强弱。

该技术的基本思想是将原来的多维问题降为低维或者单变量的问题，在得到的低维空间中，通过将变量实际变化的信号强度与不同条件下的信号强度对比来进行归因分析：若计算的某条件下变量变化的信号强度与实际变化的信号强度符号不一致，则该条件不是导致实际变量变化的原因；若计算的某条件下变量变化的信号强度与实际变化的信号强度符号一致，则说明该条件是导致实际的变量变化的原因之一，其贡献为该条件下的信号强度与导致变量实际变化的所有条件下信号强度之和的比值。

2.5 数据来源

本文中的 DEM 数据来自于美国地质调查局 EROS 数据中心的全球陆地 DEM，可从如下网址下载：http://edcdaac.usgs.gov/gtopo30/gtopo30.asp；土地利用数据来自地球系统科学数据共享平台；土壤数据来自于中国科学院南京土壤研究所的土壤数据库；叶面积指数、植被覆盖度等植被信息通过 1982～2000 年的 NOAA-AVHRR 数据反演得到；长系列实测气象数据（1961～2000 年）由中国气象局提供；主要水文站实测径流系列（1961～2000 年）由海河水利委员会提供；水利工程、灌区、取用水、社会经济等相关信息由各省（市）统计年鉴及海河水利委员会相关统计资料得到；统计降尺度模型 SDSM 可从如下网址下载得到：https://co-public.lboro.ac.uk/cocwd/SDSM/；气候模式 PCM 的相关输出数据可从如下网址下载得到：http://www.ipcc-data.org。

3 结果与分析

3.1 海河流域二元水循环模型

基于水文、气象、数字高程模型（DEM）、土地利用、土壤、植被、人口、国内生产总值（GDP）、取用水等多源信息，建立海河流域二元水循环模型，对海河流域“自然-人工”二元水循环过程进行模拟，并利用实测径流数据和地下水监测数据对模型的模拟效果进行验证。有关建模的具体过程请参考文献。

图 1 是海河流域 3 个主要控制站观台、承德、滦县模拟月径流过程与实测的对比情况，从模拟结果的统计分析来看，模拟的 1980～2000 年年均径流量误差在 7% 以内，月径流量的 Nash-Sutcliffe 效率系数在 60%～80%，模拟月径流量与实测系列的相关系数达到 80% 以上。因此，模型对径流过程的模拟具有较为满意的精度。

图 2 是海河流域 2000 年潜水水位等值线模拟结果与实测结果的对比情况，可以看出，模拟结果很好地再现了实测结果，表明二元模型在地下水流场模拟中具有较高的精度。

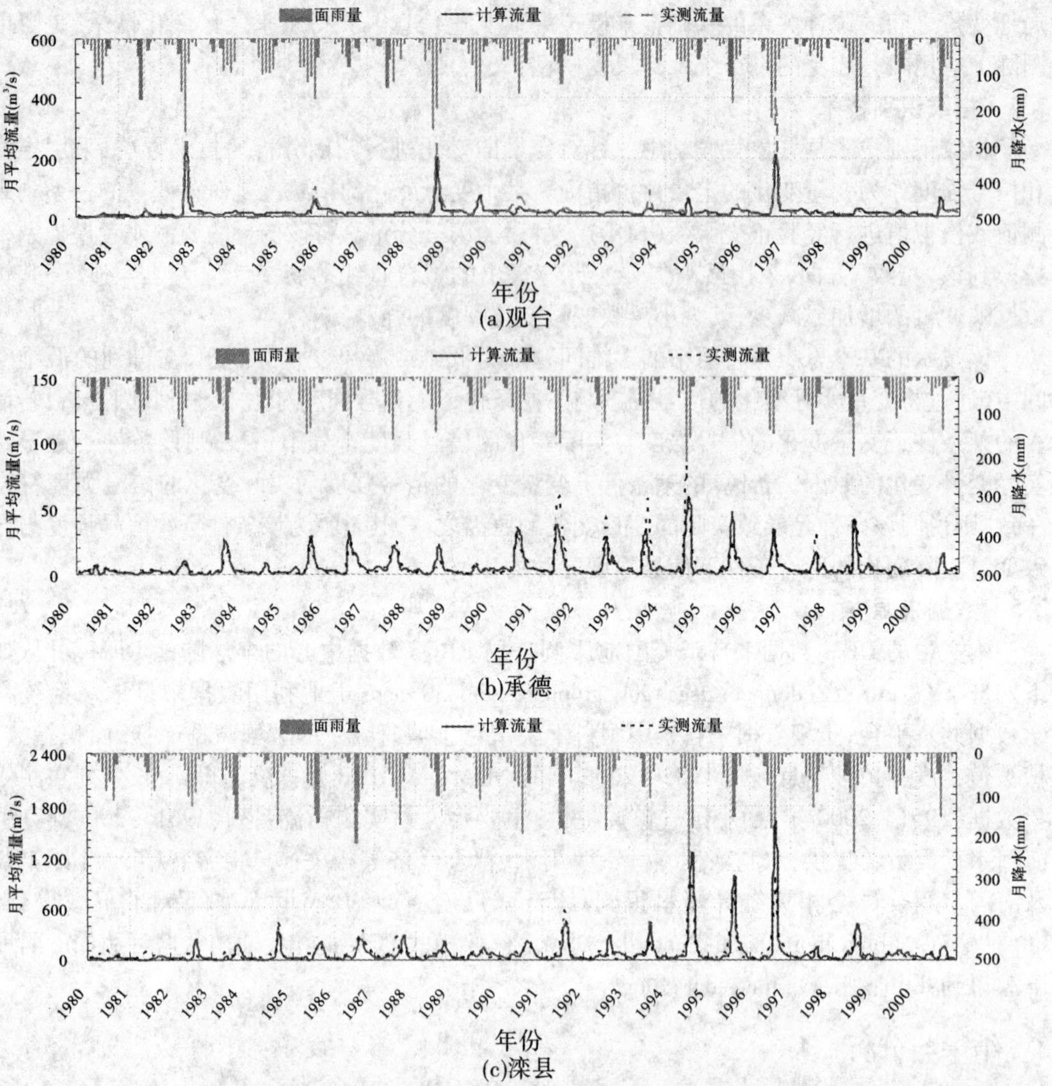

面雨量
计算流量
实测流量
月平均流量(m³/s)
月降水(mm)
年份
(a)观台
(b)承德
(c)滦县

图1 海河流域主要水文站月径流过程模拟结果与实测结果的对比

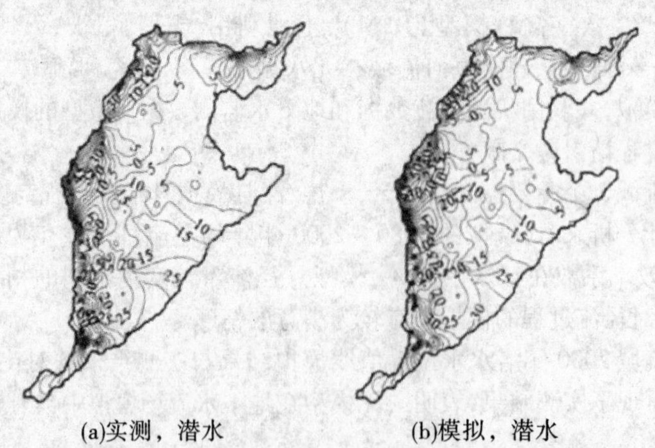

(a)实测，潜水
(b)模拟，潜水

图2 海河流域2000年潜水水位等值线模拟结果与实测值对比

3.2 统计降尺度模型的应用

在海河流域内选择了26个气象站点，基于各站点实测资料和NCEP再分析资料，选择1961～1990年为模型率定期，1991～2000年为模型验证期，分站点对SDSM模型进行率定和验证。通过对比各个站点降尺度系列和实测系列（1961～2000年）的均值、最大值、最小值、百分位数、最大五天值的总和等统计指标，二者吻合较好，降尺度结果令人满意。限于篇幅，仅以北京站为例列出降水降尺度结果和实测系列的上述统计指标的对比情况（见图3）。

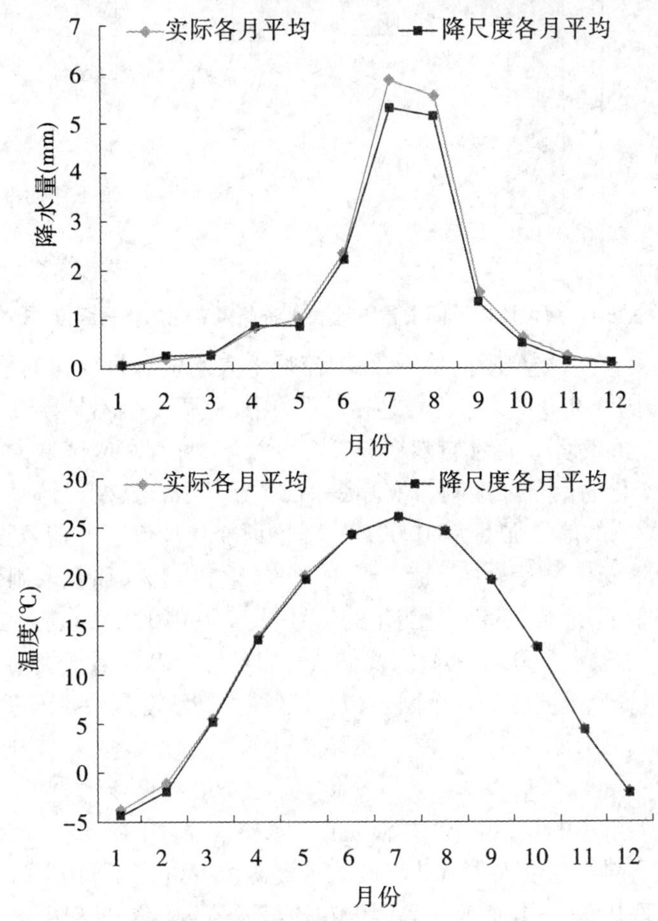

图3 北京站SDSM模型降尺度结果与实测对比

3.3 海河流域水资源演变的归因分析

影响流域水资源演变的因素较多，本文考虑了自然变异、气候变暖、取用水以及下垫面改变四个因素的影响。同时，为了深入研究区域人类活动对流域水资源演变的影响，将取用水和下垫面改变的组合作为人类活动因素进行考虑，因此设置了五个相应的情景。

对于自然变异和气候变暖这两个情景，采用气候模式提供的相应情景下的降水和温度数据，经统计降尺度和空间插值后，作为二元水循环模型的输入。同时，在二元水循环模型中不考虑取用水和下垫面改变的情况，运行二元水循环模型后可得到相应情景下的流域水资源演变情况；对于取用水、下垫面改变和人类活动这三个情景，降水和温度均采用自然变异情景下的数据，分别在二元水循环模型中设置有无取用水条件和改变下垫面条件来得到相应情景下的流域水资源演变情况。

基于得到的不同情景及历史情况下海河流域15个三级区过去40年(1961~2000年)的地表水资源量系列,利用指纹识别技术对其进行EOF分解,得到不同情景下海河流域地表水资源量变化的指纹,进而可以计算得出相应的信号强度。基于实际及不同情景下水资源量变化的信号强度计算结果(见图4),对海河流域实际的水资源演变进行归因分析。

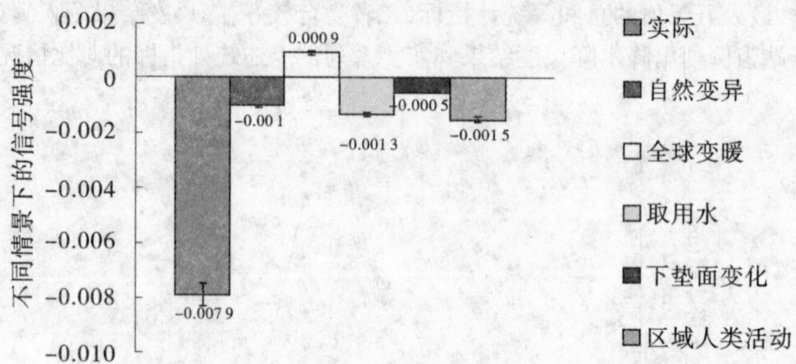

图4 海河流域不同情景下地表水资源量变化的信号强度

从图4可以看出,在气候变暖情景下,海河流域地表水资源量变化的信号强度0.000 9与实际变化的信号强度-0.007 9的符号是相反的,因此气候变暖不是导致海河流域过去40年地表水资源量变化的因素。而在自然变异、取用水、下垫面改变、区域人类活动四个情景下,地表水资源量变化的信号强度则与实际是一致的,其中自然变异情景下的信号强度为-0.001,取用水情景下的信号强度为-0.001 3,下垫面变化情景下的信号强度为-0.000 5,在影响地表水资源量变化的因素中分别占了36%、46%和18%,而在取用水和下垫面变化组合的人类活动情景下,地表水资源量变化的信号强度为-0.001 5,在影响地表水资源量变化的因素中所占比例达到了60%。因此,我们认为自然变异和区域人类活动是导致海河流域过去40年地表水资源量变化的两个因素,并且区域人类活动是主要因素。

4 结语

本文将气象气候学中的指纹识别技术与水文水资源学中的流域二元水循环模型相结合,以海河流域为例,对高强度人类活动流域的水资源演变进行了归因分析,定量区分了自然变异、气候变暖、取用水、下垫面改变等因素在流域水资源演变过程中的贡献,结果表明,气候系统的自然变异和区域人类活动是导致海河流域过去40年地表水资源量变化的两个因素,并且区域人类活动是主要因素,所占比例约为60%。本文为科学识别流域水资源演变的驱动因素和定量区分不同影响因素在流域水资源演变过程中的贡献提供了一种新的思路,相关研究成果可为未来变化环境下的流域水资源综合管理以及相关规划和措施的制定提供决策支持。

参考文献

[1] Barnett TP, et al.. Human-Induced Changes in the Hydrology of the Western United States[J]. Science, 2008(319): 1080-1083.

[2] International AD Hoc Detection and Attribution Group. Detecting and attributing external influences on the climate system: a review of recent advances[J]. Journal of climate, 2004(18): 1291-1314.

[3] 贾仰文, 王浩,周祖昊,等. 海河流域二元水循环模型开发及其应用——I. 模型开发与验证[J]. 水科学进展, 2010, 21(1):1-8.
[4] 贾仰文, 王浩, 甘泓,等. 海河流域二元水循环模型开发及其应用——II. 水资源管理战略研究应用[J]. 水科学进展, 2010, 21(1);9-15.
[5] 褚健婷. 海河流域统计降尺度方法的理论及应用研究[O]. 北京:中国科学院研究生院, 2009.

【作者简介】 丁相毅,1984 年 3 月生,2010 年 6 月于中国水利水电科学研究院水资源研究所水文水资源专业博士毕业,2010 年 8 月至今在中国水利水电科学研究院水环境研究所做博士后。

黄河上中游流域水资源利用现状分析

张　林[1]　王　婷[2]　张金慧[2]　徐立青[2]

(1. 华北水利水电学院　郑州　450011;
2. 黄河水利委员会绥德水土保持科学试验站　绥德　718000)

【摘要】 黄河是我国第二长河,贯穿九个省(区),水资源可利用总量690亿 m^3,地表水资源量580亿 m^3,地下水水资源量110亿 m^3。1987年国务院批准的正常年份黄河可供水量分水方案共分配水量为370亿 m^3;黄河上中游流域包括青海、四川、甘肃、宁夏、内蒙古、陕西、山西等七个省(区),总计分配水量224.6亿 m^3,占全流域370亿 m^3的60.7%。可见,黄河上中游水资源的利用状况及水质情况对黄河水资源开发利用有较大的影响。因此,有必要对黄河上中游流域水资源利用现状进行探讨分析,为今后水资源开发利用提出意见与建议。

【关键词】 黄河　水资源　利用　分析

1　基本情况

1.1　自然概况

黄河发源于青藏高原巴颜喀拉山北麓海拔4 500 m的约古宗列盆地,流经青海、四川、甘肃、宁夏、内蒙古、陕西、山西、河南、山东等九省(区),注入渤海。黄河的突出特点是水少沙多,水沙时空分布不均,水沙不协调,由此成为我国乃至世界上最复杂、最难治理的河流。黄河流程5 464 km,流域面积75.2万 km^2,黄河上中游是指河源至禹门口河段,流域面积49.76万 km^2,占黄河流域总面积的66.2%;干流长4 196.7 km,占黄河干流长的76.8%。

按照河道特征划分为:青甘河段:河道长2 423.2 km,高原峡谷区,河道比较窄,比降比较大,水能储量丰富;宁蒙河段:河道长1 048.4 km,河道比降平缓,部分河段河势摆动频繁,同时也是凌汛威胁比较严重的河段,受支流高含沙量洪水影响,易淤堵河道;大北干流河段:河道长725.1 km,河道比降大,坡陡流急,险滩较多,见图1。

1.2　经济社会概况

黄河上中游流域涉及青海、四川、甘肃、宁夏、内蒙古、陕西、山西等七省(区),30个地(市)和131个县(市、区)。2005年,总人口8 773万人,人口密度176人/km^2。城镇人口3 429万人,城市化率39.1%。流域人口分布极不均匀,龙羊峡至兰州最为密集,人口密度为101人/km^2,兰州至河口镇99人/km^2,河口镇至龙门78人/km^2,龙羊峡以上人口分布稀疏,人口密度仅为5人/km^2。

该区域现有耕地面积21 386万亩(1亩=1/15 hm^2),占总土地面积的28.7%,平均农村人均耕地4亩,其中人均耕地最多的是内蒙古为8.1亩,最少的是四川为1.3亩,其余省

(区)为3~5.3亩。总产粮2 604万t,平均人均粮487 kg,人均粮最多的是内蒙古980 kg,其次是宁夏762 kg、陕西490 kg、山西460 kg、甘肃338 kg、青海250 kg和四川142 kg。生产总值8 436亿元,平均人均产值9 616元,人均生产值由高到低的排列次序是内蒙古18 375元、山西10 303元、陕西9 030元、宁夏8 846元、青海7 566元、甘肃6 319元和四川4 444元。可见,各省(区)人均粮和人均产值相差很大(见表1)。

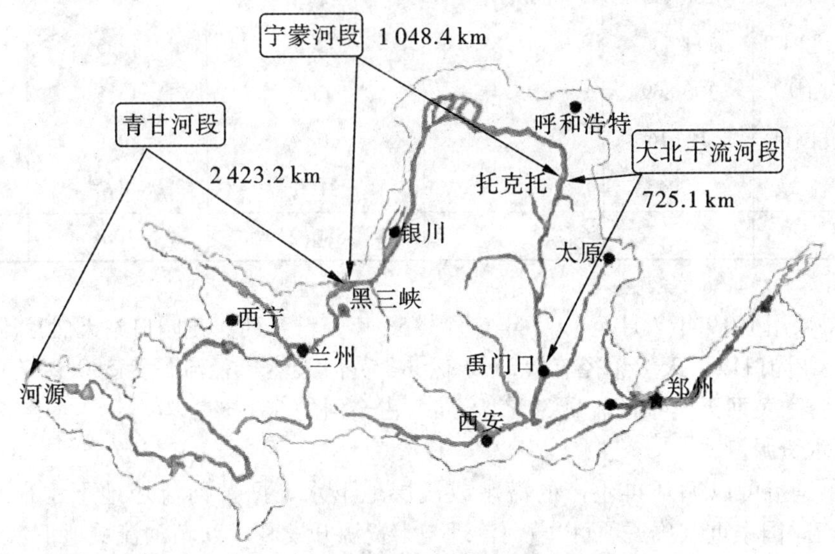

图1 黄河上中游流域分布

表1 黄河流域2005年主要经济社会指标统计

省(区)	总人口(万人)	城镇人口(万人)	生产总值(亿元)	耕地面积(万亩)	粮食产量(万t)
青海	456	177	345	837	70
四川	9	2	4	9	1
甘肃	1 839	524	1 162	5 222	444
宁夏	589	223	521	1 940	279
内蒙古	874	474	1 606	3 267	392
陕西	2 825	1 217	2 551	5 841	788
山西	2 181	812	2 247	4 270	630
小计	8 773	3 429	8 436	21 386	2 604

2 水资源概况

2.1 天然径流量

受人类活动的影响,实测径流量不能反映黄河天然径流,因此必须将实测径流量还原成天然径流量,即将实测径流量加上还原水量(生产、生活耗水和水库调节水量)。从20世纪60年代起,黄河水利委员会同有关单位开展了黄河天然年径流量的分析研究,其结果见表2。

表 2 黄河流域天然年径流量地区分布

区域	控制面积		天然径流量		年径流深
	面积(km^2)	占全河(%)	径流量(亿 m^3)	占全河(%)	(mm)
兰州	222 551	29.6	322.6	55.6	145.0
兰州—河口镇	163 415	21.4	-10.0	-1.7	
河口镇—龙门	111 586	14.8	72.5	12.5	65.0
龙门—三门峡	190 869	25.4	113.3	19.5	59.4
三门峡—花园口	41 616	5.5	60.8	10.5	146.1
下游支流			21.0	3.6	
合计			580.2	100	

注:采用 1919 ~ 1975 年 56 年系列资料。

该成果采用 1919 年 7 月至 1975 年 6 月共 56 年资料系列。花园口站天然径流量为 559 亿 m^3,加上花园口以下天然年径流量 21.0 亿 m^3,黄河天然年径流量为 580.2 亿 m^3。在河源—龙门河段,黄河天然年径流量 385.1 亿 m^3,占全河的 66.4%。

2.2 地下水资源量

地下水是指可以循环再生产的潜水或浅层地下水。据黄河流域地下水最新研究成果——"八五"国家重点攻关项目"黄河治理与水资源开发利用",黄河流域地下水天然资源量(矿化度<1 g/L 淡水总量)约 403.4 亿 m^3,其中黄河流域平原区为 154.2 亿 m^3,山丘区为 237.4 亿 m^3,合计为 391.6 亿 m^3;内流区为 11.8 亿 m^3(详见表 3)。

表 3 黄河流域及内流区地下水资源量 (单位:亿 m^3)

区域	上游	中游	下游	合计	内流区	总计
平原区	51.6	87.0	15.6	154.2	2.1	156.3
山丘区	136.2	85.3	15.9	237.4	9.7	247.1
小计	187.8	172.3	31.5	391.6	11.8	403.4

2.3 水资源可利用总量

黄河水资源可利用总量包括黄河天然河川径流量和地下水与地表水不重复部分的可开采量之和。考虑到黄河流域地下水成果受到基础资料及工作深度的限制,以及内流区零星分布的地下水难以开发利用等因素,现阶段对于黄河流域地下水与地表水不重复部分的可采量,采用 110 亿 m^3 为宜。因此,黄河流域水资源可利用总量为 690 亿 m^3,其中河川径流量可利用量为 580.2 亿 m^3,地下水与地表水不重复部分可开采量为 110 亿 m^3。

3 水资源利用现状

3.1 沿黄各省(区)取水量

根据 1998 ~2008 年《黄河水资源公报》,黄河上中游流域青海、四川、甘肃、宁夏、陕西、山西等七省(区),年平均取水总量 382.95 亿 m^3,其中取地表水量 239.44 亿 m^3,占取水总量的 62.5%;取地下水 143.51 亿 m^3,占取水总量的 37.5%。在七省(区)年平均取水总量由高到低的排列次序是内蒙古、宁夏、陕西、山西、甘肃、青海和四川,分别为 106.18 亿 m^3、

85.51 亿 m^3、69.99 亿 m^3、51.47 亿 m^3、47.94 亿 m^3、21.56 亿 m^3 和 0.29 亿 m^3；取用黄河地表水由高到低的排列次序是宁夏、内蒙古、甘肃、陕西、青海、山西和四川；取用黄河地下水由高到低的排列次序是陕西、山西、内蒙古、甘肃、宁夏、青海和四川，分别为 76.62 亿 m^3、69.90亿 m^3、37.57 亿 m^3、26.60 亿 m^3、15.78 亿 m^3、12.72 亿 m^3 和 0.24 亿 m^3（见表4）。

表4 黄河上中游各省(区)取水量与耗水量统计 （单位：亿 m^3）

省(区)	项目	1998(2002)~2008 年累计值			11(7)年平均值		
		合计	地表水	地下水	合计	地表水	地下水
青海	取水量	214.3	173.85	40.45	21.56	15.78	5.78
	耗水量	145.67	131.10	14.57	14.00	11.92	2.08
四川	取水量	2.95	2.60	0.35	0.29	0.24	0.05
	耗水量	2.62	2.37	0.25	0.26	0.22	0.04
甘肃	取水量	485.86	413.24	72.62	47.94	37.57	10.37
	耗水量	358.7	308.04	50.66	35.24	28.00	7.24
宁夏	取水量	905.06	842.8	62.26	85.51	76.62	8.89
	耗水量	451.88	421.88	30.00	42.64	38.35	4.29
内蒙古	取水量	1 022.72	768.78	253.94	106.18	69.90	36.28
	耗水量	860.76	654.38	206.38	88.97	59.49	29.48
陕西	取水量	596.32	292.61	303.71	69.99	26.6	43.39
	耗水量	477.03	248.94	228.09	55.21	22.63	32.58
山西	取水量	411.21	139.94	271.27	51.47	12.72	38.75
	耗水量	329.86	123.3	206.56	40.72	11.21	29.51
合计	取水量	3 638.42	2 633.82	1 004.60	382.95	239.44	143.51
	耗水量	2 626.52	1 890.01	736.51	277.04	171.82	105.22

注：1. 表中数据从《黄河水资源公报》1998~2008 年中获得；

2. 计算时段：地表水 1998~2008 年共 11 年，地下水 2002~2008 年共 7 年。

3.2 沿黄各省(区)耗水量

由表4 知，该区域年平均耗水总量 277.04 亿 m^3，其中地表水耗水量 171.82 亿 m^3，占耗水总量的 62.0%；地下水耗水量 105.22 亿 m^3，占耗水总量的 38.0%。年平均耗水量由高到低的排列次序是内蒙古、陕西、宁夏、山西、甘肃、青海和四川，年耗水量分别为 88.97 亿 m^3、55.21 亿 m^3、42.64 亿 m^3、40.72 亿 m^3、35.24 亿 m^3、14.00 亿 m^3 和 0.26 亿 m^3；耗用地表水由高到低的排列次序是内蒙古、宁夏、甘肃、陕西、青海、山西和四川，分别为 59.49 亿 m^3、38.35 亿 m^3、28.00 亿 m^3、22.63 亿 m^3、11.92 亿 m^3、11.21 亿 m^3 和0.22 亿 m^3；耗用地下水由高到低的排列次序是陕西、山西、内蒙古、甘肃、宁夏、青海和四川，分别为 32.58 亿 m^3、29.51 亿 m^3、29.48 亿 m^3、7.24 亿 m^3、4.29 亿 m^3、2.08 亿 m^3 和 0.04 亿 m^3。

黄河上中游沿黄七省(区)，年平均耗水量与 1987 年国务院批准的正常年份黄河可供水量分水方案规定给各省(区)的分水指示 224.60 亿 m^3相比，按各省(区)累计年平均多耗水52.44亿m^3,如果只考虑地表水耗水量,各省(区)累计年平均耗水量少耗水52.78亿m^3,只有内蒙古超指标0.89亿m^3,其余六省(区)年耗水未达到指标(详见表5)。

表5 黄河上中游沿黄各省实际耗水与指标对照 (单位:亿m^3)

省(区)	指标	11(7)年平均耗水量			指标(合计)	指标(地表水)
		合计	地表水	地下水		
青海	14.1	14.00	11.92	2.08	0.1	2.18
四川	0.4	0.26	0.22	0.04	0.14	0.18
甘肃	30.4	35.24	28.00	7.24	-4.84	2.4
宁夏	40.0	42.64	38.35	4.29	-2.64	1.65
内蒙古	58.6	88.97	59.49	29.48	-30.37	-0.89
陕西	38.0	55.21	22.63	32.58	-17.21	15.37
山西	43.1	40.72	11.21	29.51	2.38	31.89
总计	224.6	277.04	171.82	105.22	-52.44	52.78

注:指标是指1987年国务院批准的正常年份黄河可供水量分水方案规定给各省(区)的分水指示,是指地表水耗水量。

虽然其余省(区)未超指标,但是取用地下水量大,黄河地下水可供开采量仅110亿m^3,而该区域现在地下水耗水量已达到了105.22亿m^3,占黄河地下水可供开采量的95.7%。特别是陕西、山西和内蒙古等三省(区),现在地下水耗水量已达到了91.57亿m^3,占该区域地下水耗水量的87.0%,占黄河地下水可供开采量的83.2%,地下水严重超采。

4 存在的主要问题

4.1 取用地下水比较严重

在黄河上中游沿黄七省(区)中,仅有内蒙古耗水量超指标0.89亿m^3。未达到取水指标的,取用地下水比较多,年平均总耗水量(包括地表水耗水量和地下水耗水量),有四省(区)耗水量超过了1987年国务院批准的正常年份黄河可供水量分水方案规定给各省(区)的分水指示,超指标的省区有内蒙古、陕西、甘肃、宁夏,分别超指标30.37亿m^3、17.21亿m^3、4.84亿m^3和2.64亿m^3。黄河地下水可开采量110亿m^3,而黄河上中游七省(区)现已耗地下水105.22亿m^3,占黄河可供地下水开采量的95.7%。特别是陕西、山西和内蒙古,年平均耗地下水分别为30.37亿m^3、29.51亿m^3和29.48亿m^3。由此可见,该区域地下水超采严重。

4.2 水污染严重

从青海,经甘肃、宁夏,至内蒙古,黄河沿岸能源、重化工、有色金属、造纸等高污染的工业企业林立,产生出了包括COD(化学需氧量)、氨氮、重金属、高锰酸盐指数以及挥发酚等在内的大量污染物。由于环境保护设施投入大,运转成本高,沿黄重点污染源偷排现象仍比较严重,而一些"十五小"、"新五小"企业点多面广,很难根除。

2011年年初,黄河流域水资源保护局组织专家组,对黄河水污染的状况及危害进行了量化分析,发现黄河干流近40%河段的水质为劣Ⅴ类,基本丧失水体功能。随着经济发展,黄河流域废污水排放量比20世纪80年代多了一倍,达44亿m^3,污染事件不断发生。黄河上游的绝大部分支流都受到不同程度的污染，而中下游几乎所有支流水质长年处于劣Ⅴ类状态，支流变成了“排污河”。

4.3 水资源开发利用缺少可操作性规划

该流域水资源开发利用缺少可操作性的统一规划，无法指导流域内水资源的开发利用和保护，开发治理中存在重开源、轻节流和保护，重经济效益、轻生态环境保护的现象，导致国民经济与生态环境之间、地区之间、部门之间、上下游之间用水矛盾日益尖锐，水资源供需失衡。

4.4 管理体系不完善

目前，黄河上中游流域水资源管理和水资源保护流域管理的法规仍不完善，流域管理与区域管理权责划分不够具体、明确，涉河工程建设管理仍不全面；地下水开采利用缺乏系统、有效的监督、监测措施；随着经济社会发展，穿河工程、跨河工程逐渐增多，由于缺乏系统的规划，致使有些河段工程密度过大，影响防洪安全，甚至引起水事纠纷。流域管理信息化建设、科学研究支撑体系、水文化建设、公众参与流域管理的方式和制度建设等方面还很薄弱，尚需进一步加强。

5 对策

2010年黄河水利委员会提出了实施最严格的水资源管理制度，并在水资源的管理中制定了“三条红线”，即总量控制红线、节约用水红线和纳污红线。为了进一步加强该区域的水资源管理，针对目前黄河上中游水资源利用中存在的主要问题，提出解决对策如下。

5.1 提高节水意识，实施科学用水

针对目前黄河上中游水资源短缺问题，应加强宣传和引导，提高流域内全体公民的节水意识；制定和完善节水政策、法规；抓好用水管理，实行计划用水、限额供水、按方收费、超额加价等措施，大力推广经济、节水灌溉制度，灌关键水，实施优化配水；建立健全县、乡、村三级节水管理组织和节水技术推广服务体系，加强节水工程的维护管理，确保节水工程安全、高效运行。引导用水企业大力开展节水技术改造，增加污水处理设备，提高污水利用率，达到节约用水的目的。

5.2 流域与区域管理相结合

黄河上中游涉及青海、四川、甘肃、宁夏、内蒙古、陕西和山西等七省(区)，要解决好地区之间、部门之间、上下游之间的用水问题，必须加强流域与区域管理相结合的管理办法。首先，需要加快相关的管理法规和制度建设，在《中华人民共和国水法》和《中华人民共和国防洪法》等现有法律体制框架下，进一步规范流域与区域管理，流域内部和区域内部统一与分级管理的职责；其次，需要从控制力度、刚度和致密度三个方面着手，完善相关的控制机制，加强对取水活动和行为的监督检查；再次，需要加强流域管理与区域管理的相互协作，积极推进地方各级水行政主管部门建立健全相关责任制，认真履行其水资源管理职责，全面提高流域机构和地方各级水行政主管部门的管理水平。由流域机构组织制定各省(区)的取用水量方案，对黄河上中游水量实行统一分配与调度，各省(区)严格按照分配水量进行取水。只有这样，才能解决地区之间、部门之间、上下游之间的用水矛盾。

5.3 高度重视水污染防治，确保群众用水安全

由于黄河上中游流域内水资源短缺，经济社会发展用水又挤占河道生态环境用水，黄河

上中游生态水严重不足，水环境容量变得很小，治理水污染的难度很大。同时，河道湿地功能萎缩、水生态失衡、水生生物生活环境破坏、生物多样性减少，黄河上中游脆弱的水生态系统日趋恶化。随着西部大开发和黄河上中游沿岸城市化的进一步发展，黄河上中游水资源和水生态保护、沿河城镇供水安全将面临更为严峻的考验。

因此，有关部门必须高度重视黄河上中游流域内水污染的治理，重点加强工业污染源治理，建设污水处理厂，严格控制入河污染物排放总量，实现黄河上中游水资源的有效保护。根据黄河流域水资源保护和水污染防治的统一要求，按照黄河入河污染物总量控制目标，加强黄河上中游干流各项污染物入河总量的控制、监督和管理工作。

【作者简介】 张林，女，1986 年生，陕西绥德人，在读硕士研究生，专业为水力学及河流动力学。

基于南水北调西线调水的泛流域水资源系统优化研究

彭少明 王 煜 贺丽媛

（黄河勘测规划设计有限公司 郑州 450003）

【摘要】 针对南水北调西线工程的调出区和受水区水资源条件不同、生态环境状况各异、经济社会发展水平差别较大的复杂大系统问题，研究基于调水的泛流域水资源多维尺度模拟和优化方法，建立三层结构、具有水资源时空优化功能的泛流域水资源配置的模型系统，将调水区河流与黄河受水区联系在同一体系、纳入整体模型中优化。采用大系统协调技术和嵌套遗传算法动态调节机制解决泛流域水资源协调优化分配问题，通过模型求解，提出了一套系统优化的适宜调水规模、工程布局及调水量空间合理分配方案，为南水北调西线工程的宏观决策提供科学支撑。

【关键词】 南水北调西线工程 泛流域 时空优化 嵌套模型 配置方案

1 南水北调西线工程泛流域问题的提出

1.1 黄河水资源不能支持经济社会发展需求

受气候变化与人类活动影响，黄河流域水资源衰减严重，供需矛盾十分尖锐。据预测，在实施强化节水、积极挖掘供水潜力等措施下，2030 年黄河流域需水量将达到 547.33 亿 m^3，而可供水量仅为 443.18 亿 m^3，缺水量为 104.15 亿 m^3，见表 1。

表 1 黄河流域 2030 年水资源供需形势 （单位：亿 m^3）

项目	青海	四川	甘肃	宁夏	内蒙古	陕西	山西	河南	山东	黄河流域
需水量	27.67	0.36	62.61	91.16	108.85	98.08	69.87	63.26	25.47	547.33
可供水量	20.44	0.35	43.55	68.91	90.77	73.79	64.39	60.25	20.73	443.18
缺水量	7.23	0.01	19.06	22.25	18.08	24.29	5.48	3.01	4.74	104.15

由于流域水资源供需矛盾突出，黄河生态水量也将受到影响，到 2030 年减少为 185.8 亿 m^3。按照黄河河道内低限生态环境需水量 210 亿 m^3 计算，黄河生态环境缺水 24.2 亿 m^3。黄河水资源短缺不能支撑流域经济社会快速发展，供需矛盾突出、生态环境恶化，严重制约了流域健康协调发展，迫切需要实施跨流域调水补济黄河。

1.2 调水区水资源条件与可调水量

长江上游支流雅砻江、大渡河多年平均径流量分别为 600.37 亿 m^3 和 475.53 亿 m^3，水资源丰沛，与黄河源头可形成自然落差，具备调水工程条件。根据前期研究，南水北调西线工程调水河流范围包括雅砻江干流，雅砻江支流达曲、尼曲，大渡河支流色曲、杜柯河、玛柯河、阿柯河,具备调水工程建设的坝址依次为热巴、阿安、仁达、洛若、珠安达、霍那、克柯2。调水河流及河段情况见表2。

表2 南水北调西线工程调水河流及坝址径流量 (单位:亿 m^3)

调水河流	雅砻江			大渡河			
	干流	达曲	泥曲	色曲	杜柯河	玛柯河	阿柯河
调水河段	仁青里—甘孜	然充—东谷	章达—泥巴沟	色达—河西寺	年弄—壤塘	霍那—则曲口	克柯—阿坝
径流量	60.7~78.5	10~11.6	11.3~13.2	3.5~5.5	7.5~18.5	11.0~58.6	5.7~9.9

通过研究调水坝址的生态环境现状特征及其保护目标、保护对象,采用多种生态需水方法的比较计算,满足维持引水坝址下游临近河段生态流量分别为:热巴 35 m^3/s,阿达 40 m^3/s,阿安、仁达、珠安达、霍那均为5 m^3/s,洛若、克柯2均为2 m^3/s。

将各坝址的入库径流扣除下游下泄流量需求作为坝址处的最大可调水量,则南水北调西线工程热巴、阿安、仁达、洛若、珠安达、霍那、克柯2等7个坝址多年平均的可调水量为108.64亿 m^3,见表3。

表3 南水北调西线工程调水河流最大可调水量 (单位:亿 m^3)

调水河流	雅砻江			大渡河				总计
	热巴	阿安	仁达	洛若	珠安达	霍那	克柯2	
坝址径流量	64.92	10.56	12.96	5.40	17.40	13.97	12.03	137.24
最小下泄径流	17.61	2.59	1.89	0.84	1.89	1.89	1.89	28.6
最大可调水量	47.31	7.97	11.07	4.56	15.51	12.08	10.14	108.64

1.3 南水北调西线调水系统优化问题的提出

南水北调西线工程通过调水可实现雅砻江、大渡河与黄河水系的连通,形成一个泛流域水资源系统。2030年黄河流域河道内外总缺水量为128.35 m^3,而调水河流的最大可调水量仅为108.64 m^3,实施跨流域调水一方面增加受水区收益但同时会使调出区造成一定的损失,因此南水北调西线工程将面临3个层面的问题需要优化解决:①合理的调水总量问题,从泛流域的角度优化决策调多少水量;②调水工程的规模与布局问题,即7个调水工程的优化组合问题;③调水量的合理分配,最大限度地体现调水的价值。

由此可见,泛流域水资源系统是一个涉及多水源、多地区、多目标的高维、复杂的大系统工程,必须全面统筹、高效配置,平衡调水区与受水区,兼顾经济发展和河道生态环境用水,全面提高泛流域总福利水平、提升水资源承载能力。对于如此庞大的系统,此前研究仅关注给定调水量的分配问题,而对于涉及两个以上流域系统优化则显得束手无策,传统的流域水资源模型系统优化、求解面临巨大挑战。因此,需要建立基于南水北调西线调水条件下的泛流域水资源系统优化模型,将调水区和受水区联系为统一整体,实施水资源的统一优化调配。

2 泛流域优化模型的建立

2.1 泛流域水资源系统概化

南水北调西线工程从雅砻江干支流和大渡河支流引水,调水工程由7座水源水库和多段输水隧洞组成；调水从源头进入黄河，经黄河干流主要水库调节后分配到各个受水区。

水资源系统概化可将实体抽象简化为参数表达的概念性元素，并建立描述各类元素内部和相互之间水力联系和水量运移转换的框架，为建立系统的数学模型奠定基础。根据调水河流、调水工程和黄河水系的水力关系，结合工程布置及河流水系特征，将南水北调西线工程连通的泛流域水资源系统概化为调水河流、调水工程、调水渠道、受水河流、受水水库和受水单元等六类基本元素，建立南水北调西线工程泛流域水资源配置系统概化关系，见图1。

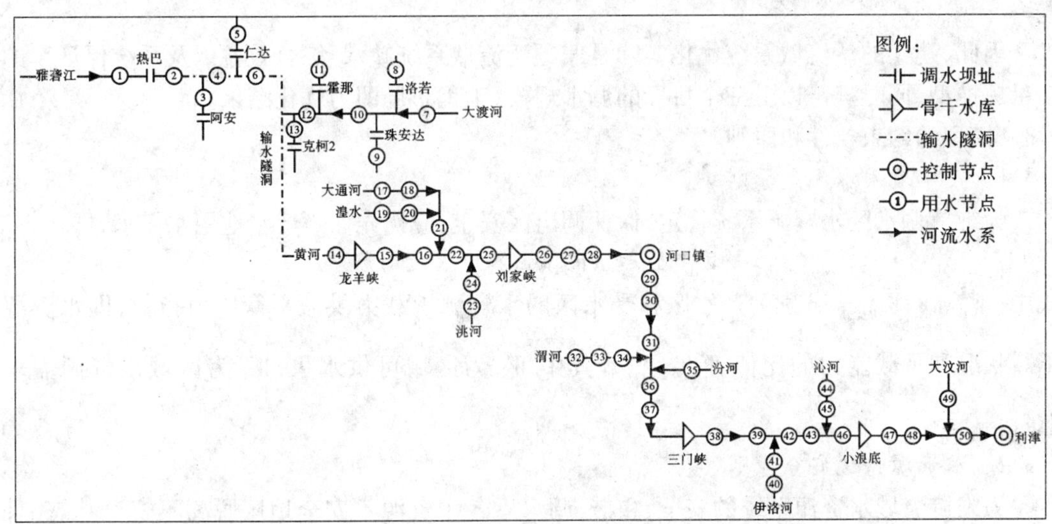

图1　南水北调西线工程泛流域水资源配置系统节点概化

2.2　模型优化目标

跨流域调水的目标：通过工程措施克服单一流域的局限性，促进不同流域间水资源互补互济和谐演进，追求系统整体最优。

2.2.1　综合缺水最小目标

协调泛流域水资源供需矛盾、解决缺水地区的水资源短缺问题，确定以泛流域系统综合缺水率最低（或综合水资源安全度最高）为目标，即

$$\min f = \sum_{i=1}^{n}\left[\omega_i\left(\frac{W_{\mathrm{d}}^{\ i} - W_{\mathrm{s}}^{\ i}}{W_{\mathrm{d}}^{\ i}}\right)^{\alpha}\right] \tag{1}$$

式中：ω_i 为 i 子区域对目标的贡献权重，以其经济发展目标、人口、经济规模、环境状况为准则，由层次分析法确定；$W_{\mathrm{d}}^{\ i}$、$W_{\mathrm{s}}^{\ i}$ 分别为 i 区域需水量和供水量；α 为幂指数，体现水资源分配原则，α 愈大则各分区缺水程度愈接近，水资源分配越公平，反之则水资源分配越高效，$0 \leqslant \alpha \leqslant 2$，在此取1.5。

2.2.2　调水综合净效益最大化

泛流域系统综合净收益最大化，保证水资源的高效利用。调水后受水区的供水增加的收益与调水区水量减少的损失（含工程成本）之差最大，即

$$B = \mathrm{Max}\left(\sum_{i=1}^{2}\sum_{j=1}^{J}\sum_{k=1}^{K} TB[Q_{\mathrm{in}}(i,j,k)] - \sum_{n=1}^{7}\sum_{m=1}^{12} TC[Q_{\mathrm{out}}(m,n)]\right] \tag{2}$$

式中：$TB[Q_{\mathrm{in}}(i,j,k)]$ 为受水区增加供水量 $Q_{\mathrm{in}}(i,j,k)$ 时的收益增加量，$i=1,2$ 指河道内外

配水，$j=1,2,\cdots,J$ 指受水区数，$k=1,2,\cdots,K$ 指配水的部门；$TC[Q_{out}(m,n)]$ 为调水量 $Q_{out}(m,n)$ 时的调水区损失总量，$m=1,2,\cdots,12$ 指调水的月份，$n=1,2,\cdots,7$ 指调水点。

2.2.3　调水保证率尽可能高

优化调水规模和工程布局，保障泛流域水资源系统稳定。通过优化选择调水规模和工程布局，满足在一定调水量情况，调水的保证率最高，即

$$Pr = \mathrm{Max}[1-(P(Q \leqslant Q_0)] \tag{3}$$

式中：Pr 为调水量满足设计调水量的概率，$P(Q \leqslant Q_0)$ 为调水量 Q 低于设计调水量 Q_0 的概率。

因此，建立的泛流域系统优化实际是兼顾泛流域系统缺水、综合效益以及系统保证率的多目标模型，同时满足以上三个目标的调水规模、工程布局即为优化结果。

2.3　模型的约束条件和原则

2.3.1　产水地用水优先原则

为维持调水区水资源系统稳定，保证调出区安全，应优先保障产水地用水需求，即

$$W_{ds调水} \leqslant KW_{ds受水} \quad 且\ K \leqslant 0.5 \tag{4}$$

式中：$W_{ds调水}$、$W_{ds受水}$ 分别为产水区和受水区的水资源承载力供需平衡压力指数，即水资源承载状况和承载能力的比值 $W_{ds}=\dfrac{W_d}{W_s}$，其中 W_s 为区域可供水量，W_d 为区域水资源需求总量。

2.3.2　水资源系统安全原则

为维持流域水资源系统的安全、和谐，调水后各水资源不安全地区脱离不安全状态，且各受水区水资源承载力供需平衡压力指数当量接近，即

$$W_{ds} = w_d / w_s \leqslant 1.2 \tag{5}$$

$$w_{ds}{}^{i} \approx w_{ds}{}^{j} \tag{6}$$

式中：$w_{ds}{}^{i}$、$w_{ds}{}^{j}$ 分别为不同区域的水资源承载力供需平衡压力指数。

2.3.3　河道内补水优先原则

长期以来，黄河流域水资源开发利用已经超过了其承载能力，引起了一系列的生态环境问题，在调水条件下，应保证生态环境具有优先补水权，即

$$w_{ds河道外} \geqslant \gamma W_{ds河道内} \tag{7}$$

式中：$w_{ds河道内}$、$w_{ds河道外}$ 分别为河道内、外水资源承载力供需平衡压力指数；γ 为倍比数，$\gamma > 1$。

河道内水资源承载力当量应不小于一定倍比的河道外水资源承载当量，河道内水资源压力不大于河道外水资源压力，否则调入水量应优先补给河道内。

2.3.4　高效用水原则

调入水量的配置按照效率优先原则，通过权重 ω_i 来实现，根据对黄河流域水资源不安全地区的缺水分析，由层次分析法制定 ω_i 的几个级别：第一层次，大城市，缺水影响特别严重，缺水影响人民生活用水；重要性9，权重 $\omega_i=1.9$；第二层次，较大城市，缺水影响严重，缺水重点产业和能源化工产业用水缺水影响工业发展以及河道内生态环境用水，重要性7，权重 $\omega_i=1.7$；第三层次，小城市，缺水影响一般工业发展用水，重要性3，权重 $\omega_i=1.3$；第四层次，农村，缺水影响农业灌溉和灌区发展，重要性1，权重 $\omega_i=1.1$。

3 模型优化求解

3.1 模型系统的实现

模型系统构建包括3层的总分结构框架，层层深入，首先解决宏观层面的调水量问题，其次解决调水工程布局及调水量部门分配等中观问题，最后解决微观层次的调水工程合理调度以及受水区水量的优化分配问题，上一层模型的解可作为内部参数输入下一层模型中，通过参数传递，层层嵌套。

表4 基于跨流域调水的泛流域水资源优化模型系统

模型	建模目标	解决问题	固定参数	求解变量
M0	总体协调模型	规模优化	α	$Q_{总量}$
M1	调水工程布局优化	布局优化	n	$Q_{out}(m,n)$
M2	受水区水量分配	分配优化	λ	$QP_{in}(i,j,k)$
M3	成本收益优化	模型求解模块	$\lambda(i,j,k)$，$\theta(m,n)$	B

M0从宏观层面解决调水总量优化问题，按照综合缺水最小目标优化，按照式(1)的目标在调水区和受水区间比较，确定调水总量 $Q_{总量}$，并作为输出实现与调水区的工程规模和布局优化模块M1及受水区的调水量分配优化模块M2之间的数据联络。

M1从调水区层面解决7个调水坝址的规模和布局优化问题；M1根据M0给定的调水规模 $Q_{总} = \sum_{n=1}^{7}\sum_{m=1}^{12} Q(m,n)$ 和调水保证率 $P(Q) > P_0$ 的要求，在满足调水点可调水量条件下提出调水区不同工程规模和布局方案及其损失影响参数传递给M3。

M2模型从受水区层面解决调入水量的时空运筹优化分配问题，按照调水分配的效益最大化，实现调水总量在时间(通过黄河干流水库的调节作用)和不同空间(在不同受水区之间)的优化分配，满足分配水量约束 $Q_{总} = \sum_{i=1}^{2}\sum_{j=1}^{J}\sum_{k=1}^{K} Q(m,n)$，$i = 1,2$ 分河道内外配水，$j = 1,2,\cdots,J$ 分地区配水，$k = 1,2,\cdots,K$ 分部门配水，地区、时间和部门均为自由变量，实质是水资源 $Q_{in}(i,j,k)$ 在三维空间中的优化，M2产生不同的分配方案和效益分析参数传递M3。

M3是连接优化模块，M3在接收M1、M2成本、收益数据基础上调用系统经济优化模块进行评估，通过分析对比各分项工程边际投入、调水区边际成本和受水区边际效益，调用分析子程序确定优化方案，并对泛流域调水系统目标实现程度进行决策。

自模型M0～M3，模型的决策变量及其自由度逐渐增加，可供优化运筹的空间增加，同时运筹难度增加。模型系统各模型之间的关系及数据传递见图2。

3.2 模型求解方法和流程

采用大系统分解协调技术及逐步优化方法和嵌套遗传算法，求解南水北调西线工程泛流域水资源优化配置这一超大系统问题。首先，模型采用大系统分解协调方法将系统分解为调水区和受水区，由嵌套的外层遗传算法生成不同的调水量传递给调水区的工程优化和受水区的配置优化；其次，模型系统设置两个并行的模型M1、M2分别计算各调水区的调水量优化和受水区的配水量优化，调水区工程优化采用逐步优化算法POA(Progressive

Optimization Algorithm)提出工程的规模、布局和运用方式等参数，受水区配置优化，采用嵌套的内层遗传算法生成不同地区的水资源配置方案及调水收益等参数；最后，系统采用系统优化的目标函数进行识别并提出优化结果，求解流程见图3。

根据调水河流最大可能调水量108.64亿m^3，设定调水量有效搜索范围为0～108.64亿m^3，模型M0中遗传算法模块，随即生成调水量的第一代种群，按照杂交、变异、演化迭代，求解式(1)，M1接收M0调水总量按照POA逐步优化，根据可调水量约束提出各调水坝址调水量为

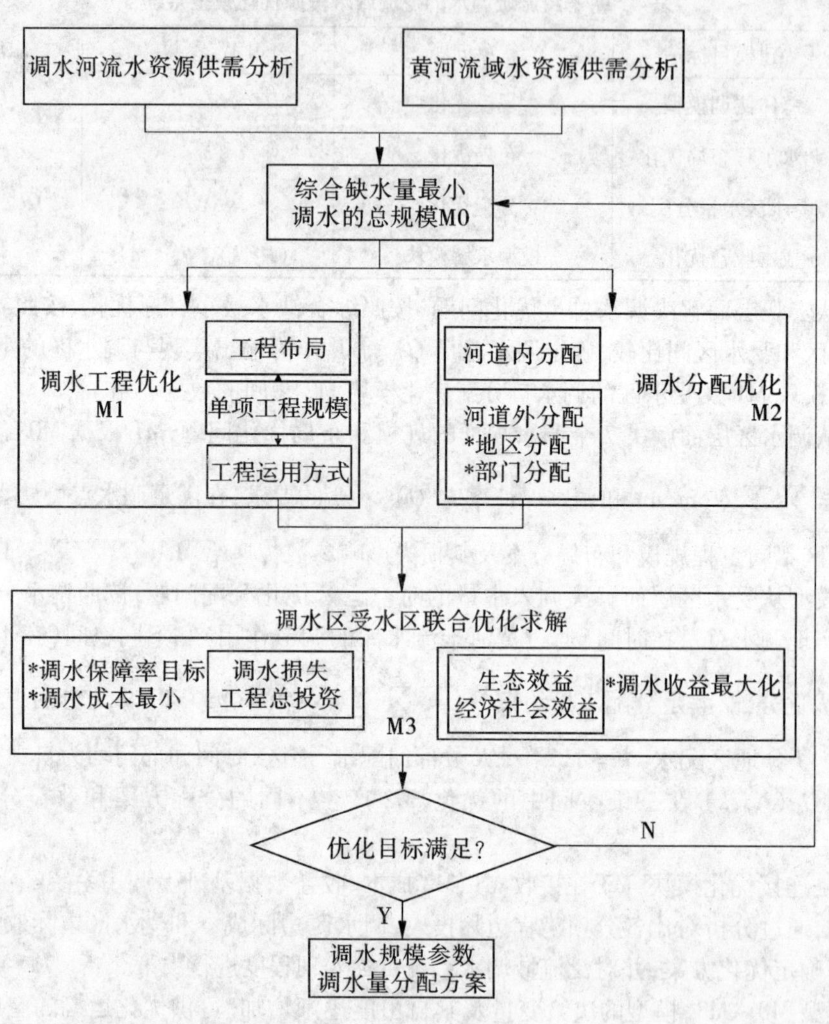

图2　模型结构与流程图

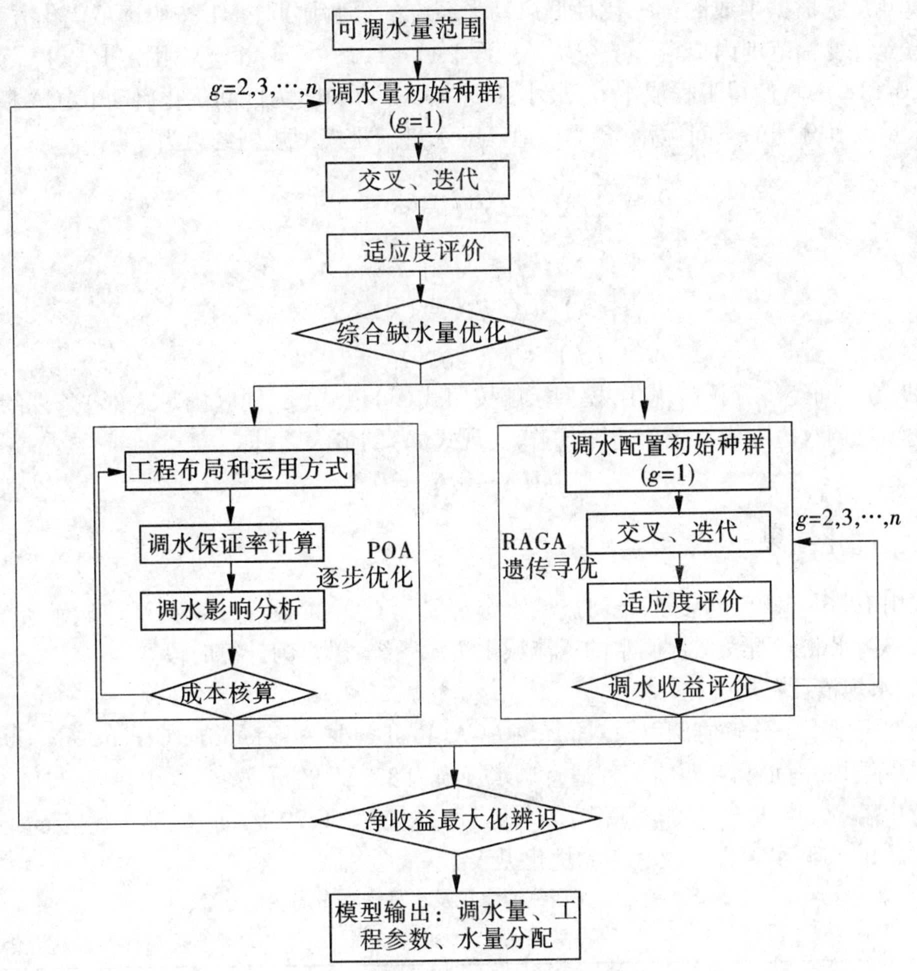

图3　南水北调西线工程泛流域水资源配置模型求解流程

$$
\begin{cases}
0 \leqslant \sum_{m=1}^{12} Q(m,1) \leqslant 47.31 \\
0 \leqslant \sum_{m=1}^{12} Q(m,2) \leqslant 7.97 \\
0 \leqslant \sum_{m=1}^{12} Q(m,3) \leqslant 11.03 \\
0 \leqslant \sum_{m=1}^{12} Q(m,4) \leqslant 4.56 \\
0 \leqslant \sum_{m=1}^{12} Q(m,5) \leqslant 15.51 \\
0 \leqslant \sum_{m=1}^{12} Q(m,6) \leqslant 12.08 \\
0 \leqslant \sum_{m=1}^{12} Q(m,7) \leqslant 10.14
\end{cases}
\tag{8}
$$

模型系统根据生成的第一代种群启动大系统的分解协调，M1 按照逐步优化方法在水库水位允许变幅范围内拟定一条初始调度线 $V_t^k(t=1,2,\cdots,n-1,n)$，固定 V_{t-1}^k、V_{t+1}^k 调整 V_t^k（采用 0.618 法），使得工程成本 C_d 最小，即求解式(4)；M2 为区间优化启动 RAGA 模块，进行调水量及其分配的空间优化，各受水地区配水量满足模型规则约束为

$$\begin{cases}0 \leqslant \sum_{i=1}^{2}\sum_{j=1}^{J}\sum_{k=1}^{K} Q(i,j,k) \leqslant Q_S(i,j,k) \\ 0 \leqslant Q(i,j,k) \leqslant Q_S(i,j,k) \\ w_d(i,j,k)/w_s(i,j,k) < 1.2 \\ w_d(1,j,k) < w_d(2,j,k)\end{cases} \tag{9}$$

M3 为空间经济分析和优化，模型求解按照式(3)～式(5)同时满足式(6)条件约束，模型优化满足调水边际成本等于边际收益，实现式(3)的最大化，即

$$MP_w - MC_w = 0 \tag{10}$$

4 系统优化结果

采用 1956～2005 年系列，黄河流域水资源量为 609.40 亿 m^3，利用泛流域水资源优化模型求解南水北调西线工程调水的泛流域水资源系统，以月为计算时段。

4.1 调水规模及布局优化

经过模型系统分解协调以及双嵌套 RAGA(Real code Accelerating Genetic Algorithm)和 POA 求解结果，南水北调西线工程适宜调水规模为 82.31 亿 m^3 方案，其中热巴43.11 亿 m^3、阿安 7.0 亿 m^3、仁达 7.5 亿 m^3、洛若 2.5 亿 m^3、珠安达 11.20 亿 m^3、霍那 7.5 亿 m^3、克柯 2 为 3.5 亿 m^3。调水工程规模及布局优化见表 5。

表 5 南水北调西线调水量及工程规模布局优化参数

（单位：流量：m^3/s，水量：亿 m^3）

调水河流	雅砻江			大渡河				总计
	干流	达曲	泥曲	色曲	杜柯河	玛柯河	阿柯河	
调水流量	153.80	25.91	27.33	19.92	36.32	27.43	12.63	303.34
调水量	43.11	7.5	8.0	2.5	10.2	7.5	3.5	82.31

4.2 调入水量的配置优化

通过泛流域模型系统配置优化计算，2030 年水平实施南水北调西线工程调水 82.31 亿 m^3，向河道外配置 55.20 亿 m^3，向河道内补水 27.11 亿 m^3，体现了经济和生态效益兼顾原则。经济社会配置的 55.20 亿 m^3，首先向重要城市配水 24.2 m^3 解决城市居民生活用水问题，体现以人为本的原则；其次向重要能源工业基地配置 22.1 m^3 解决发展问题，体现调水的效益原则；再次向水资源不安全及生态脆弱地区配水，解决了主要受水区水资源不安全问题，体现了公平原则。调水的配置方案见表 6。

4.3 调水后黄河流域水资源供需平衡分析

实施南水北调西线工程调水，2030 水平年黄河流域内多年平均缺水量减少为 23.66 亿 m^3，全流域河道外缺水率 4.0%，见表 7，可有效缓解黄河流域资源性缺水的矛盾，实现黄河流域水资源的持续协调发展。黄河河道内生态补水 25.71 亿 m^3，可满足黄河生态环境需水

量要求,实现黄河生态健康的良性维持。

表6 南水北调西线工程调水量河道外配置方案 (单位:亿 m^3)

地区	青海	甘肃	宁夏	内蒙古	陕西	山西	合计
重要城市	4.9	5.9	4.2	7.7	0.6	1.0	24.2
能源基地	0	2.8	5.9	7.8	3.5	2.1	22.1
水资源不安全区及生态脆弱地区	0.8	1.8	1.3	2.6	1.1	1.3	8.8
小计	5.7	10.5	11.3	18.1	5.1	4.5	55.2

表7 南水北调西线调水量分配及黄河流域水资源供需形势 (单位:亿 m^3)

省(区)	青海	四川	甘肃	宁夏	内蒙古	陕西	山西	河南	山东	黄河流域
调水后缺水量	1.71	0	1.43	1.59	1.72	6.22	3.24	3.01	4.74	23.66

5 结论

西线调水的优化配置是一个全局性问题,调水量、调水时机的选择,调水效益最大发挥、调水量合理分配是南水北调西线工程调水的关键技术问题。本文从流域经济社会、生态环境与水资源协调角度出发,分析流域水资源系统的耦合关系,研究跨流域超大系统、多水源的联合运用模式,提出西线调水的合理规模和优化配置方案,全面体现了调水的目标和原则,可为南水北调西线工程宏观决策提供技术支撑和参考。

参考文献

[1] 冯耀龙,练继建,王宏江,等.用水资源承载力分析跨流域调水的合理性[J].天津大学学报,2004,37(7):595-599.

[2] 王浩,游进军.水资源合理配置研究历程与进展[J].水利学报,2008,39(10):1168-1175.

[3] 雷声隆,覃强荣,郭元裕,等.自优化模拟及其在南水北调东线工程中的应用[J].水利学报,1989,(5):1-13.

[4] 王劲峰,刘昌明,于静洁.区际调水时空优化配置理论模型探讨[J].水利学报,2004,(4):7-14.

[5] 游进军,王忠静,甘泓,等.两阶段补偿式跨流域调水配置算法及应用[J].水利学报,2008,39(7):870-876.

【作者简介】 彭少明,1973年4月生,博士,2008年毕业于西安理工大学,黄河勘测规划设计有限公司,高级工程师。

塔里木河流域水权管理与对策*

陈小强　魏　强　孟栋伟

（新疆塔里木河流域管理局　库尔勒　841000）

【摘要】 实施水权管理，是塔里木河流域干旱区水资源管理的创新工作，通过阐述水权内涵及塔里木河流域初始水权，分析了流域水权管理现状。经整理相关资料，塔里木河流域超限额用水仍然严峻，影响了流域社会经济、生态的和谐发展。为了促进流域水资源的可持续利用，针对存在的问题，提出了流域水权管理的对策。

【关键词】 塔里木河流域　水权管理　对策

1　塔里木河流域水资源概况

塔里木河是我国最大的内陆河，塔里木河流域是环塔里木盆地的阿克苏河、喀什噶尔河、叶尔羌河、和田河、开都河-孔雀河、迪那河、渭干河与库车河、克里雅河和车尔臣河等九大水系144条河流的总称，流域总面积102万km^2。20世纪50年代以来，由于人类活动和塔里木流域降水稀少，蒸发强烈，温差大，多风沙等典型大陆性气候变化的影响，导致许多源流相继断流。目前，与干流有地表水联系的只有和田河、叶尔羌河、阿克苏河三条源流。此外，孔雀河通过扬水站从博斯腾湖扬水经库塔干渠向塔里木河下游灌区输水，形成现在塔里木河"四源一干"的格局。本文研究分析的区域为"四源一干"，其示意图如图1所示。四源流多年平均天然径流量256.73亿m^3（含国外入境水量57.3亿m^3），其中阿克苏河、叶尔羌河、和田河和开都河-孔雀河分别为95.33亿m^3、75.61亿m^3、45.04亿m^3和40.75亿m^3。地下水资源与河川径流不重复量约为18.15亿m^3，其中阿克苏河、叶尔羌河、和田河和开都河-孔雀河分别为11.36亿m^3、2.64亿m^3、2.34亿m^3和1.81亿m^3。水资源总量为274.88亿m^3，其中阿克苏河、叶尔羌河、和田河和开都河-孔雀河分别为106.69亿m^3、78.25亿m^3、47.38亿m^3和42.56亿m^3，见表1。由于塔里木河上中游流域无计划的土地开发，自20世纪50年代到90年代的40多年里，塔里木河从1 321 km缩短到1 001 km。塔里木河下游的生态环境急剧恶化并造成了沙漠化的扩展。

2　水权内涵及塔里木河流域初始水权的确定

2.1　水权内涵

水权一般指水资源的所有权、经营权、使用权等一系列与水资源相关权利的总称。水权

* 基金项目：水利部公益性行业科研专项经费项目（201101049）资助。

的三项基本权利,其权属主体分别为国家、企业和消费者,彼此间虽相互联系,但其实质上却是相互分离的。

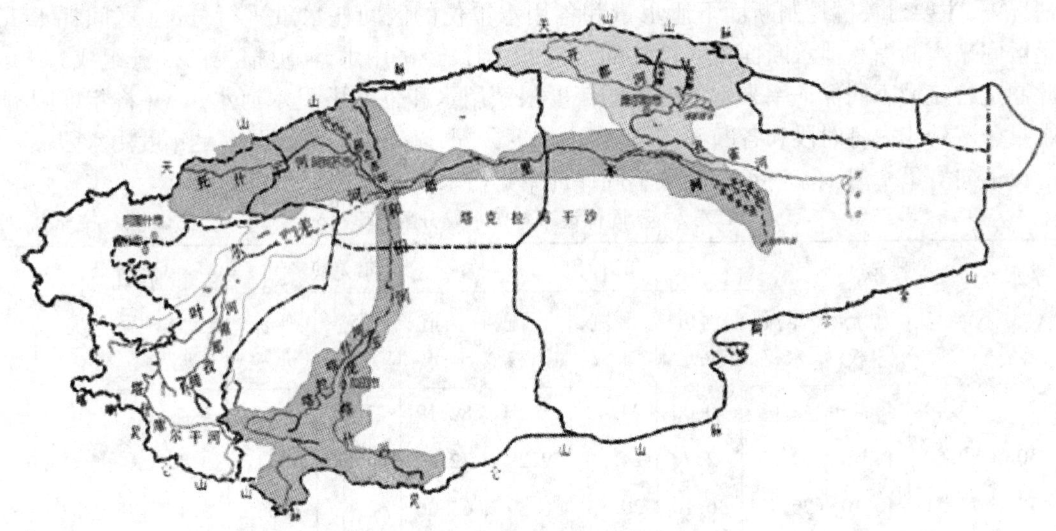

图 1 塔里木河流域“四源一干”示意图

表 1 四源流水资源总量统计 (单位:亿 m^3)

流域	地表水资源量	地下水资源量		水资源总量
		资源量	其中不重复量	
开都河-孔雀河流域	40.75	19.97	1.81	42.56
阿克苏河流域	95.33	38.12	11.36	106.69
叶尔羌河流域	75.61	45.98	2.64	78.25
和田河流域	45.04	16.11	2.34	47.38
四源流合计	256.73	120.18	18.15	274.88

注:阿克苏河流域地表水资源量含台兰河等小河区水量。

水资源的所有权是指所有者对水资源的占有、使用和收益处分的权利。《中华人民共和国水法》明确规定:水资源归国家所有。也就是说,国家是水资源的所有者;水资源的使用权是从所有权中派生并分离出来的一项权利,是水资源所有权国家所有,亦即全民所有的具体体现。换句话说,落实使用权是实现所有权的重要方式之一,所不同的是,使用权一旦从所有权中分离出来,其权属主体便发生了变化,既不是国家,也不是全民,而是具体的消费者。因此,水资源的使用权与每个消费者休戚相关,也是老百姓、各行各业和各级政府所共同关注的;水资源的经营权是架在所有权与使用权之间的一座“桥梁”,通过这座“桥梁”,抽象的国家所有权就被部分转化为具体的消费者使用权。现实中,经营权的权属主体通常是开发和经营水资源的企业,水资源的经营权包括三层含义:一是对水资源的开发权,二是对水资源开发设施的使用权,三是出售商品水的权利。当然,在多数情况下,从事水经营的企业还同时拥有对开发设施的所有权,但应当注意的是,企业并不拥有对水资源的所有权。

2.2 塔里木河流域初始水权的确定

为了缓解塔里木河流域水资源供需矛盾日益突出,流域生态环境不断恶化的局面,2003

年12月新疆维吾尔自治区人民政府下达了《关于印发塔里木河流域“四源一干”地表水水量分配方案等方案的通知》(新政函[2003]203号文),分配方案将各流域不同保证率来水情况下,主要对关键控制断面下泄水量和各用水单位的区间耗水量进行分配(区间耗水量是区间来水断面和泄水断面之间消耗的水量,由国民经济用水、河道损失两部分组成)。由此制定各流域不同保证率来水情况下的年度限额用水和下输塔里木河水量,并将年度限额水量分解到年内各时段和各断面,进行年内调度分配,以确保各源流在满足年度用水限额的前提下向塔里木河干流输水。流域不同保证率年份水量分配方案如表2所示。

表2 不同保证率年份水量分配方案 (单位:亿 m^3)

来水频率(%)	阿克苏河			和田河			叶尔羌河			开都河-孔雀河塔里木河水权	塔里木河总水权
	河道来水	源流水权	干流水权	河道来水	源流水权	干流水权	河道来水	源流水权	干流水权		
25	88.49	46.46	42.03	50.00	24.42	15.51	84.19	73.14	5.58	4.50	67.62
50	80.60	46.40	34.20	42.70	24.16	9.29	72.79	64.54	3.3	4.50	51.29
75	72.51	46.10	26.41	36.10	20.83	6.39	65.06	62.43	0	4.50	37.30
90	66.78	41.56	25.22	31.00	20.68	2.02	58.47	58.47	0	4.50	31.74

3 流域水权管理情况分析

2001年6月27日,国务院正式批准实施了《塔里木河流域近期综合治理规划报告》,提出了以生态建设和环境保护为根本的指导思想,计划用5~6年时间,投资107亿元,通过节水工程建设和加强流域水资源统一管理调度等工程措施与非工程措施,增加各源流汇入塔里木河的水量,使下游河道长期断流状况得到改善,水流到达台特玛湖,改善灌溉条件和流域生态环境,为全流域人口、资源、环境可持续协调发展创造良好的条件。

从2002年开始,在全流域实行水量统一调度。制定了《塔里木河流域水资源统一调度管理办法》,依据批准的《水量分配方案》,编制各年度水量调度方案,每年6~9月按旬、月对水量实行实时调度。通过对流域的用水监督和水量调度工作,基本保证了源流和干流各灌区限额用水量,为灌区农业生产的丰收作出了应有的贡献。但由于流域内区域管理强流域管理弱,体制不顺等多方面的原因,“四源一干”超限额用水量仍旧严峻,经统计整理相关资料,具体结果如表3所示。

表3 2001~2010年塔里木河流域“四源一干”超限额用水量统计 (单位:亿 m^3)

流域	阿克苏河流域	叶尔羌河流域	和田河流域	开都河-孔雀河流域	塔里木河干流
多年平均来水	91.23	73.99	48.1	44.67	
多年平均年度限额	59.15	72.61	25.13	3.48(生态和垦区供水)	19.31
多年平均实际引水	63.22	73.54	26.88	2.98(生态和垦区供水)	24.08
超限额引水	4.07	0.93	1.75	0.5	4.77

塔里木河近期治理临近尾声,流域水资源统一调度管理的问题与矛盾凸显。目前,流域内灌溉面积比国务院批准的规划面积新增了几百万亩,要多用水约几十亿立方米,不仅将节增出来的水全部耗尽,还挤占了各源流原来汇入干流的水量,进一步加剧了源流和干流、上

游和下游、地方和兵团、生态和生产之间的用水矛盾。

4 流域水权管理对策

4.1 转变观念,树立水权管理意识

前面的分析表明,人为因素是引起塔河流域生态问题的主因。因此,要根本性的解决出现的问题,需从改变人们陈旧的思想观念入手。以往的过度引水引起的生态环境问题已经提出了警示:绿洲生态环境是干旱沙漠地区人们赖以生存的摇篮和襁褓,要实现自我的可续发展,必须保护周边的生态环境。所以,在未来的发展中,首先得树立水权管理意识,以水资源的属有权的量制定科学合理的发展方案。

4.2 明晰水权,引进新约束机制

通过初始水权的界定,明晰了生态水权。通过各控制节点,每个用水户限引规定水量,生态用水得到了保障,不仅有助于生态环境的修复与改善,而且每个用水户都有了用水的约束。在初始水权分配以后,地区或团场若打算发展经济,增加灌溉用水或工业用水,水权不宜新增,因水权增加势必影响到其他已授予水权用户的权益以及影响生态水权的保障。借鉴东阳市和义乌市的水权实践,考虑塔里木河流域干旱缺水的特点,新增用水只能通过引进高新技术节水或从已有水权的用户手中转让得到,不仅有利于生态水权的稳定,而且可以激励各用水户节约用水、提高用水效率。

4.3 科学制定定额与适时水权

水行政主管部门要积极开展工作,进行广泛调查和研究,确定行业用水定额,同时各单位根据自身实际情况和所分配的水权确定合适的定额。具体实施初始水权方案时,并非完全一成不变,根据实际需要,可以在允许的范围内作适当的调整,包括机动水的最终使用,从而形成最终的水权分配,即适时水权。

4.4 完善法规,加强各方水权的保护力度

根椐新政函[2003]203号文的精神,各流域的初始水权得以确定,由于受利益的驱动,挤占生态用水的时有发生,塔里木河下游生态环境的修复没有保障,因此有赖于法规的完善,包括人口增长的适度控制、开垦荒地的规定、生态水权落实的保障、天然植被的保护等。通过法律法规的完善,切实做到有法可依、违法必究,打击非法垦荒破坏胡杨林与其他天然植被的行径,约束生产用水挤占生态用水,制止不符合可持续发展原则的经济增长方式。

4.5 利用水价经济杠杆,优化水资源配置

在塔里木河流域水资源短缺的情况下,通过调整水价来配置水资源,是依靠经济杠杆,通过调整人们的经济利益关系来自觉地调整用水数量和结构,实现水资源优化配置。商品水价包括三个部分,即资源水价、工程水价和环境水价。不同的用水户,在不同地区的不同时间,使用不同水源的不同量的水,其资源水价是不同的。根据水资源和经济社会发展情况主动调整资源水价,就能引导人们自觉调整用水结构和数量,实现水资源的优化配置。

4.6 加强统一管理,提高水资源利用效率,实现水资源的可持续利用

实施水权管理的前提是水资源统一管理。通过强化水资源管理,不断提高水资源使用效率,促进经济社会可持续发展。即使有了指标体系、各项制度,没有统一、权威的管理作保证,这一切都是形同虚设。水资源统一管理关系到水资源的合理、高效、可持续利用,是贯彻水权管理、建设节水型社会的体制保证,是水利发展到今天提出的一道时代命题。这种统一

管理应该包括流域管理和区域管理两个方面，当前主要是要确定流域水资源统一管理和城市水务体制两个问题。

5 结语

塔里木河流域地处内陆深处的干旱荒漠植被区，水资源的短缺已成为制约流域社会经济和生态环境可持续发展的主要因素。通过树立水权意识，使人们从片面追求经济增长转变为人与生态环境和谐发展。通过明晰水权，给每个用水户以法律形式确定一定的用水量，约束各自限额用水，保障了生态用水的稳定。通过完善法律法规，加强生态环境保护力度，制止违法垦荒、破坏生态等行径，促进塔里木河流域生态环境的良性发展。

参考文献

[1] 宋郁东，樊自立，雷志栋，等. 中国塔里木河水资源与生态问题研究[M]. 乌鲁木齐：新疆人民出版社，2000.

[2] 胡军华，唐德善. 塔里木河流域适时水权管理研究[J]. 人民长江，2006，37(11)：73-75.

[3] 朱一中，夏军. 论水权的性质及构成[J]. 地理科学进展，2006，25(1)：16-23.

[4] Jan Crouter. A Water Bank Game with Fishy Extemalities[J]. Review of Agricultural Economies，2002，25(1)：246-258.

【作者简介】 陈小强，1975 年 9 月生，2010 年 7 月毕业于新疆农业大学，工学硕士，工程师，现工作于新疆塔里木河流域管理局。

浅论地理信息数据在水利普查中的应用

杜金华[1] 夏 丽[2] 王文波[2]

(1. 赤峰市水利局;2. 赤峰市喀喇沁旗水利局 内蒙古 014400)

【摘要】 水利普查是一项重大的国情国力调查。地理信息数据的应用,对于提高水利普查工作效率和普查质量、完善水利普查信息、保障普查成果高效管理有重要作用。本文通过分析地理信息数据在国内外统计、大型普查中的应用情况,借鉴、运用成功经验,提出以地理信息数据为基础,更好地应用服务于水利普查、建立水利基础地理信息数据平台、发布水资源地图的应用设想,从而实现水利普查数据的空间化、增强普查方法的科学性、提高普查成果使用的方便性,为国家经济社会发展提供可靠的基础水信息。

【关键词】 水利普查 地理信息数据 应用

开展全国水利普查是为了查清我国江河湖泊基本情况,掌握水资源开发、利用和保护现状,摸清经济社会发展对水资源的需求,了解水利行业能力建设状况,建立国家基础水信息平台,为国家经济社会发展提供可靠的基础水信息支撑和保障。因此,在普查开展和普查成果应用中,对准确、有效的基础地理信息有着极为严格的要求。目前,测绘部门掌握的国家基础地理信息是重要的基础性、战略性信息资源,具有精度高、内容丰富等特点,是国民经济和社会信息化的基础平台与重要支撑,对完善水利普查信息、提高普查质量具有重要意义。我国目前已建成国家基础地理信息系统 1∶100 万数据库、1∶25 万数据库、1∶5 万数据库等,多数省(自治区、直辖市)已完成或正在建设本地区 1∶1 万基础地理信息数据库。国家测绘局多年来不断开展的航空摄影测量和多轨道卫星遥感资料的收集积累,为更新地理信息数据和各部门的应用奠定了基础。因此,在全国水利普查中充分应用地理信息数据,能够推动测绘成果社会化应用,更好地服务于水利普查和水利建设。

1 地理信息数据在国内外统计工作中的应用分析

1.1 国外应用情况

社会经济统计信息与地理信息的结合已不是新生事物。国际上发达国家和一部分发展中国家,在开展社会经济、人口等普查时,普遍采用地理信息、地图为辅助手段,在调查阶段和数据应用、发布阶段,多种统计专题地图发挥了直观、准确、方便理解的作用。如美国人口调查局的 TIGER 地图服务系统和 FACTFINDER 网站,TIGER 在基础线划地图上,与 1990 年、2000 年人口普查数据结合,提供相关产品和服务,并从 1995 开始提供网上地图查询;FACTFINDER 作为新的网上数据分发和制图系统,为公众提供了一种快速获取统计数据的特殊工具,使用户能够快速检索到大量的并且不断更新的人口统计数据和经济数据,这些数据包括了1亿个美国家庭和超过2 000万个商业机构。用户使用FACTFINDER,只需要简单的操作,就可以获得他们所需的在指定地理位置的准确数据,同时产生客户化地图,为用户形象地展示某一特殊区域的统计数据,极大地提高了公众获取政府数据的效率。英国国家统计局除向公众提供统计数据外,也提供打印输出统计地图的服务,并通过在线的NESS系统,提供数据查询和地图、图标显示、输出。英国国家统计局下设国家统计地理部,负责跨部门间的统计数据收集、处理、存储和集成,在网上提供PDF地图。加拿大政府统计部除发布统计数据外,还是在线的加拿大国家地图集的参与者,该地图集网站集成大量的统计地图;2001年的加拿大人口普查成果也通过专门网站发布数据和地图。从发达国家的经验可以看出,地理信息和相关技术已经在各国统计工作中得到了广泛应用。

1.2 国内应用情况

我国地理信息数据在统计部门的应用也有多年,在国家863计划课题"国家社会经济统计地理信息系统建设"支持下,以国家测绘局提供的全国1∶100万基础地理数据为平台,开发了国家社会经济统计地理信息系统,为我国统计部门进行数据整合和数据分析与应用提供了新的模式。但目前我国在开展常规的统计工作和小型的调查时,仍不能有效结合应用最新的地理信息数据,只是初步应用在人口、经济、农业等大型普查中。如2006年的第二次全国农业普查、2008年的第二次全国经济普查、2010年的第六次全国人口普查,虽然在划分普查区和绘制地图时基本实现了普查区划分与绘图电子化,但仍存在某些方面的不足:一是普查区的划分和绘图尚未统一。以往普查划区绘图常借助CAD类软件裁切出行政区边界,每个行政区再按各自的标准划分普查区,最后在普查区边界范围内绘制该区地图,这势必要造成普查区间的地域交叉、重叠和遗漏现象。二是普查区地图管理尚不规范。以往普查区地图的管理仅处于文件管理阶段,有的地方甚至处于纸质的人工管理阶段,这使原有的普查区地图在下次普查中不能重复使用,每次普查都得重起炉灶,造成了原有资源的浪费,而且无法直观地通过普查区地图反映两次或多次普查的不同变化。三是划区绘图成果应用不充分。普查区地图本应是统计地理信息服务工作的基石,目前却未能在推进统计信息服务工作中发挥它应有的作用。

2011年开展的第一次全国水利普查,积极尝试应用了最新的地理信息数据。国家测绘局向水利部门提供了1∶5万数据更新工程、西部测图工程、灾后重建测绘工程等国家重点测绘项目形成的最新1∶5万基础地理信息数据。新版1∶5万基础地理信息数据现势性强、内容丰富、精度高,包含了很多水利基础信息,如河流结构线、水流方向、堤防、水闸线、涵洞等。这是新版1∶5万数据首次在国家重大工程中的大规模全面应用。除1∶5万基础地理信息数据外,全国从中央到地方各级测绘部门还开通了地理数据提供的"绿色通道",快速提供部分1∶1万数据。水利普查以此为基础,将大幅度地减小普查的工作量,彻底改变传统作业方式,显著提高工作效率。但此次水利普查对地理信息数据的应用,也处于起步阶段,还需进一步的深化应用。

综合国内外地理信息数据应用结果,可以看出国外统计、普查工作对地理信息数据的应用已相对成熟,但国内的应用还不够充分、全面。地理信息不仅是用地图符号和图形对统计数据进行诠释,而且通过地理信息系统和地图语言等,将这些对象数据和相应的地图符号所构成的空间实体,在时空中的演化规律表现出来,使枯燥的统计数字、表格可以在时间、空间内组合,同时还具有定量、定位和可测量的特性。此外,用户通过地理信息系统交互查询、展示在地图上的直接信息和间接信息,从要素分布、相互关系以及所处的地理环境等多角度分析,还可以获得新的知识。因此,水利普查在建立和完善水利普查地理信息系统时,应全面应用现有的地理信息数据,积极开展水利普查空间数据处理,更好地开发应用水利普查成果,拓展地理信息数据在水利日常业务中的应用范围和深度。

2 地理信息数据在水利普查工作中的应用设想

此次开展的全国水利普查,根据《第一次全国水利普查实施方案》确定的技术路线,基于高分辨率遥感影像数据和最新的国家基础地理信息相结合,编制水利普查影像工作底图。利用普查工作底图对大部分水利普查对象进行图上识别和复核,不仅可以大大减小普查外业工作量,节省人力、物力投入,而且可以显著提高水利普查工作效率,为确保普查对象不重不漏目标的实现,有效提高普查数据质量奠定了良好的基础。下一步,我们应在更好地开展水利普查、建立水利基础地理信息数据平台和发布水资源地图等方面,全面应用地理信息数据,从而实现水利普查数据的空间化、增强普查方法的科学性、提高普查成果使用的方便性。

2.1 全面应用地理信息数据服务于全国水利普查

全国水利普查在采集普查对象信息要素的同时,还应从空间定量采集普查对象对资源、环境的占有量,这样做既符合中央提出的科学发展观,又有利于促进人口、资源、环境的协调发展。因此,利用已有的普查工作底图,在普查中将基本单位位置、占有资源范围、数量等标注在图上;普查后,由国土资源、测绘、建设、环保、统计等部门协助水利部门汇总、整理数据,制作水资源地图。

此外,水利普查要充分吸收测绘部门及具有测绘经验和技术的人员参与。测绘部门和人员参与并借助测绘手段开展工作,有利于快速、准确划定普查区。如通过遥感图像分析普查对象的规模、特点等影像特征信息,总结不同水利普查对象的纹理及其空间变化规律,充分考虑经济条件、统计单位分布、行业特点,人口、地形、地貌、自然状况等诸多因素,借助GIS空间分析方法和数据库技术,设计水利普查分区,辅助制订水利普查方案,研究设计切实可行、行之有效的普查路线等,从而保证普查区域的完整覆盖和地域上的不重不漏,确保以合理的精度制作普查区地图、建立数据库、提供普查员使用的普查区图、提高工作效率等。再如,基于GPS和PDA的普查数据移动采集终端,进行现场数据采集,包括建筑物、道路、控制点、地名地址等普查图要素,将采集数据与统计基础地理信息平台数据关联,更新、标注普查区图。同时也可以进行基本单位清查,录入基本单位和普查表的信息,并建立水利数据库,与普查地理单元建立关联。最终采集的信息可以传到普查GIS系统进行汇总和拼接。

2.2 建立水利基础地理信息数据平台

水利部门的各类普查数据主要是按照普查区开展调查,逐层逐级汇总、上报后进行成果发布。为此,建立统一的、标准化的、可持续更新的水利基础地理信息数据平台,作为各项水利统计或普查的基础数据之一,便于综合汇总、分析、协调和维护,减少重复性投入。在水利普查中,可以利用国家测绘局向水利部门提供的1∶5万基础地理信息数据库或更大比例尺的数据库数据,结合民政部门最新的行政区划、境界、地名资料,以及高分辨率航空、航天遥感资料等,借助GIS工具软件,加强数据采集、提交和汇总。同时,结合水利部门的实际需求,选择市、县、乡等地方,建立多尺度的水利基础地理信息背景数据库和多尺度的多种水利统计地理单元数据库,作为水利统计基础地理信息数据平台。该数据平台应该是多尺度、多时态的,可与多年度、多种统计指标相关联，支持统计信息的空间集成、分析与表达。该数据平台要素既包括基础地理信息基本要素，如地名、河流结构线、水流方向、堤防、水闸线、涵洞等，也包括省级至乡镇级等各级政区，还包括水利部门按照普查标准划分的统计地理单元等。

2.3 发布水资源地图

统计部门定期发布的统计数据通常是行政区划为单元的表格、数字，这些数据还有一种更加直观的表达方式——统计地图。水利部门可以汇合统计部门，发布水资源地图。近年国家测绘局的有关部门进行了尝试，开发了“国家动态地图网”。该网站利用 Web GIS 技术和网络技术、计算机技术和网络信息集成等相关技术，对各类与地理信息相关的国家权威部门发布或其他方式共享的社会、经济、自然等专题统计信息，以及与人民生活密切相关的空间定位型信息，实现在线或非在线发现、获取及链接。同时，对这些信息进行整理、组织和集成，开发网上 GIS 功能，建立长期稳定运行的“国家动态地图网”网站，在基础地理信息公共平台（1∶400 万及 1∶100 万比例尺数据库）上集成各类专题信息，通过万维网地理信息技术对这些数据和信息进行综合分析与利用，以多网页分组、多种表现形式的专题地图呈现给用户，这些地图随专题信息的更新而动态更新。国家动态地图网的大部分数据资源就是来自国家统计局网站，用户在阅读、了解统计数据的同时，可以使用该数据，并结合多年的数据或自己的数据、计算模型等，使用网站提供的软件工具和基础地理地图数据，制作自己需要的统计地图等。

这样的设计思想，可以应用于水利普查的数据发布，用户除了使用普查数据还有地图工具辅助，可以阅读以地图形式发布的水利普查资料。通过不同的水利专题地图，更直观、及时地展示水利信息，比较普查资料时间序列的发展变化，为各级领导提供基于电子地图的及可视化的全方位、多层面、多角度的水利数据信息，也可以为各部门、企事业单位和社会公众提供水利信息共享平台。

3 结语

实现水利普查资料的应用和转化是水利普查工作的关键，现代地理信息数据给水利普查工作在资料开发和应用方面提供了技术支持。通过地理信息数据在水利普查中的应用分析，可以看出地理空间信息和技术的应用对实现水利普查数据的空间化、增强水利普查方法的科学性以及成果使用的方便性等十分必要。水利普查成果既要包含丰富翔实的数据资料，真实反映国家改革开放与现代化建设的成果和综合国力，也要开发党政领导和社会公众关心需要的可视化、网络化创新产品，利用地理信息数据实现水利普查资料的汇集、分析、展示，向用户提供更深层的信息，拓展地理信息数据在水利日常业务中的应用范围和深度，从而为国家经济社会发展提供可靠的基础水信息支撑和保障。

参考文献

[1] 国务院第一次全国水利普查领导小组办公室. 灌区专项普查[M]. 北京：中国水利水电出版社，2010.

[2] 国务院第一次全国水利普查领导小组办公室. 水利普查空间数据采集与处理[M]. 北京：中国水利水电出版社，2011.

【作者简介】 杜金华，女，1985 年 12 月生，本科，2009 年 1 月毕业于北京工业大学。现工作于内蒙古赤峰市水利局计划财务科，助理工程师。

滦河下游区域水资源承载能力分析

仇新征　闻立芸

（水利部海委引滦工程管理局　河北　064309）

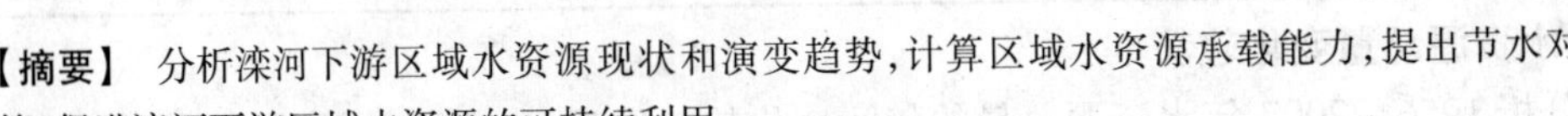

【摘要】 分析滦河下游区域水资源现状和演变趋势，计算区域水资源承载能力，提出节水对策，促进滦河下游区域水资源的可持续利用。

【关键词】 滦河下游　水资源承载力　对策　可持续利用

滦河下游区域主要包括唐山、秦皇岛两市，是京津唐经济区内的重要水源地。自1999年以来，滦河流域持续性干旱，滦河水资源经济区遭遇了空前的水源危机。滦河流域生态环境严重失衡，下游一些常年有水河流逐步向季节性河流转变，水资源量逐步减少的趋势也非常明显。水资源严重短缺已成为了长期制约滦河下游区域水生态环境与社会经济发展的瓶颈。提高滦河下游水资源承载能力，促进流域水资源可持续利用已成为我们当前需要解决的最紧要问题。

1　流域概况

滦河发源于河北省丰宁县坝上骆驼沟乡小梁山南坡大古道沟，流经内蒙古、辽宁、河北的27个市（县、区），于河北省乐亭县兜网铺注入渤海，全长888 km，流域面积44 750 km^2，其中山区占98%，平原占2%；按行政区划，河北省占总面积的81%，内蒙古自治区占15.5%，辽宁省占3.5%。

滦河下游平原河段地质区属河北拗陷冀东部分，地貌呈平原类型，由山前平原向东南逐步变为滦河三角洲平原，地面高程变化1～20 m，河道纵坡约为0.66‰，河漫滩由水流堆积作用形成，组成物质较为松散。河道呈平原游荡性河道性状，自京山铁路桥以下逐渐变宽。

随着潘家口水库、大黑汀水库、桃林口水库三大水源工程和引滦入津、引滦入唐、引青济秦等跨流域引水工程的建成运行，形成了以滦河流域为母体，辐射天津、唐山、秦皇岛三座城市的滦河水资源经济区，滦河又成为了影响京津冀区域社会经济发展的重要水源地。

2　流域水资源及演变趋势

2.1　水资源总量

根据1956～2007年水文系列资料评价成果，滦河流域多年平均降水量519.2 mm，地表水资源量39.49亿 m^3，地下水资源量19.40亿 m^3，水资源总量43.71亿 m^3，见表1。

按2007年人口计算，滦河流域人均水资源占有量为855 m^3，相当于全国平均水平的39.8%；亩均水资源占有量662 m^3，相当于全国平均水平的42.5%。人均水资源量低于国际公认的人均 1 000 m³ 紧缺标准，属重度缺水地区。

表 1　滦河流域水资源评价成果表

水资源分区		降水量（mm）	地表水资源量（亿 m³）	地下水资源量（亿 m³）	水资源总量（亿 m³）
山区	潘家口以上	476.7	21.62	11.61	23.64
	潘大区间	727.7	3.50	1.22	3.63
	桃林口以上	614.7	7.15	2.90	7.46
	大桃滦区间	668.6	6.40	2.41	7.49
	山区小计	518.0	38.67	18.14	42.22
平原区		585.0	0.82	1.26	1.49
全流域		519.2	39.49	19.40	43.71

2.2　水资源总量变化趋势

分析 1956～2007 年水资源总量的年际变化趋势发现，水资源总量的年际变化趋势与降水量、地表水资源量的年际变化趋势基本一致，但水资源总量的趋势性减少倾向比较明显，见图 1、图 2。

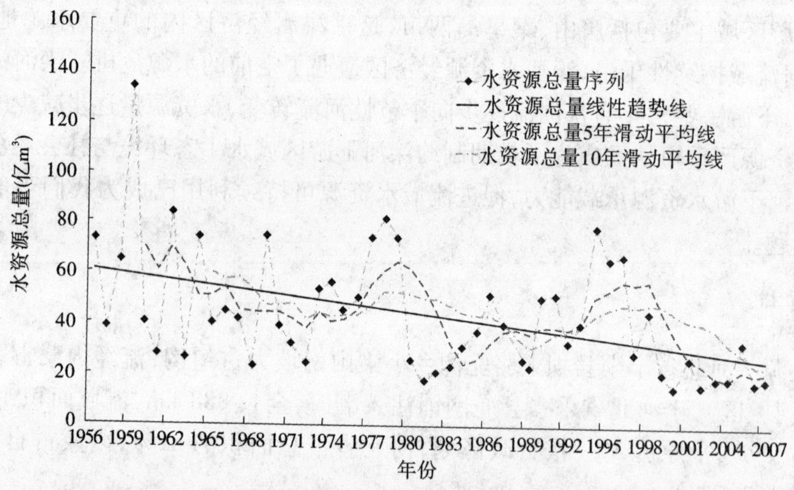

图 1　滦河流域水资源总量变化趋势图

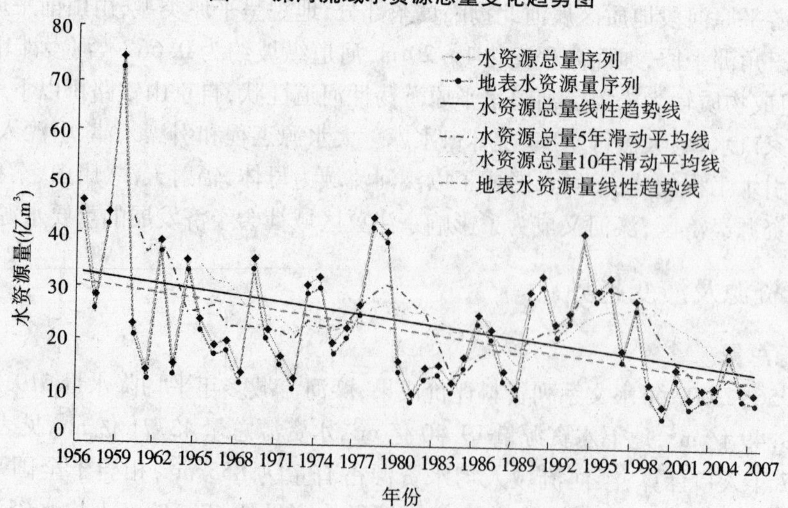

图 2　潘家口以上水资源总量变化趋势图

由于潘家口以上流域面积占滦河流域面积的75%，滦河流域水资源总量的年际变化趋势与潘家口以上山区的年际变化趋势基本一致，而潘家口以上山区水资源总量的年际变化与其地表水资源量的变化趋势一致，均呈明显的减少趋势。从趋势变化看，滦河流域水资源总量的减少趋势大于潘家口以上山区。

3 区域水资源承载能力分析

3.1 水资源承载能力概述

3.1.1 水资源承载能力的含义

滦河流域的水资源承载力可定义为：在某一社会经济发展水平下，以可预见的技术发展水平为依据，通过水资源的可持续利用，某地区的水资源能够持续支持区域经济发展规模和保证一定物质生活水平条件下的人口数量的最大能力。

3.1.2 评价指标

水资源承载力一般有两个分析评价指标：

相对承载指数(*RCI*)指现在或将来某一阶段的实际人口量与水资源承载人口的比值

=实际人口量(现在或将来某一阶段)/水资源承载人口

绝对承载指数(*ACI*)指现在或将来某一阶段的实际人口量与水资源承载人口的差值

=实际人口量(现在或将来某一阶段)-水资源承载人口

本文选取相对承载指数*RCI*来反映滦河流域各地区水资源的承载力状况。

3.1.3 水资源承载力分类系统

水资源承载力可分为基本型和亚基本型，具体标准见表2。

表2 水资源承载力的分类系统及划分标准

基本型	亚基本型	划分标准	相应人水关系
超载(Ⅰ)	强(ⅠA)	$RCI \geq 2$	人水关系是不可持续发展
	中(ⅠB)	$1.5 \leq RCI < 2$	
	弱(ⅠC)	$1 < RCI < 1.5$	
临界(Ⅱ)		$RCI = 1$	人水关系处于潜在危险中
缺载(Ⅲ)	强(ⅢA)	$RCI \leq 0.5$	人水关系呈可持续发展状态
	中(ⅢB)	$0.5 < RCI < 2/3$	
	弱(ⅢC)	$2/3 \leq RCI < 1$	

3.2 区域水资源承载能力评价

参照《滦河流域需水量预测报告》中的滦河流域不同水平、各地区的社会人口规模和经济规模(GDP)，计算出不同水平年、各地区的人均GDP，具体结果见表3。

根据流域现状发展水平下的预测人口规模和水资源可承载的人口规模，计算上述方案下流域水资源相对承载指数及其相对应的水资源承载类型，见表4。

滦河下游区域水资源承载力类型从整体上来看，唐山、秦皇岛两市在不同水平年下的国民经济发展规模和人口数量，大多超过了相应社会经济发展水平下的水资源承载能力，处于弱超载状态或临界状态。

由此可以看出，滦河下游区域的社会经济发展状况是当地水资源所不能承载的，不利于社会的可持续发展，应采取一定的措施来增加水资源的可承载能力。

表3 滦河下游区域预测的各水平年社会发展规模

水平年	地区	预测人口（万人）	预测 GDP（亿元）	预测人均 GDP（万元）	总需水（亿 m^3）	总供水（亿 m^3）	承载最大 GDP（亿元）	承载人均 GDP（亿元）	预测人均 GDP（亿元）	承载人口规模（万人）
2010	唐山	755	3 840.63	5.09	39.11	38.06	3 737.52	4.95	5.09	734.73
	秦皇岛	289.25	766.19	2.65	10.98	10.86	757.82	2.62	2.65	286.09
	滦河下游	1 044.25	4 606.82	7.74	50.09	48.92	4 495.34	7.57	7.74	1 020.82
2020	唐山	799.69	9 924.96	12.41	44.17	42.34	9 513.76	11.90	12.41	766.56
	秦皇岛	307.43	1 900.46	6.18	11.88	11.62	1 858.87	6.05	6.18	300.70
	滦河下游	1 107.12	11 825.42	18.59	56.05	53.96	11 372.63	17.95	18.59	1 067.26
2030	唐山	842.57	18 045.52	21.42	48.06	45.64	17 136.86	20.34	21.42	800.14
	秦皇岛	324.66	3 366.07	10.37	12.63	12.23	3 259.46	10.04	10.37	314.38
	滦河下游	1 167.23	21 411.59	31.79	60.69	57.87	20 396.32	30.38	31.79	1 114.52

表4 滦河流域水资源相对承载指数、承载力类型、人口超载比例

水平年	地区	相对承载指数	承载力类型	人口超载比例(%)
2010	唐山	1.0	Ⅱ	2.76
	秦皇岛	1.0	Ⅱ	1.10
	滦河下游	1.0	Ⅱ	2.31
2020	唐山	1.0	Ⅱ	4.32
	秦皇岛	1.0	Ⅱ	2.24
	滦河下游	1.0	Ⅱ	3.47
2030	唐山	1.1	ⅠC	5.30
	秦皇岛	1.0	Ⅱ	3.27
	滦河下游	1.1	ⅠC	4.36

4 滦河下游区域水资源可持续利用对策

滦河下游区域是京津冀地区重要的工业基地、粮食和副食品生产基地，近年来区域社会经济呈现出快速发展的态势，特别是南部沿海地区工业园区迅速崛起，成为经济快速增长的驱动力。随着人口增长和经济社会的快速发展，对水资源的需求量急剧增加，水资源短缺已成为严重制约滦河流域经济发展的重要因素。经济社会的稳定发展需要水资源作支撑，实现水资源的可持续利用是新时期我国水利事业的重大使命，是加速建设资源节约型和环境友好型社会的必然选择。因此，制定水资源可持续利用的对策与措施，实现水资源的合理开发利用、节约保护和优化配置，对于促进滦河流域社会经济和生态环境的协调发展具有重要意义。

为满足未来水资源可持续利用的要求，滦河流域需要从水资源的承载能力、水资源优化

配置和水资源统一管理等方面加强研究,建立和完善水资源管理机制,强化节水,提高用水效率,调整产业结构,加大水利工程建设,提高供水能力,努力构建水资源供给、水生态环境安全、防洪减灾和综合管理4大保障体系。

参考文献

[1] 温立成,鞠玉梅.滦河水资源可持续开发利用初探[J].海河水利,2003(2):26-27.

[2] 张丽,董增川.流域水资源承载能力浅析[J].中国水利,2002(10):100-104.

[3] 王友贞,施国庆,王德胜.区域水资源承载力评价指标体系的研究[J].自然资源学报,2005(4):597-604.

[5] 全海娟,许佳君,陈昌仁.我国水资源承载能力评价研究进展初探[J].水利经济,2006(6):56-58.

【作者简介】 仇新征,1977年10月生,2001年毕业于河海大学。现任水利部海河水利委员会引滦工程管理局水情调度科副科长,工程师,主要从事水库调度工作。

长沙市缺水时期水资源应急调度的研究

陈春明[1] 杨小康[1] 易瑾瑜[1] 刘东润[2]

（1. 长沙水文水资源勘测局 长沙市 410014；
2. 湖南省水文水资源勘测局 长沙市 410007）

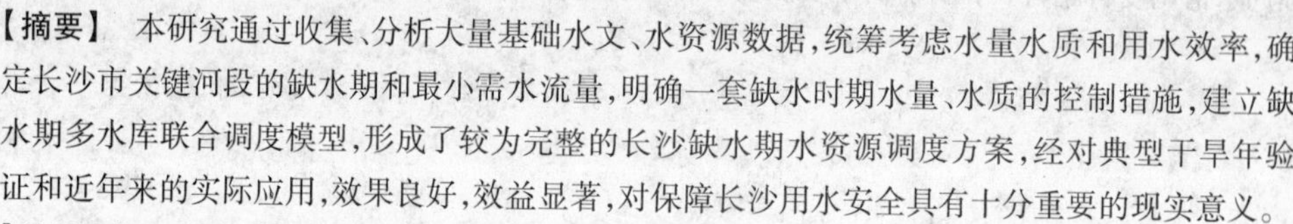

【摘要】 本研究通过收集、分析大量基础水文、水资源数据，统筹考虑水量水质和用水效率，确定长沙市关键河段的缺水期和最小需水流量，明确一套缺水时期水量、水质的控制措施，建立缺水期多水库联合调度模型，形成了较为完整的长沙缺水期水资源调度方案，经对典型干旱年验证和近年来的实际应用，效果良好，效益显著，对保障长沙用水安全具有十分重要的现实意义。

【关键词】 水资源 缺水期 应急调度

1 前言

长沙市位于湖南省东部偏北，湘江下游和长浏盆地西缘。全市土地面积 11 819.5 km^2，多年平均水资源总量为 96.19 亿 m^3，2008 年总用水量为 36.17 亿 m^3。从水资源总量来看，水资源能满足本地区的用水需求，但由于本地区水资源时空分布极不均匀，经济社会发展所需用水量不断加大，河流水环境质量恶化，缺水期水资源供需矛盾突出。

自 1998 年以来，长沙市遭遇了多年特大干旱，2003 年、2007 年出现了 50 年一遇的特大干旱。2008 年底至 2009 年初，长沙市出现了特大秋冬连旱，仅次于 1979 年，居历年第二位。2009 年底湘江长沙段刷新历史最低水位，橘子洲西面江水断流，游人可从河道中直接走上橘子洲。2008 年 10 月湘江橘子洲大桥见图 1。

图 1 2008 年 10 月湘江橘子洲大桥

多年特大干旱和湘江长沙水位数次刷新历史最低，严重影响了居民生活和工农业生产用水，使长沙用水安全受到严重威胁，省市政府对此高度重视，多次召开专门会议，研究对策，强调科学配置和调度水资源，尤其是要迫切解决缺水期水资源应急调度的问题。为防范缺水期水危机的发生，保证水资源可持续利用，因此对缺水期水资源进行合理配置和控制调度十分必要。

长沙市河流大多属湘江流域，较大的一级支流有7条，浏阳河、捞刀河、沩水河为长沙市境内三大河流，总流域面积为9 638 km^2，占全市面积的81.5%。本次研究的范围确定为湘江长沙段、浏阳河、捞刀河、沩水河。

2 河流最小需水流量的研究和确定

2.1 河流最小需水流量的定义

河段最小需水流量是指同时满足河段河道内和河道外各用水户不同用水保证率和水质目标要求的流量，它包括河道内用水量和河道外用水量。河道外用水量是指沿河两岸一定范围内的城镇及农村生活、生产、生态等用水量，河道内用水量是指生态、环境、航运等用水量。

2.2 河流最小需水流量的确定

2.2.1 河道外需水量计算

河道外需水量按生活需水、生态需水、工业需水及农业需水4部分计算。

2.2.2 河道内需水量计算

(1)生态需水量计算。目前，生态需水量的计算方法大体可分为四类，即水文学法、水力学法、栖息地法和综合法。以湿周法为代表的水力学法，对数据量要求不大，数据容易获得，而且不受人类活动对河道径流影响的限制，所以本文对已经被人类活动干扰的湘江、浏阳河、捞刀河和沩水河采用湿周法进行生态需水量估算。

(2)环境流量计算。环境流量计算主要针对所有的开发利用区和需水改善水质保护区、保留区和缓冲区进行的。对保护区、保留区和部分水质较好、用水矛盾不突出的缓冲区，其水质目标原则上是维持现状水质，故不需要环境流量。

①现状条件下的环境流量计算。代表现状条件的环境流量是采用现状纳污能力计算中的流速、水质降解系数等参数，并采用水资源综合规划的污染物入河量统计成果，利用一维模型进行计算。

②规划水平年的环境流量计算。对于入河量不大于其纳污能力的，原则上维持现状水质不变；对于入河量大于其纳污能力的，考虑规划水平年污染物的消减量，对河段进行环境流量计算。根据水资源综合规划要求，对污染物入河量采取分阶段治理方案，2020年将污染物入河控制量消减量的70%实施消减，然后根据消减后的入河排污量进行河段环境量计算。

2.3 河道最小需水流量计算成果

通过综合考虑河道外用水量和河道内用水量，计算求得代表河段最小需水流量成果见表1。

3 缺水期水资源调度的研究

本文所指的缺水期，既区别于通常所指的对应于汛期的枯水期，也不同于“无雨或小雨，造成空气干燥、土壤缺水”的气象干旱期，它是受干旱气象变化机制与土地、植被、大气系统水分散失机理影响以及人类活动等综合因素作用下,产生天然水资源不能满足生产、生活、生态环境综合用水需求(供水紧缺)所对应的时间。

表1 长沙市代表河段2020年最小需水流量成果

(单位:m^3/s)

河名	主要河段		河道外用水量	回归水量	河道内用水量	计算最小流量	采用最小流量
	起点	终点					
湘江	易家湾	长沙枢纽	43.54	34.76	593	602	600
浏阳河	双江口站以上		3.16	2.45	6.25	6.96	7.0
	双江口站	朗梨站	6.33	4.97	17.5	18.86	19.0
	朗梨站	河口			19.3	19.3	19.3
捞刀河	罗汉庄站以上		2.47	1.94	5.08	5.61	5.60
	罗汉庄站	河口			5.2	5.2	5.60
沩水河	黄材坝以上		0.57	0.44	0.50	0.63	0.63
	黄材坝	宁乡站	3.94	3.11	4.30	5.13	5.10
	宁乡站	河口	1.78	1.41	16.2	16.57	16.6

3.1 缺水期水资源调度原则与启动条件

缺水期水资源调度是以生活用水、生产用水、生态环境需水为对象,以水量、水质兼容的特定时空范围的水资源调度。它包括限污、节水和工程调水等措施。水库作为河流上的控制性工程,利用水库水资源进行统一调度,进而构建下游补水的长效机制,这已成为目前迫切需要解决的问题。

3.1.1 调度原则

根据长沙市水资源的特点,将长沙市水资源的调度分为正常调度和应急调度两部分。

(1)正常调度原则:贯彻"限污、节水、水工程调水相结合"的原则。

(2)应急调度原则:在正常调度原则基础上,贯彻"三先三后"原则,即"先生活后生产生态、先地表后地下、先节水后调引水"原则进行应急供水。必要时对城镇生活用水实行限时限量和分段片轮流供水、对第三产业(如洗车、娱乐性服务业等)原则上实行停供,对工业部门和灌溉用水实行限供、停供,确保用水安全。

根据以上确定的调度原则,长沙市主要河流河段满足最小需水流量的调度可归纳为:一是沿河两岸限制入河排污量,以减少河道内生态环境需水量;二是沿河供水区节约用水,以减少河道外的需水量;三是通过水库调水,调节河段最小需水量。

3.1.2 调度启动条件与方法

水资源的衡量标准包括水量和水质两方面,缺水期水资源调度也包括水量与水质的双重调度。当河道来水量(上断面流量)达到预警流量(1.1倍断面最小需水流量)时,及时预警,根据上游来水过程情况和气象、水文短期预报,进行综合分析判断,若预测在未来10 d内河道流量将维持在预警流量以下时,宣布进入"缺水期水资源调度状态",开始执行调度。

(1)启动条件。

当控制断面来水量等于预警流量时,根据气象水文短期预报分析和水质监测情况,预测在未来10 d内的流量都不会大于预警流量或水质不会优于Ⅲ类的条件下则开始预警。

(2)启动方法。

根据上述启动条件,缺水期水资源调度的初始启动方法则是:

当来水量等于预警流量,水质优于Ⅲ类,则发布预警,暂不启动。

当来水量等于预警流量,水质劣于Ⅲ类,则启动调度。

当来水量等于最小需水流量或水质劣于Ⅲ类,则启动调度。

一经启动调度,首先发限污指令(包括限污量和时间),若情况进一步恶化,将限污减水一并实施,同时考虑一定的传播时间后,提前做好水库电站的调度准备,实时启动水库调度程序。

3.2 缺水期水库调度模型的建立

缺水期水资源调度研究则是当前水资源调度的新课题。通过对研究区域进行概化,建立网络图(见图2)和数据整理分析,通过线段把水库、汇流、供水等结点联系起来。然后建立模拟模型,把系统中的各种未知量作为决策变量,并将其复杂关系表示成决策变量之间的约束方程组,将水资源调度目标表示成线性目标函数,进行模型模拟,并对约束函数进行调整,求出最优解。

通过系统概化构建水库群联合调度模型,优化成员水库供水任务分配因子,并结合供水水库群常规调度规则,实现共同供水任务在水库间的优化分配。采用改进粒子群算法(NSPSO)对湘江干流、支流水库群联合供水调度模型决策变量进行多目标优化,分析联合供水调度过程中目标之间的竞争关系,检验联合调度规则的合理性与有效性,在此基础上,形成较为完整的长沙缺水期水资源调度方案。

3.3 缺水期水资源调度方案

根据不同缺水性质,长沙市缺水期水资源调度可分为3个方案:

方案一:当出现以水质性缺水为主时,其调度以限制排污为主,水库调水和限制用水配合。

方案二:当出现水量缺水为主时,其调度以水库为主,限制排污和限制用水相结合。

方案三:当出现综合性缺水时,其调度以水库调水和限制排污并举,限制用水配合。

方案一的调度主要由环保部门根据排污情况去实施。方案二和方案三都要通过水库调度解决。对于水库调度,通过求解各河段缺水量,选择调水水库,提出各流域的缺水期水资源调度方案。

3.4 缺水期调度计算

湘江的调度期选择9月至次年1月,其他河流调度期选择7月至次年1月,进行调度计算。缺水期水资源调度应按洪水与水库特性,根据缺水期水资源调度原则,在必须保证生活用水的前提下,兼顾争取综合效益最大化,对水库进行缺水期水资源配置调度,如果在汛期中出现了缺水期(控制断面流量小于最小流量)的时段,也按调度程序启动水库调水。

根据缺水期多水库联合调度模型,利用坝址(断面)及各区间径流作为各结点输入条件,通过模型分解协调迭代计算,得到不同干旱年份水资源调度成果和不同缺水期水资源缺水情况,2007年(50年一遇的特大干旱)长沙市主要河流各控制河段枯水期调度成果见表2。

4 结语

本文根据长沙市经济发展需要及缺水期水资源供需矛盾突出实际,研究确定长沙市缺水期、河段及控制断面最小需水流量，运用缺水期多水库联合调度模型对长沙市主要河流重要河段缺水期进行应急调度计算。当遭遇特别干旱期，长沙市主要河流和主要水库的供水资源严重不足，不能满足工农业生产的需求，通过水库和全市水资源的合理调度可减少缺水量和缩短缺水周期。该研究成果在近年长沙市缺水期水资源调度、保障用水安全等方面得到了应用，为相关部门和政府决策提供了技术支撑和科学依据，取得了良好的社会效益和经济效益。

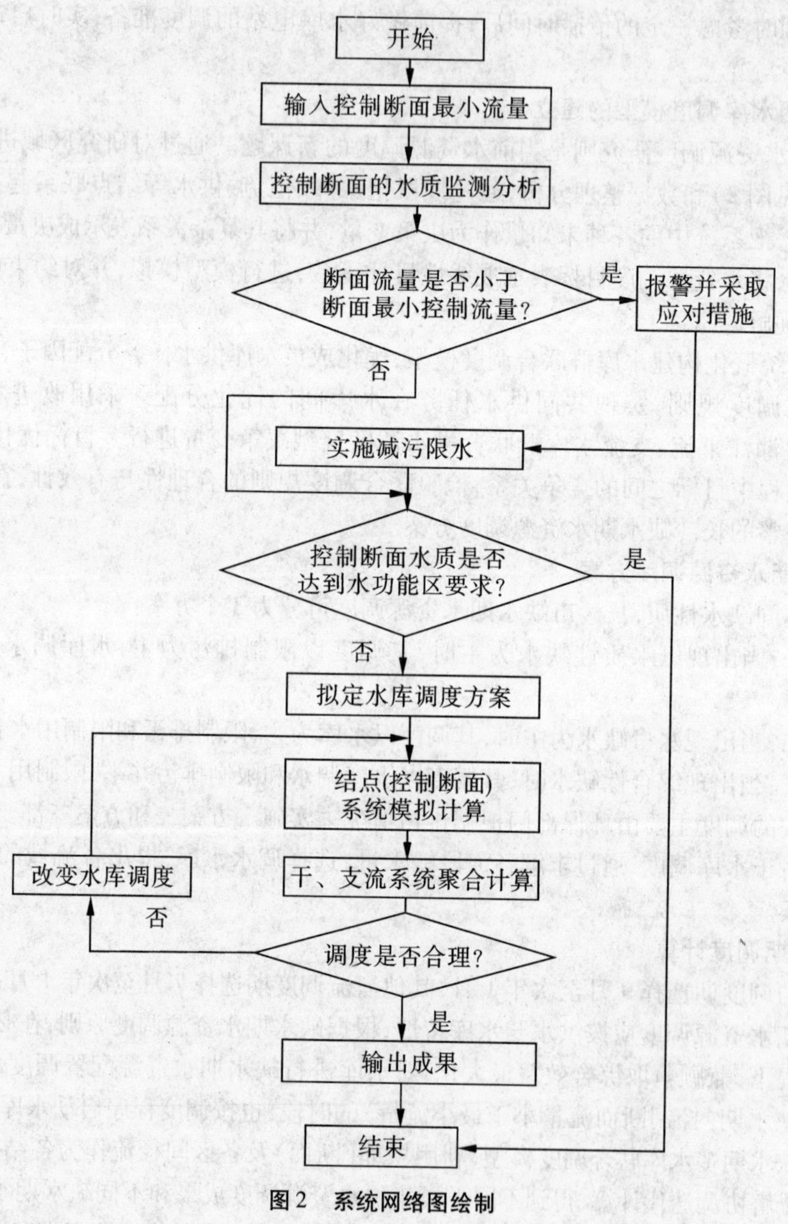

图2 系统网络图绘制

表2 2007年(50年一遇的特大干旱)长沙市主要河流各控制河段枯水期调度成果

河流	主要河段			代表性断面									
	起点	止点	河长(km)	水文站	集水面积(km^2)	断面采用最小控制流量(m^3/s) 限污后流量	断面采用最小控制流量(m^3/s) 保证率(%)	缺水量/缺水天数(万m^3/d)	97%保证率时断面流量(m^3/s)	调水量(万m^3)	各控制断面以上考虑参与应急调度的水库及调水量(万m^3)	缺水调度措施	调度后河段的缺水解决情况
湘江	易家湾	长沙航电枢纽	62	湘潭	81 638	517	87.2	95 000/114	331	95 000	东江(82 000) 大源渡(6 700) 株洲航电(6 300)	改变水库运行方式	解决、不缺水
浏阳河	双江口站以上		192.9										
	双江口站	朗梨站	110.9	双江口	2 067	11.6	85.9	544/34	7.41	>544	株树桥(550)	改变水库运行方式	解决、不缺水
	朗梨站	河口	15.1	朗梨	3 815	18.8	83.8	891/54	9.70	891	株树桥(550) 官庄(347)	改变水库运行方式	解决、不缺水
捞刀河	罗汉庄站以上		61.1										
	罗汉庄站	河口	11.5	罗汉庄	2 468	5.2	88	1 419/94	4.8	418	马尾皂(360) 关山(58)	改变水库运行方式	部分解决
沩水河	黄材坝以上		36.0										
	黄材坝	宁乡站	57.5	黄材	240.8	4.3	48.4	1 654/96	0.86	769	黄材(710) 田坪(69)	改变水库运行方式	部分解决
	宁乡站	河口	50.5	宁乡站	2 289	13.6	55.7	3 210/107	4.7	769	黄材(710) 田坪(69)	改变水库运行方式	部分解决

参考文献

[1] 杨志峰,等.生态环境需水量理论、方法与实践[M].北京:科学出版社,2003.

[2] 杨志峰,等.流域生态需水规律[M].北京:科学出版社,2006.

[3] 汪恕诚.资源水利—人与自然和谐相处[M].北京:中国水利水电出版社,2003.

[4] 冯尚友.水资源持续利用与管理导论[M].北京:科学出版社,2000.

[5] 姚荣.基于可持续发展的区域水资源合理配置研究[D].南京:河海大学,2005.

【作者简介】 陈春明,男,1979 年 10 月生,本科,2007 年毕业于河海大学,现工作于长沙水文水资源勘测局并担任水情科科长,曾获“全国水利技术能手”、“湖南省技术能手称号”、长沙市“青年岗位能手”等荣誉称号。

“ET 管理”与水资源可持续利用

徐 磊

（中国灌溉排水发展中心 北京 100054）

【摘要】 本文基于 ET 控制的水资源管理理论与实践，进一步分析了基于 ET 控制的水资源管理新理念及其基本计算方法，提出了 ET 管理的含义与基于 ET 水资源可持续利用的管理模式，指出 ET 管理就是对流域或区域水资源开发利用过程的耗水管理，就是从流域或区域整体控制总的 ET，确保流域或区域总 ET 不超过可消耗 ET，只有对 ET 进行管理和控制才能真正实现流域或区域水资源的可持续利用，通过调整 ET 在时空上和部门间的分配，通过提高各部门 ET 利用的效率，减少低效和无效 ET，增加高效 ET，才能真正实现人水和谐，促进经济发展和社会进步。

【关键词】 水资源管理 ET 管理 可持续利用

1 引言

中国属于世界上的贫水国家之一，人均淡水资源量仅为世界人均量的 1/4，居世界第 109 位，而且分布很不平衡：长江、珠江、东南和西南诸河流域，GDP 和人口约占全国的 1/2，但是水资源占了 84%；北方的海河、黄河、淮河、辽河、松花江及西北诸河流域，耕地面积占了全国的 65%，同时又是粮食的主产区，是农业的主要灌溉区，水资源却只占全国的 16%。尤其是黄河、淮河和海河 3 个流域的国土面积占全国的 15%，耕地、人口和 GDP 分别占全国的 1/3，水资源总量仅占全国的 7%，水资源供需矛盾十分突出。

目前，全国大部分地区水资源开发利用程度达到或超过国际水资源开发利用的警戒线，特别是北方地区，其水资源开发利用是掠夺性的。黄河、淮河、辽河都曾有不同程度的断流；海河流域有河皆干，有水皆污。另外，由于人类活动的加剧，造成水资源安全系统水土流失、土壤盐渍化、林草地退化、土地沙化等，使水资源安全的良性循环系统以及健康的自组织系统遭受严重的破坏，水资源安全系统面临崩溃。

这种水危机与水问题在很大程度上与水资源管理有关，来自不合时宜的水资源开发利用与水资源配置模式和管理模式。仅重视开源，即增加陆地水系统的输入和用水过程末端的节水，已无法全面提高水资源在其动态转化过程中的效用。只有重视水循环过程中的每一个环节，充分考虑水循环的每一个分过程——大气过程、地表过程、土壤过程以及地下过程的消耗转化，才能实现资源性节水的目的。因此，开展“ET 管理”为核心的水资源管理研究具有重要意义。

以 ET 为核心进行水资源管理的理念，是随着水利部世界银行贷款节水灌溉项目以及 GEF 海河流域水资源与水环境综合管理项目的实施而提出，并逐步发展完善起来的。其核心思想是在宏观区域/流域尺度开展以“ET 管理”为核心的水资源管理,减少无效或低效 ET,以实现资源性节水。目前,在我国世界银行节水灌溉项目区、海河流域、新疆吐鲁番地区等进行了试点实施。

2 “ET 管理”的含义

水资源安全的基础是健康的水循环,水循环包括天然水循环与人工水循环,或称自然水循环与社会经济系统水循环。自然水循环是指地球上的水在太阳辐射和地球重力作用下,不断地进行转化、输送、交换的连续运动过程。水通过蒸发、凝结、降水、径流的转移和交替,沿着复杂的循环路径不断运动和变化,完成水的自然循环过程。自然水循环使大气圈、水圈、岩石圈、生物圈之间相互联系起来,以水作为纽带,在各圈层之间进行能量交换。它不但改造了各个圈层,促进各圈层的发展,同时也促进了整个自然界的发展。自然水循环把三种形态的水和不同类型的水体联系起来,形成一个运动系统,水在这个系统中挟带、溶解物质和泥沙,使物质进行迁移。自然水循环使大气降水、地表水、地下水、土壤水之间相互转化,使水形成一个不断更新的统一系统。这个系统唯一的介质作用就是蒸散发(ET),也是这个系统唯一可以表述的水资源消耗项,它不仅具有区域水量平衡与能量平衡的决定性,而且对区域健康水循环、水资源的配置、水资源效用具有重要的作用。因此, ET 管理就是流域或区域水资源真实消耗的管理。

基于 ET(耗水)的水资源管理是 David 等首次提出节水中的可回收水和不可回收水概念;Keller 等提出了资源性节水系列概念,强调从整个水循环不可回收的水量中进行节水,而不仅限于用水过程中的节水,引发了人们对水资源高效利用和有效管理的新思考。这一思想起先用于农业灌溉用水管理中,后被世界银行 GEF 海河项目中广泛应用和实践。“ET 管理”的含义就是对流域或区域水资源开发利用过程的耗水管理,就是从流域/区域整体控制住总的 ET 量,确保流域/区域总 ET 不超过可消耗 ET,实现水资源的可持续利用;从流域/区域整体提高水分生产水平,提高 ET 效率,促进社会经济持续发展,为减少流域或区域 ET 消耗量,只有减少水分的蒸发蒸腾,才是区域水资源量的真正节约。

传统的水资源管理,注重取水管理,节水的效果主要由取水量的减少来衡量,取水的减少量等同于节约的水量。因此,在进行水资源开发利用时,主要在区域间和部门间分配各种可利用的水源,缺乏对 ET 总量的分配和控制。其结果是,发达地区或者强势部门通过提高水的重复利用率和消耗率,在不突破许可取水量限制的条件下,将消耗更多的水量(增加 ET),在区域/流域水资源条件(区域/流域总的可消耗 ET)基本不变的情况下,这就意味着欠发达地区或者弱势部门如农业、生态等部门可使用的水资源将被挤占。越是在水资源紧缺的地区,这种矛盾越突出。因此,按照传统水资源管理理念,水资源利用的公平性并不能真正得到保证,生态系统的安全也并不能真正得到保障。所以,只有对 ET 进行管理和控制才能真正实现流域/区域水资源的可持续利用,通过调整 ET 在时空上和部门间的分配,通过提高各部门 ET 利用的效率,减少低效和无效 ET,增加高效 ET,才能够真正实现人水和谐,促进经济发展和社会进步。

3 ET 管理模式

ET 管理模式首先综合考虑人类社会生活、生产以及自然生态环境的用水平衡。在此基础上,以流域(区域)为管理单元,从ET的分配入手统一筹划水资源的配置。通过监测的现状ET与流域可持续发展的目标ET进行比较,采取措施让现状ET不断与目标ET趋于一致,使各地的ET值与当地可利用的水资源量达到平衡。当监测的区域ET值低于目标ET时,认为区域水资源的可持续发展有了保证,可以将差额部分的ET转换,从而实现用水的转换。其重点是控制流域的实际耗水量与目标耗水量之间的平衡,以减少流域水分循环过程中无效或低效ET为目标,实现水资源的可持续利用,具体原理如下式

$$\mathrm{Target}_{\mathrm{ETACR}}=P+I-O-W_{\mathrm{CURB}}-ET_{\mathrm{ECO}}-\Delta ET_{\mathrm{AGR}}-\Delta G$$

式中:$\mathrm{Target}_{\mathrm{ETACR}}$为目标耗水量;$P$为降雨量;$I$为入流量;$O$为现状排水量;$W_{\mathrm{CURB}}$为现状工业及生活耗水量;$ET_{\mathrm{AGR}}$为现状农业耗水量;$ET_{\mathrm{ECO}}$为现状生态耗水量;$\Delta ET_{\mathrm{AGR}}$为农业耗水量的年度减少量;$\Delta G$为流域内水资源变化量。

ET管理不是从水量的直接分配出发,而是综合考虑经济、社会、生态目标,首先根据流域水量平衡核定流域总可消耗ET,确定流域可利用水资源量。在此基础上,对流域内上下游各行政区域进行目标ET分配,各行政区域再将目标ET层层分配到下级行政单位并实施管理。同时需要考虑不同水文年份下ET分配的调整。具有运行模式见图1。

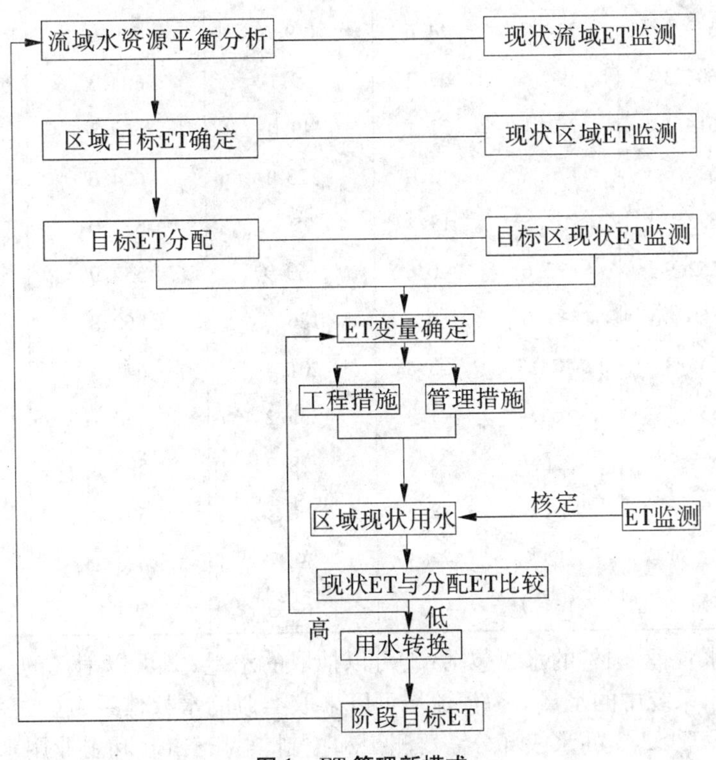

图1 ET管理新模式

4 基于ET的海河流域地下水管理

海河流域在我国政治、经济和文化领域具有重要的战略地位,也是我国水资源短缺和地下水超采最为严重地区。基于ET的海河流域水资源管理就是以ET管理作为技术手段,提出适合海河流域特点的地下水水权及打井许可管理体系,为逐步实现地下水"零"超采、逐步恢复地下水,实现基于ET的采补平衡,保证水资源的可持续利用与社会经济的可持续

发展。

依据中国科学院遥感应用所2007年9月29日提供的《海河流域ET生产结果2002～2006》和《海河流域地下水可持续开发利用水权及取水许可管理战略研究》，海河流域2005年平原区19地市现状ET水平、地下水净使用量及水权分析见表1。

表1 海河流域2005年平原区19地市现状ET水平、地下水净使用量及水权分析

行政区	实际ET（mm）	实际地下水净使用量		水权分析结果（mm）		
		mm	亿 m^3	地下水开采量	消耗量ET	回归本系统水量
北京	518.7	83.6	5.4	270.6	518.7	270.6
天津	549.1	-35.6	-4.0	42.3	549.1	42.3
石家庄	636.9	212.3	14.3	142.2	636.9	142.2
唐山	552.0	-48.0	-4.2	93.7	552.0	93.7
秦皇岛	538.2	-70.7	-2.9	108.8	538.2	108.8
廊坊	591.8	185.8	11.9	79.6	591.8	79.6
保定	629.9	224.7	24.6	119.6	629.9	119.6
沧州	604.3	194.6	27.3	39.1	604.3	39.1
衡水	619.1	251.1	22.1	49.0	619.1	49.0
邯郸	624.6	176.5	13.0	123.1	624.6	123.1
邢台	622.0	203.5	18.4	95.8	622.0	95.8
安阳	552.9	-22.6	-0.6	160.9	552.9	160.9
鹤壁	560.8	-37.0	-0.5	186.0	560.8	186.0
新乡	645.8	-174.3	-3.8	201.4	645.8	201.4
焦作	558.1	-164.4	-1.9	235.2	558.1	235.2
濮阳	598.3	-226.6	-4.3	228.9	598.3	228.9
德州	597.8	-79.6	-8.2	98.6	597.8	98.6
滨州	623.1	-32.9	-2.3	39.3	623.1	39.3
聊城	594.4	-161.7	-13.7	167.3	594.4	167.3

对于地下水管理，国外的水权包括：①可以抽取的水量；②可以消耗的水量（ET）；③必须回归到当地水系统中的水量。海河流域可以采取类似的水权体系，这样会增加水权管理的复杂性，因为至少2/3的水权部分需要测量和控制。对于市政和工业用水者来说他们的取水总量和排放总量可以测量与控制。伴随着利用遥感估算真正ET值技术的出现，对于灌溉水使用者来说，其取水量和ET量都能得到测量。同时，水权的三个部分，对于一个水市场的良好运行也是非常重要的。如果没有这些方面的内容，一个用水者可以将原本只有很低ET值的一个水权出卖给将会消耗很高ET值的某个用水者。依据现状水资源条件下，考虑适宜的入海水量、控制地下水超采及其他因素条件下，确定的ET值和现状目标ET条件下地下水净使用量，具体见表2。

表2 海河流域现状基准年平原区19地市目标ET、净使用量及水权

行政区	目标ET	地下水目标净使用量		开采量	消耗量（目标ET）	回补量
	（mm）	mm	亿 m^3	（mm）	（mm）	（mm）
北京	535.9	-48.5	-5.4	414.5	535.9	357.3
天津	576.0	75.9	5.1	2.3	576.0	50.8
石家庄	586.1	57.4	5.0	260.5	586.1	184.6
唐山	699.7	68.2	2.8	155.7	699.7	98.3
秦皇岛	703.0	71.9	4.6	163.2	703.0	95.0
廊坊	609.8	83.8	9.2	168.2	609.8	96.3
保定	553.9	110.4	15.5	236.9	553.9	153.2
沧州	590.7	120.4	10.6	160.6	590.7	50.2
衡水	598.8	35.0	2.6	188.6	598.8	68.2
邯郸	631.1	46.0	4.2	151.5	631.1	116.5
邢台	600.0	137.9	3.7	156.2	600.0	110.2
安阳	663.2	149.5	2.0	263.8	663.2	125.8
鹤壁	672.6	86.1	1.9	253.3	672.6	103.8
新乡	774.6	38.6	0.5	272.4	774.6	186.3
焦作	669.4	93.2	1.8	245.0	669.4	206.4
濮阳	689.0	-37.5	-3.8	190.8	689.0	97.6
德州	619.4	-39.8	-2.8	101.6	619.4	139.1
滨州	645.6	-35.6	-3.0	26.0	645.6	65.8
聊城	615.9	57.2	3.7	130.9	615.9	166.4

根据流域国民经济的发展，2020年的目标ET考虑南水北调（东、中线）通水，入海水量标准进一步提高，具体见表3。

5 结语

本文从我国水资源利用现状的具体问题入手，阐述了水资源"ET管理"的含义，并将其与传统水资源管理的差异进行了对比分析，提出了进行水资源"ET管理"的具体模式，即以人类社会生活、生产以及自然生态环境的用水平衡为基础，以流域（区域）为管理单元，统筹分配ET，并将现状ET与目标ET相对比，通过不断调整具体措施，最终实现水资源的可持续发展。

以"ET管理"为核心的水资源管理理念是一种全新的理念，它是从水资源消耗的效率角度出发，强调尽量减少水分循环过程中低效或无效ET，同时兼顾生态、环境用水，综合考虑水分循环的全过程，使有限的水资源利用效率最大化，对传统水资源管理进行了有意义的

补充。这也与我国"实施最严格的水资源管理制度"思想是一致的，是水资源管理的"三条红线"，在实践中的具体应用将有助于实现我国水资源的可持续利用。

表3　海河流域规划年平原区19地市目标ET、净使用量及水权

行政区	目标ET (mm)	地下水目标净使用量		开采量 (mm)	消耗量（目标ET） (mm)	回补量 (mm)
		mm	亿 m^3			
北京	557.1	-44.9	-2.9	20.0	557.1	357.3
天津	626.6	-49.7	-5.6	0.1	626.6	50.8
石家庄	574.0	81.3	5.5	17.9	574.0	184.6
唐山	649.6	31.4	2.7	11.4	649.6	98.3
秦皇岛	648.0	23.6	1.0	4.8	648.0	95.0
廊坊	565.2	64.6	4.2	10.3	565.2	96.3
保定	563.3	42.7	4.7	21.5	563.3	153.2
沧州	545.1	51.2	7.2	14.3	545.1	50.2
衡水	562.5	63.8	5.6	11.6	562.5	68.2
邯郸	594.2	89.4	6.6	15.2	594.2	116.5
邢台	565.6	79.9	7.2	17.2	565.6	110.2
安阳	654.6	71.7	1.9	5.3	654.6	125.8
鹤壁	663.9	34.1	0.5	1.9	663.9	103.8
新乡	764.6	-106.6	-2.3	1.7	764.6	186.3
焦作	660.7	-187.7	-2.2	0.2	660.7	206.4
濮阳	692.0	89.2	1.7	3.6	692.0	97.6
德州	615.5	-79.4	-8.2	6.1	615.5	139.1
滨州	641.5	-48.1	-3.4	1.3	641.5	65.8
聊城	612.0	-37.7	-3.2	10.9	612.0	166.4

参考文献

[1] 王浩，杨贵羽，贾仰文，等. 以黄河流域土壤水资源为例说明以"ET管理"为核心的现代水资源管理的必要性和可行性[J]. 中国科学，2009，39(10)：1691-1701.

[2] David C D, Robert M H. Agriculture and water conservation in California, with emphasis on the San Joaquin valley[M]. Technical report, Department of Land, Air and Water Resources University of California, Davis, October, 1982.

[3] Keller A, Keller J, Secker D. Integrated water resources system: Theory and policy implications[M]. Research Reports, Colombo, Sri Lanka. International Water Management Institute, 1996.

【作者简介】 徐磊，1981年7月17日生，博士，2008年6月毕业于中国农业大学，现于中国灌溉排水发展中心工作，工程师。

内陆核电建设对水资源管理的新挑战及对策的初步研究

刘 达[1,2,3] 黄本胜[1,2,3] 邱 静[1,2,3] 马 瑞[4]

(1.广东省水利水电科学研究院 广州 510610;
2.河口水利技术国家地方联合工程实验室 广州 510610;
3.广东省水动力学应用研究重点实验室 广州 510610;
4.水利部珠江水利委员会 广州 510610)

【摘要】 本文分析了内陆核电建设可能给水资源管理带来的新挑战,在深入研究国内外相关管理经验和制度的基础上,提出了内陆核电站水资源的管理应分为三个层次,即前期的规划设计阶段的管理、正常运行期的日常管理及突发性事故的水安全应急管理。对前期规划设计阶段的管理,论述了水行政主管部门提前介入内陆核电站前期工作的必要性及介入形式;内陆核电正常运行期的水资源日常管理方面,主要阐述了取用水要求、水处理及水质监测要求、专业人员培训与专项课题研究等;内陆核电水资源的应急管理上,分析了水行政部门对核电水资源安全应急管理的合理介入方式。

【关键词】 内陆核电 水资源管理 低放射性废液 核事故 水资源条件

1 内陆核电建设对水资源管理的新挑战

传统的以煤炭发电为主的能源结构由于其高能耗、高污染、高二氧化碳排放将受到越来越多的环境保护的制约,此外煤炭储量也越来越难以满足我国国民经济和社会高速发展的巨大需求,而核电由于其能提供大量的清洁能源,因此在原国家的能源规划中拟大力发展,并积极谋划大力发展内陆核电,计划到2020年,核电运行和在建装机容量要达到1亿kW,核电机组将达到100多台,力争核电占电力总装机的比例达到5%以上。国内三大核电公司此前已初步普选或确定了70多个内陆核电站址,分布在十几个省(市、区),其中部分厂址已不同程度地开展了前期工作。

虽然福岛核电事故对中国乃至世界核电建设进程产生了深远的影响,致使我国调整了核电发展中长期规划,并规定在核安全规划批准前,暂停审批核电项目,包括开展前期工作的项目。但是,从中国经济和能源需求的长期趋势来看,发展核电总的政策趋势应该不会有太大的改变,但核电安全审查的力度会不断加强。

那么内陆核电的大规模开发将给区域水资源安全带来何种影响,其低放射性废液的排放是否会对饮用水源地及作为饮用水源的水库产生负面影响,影响有多大,是否采取了可靠、有效的舒缓措施予以消解,低放射性废液排放在平原河网区、河口或是库区可能存在的累积影响,对水资源安全的影响程度等都是水行政主管部门取水许可审批需要回答的问题。

水资源管理将面临新的挑战,水行政主管部门对核电水资源的管理宜做好充分准备。

内陆核电项目用水量大且要求的保证率高,目前我国规划的单个内陆核电厂址中往往拟布局 4 台百万千瓦级核电机组,即使采用循环冷却方式,设计年需新水量也达约 1.6 亿 m^3。同时,由于低放射性废液是核电项目常规退水的组成部分之一,并存在高放、中放物质的浓缩、储存、处置问题,内陆核电项目放射性流出物常规和事故排放对水资源,尤其是水源的影响、应急和控制问题始终是内陆核电发展广受关注、不容回避的关键问题,福岛核事故之后更将水安全、核安全并重的理念进一步推上显要位置。但由于内陆核电建设是件新鲜事物,水资源综合规划并未专门考虑内陆核电发展的水资源配置、低放废液受纳稀释扩散、核事故应急的水源保障等重要问题,亦未专门编制核电发展水资源保障、保护、管理相关的专项规划,尚无符合我国国民经济和社会发展阶段特征和具体国情的水安全、核安全保障控制和管理的完整政策和技术体系;目前仅有于可研阶段开展的建设项目水资源论证探讨核电建设水资源保障及安全的相关问题,但该阶段除重大的颠覆性因素存在,内陆核电项目选址和建设的总体趋势已不可逆转,从而可能将水安全、核安全双重保障与国家经济资源投入效益保障置于矛盾不可调和的两难境地。

目前,我国的内陆核电厂址主要集中在我国东南部长江、珠江等大流域,虽然前述两个流域的水量较大,水体扩散条件较好,但东南地区气候具有明显的大陆性气候特点,水资源条件在丰枯期的差异特别明显,因此流域内地表水的低放废液稀释、承载能力的稳定性、持续性也应该成为内陆核电站址布局所要考虑的一个重要问题。在国外建设一个核电站往往只有一两台机组,而我国最多有 8 台机组,机组台数的倍增必然带来更大的运行风险。虽然核电站的规划建设都有详细的突发事件应急预案,但是毕竟长三角、珠三角地区是我国经济最为发达的区域,人口密集,用水群体大,潜在厂址往往有相当大的人群在周围居住,公众对水安全、水源保障等问题比较敏感,这两个地区对我国来说也经受不起任何的突发事件带来的影响,不容任何闪失。

2 内陆核电水资源管理对策的初步研究

面对内陆核电的长期发展趋势,水行政主管部门应提前做好应对准备,必须及早介入选址工作,保障内陆核电项目布局选址的合理性,从前期工作的开端就发挥更重要的作用、提供更专业和审慎的政策与技术要求。与此同时,要做好内陆核电运行期的水资源管理工作的相关研究。

对于内陆核电站水资源的管理应分为三个层次的管理,即前期的规划设计阶段的管理、正常运行的日常管理及突发性事故的水安全应急管理。

2.1 内陆核电规划设计阶段的管理

保障水源安全是水利部门的重大管理责任,水行政主管部门应该介入内陆核电站的各个工作阶段和论证环节,从开始就把其建设可能涉及的水资源安全问题抓住,水利部门以整个流域为整体的水体承载能力对内陆厂址布局提出合理建议,对不符合水资源安全要求的厂址要坚决向上级主管部门提出意见。对于确定选址的内陆核电项目,水利部门从水资源安全的角度对内陆核电站的管理应该贯穿其建设的全过程,其规划建设方案主要通过水资源论证的形式报送水行政主管部门审批,对其取、用、退水方案严格把关。

核电站规划设计阶段的水资源管理主要体现在对内陆核电建设项目的水资源论证的管理

和评审把关上，具体对应核电规划设计的各个阶段而言，就是在规划选址阶段进行《规划选址阶段水资源初步论证》，在初可研阶段进行《初可研阶段水资源论证》，可研阶段进行《可研阶段水资源论证》。三个阶段的水资源论证目的和研究重点各有不同，联系紧密，逐步深入。

水行政主管部门重视和加强对内陆核电水资源的管理、提前介入内陆核电站前期工作是贯彻《水法》、落实最严格水资源管理制度的要求。原因如下：

(1)水资源条件是内陆核电选址和建设的关键因素。内陆核电项目用水量大，要求保证率高，需要充足稳定的水源保障。内陆核电一个厂址一般布局4台百万千瓦核电机组，即使采用循环冷却方式，年需新水量也要1.6亿 m^3。但由于内陆核电建设是件新鲜事物，水资源综合规划并未考虑内陆核电大规模发展的水量配置问题，因此核电建设将对现有流域、区域水资源配置格局和调度产生较大的影响。作为水资源技术管理部门，水行政主管部门在保障内陆核电项目布局合理性方面，从前期工作的开端就有其不可或缺的重要作用。

(2)核电事故和突发事件将极大地影响水体安全。核电项目使用放射性元素作为热动力源，核岛内的核素一旦外泄并进入下游水体，将对下游供水水质安全产生重大影响。内陆核电多处于人口稠密、经济发达地区，一旦因核安全事故致使核素排入江河或地下，将对广大地区水体和人员安全产生难以估量的影响和伤害，内陆核电项目的水安全应急管理的重要性非同一般，一遇事故必然造成无法挽回的重大损失，水行政主管部门作为水域纳污控制和监管的责任部门，要在事故应急预案的制订过程中做好指导、审查、管理工作。

2.2 内陆核电水资源的日常管理

核电站退水中的低放废液的长期排放会造成在水流交换弱的地区形成放射性元素的富集，还可以通过底泥吸附、食物链传递、地下水侧渗等方式形成放射性元素的富集，影响下游水质，会对水资源安全构成威胁。因此，内陆核电水资源保护日常管理的重要性也非同一般，水行政主管部门作为水域纳污控制和监管的责任部门，要做好内陆核电水资源的日常管理工作。

2.2.1 内陆核电站低放废液排放的日常监管

(1)项目施工前，按照节水“三同时”管理的要求对土建及调试用水方案涉及的取水、用水、水处理、排水工程进行统一验收，合格后方予发放施工用水许可证。

(2)当来水条件发生较大改变而需要调整低放废液的排放方案时，需要及时向水行政主管部门提出书面申请，得到批准后方可调整排放方案。

(3)严格监测要求。

编制详细的水质监测和退水排放方案，报水行政主管部门审批同意后方可实施。建设在线实时来水量、核电站取退水计量监测系统。

2.2.2 加强水利(水务)部门的核电专业技术队伍的建设

做好管理和技术队伍的能力建设，对内陆拟建核电站地区的水利(水务)部门领导开展核电水资源管理的培训班，提高其核电水资源的管理水平。大力开展核电项目水资源管理调研、技术交流、国际研讨；开展核电水资源管理和论证从业人员的培训，建立规范、完备的培训体系。

2.2.3 加大核电水资源管理的研究投入

核电水资源管理是一项全新的业务，需要财政加大支持力度。同时，为提高核电水资源管理工作的前瞻性、科学性、严谨性，使核电水资源管理和技术工作有章可循，急需开展相关政策和技术要求及其框架的研究制定、核电选址和布局的水资源论证、低放废液排放影响等专题研究、人员培训等工作,并需加强监督检查力度,这些都需要大量的经费予以支持,以保障这些工作的顺利开展。

与此同时,水利部门还应加强与核电主管部门、安全监督部门、环保部门及相关设计研究和生产部门的协调、沟通与合作。

2.3 内陆核电突发性事故的水资源安全应急管理

由于我国人口密集,即便发生小的事故,其可能对水资源安全带来的影响都是很大的,具有高度的敏感性。因此,对于核安全事故来说,水利部门也应该介入核电站突发性事故应急预案的编制,提出水利部门的考量和意见。

此外,对于核电站正常运行时外部条件的突发性变化来说,也应该有应急预案。例如,当核电最不利工况或非正常工况和连续枯水年叠加时,低放废液的排放可能会出现异常,也会对水资源安全构成威胁。从美国大旱年份看核电运行,发现核电比火电对水的依赖性要大得多。2003 年法国遇到连续枯水状况,1/4 核电站停止生产运行。

内陆核电站的水资源安全应急管理必须对突发事件考虑全面,对风险进行充分的估计,既要考虑诸如发生突发性放射性物质外溢的核电站安全的突发性事故情况下的应急措施,又要考虑诸如连续枯水状态等外部条件的突发性变化的应急对策。此外,还要考虑如福岛核事故时大量辐射冷却水的排放问题,福岛核电站最近不得不再次将大量辐射冷却水排入大海,而内陆核电站不具备此条件,该如何应对也提前予以考虑并提出切实可行的解决方案。

因此,必须要制定包含这两个方面风险因素的突发性事件应急预案,不但包括技术上的应急对策,还包含水利部门在管理上的应急对策,可以参考水利三防管理机制构建水资源安全应急管理机制。要求内陆核电建设项目业主制订合理可行的突发事故应急预案,应急预案必须通过专题论证,取得水行政主管部门的审批同意方可实施。在应急预案中必须明确对由此而造成的水污染事件的应急处置措施,并报水行政主管部门审查、备案;事故排放发生时必须及时报告水行政管理部门,并启动应急预案。

3 结语

内陆核电的潜在厂址往往有相当大的人群在周围居住,公众对水安全、水源保障等问题比较敏感,核电水安全问题不容任何闪失,内陆核电的建设将给水资源管理工作带来新的挑战。面对内陆核电的发展趋势,水行政主管部门应做好充分的应对准备,及早介入选址工作,保障内陆核电项目布局选址合理性,与此同时,要启动内陆核电运行期及事故突发期的水资源管理工作的相关研究。

参考文献

[1] 刘永叶,刘森林. 我国内陆核电建设的环境问题. 全国内陆核电站核与辐射安全学术讨论会,2009.

[2] 马名杰. 对在内陆发展核电的几点思考[J]. 国防科技工业,2006(4):5-10.

[3] 田杰棠. 借鉴国外经验,发展内陆核电[J]. 国防科技工业,2006(4):22-26.

[4] 徐东武,潘吕法. 浙江龙游核电厂液态放射性流出物对饮用水源的影响. 全国内陆核电站核与辐射安全学术讨论会,2009(10):38-42.

[5] 刘达,黄本胜,邱静,等. 内陆核电站建设对水资源安全影响的问题及研究现状[J]. 广东水利水电,2010(10):28-33.

【作者简介】 刘达,男,1981 年 11 月生,吉林延边人,四川大学工学硕士(2007 年毕业),四川大学在读博士研究生(2009 级),现工作于广东省水利水电科学研究院水资源与水生态环境工程研究所,工程师。

乐昌峡水利枢纽工程一、二维水流洪水演进数值模拟研究

徐林春[1,2] 黄 东[2] 郑国栋[2] 刘画眉[2]
郑 泳[2] 蔡素芳[2] 林美兰[2]
(1. 武汉大学 水资源与水电工程科学国家重点实验室 武汉 430072;
2. 广东省水利水电科学研究院 广州 510635)

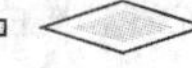

【摘要】 本文以乐昌峡水利枢纽工程坝下防护范围为研究对象,建立了水库下游河道及两岸淹没区一、二维联解洪水演进水流数学模型。采用室内试验数据对模型数值计算结果进行率定和验证,结果表明,所建模型具有较高的计算精度和可靠度。针对工程动库调洪演算成果,选取典型洪水下的不同防洪调度方案进行了模拟计算,给出了各工况下水库下游河道沿程各特征断面的洪水风险信息,为进行淹没区的洪灾损失评估及相应的防洪调度决策提供了必要的参数和依据。

【关键词】 乐昌峡 一、二维联解 洪水演进 数值模拟 率定验证 防洪调度 洪水风险信息

为建设现代化水利,科学估算通过水利枢纽工程的优化调度而产生的防洪效益,有必要建立一套全面、完整的枢纽工程防洪减灾评估体系,准确评估工程抵御每场洪水能够减免的洪灾损失情况,并在此基础上根据社会经济发展情况,预测未来的损失,提出切实可行的防洪减灾措施和对策。其中,洪水演进模型是整个防洪减灾评估体系的核心内容。该模型根据水库的防洪调度方式,可以预先模拟洪水运动情况,以及评估防洪调度措施的作用和效果,寻找最佳调度方案,使防洪调度趋于科学合理,以达到减轻洪灾损失、降低洪水风险的目的。并且通过模拟历史洪水、实时洪水和未来洪水,把有限的实测水情信息推演到全流域,可达到对流域洪水水情、灾情进行事先评估、实时评估和事后评估的目的。

1 数学模型

1.1 一维网河数学模型

一维非恒定流网河数学模型的基本方程采用圣维南方程组

$$\begin{cases} \dfrac{\partial Q}{\partial x} + B_{\mathrm{T}} \dfrac{\partial z}{\partial t} = q_1 \\ \dfrac{\partial Q}{\partial t} + 2u \dfrac{\partial Q}{\partial x} + (gA - Bu^2) \dfrac{\partial z}{\partial x} - u^2 \dfrac{\partial A}{\partial x} + g \dfrac{n^2 |u| Q}{R^{3/4}} = 0 \end{cases} \tag{1}$$

式中:Q 为流量;z 为水位;R 为水力半径;u 为流速;q_1 为旁侧入流;n 为糙率;B_{T} 为包括主河道泄流宽度和仅起调蓄作用的附加宽度;B 为过流河宽;A 为过水面积;g 为重力加速度;

x 、t 为空间和时间坐标。

式(1)再结合汉点连接方程即构成一维网河水流数学模型,可采用网河三级联解方法求解。具体求解方法可见参考文献[1]~[3]。

1.2 二维水流数学模型

平面二维水流基本方程包括水流连续方程和水流运动方程,即

$$\frac{\partial z}{\partial t} + \frac{\partial M}{\partial x} + \frac{\partial N}{\partial y} = 0 \tag{2}$$

$$\frac{\partial M}{\partial t} + \frac{\partial uM}{\partial x} + \frac{\partial vM}{\partial y} = - gh\frac{\partial(h + z_{\mathrm{b}})}{\partial x} - \frac{gn^2 u\sqrt{u^2 + v^2}}{h^{\frac{1}{3}}} + \gamma_{\mathrm{t}}\left(\frac{\partial^2 M}{\partial x^2} + \frac{\partial^2 M}{\partial y^2}\right) \tag{3}$$

$$\frac{\partial N}{\partial t} + \frac{\partial uN}{\partial x} + \frac{\partial vN}{\partial y} = - gh\frac{\partial(h + z_{\mathrm{b}})}{\partial y} - \frac{gn^2 v\sqrt{u^2 + v^2}}{h^{\frac{1}{3}}} + \gamma_{\mathrm{t}}\left(\frac{\partial^2 N}{\partial x^2} + \frac{\partial^2 N}{\partial y^2}\right) \tag{4}$$

式中:h 为水深;u 、v 分别为 x 、y 方向的流速;$M = uh$;$N = Vh$;z_{b} 为河床高程;n 为曼宁糙率系数;γ_{t} 为紊动黏滞系数。

方程离散时空间采用有限体积法,运用守恒格式对水流连续方程进行离散,保证计算域内水量守恒,时间采用蛙跳法,计算物理量使用交错网格。

1.3 模型处理

1.3.1 动边界的处理

令闭边界的法向流速为0,而沿切线方向的流速为非 0 值,即 $V_n|_{\Gamma} = 0, V_l|_{\Gamma} \neq 0$。随着水位的变化,动边界位置不断变化,根据计算水位和底部高程判断网格单元是否露出水面,定义临界水深 $h = 0.005 \sim 0.01$ m,当水深 $h > \Delta h$ 时,糙率取正常值,反之糙率取一大值(如 10^{10} 量级),经计算验证,取 $\Delta h = 0.01$ m。在河网方程组中增加一组流量为零的河段方程,将该组河段方程嵌入到河网节点方程组中参与河网联解计算,这样就可将一个动边界问题转化为固定边界问题来处理,以此处理干河床上的洪水演进问题。

1.3.2 一、二维联解

一维河道和二维淹没区之间以溃口作为连接点,连接点每个时刻的水位和流量是作为计算过程中的内边界条件耦合自动算出,实现了一维模型和二维模型交界面上水位、流量、能量、阻力的平顺衔接,计算稳定并反映实际情况。其中,溃口假设堤防瞬间按预定宽度全部溃决,或在一定时间内决口线性扩展达到预定最大宽度,按平底宽顶堰自由或淹没出流计算流入淹没区的流量过程,淹没区内按干河床进行洪水演进计算。

1.3.3 水流阻力项计算

在洪水演进计算中,河道水流阻力的主要影响因素有洪水量级和涨落变化、回水影响以及河床冲淤变化。为了恰当地反映这些因素对水流阻力的影响,关键在于把握区域影响水流阻力的主要因素的变化规律,建立它们与 n 值的相关关系。

在整个计算域内糙率 n 随河床底质的不同分布及水流强度的变化而变化,依下式进行计算

$$n = A_0 g^{-0.5} E_{*\mathrm{bc}} E^{-0.5} h^{0.167} \tag{5}$$

式中:E 为合流速,$E = \sqrt{U^2 + V^2}$;$E_{*\mathrm{bc}}$ 为床面不冲摩阻流速。

2 模型验证

2.1 一维模型计算精度分析

采用室内试验与模型数值计算结果进行比较，分析和判别模型的计算精度与可靠性。

试验水槽平面形状示意图如图1所示。水槽中部有一个收缩扩散段，闸门位于其中最窄处，瞬时打开。所选的计算条件是上游水深30 cm，下游水深10.1 cm，底坡为0。计算糙率取0.012。下游末端的边界条件为矩形薄壁堰的水位—流量关系。图2和图3为水槽中线上几个测点水位过程线的计算与实测值的对比，其中负号表示闸门上游，$x=-0$ 指上游紧靠闸门处的测点。由图2、图3可见，各测点的计算值与实测值符合良好。

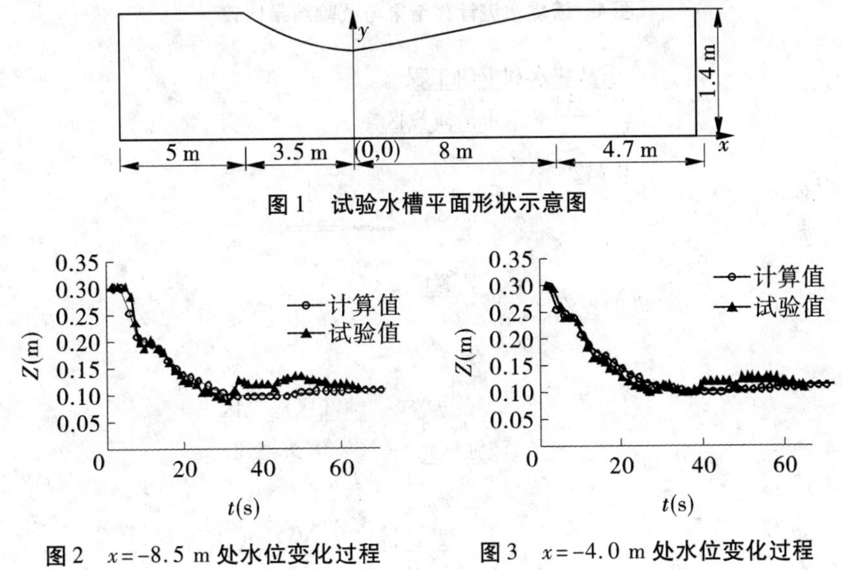

图1 试验水槽平面形状示意图

图2 $x=-8.5$ m 处水位变化过程

图3 $x=-4.0$ m 处水位变化过程

另外，还将模型结果与Rittor解、斯托克(Stoker)解等溃坝洪水理论解进行了比较，计算结果与理论解吻合良好，表明一维模型模拟水流运动具有较高精度。

2.2 二维模型计算精度分析

二维试验结果为一堤防决口后水流在堤后地面上流动的情况。地面为水平，边长1.84 m。平地的一侧为水箱，中间用堤防隔开。堤防决口处的口门宽为0.2 m，位于堤防的中间，决口后水流的前端位置通过摄像机记录，同时在与口门垂直的中心线上设置若干台水位仪记录水深。水流前端位置的数值计算结果和试验结果如图4所示，可见二者基本吻合。

3 模型应用

3.1 模型范围的确定

本文采用大范围的北江干流及其支流的一维网河水流数学模型与枢纽工程下游两岸淹没区二维水流数学模型进行耦合联解，以此进行淹没区的洪水演进模拟。一维网河数学模型计算范围的上边界从武江乐昌峡水利枢纽坝址、浈江湾头水利枢纽坝址开始，下边界至孟州坝水利枢纽坝址处，根据两岸地形条件，于乐昌、乳源及浈江区各选定一处淹没区进行二维洪水演进计算。模型计算范围见图 5。

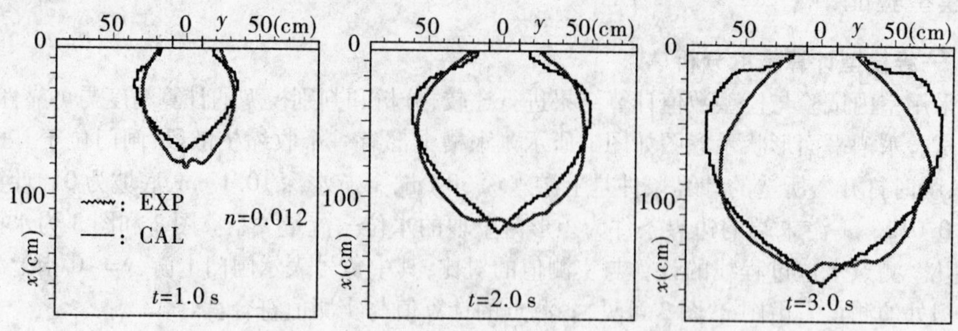

图 4　溃堤水流计算结果与试验结果比较

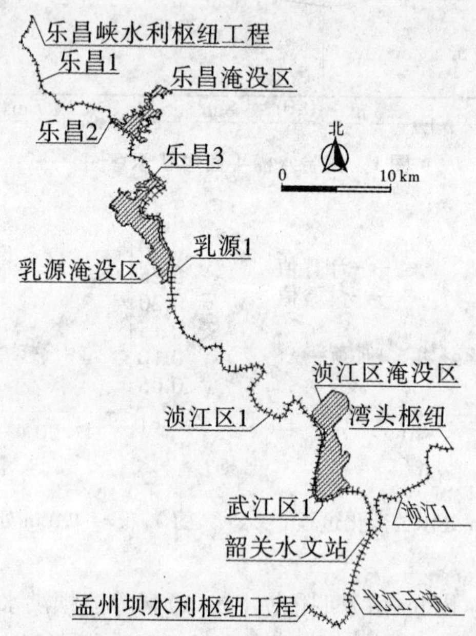

图 5　一维网河数学模型计算断面布置及二维淹没区范围示意图

3.2　边界条件的确定

采用上述一、二维洪水演进数学模型，直接利用《广东省乐昌峡水利枢纽库区动库容调洪研究报告》中“06.7”典型洪水放大后频率为 2%、1% 和 0.1% 的三组频率在两种调洪方案下的调洪演算成果作为上边界条件，下边界孟州坝水利枢纽工程则结合枢纽调度原则给定坝址相应水位—流量过程，针对乐昌峡水利枢纽工程坝下防护范围进行模拟计算。

3.3　一维计算结果

表 1 和表 2 分别给出了“06 · 7”型 50 年、100 年和 1 000 年一遇洪水时，坝址天然来流和建库后两个调洪方案共三组水文组合条件下，水库下游河道沿程部分特征断面的最大流量、最高水位和相应出现时间。

表1 不同工况和“06·7”型不同计算水文组合下最大流量计算结果统计

断面位置	工况	P=2%		P=1%		P=0.1%	
		最大流量 (m^3/s)	出现时刻 (h)	最大流量 (m^3/s)	出现时刻 (h)	最大流量 (m^3/s)	出现时刻 (h)
乐昌1	天然来流	5 326	38	6 024	38	8 464	38
	调洪方案一	3 190	26	3 928	49	8 450	39
	调洪方案二	—	—	3 884	51	8 246	41
乐昌2	天然来流	5 319	39	6 017	39	8 461	39
	调洪方案一	3 190	29	3 909	50	8 448	39
	调洪方案二	—	—	3 876	52	8 240	41
乐昌3	天然来流	5 311	39	6 009	39	8 456	39
	调洪方案一	3 190	33	3 889	51	8 445	39
	调洪方案二	—	—	3 870	52	8 235	42
武江区1	天然来流	5 123	43	5 808	43	8 349	42
	调洪方案一	3 184	40	3 689	56	8 336	43
	调洪方案二	—	—	3 742	57	8 085	44
韶关水文站	天然来流	9 390	43	10 423	43	13 346	42
	调洪方案一	7 416	39	8 446	75	12 160	47
	调洪方案二	—	—	8 446	75	12 204	47

表2 不同工况和“06·7”型不同计算水文组合下最高水位计算结果统计

断面位置	工况	P=2%		P=1%		P=0.1%	
		最高水位 (m)	出现时刻 (h)	最高水位 (m)	出现时刻 (h)	最高水位 (m)	出现时刻 (h)
乐昌1	天然来流	108.47	38	109.6	38	113.13	38
	调洪方案一	104.52	25	105.94	50	113.11	39
	调洪方案二	—	—	105.86	51	112.82	41
乐昌2	天然来流	93	39	93.55	39	95.22	39
	调洪方案一	90.38	30	91.21	50	95.21	39
	调洪方案二	—	—	91.18	52	95.08	42
乐昌3	天然来流	86.92	39	87.36	39	88.83	39
	调洪方案一	85.27	27	85.83	51	88.83	40
	调洪方案二	—	—	85.82	52	88.7	42
武江区1	天然来流	60.54	44	61.27	43	63.63	43
	调洪方案一	58.41	40	58.77	76	62.86	47
	调洪方案二	—	—	58.77	76	62.95	48
韶关水文站	天然来流	57.85	44	58.67	44	61.11	43
	调洪方案一	56.19	40	57.04	75	60.01	48
	调洪方案二	—	—	57.04	75	60.07	49

注：“—”表示50年一遇洪水下的调洪方案二与调洪方案一相同，未重复计算。

结果表明,对于100年一遇设计标准洪水及其以下的频率洪水,乐昌峡水利枢纽能够较好地发挥水库削峰滞洪的防洪效益,其中对武江的削峰作用尤为显著,且洪峰出现时间在水库调洪后均明显滞后,各特征断面的最大洪峰流量和最高水位均有所减小,可以有效缓解水库下游两岸堤围的防洪压力,为确保人民的生命财产安全提供了一个重要保障;但对于1 000年一遇的超标准大洪水,乐昌峡水库仅能做到短时滞洪的作用,无法削减洪峰水位或流量。

3.4　二维计算结果

由于篇幅所限,本文仅针对乐昌淹没区的洪水演进和淹没情况进行简单分析。

对于"06 · 7"型50年一遇洪水,经过上游乐昌峡水利枢纽工程的调洪作用,可利用水库有效库容,完全发挥水库的削峰滞洪作用,确保乐昌淹没区的防洪安全,不致发生淹没。

对于"06 · 7"型100年一遇洪水,经过上游乐昌峡水利枢纽工程的削峰滞洪作用,乐昌淹没区的漫堤进洪时间被大大推迟,淹没历时大幅缩短,进洪量也大幅降低,淹没区内的最大淹没面积和最大淹没水深较建库前天然情况下均有大幅减小和降低,乐昌峡水利枢纽工程的防洪作用十分明显,防洪减灾效益显著;对于该水文组合,调洪方案二的最大淹没面积和淹没水深基本略大于调洪方案一的成果。

对于"06 · 7"型1 000年一遇的超标准大洪水,通过乐昌峡水利枢纽工程的调洪作用,仅可推迟下游乐昌淹没区的漫堤分洪时间,并缩短淹没历时,但削峰作用并不明显,建库前后乐昌淹没区的最大淹没面积和最大淹没水深相差不大;对于该水文组合,调洪方案二的最大淹没面积和淹没水深均略小于调洪方案一的成果。

表3~表5分别统计了乐昌淹没区100年一遇洪水条件下各工况部分特征时刻淹没面积及淹没水深等参数,图6~图8分别给出了相应各工况下的最大淹没范围。

表3　乐昌淹没区天然来流情况下各特征时刻淹没面积成果统计($P=1\%$)

不同淹没水深(m)	不同漫堤分洪时间(h)淹没面积(km^2)			
	5	15	23	36
≥12	—	—	0.002 5	0.002 5
≥10	—	—	0.007 5	0.005 0
≥8	—	—	0.100 0	0.042 5
≥6	—	0.055 0	0.412 5	0.260 0
≥4	0.010 0	0.390 0	1.025 0	0.637 5
≥2	0.115 0	1.002 5	2.107 5	1.510 0
≥1	0.220 0	1.342 5	2.690 0	2.090 0
最大淹没面积(km^2)	0.370 0	1.692 5	3.267 5	2.677 5
最大淹没水深(m)	4.94	7.84	13.97	13.01

表 4　乐昌淹没区调洪方案一情况下各特征时刻淹没面积成果统计（$P=1\%$）

不同淹没水深（m）	不同漫堤分洪时间（h）淹没面积（km^2）			
	2	5	7	9
≥4	—	0.010 0	0.010 0	0.010 0
≥3	0.007 5	0.042 5	0.042 5	0.042 5
≥2	0.042 5	0.112 5	0.112 5	0.112 5
≥1	0.110 0	0.192 5	0.192 5	0.192 5
最大淹没面积（km^2）	0.210 0	0.342 5	0.342 5	0.342 5
最大淹没水深（m）	3.82	4.96	4.72	4.59

表 5　乐昌淹没区调洪方案二情况下各特征时刻淹没面积成果统计（$P=1\%$）

不同淹没水深（m）	不同漫堤分洪时间（h）淹没面积（km^2）			
	2	5	7	11
≥5	—	0.010 0	—	—
≥4	—	0.042 5	0.010 0	0.010 0
≥3	0.007 5	0.112 5	0.042 5	0.042 5
≥2	0.037 5	0.192 5	0.112 5	0.112 5
≥1	0.097 5	0.312 5	0.195 0	0.195 0
最大淹没面积（km^2）	0.187 5	0.425 0	0.345 0	0.345 0
最大淹没水深（m）	3.60	5.08	4.90	4.58

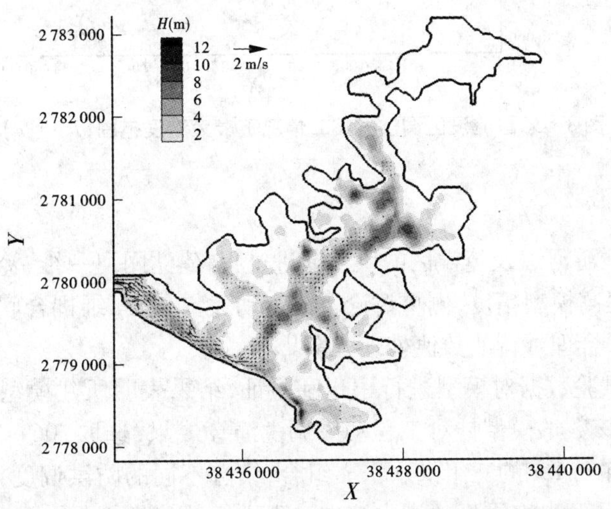

图 6　乐昌淹没区天然来流情况下最大淹没范围（$P=1\%$）

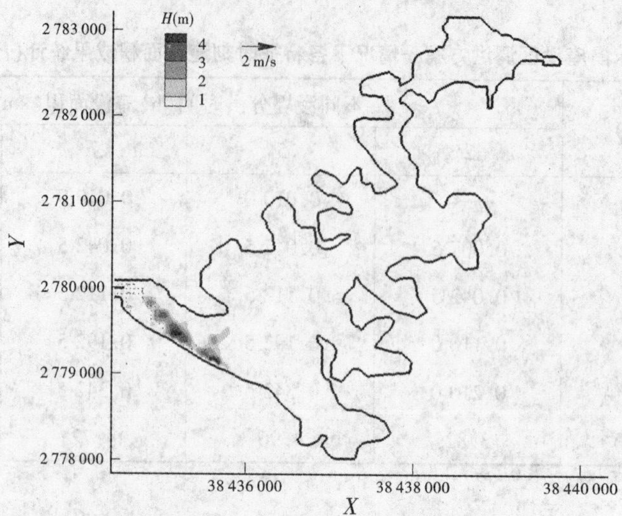

图7 乐昌淹没区调洪方案一情况下最大淹没范围($P=1\%$)

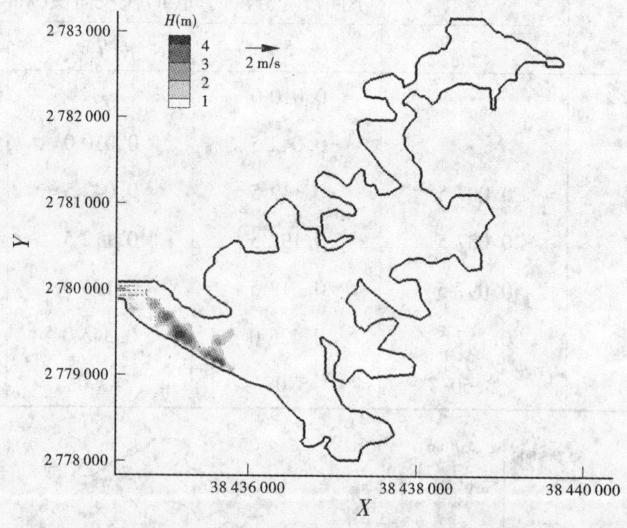

图8 乐昌淹没区调洪方案二情况下最大淹没范围($P=1\%$)

4 结语

(1)针对水库下游河道及淹没区的洪水演进过程，建立网河一维-淹没区二维联解的水动力计算模型，为提高模型精度，侧重讨论了模型动边界、一二维耦合联解以及水流阻力项计算等问题，提出了合理可行的数值处理方案。

(2)采用室内试验数据对模型进行率定和验证，结果表明所建模型具有较高的计算精度和可靠度。利用该数学模型针对工程动库调洪演算成果，选取“06·7”型典型洪水放大后频率为2%、1%和0.1%的三组洪水水文组合，结合不同的防洪调度方案，对乐昌峡水利枢纽工程下游河道及两岸淹没区的洪水演进过程进行数值模拟，给出了各工况下水库下游河道沿程各特征断面的最大流量、最高水位和相应出现时间以及淹没区的淹没面积、淹没水深和淹没历时等洪水风险信息，为进行淹没区的洪灾损失评估及相应的防洪调度决策提供了必要的参数和依据。

(3)计算结果表明,对于100年一遇设计标准洪水及其以下的频率洪水,乐昌峡水利枢纽能够较好地发挥水库削峰滞洪的防洪效益,其中对武江的削峰作用尤为显著,且洪峰出现时间在水库调洪后均明显滞后,各特征断面的最大洪峰流量和最高水位均有所减小,可以有效缓解水库下游两岸堤围的防洪压力,淹没区的漫堤进洪时间亦被大大推迟,淹没历时大幅缩短,最大淹没面积和最大淹没水深较建库前天然情况下均有大幅减小和降低,为确保人民的生命财产安全提供了一个重要保障。该模型具有较好的实用价值,可在同类工程洪水演进计算中推广。

参考文献

[1] 张二骏,张东生,等. 河网非恒定流三级联合算法[J]. 华东水利学院学报. 1982(1):1-13.

[2] 李义天. 河网非恒定流隐式方程组的汊点分组解法[J],水利学报. 1997(3):49-51.

[3] Cunge J A, et al. Practical aspects of computational river hydraulics[M]. Pitman Advanced Publishing Program. 1980.

【作者简介】 徐林春,1979年生,硕士研究生,2004年6月毕业于武汉大学,水力学与河流动力学专业,在职博士生,目前就职于广东省水利水电科学研究院,副总工,高级工程师。

全国城市饮用水水源地分区安全评价与措施布局

张建永　朱党生　曾肇京　刘卓颖

（水利部水利水电规划设计总院　北京　100120）

【摘要】 本文结合我国东、中、西三大区经济社会发展布局，在综合分析我国各省级行政区城市供水组成、城镇化水平及综合生活人均用水量情况基础上，将全国划分为8个规划分区。根据全国城市饮用水水源地调查评价成果，对各规划分区城市饮用水水源地分布状况进行了说明，对城市饮用水水源地水质、水量安全状况进行了综合评价。总体评价表明，我国城市饮用水水源地安全形势仍十分严峻，各规划分区的饮用水安全特点和主要问题具有明显的区域差异性。结合城市饮用水水源地安全分区评价成果，提出了各规划分区水源地安全保障措施总体布局。

【关键词】 城市　饮用水水源地　安全评价　措施布局　分区

饮用水安全既涉及人民群众的生命健康，又涉及经济社会的可持续发展，成为全面建设小康社会的重要支撑条件，更是体现以人为本，构建社会主义和谐社会的客观需要。近年来，中央和地方加大了城乡饮用水安全保障工作的力度，采取了一系列工程和管理措施，解决了一些城乡居民的饮用水安全问题。随着经济社会的发展、人口增长和城市化进程的加快，我国对饮用水水源的水质、水量安全保障也提出了更高的要求。为此，水利部于2005年7月在全国开展了城市饮用水水源地安全保障规划工作。规划与全国农村饮用水安全工程规划相衔接，涉及全国661个设市城市和1 746个县级城镇的4 555个集中式饮用水水源地，在对饮用水水源地安全状况进行全面调查和评价基础上，针对水源地主要安全问题和因素，制定了水源地保护和综合治理工程、水源地建设工程（包括改扩建和新建水源）、水源地监控系统建设、水源地监督管理和应急预案等对策措施。

我国各地自然地理条件，特别是水资源条件差异显著，经济社会发展状况、城市化进程及供水组成有明显的区域差别，不同区域的城市饮用水安全状况也呈现不同的特征。根据各省（自治区、直辖市）城市饮用水水源地安全评价成果，将全国划分为若干规划分区，分析各规划分区水源地安全态势，提出安全保障措施总体布局，对指导和推动全国城市饮用水安全保障工作具有重要意义。

1　规划分区划分

为既能反映各地区饮用水安全保障的不同特点，又能反映相似地区的共同规律，在宏观自然条件背景下（如气候、水资源条件等），结合我国东、中、西经济社会发展布局，以省区为基本单位（保证行政区的相对完整），对全国各省级行政区的城市供水组成、城镇化水平及综合生活人均用水量等因素进行分析,将位置临近的省区进行整合,划分若干规划分区。其中,城市供水组成为城市河湖引水、水库、地下水水源地综合生活供水量分别占城市综合生活总供水量的比例;城镇化水平为城镇人口占总人口的比例;人均综合生活用水量为城市居民日常生活用水和公共建筑用水之和。

分析表明,全国各省级行政区中,地下水水源地供水比例相对较大的有青海、宁夏、山西、西藏、内蒙古、河北、新疆、陕西、河南、北京、山东、黑龙江、辽宁、甘肃、吉林、安徽等,除西藏外,主要分布在北方;河道型和水库型水源地供水相对较大的有天津、上海、重庆、浙江、广东、江西、湖北、海南、江苏、福建、湖南、广西、四川、云南、贵州等,除天津外,主要分布在南方,见图1。其次,我国的城镇化水平呈自东向西梯度递减态势,其中东北及东部沿海地带是中国地区经济发达地区,也是我国人口及城镇分布密集区,而西部地区的人口和城镇分布则相对稀疏。由于气候和水资源条件的不同,我国各省级行政区城市综合生活人均用水量也存在明显的区域差异,南方城市的综合生活人均用水量明显高于北方城市,同时城市综合生活用水量总体上由东南向西北逐渐递减。考虑省区临近、保持行政区划完整的要求,将全国大致划分为8个规划区域,即一区(北京、天津、河北、山东),二区(江苏、浙江、上海),三区(广东、福建、海南),四区(黑龙江、吉林、辽宁),五区(山西、河南、安徽),六区(湖北、湖南、江西),七区(陕西、甘肃、宁夏、青海、新疆、内蒙古),八区(重庆、四川、云南、贵州、广西、西藏)。

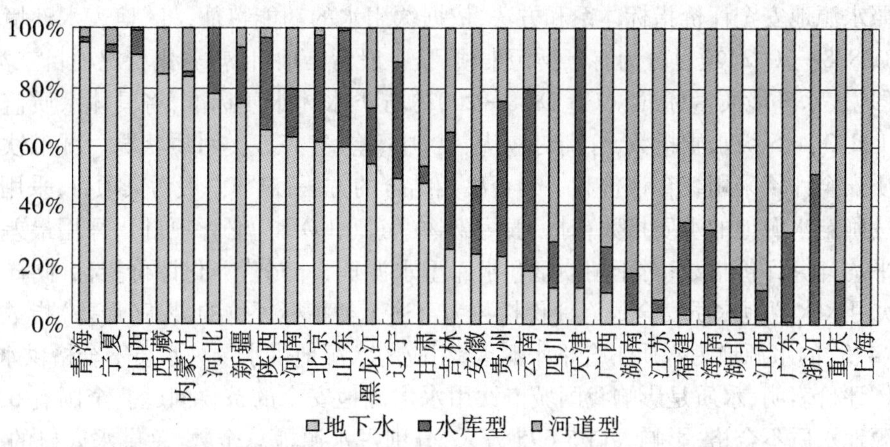

图1　全国各省级行政区城市供水组成

2　规划分区水源地分布状况

调查表明,2004年全国661个设市城市和1 746个县级城镇(含县城和其他县镇)集中式饮用水水源地共4 555个,综合生活总供水量275.76亿 m^3,水源地供水总人口为37 984万人。我国北方地区以地下水水源地供水为主,如七区、五区、一区的地下水供水人口比例分别为72.2%、61.3%和58.0%,南方以地表水水源地供水为主,如二区、三区、六区、八区的地表水供水人口比例分别为97.1%、98.0%、96.7%和86.3%。

各规划分区水源地分布状况分析如下:一区位于华北地区东北部,除少数河道型水源地供水外,主要以水库型和地下水型水源地供水为主,二区以长江口地区为主,处于太湖流域及淮河沂沭泗地区,主要以地表水水源地供水为主,其湖泊型供水人口全国最多,占全国湖泊型水源地供水人口的71.0%；三区以珠江口地区为主，主要靠河道型和水库型水源地供水，其中河道型供水人口最多，本区无湖泊型水源地供水；四区为东北三省，主要以地下水型、河道型和水库型水源地供水为主；五区位于我国中部地区的北部，供水组成与四区类似；六区的湖北、湖南和江西，反映中部地区特点，主要以河道型和水库型水源地供水为主；七区位于西北地区，主要以地下水型水源地供水为主；八区位于西南地区，主要以河道型、水库型等地表水水源地供水人口为主。各规划分区不同类型城市饮用水水源地供水人口见图2。

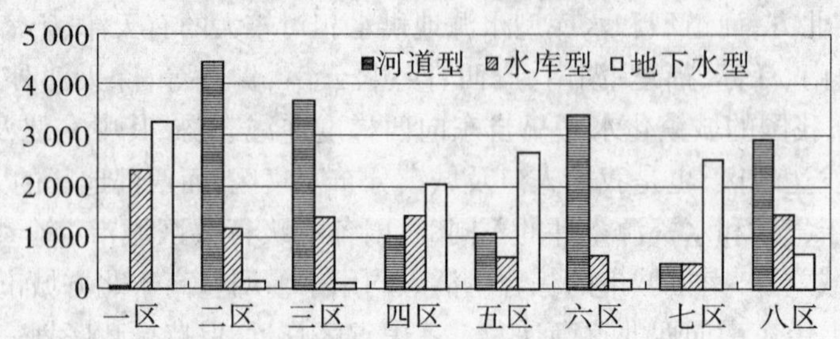

图2 各规划分区不同类型城市饮用水水源地供水人口 （单位：万人）

3 规划分区水源地安全态势

3.1 规划分区水源地水质安全状况

按照水源地安全评价指标体系和方法，依据饮用水的功能特征，将《地表水环境质量标准》(GB 3838—2002)项目分为有毒类污染项目、一般污染项目、营养化状况项目三类指标，并对应《地表水环境质量标准》(GB 3838—2002)五类水体水质标准，将具体水质监测指标换算为1、2、3、4、5级水质指数，分别对应优、良、中、差、劣等五类水质状况。城市饮用水水源地水质安全评价采用综合评价和一票否决相结合的方法：对非一般污染项目，采用一票否决的方法确定水源地的水质指数；对一般污染物(氨氮、COD为必评项目)，采用最差5项进行算术平均计算评价指数；对于湖泊型和水库型水源地，补充营养化评价指标，同样划分为5级。饮用水水源地水质安全最终评价指数为上述3类指标评价的最高(最差)指数。水质指数为“4、5”的水源地为水质不合格水源地，相应的水源地供水量为水质不合格供水量。

总体评价表明，水质是影响我国城市饮用水水源地安全的首要问题。全国有638个饮用水水源地水质不合格，影响人口5 695万人，分别占水源地总个数、总供水人口的14.0%和15.0%。其中四区、五区、七区、二区水质不合格影响人口比例较高，分别为26.0%、19.1%、19.0%和17.0%，而六区和八区水源地水质相对较好，水质不合格影响人口比例分别为5.2%和8.7%。

从水源地类型分析，河道型水质不合格水源地113个，影响人口2 238万人，除一区外，在其他各区具有分布，其中二区、三区、四区影响人口较多，分别占河道型水质不合格影响人口总数的28.2%、27.0%和13.9%。地下水型水质不合格水源地411个，影响人口为2 260万人，主要分布在一区、五区、七区和四区，影响人口分别占地下水型水质不合格影响人口总数的30.9%、25.9%、22.0%和18.0%。水库型水质不合格水源地102个，影响人口773万人，主要分布在四区、八区和一区，影响人口分别占水库型水质不合格影响人口总数的52.5%、24.0%和10.7%；湖泊型水质不合格水源地12个，影响人口相对较少，为424万人，主要分布在二区、八区和六区，其中二区水质不合格影响人口最多，占湖泊型水质不合格影响人口总数的82.5%。各规划分区不同水源地类型水质不合格影响人口见图3。

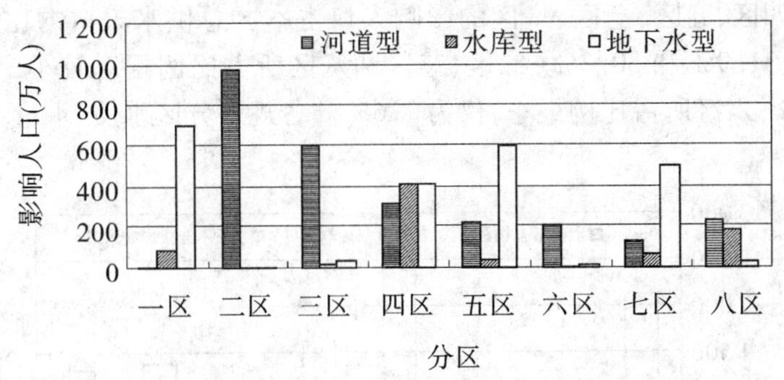

图3 各规划分区不同水源地类型水质不合格影响人口

从污染类型分析,638个水质不合格的水源地中,仅一般污染指标不合格的水源地有170个,影响人口1 956万人,主要分布在四区、二区、五区和三区,影响人口分别占该类型影响人口总数的32.9%、18.2%、14.8%和14.5%;仅非一般污染指标不合格的水源地363个,影响人口2 202万人,主要分布在一区、五区、七区,影响人口分别占该类型影响人口总数的30.8%、23.1%和15.6%;仅存在富营养化的水源地37个,影响人口628万人,主要分布在四区、二区和八区,影响人口分别占该类型影响人口总数43.1%、39.8%和15.0%;存在一般污染、非一般污染、富营养化两项及以上不合格的复合型污染水源地68个,影响人口910万人,主要分布在二区、三区、四区和一区,影响人口分别占该类型影响人口总数的34.2%、33.5%、12.2%和10.5%。

3.2 规划分区水源地水量安全状况

2004年,全国规划范围内供水人口4.18亿人(包括集中供水水源覆盖范围内的3.80亿人及部分自备、零星水源供水人口0.38亿人),综合生活供水量293.91亿m^3(包括居民生活用水和城市公共用水),综合生活人均用水量193 L/d。由于气候和水资源条件不同,南方的用水指标明显高于北方城市,如二区、三区、六区、八区综合生活人均用水量相对较高,一区、四区、五区、七区等相对较低。按照水源地安全评价指标体系和方法,水源地水量评价主要针对工程供水能力、枯水年来水量保证率或地下水开采率,反映水量是否满足水源设计水量要求,任一评价指标为不合格,则水源地水量安全评价为不合格。评价结果表明,水量评价为不合格的水源地1 233个,影响人口4 875万人。从规划分区分析,水量不合格水源地主要分布在一区、四区、五区、七区和八区,占本区水源地供水人口的比例分别为20.3%、23.3%、16.5%、16.7%和16.2%。全国地下水水源地综合生活供水量为66.79亿m^3,其中地下水超采量为11.9亿m^3,占地下水综合生活供水量的17.8%。超采水源地主要分布在我国北方地区,其中一区、七区、五区和四区的超采量较大,分别占全国综合生活总超采量的比例分别为44.0%、20.2%、16.5%和16.4%。

3.3 规划分区水源地水质水量综合评价

根据水源地水质和水量安全评价情况,对水源地安全状况进行总体评价。结果表明,水质不合格影响人口共计5 695万人,水量不合格影响人口共计4 875万人,扣除重复量后总的影响人口为9 480万人,占水源地总供水人口的25%。从规划分区分析,一区水质、水量不合格问题并存,以水量不合格影响为主;二区以水质不合格问题为主;三区、四区、五区和七区水质、水量不合格问题并重；六区水质、水量不合格问题相对较小；八区以水量不安全问题为主。其中，四区、五区、七区、一区的影响人口占本区总供水人口的比例较大，分别为41.0%、33.3%、31.9%和30.3%；二区、三区和八区所占比例相对较小，分别为17.5%、17.8%和23.6%；六区所占比例最小，仅为7.6%。各规划分区水质、水量不合格影响人口见图4。

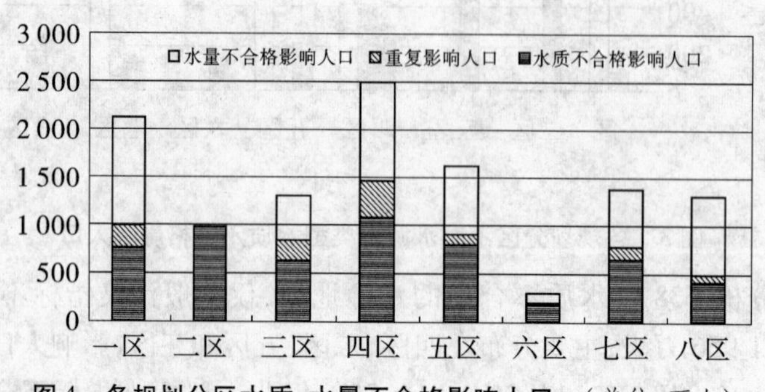

图4 各规划分区水质、水量不合格影响人口 （单位：万人）

4 水源地安全保障总体布局

我国水资源地区分布不均，水资源分布与城市经济社会发展布局、城市化进程也不匹配。为全面解决设市城市和县级城镇的集中式饮用水水源地安全保障问题，满足2020年全面实现小康社会目标对饮用水安全的要求，应在综合分析城市饮用水水源安全调查评价成果上，提出城市饮用水水源地安全保障对策措施。为指导各地城市饮用水水源地安全保障工作，结合各规划分区水源地安全态势及需水预测分析成果，提出各规划分区水源地安全保障措施总体布局。

（1）一区主要以地下水型和水库型水源地供水为主，地下水超采严重（占本区地下水供水量的28%），为资源型缺水地区，现状人均综合生活用水量低于全国平均水平。水质不合格水量主要以地下水水源地和非一般污染为主，不安全设市城市和县城个数最多，分别占本区设市城市和县城总数的47%和51%。区内25%的供水人口受到水质、水量不合格问题影响。由于不合格原因基本是地下水超采导致的，因此规划重点是加大地表水供水，主要是通过水库和南水北调跨流域调水等工程措施增加供水量。

（2）二区用水水平高，几乎全部为地表水供水（占本区总供水量的98%），湖泊型水源地供水量全国最多，人均综合生活用水量高。水质问题较多，不合格水源地主要为河道型水源地和湖泊型水源地，其中河道型水源地影响人口占本区影响人口总数的64%。水质不合格以一般污染、富营养化和复合污染为主，水质不合格水量占本区水源地供水量的16%，其绝对量大，占全国水质不合格总水量的21%。由于本区地表水水量丰富，因此应充分利用长江、钱塘江等江河水量，加大环境治理力度，减轻海水入侵影响，完善江苏江水北调、引江济太、实施太湖流域水环境综合治理、新建上海市城市水源地等。

（3）三区主要以河道型和水库型水源地供水为主（占本区总供水量的98%），用水水平最高。水污染严重，水质不合格水量占本区总供水量的13%，且以河道型水源地为主（占本区水质不合格总水量的92%），主要是一般污染和复合污染，并影响到对港、澳地区供水。由于本区水资源丰富，规划应充分利用西江、北江、闽江等江河水量，加大水环境容量，减轻咸潮影响，调整河道水源地进水口布局，工程措施除适当兴建供水水库外，区内主要以改扩建供水工程为主。

(4)四区地下水超采水量较多(占本区地下水供水量的16%)，加之地表水污染严重，资源型缺水和水质型缺水共存。区内26%的供水人口为水质不合格影响人口，为全国八大区之首。本区河道型、水库型、地下水型水源地均受到一定污染，水质不合格水量占本区总供水量的25%，污染类型主要以一般污染和富营养化为主。不安全设市城市占本区设市城市总数的34%。本区水资源并不丰富，因此应特别加大已有供水水源地的保护治理力度，新建水源地应以水库和引调水工程为主，如辽河大伙房引水、吉林中部供水、黑龙江引嫩工程等。

(5)五区与四区问题相当，既有地下水超采(占本区地下水供水量的14%)，又有地表水污染问题，人均综合生活用水量较低。水质不合格水量占本区总供水量的18%，污染类型主要以一般污染和非一般污染为主。不安全设市城市占本区城市总数的35%。区内33%的供水人口受到水质、水量不合格问题影响。由于当地水环境容量不足，要重视利用长江、黄河、淮河等水量，加强水环境治理，在南水北调、引黄入晋、引江济淮等工程基础上，重点建设水源地改扩建及配套工程。

(6)六区地表水相当丰富，一般无地下水超采问题，用水定额居中，水污染相对较轻，以非一般污染为主。本区是全国八大区中问题最小的一个区，区内水资源丰富，水质较好，规划要重视对已有水源地的保护，主要以改扩建工程和水源调配增加供水量。

(7)七区以地下水供水为主(占本区供水量的70%)，地下水超采相对严重(占本区地下水供水量的17%)，人均综合生活用水量偏低。水质不合格水源地也以地下水为主，中小城市水质相对较差，供水分散。不安全设市城市比例较高，占本区设市城市总数的46%，区内32%的供水人口受到水质、水量不合格问题影响。本区水资源短缺，地表水和地下水转换多，因此要重视水资源保护和治理工作，同时加大水库和饮水工程建设，保障供水安全。

(8)八区以地表水供水为主(占本区总供水量的87%)，水质问题总体较轻，水质不合格水量仅占总供水量的9%，以非一般污染(57%)为主，个别地区存在背景值高的问题严重。本区水资源丰富，由于地形原因使水资源开发利用相当困难，属工程型缺水地区。规划应在小型、分散的基础上，建设完善一些调水、引水工程，由于新建工程单方水成本较高，因此要坚持改扩建和新建工程并举，充分利用已建工程潜力。

5 结语

随着城市化进程的加快，我国城市饮用水水源安全面临水质不断恶化，部分水源地丧失功能；水量不足、保证率不高；地下水污染、超采严重；水源地安全管理和保障措施薄弱等问题，饮用水安全形势仍十分严峻。城市饮用水安全既受区域水资源条件等自然因素制约，又与城市布局及人口发展、水资源开发利用和水污染防治水平等经济社会因素密切相关，不同区域的城市饮用水安全状况和保障措施具有明显的区域差异性。以规划分区为指导，以城市为对象，以饮用水水源地为基本单元，找出城市饮用水水源地安全存在的主要问题，分析其综合生活用水需求和配置方案，合理提出饮用水水源地安全保障的工程和非工程措施，可为各省(自治区、直辖市)以及各城市的饮用水水源地安全保障提供指导，也可为跨区域、跨流域的水资源配置提供科学依据。

参考文献

[1] 翟浩辉.把握重点,统筹规划,保障城市饮用水水源地安全[J].南水北调与水利科技,2006,4(5):1-3.

[2] 国务院办公厅.国务院办公厅关于加强饮用水安全保障工作的通知(国办发[2005]45 号),北京,2005.

[3] 水利部水利水电规划设计总院.全国城市饮用水水源地安全保障规划技术大纲[R].北京:水利部水利水电规划设计总院,2005.

[4] 刘勇.中国新三大地带宏观区域格局的划分[J].地理学报,2005(5):361-369.

[5] 水利部.全国城市饮用水水源地安全保障规划[R].北京:水利部水利水电规划设计总院,2006.

【作者简介】 张建永,山东寿光人,北京师范大学环境科学专业,硕士,高级工程师,工作单位为水利部水利水电规划设计总院环境与移民处工作。

二、水环境与水生态

东江与水库联网供水水源工程污染控制措施浅议

肖许沐　季　冰　王丽影　孙　浩　胡和平　莫妙兴

（中水珠江规划勘测设计有限公司水环境与生态修复研究所　广州　510610）

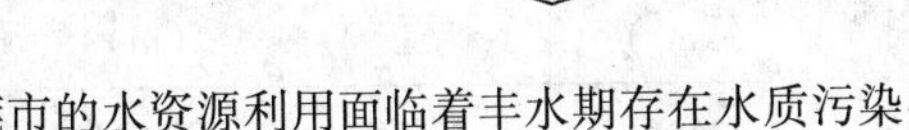

【摘要】 目前，东莞市的水资源利用面临着丰水期存在水质污染导致的水质性缺水威胁，枯水期存在水量缺水等问题，水库联网工程是东莞市水资源可持续利用的一项战略措施，是市政府督办的重点民生项目，本工程从分析联网水库环境现状和水环境容量入手，提出联网水库流域范围内水环境治理的总体思路及主要措施，以供类似工程借鉴和参考。

【关键词】 水资源利用　水环境容量　水环境治理　主要措施

目前，东莞市90%以上的供水水源直接取自东江干支流，由于东江流域来水年际年内分布不均，年际经常出现连丰连枯的现象，年内70%～80%以上的径流量集中在汛期，致使东莞市的枯季供水安全常常受到威胁；因此，2000年以来，东莞市政府和各有关部门按照水利部、国家发展和改革委员会的统一部署，先后开展了《东莞市水资源综合规划》和《东莞市饮用水源地安全保障规划》，为了在水质、水量上满足供水安全要求，规划提出了东江与水库联网工程方案，在考虑入境水为主要水源的同时，充分利用当地水资源及已有蓄水工程的调蓄能力，在东江沙角设取水口，抽取东江优质雨洪资源入联网水库以提高供水的保证程度，减少咸潮对东莞市供水的影响，缓解日趋突出的供需矛盾，尤其要解决水资源供需矛盾极为突出的中部及沿海片的供水问题，确保东莞市经济建设的可持续发展。考虑地形地貌和线路走向等各方面因素，拟定将松木山、同沙、横岗、水濂山、莲花山、芦花坑、白坑、马尾及五点梅等9座水库联网供水。

1　联网供水水源工程概况

东江与水库联网供水水源工程位于东莞市境内，是为东莞市提供原水、充分利用当地水资源、最大限度地发挥现有水库的调蓄作用而建设的。工程是针对目前东莞市中部及沿海片供水保证率低、供水设施调节能力缺乏和抗风险能力低的突出问题而提出的。

供水工程在东莞市石排镇东江左岸沙角村设有取水泵站，取水流量为27 m^3/s，通过30.336 km的有压管道和泵站将水送入松木山水库，沿线在泰岗圩设加压泵站，大江头设有调压塔，松木山水库旁设有自动化控制大楼（调度中心）。水库联网分东西线：东线为松木山—莲花山—马尾—五点梅—芦花坑。西线为松木山—同沙，在同沙再设一加压泵站至横岗，由横岗分别至水濂山、白坑及芦花坑，形成9个水库的环状闭合联网。联合线路长40.081 km，工程全长70.417 km，贯通15个镇区。

2 联网水库环境现状

2.1 水库概况

本联网工程共涉及9座水库，其中横岗、松木山和同沙水库为中型水库，总兴利库容为9 108万 m^3，占9座水库总兴利库容11 287万 m^3 的81%，建成于1959～1960年，其余均为小(1)型水库，建成于1958～1960年。9座水库总集水面积为241.37 km^2，采用1956～2000年系列，按径流系数法分析，联网水库多年平均径流量为1.95亿 m^3，兴利库容为1.13亿 m^3，各联网水库现状特征值见表1。

表1 各联网水库现状特征值

序号	水库名称	类型	镇区	集水面积（km^2）	平均径流量（万 m^3）	正常蓄水位（m）	兴利库容（万 m^3）	死库容（万 m^3）
1	横岗	中型	厚街	44.6	3 468	22.74	2 094	146
2	松木山	中型	大朗	54.2	4 341	24.74	3 798	172
3	同沙	中型	东城	100	8 326	19.74	3 216	166
4	莲花山	小(1)型	长安	8.5	760	19.74	419	0.32
5	马尾	小(1)型	长安	6.57	441	10.99	261	0.05
6	五点梅	小(1)型	长安	3.5	286	10.99	543	51.5
7	白坑	小(1)型	虎门	5.8	474	12.24	319	30.5
8	芦花坑	小(1)型	虎门	6	490	10.99	160	34.22
9	水濂山	小(1)型	南城	12.2	881	21.74	477	140
	合计			241.37	19 467		11 287	740.59

注：水位以国家85高程系为基准。

2.2 水质现状分析

根据9座联网水库水质现状监测结果，采用单因子评价法进行评价，各水库水质受到不同程度的污染，除水濂山水库水质为Ⅲ类外，其余8座水库水质均为Ⅳ类至劣Ⅴ类，其中同沙、松木山、莲花山水库3座水库库区水质最差，水质评价结果为劣Ⅴ类，主要超标项目为总氮、粪大肠菌群、五日生化需氧量等；白坑、马尾2座水库水质评价结果为Ⅴ类，主要超标项目为粪大肠菌群、五日生化需氧量、高锰酸盐指数等。

2.3 底质现状分析

在水库的污染源中，污染底泥属于内源污染源，底泥中的污染物会不断地向水体中迁移释放，造成水库水质恶化，因此必须对底质进行调查评价。同沙水库底质监测，共监测11项指标，分别为pH值、铜、锌、铅、镉、总铬、镍、砷、总氮、总磷及有机碳。每个点根据底泥的厚度分层取样。评价以《土壤环境质量标准》（GB 15618—1995）中集中式水源地土壤标准为准，9座联网水库的重金属指标总体上不能满足《土壤环境质量标准》（GB 15618—1995）中的Ⅰ类集中式水源地底质的相关要求，铜、锌、总铬、铅和镍等普遍超标，重金属污染底泥释放的污染物会对饮用水安全带来非常大的风险，从环境的角度出发，必须对水库进行环境清淤，以清除水库内受污染底泥，并在清淤过程中防止重金属的释放。

3 联网水库水环境容量分析

3.1 有机物水环境容量模型

由于九座水库与东江联网后，各水库类似于水缸，可以被看做一个均匀混合的水体，因此 COD_{Mn} 和 BOD_5 负荷量可以采用完全混合衰减模型，可以采用下述计算方程

$$V\frac{dC_t}{dt}=W-QC_t-KVC_t \tag{1}$$

式中：W 为 COD_{Mn}、BOD_5 环境容量，kg/a；Q 为水库年排水量，万 m^3/a；K 为 COD_{Mn}、BOD_5 降解系数，a^{-1}；V 为水库容积，万 m^3；C_t 为水库中 COD_{Mn}、BOD_5 瞬时浓度，kg/万 m^3；T 为计算时段长度，a。

求解式(1)得环境容量表达式为

$$W=\frac{C_s-C_0\exp\left[-\left(K+\frac{Q}{V}\right)t\right]}{1-\exp\left[-\left(K+\frac{Q}{V}\right)t\right]}(Q+KV) \tag{2}$$

式中：C_0 为 COD_{Mn}、BOD_5 初始浓度，kg/万 m^3；C_s 为 COD_{Mn}、BOD_5 目标浓度，kg/万 m^3；其余符号意义同前。

因为九大联网水库之间存在调水问题，上式计算得到的 W 并非最大允许排放量，应扣除调水过程中代入目标水库的污染物的量，所以最大允许排放量应修正如下

$$W^*=W-Q_{in}C_{in} \tag{3}$$

式中：W^* 为 COD_{Mn}、BOD_5 最大允许排放量，kg/a；Q_{in} 为从联网水库调水的量，万 m^3；C_{in} 为调水中的 COD_{Mn}、BOD_5 浓度，kg/万 m^3。

3.2 总氮、总磷水环境容量模型

在富营养化水库中 T-N 和 T-P 水环境容量计算根据《水域纳污能力计算规程》(GB/T 25173—2010)推荐采用狄龙模型，即

$$L=\frac{C_s\overline{Z}(Q_{\lambda}/V)}{1-R} \tag{4}$$

式中：L 为总磷、总氮单位允许负荷量，g/(m^2·a)；C_s 为总磷、总氮的水环境标准，或者氮磷的入库浓度，mg/L；$\overline{Z}$ 为平均水深，m；R 为氮磷的滞留系数，a^{-1}，$R=0.426e^{-0.271q}+0.574e^{-0.00949q}$；$Q_{入}$ 为年入库水量，为本地来水量和调入水量之和，m^3/a；V 为水库库容，m^3。

由于存在水库间的调水，且一般年均调水量相对于库容较大，计算出的总氮总磷容许负荷需减去因调水代入的总氮总磷量才得到总氮总磷的实际环境容量，即

$$M=LA-Q_{in}C_{in} \tag{5}$$

式中：A 为水库水面面积，m^2；Q_{in} 为从联网水库调水的量，万 m^3；C_{in} 为调水中的 T-N 和 T-P，kg/万 m^3。

3.3 水环境容量计算

对以上公式中参数进行识别(其中降解系数 K 参考鄱阳湖等有关水库湖泊的研究成果)，得出 9 座水库水质各指标的水环境容量值，见表 2。

表2 水环境容量值

水库	COD_{Cr}	BOD_5	T-N	T-P
同沙水库	4 646	808	231	68
横岗	2 026	349	139	46
松木山水库	3 828.89	1 052.68	224.8	72.24
水濂山	505	87	22	6.6
白坑	322	56	13.3	3.6
莲花山水库	414.23	111.03	22.92	4.16
马尾水库	281.93	74.55	25.56	4.2
五点梅水库	444.8	126.4	13.34	1.82
芦花坑水库	247.99	65.49	8.88	0.6

4 水源污染控制措施分析

通过对9座联网水库入库污染物量、水环境容量和入库消减量的计算分析，提出工程措施和非工程措施的综合治理思路，其中工程措施是减少入库污染物的关键步骤，主要是针对点源污染、面源污染及内源污染问题；非工程措施主要是针对各类污染源的管理和限制，在治污工作中起根本的作用，但对于目前的环境管理状况，非工程措施在今后相当长的一段时间内逐步完善，最终发挥效力。要达到既定的水质目标，近期内起主要作用的是工程措施。

4.1 工程措施

4.1.1 截污及治污工程

作为饮用水源地，在保护区内的污染源应以正本清源为目标，将点源污染截到污水处理厂，逐步实现雨、污分流，2002年以来，在水环境综合整治中，东莞市及各镇区环保部门进行了大量的工作，全市计划完成36座污水处理厂及其配套截污干管工程，在这36项截污治污工程中，9座联网水库汇水区域涉及到大岭山、长安、虎门、松山湖科技产业园、南城、东城、厚街7个镇区，涉及的污水处理厂及其配套截污主干管网工程共6项。

4.1.2 雨季面源污染治理工程

为了经济有效地削减分散点源和非点源污染物入库量，在加强点源治理的同时，有计划、因地制宜地运用生态学的原理和方法，采取工程措施，进行生态修复，以期达到最大限度地控制污染物入库量的目的。雨季面源污染治理工程采用人工湿地治污方法。

雨季面源污水先经过调节前置库（构造成氧化塘形式），经过预处理后，再经过人工湿地处理达到地表水Ⅲ类标准后排入水库，大大减少了入库污染物，为水库提供清洁的饮用水源补充水，人工湿地工艺流程图见图1，该污水系统具有高效率、低投资、低运转费、低维持费、低能耗等特点。

4.1.3 内源污染治理工程

在外源污染负荷得到有效控制的情况下，污染底泥的内源污染负荷往往具有等效的潜在性污染影响，而清淤是解决内源污染有效的措施。

目前，对水库的清淤方法主要有直接挖运法、动态清淤法和静态清淤法，其中直接挖运法有排干直接挖运法、抓斗船开挖法，动态清淤法有绞吸式挖泥船、射流式开挖船和潜水式开挖船等开挖方法，静态清淤法有气力泵系统船开挖方法等，针对本工程特点，主要对常用的抓斗船开挖法（直接挖运法代表）、绞吸式挖泥船开挖法（动态清淤法代表）及气力泵船开挖法（静态清淤法代表）进行比较。

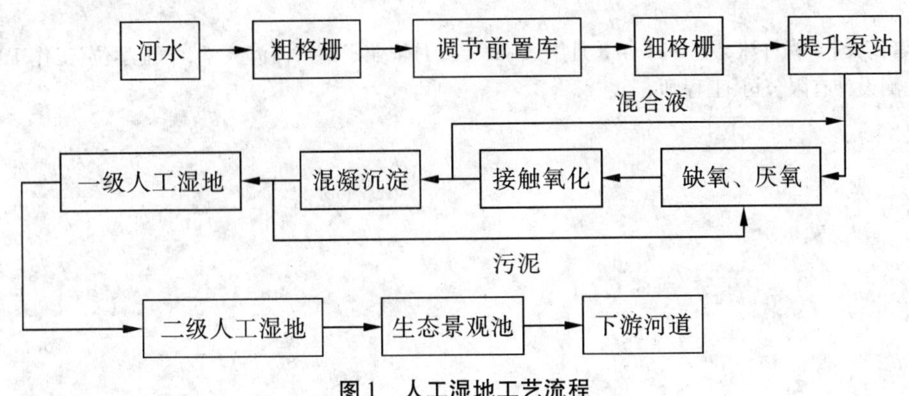

图1　人工湿地工艺流程

污染淤泥处置方法有物理晾晒、机械脱水、化学固化、热处理方法，物理晾晒虽施工简单、处理成本低，但其缺陷也较明显，如晾晒需占用大面积场地，而且易受天气影响；机械脱水设备价格较高，处理效率低，难以满足大型疏浚工程要求，而且机械脱水只是进行了淤泥减量化，无法对其中的重金属等污染物进行稳定化；采用热处理方法处置淤泥，能产生较高附加值的产品，但它对淤泥性质有一定要求，而且处理费用高、处理淤泥量有限，应用于处理大量发生的淤泥也不大适宜；化学固化方法对淤泥进行处置，处理效率高，施工方便，易于进行推广应用，该技术已在无锡长广溪清淤、管社山湖底清淤得到成功应用，最终采用化学固化方法，但化学方法处理淤泥的投资稍大。

4.2　非工程措施

非工程措施中至关重要的是进行水库水源保护区的划分，制定水库饮用水水源保护管理办法，为水源保护提供法律依据，也为规范保护区内水事行为提供制度保障。划定饮用水水源保护区是保护饮用水水源地的执法依据，也是实施各级饮用水水源地环境保护规划的工作基础。各水库水源保护区包括水域范围和陆域范围，饮用水源保护区陆域范围分一级保护区、二级保护区和准保护区三个级别，其中大中型水库划分为一级保护区、二级保护区和准保护区，小型水库一般不设准保护区。各级别的划分原则上是以正常蓄水位或岸线为基线向陆域纵深200 m、2 000 m等为界线。

5　结语

（1）东江与水库联网供水水源工程从战略的高度，整体研究区域的供水布局，协调好入境水与当地水的关系，在以入境水为主要水源的同时，充分利用当地水资源及已有蓄水工程的调节能力，研究东江丰水期的水资源利用，实现水资源的优化配置，缓解日趋突出的供需矛盾，确保东莞市经济建设的可持续发展。

（2）水源保护工程措施包括截污及治污工程、雨季面源污染治理工程、内源污染治理工程。通过合理的工程措施将入库污染物量降至最低，工程措施是减少入库污染物的关键步骤，目前工程措施在治污工作中起主要作用，远期非工程措施对保护联网水库饮用水源起到根本性的作用，通过工程措施和非工程措施，联网水库水质才能达到预期的目标。

参考文献

[1] 姜平. 贵州省湖库污染现状和湖库富营养化控制对策[J]. 贵州环保科技，2004，1-6.

【作者简介】 肖许沐，男，1981 年 8 月生，2005 年 6 月毕业于南京河海大学，学士学位，工作于中水珠江规划勘测设计有限公司，工程师。

生物完整性指标(IBI)评价方法在河流健康管理中的应用*

朱 迪 杨 志 常剑波

(水利部中国科学院水工程生态研究所,
水利部水工程生态效应与生态修复重点实验室 武汉 430079)

【摘要】 本文探讨了生物完整性评价指标的研究进展及适应性 IBI 指数在河流健康管理中应用,列举了长江中上游地区 IBI 指标评价指标体系构建,并对中游浅水湖泊和上游三峡库区的生物完整性演变趋势作了评价。结果表明长江流域多年来(1964~1998 年、1997~2002 年、2003~2008 年、2010 年)鱼类生物完整性的变化呈明显下降趋势。中游浅水湖泊受人类干扰较大的湖泊或者年份 IBI 得分较低;在长江上游三峡库区江段,随着从坝前向库尾延伸,IBI 指数值呈现递增趋势。库尾江段的健康状况明显优于库区中下游江段,同时支流江段的健康状况明显优于库区干流水域;此外,受工程影响较小的巫山江段的健康状况也优于受电站影响较大的北碚江段。但总体而言,各个江段的生物完整性等级均为差(IBI 值<42)。上述案例所呈现的生物完整性的时空变化趋势与其他相关研究所反映的长江流域水环境质量互为补充验证,可以为湖泊管理、持续发展和利用以及水环境保护提供更为充分的科学依据。

【关键词】 河流健康 生物完整性指数(IBI) 长江 三峡大坝(TGD)库区 流域管理

1 河流健康的概念与内涵

河流健康的概念是人们在过度开发利用河流系统,使得河流环境不断恶化,需要修复河流的自然状态下提出来的,从这个角度看河流健康的本质是河流生态系统健康,也是目前人们关注最多的。Schofield 和 Davies(1996)把健康定义为自然性:“河流健康就是指与相同类型的未受干扰的(原始的)河流的相似程度,尤其是在生物完整性和生态功能方面。”Simpson 等(1999)则把其定义为河流生态系统支持与维持的主要生态过程,以及具有一定种类组成、多样性和功能组织的生物群落尽可能接近首受扰前的状态的能力。这些定义提供了这样一种思路:通过与基准状态的河流相比较来监测和评价河流健康。

传统的河流环境评价是以物理、化学指标为基础,通过对物理、化学指标的分析来反映河流系统所处的环境条件状况(赵彦伟,2005)。这种评价并不能有效地反映河流生态系统的健康状态,因为它实质上说明的是生态系统所面临的环境压力,而不是生态系统对环境条件变化的反应及受到的影响。从现今发展来看,采用多种评价指标进行河流健康评价

* 资助项目:水利部公益性行业科研专项(200901013),科技部国际合作项目(DFA31550)。

(Smith,1999;Karr 和 Chu,2000;林木隆等,2006)已成为了一种趋势(见表1)。

表1 国外主要的河流健康评价方法(林木隆等,2006)

评价方法	提出者	内容简介	主要特点
RIVPACS	Wright 1984	利用区域特征预测河流自然状况下应存在的大型无脊椎动物,并将预测值与该河流大型无脊椎动物的实际监测值相比较,从而评价河流健康状况	能较为精确地预测某地理论上应该存在的生物量;但该方法基于河流任何变化都会影响大型无脊椎动物这一假设,具有一定片面性
AUSRIVAS	Simpson 1994	针对澳大利亚河流特点,在评价数据的采集和分析方面对 RIVPACS 方法进行了修改,使得模型能够广泛用于澳大利亚河流健康状况的评价	能预测河流理论上应该存在的生物量,结果易于被管理者理解;但该方法仅考虑了大型无脊椎动物,并且未能将水质及生境退化与生物条件相联系
IBI	Karr 1981	着眼于水域生物群落结构和功能,用12项指标(河流鱼类物种丰富度、指示种类别、营养类型等)评价河流健康状况	包含一系列对环境状况改变较敏感的指标,从而对所研究河流的健康状况作出全面评价;但对分析人员专业性要求较高
RCE	Petersen 1992	用于快速评价农业地区河流状况,包括河岸带完整性、河道宽/深结构、河岸结构、河床条件、水生植被、鱼类等16个指标,将河流健康状况划分为5个等级	能够在短时间内快速评价河流的健康状况;但该方法主要适用于农业地区,如用于评价城市化地区河流的健康状况,则需要进行一定程度的改进
ISC	Ladson 1999	构建了基于河流水文学、形态特征、河岸带状况、水质及水生生物5方面的指标体系,将每条河流的每项指标与参照点对比评分,总分作为评价的综合指数	将河流状态的主要表征因子融合在一起,而为科学管理提供指导;但缺乏对单个指标相应变化的反映,参考河段的选择较为主观
RHS	Raven 1997	通过调查背景信息、河道数据、沉积物特征、植被类型、河岸侵蚀、河岸带特征以及土地利用等指标来评价河流生境的自然特征和质量	较好地将生境指标与河流形态、生物组成相联系;但选用的某些指标与生物的内在联系未能明确,部分用于评价的数据以定性为主,使得数理统计较为困难
RHP	Rowntree 1994	选用河流无脊椎动物、鱼类、河岸植被、生境完整性、水质、水文、形态等7类指标评价河流的健康状况	较好地运用生物群落指标来表征河流系统对各种外界干扰的响应;但在实际应用中,部分指标的获取存在一定困难

近年来,我国已经开始关注从河流健康状况视角保护河流,在河流健康状况评价方法学、河流的可持续管理等方面陆续开展了一定的工作(唐涛,2002)。河流健康的概念已经得到了我国政府部门的认可,前水利部部长汪恕诚在第二届黄河国际论坛上指出,中国河流流域管理机构的首要任务是“维护河流的健康生命”。各大流域机构都已将维护河流健康作为主要的工作目标,如长江流域健康长江(蔡其华,2005)、珠江流域健康指标体系(林木隆等,2006)和黄河健康评价与修复基本框架(赵彦伟等,2005)。本文通过选择符合地区特

征的受人为干扰相对较小的河流作为参照基准,以鱼类完整性指数为评价指标,从河流健康角度评价长江不同区域水体的生态环境质量,为广泛开展的河流恢复项目提供基础数据和决策依据。

2 研究方法和材料

2.1 生物完整性指数法(IBI)

生物完整性指数(IBI)首先是由 Karr(1981)提出的一种水域生态系统健康状况评价指标,其定义为:"一个地区的天然栖息地中的群落所具有的种类组成、多样性和功能结构特征,以及该群落所具有的维持自身平衡、保持结构完整和适应环境变化的能力。"IBI 由 12 个指标 3 大类别组成(Karr,1981;Karr 和 Chu,1999):①种类丰度和组成;②食性结构;③鱼类丰度和健康状况。每一个指标被赋值 5 或 3 或 1,若该指标的原始数据接近期望值即被赋值为 5,若该指标的数据严重偏离期望值就赋值为 1,若处于两者之间则赋值为 3。所有指标赋值的总和表示实测数据和期望鱼类群落数据的偏离程度。

2.2 鱼类群落调查方法

本文数据获得均采用商业捕捞渔获物调查法。调查网具为三层刺网和定置沿绳钼钩等,现场对象为专业捕捞的渔民。在调查期间逐日进行统计,分析调查起始日至该调查日的鱼类种类变化特征,若连续 7 个调查日中渔获物的种类组成没有差异,则表明调查已经满足采样统计要求,此次采样工作结束。对所获得的渔获物样本,现场鉴定渔获物种类,逐尾测量全长、体长、体重,并记录鱼体的受伤、寄生虫等个体情况。对个别没有当时确定到种的鱼类标本,用5% ~10% 的福尔马林浸泡,运回实验室进行二次鉴定和测量。种类鉴定主要依据《四川鱼类志》、《中国动物志 硬骨鱼纲 鲤形目(中卷)》、《中国动物志 硬骨鱼纲 鲤形目(下卷)》等文献。

3 IBI 在长江流域健康评价中的应用

3.1 长江中游浅水湖泊 IBI 指标体系及评价结果

根据历史基准资料,建立了适合长江中浅水湖泊的 IBI 指标体系,其指标体系以及生物完整性等级划分与特征如表 2、表 3 所示。IBI 的结果表明,1964 ~ 1998 年,洪湖生物完整性逐年降低;以 1992 年前后的资料进行横向比较,东湖、洪湖、保安湖和三湖连江水库的 IBI 值分别为 22、28、38 和 42,其中东湖的生物完整性表现极差、洪湖为差、保安湖和三湖连江水库为一般(见表 4),表明各个湖泊均受到一定程度的人工干扰(朱迪,常剑波,2004)。

3.2 长江上游三峡库区鱼类生物完整性变化

3.2.1 适合长江上游三峡库区江段的鱼类生物完整性指标体系

根据长江上游宜宾至宜昌江段历史数据和资料(1997 ~ 2002 年),分析鱼类群聚结构和组成特征,构建适合长江上游江段的鱼类生物完整性指标体系(见表 5)。

3.2.2 1997 ~ 2002 年三峡库区江段的鱼类生物完整性指标评价

对长江上游宜宾至宜昌干流河段初步评价,其结果表明:根据计算结果(见表 6),三峡库区江段的多数的生物完整性处于一般和良好等级。其中,宜宾、合江、木洞和宜昌在 1997 年的生物完整性分别为"极好"、"好"、"好"和"一般";1998 年均为"好";1999 年和 2000 年都是分别为 "好"和"一般";在 2001 年和 2002 年的生物完整性都是"一般"。在统计上,IBI

的得分在6年的范围内降低了,多数IBI值被划分为好和一般两个等级(Zhu Di, Chang Jianbo,2008)。

表2 适合长江中游浅水湖泊的IBI指标体系

属性(Attribute)	指标(Metrics)	评分标准(Metrics Scores Criteria)		
		5	3	1
种类丰度和组成 Species richness and composition	种类数占期望值的比例	>60%	30% ~60%	<30%
	鲤科鱼类种类数百分比	<45%	45% ~60%	>60%
	鳅科鱼类种类百分比	2% ~4%	4% ~6%	6% ~8%
	鲿科鱼类种类百分比	2% ~5%	5% ~8%	9% ~12%
	商业捕捞获得的鱼类科数	>18	12 ~18	<12
	鲫鱼(放养鱼类)比例	7% ~22%	23% ~38%	39% ~54%
营养结构 Trophic composition	杂食性鱼类的数量比例	<10%	10% ~40%	>40%
	底栖动物食性鱼类的数量比例	>45%	20% ~45%	<20%
	鱼食性鱼类的数量比例	>10%	5% ~10%	<5%
丰富度和个体健康状况 Abundance and individual conditions	单位渔产量(kg/ha)	>100	80 ~40	<40
	外来种所占比例	0	0% ~1%	>1%
	感染疾病和外形异常个体比例	0% ~2%	2% ~5%	>5%

表3 IBI值、生物完整性等级划分及特征

IBI数值(IBI Values)	完整性等级(Integrity Classes)
53 ~60	极好(Excellent)
43 ~52	好(Good)
33 ~42	一般(Fair)
23 ~32	差(Poor)
13 ~22	极差(Very Poor)
NF	没有鱼(No fish)

表4 长江中游浅水湖泊生物完整性指数(IBI)评价赋分

指标 Metrics	洪湖 Honghu				东湖 Donghu	保安湖 Baoanhu	三湖连江 Sanhulianjian
	1964	1981	1993	1998	1992	1992	1992
IBI值(IBI value)	52	40	28	26	22	38	42
生物完整性等级(Integrity Classes)	好(Good)	一般(Fair)	差(Poor)	差(Poor)	极差(Very poor)	一般(Fair)	一般(Fair)

表 5 适合长江上游的 IBI 指标和赋值标准

属性 (Attributes)	指标 (Metrics)	赋值(IBI Scoring) 5	3	1
种类丰度和组成 (Species richness and composition)	所有本地种数占期望值的比例	>50%	30% ~ 50%	<30%
	鲤科鱼类所占比例	>50%	25% ~ 50%	<25%
	鳅科鱼类所占比例	>15%	5% ~ 15%	<5%
	鲿科鱼类所占比例	>20%	5% ~ 20%	<5%
非耐受/耐受性种类 (Intolerance and tolerance species)	耐受性个体所占比例	< 6%	6% ~ 12%	>12%
	渔获物中出现的科数	>18	12 ~ 18	<12
营养结构 (Trophic composition)	杂食性鱼类所占比例	<8%	8% ~ 15%	>15%
	底栖动物食性鱼类所占比例	>40%	20% ~ 40%	<20%
	顶级肉食性鱼类所占比例	>15%	5% ~ 15%	<5%
丰富度指标(Abundance)	单位努力捕捞量(CPUE, kg/船)	>2	1% ~ 2%	<1
个体健康状况 (Individual condition)	非本地种所占的比例	<1%	1% ~ 2%	>2%
	DELT 个体所占比例	<2%	2% ~ 5%	>5%

注: * DELT 指外表畸形、鳍损伤、鱼体受损、肿瘤、疾病和寄生虫的个体。

表 6 1997 ~ 2002 年四个监测江段的 IBI 指标评价

年代(Years)	宜昌 1997	1998	1999	2000	2001	2002	合江 1997	1998	1999	2000	2001	2002
IBI 得分 (IBI value)	46	52	42	40	42	44	38	32	50	48	44	46
完整性等级 (Integrity Classes)	极好	好	一般	一般	一般	好	一般	差	好	好	好	好

年代(Years)	宜宾 1997	1998	1999	2000	2001	2002	木洞 1997	1998	1999	2000	2001	2002
IBI 得分 (IBI value)	56	42	50	42	44	42	40	52	48	50	44	42
完整性等级 (Integrity Classes)	极好	一般	好	一般	好	一般	一般	好	好	好	好	一般

3.2.3 2003 ~ 2008 三峡库区鱼类生物完整性指标评价

根据 2003 ~ 2008 年对长江上游三峡库区的鱼类调查数据,采取表 5 建立的适合长江库区的 IBI 指标和赋值标准及等级评价标准,分别整理出三峡库区不同江段的鱼类分布、群落数量结构和鱼类食性数据,计算各个区域相应指标的得分。根据计算结果,长江上游鱼类生物完整性处于一般等级,三峡库区万州站个别年份(2006 ~ 2007 年)鱼类生物完整性等级为差。在统计上,长江上游 IBI 得分在 6 年(2003 ~ 2008 年)内降低,多数 IBI 值被划分为一般等级(见图 1)。

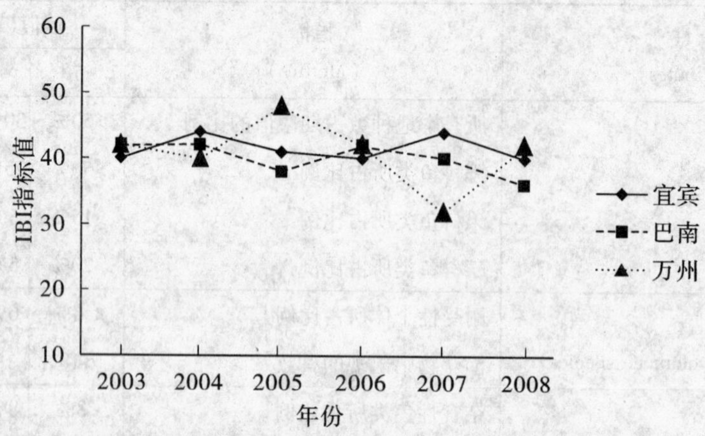

图 1　2003～2008 年间长江三峡库区宜宾、巴南和万州段 IBI 值变化

3.2.4　2010 年三峡库区鱼类生物完整性评价

对 2010 年三峡库区及四条支流香溪河、九畹溪、大宁河、磨刀溪的生物完整性进行评价,其结果为:香溪河、九畹溪、大宁河、磨刀溪的生物完整性得分分别为 40、38、40、44,表明香溪河、九畹溪、大宁河生物完整性等级均为“一般”,磨刀溪为“好”。由结果可以看出,在库区干流江段,随着从坝前向库尾延伸,健康指数呈现递增趋势。库尾江段的健康状况明显优于库区中下游江段;同时,支流江段的健康状况明显优于库区干流水域;此外,受工程影响较小的巫山江段的健康状况也优于受电站影响较大的北碚江段。但总体而言,各个江段的生物完整性等级均为差(IBI 值<42)(见图 2),表明库区整体水域受人工干扰的程度较大。

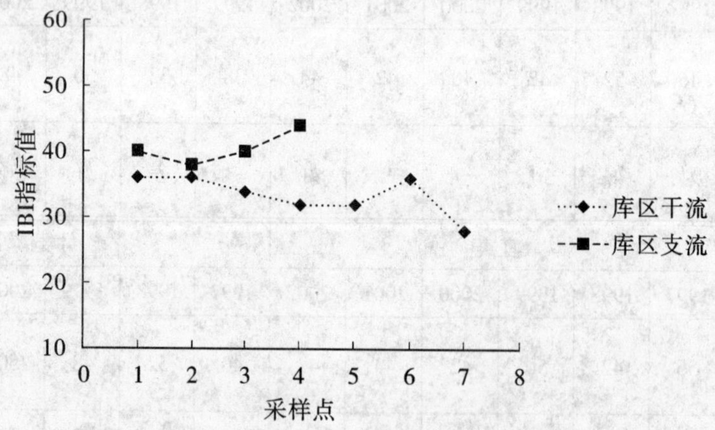

图 2　2010 年三峡库区各个监测断面及四条支流 IBI 得分

4　讨论

我国有丰富的水资源,随着人类活动的加剧,围湖造田、污染、江湖阻隔以及渔业的过度开发利用等已成为长江流域水体面临的几个主要环境问题,严重干扰了水体的生物完整性,如文中所示,三峡库区各个江段的生物完整性等级均处于“差”到“一般”等级;在库区干流江段,随着从坝前向库尾延伸,IBI 值呈现递增趋势,其中库尾江段的健康状况明显优于库

区中下游江段,支流江段的健康状况明显优于库区干流水域,表明库区整体水域受人工干扰的程度较大。同时,人类活动影响如大型水利工程建设、过度捕捞、环境污染等不同的环境胁迫,对水域生物完整性所造成的后果也是不一样的,应该进行分类和界定,以期找到最有效的解决方案。逝者已矣,来者可追。水生生态环境过去所遭受的破坏已不能完全恢复,但是我们不能使已遭受破坏的更加恶化,而且更应花大力气去保护那些尚未遭受污染的水域环境。

在我国的河流评价和管理中,人们习惯于用单一评价指标作为关键控制性指标(赵彦伟,2005),认为IBI评价指标体系过于专业和复杂,难以在实际管理中推广使用,甚至有人建议在评价健康长江的指标体系中用鱼类损失指数作为鱼类群落状况指标。在国外相关文献和记载中,一是专业技术人员参与河流管理,包括生态监测、数据分析乃至管理策略制定等;另外,长期的监测积累了大量数据,并制定细化了相关操作流程、技术方案、评价划分标准,这就使得执行者有章可循,如美国环境管理局(EPA)制订的监测评价流程,使得IBI评价法已成为其水体生物监测的基础等。因此,IBI在我国流域健康管理中应用时,须进行属地化研究,制订一套适合我国流域实际情况的、与美国EPA标准有差别的、国际认可的评价程序和是势在必行。

参考文献

[1] Schofield N J,Davies PE. Measuring the health of our rivers[J], Water, 1996,May/June,39-43.

[2] 赵彦伟,杨志峰,姚长青. 黄河健康评价与修复基本框架[J]. 水土保持学报. 2005,5(19):131-134.

[3] Karr JR. Assessment of biotic integrity using fish communities[J]. Fisheries,1981, 6(6): 21-27.

[4] Smith M J, Kay W R, Edward D H D, et al. AusRivAS: Using macroinvertebrates to assess ecological condition of rivers inWestern Australia [J]. Freshwater Biology,1999, 41: 269-282.

[5] 林木隆,李向阳,杨明海. 珠江流域河流健康评价指标体系初探[J]. 人民珠江,2006(4):1-3.

[6] 唐涛,蔡庆华,刘健康. 河流生态系统健康及其评价[J]. 应用生态学报,2002,13(9):1191- 1194.

[7] 蔡其华. 维护健康长江,促进人水和谐. 中国水利,2005(8):7-9.

[8] 朱迪,常剑波. 长江中游浅水湖泊生物完整性评价. 生态学报. 2004, 24 (12):2761-2767.

[9] Di Zhu, Jianbo Chang. Annual variations of biotic integrity in the upper Yangtze River using an adapted index of biotic integrity(IBI)[J]. Ecological Indicators,2008,8(15):564-572.

[10] Karr, J R,E W Chu. Sustaining Living Rivers[J]. Hydrobiologia, 2000, 422/423: 1-14.

【作者简介】 朱迪,女,1978年12月生,毕业于中国科学院水生生物研究所,博士,现工作于水利部中国科学院水工程生态研究所,助理研究员。

淮河流域防污体系与防污标准探讨*

张　翔[1]　李　良[1]　高仕春[1]　冉啟香[2]

(1. 武汉大学水资源与水电工程科学国家重点实验室　武汉　430072;
2. 长江水利委员长江上游水文水资源勘测局　重庆　400014)

【摘要】 流域水污染综合防治需要从水的自然循环和社会循环两方面入手。河流防污体系由污水处理厂、面源污染控制等工程措施和水量水质联合调度、风险管理等非工程措施组成,河流防污标准指在上述工程措施和非工程措施作用下,流域防污能力,特别是防止发生重大突发水污染事件的能力。本文从防污调度等非工程措施角度,介绍淮河流域防污体系中防污调度的组成和调度模型,对淮河流域防污调度的经验方案进行分析和模拟,应用Copula函数分析水量和水质的联合概率分布,探讨Copula函数在评价防污调度风险和确定流域防污标准方面的可行性。

【关键词】 防污体系　防污标准　防污调度　Copula函数　联合概率分布

1　引言

水资源严重短缺与水环境日益恶化是制约我国社会经济发展的瓶颈问题,特别是由污染造成的水质型缺水现象日益普遍,如何有效地遏制水环境恶化趋势,在受到人类活动剧烈影响的地区,重新建立起水的健康循环,是当前水资源可持续利用与人类永续发展的根本性问题。从流域水循环角度构筑水污染防治和水环境恢复体系,包括工程措施和非工程措施,以污水处理和污染源控制为主的水环境治理多采用工程措施,而非工程措施则主要是从水循环角度,充分利用水的可再生性和自净能力,通过水量的合理调配和水质预测预警,有效预防水污染事故发生和减少水污染事故损失,流域水污染防治的工程措施和非工程措施能够为流域水质安全提供有力的保障,也是流域水环境生态恢复的重要保障。

淮河流域闸坝众多,闸坝上游经常蓄积大量工业废水和生活污水,当闸上污水集中下泄时,特别是汛期第一场洪水之前,闸坝通常在较短时段内大流量下泄,常常导致高浓度污水团集中下泄,造成下游严重的水污染事件。淮河流域特殊的水文气象条件和剧烈的人类活动,使得流域水污染防治工作在加强污染源控制和污水处理的工程措施的同时,必须综合考虑流域水循环、上下游和干支流的关系,加强水文预报、水质预测预警和水量的合理调配等非工程措施,通过闸坝群防污调度,提高淮河流域的综合防污能力。本文主要从防污调度等非工程措施的角度,介绍淮河流域防污体系中防污调度的组成和调度模型,提出流域防污标

* 资助项目:国家水体污染控制与治理科技重大专项(2009ZX07210-006),教育部博士学科点基金(20100141110 03)和国家自然科学基金面上项目(71073115)。

准的概念,应用 Copula 函数分析水量和水质的联合概率分布,探讨 Copula 函数在评价防污调度风险和确定流域防污标准方面的可行性。

2 淮河-沙颍河-涡河防污调度体系

淮河流域水系发达,沙颍河、涡河是淮河两条最大的支流,也是流域污染最为严重的河流之一。根据监测,2004 年沙颍河、涡河年接纳污水总量达到 10.4 亿 m^3,主要污染物化学需氧量 COD 为 29.8 万 t,分别占淮河水系总量的 42.6% 和 52.3%;COD 排放量超过国家确定的“十一五”水污染防治目标的 1.2 倍。排污量远远超过水环境承载能力,导致沙颍河、涡河几乎常年处在严重污染状态,对淮河干流构成了严重的污染威胁。为减轻沙颍河、涡河污染水体随洪水集中下泄对淮河干流的污染威胁,水利部淮河水利委员会和流域水利部门制订污染联防预案,在保证防汛安全和工程供水的前提下,科学调度水闸,发挥水闸等水利工程的调蓄和控制作用,进行污染联防调度。在多年的实际应用中表明水闸污染联防有效减轻了水污染影响,避免了淮河中下游重大水污染事故的发生。

以淮河干流蚌埠闸为控制断面,淮河-沙颍河-涡河防污调度体系如图 1 所示。淮河-沙颍河-涡河防污调度体系是由多条河流和众多闸坝组成的复杂大系统,针对系统的空间特性,将此大系统分为三个子系统:子系统 1——沙颍河流域,子系统 2——涡河流域,子系统 3——淮河干流。沙颍河流域防污调度的闸坝分别为:周口闸、郑埠口闸、槐店闸、耿楼闸、李坟闸、阜阳闸、颍上闸,其中颍上闸特别重要,控制着沙颍河进入淮河干流的水量和水质;涡河流域则主要考虑蒙城闸,蒙城闸控制着涡河流域进入到淮河干流的水量和水质;淮河干流选取三个断面,即上断面王家坝、中间控制断面鲁台子和下断面蚌埠。王家坝为淮河干流上游来水控制断面,鲁台子为沙颍河和淮河干流汇合之后的控制断面,蚌埠为整个系统的控制断面。

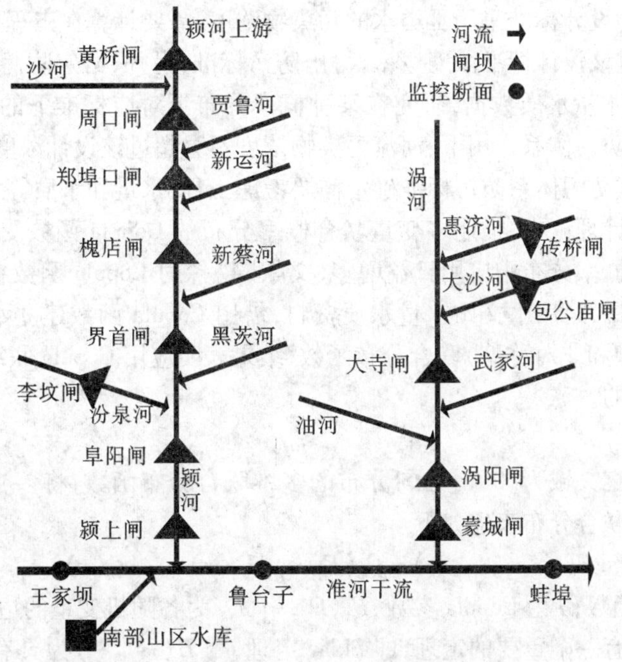

图 1 淮河-沙颍河-涡河防污调度体系分布图

分析淮河干流过去多次发生的较大水污染事件，发现其均与沙颍河、涡河与淮河流量与水质的对比关系有关，发生时间主要集中在春汛与夏汛。因此，合理利用沙颍河和涡河的闸坝，调整有关闸上的蓄水水质浓度，明确主要闸坝开闸放水的限制浓度，并根据淮干流量和水质，科学调度支流入淮干的流量和南部山区水库的防污应急调度，是淮河流域防污体系非工程措施的重要组成部分。

3　流域防污体系的防污标准与 Copula 函数

3.1　流域防污标准

河流防污标准指在水污染防治的工程措施和非工程措施的作用下，流域的防污能力，特别是防止发生重大突发水污染事件的能力。防污标准越高，污水处理能力和防止污染事件发生的能力越强。制定流域防污体系的防污标准，需要综合考虑污水防治的工程措施的规模和可靠性、河流水量水质联合调度的合理性和风险、水质预测预警的准确性等，可见流域防污体系的安全性受到工程不确定性、水文不确定性、水质不确定性等多因素的影响。合理确定流域防污体系的防污标准需要综合考虑流域防污体系的规模、保护对象的重要性和对水质安全的需求，采用概率统计的方法，给出定量的可供流域防污规划设计使用的标准。

在具体研究中，面临两方面的挑战：①工程措施的评价；②水环境可恢复能力的评价。工程措施标准高，污水处理能力大，污染源控制措施健全，则工程防污标准高。从流域水循环、上下游和干支流的关系来看，通过水量水质的合理调度，使入河道污水能够得到迅速的稀释降解，河流具有较强水环境恢复能力，具有准确的和较长预见期的水质预测预警系统，说明防污非工程措施标准较高。实际评价中，可以采用控制断面在一定水量条件下，污染物浓度大于某一标准值出现的可能性来评价防污标准的高低。防污标准越高，河流控制断面污染物浓度超过某一设计标准可能性就越小。在采用用水功能区划水质保护目标作为设计标准，则该河段防污设计标准采用Ⅲ类水的污染物浓度，需要计算在工程措施和非工程措施下，污染物浓度超过该设计标准浓度概率，得出防污标准；另外，给定防污标准(以概率形式给出)，通过实测的水量水质数据，研究在某种概率分布下对应标准下的污染物浓度，即达到防污标准的流域防污体系作用下，河流污染物浓度不会超过该设计浓度。

综上所述，流域防污体系防污标准确定需要考虑工程、水量和水质等多变量联合概率分布。近年来概率统计理论中，描述多变量联合概率分布的 Copula 函数受到了广泛的关注，在水文多变量联合概率分布中应用日益增多，文献[4]探讨 Copula 函数在水量水质联合概率分布模拟中的应用，运用水文站水量水质资料，借用 Copula 函数建立水量和水质的二维和三维(分为对称型和非对称型)联合分布函数，结果表明应用 Copula 联结函数模拟水量水质联合分布是可行的。

3.2　Copula 函数

Copula 函数是定义域为[0,1]均匀分布的多维联合分布函数，将多个随机变量的边际分布连接起来构造联合分布表述如下

$$F(x_1,x_2,\cdots,x_n)=C_\theta(F_1(x),F_2(x),\cdots,F_n(x)) \tag{1}$$

式中：C 为 Copula 函数；θ 为 Copula 参数；$F_1,F_2,\cdots,F_n$ 为各随机变量的边际分布。

Copula 函数具有一个重要的定理，即 Sklar 定理，令 $H(\cdot,\cdot)$ 为具有边缘分布 $F(\cdot)$ 和 $G(\cdot)$ 的联合分布函数，那么存在一个 Copula 函数 $C(\cdot,\cdot)$，满足

$$H(x,y)=C(F(x),G(y)) \tag{2}$$

若 $F(\cdot)$、$G(\cdot)$ 连续,则 $C(\cdot,\cdot)$ 唯一确定;反之,若 $F(\cdot)$、$G(\cdot)$ 为一元分布函数,$C(\cdot,\cdot)$ 为相应的 Copula 函数,那么有上式定义的 $H(\cdot,\cdot)$ 是具有边缘分布 $F(\cdot)$、$G(\cdot)$ 的联合分布函数。

目前来说,Copula 函数有很多种类型,总体上可以划分为三种类型:椭圆型(包括正态 Copula 函数和学生 t-Copula 函数)、二次型、Archimedean 型,水文领域中常见的是 Archimedean 型,Archimedean 型又分为对称型 Archimedean 型和非对称型 Archimedean 型两种型式。

在众多水文事件中,往往包含多个相互关联的特征变量,在建立联合分布时需要考虑变量之间的相关关系。Copula 函数用来描述多个水文变量之间的相关性结构,能够灵活地构造边缘分布为任意分布的多个水文变量联合分布。

4 实例分析

图 1 为淮河-沙颍河-涡河防污调度体系分布图,包括众多的闸坝和南部山区的水库。为便于探讨防污体系的防污标准问题,对实例进行简化,研究范围主要考虑淮河-沙颍河,并以沙颍河入淮河干流的控制闸颍上闸的调度为研究对象,以淮河干流王家坝为上断面,考虑淮南较大支流史河、淠河,以蒋家集水文站和横排头水文站来水为区间入流,以鲁台子断面为水质达标控制断面;研究对象仅是非工程措施。

4.1 经验调度方案与模拟

依据水闸防污调度具体实施途径,参考淮河水闸联防常规调度方案,制订颍上闸调度运用规则。具体如下:①实施小流量分段放流;②依据淮河干流王家坝流量和颍上闸闸上水质相机放流。在保证防洪安全的前提下,以尽可能降低沙颍河污染水体对淮河干流水环境质量的影响为目标,选取淮河主要水质污染指标——氨氮和高锰酸盐指数作为水质指标,以颍上闸闸上水质和淮河干流王家坝来水为调度依据,并以颍上闸为调度对象,按泄流倍比调控颍上闸泄流量,设置六个调度方案。泄流倍比是颍上闸下泄流量与淮河干流王家坝流量的比值,方案一至方案五未考虑颍上闸闸上水质的变化,方案六根据颍上闸闸上不同的水质状况,采用动态变泄流倍比方案,见表 1 和表 2。

表 1 颍上闸调度方案集

方案	一	二	三	四	五
泄流倍比	1.0	0.6	0.4	0.3	0.2

表 2 方案六泄流倍比

闸上水质指标浓度(mg/L)	高锰酸盐指数	<15	15~20	20~25	25~30	>30
	氨氮	<2	2~3	3~4	4~5	>5
泄流倍比		1.0	0.6	0.4	0.3	0.2

4.2 调度方案的防污标准

采用蒙特卡洛方法生成不同流量和水质条件的调度情景,采用 BP 人工神经网络模型对不同调度方案进行模拟,得到调度方案下鲁台子控制断面的水量水质过程。本文应用 Copula 函数,对水量水质联合重现期进行了研究。如果 x 和 y 是两个相互独立的随机变量,

其重现期可分别写为

$$T(x) = 1/[1 - F_X(x)] \qquad T(y) = 1/[1 - F_Y(y)] \tag{3}$$

式中：$T(x)$、$T(y)$ 分别为 x 和 y 的重现期；$F_X(x)$、$F_Y(y)$ 分别为 x 和 y 的概率分布。

如果 x 和 y 是两个相互关联的随机变量，则两变量的联合重现期为

$$T(x,y) = 1/[1 - F(x,y)] \tag{4}$$

式中：$T(x,y)$ 为两变量的联合重现期；$F(x,y)$ 为两变量的联合概率分布。

采用鲁台子控制断面未经过调度（记为方案0）的水量水质实测过程和经过调度模拟的水量水质过程，应用 Copula 函数分别建立水量和水质的联合概率分布函数，应用式(4)计算未经过调度和经过6个不同方案调度下鲁台子控制断面的水量水质联合重现期。图2、图3为实施表1中调度方案1后，鲁台子控制断面模拟的高锰酸盐指数、氨氮与水量的联合重现期等值线图，限于篇幅，表1和表2中的其他调度方案的计算结果未列出。从联合重现期图中，看出调度后高锰酸盐指数和水量的联合重现期增大了，而氨氮指数和水量的联合重现期没有明显增大，说明调度方案对高锰酸盐指数起到了很好调度作用，对氨氮指数影响不大，总体上水量水质联合调度提高流域的防污标准和防污能力，降低水污染超标事件的发生概率。从等值线图中可以查出水量和水质组合的联合重现期和概率值，也可以根据给定的联合重现期获得不同变量的组合。

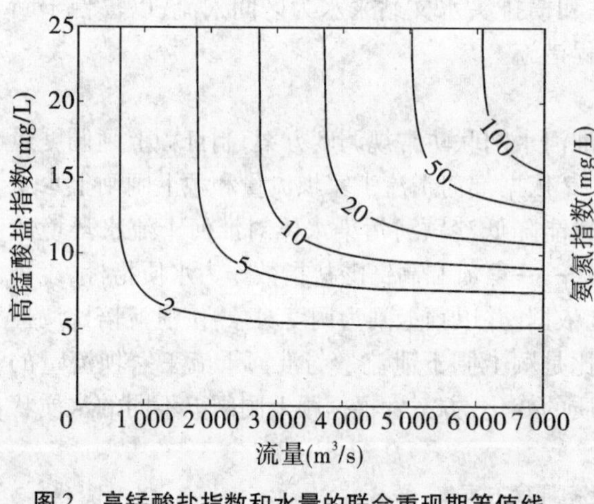

图2 高锰酸盐指数和水量的联合重现期等值线

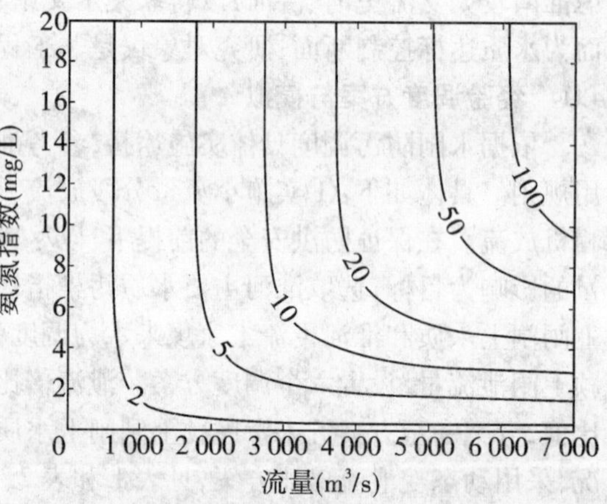

图3 氨氮指数和水量的联合重现期等值线

5 结论

流域防污体系由污水处理和污染源控制等工程措施和水量的合理调配等非工程措施构成。从流域水循环角度，流域防污体系是进行水污染防治、恢复水环境、预防水污染事故的发生和减少水污染事故所带来的损失的综合保障体系，对提高流域水质安全具有重要作用。防污标准是流域防污体系防污能力高低的评价指标，一方面防污标准越高，表明流域防止污染事件发生的能力越大，另一方面表明流域水环境恢复能力越强，合理制定流域的防污标准是流域水环境综合管理研究的创新和挑战。

流域防污调度是流域防污体系中重要的非工程措施之一。流域防污调度通过水量的合理调配，充分利用水体的纳污和自净能力，降低污染物浓度，减少发生水污染事故，特别是突发水污染事故的可能性。研究表明，流域防污调度水量与水质的联合分布可以用 Copula 函

数来建立，Copula函数也是评价流域防污调度体系防污标准的有力工具。在淮河流域防污调度中的应用表明，淮河流域防污调度的经验方案能够有效降低突发水污染事件的发生概率，提高流域防污体系的防污标准和防污能力。同时，Copula函数在评价流域防污体系防污标准方面是可行的，如何综合考虑流域防污的工程措施和非工程措施，确定流域防污体系的防污标准是今后的研究重点。

参考文献

[1] 熊立华，郭生练，肖义，等. Copula联结函数在多变量水文频率分析中的应用[J]. 武汉大学学报：工学版，2005，38(6)：16-19.

[2] 肖义. 基于Copula函数的多变量水文分析计算研究[D]. 武汉：武汉大学，2007.

[3] 郭生练，闫宝伟，肖义，等. Copula函数在多变量水文分析计算中的应用及研究进展[J]. 水文，2008，28(3)：1-7.

[4] 张翔，冉啟香，夏军，等. 基于Copula函数的水量水质联合分布频率分析[J]. 水利学报，2011，42(4)：483-489.

[5] Abe Sklar. Random variables, joint distribution functions, and copulas[J], Kybernetika, 1973, 9(6): 449-460.

【作者简介】 张翔，男，1997年毕业于四川大学，获水文水资源专业博士学位。现任武汉大学水资源与水电工程科学国家重点实验室教授，博士生导师，水文水资源实验研究中心主任，兼任中国自然资源学会理事和水资源专业委员会委员、中国可持续发展研究会水问题委员会委员、中国水利学会青年委员会委员，主要研究方向为基于生态水文的可持续水资源管理。

冻融侵蚀黑土导致农业非点源污染研究*

赵显波[1,2]　许士国[2]　刘振平[2]　刘铁军[3]

（1.黑龙江省水利科学研究院　哈尔滨　150080；
2.大连理工大学　大连　116023；
3.水利部牧区水利科学研究所　呼和浩特　010010）

【摘要】 冻融侵蚀是典型黑土区的全概率事件。在对国内外冻融侵蚀黑土导致的非点源污染研究现状与发展动态分析的基础上，从黑土区气候、黑土及冻土性质、地形条件、环境及水文学等不同角度定性分析冻融侵蚀及冻融侵蚀黑土导致农业非点源污染产生的机理，需要进一步定量研究冻融侵蚀引起的黑土流失量及定质研究冻融侵蚀引起农业非点源污染物进入水体的量。

【关键词】 冻融侵蚀　水土流失　黑土　冻土　农业非点源污染

根据《第一次全国污染源普查公报》显示，非点源污染物排放对水环境影响较大，是总氮（TN）、总磷（TP）排放的主要来源。随着点源污染控制能力的提高，非点源污染的严重性逐渐凸显，非点源污染问题正成为当前科学研究和社会关注的焦点。黑龙江是农业大省，为在有限的农田上实现农民增收、农业增效、农村致富，近30年来化肥、农药施用量急剧增加，导致由土壤侵蚀引起的农业非点源污染问题日益突出。黑龙江省内河流存在春夏两季径流、泥沙及污染高峰现象，且黑龙江省处在冰封期长、冬季初春无植被生长的生态系统脆弱区，初春植被缺失无法实现污染物入河前植物拦截消减作用。因此，黑龙江省特有的由冻融侵蚀黑土（冻融面积5.76万km^2）导致的农业非点源污染问题引起环境、水利、农业等部门的高度关注。

1　国内外研究现状及发展动态分析

长期以来，人们过度关注施用化肥带来的丰厚社会经济效益，而忽略了它对环境的影响。在点源污染逐步得到控制的情况下，农业非点源污染引发相关领域人员的高度关注，寻求科学合理的农业非点源污染防治的理论方法，有效预防或控制水肥土流失，减少非点源污染，是东北地区冻融侵蚀黑土农业非点源污染研究领域的热点和前沿问题。

1.1　非点源污染的研究进展

农业非点源污染是导致水体水质下降的最主要因素，农业非点源是总氮、总磷排放的主要来源，导致了一系列环境灾害问题如湖泊水库富营养化程度加重、水域功能下降或丧失、湖库淤积影响兴利、河床淤积抬高加大洪水风险等。对于非点源污染的环境效应的结果已

* 基金项目：黑龙江省青年基金（QC2010099），国家自然基金（40901136）。

有较多研究,但多是定性分析、评价,而对农业非点源污染形成的过程问题有待进一步研究。

1.2 土壤侵蚀的研究进展

非点源污染源自水土流失,表现形式是土壤侵蚀。特别是东北黑土区作为世界范围仅有的三大块黑土带之一,相对狭小但肥力最好。黑土是珍贵的自然资源,在自然气候生物适宜条件下每300~500年才能生成1 cm。东北黑土区一年一季的种植,化肥施用量大,垦殖指数高,坡耕地比例大特点致使黑土区水土流失面积大、范围广、速度快,导致耕地面积减少,土壤性质改变,肥力下降。土壤结构恶化、黑土层变薄对东北地区粮食稳产增产构成严重威胁。

已有研究主要是土壤侵蚀的定性评述,如黑土变薄的定量监测、侵蚀沟数量增多监测等,据此推出的土壤侵蚀模数,少有考虑土壤侵蚀水土流失对河流水环境的造成的影响,以及由土壤侵蚀水土流失给水体系统带来的累积的环境效应。因此,综合考虑土壤侵蚀引起的农业非点源污染,考虑从肥料到污染物的转化累积,从黑土地的黑土到河道泥沙的转化累积等是需要进一步解决的问题。

1.3 冻融侵蚀研究进展

我国的冻土约占全国总面积的3/4,其中:多年冻土面积占国土面积的22.3%,季节性冻土占国土面积的53.5%。已有的研究从冻融循环对土壤水热盐迁移机制及试验研究得出温度、未冻水含量和土水势是冻融土壤中水盐运移的三大基本因素,盐随水走是盐分在土壤中迁移的主要形式,影响水分迁移的因素也会影响盐分的迁移。冻融循环对土壤水、热、盐运动的定量研究包括室内土柱冻融试验、冻土水热耦合方、冻融循环对土壤结构改变的研究等。

黑龙江省处于中国东北寒带中—深季节冻土区的东北黑土区,冬季较长,积雪量大。综合关于冻融侵蚀黑土的研究,从黑土区土壤的特性和自然环境来看,黑土的土体完全具备产生冻胀的条件:地面表土主要为黑土(母质层多为深厚的黄土性黏土),土壤颗粒细碎,具有冻胀敏感性;土体中有初始水分供给,供给水主要为秋季降水;初冬缓慢而持续降低的大气温度及漫长的冬季负温持续期,为土体产生冻胀(土壤膨胀变形)提供了适宜的冻结条件,且有以雪为形式的降水;在春季融化期内,融化层大致沿冻结线逐渐增厚,融化速度较快,融化作用从冻土体中释放出来的迁移水和雪水,由于很难在透水性能极差的未融土层上下渗,于是部分融土便与水一起,沿着融冻界面顺坡下滑;在一年一度的冻融过程中,土体始终处于较为疏松的状态,因而冻融侵蚀黑土流失严重。省内河流存在春季与夏季径流、泥沙及污染双高峰,更进一步说明省内漫川漫岗坡耕地普遍受冻融侵蚀影响。松辽委调查报告指出,部分东北黑土侵蚀区土壤冻融侵蚀的速度已不亚于暴雨造成的水土流失。

从当前冻融侵蚀土壤的问题研究来看:多是通过室内填土柱做冻融试验对冻融侵蚀土壤的机制进行定性描述;少有的定量研究或数值模拟也是模拟研究室内的土柱中的土壤冻融过程,就一项或几项指标定性或定量分析;由于室内模拟冻融试验土柱的边界条件比较简单,与自然条件下的冻融过程差异大,很难直接应用于生产实际。因此,野外的实际监测与实验室内模拟结合关于冻融侵蚀黑土水热盐运移问题有待进一步探讨。

2 研究区自然状况

冻融侵蚀引起的黑土流失与非点源污染研究是在中—深季节冻土区、黑土区、国家重点治理的水污染控制区以及生态脆弱区的自然条件大背景下进行研究的。

2.1 冻土区

黑龙江省地处高纬度，受温带、寒温带大陆性季风气候影响，年降水总量小年平均降水量多介于400~650 mm，降水分布不均匀，80%降水量集中在6~9月，10%左右降水量以降雪的形式在霜冻期降下，春旱夏涝，径流过程的年内、年际变化剧烈。年平均气温为5~-4 ℃，气温昼夜温差大，处中—深季节冻土区。

2.2 黑土地

黑龙江省作为国家重要的商品农业基地和粮食安全保障基地，拥有世界范围仅有三大块黑土带之一的我国东北黑土。稀缺的黑土还具有珍贵性和再生性差的特点，珍贵性具体表现为：黑土地是一种性状好、肥力高、适宜植物生长的土壤类型，与国内黄土地、棕土地、红土地等其他主要耕地土壤相比较，黑土具有腐殖质含量高、结构疏松、容重低、水稳性团粒结构比重大、持水能力强、通透性良好、微生物活性强等特点，具有良好的保肥、保水与通气性能；黑土再生性差表现为在自然气候生物适宜条件下每300~500年生成1 cm，黑土区每2~3年侵蚀流失掉1 cm，黑土易蚀特点使目前珍贵黑土资源退化严重。黑龙江省作为农牧业生产基地，农业主产玉米、大豆、水稻，一年一熟，农业耕作区地形特点是漫川漫岗形。

2.3 水体污染严重

水资源面临水多、水少和水脏的三大问题。松花江流域在"十一五"期间被列为国家重点治理的水污染控制区。在黑龙江省由冻融侵蚀引起的黑土流失，黑土层变薄、肥力减弱成为制约黑龙江省农业可持续发展的主要因素，同时黑土带着营养盐进入水体，影响水环境的质量，为水体富营养化留下隐患，泥土淤积河道为水体行洪、兴利留下隐患。由冻融侵蚀带来的水土流失问题给土地与水带来了双重危害。

2.4 生态脆弱区

黑龙江省四季分明，夏季酷暑，农作物一年一熟，秋季收获后翻地无植被覆盖，冬季严寒漫长冰雪覆盖，春季干旱少雨，初春冰雪开始融化产生冻融侵蚀，冬季及初春无植被生长属生态脆弱区的生态脆弱期。

3 不同角度分析冻融侵蚀黑土及其导致非点源污染产生机制

3.1 水的物理性质分析冻融侵蚀黑土产生机制

水的密度1.0 g/cm^3，冰点0 ℃，冰的密度0.917 g/cm^3，单位质量的水变成冰时，体积将增大9%。当温度降低到冰点以后，黑土冻结过程中，土壤水分迁移，使已冻土体的含水量大增，冻融循环次数增加，表层土空隙变得更大，松散表层土壤随融化的水流走，形成冻融土壤侵蚀。

3.2 气候主导因素分析冻融侵蚀黑土产生机制

黑土区地处高纬度，受温带、寒温带大陆性季风气候影响，四季分明，一年一季的种植，一般秋季收获后翻地，黑土耕作区无植被覆盖，土壤颗粒细碎，黑土表土较松，秋季有降水，蒸发少，土壤含水量增加；冬季寒冷漫长，有以雪为形式的降水，黑土耕作区被积雪覆盖；春季干旱少雨，冰雪开始融化，气温昼夜温差大，土壤反复冻融，可蚀性增强，加之黑土耕作区无植被覆盖，融雪径流对解冻的黑土形成侵蚀。

土物理性质分析冻融侵蚀黑土产生机制

壤是冻融侵蚀的对象,由于黑土受其质地、结构、孔隙度及透水性的影响,黑土易受侵土的成土母质比较单纯,主要是第四纪更新世砂砾黏土层,其粗粉砂含量占30%左明黑土具有黄土特征,黑土最大特征是上部具有腐殖质层,该层中大部为粒状及团状构,土层疏松多孔,孔隙大,持水量大,而黑土底层质地黏重,结构性较差,透水性较小。这性质的差异,使黑土表层极易形成径流,造成表层土壤流失。

3.4 典型黑土区地形条件分析冻融侵蚀黑土产生机制

典型黑土区以松嫩平原及四周台地低丘区为中心,是漫川漫岗形的土地,黑土区耕地坡度一般小于10°,大于15°的不多,3°~7°的占绝大部分,坡长500~2 000 m,有的达4 000 m。坡度、坡长都是影响冻融侵蚀与坡面径流过程的重要地貌因素。坡度越大,冻融侵蚀越严重,水土流失越严重。坡面越长,冻融面积越大,水流流程长,水流对坡面侵蚀的概率高,单位径流量所引起的侵蚀量越大。坡面越长,所以相应于同样径流量,坡面越长土壤侵蚀量就越多。

3.5 概率的角度分析冻融侵蚀黑土产生机制

典型黑土区夏季作物生长旺盛期,植被覆盖度高,只是场次暴雨侵蚀量,且处于高纬度,受温带、寒温带大陆性季风气候影响,形成暴雨的概率小(小概率事件),而冬季漫长冰雪覆盖,黑土耕作区无植被覆盖,初春解冻期无植物生长,且黑土区冬冻春融是自然规律,加之春季解冻期气温日较差大,在0 ℃上下反复,昼融夜冻,反复冻融,冻融是典型黑土区的全概率事件,使得黑土区春季冻融侵蚀对水土流失的贡献更大。

3.6 冻土的物理、力学性质角度分析冻融侵蚀黑土产生机制

典型黑土区冬季严寒,表层土体的水分因冻结而体积膨胀,对土体产生巨大的压力,春季表层冰雪融化,而下层冻层传热慢融化也慢,形成不透水层,因此产生地表径流,造成水土流失,加之温度周期性地发生正负变化,冻土层中的地下冰地下水不断发生相变和位移(物理角度),使冻土层产生冻胀、融沉、流变等一系列的应力变形,当冻胀力大于土体内聚力时(力学角度),产生土壤冻融侵蚀。

3.7 水文学角度分析冻融侵蚀黑土产生机制

典型黑土区春季积雪和表层土壤开始融化,而下层冻土传热慢,融化也慢,因此在土壤中形成一个不透水层,随着冰雪融化,不透水层上部的土壤达到饱和状态后,多余的水分就不能及时入渗到下层土壤中,因而产生融雪径流(水文学角度),此时解冻的表层土壤在反复的冻融过程中其团聚体大小及稳定被破坏,极容易遭受融雪径流冲刷表层土壤,形成冻融侵蚀水土流失。

当水分供给充足时,冻融循环次数增加,土壤表层含水量成幂次增加$(\rho_w/\rho_i)^n$(其中:n为冻融循环次数,ρ_w为水的密度1.0 g/cm^3,ρ_i为冰的密度0.917 g/cm^3),表层土体含水量增大,表层土壤体积随之增大即表层土壤变的松散,土壤表层水过饱和后,汇满产流(水文学角度),松散的表层土随着水流失,形成冻融侵蚀。

3.8 物理化学角度分析冻融侵蚀黑土导致非点源污染产生机制

在土壤冻融过程中,由于受相变的影响,土壤中冰、水共存,水、热、盐分的迁移具有更强烈的耦合性。土壤冻结过程中,冻结区水分冻结成冰,使土壤水热状况发生变化,土水系统的动态平衡遭到破坏,水分在土水势梯度作用下不断从未冻区向冻结区迁移。结果是冻结区冰含量不断增加,未冻区水中盐分不断积累,并引起土体冻胀。随着气候逐渐变暖……从表层开始融化,而其下冻结层阻碍上层融水下渗,致使上层土壤含水率增加,汇满……走表层黑土及表层黑土中营养盐,土壤冻融侵蚀产生农业非点源污染。

3.9 环境角度分析冻融侵蚀黑土导致非点源污染产生机制

黑土区的黑龙江省每年流失掉的黑土在2亿~3亿t,其中,流失黑土中含的营养盐……合标准化肥为500万~600万t,相当于黑龙江省施用化肥的总量。从水环境角度说明农业非点源污染严重,对水体构成富营养化威胁。水土流失是各种生态环境退化的集中反映,是导致生态环境进一步恶化的根源。黑土区由冻融侵蚀导致的水环境非点源污染问题亦不容忽视。

4 结论与展望

综上所述,对于水体非点源污染已有较多研究,但多是针对水体非点源污染环境效应研究;对于土壤侵蚀水土流失亦有较多研究,都是针对局部水土流失定性分析的;对于冻融侵蚀研究,多是机制定性评述,少有定量研究还是基于室内填土柱的冻融试验。在东北黑土区综合考虑自然条件下冻融侵蚀黑土量和质,即冻融侵蚀黑土导致的农业非点源污染定量、定质过程有待于进一步研究。

黑龙江省特有冻融侵蚀是春季融化一层,融雪径流侵蚀冲刷一层,带走黑土,同时带走黑土中肥料,污染物通过融雪径流进入水体,形成春季非点源污染高峰。研究冻融侵蚀黑土的环境影响因子间相互影响,以及污染物在冻融条件下的随时间、空间变化迁移转化规律,将对遏制黑土流失、缓解水质性水资源短缺压力、建设黑龙江成为国家商品粮基地和重要生态安全保障区具有重要意义。

参考文献

[1] 赵显波,许士国,刘振平.冻融侵蚀引起的黑土流失与非点源污染研究[C]//第八届中国水论坛,2010.12:268-272.

[2] 张维理,武淑霞,冀宏杰,等.中国农业面源污染形势估计及控制对策 I.21世纪初期中国农业面源污染的形势估计[J].中国农业科学,2004,37(7):1008-1017.

[3] 范昊明,蔡强国,陈光,等.世界三大黑土区水土流失与防治比较分析[J].自然资源学报,2005,20(03):387-393.

[4] 刘兴土,阎百兴.东北黑土区水土流失与粮食安全[J].中国水土保持,2009(01).17-19.

[5] 徐学祖,王家澄,张立新.冻土物理学[M].北京:科学出版社,2010.

【作者简介】 赵显波,女,1980年2月生,博士,毕业于大连理工大学,现在黑龙江省水利科学研究院工作,工程师,从事以水资源为基础的水环境研究。

呼伦湖游憩功能损害程度评价*

郝伟罡[1]　延文龙[2]　庄　健[3]　李振刚[1]　王丽霞[1]

（1. 水利部牧区水利科学研究所　呼和浩特　010020；
2. 内蒙古自治区乌海市　乌海　016000；
3. 达茂旗水务局　包头　014300）

【摘要】 呼伦湖是著名的旅游风景区，近年来呼伦湖水量骤减，湖泊富营养化程度加剧，对其湖泊旅游功能造成很大破坏。选取人类对呼伦湖影响最为显著的区域小河口一带2007年水质实测数据，运用水污染损失率为基础的评价方法，对呼伦湖旅游功能损害程度进行评价。结果表明，呼伦湖水污染对该湖旅游功能的损害程度已接近中等水平。

【关键词】 呼伦湖　功能损害　污染损失率

1　引言

呼伦湖是中国第五大淡水湖，也是东北地区第一大湖，是内蒙古十大旅游景点之一。呼伦湖不仅是众多鸟类栖息、繁殖的理想场所，而且是东北亚鸟类迁徙的重要集散地，为众多的野生动物提供了优良的栖息和繁殖的场所。呼伦湖畔景色绚丽，气候宜人，是开展湿地观鸟、湖岛观景等观光旅游、生态旅游的理想场所，目前已开发的景点有小河口、呼伦日出、呼伦蜃景、乌兰布冷玛瑙滩、成吉思汉栓马桩、鸥岛等。旅游业是一项综合性产业，呼伦湖旅游业发展必然带动地区经济的发展，促进经济落后地区脱贫致富，产生巨大的经济效益。

2　研究区概况

呼伦湖地处呼伦贝尔草原中部新巴尔虎左旗、新巴尔虎右旗和满洲里市间，东经117°00′10″～117°41′40″，北纬48°30′40″～49°20′40″（见图1）。当湖水位在545.55 m时，蓄水量约138亿m^3，水面约2 339 km^2，最大水深8 m，平均水深5.7 m。湖面呈不规则的斜长型，湖面长93 km，最大宽度41 km，平均宽度25 km。呼伦湖及周边7 400 km^2于1990年被划定为自治区级保护区，1992年被批准为国家级自然保护区，2002年1月被列入国际重要湿地名录，2002年11月被联合国教科文组织人与生物圈计划吸收为世界生物圈保护区网络成员。呼伦湖及其周边湿地被称为呼伦贝尔“草原的肾”，其水域与湿地为呼伦贝尔草原的生态保护和地区社会与经济发展，发挥着不可替代的重要作用。

---

* 基金项目：国家自然科学基金资助项目（51009098，50779040），中国水科院科研专项基金资助项目（KW2010010701）。

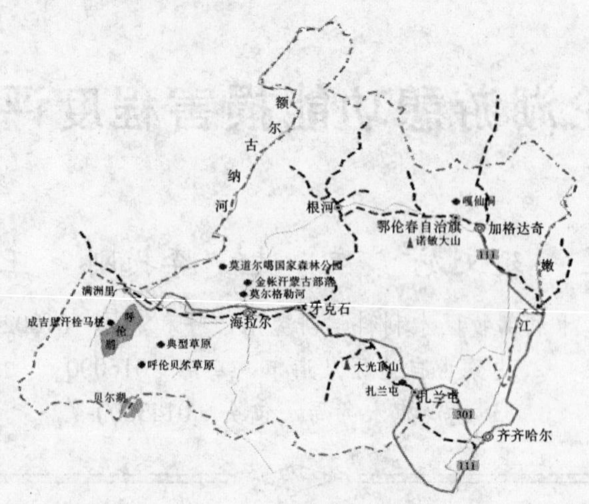

图 1　呼伦湖地理位置

呼伦湖拥有丰富的水产资源，成为内蒙古最大的商品鱼基地和绿色食品基地。据统计，湖区共有鱼类 33 种；浮游植物 8 门 21 目 38 科，共 187 种属；浮游动物 59 种；野生植物 74 科 292 属 653 种，占呼伦贝尔市野生植物总数的 48.3%；鸟类 17 目 40 科 241 种，其中，属国家一级 5 种，二级 14 种，占重点保护鸟类的 19.6%。

3　理论与计量模型

L. D 詹姆斯注意到水污染造成的经济损失与污染物浓度有关，提出了损失—浓度曲线（见图 2）。曲线是 S 形，随浓度上升，损失在一定范围内迅速增加，水质污染到一定程度造成的损失趋于极限。

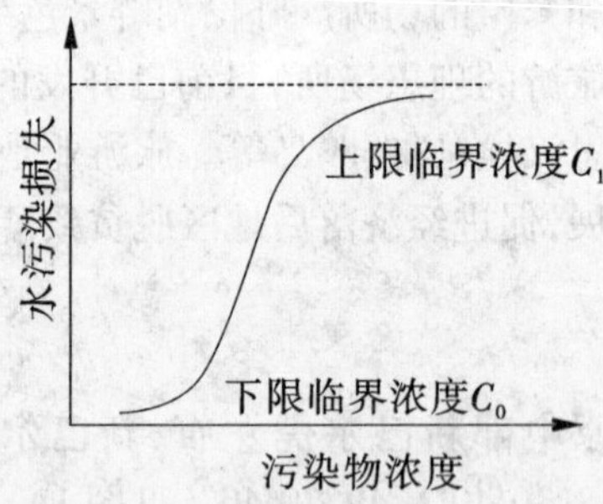

图 2　污染损失—浓度曲线

3.1　湖泊功能损害程度评价

设湖泊的某一功能 i，有 n 种污染物。某种污染物对水体造成的损失，可用污染损失—浓度曲线（见图 2）加以描述，公式如下

$$S = \frac{K}{1 + a\exp(-bc)} \tag{1}$$

式中：S 为某污染物对水体造成的经济损失；c 为某污染物浓度，mg/L；a、b 分别为由该污染物特征所决定的参数，可通过毒理试验或损害试验确定；K 为该功能资源量或价值量。

定义 R 为某种浓度 c 下，污染物引起的损失与功能价值 K_i 之比，称为该污染物对该功能的损失率，简称污染损失率。由式（1）得

$$R = \frac{S}{K} = \frac{1}{1 + a\exp(-bc)} \tag{2}$$

一般地,第 j 种污染物对湖泊第 i 项功能的污染损失率可写为

$$R_{ij} = \frac{1}{1 + a_{ij}\exp(-b_{ij}c_j)} \tag{3}$$

由于 a_{ij}、b_{ij} 是仅与污染物自身特性和水体使用功能有关的参数,不能随意选取,因而具有客观性。表 1 是几种常见污染物的 a_{ij}、b_{ij} 参考值。

表 1　参数 a、b 的参考值

使用功能	参数	BOD_5	TN	TP
旅游	a	18 895.6	799.4	799.4
	b	5.25	2.09	2.09

当水中污染物多于一种时,其综合损失率不是各项损失率简单求和。有 n 种影响相互独立的污染物存在时,减去交互影响部分后的综合损失率 $R_i^{(n)}$ 为

$$R_i^{(n)} = R_i^{(n-1)} + (1 - R_i^{(n-1)})R_{in} \tag{4}$$

这是一个递推公式,其中 $R_i^{(1)} = R_{i1}$。计算时,先将各污染物浓度代入式(3),求出单项污染损失率 R_{ij},然后用式(4)算出综合污染损失率 $R_i^{(n)}$。计算 $R_i^{(n)}$ 时,先算两种污染物综合损失率 $R_i^{(2)}$,再计算 $R_i^{(3)} = R_i^{(2)} + [1 - R_i^{(2)}]R_{i3}$。依次类推,直到算出 $R_i^{(n)}$。

单项污染损失率 R_{ij} 表明了 j 污染物对 i 功能的损害程度。以此作为该污染物对 i 功能损害程度的单项评价值。综合污染损失率 $R_i^{(n)}$ 表明 n 种污染物共存时对 i 功能的损害程度,以此作为功能损害程度的综合评价值。

3.2　分级标准的划分

为了得出更加直观、形象的结果,可以根据综合损失率大小,依水质评价惯例将湖泊功能损害或水质划分成若干级,见表 2

表 2　分级标准的划分

综合损失率 R_i(%)	<1	1~5	5~20	20~50	50~90	>90
污染程度	优良	尚清洁	轻污染	中污染	重污染	严重污染
损害程度	不损害	微损害	轻损害	中损害	重损害	功能丧失

3.3　湖泊损害程度与湖泊水质评价的关系

环境质量是与环境损害程度直接相连的。因此,可以把污染损失率同时作为湖泊水质的评价结果。不仅克服以往各类水质评价模式评价指数缺乏明确物理意义,掩盖或夸大某些污染物影响的弊端,而且将湖泊损害程度与湖泊水质评价统一起来,使水质评价结果也具有了明确的物理意义。

4　呼伦湖旅游功能损害程度评价

近些年来,由于受气候暖干化与人类活动的双重影响,呼伦湖水量骤减,水位逐年下降,各种盐分大量蓄积,湖泊富营养化程度加剧。导致旅游景观质量受损,可观赏性下降。本文运用上述模型对呼伦湖旅游功能损害作量化评价,以期为该湖合理开发利用提供依据。选

取人类对呼伦湖影响最为显著的区域小河口一带 2007 年水质实测数据，参照国家《地表水环境质量标准》(GB 3838—2002)，选取 BOD_5、总氮、总磷作为评价因子，分别用式(3)、式(4)计算呼伦湖 2007 年旅游功能的单项指标污染损失率和综合污染损失率，计算结果见表 3。

表 3　呼伦湖不同使用功能的污染损失率

损失率 R(%)	BOD_5	TN	TP	综合损失率(%)	污染程度	损害程度
旅游	0.90	15.87	2.06	18.34	轻污染	轻损害

由表 3 可见，呼伦湖水作为人体非直接接触的旅游娱乐用水已受到一定程度污染(轻污染，轻损害)，且接近中等损害程度(综合污染损失率距中污染标准下限只差 1.66 个百分点)。呼伦湖旅游功能损害的主要问题是水体中总氮含量高，营养物质过剩。

5　结论

通过污染损失率评价法将呼伦湖旅游功能损害程度与水质评价统一起来，使评价结果具有了明确的物理意义。呼伦湖是世界范围内干旱半干旱地区极为少见的具有很高生态效益的大型多功能湖泊，是著名的旅游风景区，其湖水质量好坏必然对该地区经济发展产生较大影响。从分析结果来看，呼伦湖水污染对该湖旅游功能的损害程度已接近中等水平，主要超标污染物指标为总氮，这与呼伦湖以草原牧业型污染为主有关，应加强流域内畜牧业的管理，以及在旅游区内加强游船的管理，根据环境容量，控制船只数量，严禁在水域内排放油污。积极推广电瓶船，减少漏油造成的危害。针对湖区不同区域水环境差异作好调查研究，在科学论证的基础上尽快实施湖泊综合治理工程。

参考文献

[1] 韩向红，杨持. 呼伦湖自净功能及其在区域环境保护中的作用分析[J]. 自然资源学报，2002，17(6)：684-690.

[2] L. D 詹姆斯，等. 水资源规划经济学[M]. 北京：水利电力出版社，1984.

[3] 朱发庆，等. 东湖水污染经济损失研究[J]. 环境科学学报，1993，13(2)：214-222.

[4] 朱发庆，吕斌. 湖泊使用功能损害程度评价[J]. 上海环境科学，1996，15(3)：4-6，12.

[5] 郝伟罡，郭中小，苗澍，等. 干旱区草型湖泊使用功能损害程度量化评估[J]. 中国水利，2006，562(16)：11-14.

【作者简介】 郝伟罡，1979 年 10 月生，2006 年毕业于内蒙古农业大学，工学硕士，现工作于水利部牧区水利科学研究所，工程师。

西北干旱区内陆河流域生态干旱及其评估指标*

梁犁丽[1]　冶运涛[1,2]　胡宇丰[1]　郝春沣[1]

(1. 中国水利水电科学研究院　北京　100038；
2. 南京水利科学研究院水文水资源与水利工程科学国家重点实验室　南京　210029)

【摘要】 西北干旱区内陆河流域生态环境脆弱，水资源是影响其绿洲生态系统稳定的关键因素。近年来，人类活动对西北内陆河流域水资源的影响加剧，生态用水被大量挤占，干旱对天然绿洲生态系统的影响作用日趋显著。基于内陆河流域水资源与绿洲稳定的相互作用关系，本文主要探讨了基于水分驱动的天然绿洲生态干旱及其演变特征，并根据生态干旱的内在和外在表现，提出了以 NDVI、潜水埋深与植被生存临界水位、河流断流长度等为代表的内陆河流域绿洲生态干旱若干评估指标。

【关键词】 生态干旱　绿洲演化　评估指标　NDVI　内陆河流域

西北干旱区内陆河流域降水量稀少，生态环境脆弱，极易受干旱威胁。近年来，由于受全球气候变化和人类活动双重因素交织影响，其生态水文过程正在发生显著改变，干旱发生频率增加，危害程度加深，不仅严重制约了社会经济的发展，而且加重了生态环境系统恶化趋势，对绿洲生态系统产生了不可忽视的影响。

对于"生态干旱"目前尚无明确定义，多表述为"干旱引起的生态环境变化"或归为农业干旱中，仅有少数研究者在生态风险评价研究中有所提及；国外不少研究者认为气候变化和干旱，尤其是水资源短缺，对植被生存产生重要影响，进而影响整个区域的生态系统平衡。在干旱对生态系统的影响评估方面，有研究者指出，植被指数能够精确反映绿洲植被变化情况，地下水位的高低直接影响植被的生存状况，绿洲开发和水资源利用是造成下游河道断流、尾闾湖泊干涸的主要原因之一，但尚缺乏系统的生态干旱评估指标。本文针对内陆河流域绿洲生态系统的干旱问题，在分析绿洲稳定影响因素的基础上，基于水资源与绿洲生态系统的相互作用关系，提出天然绿洲生态干旱的概念，阐述基于水分驱动的内陆河流域天然绿洲生态干旱及其特征，初步建立生态干旱评估指标体系，引发国内学者对流域"生态干旱"的概念、内涵、评价、预防措施等相关问题的思考和重视。

* **基金项目**：国家自然科学基金（50709042），水文水资源与水利工程科学国家重点实验室开放研究基金项目（2011491911）。

1　内陆河流域绿洲稳定性研究

绿洲(Oasis)指存在于干旱区、以植被为主体、具有明显高于其周边环境的第一性生产力、依赖外源性自然水源存在的生态系统。干旱区内陆河流域的绿洲包含了存在于其中的所有子生态系统,既包括林草、水域、湿地等天然生态系统,也包括农田系统、水库坑塘和城镇村庄等人工生态系统。以人类活动为标志,可将绿洲大致分为天然绿洲和人工绿洲。本文所探讨的绿洲生态干旱仅指天然绿洲生态系统,人工绿洲干旱可归于农业干旱中。

20 世纪 70 年代 May 指出生态系统的特性、功能等具有多个稳定态,稳定态之间存在"阈值和断点(Thresholds 和 Breakpoints)"。多数研究者认为,稳定性包含了两个方面内容:一是系统保持现状的能力,即抗干扰的能力;二是系统受到干扰后的恢复能力。基于这一共性认识,柳新伟等进一步给出生态系统稳定性的定义,即不超过生态阈值的生态系统的敏感性和恢复力。

绿洲的稳定与局地气候、植被覆盖、作物生理生态指标、下垫面的相互作用密切,是具体尺度下的相对概念,水资源是内陆河流域绿洲生态系统稳定的关键制约因子,水资源的变迁对绿洲稳定具有重要影响,绿洲随着水源的变化而迁移,若水资源量时空变化不大,则绿洲的规模不会发生大的改变,且由于水分条件的限制,绿洲规模存在阈值。但由于流域水资源量具有时空不确定性,人为和自然因素引起的长期持续水资源短缺足以导致绿洲生态系统水分收支的严重不平衡,出现植被、河流、湖泊等天然绿洲生态系统严重缺水的现象,如连续水分亏缺引起的天然植被退化、地下水位下降、河流断流、湖泊萎缩、水域生物栖息地破坏等,本文称其为天然绿洲生态系统的"生态干旱"。

2　基于水分驱动的内陆河流域天然绿洲生态干旱及其特征

2.1　天然绿洲生态干旱

随着人类活动范围的扩大和水资源开发程度的提高,在内陆河流域平原绿洲区自然水循环的基础上,社会水循环效应不断增强。平原绿洲区社会水循环总的水文效应是:地表水通量、地下水的补给和排泄条件变化大大改变了水资源量及其时空分布。在绿洲中游区,水资源大量流入社会水循环系统且用水消耗不断加大,造成绿洲下游区自然水循环中水通量持续减少,导致在连续降水偏少年份或季节,干旱对天然绿洲生态系统产生巨大影响:河流径流量减少,断流长度增加,尾闾湖泊或湿地面积不断萎缩甚至干涸;河流中下游地表和地下径流的水平运动减弱,地下水位下降,依赖潜水维持生命活动的地表植被因地下水位不能达到临界水深而退化,引起分布于河流沿岸的河谷林带变窄,整体功能减弱,河流尾闾及湖沼湿地植被退化;在地下水位降低的同时,地下水矿化度升高,水质恶化,根系层土壤含盐量升高,引起平原区依赖地下径流或降水生存的地带性植被类型的演化,导致地表植被覆盖度降低,进而造成土地沙化。

以新疆内陆河流域为例,在干旱季节,博斯腾湖及其周围湿地大面积萎缩,补给塔里木河的水量减少,塔河断流河长上移,两岸靠塔河径流补给的胡杨大面积死亡;北疆博尔塔拉河、精河、奎屯河的尾闾艾比湖水面在 20 世纪五六十年代有 1 200 km^2,而在干旱的 2006 年只有 483 km^2,2011 年持续干旱致使湖面面积不足 400 km^2,艾比湖湿地国家级保护区也大面积萎缩,湖滨地区荒漠化程度加剧,如果旱情继续,艾比湖则可能成为季节湖或最终消失,成为中国西部沙尘暴主要策源地之一。

2.2 水分变化条件下的绿洲植被干旱及其演化特征

在西北干旱区，绿洲植被的分布、长势及演替与地下水位、土壤水分和盐分的变化关系密切。若天然绿洲区缺水，则地下水位降低，毛管上升水不易到达植物根系层，一方面上层土壤发生干旱，植物生长受到水分胁迫；另一方面土壤水盐度升高，植被细胞失水可能性增加，促使植被退化，甚至因发生干旱而死亡，覆盖度降低。地下水位与植被生长的关系如图1所示。

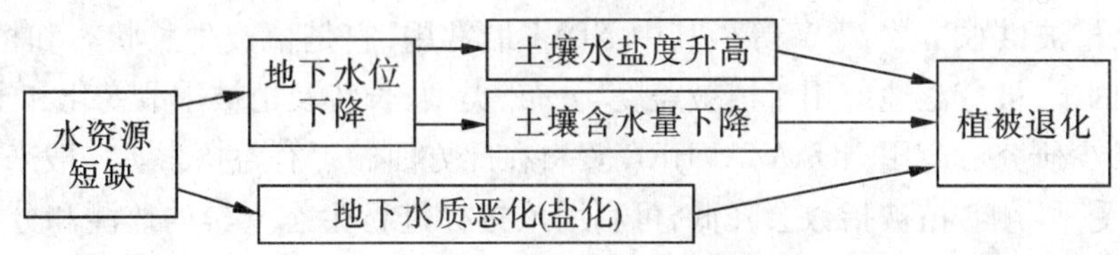

图1 地下水位与植被生长的关系

人类活动和气候变化，特别是降水和温度波动是影响水资源丰枯变化和天然绿洲演变的驱动力，水资源的循环转化过程决定了绿洲的水分状况。社会水循环表现了水资源开发利用与平原绿洲生态系统演化的动态关系，水源条件的微小变化即可引起生态环境的小幅波动，造成绿洲生态系统结构和组成的变化，促使天然绿洲与荒漠生态系统的相互演变，图2所示的河岸林→灌丛草地→盐化草甸植物群落→盐生植被→荒漠植被→裸地（或沙化的植被）的演替过程在干旱区十分普遍。

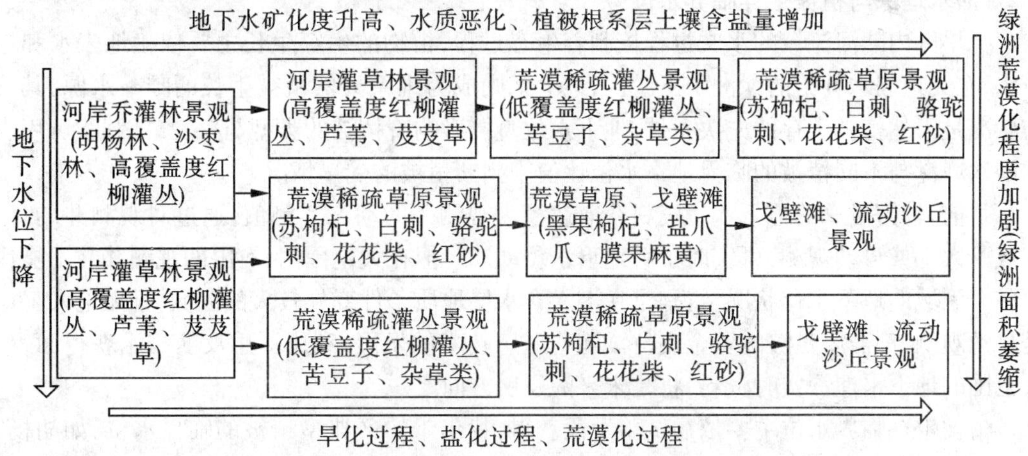

图2 天然绿洲各景观生态类型之间的演替规律

2.3 天然绿洲生态干旱与其他干旱类型的相互关系

根据研究对象的不同，干旱类型可分为气象干旱、水文干旱和农业干旱，本文所提出的生态干旱是指内陆河流域天然绿洲生态系统内的水分短缺现象，除人为因素外，降水和径流是影响生态干旱的决定性因素，即气象干旱和水文干旱持续一段时间后，肯定会引起生态干旱；农业干旱出现后，为缓解农业干旱则可能牺牲生态环境，进而引发生态干旱；即使未出现农业干旱，灌溉农业用水引走了大量的天然径流，挤占生态用水，也可能引起生态干旱；人类城市化进程的加剧使工业和生活用水增加，减少了天然径流或部分抽取地下水，加剧了生态干旱的发生发展。除此之外，突发性污染事件和生产生活排出的污水可能在短期内引起社会经济和农业干旱,进而引起生态干旱。

3　生态干旱评估指标

3.1　植被指数

植被指数是遥感监测地面植被生长状况的一个指数,它由卫星传感器可见光和近红外通道探测数据的线性或非线性组合形成,可以较好地反映地表绿色植被的生长和分布状况。常用的植被指数有归一化植被指数(NDVI)、条件植被指数 VCI、条件温度指数 TCI,距平植被指数 AVI,植被供水指数 VSW 等。其中 NDVI 最常用,它是植被生长状态和植被覆盖度的最佳指示因子,被广泛地应用于植被盖度、分布、类型、长势及植被季节变化和土地覆盖研究,国内外不少研究者应用 NOAA/AVHRR 资料在此方面做了有益探索。一般来说,缺水时作物生长将受到影响,植被指数会降低,可利用该指标进行生态干旱的监测和分析,其计算公式如下

$$NDVI = (NIR - RED)/(NIR + RED) \tag{1}$$

式中:NIR 和 RED 分别为可见光和红外光波段灰度值。

计算此指标需要流域遥感图像资料,随着遥感技术的发展和广泛应用,以遥感卫星获得长时间序列各种精度的观测数据并进行一定的处理和加工已经成为许多国际组织和机构数据集计划的重要组成部分。使用 1 km 空间分辨率的遥感数据具有时间序列上的高分辨率,特别是 *NDVI* 合成数据能很好地反映地表植被的季候特征与变化。

3.2　潜水埋深与植被生存临界水位

干旱区内陆河流域的水文过程控制着生态过程,植物的生存生长主要依赖地表水和地下水,在无地表水补给的绿洲中下游区,地下水则成为维系天然植被生长的唯一水源,其分布状况是天然植被生存的先决条件,地下水位直接影响着植被的类型与存亡。因此,可用当前潜水埋深与不同植被的临界地下水位比值来判断植被的受旱情况。

目前,较为可靠的地下水位资料获取途径为观测井的动态实测值,通过对观测井的日、旬或月人工或自动观测,可实时了解地下水位动态。但在大范围内,有限的观测井并不能代表整个区域的地下水位状况,特别是在潜水含水层地质条件变异较大的区域,且无论人工还是自动观测,都需要足够的资金支持。随着遥感技术的发展,利用遥感反演技术获得面上、大范围的地下水埋深实时资料,将是未来的发展方向之一。

植被生存临界水位主要依靠试验手段获得,找出干旱区典型植被的临界水位,如胡杨、沙枣、沙柳、柽柳及芦苇、芨芨草等。

3.3　河流断流长度和水域萎缩面积

内陆河流域河流补给来源主要为高山融水和降水,多为季节性河流。受干旱气候和水资源开发利用影响,枯水期或干旱期河流生态系统旱情主要表现在断流长度不断增加,断流时间持续加长,故河流断流长度可作为生态干旱评估指标之一。内陆河流域的湖泊、湿地一般为河流的尾闾,由于干旱造成的河流来水量减少直接影响湖泊湿地的水域面积。因此,水域萎缩面积可作为湖泊、湿地生态系统干旱的表现指标。

河流断流长度与水域面积指标均可实测或根据遥感影像确定,资料易获得,评估方法简单易行,在具体评估时可将两个指标转化为无量纲的河流断流比率和面积萎缩率。

4 结论与讨论

本文在探讨干旱区内陆河流域生态系统稳定性的基础上,指出水资源是影响绿洲生态系统稳定的关键制约因子;基于内陆河流域自然和社会水循环特点,阐述了水分驱动下的天然绿洲生态干旱及其特征,分析了在水分条件变化下,内陆河流域中下游绿洲植被干旱及其演化规律,并分析了天然绿洲生态干旱与其他干旱类型的相互关系;总结归纳了以NDVI、潜水埋深/植被生存临界水位、河流断流比率及湖泊湿地面积萎缩率为主要指标的生态干旱评估指标体系,指出了指标获取途径。由于本文为绿洲生态系统生态干旱方面的尝试性研究,还存在以下有待商榷或不足之处:

(1)在生态干旱理论探索方面,水分短缺是造成生态系统演化的不稳定因素之一,近期,不少研究者开始采用更加具有物理机制的模型来研究干旱缺水对生态系统的影响,例如吸取水文学的方法,将干旱现象与水文循环过程相结合,更加真实客观地反映干旱状况。

(2)在生态干旱研究内容和评估方法方面,目前研究内容仅限于湖泊生态系统,且生态干旱评估方法研究尚不多见,少数涉及生态干旱的研究也是借鉴生态风险评价方法构建评价指标体系。随着对生态环境和旱灾风险管理的重视,生态干旱研究将越来越深入,将可能成为独立的干旱类型之一,研究内容将扩展至河流、植被等生态系统,研究手段和方法将更加丰富。

(3)本文提出的生态干旱评估指标还存在以下问题:①评估指标归因的不确定性。由于流域来水具有随机性,评估指标在人为开发利用条件一定的情况下适用,如潜水埋深不仅受干旱状况的影响,更多地受水资源开发利用的人为因素影响。②生态干旱评价指标计算参数之间可能存在一定的联系,如潜水位和河流断流长度、湖泊湿地萎缩面积都与地下径流补给量有关,潜水位的下降可能引起河流断流和湖泊湿地的萎缩。

(4)在生态干旱与其他干旱类型的相关性方面,生态干旱是干旱发展到一定阶段的产物,和其他干旱类型关系密切,它在受自然因素影响的同时,也受人为因素的严重影响。在内陆河流域绿洲区,社会经济用水严重挤占生态用水,造成生态干旱的发生和发展;与农业干旱和水文干旱相比,生态干旱时间尺度较大,评估指标表现可能有所滞后,但生态旱情一旦形成,后果往往比较严重。

参考文献

[1] 贡璐,鞠强,潘晓玲. 博斯腾湖区域景观生态风险评价研究[J]. 干旱区资源与环境,2007,21(1):27-31.

[2] 张丽丽,殷峻暹,侯召成. 基于模糊隶属度的白洋淀生态干旱评价函数研究[J]. 河海大学学报:自然科学版,2010,38(3):252-257.

[3] Juan C. Linares, Lahcen Taïqui, Jesús Julio Camarero. Increasing Drought Sensitivity and Decline of Atlas Cedar (Cedrus atlantica) in the Moroccan Middle Atlas Forests[J]. Forests, 2011, 2: 777-796.

[4] Dobbertin, M. Tree growth as indicator of tree vitality and of tree reaction to environmental stress: A review[J]. European Journal of Forest Research, 2005, 124: 319-333.

[5] Allen, C. D., Macalady, A. K., Chenchouni, H. et al. A global overview of drought and heat-induced tree mortality reveals emerging climate change risks for forests [J]. Forest Ecology and Management, 2010, 259: 660-684.

[6] Sarris, D., Christodoulakis, D., Körner, C. Impact of recent climatic change on growth of low elevation eastern Mediterranean forest trees [J]. Climate Change, 2010, 106: 203-223.

[7] Condit, R. Ecological implications of changes in drought patterns: Shifts in forest composition in Panama[J]. Climate

Change, 1998, 39: 413-427.

[8] 郭铌,管晓丹. 植被状况指数的改进及在西北干旱监测中的应用[J]. 地球科学进展,2007,22(11):1160-1168.

[9] 郝兴明,李卫红,陈亚宁. 新疆塔里木河下游荒漠河岸(林)植被合理生态水位[J]. 植物生态学报,2008,32(4):838-847.

[10] 程维明,周成虎,刘海江,等. 玛纳斯河流域50年绿洲扩张及生态环境演变研究[J]. 中国科学 D辑地球科学,2005,35(11):1074-1086.

[11] Robert M May. Thresholds and breakpoints in ecosystems with a multiplicity of stable states [J]. Nature, 1977,269(6):471-477.

[12] 马风云. 生态系统稳定性若干问题研究评述[J]. 中国沙漠,2002,22(4):401-407.

[13] Stephen H Roxburgh, J. Bastow Wilson. Stability and coexistence in a lawn community: experimental assessment of the stability of the actual community [J]. Oikos,2000, 88(2): 409-423.

[14] David A Wardle, Karen I. Bonner, Gary M. Barker. Stability of ecosystem properties in response to above ground functional group riches and composition [J]. Oikos, 2000, 89(1): 11-23.

[15] 柳新伟,周厚诚,李萍,等. 生态系统稳定性定义剖析[J]. 生态学报,2004,24(11):2635-2640.

[16] 曹宇,肖笃宁,欧阳华,等. 额济纳天然绿洲景观演化驱动因子分析[J]. 生态学报,2004,24(9):1895-1902.

[17] 袁文平,周广胜. 干旱指标的理论分析与研究展望[J]. 地球科学进展,2004,19(6):982-990.

【作者简介】 梁犁丽,女,1982年9月生,博士,2011年6月毕业于中国水利水电科学研究院水资源所,获工学博士学位。现工作于中国水利水电科学研究院中水科技水情水调事业部,工程师,主要研究方向为生态水文、水文预报、干旱评估。

基于景观格局的河流廊道景观异质性分析方法研究*

赵进勇[1]　董哲仁[1]　游文荪[2]　邢乃春[3]　翟正丽[1]

(1. 中国水利水电科学研究院　北京　100048；
2. 江西省水利科学研究院　南昌　330029；
3. 水利部水利水电规划设计总院　北京　100011)

【摘要】 本文建立了土地分类系统与河流地貌单元特征相结合的河流廊道尺度景观类型分类体系。利用景观格局分析方法和3S技术，发展了河流廊道尺度下考虑边滩、江心洲等河流地貌单元特征的景观异质性定量分析方法，可在河流廊道尺度上进行景观格局分析，从而反映河流廊道景观的空间异质性和时间异质性。景观格局研究主要集中于两个方面：景观格局的空间特征分析和变化分析，并利用河流廊道尺度下的景观异质性定量分析方法对瓯江目标河段的景观异质性进行了分析。

【关键词】 河流廊道　景观格局　异质性

景观空间格局分析是景观生态学研究的重要组成部分。对特定区域的景观生态空间格局进行研究，是揭示该区域生态状况及空间变异特征的有效手段。随着计算机技术、信息技术及3S技术的发展，景观格局变化研究逐步发展到定量化阶段，主要表现为运用不同时期遥感影像对区域景观格局及其变化过程进行研究。河流廊道是陆地生态景观中最重要的廊道，对于生态系统和人类社会都具有生命源泉的功能。河流廊道范围可以定义为河流及其两岸水陆交错区植被带，或者定义为河流及其某一洪水频率下的洪泛区的带状地区，广义的河流廊道还应包括由河流连接的湖泊、水库、池塘、湿地、河汊、蓄滞洪区以及河口地区。把河流廊道作为一个整体研究时，景观生态学具有很大的优势，能够将河流廊道的结构、功能和动态有机地结合起来。同时，非生物环境的空间异质性与生物群落多样性的关系反映了非生命系统与生命系统之间的依存和耦合关系。一个地区的生境空间异质性越高，就意味着创造了多样的小生境，能够允许更多的物种共存。

为了突出河流景观类型的功能特点，反映河流廊道尺度下的景观异质性特征，本文尝试在河流廊道尺度上进行景观格局分析，以反映其景观空间异质性和时间异质性。本文以瓯江流域开潭-玉溪河段高分辨率遥感影像、野外实地调查资料和各种相关的统计数据为数据源，利用ENVI4.4软件解译得到该区土地利用现状图，然后利用景观格局定量化分析软

* 基金项目：水利部公益性行业科研专项经费资助项目(200801023、201101034)，中国水科院科研专项(防集1019、防集05ZD01、防集KF0703)。

件 Fragstats3.3 计算景观指数,对该区进行了大比例尺的景观异质性分析,以期为该区的水土资源开发利用提供一定的参考依据。

1 研究区概况

本文研究的目标河段地处瓯江中、上游,位于丽水市莲都区境内,介于东经 119°42′~119°57′、北纬 28°17′~28°25′,全长为 41.7 km,河道比降为 0.7‰~0.8‰。东邻青田县,西接云和县、松阳县,南与景宁畲族自治县交界,西北与武义县相依,东北与缙云县连靠。见图 1。

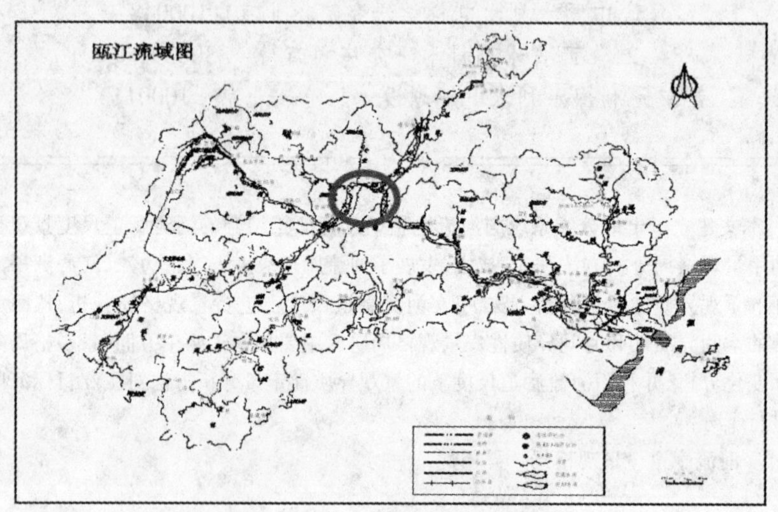

图 1 研究区域河流廊道示意图

2 河流景观异质性定量分类及指标体系

首先以瓯江流域 1∶50 000 地形图为基准,分别对 2004 年和 2009 年遥感影像进行分类校正。对 ALOS、SPOT 的同时相全色波段和多光谱波段进行融合处理,获得了研究区域精度为 2.5 m 的影像。此外,还应用了瓯江开潭至玉溪段 1∶1 000 比例尺地形图、土地利用现状、统计年鉴等数据。

2.1 分类体系的建立

景观分类体系的建立和科学合理的分类是进行土地利用分类信息提取和景观格局分析的必要前提。江心洲、边滩、牛轭湖、季节性干河床和冲洪积扇等景观特征在河流廊道尺度下的栖息地营造中有重要作用,所以本文尝试建立土地分类系统与河流地貌单元特征相结合的研究区域河流廊道尺度景观类型分类体系。

根据研究区域河流廊道内的自然环境、生物系统和人文社会特征,考虑山区河流的形态特征、河流地貌单元特点,并结合土地利用现状分类国家标准,同时参考当地的土地利用状况、人类活动影响等因素,以及遥感图像的可判程度等将景观类型分为耕地、有林地、疏林地、灌木林、草地、建设用地、水体、江心洲、边滩九大类别。研究区域河流廊道尺度景观类型分类体系见表 1。2004 年和 2009 年景观类型示意图见 1 图 2。

表1 研究区域河流廊道尺度景观类型分类体系

一级		二级		含义
编号	名称	编号	名称	
1	耕地	11	耕地	指无灌溉水源及设施，靠天然降水生长作物的耕地；有水源和浇灌设施，在一般年景下能正常灌溉的旱作物耕地；以种菜为主的耕地，正常轮作的休闲地和轮歇地
2	林地	21	有林地	指郁闭度>30%的天然林和人工林。包括用材林、经济林、防护林等成片林地
		22	疏林地	郁闭度<30%的稀疏林地
		23	灌木林	指郁闭度>40%、高度在2 m以下的灌木丛
3	草地	31	草地	城镇绿化草地、岸边草地
4	水体	41	水体	指天然形成或人工开挖的河流水域
5	边滩	51	边滩	天然卵石边滩、采砂废弃卵石岸边堆积体
6	江心洲	61	江心洲	采砂废弃卵石河中堆积体
7	建设用地	71	建设用地	指大、中、小城市、县、镇以上建成区用地及农村居民点

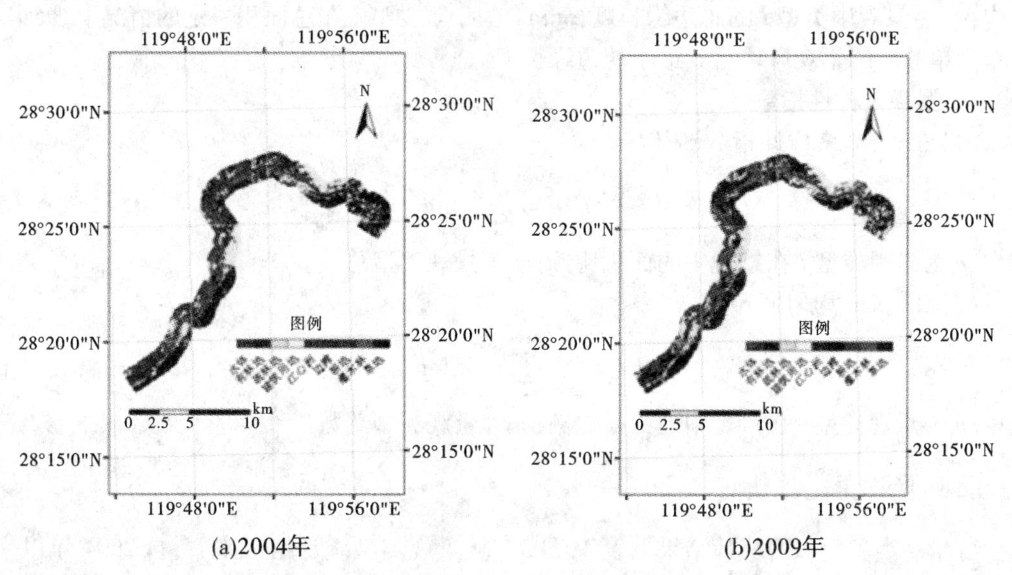

图2 2004年和2009年景观类型示意图

2.2 景观格局分析与评价指标体系

综合考虑各景观格局指数的生态学意义、目标河段的栖息地特点、前人的研究成果等因素，本着简单、有效、实用的原则，选取了三类指标：景观破碎化指数（斑块密度、最大斑块指数）、景观形状指数（边缘密度、景观形状指数）和景观多样性指数（景观蔓延度指数、Shannon景观多样性指数、Shannon景观均匀度指数），分析各景观要素变化的空间结构规律，并据此对评价区域的景观异质性进行分析。

2.2.1 景观斑块指数

(1)斑块密度 PD:

$$PD = n_i / A \tag{1}$$

式中:n_i 为第 i 类景观类型的斑块数量或研究区域斑块总数;A 为研究区域总面积。

(2)最大斑块指数 LPI:

$$LPI = \frac{\max\limits_{j=1}^{n} a_{ij}}{A} \tag{2}$$

式中:a_{ij}为第 i 类景观类型的最大斑块面积或研究区域内最大斑块面积;A 为区域总面积。

2.2.2 景观形状指数

(1)边缘密度 ED:

$$ED = \frac{1}{A}\sum_{i=1}^{m}\sum_{j=1}^{n} p_{ij} \tag{3}$$

式中:m 为研究范围内某一空间分辨率上景观要素类型总数,p_{ij}为景观中第 i 类景观要素斑块与相邻第 j 类景观要素斑块间的边界长度;A 为区域总面积。

(2)景观形状指数 LSI:

该指数表示景观空间的聚集程度,也可以表示景观形状的复杂程度,其表达式如下:

$$LSI = e_i / \mathrm{min}e_i \tag{4}$$

式中:e_i 是景观中类型 i 的总边缘长度;$\mathrm{min}e_i$ 为景观类型 i 在总面积一定的情况下,聚集成一个简单紧凑的景观斑块后其最小的边缘长度。

2.2.3 景观多样性指数

(1)Shannon 多样性指数 $SHDI$

$$SHDI = \sum_{i=1}^{m} p_i \ln p_i \tag{5}$$

式中:p_i为景观类型 i 在景观中的面积比例;m 为景观类型总数。

(2)Shannon 均匀度 $SHEI$

$$SHEI = (-\sum_{i=1}^{m} p_i \times \ln p_i)/\ln m \tag{6}$$

式中:p_i为景观类型 i 在景观中的面积比例;m 为景观类型总数。

3 景观类型变化分析

各景观类型面积的统计结果见表 2、图 3。从景观类型统计结果来看,自 2004 年以来,研究区域景观类型面积变化中,水体、建设用地、疏林地、灌木林和草地面积总体上有所增加,但有林地、边滩、江心洲和耕地的面积有所减少。在面积增加的土地利用类型中,水体面积由 2004 年的 6.58 km^2增加到 2009 年的 9.37 km^2,面积净增加 2.789 km^2,是面积增加最大的景观类型;建设用地面积由 2004 年的 13.92 km^2增加到 2009 年的 15.86 km^2,面积净增加 1.94 km^2,是面积增加第二大的景观类型;草地、疏林地和灌木林地所占面积变化不大,在 2004~2009 年间面积有一定的增加。在面积减少的景观类型中,边滩的面积由 5.52 km^2 减少到 2009 年的 2.55 km^2,减少了约 53.8%;江心洲的面积由 0.83 km^2 减少到 2009 年的 0.56 km^2,减少了约 32.5%;有林地和耕地景观类型的面积稍有减少。

表2 两个时期研究区域景观类型统计

景观类型	2004年		2009年	
	面积(km^2)	比例(%)	面积(km^2)	比例(%)
水体	6.58	8.68	9.37	12.36
有林地	24.44	32.25	22.19	29.28
疏林地	1.85	2.44	2.4	3.17
建设用地	13.92	18.38	15.86	20.93
江心洲	0.83	1.09	0.56	0.74
边滩	5.52	7.28	2.55	3.36
耕地	15.68	20.69	15.39	20.31
灌木林	5.73	7.57	6.16	8.13
草地	1.23	1.62	1.30	1.72
合计	75.78	100	75.78	100

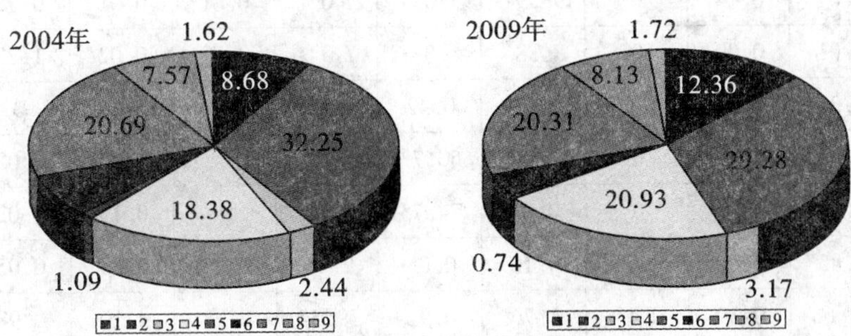

1—水体景观；2—有林地景观；3—疏林地景观；4—建设用地景观；5—江心洲景观；6—边滩景观；7—耕地景观；8—灌木林地景观；9—草地景观

图3 2004年和2009年瓯江流域景观结构图

为了进一步研究瓯江流域景观格局的动态演变过程，利用ArcGIS软件工具箱的INTERSECT命令获取了2004~2009年的景观要素变化转移矩阵，通过该矩阵，得出4年多来各种景观要素相互转移变化情况，如表3所示。由景观要素面积转移分析可得到以下结论：

(1)水体、边滩、江心洲、灌木林地等景观类型之间互相转化。水体景观类型有一部分转化为边滩、江心洲和灌木林，这主要是因为在河边采砂废弃卵石而形成新的边滩，在河床采砂产生的废弃卵石形成新的江心洲，并且其中部分边滩上面生长形成灌木林；边滩、江心洲和灌木林景观类型的一部分转化为水体，主要是因为水位上涨而淹没边滩、江心洲和灌木林。边滩和江心洲之间的互相转化主要是因为：原来的自然边滩或采砂废弃卵石边滩被二次开挖，形成新的江心洲；在河边采砂产生的废弃卵石形成新的边滩，原来的因采砂废弃卵石而形成的江心洲与新形成的边滩合并，形成新的大面积边滩。

(2)有林地、疏林地、灌木林、草地、耕地、建设用地等各种景观类型相互转化，既与植被演替相关，也与人类活动影响密切相关。

表3　2004～2009年景观要素面积转移矩阵(km²,%)

景观要素类型	水体(%)	有林地(%)	疏林地(%)	建设用地(%)	江心洲(%)	边滩(%)	耕地(%)	灌木林(%)	草地(%)	2004年总面积(km²)
水体	6.07	0	0	0.02	0.09	0.27	0.01	0.12	0	6.58
转出率	92.19	0.03	0.04	0.33	1.39	4.15	0.08	1.77	0.02	比例
转入率	64.75	0.01	0.11	0.14	16.24	10.71	0.03	1.89	0.10	8.68
有林地	0.02	21.52	1.17	1.03	0	0.02	0.44	0.22	0.02	24.44
转出率	0.10	88.05	4.77	4.20	0	0.08	1.80	0.90	0.10	比例
转入率	0.26	96.95	48.52	6.48	0	0.77	2.86	3.57	1.89	32.26
疏林地	0	0.61	1.22	0.02	0	0	0	0	0	1.85
转出率	0	33.01	65.83	1.01	0	0	0.07	0	0.08	比例
转入率	0	2.75	50.69	0.12	0	0	0.01	0	0.11	2.44
建设用地	0	0.02	0	13.77	0	0	0.11	0	0.02	13.92
转出率	0.01	0.14	0	98.91	0	0	0.81	0.01	0.12	比例
转入率	0.01	0.09	0	86.86	0	0	0.73	0.02	1.29	18.38
江心洲	0.35	0	0	0	0.42	0.05	0	0.01	0	0.83
转出率	41.91	0	0	0.40	50.73	6.00	0	0.93	0.03	比例
转入率	3.71	0	0	0.02	74.75	1.95	0	0.13	0.02	1.09
边滩	2.85	0.02	0.01	0.10	0.05	2.18	0.02	0.23	0.05	5.52
转出率	51.66	0.39	0.18	1.76	0.92	39.54	0.38	4.19	0.98	比例
转入率	30.43	0.10	0.41	0.61	9.02	85.57	0.14	3.76	4.17	7.28
耕地	0	0.01	0.01	0.84	0	0	14.79	0	0.02	15.67
转出率	0	0.06	0.04	5.37	0	0.03	94.37	0	0.13	比例
转入率	0	0.04	0.26	5.31	0	0.18	96.11	0	1.57	20.68
灌木林	0.06	0	0	0.06	0	0.01	0.01	5.58	0.01	5.73
转出率	0.96	0.06	0	1.05	0	0.18	0.26	97.35	0.14	比例
转入率	0.59	0.02	0	0.38	0	0.40	0.10	90.62	0.62	7.57
草地	0.02	0.01	0	0.01	0	0.01	0	0	1.17	1.23
转出率	1.86	0.78	0	1.09	0	0.86	0.27	0.02	95.12	比例
转入率	0.24	0.04	0	0.08	0	0.41	0.02	0	90.23	1.62
2009年总面积(km²)	9.37	22.19	2.40	15.85	0.56	2.55	15.39	6.15	1.30	75.77
比例(%)	12.36	29.29	3.17	20.93	0.74	3.36	20.31	8.13	1.71	
变化率(%)	3.68	-2.96	0.73	2.55	-0.35	-3.92	-0.37	0.56	0.09	

注:转出率表示2004年第 i 种景观类型转变为2009年第 j 种景观类型面积的比例,为横向数据;转入率表示2009年第 j 种景观类型由2004年第 i 种景观类型转化来的面积比例,为纵向数据。

4 景观格局变化分析

4.1 河流廊道尺度景观斑块性质分析

2004 ~2009 年研究区总体景观格局斑块指数的变化情况见表4,结果表明,自2004 年以来,研究区景观斑块密度减小,最大斑块指数减小。在这个时期内,研究区域的空间异质性程度逐渐缩小。原因主要是受人类活动的干扰,许多异质性的小斑块根据人们的活动目的,逐渐被合并成面积较大相对同质的斑块。大面积结构复杂自然生态系统被结构单一的人工生态系统所取代,会导致原有栖息地质量和生物多样性的下降。

表4 2004 ~2009 年研究区域景观格局指数表

指数类型	2004 年	2009 年
PD	20.4	16.8
LPI	9.7	7.9
ED	118.9	114
LSI	28.8	27.8
SHDI	1.807	1.802
SHEI	0.822 5	0.820 3

研究区2004 年和2009 年不同景观类型的斑块指数变化情况见图4。从中可以看出,只有水体景观的 *PD* 指数呈增加的趋势。其他的8 类景观的PD 指数都呈减少的趋势,这表明这几种景观的异质性在逐年减小。疏林地和边滩的 *PD* 指数在这一时期内变化较为剧烈,这与水体和边滩之间的大面积转化相吻合。

9 种景观类型中,有林地景观、水体景观、耕地景观和建设用地景观的 *LPI* 指数明显高于其他景观类型,这说明林地、水体、耕地和建设用地景观在整个区域内占绝对优势地位,其斑块丰度远高于其他景观。2004 ~2009 年,江心洲和边滩的异质性程度减小,说明采砂活动对河床和岸边区域的扰动呈减小趋势。

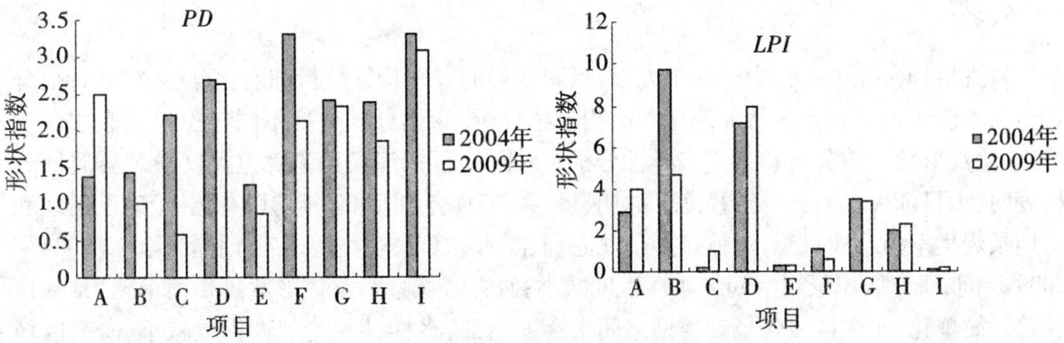

A—水体景观; B—有林地景观; C—疏林地景观; D—建设用地景观; E—江心洲景观; F—边滩景观; G—耕地景观; H—灌木林地景观; I—草地景观

图4 2004 ~2009 年研究区域各景观类型斑块指数变化

4.2 河流廊道景观形状分析

研究区总体景观格局形状指数的变化情况见图4,从结果上来看,2004 ~2009 年,区域景观的边缘密度与景观形状指数均呈减小的趋势,研究区域景观整体形状趋于简单化。

研究区2004年和2009年不同景观类型的形状指数变化情况见图5。从总体上看,与其他景观类型相比,建设用地、水体、有林地和耕地景观的ED较高。耕地、灌木林和草地景观的边缘密度变化不明显。水体和建设用地景观的边缘密度增加,有林地、疏林地、江心洲和边滩景观的边缘密度减小,这是由于水体和建设用地扩张,其边界长度增加,从而边缘密度增大。有林地、疏林地、江心洲和边滩景观类型的斑块个数减少,边缘密度减小。有林地、疏林地、江心洲的形状指数在2004~2009年间减小,说明景观形状趋向规则化,所受扰动减小。但边滩因受采砂扰动,形状指数增大,其他景观的形状指数变化不大。

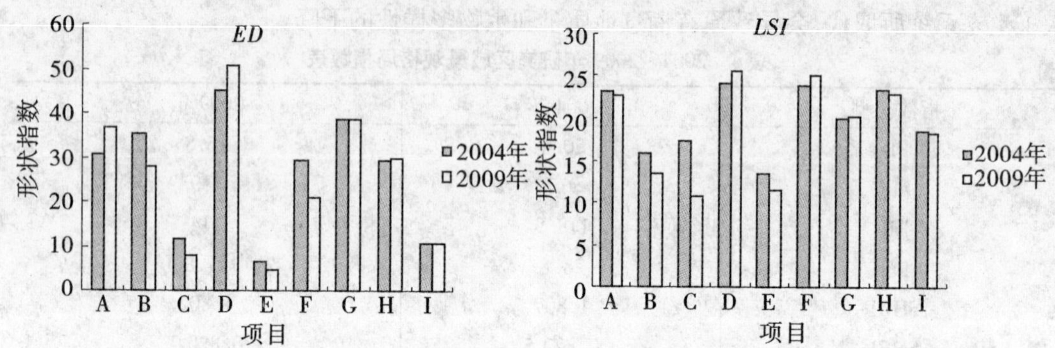

A—水体景观; B—有林地景观; C—疏林地景观; D—建设用地景观; E—江心洲景观;
F—边滩景观; G—耕地景观; H—灌木林地景观; I—草地景观

图5 2004~2009年研究区域各景观类型斑块指数变化

4.3 景观多样性分析

研究区总体土地利用景观格局多样性指数的变化趋势见表4。从结果上来看,2004~2009年间,景观多样性指数都减小了。景观多样性和均匀性指数减小,说明2004~2009年间区域内景观异质性减小,景观多样性减小,说明在区域经济不断发展的压力下,目标河段的栖息地质量有退化趋势。

5 结语

景观格局研究主要集中于两个方面:景观格局的空间特征分析和变化分析,空间特征分析包括不同年份景观要素特征统计和面积转移分析,变化分析包括不同年份景观斑块性质分析、景观形状指数分析和景观多样性分析。对瓯江目标河段河流廊道尺度景观异质性分析表明:①目标河段河流廊道尺度下的栖息地在2004年和2009年均以有林地、建设用地和耕地景观类型为基质景观,水体、边滩、江心洲、灌木林等景观类型之间互相转化。②2004~2009年间,研究区域景观空间异质性程度减小;研究区域景观整体形状趋于简单化,景观格局趋于简单化;研究区域各景观类型之间聚合度增加,多样性和均匀性指数减小,研究区域内栖息地质量有退化趋势。

参考文献

[1] Fu B J. The spatial pattern analysis of agricultural landscape in the loess area[J]. Acta Ecologica Sinica, 1995, 15(2): 113-120.

[2] 韩海辉,杨太保,王艺霖. 近30年青海贵南县土地利用与景观格局变化[J]. 地理科学进展, 2009, 28(2):207-215.

[3] 周华荣. 干旱区河流廊道景观生态学研究——以新疆塔里木河中下游区域为例[M]. 北京:科学出版社,2007.

[4] 董哲仁. 河流生态修复的尺度、格局和模型[J]. 水利学报, 2007, 37(1):1476-1481.
[5] Kaiyu song, Jinyong Zhao, Wei Ouyang, Yuan Zhang and Fanghua Hao. LUCC and landscape pattern variation of a wetland in warm Southern China over three decades [J]. Procedia environmental sciences, 2010, 2: 1296-1306.

【作者简介】 赵进勇,1976 年 10 月生,2001 年河海大学毕业,博士,现就职于中国水利水电科学研究院,高级工程师,亚洲河流生态修复协作网络(中国)总协调人,IAHR 会员,水利部水工程生态效应与生态修复重点实验室客座研究员,研究领域为生态水工学。

南桠河冶勒水电站工程对区域生态影响分析

王世岩　刘　畅　毛战坡　杨素珍

（中国水利水电科学研究院 水环境研究所　北京　100038）

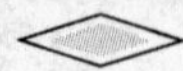

【摘要】 冶勒水电站工程为南桠河流域梯级规划“一库六级”的第六级龙头水库电站，工程区域位于处四川盆地西侧，属于青藏高原向四川盆地过渡地带，区域生态环境相对敏感和脆弱。本文在阐述冶勒水电站工程区域生态环境现状的基础上，通过遥感专题解译、GIS空间分析以及野外实地调查，对工程建设前后生态景观格局、景观多样性以及区域生物多样性变化进行了分析。结果表明：南桠河冶勒水电站工程建设对于区域生态景观格局影响不大，但受大坝阻隔影响，对于河流水生生物尤其洄游性鱼类栖息活动等造成较大影响。为此，提出了减缓水生态影响的补偿措施等建议。

【关键词】 南桠河　冶勒水电站　生态影响　GIS

水电资源作为一种清洁的可再生能源，对促进经济社会发展、优化能源结构、减少温室气体排放具有重要意义。但同时水电开发对流域生态环境是一种较大规模的扰动，尤其水利水电工程大坝的修建常使原本完整的流域生态系统被分隔成上游、水库和下游三部分，致使局地气候、水文条件以及物质循环和能量流动等发生变化，产生诸多的生态问题，导致流域生态系统某些不可逆转的改变。目前，随着最新修订的《环境影响评价技术导则生态导则》（HJ 19—2011）的颁布实施，生态影响评价已经成为水利水电工程环境影响评价的重要组成部分，对于水电开发工程区域生态环境保护有着重要意义。本研究以南桠河冶勒水电站工程为例，通过野外实地调查、遥感影像专题解译与GIS空间叠加等分析，研究水电站工程对区域生态景观格局、景观多样性、区域植被覆盖度以及河流水生生态的影响，为区域生态环境保护提供一定参考。

1　冶勒水电站工程区域生态环境基本特征

冶勒水电站工程位于四川省凉山州冕宁县和雅安市石棉县境内，为南桠河流域梯级规划“一库六级”的第六级龙头水库电站。电站以发电为主，无航运、漂木、防洪、灌溉等综合利用要求。电站采用混合式开发，坝前水库正常蓄水位2 650 m，水库总库容2.98亿m^3、调节库容2.76亿m^3，具有多年调节能力。该水电站主体工程2000年末开工，2006年8月，主体工程全面完工。

南桠河是大渡河的一级支流，发源于四川省甘孜藏族自治州九龙县牦牛山东麓，分南、北两源，北源为勒丫河，南源为石灰窑河，两源于两岔河汇合后始称南桠河。南桠河干流大致自西南流向东北，在栗子坪乡先后从右岸纳入孟获城河和阿鲁伦底河，在回隆乡凉桥右岸处有竹马河加入,于石棉县城注入大渡河。全长78 km,集水面积1 200 km^2。流域地形属高山区,地势西高东低。西北、西南、东北三面的分水岭高程均在4 000 m以上,东南面分水岭高程相对较低,一般为3 200~3 500 m。

区域气候属于川西高原气候区,仅在4 000 m以上为高原亚寒带。区域地势高差悬殊,气候差异大,降水和湿度从上游至下游逐渐减小,气温和蒸发量则逐渐增大。南瓜桥以下气温较高,雨量较少。石棉气象站海拔874 m,多年平均气温17 ℃,极端最高气温38.1 ℃,极端最低气温-3.9 ℃。多年平均降雨量和降雨天数分别为801.2 mm和143 d,多年平均相对湿度68%。南瓜桥至冶勒地区气温明显降低,雨量增大。

该区的植被主要由针阔混交林、亚高山寒温带针叶林、亚高山灌丛以及高山灌丛草甸组成,由于长期以来人为的影响,山地暖温带针阔混交林和亚高山寒温带针叶林受到较大破坏,逐步退化为较大面积的亚高山灌丛草甸。本文开展冶勒水电站工程对区域生态环境的影响分析,将对保护区域脆弱的亚高山灌丛、高山灌丛草甸生态系统以及河流水生态等方面具有重要意义。

2 冶勒水电站对区域生态景观结构影响

2.1 调查分析区域确定

为了分析冶勒水电站对区域生态景观格局的影响,首先需要框定调查范围,在本文中框定原则主要有3个方面:①包括整个冶勒水电站工程区域;②为反映工程建设后对下游区域的影响,调查范围应包括工程坝下部分一定区域;③将移民安置区也包含在生态调查范围之内。按照上述范围框定原则,在ArcGIS地理信息系统软件中标出该区域,为东西长约32 km,南北宽约22 km,面积约705 km^2的矩形框(见图1)。

图1 生态调查评价区域范围示意图

在所划定的陆生生态调查区域范围内,为了解冶勒水电站工程建设对区域土地利用、生态景观格局等方面的影响,本研究采用了卫星遥感的方法,分别选择工程建设前和建设后的卫星遥感数据对工程周边的生态环境进行了调查。工程建设前遥感影像数据采用1999年

8 月的 TM 陆地资源卫星数据，空间分辨率为 30 m；工程建设后采用 2010 年 9 月中巴卫星影像数据(CBERS-2)，空间分辨率为 20 m。通过遥感影像大气校正、几何精校正、图像增强等专题处理和解译，获得工程建设前后区域土地利用类型图及相应专题数据。

2.2 工程建设前后生态景观格局变化

景观生态系统质量现状由生态评价区域内自然环境、各种生物以及人类社会之间复杂的相互作用决定。从景观生态学结构与功能相匹配理论来说，结构是否合理决定景观功能的优劣，在组成景观生态系统各类组分中，拼块、廊道和模地是关键的 3 个组成部分，其中模地是景观背景区域，很大程度上决定景观性质，对景观动态起着主导作用。

通过 ArcGIS 地理信息系统软件支持，对冶勒水电站工程区域遥感解译的生态景观空间数据进行分析处理。将各生态景观类型斑块面积、斑块数、景观斑块频率指标按照景观格局参数的计算方法，计算拼块密度(R_d)、频率(R_f)、景观比例(L_p)，确定拼块优势度值(D_o)。表 1 为工程建设前后区域生态景观对比情况。

表 1 工程建设前后区域生态景观对比情况

类型	时期	拼块数	密度(%)	频度(%)	景观比例(%)	优势度值(%)
耕地	建设前	35	13.41	22.04	3.61	10.67
	建设后	33	12.41	21.35	3.50	10.19
林地	建设前	122	46.74	100.00	58.02	81.64
	建设后	123	46.24	100.00	58.06	81.60
草地	建设前	92	35.25	99.65	36.48	53.40
	建设后	94	35.34	99.60	35.28	53.13
水域	建设前	2	0.77	3.58	0.17	1.17
	建设后	6	2.26	7.02	1.44	3.04
居民地	建设前	1	0.38	0.14	0.01	0.13
	建设后	1	0.38	0.14	0.01	0.13
未利用地	建设前	9	3.45	6.06	1.71	3.23
	建设后	9	3.38	6.06	1.71	3.22

可以看出，工程建设后区域耕地拼块的优势度值下降 0.48%，林地景观优势度下降 0.04%，草地下降 0.27%，未利用地下降 0.01%，而水域景观优势度则增加了 1.87%。这说明工程建设后，区域生态景观除旱田景观、林地景观及部分未利用地出现减少变化外，其余类型的生态景观均呈现出增加变化特征，尤其水域景观增加较明显，这与冶勒水库工程建成后形成较大的库区水面有直接关系。而林地和草地景观面积的减少会对区域生态环境产生一定的负面影响，但其优势度值变化幅度较小，林地模地的地位没有发生改变。

2.3 工程建设前后景观多样性变化

多样性指数一般用来衡量生态景观体系的复杂程度，在区域景观多样性评价中，采用香农-威纳指数和 Evenness 指数进行分析。香农-威纳指数来源于信息理论，既考虑景观内斑块数目，又考虑斑块相对多度。景观中斑块种类增多代表景观的复杂程度增高，景观所含的信息量越大。

按照已经计算得到景观斑块数,采用香农-威纳指数和 Evenness 指数公式进行计算,得到冶勒水电站工程评价区域景观多样性情况。工程建设前调查评价区域香农-威纳多样性指数为 1.418,Evenness 均匀度指数为 0.255;工程建设后区域香农-威纳多样性指数为 1.469,Evenness 均匀度指数为 0.263。对比工程建设前后区域景观多样性变化值,工程建成后区域景观多样性和均匀性变化幅度很小,说明工程建设对于区域生态景观多样性影响不大。

3 冶勒水电站建设对区域生物多样性的影响

3.1 对植被覆盖度的影响

为了分析冶勒水电站工程建设后对区域植被覆盖度的影响,本文采用了归一化植被指数计算区域植被覆盖度的方法。归一化植被指数(the Normalized Difference Vegetation Index, NDVI)是目前应用最广泛的一种植被指数,NDVI 的变化在一定程度上能够代表地表覆盖度的变化,是植物生长状况及植被空间分布密度的最佳指示因子,与地表植被覆盖分布密度呈线性相关。因此,NDVI 在使用遥感影像进行植被覆盖度变化的研究中得到了广泛应用。而且 NDVI 通过比值可以消除大部分与太阳角、地形、云/暗影和大气条件有关的辐射照度条件的变化,增强了 NDVI 对植被的响应能力。植被覆盖度的计算见下式:

$$f=\frac{NDVI-NDVI_{min}}{NDVI_{max}-NDVI_{min}}$$

式中:$NDVI$ 为所求遥感影像像元的归一化植被指数;$NDVI_{min}$、$NDVI_{max}$ 分别为非植被覆盖度部分(裸地和未利用地)和植被覆盖部分(林地、草地等)归一化植被指数值的最小值和最大值。

利用 ArcGIS 空间分析中地图代数、图层叠加等功能,结合区域土地利用类型数据,对工程建设前后区域遥感影像进行 *NDVI* 数据运算,并对计算结果进行空间拉伸等处理,获得工程建设前后区域植被覆盖度分布图(见图 2)。

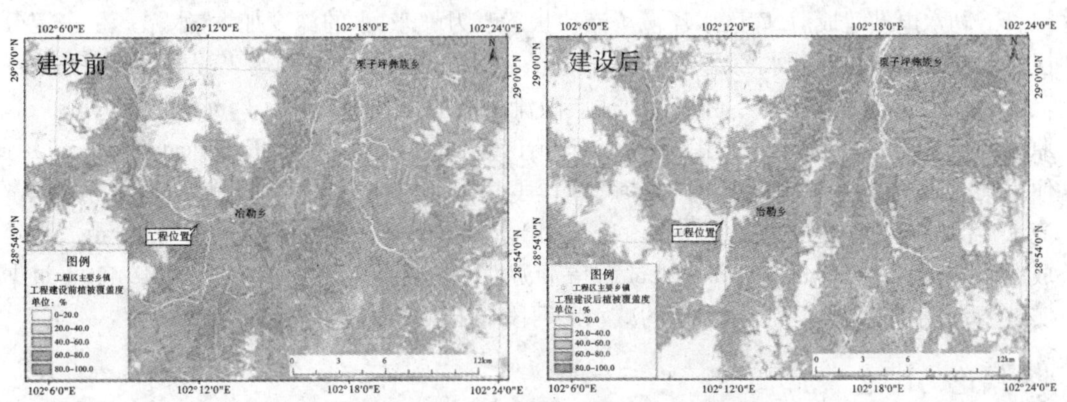

图 2 冶勒水电站工程建设前后区域植被覆盖度

通过利用 ArcGIS 中 Zonal Statistics 的分析功能,对于工程调查区域在建设前后的植被覆盖度最小值、最大值、平均值以及变化幅度、标准差等进行统计,结果见表 2。可以看出,工程建成后区域植被覆盖度略有下降,植被覆盖度平均值较工程建设前下降了约 3 个百分点,下降的原因也反映出工程建成后冶勒水库压占较大面积的草地等导致区域植被覆盖度略有所降低。

表 2 工程建设前后区域植被覆盖度比较

时期	最小值	最大值(%)	变化幅度(%)	年均值(%)	标准差(%)
工程建设前	0	100	100	79.51	11.29
工程建成后	0	98.10	98.10	76.39	14.53

3.2 对陆生动物的影响

根据相关调查资料,工程区域有兽类 58 种,隶属 6 目 22 科。其中,食肉目、偶蹄目、啮齿目共占兽类科数、种数 77.27% 和 81.04%。动物种类繁多,其中东洋界种类 43 种,古北界种类 12 种,广布种 3 种;鸟类 153 种,隶属 12 目 39 科,其中繁殖鸟 147 种。东洋界种类 63 种,古北界种类 60 种,广布种 24 种。爬行动物 9 种,隶属 1 目 4 科 9 属。其中,东洋界物种 8 种;两栖类 12 种,隶属 2 目 6 科 11 属,依其地理分布范围均属东洋界物种。

通过工程建设监理期间记录以及走访石棉林业部门、自然保护区管理部门等调查,工程施工期对区域的鸟类以及两栖类产生一定的影响,但没有出现区系、种群数量大的变化。工程建设对区域分布的国家保护动物中的鸟类产生一定的影响,但由于区域水域面积的增大,会有利于涉水的水禽类生存;工程区域大型珍稀保护动物多栖息活动于深山远离人为活动的区域,工程建设对于珍稀特产兽类——大熊猫(Ailuropoda Melanoleuca)、云豹(Clouded Leopard)、扭角羚(Budorcas Taxicolor)的影响不大。

3.3 对水生生物的影响

通过对冶勒水库库区水生生物样品采样和分析,水库形成以后,原有的适宜流水的硅藻类的数量减少,适宜静水蓝藻门、绿藻门的种类增加并变成为优势种,门类反而减少。浮游植物总体密度变化表现为工程竣工前较大、工程竣工后减小、随后再度增大,且大大超过工程竣工前水平。由于具有较为丰富的饵料基础,为浮游动物的生长繁殖提供了适宜的条件,2010 年浮游动物的密度及生物量接近工程建设前的水平。

冶勒水电站建成后,库区水环境从流动状态改为静态,水的深度加大,水温产生一定分层。同时由于水位随蓄水、发电而经常性的改变,水的物理化学性质与工程建设前的激流环境已完全不同,在此情形下,原河流中适应激流生活的鱼类,如副鳅(Paracobis Auratus)、高原鳅(TriplopHysa Steliczkae)、山鳅(Oreiagabryi Sauvage)、石爬鱼兆(Euchiloglans Davidi)等便很难适应,上溯到上游支流石灰窑河、勒丫河。剩下种类较少的是属于那些适宜静水或微流水生活的鱼类如齐口裂腹鱼(Scguztgirax (s) Prenanti)和一些小杂鱼。另外,由于大坝建成后,对松潘裸鲤(Gymnocypris Potanini)等洄游性的鱼类直接产生了生态阻隔影响,不过据实地调查,水库区及坝下一定长度的减水区河段却出现相对较多的该鱼类资源。

4 结论与讨论

(1)冶勒水电站工程建设前后区域生态景观除旱田景观、林地景观及部分未利用地出现减少变化外,其余类型的生态景观均呈现出增加变化特征,尤其水域景观增加较明显,这与冶勒水库工程建成后形成较大的库区水面有直接关系。而林地和草地景观面积的减少会对区域生态环境产生一定的负面影响,但其优势度值变化幅度较小,林地的模地地位没有发生改变,表明对区域生态景观格局影响相对较小。

(2)通过对冶勒水电站工程建设前后区域香农-威纳多样性指数、均匀性指数的对比分析表明，水电站工程建成后区域景观多样性和均匀性变化幅度较小，说明工程建设对于区域生态景观多样性影响不大。

(3)冶勒水电站工程对区域生物多样性分析表明：工程形成的大面积水库水面造成区域植被覆盖度略有下降；区域珍稀动物由于栖息活动于深山远离人为活动的区域，基本没有受到水电工程的影响；而由于工程建成后，原河道的激流环境变为水流较缓的库区静态环境，水生生物种类和数量受到一定影响，尤其洄游性鱼类受到大坝的生态阻隔影响较大。因此，需采取一定的水生生物补偿措施，如建设鱼类增殖放流站等，以恢复和增加鱼类资源，减缓水电工程大坝阻隔以及减水河段对于水生态的影响。

参考文献

[1] 董哲仁. 怒江水电开发的生态影响[J]. 生态学报，2006，26(5)：1591-1596.

[2] 黄岁，Onyx W H W. 泥沙运动引起的环境问题及环境泥沙学[J]. 水科学进展，1998，(4)：313-318.

[3] 虞泽荪，秦自生，郭延蜀，等. 二滩电站工程对陆生植物和植被的影响与对策[J]. 四川师范学院学报：自然科学版，1998，(1)：60-64.

[4] 李正霞. 水利工程与生态环境[J]. 陕西水力发电，2000(3)：31-32.

[5] 王世岩，陈建，刘畅，等. 西霞院水库工程区土地利用/覆盖变化特征[J]. 人民黄河，2011，33(10)：59-62.

[6] 锦矗. 四川冶勒自然保护区综合科学考察报告[M]. 成都：四川科学技术出版社，2004.

[7] 国家环境保护总局自然生态保护司. 非污染生态影响评价技术导则培训教材[M]. 北京：中国环境科学出版社，1999.

【作者简介】 王世岩，1974 年 5 月生，男，博士，高级工程师，主要从事流域水生态区划、环境评价、湿地生态等方面的研究工作，现工作于中国水利水电科学研究院水环境研究所。

三峡库区表层沉积物中铅污染特征与同位素示踪初步研究*

高 博 周怀东 殷淑华 郝 红

（中国水利水电科学研究院水环境研究所 北京 100038）

【摘要】 为研究三峡库区铅污染特征及其来源，于2008年10月采集了24个表层沉积物样品，测定了沉积物中重金属元素铅的含量，采用潜在生态风险评价法对其污染进行了评价，并利用铅同位素示踪法对污染源进行了识别。结果表明，库区重金属铅元素含量范围为21.5～93.0 mg/kg（平均值为59.4 mg/kg），高于库区土壤和沉积物背景值。潜在生态风险指数法评价结果表明，三峡库区沉积物中存在中等程度的铅富集，但仍处于轻微生态危害等级，并未对库区构成明显的潜在生态危害。库区沉积物铅同位素比值主要分布在矿产开采和自然源之间，这表明库区中铅污染可能主要受该区域采矿活动和自然源的影响。

【关键词】 铅污染 沉积物 铅同位素 潜在生态风险评价法 三峡库区

水体沉积物作为水环境中重金属的主要蓄积库，可以反映河流受重金属污染的状况，其污染状况是全面衡量水环境质量状况的重要因素。随着水环境条件的变化，沉积物中的重金属可能成为二次污染源再次释放到水体中，从而威胁库区水体的水质安全。

三峡水库设计蓄水位为175 m，2008年11月三峡水库首次试验性蓄水至172 m，库区水文情势和水环境条件也随之会出现相应变化，这将会影响到库区重金属元素的环境行为。为全面了解这一特殊时期（蓄水运用期）三峡库区沉积物中铅的污染特征及其生态环境风险，本文以三峡库区干流及其主要支流为研究区域，于2008年10月采集了库区干支流表层沉积物样品，分析测定了沉积物中铅元素含量及其分布特征，并采用瑞典学者Häkanson提出的潜在生态危害法对库区水体沉积物中铅污染及潜在生态风险进行定量评价。另外，本文采用铅同位素示踪技术对库区铅污染来源进行初步的识别，以期为库区环境中铅污染控制提供基础数据和科学依据。

1 材料与方法

1.1 样品的采集

为了研究三峡水库蓄水运用期沉积物中重金属污染状况，于2008年10月在三峡库区干支流共采集表层沉积物样品24个。由于三峡水库干流中上游水流速度较快和水深较深，且中上游表层沉积物多为沙粒，沉积物样品难以采集，因此本研究样品主要为库区中下游干

* 基金项目：国家水体污染控制与治理科技重大专项（2009ZX07104-001；2009ZX07104-002-03）。

流及支流。其中,干流包括太平溪和官渡口,干流断面采用左、中、右断面分别采样;支流包括13条库区一级支流,主要包括下岩寺、黄金河、汝溪河、澎溪河、汤溪河、磨刀溪、长滩河、梅溪河、草塘河、大宁河、神农溪、童庄河、香溪河。支流采样点的布设一般选取支流河口(距干流3 km处)以及支流中游。

1.2 铅含量测定

所有试验用水由Milli-Q高纯水发生器制得(>18.2 MΩ · cm)。HCl、HNO_3和HF为微电子级(BV-Ⅲ级,北京化学试剂研究所)。试验过程中所用器皿均采用20% HNO_3浸泡过夜,并用高纯水冲洗干净后备用。

准确称取样品40 mg,置于容量为10 mL的聚四氟乙烯消解罐中。然后加入2 mL HNO_3和0.2 mL H_2O_2,超声1 h后在电热盘上于60 ℃保温24 h。蒸干样品,加入2 mL 6 mol/L的HNO_3,超声1 h后保温过夜,然后加入2 mL HF放在电热盘上,再于60 ℃保温24 h。蒸干样品,加入1 mL 6 mol/L HNO_3和1 mL HF后,放入高压釜中190 ℃消解48 h。此消解程序可以保证沉积物样品完全消解并得到澄清的溶液。稀释后,加入内标,采用Elan DRC-e型电感耦合等离子体质谱仪(美国Perkin-Elmer公司)测定了样品中铅的含量。在分析样品的同时,采用相同的分析程序分析了沉积物标准样品GBW07312 (GSD-10)(中国地质科学院地球物理地球化学勘查研究所)重金属元素Pb含量,测定值与标准值吻合,回收率为95%。

1.3 铅同位素比值测定

铅同位素的测定采用电感耦合等离子体质谱仪直接测定,具体参数见表1。考虑到^{204}Pb丰度较低,本研究只讨论$^{206}Pb/^{207}Pb$和$^{208}Pb/^{207}Pb$丰度比数据。采用美国NIST标准物质SRM-981溶液校正质量歧视效应和仪器参数的漂移。NIST SRM-981标准样品$^{206}Pb/^{207}Pb$和$^{208}Pb/^{207}Pb$测定结果分别为1.092 6和2.374 3,与标准值(1.093 3和2.370 4)相吻合,同位素比值测定精度<0.5%。

表1 Pb同位素比值测定的仪器工作参数

工作参数	灵敏度	CeO/Ce	Ba^{++}/Ba	Sweeps	Readings	reduplicates	采样锥	样品浓度
设定值	10 000cps/1ppb	<3%	<3%	250	5	5	Ni	10~30 μg/kg

1.4 评价方法

为了定量表达北江沉积物中铅的潜在生态风险,应用瑞典学者H · kanson在1980年提出的沉积物潜在生态危害评价方法计算了沉积物中铅的单因子污染参数和潜在生态风险参数,计算公式如下:

$$C_r = \frac{C_x}{C_0} \tag{1}$$

$$E_r = T_r C_r \tag{2}$$

式中:E_r为潜在生态风险参数;C_r为铅的污染参数;C_x为标称沉积物铅含量的实测值;C_0为参比值,目前各国学者对参比值的选择各不相同,为了更好地反映库区目前铅污染的情况,本文以中国长江沉积物背景值为参比值;T_r为铅的毒性响应系数,其值为5。

单因子污染参数C_r和单因子潜在生态风险参数E_r值所对应的污染程度以及潜在生态风险程度如表2所示。

表 2　C_r和 E_r值所对应的污染程度以及潜在生态风险程度

污染参数范围	污染程度	潜在生态风险参数	潜在生态风险程度
$C_r<1$	轻微	$E_r<40$	轻微
$1\leqslant C_r<3$	中度	$40\leqslant E_r<80$	中度
$3\leqslant C_r<6$	较高	$80\leqslant E_r<160$	较高
$C_r\geqslant 6$	高度	$160\leqslant E_r<320$	高
		$E_r\geqslant 320$	极高

2　结果与讨论

2.1　沉积物中铅的含量及分布特征

三峡库区表层沉积物中铅含量及同位素比值测定结果见表 3。由表 3 可见,沉积物中铅元素的含量范围分别为:21.5 ~93.0 mg/kg,平均值为 59.4 mg/kg,分别是三峡库区重庆市土壤重金属背景值(23.88 mg/kg)和长江流域沉积物背景值(27.0 mg/kg)的 2.5 和 2.2 倍。另外,蓄水运用期沉积物中铅含量高于 1999 年和 2001 年监测结果(分别为36.9 mg/kg 和 53.4 mg/kg)。由此可见,库区沉积物中存在一定程度的铅累积,而且近年来库区沉积物中铅有增加的趋势。其中,白帝城采样点的铅含量最高,最低点在神农溪支流。库区沉积物中铅浓度存在较大的变异系数(*RSD* 为 44.0%),说明沉积物中铅元素含量受到人为活动的干扰。干流官渡口和太平溪断面采样点的铅元素平均含量均大于支流,支流河口高于支流中游。

表 3　三峡库区表层沉积物中 Pb 含量及同位素比值

采样点	Pb 含量(mg/kg)	$^{206}Pb/^{207}Pb$	$^{208}Pb/^{207}Pb$	采样点	Pb 含量(mg/kg)	$^{206}Pb/^{207}Pb$	$^{208}Pb/^{207}Pb$
黄金河河口	80.4	1.173	2.469	官渡口	77.3	1.173	2.462
大宁河河口	29.5	1.199	2.485	香溪河河口	72.8	1.173	2.468
汤溪河河口	23.1	1.191	2.483	汝溪河河口	77.6	1.174	2.465
太平溪(左)	83.6	1.176	2.466	磨刀溪河口	68.5	1.178	2.490
太平溪(中)	83.9	1.175	2.477	磨刀溪(中)	26.3	1.207	2.509
太平溪(右)	83.2	1.175	2.461	汤溪河(中)	24.4	1.200	2.502
磨刀溪中	23.7	1.177	2.483	梅溪河(中)	34.3	1.189	2.494
长滩河河口	18.0	1.210	2.495	下岩寺	63.5	1.177	2.484
澎溪河河口	85.2	1.177	2.468	神农溪(中)	21.5	1.210	2.492
童庄河河口	83.2	1.175	2.469	神农溪河口	61.0	1.181	2.470
童庄河中	59.9	1.181	2.475	白帝城	93.0	1.175	2.464
童庄河(左)	82.1	1.175	2.463	梅溪河河口	69.7	1.178	2.468

2.2　沉积物中 Pb 污染程度评价

三峡库区表层沉积物中铅的单因子污染参数(C_r)及其潜在生态风险参数(E_r)的计算

结果见表4。从各采样点铅的单因子污染参数(C_r)来看,三峡库区表层沉积物的铅污染参数变化范围为0.7~3.4,平均值为2.2,属于中度铅污染水平,说明库区沉积物中存在中等程度的铅积累。在库区24个采样点中,有6个采样点的沉积物中铅污染程度达到较高水平,占总样品量的25%。15个采样点污染水平为中等。另外,汤溪河等5个采样点的污染参数小于1.0,属于轻微污染级别。

根据本文中所计算的潜在生态风险指数E_r值表明,三峡库区沉积物中铅的潜在生态风险为低度。铅潜在风险参数变化范围为3.3~21.3,平均值达到17.2。结果同时表明,虽然沉积物中铅具有中等污染程度,但其并不具有较高的生态风险,这主要是因为铅的毒性系数较低。但是,考虑到库区作为饮用水保障以及库区生物富集作用,因此并不能忽视库区铅污染现状。

表4 三峡库区沉积物潜在生态风险指数评价结果

采样点	污染参数(C_r)	污染程度	潜在生态风险参数(E_r)	生态风险程度
黄金河河口	3.0	中度	14.9	低度
大宁河河口	1.1	中度	5.5	低度
汤溪河河口	0.9	轻微	4.3	低度
太平溪(左)	3.1	较高	15.5	低度
太平溪(中)	3.1	较高	15.5	低度
太平溪(右)	3.1	较高	15.4	低度
磨刀溪中	0.9	轻微	4.4	低度
长滩河河口	0.7	轻微	3.3	低度
澎溪河河口	3.2	较高	15.8	低度
童庄河河口	3.1	较高	15.4	低度
童庄河中	2.2	中度	11.1	低度
童庄河(左)	3.0	中度	15.2	低度
官渡口	2.9	中度	14.3	低度
香溪河河口	2.7	中度	13.5	低度
汝溪河河口	2.9	中度	14.4	低度
磨刀溪河口	2.5	中度	12.7	低度
磨刀溪(中)	1.0	中度	4.9	低度
汤溪河(中)	0.9	轻微	4.5	低度
梅溪河(中)	1.3	中度	6.3	低度
下岩寺	2.4	中度	11.8	低度
神农溪(中)	0.8	中度	4.0	低度
神农溪河口	2.3	中度	11.3	低度
白帝城	3.4	较高	17.2	低度
梅溪河河口	2.6	中度	12.9	低度
平均值	2.2	中度	11.0	低度

2.3 沉积物 Pb 同位素组成及示踪

三峡库区表层沉积物中铅同位素比值测定结果见表 3。结果表明,沉积物中$^{206}Pb/^{207}Pb$与$^{208}Pb/^{207}Pb$比值的范围分别是 1.173 ~ 1.210 和 2.462 ~ 2.509,平均值分别为 1.183 和 2.477。对于自然产生的 Pb 而言,$^{206}Pb/^{207}Pb$一般较高(>1.20),人为因素产生的 Pb,$^{206}Pb/^{207}Pb$比值较低,在 0.96 ~ 1.20。因此,库区沉积物中铅污染已受到人为因素影响。为了进一步识别库区沉积物中的铅来源,对库区沉积物中的铅同位素比值和 1/Pb 进行分析。一般而言,若 1/Pb 与$^{206}Pb/^{207}Pb$呈显著的线性正相关关系,这种关系可以用二元混合模型(Binary Mixing)加以解释。本文中 1/Pb 与$^{206}Pb/^{207}Pb$呈显著的线性正相关关系(r^2为 0.763 9),说明库区沉积物中的铅污染主要是两种不同铅源共同作用的结果。

环境中,铅主要来源于金属冶炼等工业活动、汽车尾气的排放以及煤炭的燃烧。因此,为了进一步更准确的识别库区重金属的来源,需对库区沉积物中的铅同位素进行进一步分析。利用铅同位素比值进行污染源识别时,需分析各种端元物质(各类源的代表性物质)的铅同位素组成特征,继而把库区沉积物铅同位素比值投影到端元物质铅同位素比值关系图上($^{206}Pb/^{207}Pb$和$^{208}Pb/^{207}Pb$),根据样品和端元物质的铅同位素比值的分布状况来确定其污染源。为了识别库区沉积物中铅来源,将矿产开发和大气沉降的铅同位素比值也同时投影到图 1 中。库区沉积物中$^{206}Pb/^{207}Pb$与$^{208}Pb/^{207}Pb$比值的关系见图 1。结果表明,$^{206}Pb/^{207}Pb$和$^{208}Pb/^{207}Pb$存在显著的线性正相关关系(r^2 = 0.658 3),这也证明了库区沉积物中受两种铅来源的共同作用。根据二元混合模型可以估算出两类铅源对沉积物的相对贡献比例。本文假设铅主要受人为源和自然源($^{206}Pb/^{207}Pb$ = 1.200)共同作用,假设人为源主要为采矿活动,其$^{206}Pb/^{207}Pb$为 1.166,则计算出的矿产开采排放铅污染对库区沉积物的贡献率大约为 42.5%。

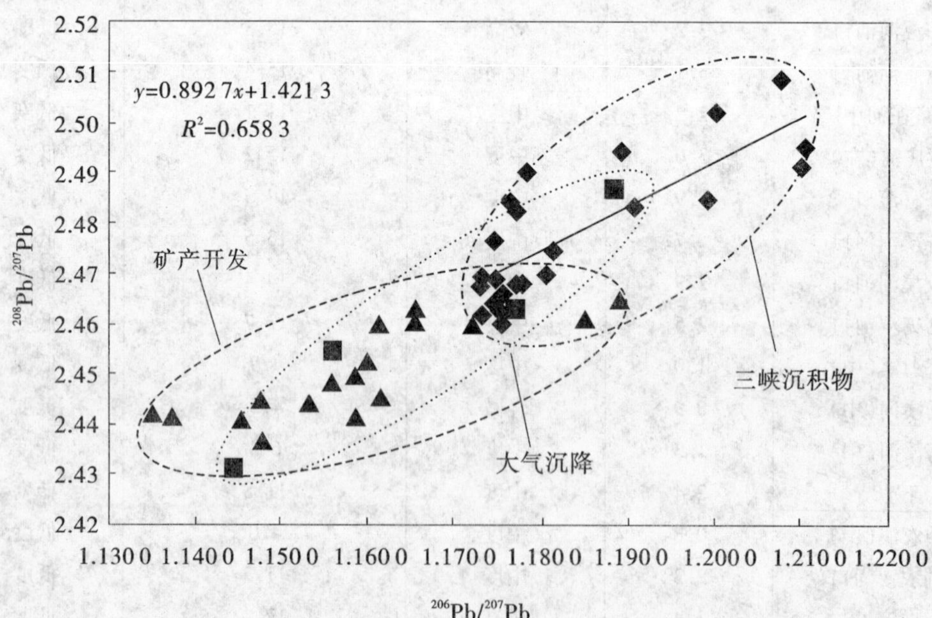

图 1 三峡库区沉积物中$^{206}Pb/^{207}Pb$与$^{208}Pb/^{207}Pb$关系图

库区沉积物铅同位素比值主要分布在矿产开采和自然源之间,这可能表明库区中铅污染主要受该区域采矿活动和自然源的共同控制。进一步分析可知,大宁河、长滩河和神农溪沉积物中的铅主要是由自然源控制,基本没有受到人为污染,而其他采样点的铅同位素比值$^{206}Pb/^{207}Pb$明显低于1.20,说明主要是受人为和自然源共同控制,并且接近矿产开采排放特征。另外,值得注意的是,通过大气传输及沉降作用,库区周边地区的外源铅(包括机动车排放)可能会对水体造成一定的影响。

由于缺乏库区不同端元物质的铅同位素信息(各类源的代表物质),目前只能够鉴别出库区的沉积物的两种铅的来源。参考相关研究发现,矿产开采可能是主要的污染源。针对库区流域内重要的铅排放源,采集具有代表性的端元物质将是后续研究工作的重要内容,以便于准确地对库区地重金属铅污染进行源解析。

3 结语

(1)三峡库区沉积物中铅元素的含量范围分别为:21.5~93.0 mg/kg,平均值为59.4 mg/kg,高于库区重庆市土壤环境背景值和长江沉积物背景值,说明沉积物中铅受到人为活动的干扰。

(2)潜在生态风险指数评价结果表明,三峡库区沉积物中存在中等程度的铅污染,但其潜在生态风险为轻微水平。

(3)铅同位素示踪研究结果表明,库区沉积物中铅污染主要受两种不同铅源的共同作用。目前大致判断库区沉积物铅污染主要受人为源(矿产开采)和自然源的控制。

参考文献

[1] Förstner U. Contaminated Sediments: Lectures on Environmental Aspects of Particle-Assciated Chemicals in Aquatic Systems. Berlin: Springer-Verlag,1989:110-192.

[2] Häkanson L. An ecological risk index for aquatic pollution control-A sedimentological approach[J]. Water Research, 1980, 14(8):975-1001.

[3] 赵一阳,鄢明才. 中国浅海沉积物地球化学[M]. 北京:科学出版社, 1994.

[4] Zhu B Q. The mapping of geochemical provinces in China based on Pb isotopes[J]. Journal of Geochemical Exploration, 1995,55:171-181.

[5] Mukai H, Tanaka A, Fujii T, Zeng Y Q, Hong Y T, Tang J, Guo S, Xue H S, Sun, Z L, Zhou J T, Xue D M, Zhao J, Zhao G H, Gu J L, Zhai P Y. Regional Characteristics of Sulfur and Lead Isotope Ratios in the Atmosphere at Several Chinese Urban Sites[J]. Environmental Science and Technology, 2001, 35, 1064-1071.

【作者简介】 高博,男,1978年6月生,2008年7月毕业于中国科学院广州地球化学研究所,获环境科学专业博士学位,目前工作于中国水利水电科学研究院,工程师。

西南黄海海滨湿地磷化氢释放通量与“耦合温室效应”研究

洪宇宁[1,2] 张 蕤[3] 耿金菊[2] 丁丽丽[2] 左 平[4] 王晓蓉[2] 任洪强[2]

（1. 中国水利水电科学研究院水环境研究所 北京 100038；
2. 南京大学环境学院污染控制与资源化研究国家重点实验室 南京 210093；
3. 南京师范大学化学与环境科学学院 南京 210097；
4. 南京大学地理与海洋学院 南京 210093）

【摘要】 磷化氢(PH_3)是磷的生物地球化学循环的气态载体，在光照条件下消耗，可能存在“耦合温室效应”。本文研究了西南黄海海滨湿地芦苇带、盐蒿带、米草带和光滩带4个典型植被带磷化氢气体的释放通量，并测定了气体样品中的温室气体(CO_2、CH_4、N_2O)浓度。结果表明，海滨湿地大气中普遍存在 PH_3 气体，盐城温带海滨湿地的 PH_3 释放通量为 -230 ~ 276 ng/(m^2·h)，其释放过程可能是土壤释放与吸收两种作用动态竞争的过程；盐城海滨湿地地表植被对 PH_3 分布存在显著影响，光滩带与盐蒿带沉积物-大气界面气态 PH_3 释放通量均高于米草带与芦苇带；气态 PH_3 对大气中含量较高的还原性气体存在同步变化关系，与大气中含量较低的还原性气体存在竞争氧化性物质的关系，对于痕量温室气体存在显著的“耦合温室效应”。

【关键词】 海滨湿地 磷化氢 温室气体 植被带

传统理论认为，在磷的生物地球化学循环中不存在磷的气态化合物。研究者改进了分析手段，发现PH_3在各种环境中普遍存在，并在港口和近海检出了较高含量的气态和基质结合态磷化氢。目前，对包括潮间带在内的海滨湿地的磷化氢分布与释放情况尚未有系统研究。西南黄海地区作为我国传统渔业养殖地区，每年人为原因输入海滨湿地的磷较多，但是尚未得到系统研究。

大气中磷化氢在光照条件下消耗速率明显快于避光条件，影响因子可能是光照生成的自由基。考虑到自由基对温室气体的清除作用，磷化氢可能与温室气体竞争大气中的自由基，影响温室气体的归趋，从而间接地促进了温室气体在大气中的积累，增强温室效应，造成“耦合温室效应”。但是自然条件下磷化氢与温室气体的释放过程事实上存在怎样的关系，目前尚未见报道。

本文主要以西南黄海海滨湿地为研究对象，考察了3月初不同植被带（芦苇带、盐蒿带、米草带、光滩带）磷化氢气体与主要温室气体(CO_2、CH_4、N_2O)的释放情况。

1 材料和方法

分别于2007年8月、11月，2008年3月、5月，于江苏盐城珍禽国家自然保护区，根据地表植被类型的变化，选取4个代表性的点位采样，采样时间为上午7时。采样地的地表植被类型的过渡，反映了地表水体盐度逐渐升高，逐渐由淡水过渡为海水，采样信息见表1。在采样点上用静态箱法以10 min为间隔采集箱内空气样品，封存于聚四氟乙烯气袋中，避光保存。采样时间共计60 min。每个采样点均取两个平行样品。

表1 采样地点信息

植被优势种	经纬度	不同季节采样过程平均气温(℃)							
		2007年8月		2007年11月		2008年3月		2008年5月	
无(光滩)	33°36′26.4″N 120°37′1.2″E	29.2	31.0	12.2	14.8	11.1	13.0	17.0	18.0
米草	33°36′22.2″N 120°36′37.2″E	28.5	29.5	17.3	18.5	6.1	10.3	19.0	24.5
盐蒿	33°35′55.2″N 120°35′24.6″E	34.0	33.0	20.1	21.5	18.0	18.2	23.0	25.0
芦苇	33°34′30.6″N 120°32′52.8″E	28.0	29.5	18.0	18.5	18.2	18.3	26.0	28.5

空气中PH_3含量测定方法为改进的柱前二次低温冷阱富集和气相色谱/氮磷监测器(NPD)联用技术，N_2O的测定方法为气相色谱(Shimadzu GC-14B)/^{63}Ni电子捕获检测器(ECD)联用技术，CO_2的测定方法为气相色谱(Shimadzu GC-14B)/热导检测器(TCD)联用技术，CH_4的测定方法为气相色谱(Shimadzu GC-12A)/氢火焰检测器(FID)联用技术。

PH_3释放通量的计算公式为$F = H\Delta c/\Delta t$（H为静态箱有效高度，m，c为磷化氢质量浓度，ng/m^3，t为采样时间，h），根据采样间隔始末时刻磷化氢浓度差计算通量。本文将采样间隔设定为10 min，因此同样以10 min为Δt计算通量，以采样的起点、终点的磷化氢浓度计算总的通量。由于采样过程中磷化氢浓度波动较大，在通量上反映为正负通量交互出现。

2 结果与讨论

2.1 黄海海滨湿地磷化氢释放通量分布

如图1所示，盐城温带海滨湿地土壤-大气界面磷化氢的释放通量在2008年5月的米草带达最低值-230 ng/(m^2·h)，在2007年8月的盐蒿带达最高值276 ng/(m^2·h)。由图1可知，磷化氢正负通量在不同季节不同植被带均有出现，海滨湿地并非固定的源或汇。

按采样地区划分，磷化氢释放通量排序为：盐蒿带>光滩带>米草带>芦苇带。

其中，芦苇带的磷化氢释放通量最小为(-45.9±122.3) ng/(m^2·h)。考虑到采样过程中芦苇带除冬季外均有上覆水，其他采样点则没有上覆水或者上覆水极浅（水深小于1 cm），因此芦苇带的通量较低可能是由两个原因引起的：一是芦苇带上覆水较深（一般大于10 cm），由于磷化氢在水相扩散过程中遇到的传质阻力较大，因此释放通量下降；二是上覆水中可能存在较多氧化性物质，导致磷化氢在水相扩散过程中发生氧化，以致磷化氢通量下降。

盐城温带海滨湿地既是磷化氢的源，也是磷化氢的汇，但是不同植被带间的区别较大。

盐蒿带与光滩带是盐城温带海滨湿地磷化氢释放的主要源,米草带和芦苇带主要为汇。米草带、芦苇带的植被主要由植株较高、根系发达的大米草和芦苇构成,盐蒿带植被主要由植株低矮、根系不发达的盐蒿构成,光滩带地表裸露,基本无植被。由于海滨土壤土质黏重,透气性差,因此植株根系的发育有助于改善土壤的透气性,提高环境的氧化还原电位。因此,植株发达的环境中,氧化还原电位较高,磷化氢易被氧化消耗,磷化氢释放倾向于负通量;植株不发达的环境中,氧化还原电位较低,磷化氢不易被消耗而易于在原位积累,因此磷化氢释放倾向于正通量。

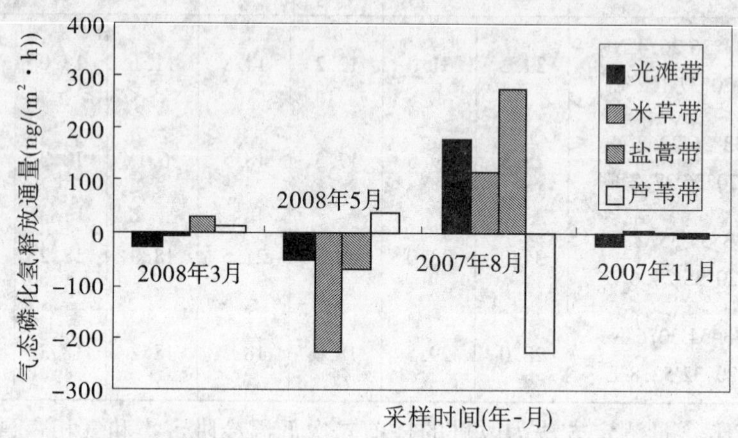

图 1　盐城温带海滨湿地不同季节和植被带的水/土壤-大气界面磷化氢通量

同时,由于在盐城温带海滨湿地,随着与海岸距离的增大,地表水体的盐度下降,表现为地表植被优势种逐步由耐盐的大米草过渡为盐蒿,继而变为适宜淡水环境的芦苇。但是磷化氢释放通量并未呈现出类似的变化趋势,未能观察到地表水体盐度对磷化氢释放通量存在影响。

按采样时间划分,磷化氢释放通量排序为 2007 年 8 月[(86.3±219.1) ng/(m^2·h),采样时平均气温 29.9 ℃)]>2008 年 3 月[(2.22±25.62) ng/(m^2·h),采样时平均气温 13.4 ℃)]>2007 年 11 月[(-4.42±12.04) ng/(m^2·h),采样时平均气温 16.9 ℃)]>2008 年 5 月[(-77.9±111.9) ng/(m^2·h),采样时平均气温 21.3 ℃)]。

2.2　磷化氢的"耦合温室效应"

盐城温带海滨湿地温室气体与气态磷化氢间的 Pearson 相关系数见表 2。由表 2 可知,在盐城温带海滨湿地大气中,气态磷化氢浓度与甲烷、二氧化碳浓度呈显著正相关,与氧化亚氮呈显著负相关。

气态磷化氢与二氧化碳间存在显著正相关。Roels 和 Verstraete 的研究结果则表明气态磷化氢与二氧化碳不存在显著相关性。由于二氧化碳是生物呼吸作用的氧化产物,与磷化氢生成过程截然相反,因此两者间的显著正相关不可能是由两者生成过程的相关所致。因此,两者的显著正相关主要原因是二氧化碳浓度升高反映了土壤呼吸作用强度大,微生物代谢活跃,有利于磷化氢的生成。

气态磷化氢与甲烷间存在显著正相关的原因可能有两个:一是两者的生成和保存都需要相似的环境条件,都需要还原性环境;二是生态系统中甲烷浓度升高往往表明微生物代谢活跃,而微生物代谢增强可能有助于气态磷化氢的生成。Gassmann 等发现生物圈中磷化氢总是伴随着甲烷存在,并根据 Dévai 等人测定的磷化氢与甲烷的释放比例推算出全球磷化氢释放规模为 40 000 t。Eismann 等发现在猪粪发酵过程中,甲烷与磷化氢在生成过程中存在显著正相关,但是没有发现产甲烷菌参与磷化氢生成的证据。相反,Eismann 等还发现磷化氢对产甲烷过程存在抑制。Rutishauser 等、Jenkins、Han 等的试验结果中,产甲烷过程与产磷化氢过程并不存在显著相关。那么,本试验中磷化氢与甲烷间的显著正相关是否是产甲烷过程与产磷化氢过程的耦合所致?

表 2 盐城温带海滨湿地温室气体与气态磷化氢间的 Pearson 相关系数

项目	磷化氢	甲烷	氧化亚氮	二氧化碳
磷化氢	1			
甲烷	0.207*	1		
氧化亚氮	-0.231*	0.059	1	
二氧化碳	0.356**	0.483**	0.113	1

注:表中带*的表示在95%置信区间内显著相关,$n=99$;
带**的表示在99%置信区间内显著相关,$n=99$。

为了探明这一问题,本文进一步以甲烷浓度为控制变量,对气态磷化氢浓度与二氧化碳浓度做偏相关分析,得到偏相关系数为 0.299 3,偏相关检验的双尾概率 $P=0.03$,在显著性水平为95%时呈显著正相关。相反地,当以二氧化碳浓度为控制变量,对气态磷化氢浓度与甲烷浓度做偏相关分析时,得到偏相关系数为 0.042 5,偏相关检验的双尾概率 $P=0.678$,不存在显著相关。因此,可以认为,甲烷浓度与气态磷化氢浓度存在虚假相关,二氧化碳浓度与气态磷化氢浓度存在真实相关。二氧化碳浓度高低可以体现自然环境的还原性强弱。还原性环境同时影响了磷化氢浓度和甲烷浓度,促使磷化氢浓度和甲烷浓度同时升高,使甲烷浓度与气态磷化氢浓度表现为存在显著正相关,但是甲烷浓度本身与磷化氢浓度不存在直接相关,两者的生成过程也不相关。

气态磷化氢浓度与氧化亚氮浓度间存在显著负相关,可能是磷化氢与氧化亚氮竞争自由基所致。因此,气态磷化氢对氧化亚氮存在“耦合温室效应”。但是,为何同样还原性弱于磷化氢的甲烷却表现为与磷化氢呈显著正相关,而非如氧化亚氮那样呈显著负相关?本文认为,磷化氢与还原性气体存在两种关系:一是由于同为还原性气体,所以磷化氢与还原性气体对生成和保存环境具有相似的要求,随着环境还原性的强弱变化体现为同步变化关系;二是在氧化性物质不足的情况下,磷化氢与还原性气体将竞争氧化性物质,存在竞争关系。由于磷化氢为痕量气体,所以磷化氢的竞争作用对在大气中含量较高的气体(如甲烷)影响极小,而主要表现为随环境还原性的强弱而发生的浓度同步增减,表现为同步变化关系;相反地,对于在大气中含量较低的气体(如氧化亚氮),磷化氢的竞争作用能够显著影响其浓度增减,因此磷化氢与其的关系最终表现为竞争关系。

3 结语

(1)盐城温带海滨湿地的磷化氢释放通量为-230~276 ng/(m^2·h)。

(2)盐城海滨湿地地表植被对磷化氢分布存在显著影响,光滩带与盐蒿带沉积物-大气界面气态磷化氢释放通量均高于米草带与芦苇带。

(3)气态磷化氢对大气中含量较高的还原性气体存在同步变化关系,与大气中含量较低的还原性气体存在竞争氧化性物质的关系,对于痕量温室气体存在显著的"耦合温室效应"。

参考文献

[1] Roels J, Verstraete W. Occurrence and origin of phosphine in landfill gas [J]. Science of the Total Environment, 2004, 327: 185-196.

[2] Gassmann G, Glindemann D. Phosphine (PH_3) in the biosphere [J]. Angewandte Chemie: International Edition in English, 1993, 32(5): 761-763.

[3] Dévai I, Felföldy L, Wittner I, et al. Detection of phosphine: new aspects of the phosphorus cycle in the hydrosphere [J]. Nature, 1988, 333: 343-345.

[4] Eismann F, Glindemann D, Bergmann A, et al. Balancing phosphine in manure fermentation [J]. Journal of Environmental Science and Health, 1997, 32(6): 955-968.

[5] Eismann F, Glindemann D, Bergmann A, et al. Effect of free phosphine on anaerobic digestion [J]. Water Research, 1997, 31(11): 2771-2774.

[6] Metcalf W W, Wolfe R S. Molecular genetic analysis of phosphite and hypophosphite oxidation by Pseudomonas stutzeri WM88 [J]. Journal of Bacteriology, 1998, 180(2): 5547-558.

[7] Jenkins RO, Morris TA, Craig PJ, et al. Phosphine generation by mixed- and monoseptic-cultures of anaerobic bacteria [J]. The Science of the Total Environment, 2000, 250: 73-81.

[8] Han SH, Zhuang YH, Zhang HX, et al. Phosphine and methane generation by the addition of organic compounds containing carbon-phosphorus bonds into incubated soil [J]. Chemosphere, 2002, 49: 651-657.

【作者简介】 洪宇宁,1984年4月生,博士研究生,工程师,2010年于南京大学获得博士学位,工作单位为中国水利水电科学研究院。

三、防洪减灾与水文学

基于水文分割方法的鄱阳湖流域入湖非点源污染负荷研究

涂安国[1] 李 英[2] 莫明浩[1] 杨 洁[1]

(1. 江西省水土保持科学研究所 南昌 330029;
2. 江西省赣西土木工程勘测设计院 宜春 336000)

【摘要】 非点源是影响地表水环境的主要污染源之一。为研究鄱阳湖流域非点源污染情况,选取鄱阳湖入湖主要河流为研究对象,在主要河流入湖控制站对流域径流量及水质进行同步监测,在基流分割基础上,应用非点源污染负荷估算公式,探求氮磷径流入湖负荷流失规律。结果表明:2008 年赣江、抚河、信江、饶河(昌江和乐安江之和)、修水等五大河及博阳河、西河入湖氨氮和总磷的污染负荷(溶解态)分别为 41 074.81 t 和 9 114.94 t;在总量组成中,非点源是入湖污染物主要来源,占总入湖负荷的 68% ~76%;在空间分布上,入湖五河中赣江对入湖非点源污染负荷的贡献占绝对优势,其他入湖河流的非点源污染总贡献率仅占 1.35% ~11.20%。

【关键词】 鄱阳湖流域 非点源 污染负荷估算 水文分割

非点源污染已渐成世界湖泊水质问题主导因素。鄱阳湖是世界重要湿地和候鸟栖息地,是我国第一大淡水湖,汇纳江西省境内赣江、抚河、信江、饶河、修河等五大河流及环湖小流域来水,流域面积占全省国土面积的 92.6%,是江西省农业非点源污染的最终受纳水体。近年来,湖体氨氮和总磷超标严重,鄱阳湖水体富营养化程度加剧,流域农业非点源污染呈恶化趋势。目前,虽然学者们就鄱阳湖环湖区非点源污染现状、污染负荷估算、氮磷输出特征开展了一些研究,但研究都不够深入,尤其是对整个鄱阳湖流域的非点源污染负荷估算的相关报道仍未见。

非点源污染负荷量化是了解、掌握非点源污染状况和管理水环境质量的基础,也是非点源污染研究的重点和热点。随着水环境实施总量控制的管理,非点源污染定量化尤为重要。我国非点源污染负荷研究多为小流域或小区尺度的负荷研究,计算结果为流域不同土地类型汇入地表水的非点源污染负荷之和,而流域出口非点源污染负荷仅占输入河流总负荷的一部分,因此寻求准确估算较大流域非点源污染输出负荷的方法十分迫切和必要。对具有较长系列水质、水量同步监测数据的大流域,流域水文分割法为非点源污染负荷的定量化研究提供了一种有效途径。本文采用流域水文分割法原理,以鄱阳湖流域主要入湖河流为研究对象,开展入湖非点源污染负荷研究,以期为鄱阳湖流域非点源污染防治提供决策支持。

1 研究区背景

鄱阳湖流域位于长江中下游南岸,与江西省行政辖区基本重叠,集水总面积 16.22 万

km^2。鄱阳湖承接赣江、抚河、信江、饶河、修水五大河及博阳河、西河等河流的来水，调蓄后经湖口汇入长江（见图1）。流域内山区丘陵约占总面积的78%，水面占10%；地带性土壤是红壤，占总面积的56%；土地利用类型以林地为主，约占总面积的61.1%。鄱阳湖流域属亚热带湿润季风气候，降水丰富，多年平均降水量为1 620 mm，蒸发量为700～800 mm。流域内降雨径流时空分布不均，年内、年际变化明显，具有明显的季节性和区域性，且洪旱灾害频繁。鄱阳湖平原及入湖五河两侧平坦之地均发展为城市、村镇或开垦为良田，农业经济发达，是国家重要商品粮基地之一。

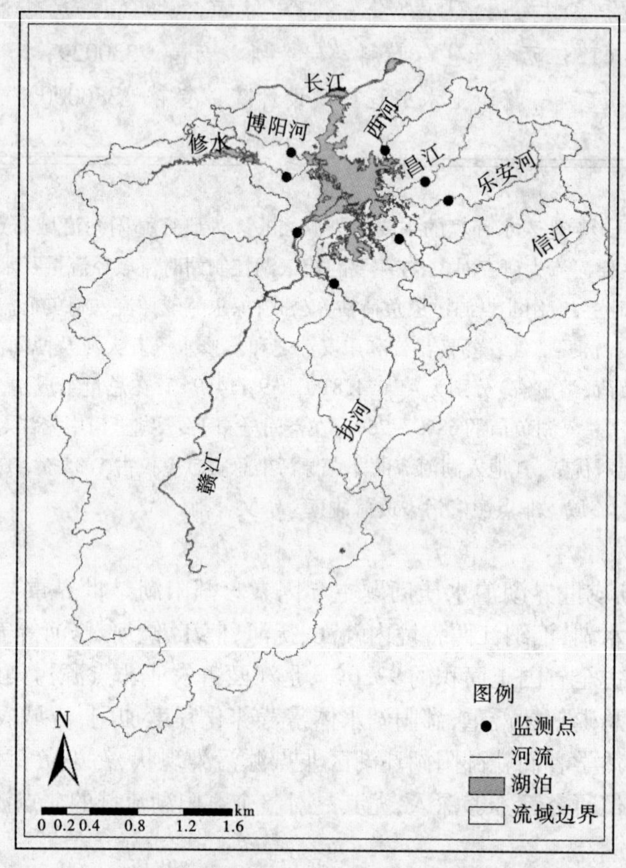

图1 鄱阳湖流域水系和主要水文站分布

近年来，随着流域经济社会的快速发展，鄱阳湖水质总体呈现下降趋势，由水土流失等原因引起的非点源污染是其水质恶化的重要原因。鄱阳湖水体中氮磷含量在不断增加，水质恶化风险较大，已具备了较大规模“水华”暴发的营养条件和风险。

2 研究方法

2.1 基流分割

借鉴应用水文学的处理方式，将总径流分为地表径流和基流两部分。基流为流域出口断面的流量过程线中除去地表径流外的部分，在径流过程线中表现基本稳定，大小主要受流域土壤、植被、地形、地质及气候等的影响。目前的基流分割方法主要有电子滤波法、环境同位素法、水文模型法，加里宁水量平衡法、直线分割法、综合退水曲线法等。其中，直线分割法具有直观、易于操作等特点,应用较多,可细分为平割法和斜割法 2 种。平割法又称枯季最小流量法,即将枯季最小流量作为地下水流出量进行水文分割,实际操作中可采用最枯日平均流量、月平均流量或枯水期 3 个月平均流量等。本文将采用直线平割法对入湖河流基流进行分割。

2.2 估算方法

降雨径流冲刷是产生非点源污染的原动力,也是非点源污染物的载体。对水体来说,只有进入水体的非点源才是有效的,其负荷计算对水体环境质量改善才有意义。研究表明,非点源污染主要是由汛期地表径流引起的,枯水季节的水质污染主要是由点源排放引起的。鄱阳湖流域每年 11 月下旬至次年 2 月中下旬(即枯水期)降水只占全年降水量的 7.2% ~ 19.5%,且雨强小,几乎没有地表径流和地表径流携带非点源污染物进入地表水体。因此,可以认为汛期水环境质量是由点源和非点源共同引起的,而枯水期水质主要受点源影响。

流域水文分割法运用水文学原理,考虑到点源、非点源的形成和输移规律,认为基流输送物质代表点源负荷和自然背景值,而地表径流输送的物质为非点源负荷,采用径流组分分割方法实现非点源污染负荷估算,具有方法简便、易行、数据要求较低的特点,并已在部分流域有较好的运用。综合考虑鄱阳湖流域的水文、气象特征,本文将采用此方法进行鄱阳湖流域入湖非点源污染负荷估算。

依据流域水文估算法点源可根据河流的水文特点由河川基流(非汛期径流)推求,溶解态非点源负荷可根据暴雨径流推求,则河流年入库污染总负荷可表示为

$$W_{\mathrm{t}} = \sum_{i=1}^{n} C_{\mathrm{p}i} Q_{\mathrm{p}i} \Delta t + \sum_{i=1}^{n} C_{n\mathrm{p}i} Q_{n\mathrm{p}i} \Delta t$$

式中:W_{t} 为入湖总污染负荷;$C_{\mathrm{p}i}$ 为第 i 时刻入湖点源污染物浓度;$Q_{\mathrm{p}i}$ 为第 i 时刻入湖点源径流量;$C_{n\mathrm{p}i}$ 为第 i 时刻入湖非点源污染物浓度;$Q_{\mathrm{p}i}$ 为第 i 时刻入湖非点源径流量;Δt 为第 i 次监测所代表的时间段。

其中,W_{t} 可由监测断面的水质、水量数据直接求出,即

$$W_{\mathrm{t}} = \sum_{i=1}^{n} C_i Q_i \Delta t$$

式中: C_i 为第 i 次监测的污染物质量浓度; Q_i 为第 i 次监测的流量; Δt 为第 i 次监测所代表的时间段。

由此,溶解态非点源污染负荷可表示为 $W_{n\mathrm{p}} = W_{\mathrm{t}} - W_{\mathrm{p}}$,即

$$\sum_{i=1}^{n} C_{n\mathrm{p}i} Q_{n\mathrm{p}i} \Delta t = \sum_{i=1}^{n} C_i Q_i \Delta t - \sum_{i=1}^{n} C_{\mathrm{p}i} Q_{\mathrm{p}i} \Delta t$$

式中:溶解态非点源污染负荷 $W_{n\mathrm{p}}$ 通过溶解态总污染负荷 W_{t} 与点源污染负荷 W_{p} 之差估算;点源污染负荷 W_{p} 通过枯水期实测污染物质量浓度计算;河川基流由径流分割得到。

2.3 断面选取和数据来源

流域水文分割法是基于详细的水文资料和水质监测数据建立的计算方法,因此对入湖断面的要求一是距离入湖口近,二是具有多年连续的水文和水质监测数据。自 2007 年 10 月开始,江西省水环境监测中心每月都对赣江(外洲站)、抚河(李家渡站)、信江(梅港站)、昌江(渡峰坑站)、乐安江(石镇街站)、修河(永修站)、西河(石门街站)及博阳河(梓坊站)等河流入湖控制断面进行水质水量同步监测,研究选取以上入湖断面及其 2008 年水质水量

监测数据为基础进行入湖非点源污染负荷估算。

3 结果分析与讨论

3.1 入湖河流基流分割

采用2008年外州(赣江)、梅港(信江)、李家渡(抚河)、永修(修水)、渡峰坑(昌江)、石镇街(乐安江)站逐月流量观测数据(昌江和乐安江为饶河入湖2支流),绘出流量线(见图2)。分析可知,入湖河流流量呈多峰特性,4~11月的汛期较大,1~3月及12月一般为枯水月份。

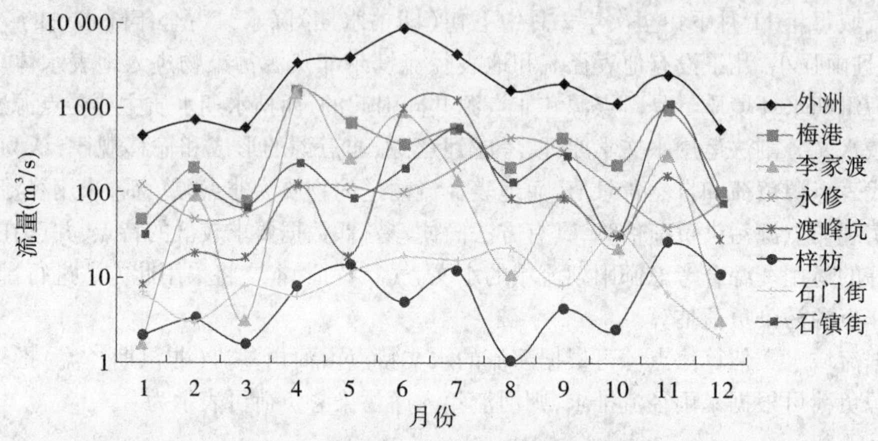

图2 2008年入湖河流控制站流量变化

根据水文分割原理,选取直线平割法,以2008年各河流最枯3个月平均流量为基流进行水文分割,得到鄱阳湖流域入湖河流基流、地表径流量(见表1)。2008年入湖水量中,赣江比重第一,占63.96%;信江第二,占12.01%,除抚河和昌江外,各河流基流量占径流量的16.99%~32.81%。

表1 2008年鄱阳湖流域入湖河流基流分割

流域	总径流量(亿 m^3)	基流流量(m^3/s)	基流量(亿 m^3)	地表径流量(亿 m^3)
赣江	782.64	554.67	175.40	607.24
信江	146.98	78.97	24.97	122.00
抚河	86.33	2.65	0.84	85.49
修水	47.10	50.00	15.81	31.29
昌江	73.30	15.27	4.83	68.47
博阳河	2.47	2.44	0.77	1.70
乐安江	81.06	58.67	18.55	62.51
西河	3.84	3.99	1.26	2.58

3.2 入湖非点源污染负荷计算及特征分析

3.2.1 污染负荷估算

根据上述河流月水质和水文监测数据,应用上述公式计算河流入湖的氮磷负荷(见表2)。结果显示,鄱阳湖流域入湖氮磷负荷年内分配不均,变异系数平均分别达0.80和0.76,主要集中在4~9月,入湖负荷可占全年总量的55.45%~94.39%,平均达76.39%。

入湖氮磷负荷和入湖径流量的年内分配基本一致,河流营养盐负荷和径流量在0.05水平基本呈显著性正相关,相关系数最大可达0.96以上。

表2 鄱阳湖流域入湖河流氮磷负荷年内分配情况

河流	项目	1月	2月	3月	4月	5月	6月	7月	8月	9月	10月	11月	12月
赣江	氨氮	0.64	3.21	1.74	22.43	19.78	19.35	11.26	4.36	3.63	4.5	6.13	2.97
	总磷	1.88	1.61	0.58	1.54	3.06	27.61	13.36	13.62	7.72	11.54	15.83	1.64
信江	氨氮	1.37	2.42	4.29	40.9	9.17	3.88	1.56	4.79	18.87	6.43	4.41	1.9
	总磷	2.72	7.69	0.75	23.2	11.68	9.05	1.9	2.7	11.49	5.65	16.4	6.77
抚河	氨氮	0.11	1.67	0.18	49.13	25.33	16.25	3.07	0.06	0.55	0.27	3.33	0.05
	总磷	0.09	4.7	0.1	13.41	2.48	30.76	10.17	0.21	1.56	0.64	35.85	0.02
修水	氨氮	8.36	2.49	8.63	7.82	1.65	5.6	14.78	15.15	14.46	8.42	2.84	9.8
	总磷	13.7	3.89	2.41	9.07	6.46	11.67	7.46	11.86	23.06	4.28	2.84	3.3
昌江	氨氮	1.2	1.22	2.26	3.03	0.96	42.2	35.6	6.53	2.01	0.85	2.37	1.76
	总磷	1.18	1.32	1.86	11.68	1.05	61.17	7.34	0.56	1.66	2.14	7.47	2.56
博阳河	氨氮	2.76	3.44	2.03	11.85	6.44	14.1	15.4	1.12	5.75	3.64	23.98	9.48
	总磷	2.78	4.8	2.3	11.67	11.99	9.6	14.65	1.36	4.45	3.02	26.42	6.97
乐安江	氨氮	3.85	8.9	8.21	11.09	8.13	14.24	12.99	5.28	3.73	7.79	8.6	7.19
	总磷	1.67	2.32	5.7	9.82	4.14	10.66	20.87	2.62	6.65	1.83	28.93	4.78
西河	氨氮	1.24	1.61	2.78	7.49	5.75	6.72	10.98	10.63	27.11	2.7	17.98	5.01
	总磷	1.86	4.97	4.2	4.53	4.58	10.5	7.31	9.78	14.17	5.64	25.2	7.25

采用流域水文分割法估算方法计算得到鄱阳湖流域的入湖污染负荷见表3。2008年,通过五河和西河、博阳河等河流携带的溶解性氨氮、总磷分别为36.67~24 786.9 t和9.52~5 220.56 t,其中来自非点源的氨氮、总磷分别占26.51%~98.66%和53.63%~99.19%。可见,鄱阳湖流域非点源污染严重,非点源是鄱阳湖入湖污染物主要来源,对氨氮贡献率为68.36%,总氮贡献率为75.29%,远远高于点源污染贡献率。

表3 2008年鄱阳湖流域非点源污染负荷计算结果

流域	总质量(t)		点源(t)		溶解态非点源(t)		非点源贡献率(%)	
	氨氮	总磷	氨氮	总磷	氨氮	总磷	氨氮	总磷
赣江	24 786.90	5 220.56	5 220.89	842.74	19 566.01	4 377.82	78.94	83.86
信江	2 995.59	1 881.97	891.11	759.16	2 104.48	1 122.82	70.25	59.66
抚河	2 201.02	416.50	29.40	3.37	2 171.62	413.13	98.66	99.19
修水	874.99	233.81	493.27	84.11	381.72	149.7	43.63	64.03
昌江	1 986.16	218.84	359.71	36.61	1 626.45	182.23	81.89	83.27
博阳河	36.67	9.52	8.53	2.41	28.14	7.11	76.74	74.7
乐安江	8 111.37	1 094.01	5 961.01	507.35	2 150.36	586.66	26.51	53.63
西河	82.11	39.73	31.39	16.56	50.72	23.17	61.77	58.31
合计	41 074.81	9 114.94	12 995.31	2 252.31	28 079.50	6 862.64	68.36	75.29

3.2.2　非点源污染空间分布特征

图3显示鄱阳湖入湖河流非点源污染负荷贡献率对比情况。空间上,赣江对入湖非点源污染负荷的贡献占绝对优势,氨氮、总磷贡献率分别为69.68%和63.79%,其次为信江和抚河,其他入湖河流非点源贡献率都较小,博阳河和西河污染贡献率均小于0.34%。

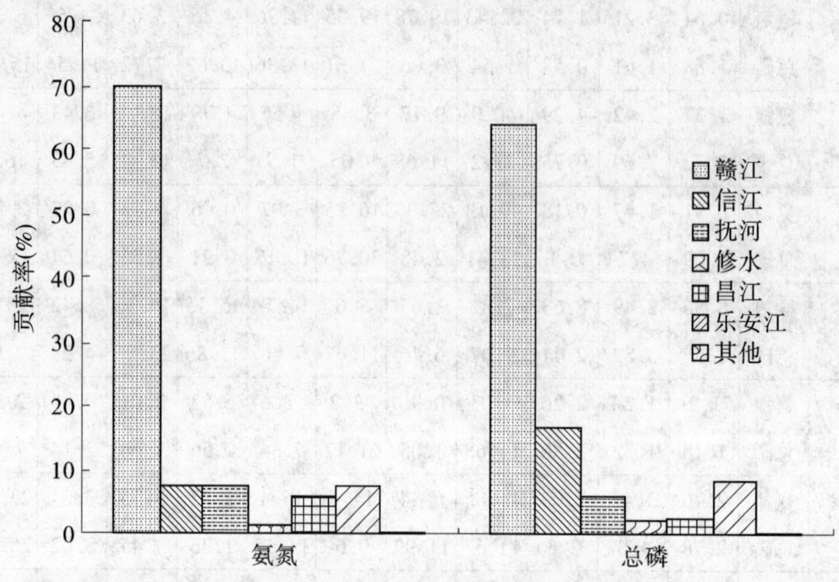

图3　鄱阳湖入湖河流非点源污染负荷贡献率对比

3.3　讨论

2008年鄱阳湖流域平均降水量1 536.0 mm,属平水年份(频率$P=62\%$)。鄱阳湖流域入湖非点源污染负荷占入湖总负荷的68%~76%,这与国内相关研究进行对比,比例较合理,表明应用水文分割法估算鄱阳湖流域入湖河流非点源污染负荷方法可行。但是,由于河流水文过程复杂,水文分割方法的不成熟性和水文分割法假定条件不确定性,估算结果可能有偏差,值得重视和讨论。

比较表2和图2可知,抚河非点源贡献率虽达98%以上,但对鄱阳湖入湖总量贡献率并非最大,而小于赣江、信江和乐安江。研究发现,入湖河流非点源污染氨氮负荷和径流量、流域面积相关系数分别为0.987和0.984,总磷负荷和基流、流域面积相关系数分别为0.987和0.981,显著性水平达0.01,呈显著正相关;非点源氨氮、总磷污染负荷与基流占总流量比例呈负相关。流域径流量和面积大小直接影响污染物负荷,抚河流域是基流比例最小的流域,其流域面积小于赣江和信江。

4　结论

(1)2008年,通过“五河”和西河、博阳河等河流挟带进入鄱阳湖的溶解性氨氮、总磷分别为36.67~24 786.9 t和9.52~5 220.56 t,其中非点源氨氮、总磷分别占26.51%~98.66%和53.63%~99.19%。总量组成中,非点源是鄱阳湖入湖污染物主要来源,占总入湖负荷的68%~76%。

(2)空间分布上,主要河流污染贡献率不同,以及污染特征也不相同。赣江流域对鄱阳湖入湖污染贡献率最高,尤其是非点源污染,是流域污染控制重点。五河中修水的综合贡献率最小,氨氮占总量的43.62%,低于点源,非点源总磷稍占优势。水土流失是非点源的载体,因此加大对赣江流域生态建设是控制非点源污染的有效途径。

参考文献

[1] 王晓鸿,樊哲文,崔丽娟,等.鄱阳湖湿地生态系统评估[M].北京:科学出版社,2004.

[2] L F Leon, D C Lam, D A Swayne, et al. Integration of a nonpoint source pollution model with a decision support system [J]. Environmental Modeling and Software, 2000(15): 249-255.

[3] 赵其国,黄国勤,钱海燕.鄱阳湖生态环境与可持续发展[J].土壤学报,2007,44(3):318-326.

[4] 蔡明,李怀恩,刘晓军.非点源污染负荷估算方法研究[J].人民黄河,2007,29(7): 36-37, 39

[5] 郑丙辉,王丽婧,龚斌.三峡水库上游河流入库面源污染负荷研究[J].环境科学研究,2009,22(2):125-131.

【作者简介】 涂安国,男,1983年11月生,2010年6月于河海大学硕士研究生毕业,获硕士学位。现在江西省水土保持科学研究所工作。

北疆供水工程长距离供水调度模型研究

李海涛[1]　江　浩[2]　谢小燕[2]

（1. 新疆额尔齐斯河流域开发工程建设管理局　乌鲁木齐　830000）；
2. 南瑞集团公司　南京　210003）

【摘要】 北疆供水工程渠线距离长，跨流域、跨地域，供水调度复杂。本文根据北疆供水工程调度特点，对如何建立数学模型，制定科学的供水调度预案进行了研究。

【关键词】 北疆供水　调度模型　研究

1　概述

北疆供水工程的基本任务是从“635”水库取水，为受水区提供城市生活和工业用水，同时兼顾农业和生态及环境的补水任务，渠线总长度 510 km。供水调度既受水文气象的影响，又受水力控制等工程因素的影响，同时还涉及干渠沿线、水源地、受水区等多方面利益的协调，从调度技术上讲，北疆供水调度是一个极其复杂的问题，国内外几乎没有直接现成例子可供借鉴。

2　调度特点分析

2.1　可供水量的不确定性

由于水文气象的不确定性和随机性，现有的水文预报技术还达不到长时间准确预报的程度，因此在实际调度中水库全年可供水量是个不确定因素，很难给用水户一个明确的供水信号。由于总干渠上没有在线的调节水库，整个渠道供水完全依赖“635”水库的调蓄能力，“635”水库来水的不确定性为渠道供水调度增加了难度。可供水量不确定性带来的是调度计划编制的不精确，不利于资源的合理分配和管理。

2.2　需水量的不确定性

受水区需水量也是个不确定因素，北疆供水长度为 500 多 km，沿途共有 12 个分水口，跨越戈壁、沙漠、草原和耕地，影响受水区水文气象的因素极其复杂，受水区的需水量是一个不确定的随机量。同时，政策、价格和调度体制等因素，也会影响受水区需外调水量。

2.3　渠道运行的特殊性

北疆供水工程途经多个行政区域并跨越多个流域，给多个农牧灌区供水，若对引水系统采用常规方法分散管理和调度，则在运行期间当发生分水、停(启动)泵等情况时，由于不能及时进行有效的控制，将会引起水位突涨突落，导致边坡破坏，水量流失，诱发事故。此外，还会增加年运行管理费用，甚至引起区域用水纠纷。因此，北疆供水工程渠道的运行必须满足以下条件：

(1)稳。即稳定流量运行。在运行中应尽量保持流量和水位的稳定，使渠水按照近似明渠均匀流运行，流态稳定。尽量避免因流量、水位的变幅过大而导致水流流态紊乱，对渠道产生淘蚀和破坏。

(2)控。即适时调控。按照运行计划的要求，适时调节闸门开度，控制运行流量。总干渠运行流量的控制，主要依靠总干进水闸的合理调节、科学匹配来实现。总干进水闸应根据指令要求和闸前库水位的变化适时调节闸门的开度，控制进水流量的稳定。

(3)缓。即流量的增加或减小、水位的抬高或降低应缓慢进行，避免水位的骤升、骤降。水位骤升时，水头流速过大，容易对渠道造成冲刷破坏。水位骤降，渗入膜后及渠堤内的水不能随干渠水位的降低而及时逸出，产生扬压力，对衬砌板块产生顶托破坏，同时容易造成渠坡失稳滑坡。

基于上述特点对供水调度进行了研究，采用分水口权重的最大经济效益为目标函数，建立数学模型，用相应算法求解得出最佳配置。

3 调度模型

调度目标是使得水库运行期内供水收益最大，即

$$\text{Obj}:\text{Max} \quad E = \sum_{i=1}^{I}\sum_{t=1}^{T}(Q_{\text{ws}i,t}P_i w_i \Delta t)$$

式中：E 为所有供水对象的总收益；Δt 为时段数；$Q_{\text{ws}i,t}$ 为 t 时段，i 分水口所需流量；P_i 为 i 分水口的供水价格；w_i 为 i 分水口的权重系数。

约束条件：

(1)进水口最大可供流量约束。

$$\sum_{i=1}^{I} Q_{\text{ws}i,t}(t-\tau)\sigma_i \leqslant Q_{\inf\max}$$

式中：$Q_{\text{ws}i,t}(t-\tau)$ 为 t 时段，i 分水口的需水流量；τ 为 i 分水口的水流传播时间；$Q_{\inf\max}$ 为进水口最大可供流量；σ_i 为分水口的渠道水利用系数。

(2)渠道水位变幅。

$$|(Z_{i,t}-Z_{i,t-1})| \leqslant \Delta Z_i$$

式中：$Z_{i,t}$ 为 i 分水口 t 时段的水位；$Z_{i,t-1}$ 为 i 分水口 t-1 时段的水位；ΔZ_i 为 i 分水口的水位变幅。

(3)节制分水闸运行要求。

$$Q_{\inf,t} - \sum_{i=1}^{i} Q_{\text{ws}\,i,t}(t-\tau)\sigma_i \geqslant Q_{\lim i}$$

式中：$Q_{\inf,t}$ 为 t 时段进水口的进水流量；$\sum_{i=1}^{i} Q_{\text{ws}\,i,t}(t-\tau)$ 为前 i 个分水口 t 时段的分水流量；$Q_{\lim i}$ 为 i 分水口节制运行所需最小流量。

(4)分水口分水流量变幅。

$$|(Q_{i,t}-Q_{i,t-1})| \leqslant \Delta Q_i$$

式中：$Q_{i,t}$ 为 i 供水口 t 时段的水位；$Q_{i,t-1}$ 为 i 供水口 $t-1$ 时段的水位；ΔQ_i 为 i 供水口的水位变幅。

(5)分水口最大分水流量约束。

$$Q_{i,t} \leqslant Q_{\max i}$$

式中：$Q_{\max i}$ 为 i 供水口最大供水流量

(6)倒虹吸运行条件。

$$Q_{\text{inv},t} \geqslant Q_{\min,\text{inv}}$$
$$Q_{\text{inv},t} \leqslant Q_{\max,\text{inv}}$$

式中：$Q_{\text{inv},t}$ 为 t 时段倒虹吸的进水流量；$Q_{\min,\text{inv}}$ 为倒虹吸的最小进水流量；$Q_{\max,\text{inv}}$ 为倒虹吸的最大进水流量。

(7)供水水量约束。

$$W_i \leqslant W_{i,\text{need}}$$

式中：W_i 为 i 分水口的分水量；$W_{i,\text{need}}$ 为 i 分水口的需水量。

边界条件：

模型计算边界条件为各个水位站的初始水位和各个闸门的初始开度。

4 模型求解

模型求解采用目标规划法。

4.1 目标规划模型

对于 L 个目标，K 个优先等级（$K \leqslant L$）的一般目标规划问题。对于同一个优先级别的不同目标，它们的正负偏差变量的重要程度还可以有差别。如对于第 $k(k=1,2,\cdots,K)$ 级目标的正负偏差变量分别赋予不同的权系数 w_{kl}^{+} 和 w_{kl}^{-}（$l=1,2,\cdots,L$），则目标规划问题的一般数学模型可表述为

$$\min z = \sum_{k=1}^{K} P_k \sum_{l=1}^{L} (w_{kl}^{-} d_l^{-} + w_{kl}^{+} d_l^{+})$$

$$\text{s.t.} \begin{cases} \sum_{j=1}^{n} c_{ij} x_j + d_l^{-} - d_l^{+} = q_l & (l=1,2,\cdots,L) \\ \sum_{j=1}^{n} a_{ij} x_j = b_i & (i=1,2,\cdots,m; j=1,2,\cdots,n) \\ x_j \geqslant 0 & (j=1,2,\cdots,n) \\ d_l^{-}, d_l^{+} \geqslant 0 & (l=1,2,\cdots,L) \end{cases}$$

4.2 建立模型的步骤

(1)根据问题所提出的各个目标与条件，确定目标值，列出目标约束与绝对约束。

(2)根据决策者的需要将某些或全部绝对约束转化为目标约束，这时只需要给绝对约束加上负偏差变量和减去正偏差变量。

(3)给各个目标赋予相应的优先因子 $P_k(k=1,2,\cdots,K)$。

(4)对同一优先等级中的各偏差变量，根据需要可按其重要程度不同，赋予相应的权系数 w_{kl}^{+} 和 w_{kl}^{-}（$l=1,2,\cdots,L$）。

(5)根据决策者需求，按下列三种情况：

①恰好达到目标值,取 $d_l^- + d_l^+$;

②允许超过目标值,取 d_l^- ;

③不允许超过目标值,取 d_l^+ 。

构造一个由优先因子和权系数相对应的偏差变量组成的,要求实现极小化的目标函数。

4.3 优先因子与权系数

在一个多目标决策问题中,要找出使所有目标都达到最优的解是很不容易的;在有些情况下,这样的解根本不存在(当这些目标是互相矛盾时)。实际作法是:决策者将这些目标分出主次,或根据这些目标的轻重缓急不同,区别对待,也就是说,将这些目标按其重要程度排序,并用优先因子 $P_k(k = 1,2,\cdots,K)$ 来标记,即要求第一位达到的目标赋予优先因子 P_1,要求第二位达到的目标赋予优先因子 P_2,…要求第 K 位达到的目标赋予优先因子 P_K。规定

$$P_1 \gg P_2 \gg \cdots \gg P_K$$

符号"$\gg$"表示"远大于";$P_K \gg P_{K+1}$ 表示 P_K 与 P_{K+1} 不是同一级别的量,即 P_K 比 P_{K+1} 有更大的优先权。这些目标优先等级因子也可以理解为一种特殊的系数,可以量化,但必须满足

$$P_k > MP_{K+1}(k = 1,2,\cdots,K - 1)$$

其中,$M > 0$ 是一个充分大的数。

在同一优先级别中,可能包含有两个或多个目标,它们的正负偏差变量的重要程度还可以有差别,这时还可以给处于同一优先级别的目标赋予不同的权系数 w_j,这些都由决策者按具体情况而定。

5 调度流程

调度流程可以概括为制定用水协议、编制年调度预案和滚动计划三步。

(1)制订用水计划。根据相关资料,制订各用水单位当年用水计划,与需水预测模型预测结果进行比较分析并作协调处理,最后由额管局与各用水单位签订用水协议,保证全年用水不超过协议中规定的用水限额。

(2)编制年度调度预案。运行供水调度模型和各用水单位提出的需水计划,制订下一年度的年度供水调度预案,保证各用户年用水总量不超过需水计划。

(3)滚动计划。在实际调度运行中,根据来水信息的变化和各用水单位实际用水信息,运用调度模型滚动修正年度调度预案。

6 实例

以北疆供水2009年沿线12个需水用户的需水要求和"635"水库的可供水量为依据,通过模型进行供水调度计算,可以输出总干渠进水闸的模拟调度过程,对比当年总干渠人工调度的实际过程可以发现:模拟调度过程与实际调度过程相吻合,见图1。

7 结论

在充分考虑北疆供水调度特点的基础上提出的供水调度模型针对性强,在此基础上开发的供水调度模型已经嵌入北疆供水调度决策支持系统,现已进入试运行阶段,并取得了良

好的效果。模型能很好的协调好沿线各分水口需水计划矛盾，借助系统理论、数据库技术以及数据采集传输等现代化手段，提出了供水调度方法，从而为实现长期调水计划制作、滚动计划制作和精确调水的目标奠定了基础。

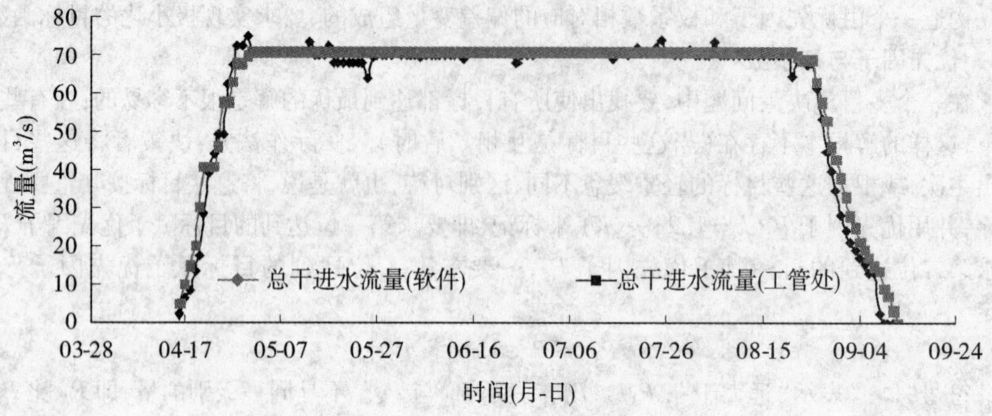

图 1 2009 年供水调度方案总干进水流量对比

参考文献

[1] 冯尚有. 多目标决策理论方法与应用[M]. 武汉：华中理工大学出版社，1990.

【作者简介】 李海涛，1975 年 8 月生，2005 年 7 月毕业于新疆农业大学水利水电工程专业，本科，现就职于新疆额尔齐斯河流域开发工程建设管理局，任西水东引枢纽项目部副主任，高级工程师。

施工期遭遇超标准洪水的应对实例

朱国建

（新疆额尔齐斯河流域开发工程建设管理局　乌鲁木齐　830000）

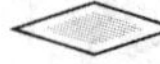

【摘要】 由于自然条件突变，水利枢纽工程在施工建设期间会遭遇到超设计标准洪水，根据实际情况，进行认真分析比对，采取最为有利的防洪度汛方案，能够最大程度减少各项经济损失，并能够安全度汛。本文通过实例，对类似问题，提供经验参考。

【关键词】 超设计标准洪水　分析比对　安全度汛

1　工程简介

某水利枢纽工程等别为Ⅱ等工程，工程规模为大(2)型，施工导流标准为20年一遇洪水(P=5%、洪峰流量为1 930 m^3/s)。导流洞属临时工程，布置在河左岸，总长度679.233 m，城门洞断面型式，断面几何尺寸为11.5 m×13.5 m，底坡为8/1 000，导流洞设计过流能力为1 594 m/s。依据开挖揭示的地质条件，导流洞全断面仅衬砌了125 m，其余段边顶拱为厚0.15 m喷射混凝土支护、底板为厚0.3 m混凝土护砌，导流洞于2009年9月建成过水。

上下游围堰堰体防渗采用土工膜斜墙防渗，堰体采用砂砾料及开挖料填筑，上游围堰堰顶高程为604.0 m，堰顶宽度为10 m，最大堰高34.67 m；下游围堰堰顶高程为575.5 m，堰顶宽度为8 m，最大堰高10.7 m。上游围堰于2009年11月填筑至597.5 m高程(达到10年一遇挡水度汛高程)，下游围堰也于当年填筑至设计高程。

2　2010年洪水预估分析

根据当地气象部门通报，2009年冬季至2010年春季以来，水利枢纽所在流域遭遇60年一遇特大暴风雪，累计降水量达到110 mm，平原区平均积雪厚52 cm，浅山地带积雪1.0 m，深山积雪厚达2.5 m。设计单位对枢纽所在流域融雪洪水及融雪成分较大的混合型洪水的年最大洪峰进行了分析预估，经采用5种方案(重现期类比、雪量距平类比、相同背景情况类比、前期降水量与年最大洪峰相关分析、前期最大积雪深度与年最大洪峰相关分析)初步估算，枢纽工程坝址2010年年最大洪峰为885～3 102 m/s，比较可能的范围为2 040～2 400 m/s。另据分析预估，该流域年最大洪峰最早出现在5月1日，5月31日前发生年最大洪水的概率为41%、6月中旬前发生年最大洪水的概率为84%。施工期导流度汛特征值见表1。

3　防洪度汛的基本思路

通过分析，预计2010年春汛洪水标准可能超过设计导流标准(可能的洪峰流量为20～

50 年一遇），因此防洪度汛坚持的原则是：①保证施工安全；②减小超标准洪水对已完建项目造成较大规模的破坏；③确保当遭遇超标准洪水时，因工程措施不当，致使上游围堰失事产生溃坝洪水时，不对下游两乡（坝址距两个乡约 15 km）一镇（坝址距县城约 30 km）人民生命财产造成威胁或损失。

表 1　工程施工期导流度汛特性

洪水标准（%）	洪峰流量（m^3/s）	水位（m）	堰顶高程（m）	泄水建筑物		
				泄洪方式	孔口尺寸（宽×高）（m×m）	下泄流量（m^3/s）
10	1 594.0	596.32	597.5	导流洞	9.5×10	1 444.77
5	1 930.0	602.32	604.0		9.5×10	1 689.74
2	2 331.0	611.47	604.0		9.5×10	2 006.49
1	2 740.0	618.56	604.0		9.5×10	2 221.15

4　防洪度汛的方案比选

根据洪峰流量预估和当前工程建设情况，初步拟选了四个方案进行比选。

4.1　方案一

按照项目审查部门批准的 20 年一遇度汛标准，围堰按 20 年一遇洪水度汛标准提前一年实施（原设计施工期第一年度汛标准为 10 年一遇洪水、即堰顶高程为 597.5 m），围堰提前填至 604.0 m 高程，在围堰左侧增设非常溢洪道、下游设置过水保护措施，非常溢洪道在堰顶处的底板高程 598.0 m，总长 111 m，采用矩形断面，底宽 20 m，墙高 4 m，由进口引渠、堰顶控制段、泄槽段、出口消能段，基本形式塑膜防渗体加 1 m 厚铅丝石笼或格宾网（TRW 网装石）的结合（具体方案由设计单位提出），建设费用计 250 万～300 万元，超过 10 年标准的洪水由导流洞、围堰上非常溢洪道联合下泄，经调洪演算分析（见表 2），围堰可挡 30 年一遇洪水。

表 2　不同频率调洪成果

标准（年）	洪峰流量（m^3/s）	下泄流量总量（m^3/s）	导流洞下泄流量（m^3/s）	溢洪道下泄流量（m^3/s）	溢洪道底高（m）	水位高程（m）	围堰顶高程（m）
10	1 594	1 445	1 445	0	598	596.32	604
15	1 790	1 652	1 567	85	598	599.21	604
20	1 930	1 753	1 593	160	598	599.84	604
30	2 132	1 957	1 657	300	598	601.46	604

4.2　方案二

按审查部门批准的度汛标准，围堰按 10 年一遇洪水度汛标准实施，堰顶高程 597.5 m，在围堰左侧增设非常溢洪道、下游设置过水保护措施，非常溢洪道在堰顶处的底板高程 591.0 m，总长 98 m，采用矩形断面，底宽 20 m，墙高 6.5～4.0 m，由进口引渠、堰顶控制段、泄槽段、出口消能段，基本形式塑膜防渗体加 1 m 厚铅丝石笼或格宾网（TRW 网装石）的结合（具体方案由设计单位提出），建设费用与方案一相当，超过 10 年标准的洪水由导流洞、围堰上非常溢洪道联合下泄,经调洪演算分析(见表3),围堰可挡20年一遇洪水。

表3 不同频率调洪成果

标准(年)	洪峰流量(m^3/s)	下泄流量总量(m^3/s)	导流洞下泄流量(m^3/s)	溢洪道下泄流量(m^3/s)	溢洪道底高(m)	水位高程(m)	围堰顶高程(m)
10	1 594	1 474	1 327	147	591	593.78	597.5
15	1 790	1 675	1 396	279	591	595.24	597.5
20	1 930	1 796	1 433	363	591	596.06	597.5

4.3 方案三

按审查部门批准的度汛标准,围堰按10年一遇洪水度汛标准实施,堰顶高程598.2 m,在围堰左侧增设非常溢洪道、下游设置过水保护措施,非常溢洪道在堰顶处的底板高程593.0 m,总长98 m,采用矩形断面,底宽25 m,墙高4~5.2 m,由进口引渠、堰顶控制段、泄槽段、出口消能段,基本形式塑膜防渗体加1 m厚铅丝石笼或格宾网(TRW网装石)的结合(具体方案由设计单位提出),建设费用与方案二相当,超过10年标准的洪水由导流洞、围堰上非常溢洪道联合下泄,经调洪演算分析(见表4),围堰可挡20年一遇洪水。

表4 不同频率调洪成果

标准(年)	洪峰流量(m^3/s)	下泄流量总量(m^3/s)	导流洞下泄流量(m^3/s)	溢洪道下泄流量(m^3/s)	溢洪道底高(m)	水位高程(m)	围堰顶高程(m)
10	1 594	1 445	1 445	0	593	596.32	598.2
15	1 790	1 665	1 439	226	593	596.18	598.2
20	1 930	1 794	1 475	319	593	597.0	598.2

4.4 方案四

降低围堰高度。因围堰基础防渗墙、上游面防渗土工膜、护坡混凝土已施工至575.0 m高程,故将上游围堰由已填筑至的597.5 m高程拆除至575.0 m高程,575.0 m(此高程较导流洞进口底板高5 m)高程以下堰体采取铅丝石笼包护,之上再覆薄层混凝土护面,形成可过水的低堰。经计算,围堰拆填(拆除、再填)18.3万m^3、铅丝石笼650个,建设费用约350万元。

5 确定防洪度汛方案

方案一围堰按20年一遇洪水标准填筑,堰顶高程604 m,导流洞加非常溢洪道过洪能力可以达到1 957 m^3/s,即30年一遇洪水标准,汛期拦蓄库容2 145万m^3(中型水库规模),围前水位为601.46 m,拦蓄的库容量大,如遇更大洪水致使围堰失事,将给下游带来严重危害。此种条件下导流洞承担的泄量较大,为1 445~1 657 m^3/s,洞内流速大,对未进行全断面混凝土衬砌段产生冲刷破坏的可能性大,对导流洞的安全运行不利。另外,工程建设区位于北方严寒地区,适宜混凝土施工的最早时间为4月上旬,当地4月大气环境温度极不稳定,常伴有大风、降雪、降温等恶劣天气条件,而5月初洪峰可能来临,因此可利用的有效施工时段很短，设计方案所要求的工程项目不一定能如期完成，风险很大。

方案二围堰按 10 年一遇洪水标准填筑，堰顶高程 597.5 m，导流洞加非常溢洪道过洪能力可以达到 1 796 m^3/s，即 20 年一遇洪水标准，汛期拦蓄库容 1 285 万 m^3（中型水库规模），比方案一少拦蓄 860 万 m^3，围堰前水位 596 m，堰前水位相对较低，如遭遇超 20 年一遇洪水致使围堰失事，仍会对下游造成危害。此方案的优点是导流洞承担的泄量较小，为 1 327 ~1 433 m^3/s，洞内流速较小，对导流洞不会冲刷破坏，利于导流洞的安全运行。施工时段、施工环境与方案一相同。

方案三基本按 10 年一遇洪水标准填筑，堰顶高程仅较方案二填高 0.7 m，为 598.2 m，在遭遇 20 年一遇洪水时，导流洞加非常溢洪道泄量可达 1 794 m^3/s，汛期拦蓄库容、堰前水位与方案二相当，其他有利条件与不利条件与方案二相当。

方案四仅保留了高 8.0 m 的围堰体，并采取工程措施加固保护了围堰已施工完毕的基础和上游面防渗体。此方案堰前水位很低，拦蓄库容仅约 12 万 m^3，优点是：①因堰前水位低、拦蓄库容小，即便失事对下游不会形成危害；②大部分洪水将由围堰顶通过，导流洞承担泄量很小，对导流洞不会构成任何威胁；③施工简单、工程量小，完全有把握在有限施工时段内完成。不足是需建设费用约 350 万元，较前三个方案多；可能给设计总工期的实现带来一定困难，尤其是可能对坝基开挖产生直接影响。

综合分析比较认为，洪水具有随机性、不确定性，一旦围堰失事形成溃坝洪水给下游人民生命财产造成的损失不可估量，方案四最不利因素是可能给工程后续施工带来一定困难，但当前施工进展情况是：枢纽位于高山峡谷区，坝肩两岸临时施工道路具备通车条件尚需 2 个月，两坝肩石方开挖最少也需要 3 个月，主河槽开挖时间最早也要到 8 月初，而此时汛期已过。因此，最终确定按方案四实施。

6 实施方案，安全度汛

2010 年 4 月初开始进行降低围堰高度、加固防护围堰基础和围堰上游面防渗体等施工内容，因 4 月大气环境温度极不稳定，经历了 3 次寒潮降温天气过程，有时夜间温度在零下十几度，施工并不顺利。到 4 月 29 日，设计要求的工作量全部完成，计拆除围堰体18 万 m^3、铅丝石笼 547 个、浇筑混凝土 610 m^3。

5 月 1 日，堰前水位上涨至 575 m 高程，堰顶开始过流，此后河道来水逐渐增加，6 月 2 日，坝址上游（距坝址 24 km）水文站实测最大洪峰流量为 1 600 m^3/s，到 7 月末，河道来水回落至正常水平，主汛期结束。因水文站断面至坝址断面间尚有几条小支流，主汛期过后经水文部门分析还原计算，坝址断面最大洪峰流量为 1 700 ~2 000 m^3/s，相当于 15 ~25 年一遇洪水，接近前期预测结果。

汛期期间，主河道、导流洞水流平稳，退水后经检查，过水低堰堰体完好无损。

对坝基开挖的影响：当年 9 月底两坝肩才开挖至主河槽，因此对坝基开挖未产生任何影响。

【作者简介】 朱国建，1979 年 4 月生，2002 年 7 月毕业于新疆农业大学，学士学位，在新疆额尔齐斯河流域开发工程建设管理局从事水利工程建设管理工作，工程师。2006 年、2007 年、2010 年、2011 年荣获先进工作者和优秀共产党员称号。

1956~2005年期间密云水库入库径流减少的原因分析

马　欢

（水利部海河水利委员会　天津　300170）

【摘要】 密云水库入库径流在1956~2005年期间明显减少，本文以定量分析径流减少的原因为目标，通过数据分析和分布式水文模型，区分了不同因素对径流减少的贡献率，认为气候变化是密云水库入库径流减少的主要原因，其次为人工取水和下垫面变化。

【关键词】 密云水库　径流减少　气候变化　人类活动

海河流域属资源型严重缺水地区，20世纪80年代以来海河流域地表水资源量进一步减少，加剧了水资源供需矛盾，气候变化和人类活动是导致海河流域地表径流减少的主要原因。气候变化对径流的影响主要体现在降水量和降雨强度的变化对地表径流的影响，气温等其他气候要素也会通过影响蒸散发等过程而影响径流量；人类活动对径流的影响包括人工取用水量的增加，以及人类活动引起的下垫面变化等方面。

明确径流变化的原因是进行水资源规划与管理的基础，采用分布式水文模型评价气象条件和下垫面变化对径流的影响，是当前的研究热点之一，我国学者在海河流域也开展了一些相关研究。

本文选取密云水库上游流域作为研究对象，对1956~2005这50年期间密云水库入库径流减少的原因进行分析，采用数据分析和分布式水文模型，定量区分气候变化和人类活动对径流减少的贡献率。

1　研究区概况

密云水库建成于1960年，总库容44亿m^3，是北京市的主要供水水源。水库坝址位于潮河与白河交汇处（见图1），总集水面积约15 800 km^2（其中6 700 km^2位于潮河流域，9 100 km^2位于白河流域），占潮白河流域面积的近90%。20世纪80年代以来，密云水库入库流量明显减少，威胁到北京市供水安全。

密云水库上游流域属典型山区流域，高程变化大，平均海拔933 m（见图2）；土壤类型以褐土和棕壤为主（见图3）。流域内有多个中小型水库（见图4），提供流域内的生产生活用水，人工取用水直接影响密云水库的入库径流。

流域内土地利用类型以林地、灌木和草地为主，由于1985年以来，水库上游开展了植树造林等水源地保护工程，20世纪90年代的土地利用情况较80年代初发生了较大变化，两时期土地利用情况对比如图5所示。

图 1　密云水库上游流域位置

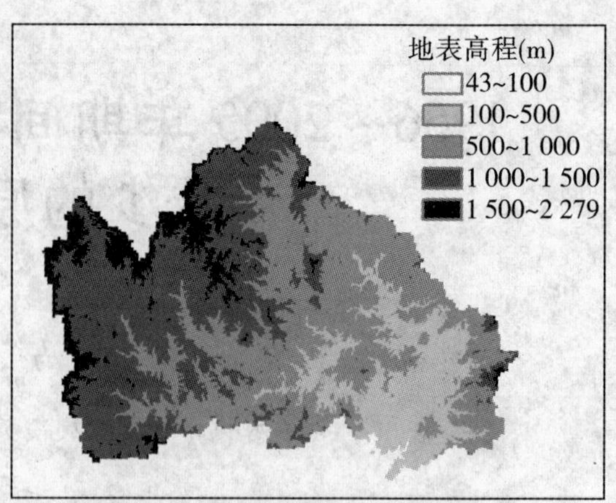

图 2　密云水库上游流域地表高程分布

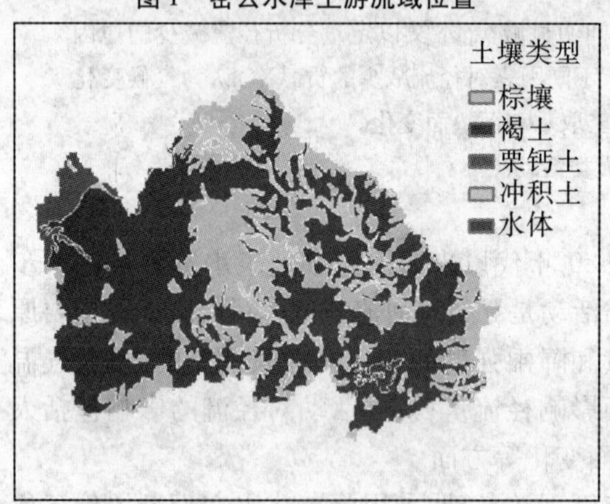

图 3　密云水库上游流域土壤类型分布

图 4　密云水库上游流域主要水库分布

（a）　80 年代中期

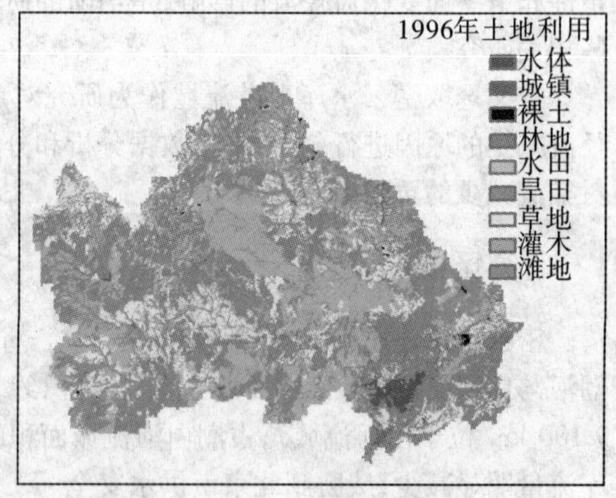

（b）　90 年代中期

图 5　两时期土地利用情况对比

从图 5 可以大致看出，与 20 世纪 80 年代相比，90 年代的土地利用类型中，林地和灌木面积比例有所增加，草地比例有所下降。流域内主要土地利用类型的具体变化情况如表 1 所示。

表 1 20 世纪 80 年代与 90 年代主要土地利用类型所占面积比例

土地利用类型	所占面积比例(%)	
	80 年代	90 年代
林地	35.1	38.1
草地	26.9	15.9
旱田	21.5	16.6
灌木	13.8	26.4

2 密云水库入库径流变化及主要原因分析

2.1 资料收集

为了分析密云水库入库径流减少的原因,本文收集了流域内的气象水文资料和空间地理信息。

气象资料来自流域内及周边共 8 个气象站的逐日数据,观测项目包括降水,最高、最低气温,最大、最小相对湿度,风速及日照时数。降水资料除气象站的降水数据外,还包括流域内及周边共 17 个雨量站的逐日降水量。气象数据(包括雨量站数据)的系列长度为 1956 ~ 2005 年。径流数据为密云水库逐月径流数据,系列长度为 1956 ~ 2005 年。流域内逐月取用水数据采用第二次水资源评价的“还原水量”统计资料进行估计,系列长度为 1956 ~ 2000 年。

地理信息数据包括地表高程、土壤类型、土地利用以及植被覆盖等,是分布式水文模型所需基础数据。地表高程数据来自全球地形数据库,空间分辨率为 3 s(约 90 m);土壤类型数据来自中国科学院南京土壤所 1∶100 万中国土壤类型图;20 世纪 80 年代和 90 年代两期土地利用数据来自中国科学院资源环境数据中心 1∶10 万中国土地利用图;植被覆盖情况通过归一化植被指数(Normalized Difference Vegetation Index, NDVI)数据反映,逐月 NDVI 数据来自 NOAA/AVHRR 全球数据产品,数据系列长度为 1981 ~ 2005 年,空间分辨率为 8 km。

2.2 入库径流减少原因初步分析与减少程度评价

根据密云水库 1956 ~ 2005 年的逐月入库流量数据,可得到过去 50 年的逐年径流,其变化过程如图 6 所示,流域逐年平均气温变化过程如图 7 所示。可以看出,密云水库入库径流在过去 50 年中明显降低,年平均气温在过去 50 年间明显升高。

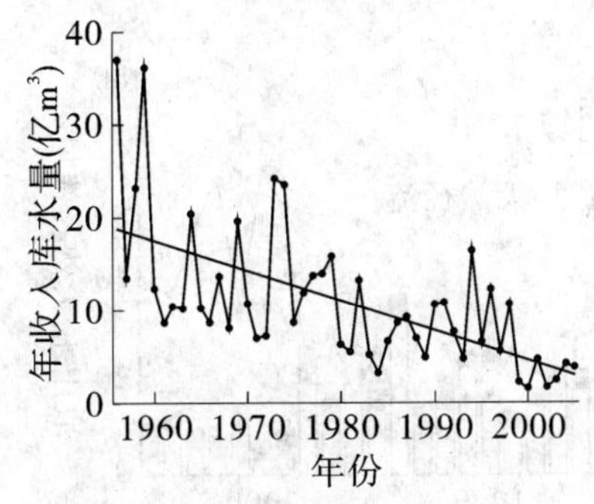

图 6 密云水库逐年入库径流量

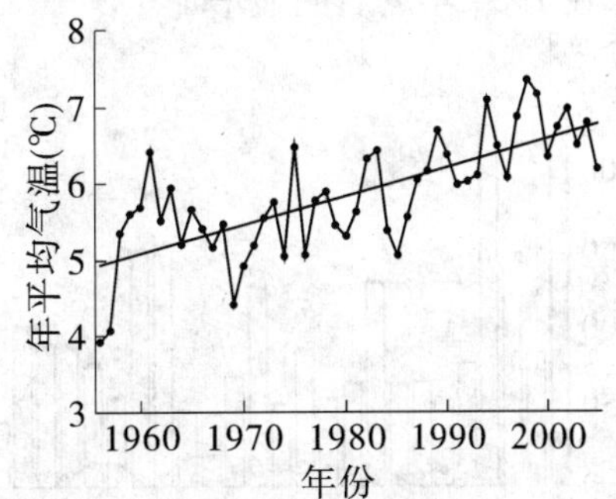

图 7 密云水库上游流域逐年平均气温

将入库径流量折算为流域面平均径流深(流域面积以 15 800 km^2 计),与降水量进行对比,如图 8 所示。可以看出,密云水库上游流域的年降水呈一定的减少趋势,同时径流系数(年径流/年降水)在过去 50 年间明显下降。

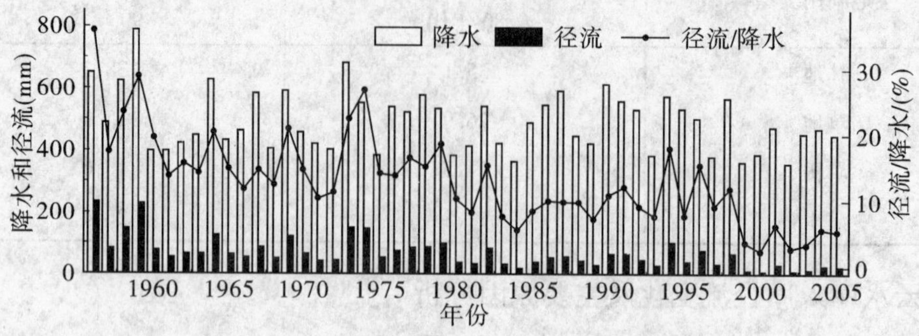

图 8　密云水库上游流域逐年降水、径流与径流系数

根据 1956 ~ 2000 年密云水库"还原水量"统计资料估计逐年人工取水量,如图 9 所示,可以看出,流域内人工取水量在 1983 年以前比较稳定,而在 1984 年后明显增加,这可能与 1983 年白河堡等水库建成有关。人工取水量在 1996 年以后趋于稳定,主要是由于近年来密云水库蓄水量急剧减少,为保证水库水位而限制上游取水,因而取用水量的上升趋势得到控制,后文分析时采用 1996 ~ 2000 年的年均人工取水量作为 2001 ~ 2005 年的逐年人工取水量。

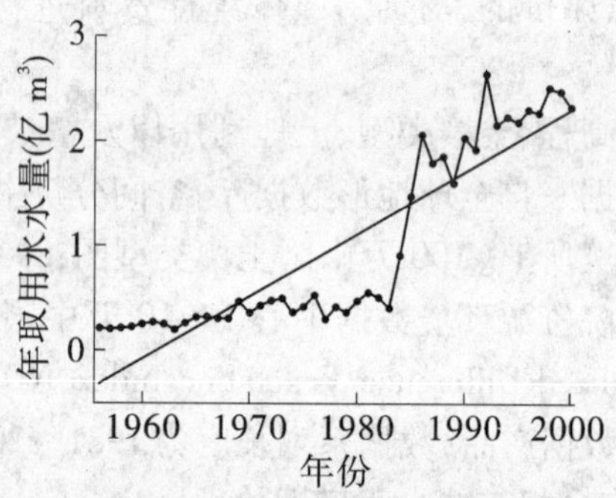

图 9　密云水库上游流域逐年人工取水

"还原水量"与实测入库径流量之和可以理解为还原后的"天然"径流量。将逐年实测径流与人工取水量均折算为面平均水量进行对比,如图 10 所示,可以看出人工取水占天然径流的比重在 1984 年之后明显增加。

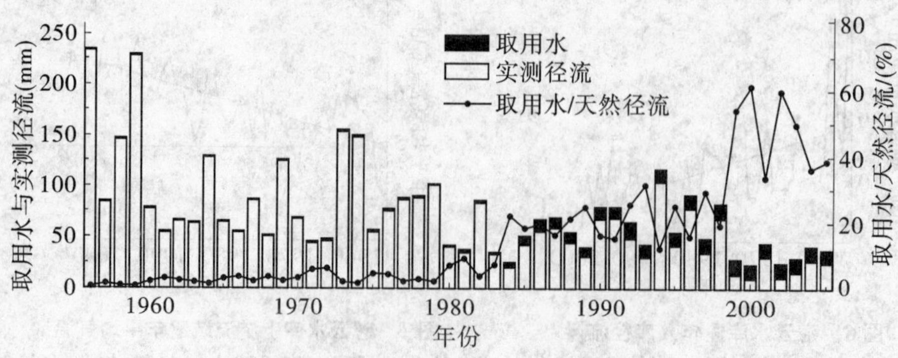

图 10　密云水库逐年入库水量与人工取水量

为定量评价径流的减少程度,本文将整个研究期划分为两个时段,通过比较两个时段年平均径流的差异,评价径流的减少量。在选取两个时段的分界点时,综合考虑入库径流、径流系数以及人类活动等因素的变化情况。从前分析可知,1984 年以来人工取水量占天然径流量的比例明显提高,径流系数则明显下降,流域内大规模植树造林工作始于 20 世纪 80 年代中期。综合以上因素,本研究以 1984 年作为分界点,将研究期划分为 1956~1983 年和 1984~2005 年两个时段。表 2 给出了两个时段的水文气象及土地利用情况对比,这里分别采用 20 世纪 80 年代与 90 年代的土地利用数据反映两个时段的土地利用情况。

表 2　不同时段年均入库径流变化及气候要素与下垫面变化情况

时段	年均入库径流(mm)	年均人工取水(mm)	气候要素年平均值		面积比例(%)	
			降水(mm)	平均气温(℃)	林地与灌木	草地
1956~1983 年	90.3	2.2	506.2	5.5	49	27
1984~2005 年	41.8	13.4	475.7	6.4	65	16
变化量	-48.5	11.2	-30.5	0.9	16	-11

从表 2 可以看出,第二时段的年均径流相对第一时段减少了 48.5 mm(7.7 亿 m^3),减少程度为 54%;两时段人工取水、气候条件及下垫面均有较大变化,这些变化都可能导致径流的减少。下文将分别就不同因素对径流减少的贡献进行定量评价。

3　不同因素对密云水库上游流域径流减少的贡献

3.1　分析方法

前文已将整个研究期划分为 1956~1983 年和 1984~2005 年两个时段,并将两时段年均入库径流的减少量作为径流减少程度的评价标准。将以这两个时段为分析对象,估计不同因素引起的两时段年均径流的减少量,从而得到不同因素对径流减少的贡献率。

人工取水对径流减少的贡献直接通过数据分析得到,人工取水的增加将直接造成入库径流减少,本文将两时段年均人工取水的增加量作为人工取水引起的入库径流减少量。

对于气候变化和下垫面改变对径流减少的影响,很难直接通过数据分析的方法得到,本文借助分布式水文模型来识别二者对径流减少的贡献率。

在分布式水文模型中输入 1956~2005 年的气象数据,且保持模型中的下垫面条件不变,此时模型输出的径流仅反映气候变化的影响。两时段模拟径流年平均值之差即为气候变化引起的径流减少量。

在保持气象条件一致的情况下,通过在分布式水文模型中采用不同时期的下垫面数据,反映下垫面变化对径流的影响。以 20 世纪 80 年代的土地利用和 1981~1983 年的平均逐月 NDVI 作为变化前的下垫面条件(后文以“下垫面 A”表示);以 20 世纪 90 年代的土地利用和 1984~2005 年的实际逐月 NDVI 作为变化后的下垫面条件(后文以“下垫面 B”表示)。以 1984~2005 年的气象数据作为输入,采用下垫面 A 时,模拟径流反映的是下垫面不变假设下,该气象条件可产生的径流量;采用下垫面 B 时,模拟径流即为下垫面发生变化后,同一气象条件可产生的径流量。两种情况下的模拟径流年均值之差,即为下垫面变化对径流的影响。

3.2 模型率定与验证

本文采用的分布式水文模型为 GBHM,它对水文过程的描述包括冠层截留、蒸散发、入渗、地表产流、河道汇流等,模型用到的参数包括植被参数、土壤参数、河道参数等,参数取值根据地理信息数据确定。

由于模型中未考虑人工取水的影响,本文采用还原后的密云水库逐月天然径流(实测径流与还原水量之和)对模型进行率定和验证。以 1956 ~ 1965 年为率定期,1966 ~ 1975 年为验证期,模拟结果如图 11 所示。

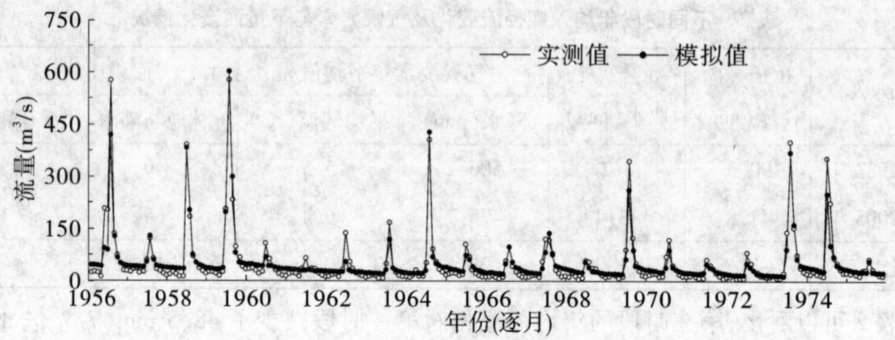

图 11 密云水库逐月入库流量模拟结果

采用 Nash 效率系数(Nash-Sutcliffe Efficiency,*NSE*)和相对误差(Relative Error, *RE*)两个指标对模拟结果进行评价,计算方法分别如下

$$NSE = 1 - \frac{\sum (Q_{\mathrm{nat},i} - Q_{\mathrm{sim},i})^2}{\sum (Q_{\mathrm{nat},i} - \overline{Q}_{\mathrm{nat}})^2} \tag{1}$$

$$RE = \frac{V_{\mathrm{sim}} - V_{\mathrm{nat}}}{V_{\mathrm{nat}}} \tag{2}$$

式中:$Q_{\mathrm{nat},i}$ 与 $Q_{\mathrm{sim},i}$ 分别为逐月天然径流和模拟径流中的第 i 个样本值;$\overline{Q}_{\mathrm{nat}}$ 为天然径流的平均值,V_{nat} 和 V_{sim} 分别为模拟时段内天然与模拟径流总量。模型的 *NSE* 和 *RE* 在率定期分别为 0.86 和 4.4%,验证期分别为 0.83 和 1.9%,表明模型模拟精度较好。

3.3 不同因素对径流减少的贡献

3.3.1 人工取水对径流减少的贡献

如前所述,人工取水对径流的影响采用数据分析的方法进行评价。从表 2 可知,两时段入库径流年均值减少了 48.5 mm,而年均人工取水量增加了 11.2 mm,即人工取水增加对入库径流减少量的贡献率为 23%(11.2 mm/48.5 mm)。

3.3.2 气候变化对径流减少的贡献

采用 GBHM 模型评价气候变化对径流减少的贡献。模型采用下垫面 A,以 1956 ~ 2005 年实测气象数据作为输入,这样模拟径流仅反映气候变化的影响。两时段年均径流模拟值以及气候变化对入库径流减少的贡献率如表 3 所示。

3.3.3 下垫面变化对径流减少的贡献

在模型中采用下垫面 B,以 1984 ~ 2005 年的气象数据作为输入,此时模拟径流反映的是下垫面发生变化后,该气象条件下可产生的径流量。采用下垫面 B 计算得到的 1984 ~ 2005 年时段年均径流量为 59.6 mm,比采用下垫面 A 时的年均径流量(68.3 mm)减少了

8.7 mm,该减少量可认为是下垫面变化引起的径流减少量,占实测径流减少量(48.5 mm)的18%。

表3 气候变化对径流减少的贡献率

时段	实测径流(mm)	模拟径流(mm)
1956~1983年	90.3	94.9
1984~2005年	41.8	68.3
径流变化量	-48.5	-26.6
贡献率		55%

3.3.4 不同因素对径流减少的贡献率汇总

根据前研究结果,人工取水、气候变化和下垫面变化对径流减少的贡献率分别为23%、55%和18%,三者之和(96%)不等于100%,这主要是由于数据资料和水文模型都不可避免的存在一定误差。假设三者误差程度相同,可对贡献率进行归一化处理,如表4所示。由表4可知,气候变化对1956~1983年与1984~2005年两个时段径流减少的贡献最大,其次为人工取水和下垫面变化。将人工取水和下垫面变化归为人类活动,其贡献率为43%,略小于气候变化的贡献率(57%)。

表4 不同因素对密云水库入库径流减少的贡献率

因素		引起的径流减少量		对径流减少的贡献率(%)	归一化后的贡献率(%)
		mm	亿 m^3		
人类活动	人工取水	11.2	1.8	23	24
	下垫面变化	8.7	1.4	18	19
气候变化		26.2	3.9	55	57

4 结论与讨论

本文从密云水库入库径流减少的原因出发,通过数据分析和水文模型,定量分析了不同因素对入库径流减小的贡献。研究结果表明,1984~2005年时段的年均入库径流相对于1956~1983年减小了48.5 mm(7.7亿 m^3),减少程度为54%。气候变化对径流减少的贡献率为57%,人工取水和下垫面变化对径流减少的贡献率分别为24%和19%。

本文通过把整个研究期划分为两个时段进行对比,来评价径流的减少程度和减少原因,划分节点选取的不同会导致结论的差异。本文选取节点时考虑了入库径流、人工取水、下垫面和气候条件发生的变化,研究结论可作为评价密云水库入库径流减少原因的一种参考。由于资料限制,本文仅通过两期土地利用和逐月NDVI数据来反映下垫面变化,有待进一步深入研究。

参考文献

[1] 任宪韶.海河流域水资源评价[M].北京:中国水利水电出版社,2007.

[2] 王忠静,杨芬,赵建世,等.基于分布式水文模型的水资源评价新方法[J].水利学报,2008,39(12):1279-1285.

[3] 贾仰文,王浩,甘泓,等.海河流域二元水循环模型开发及其应用——Ⅱ水资源管理战略研究应用[J].水科学进展,

2010, 21(1): 9-15.

[4] Wang G S, Xia J, Chen J. Quantification of the effects of climate variations and human activities on runoff by a monthly water balance model: a case study of the Chaobai River basin in northern China[J]. Water Resources Research, 2009, 45.

[5] 庞靖鹏.非点源污染分布式模拟——以密云水库水源地保护为例[D].北京:北京师范大学, 2007.

[6] 李亚光.密云水库上游水源保护林水源涵养与防止土壤侵蚀效益研究报告[R].北京:北京林业大学, 1995.

[7] 北京市潮白河管理处.潮白河水旱灾害[M].北京:中国水利水电出版社, 2004.

[8] Yang D W, Herath S, Musiake K. Development of a geomorphology-based hydrological model for large catchments[J]. Annual Journal of Hydraulic Engineering, JSCE, 1998, 42: 169-174.

【作者简介】 马欢,女,1983 年 10 月生,研究生学历,2011 年 7 月获清华大学博士学位。

浅析滦河潘家口以上流域降雨径流关系变化原因

于宝军　全黎熙

（水利部海委引滦工程管理局　河北　064309）

【摘要】 水文系统的统计规律性并不意味着水文时间系列是稳定的，只是它的状态分布依赖于时间系列。事实上由于受地理、气候、环境等自然系统的影响，水文系统的演变一直存在着。近年来，滦河潘家口以上流域的承德地区现代工农业飞速发展，引用地表水急剧增加，其用水量已经成为影响潘家口水库水文站水文要素系列稳定的重要因素。本文从气象、水工建筑物、流域内工农业活动以及水土保持措施等几个方面，分析了滦河潘家口以上流域降雨径流减少的原因，并提出了降雨径流一致性的修正问题，以期在今后的系列统计应用中避免造成系统误差。

【关键词】 径流量　降水径流关系　一致性

1　滦河潘家口以上流域概况

滦河发源于河北省丰宁县巴彦图古尔山麓，流经坝上草原，穿过燕山山脉，于乐亭县注入渤海，全长877 km，流域面积44 750 km^2，其中潘家口以上流域面积33 700 km^2，流域地势由西北向东南倾斜，山地占绝大部分，多伦以上为内蒙古高原，海拔1 300～1 400 m，地势平坦。

滦河流域地处副热带季风区，夏季短，炎热多雨，冬季长，寒冷干燥。在时间分布上降水主要集中在汛期，在空间分布上降水多发生在北部山区，多年平均值为720 mm，降水年际变化大，丰水年1997年降水多达1 040 mm，枯水年1982年仅为410 mm，两年相差1.5倍。

20世纪90年代中期以前滦河水量较为丰沛，滦县站多年平均径流量为46.3亿m^3，潘家口多年平均径流量为24.5亿m^3，径流量年内变化很大。汛期7月、8月水量较多，占年总量的50%以上；枯季1月、2月最少，两月水量之和不足全年的10%。

2　问题的提出

水文资料的一致性，是指产生各年水文资料的流域或河道的产、汇流条件在评价期内无根本变化。目前，主要通过降雨径流关系来评价系列的一致性。如果不同年份的点据基本呈单一线，则表明径流系列具有较好的一致性；反之，则表明径流系列的一致性受到了破坏，需要一致性修正。

滦河干流乌龙矶水文站和其下游石佛水文站是滦河流域引滦工程潘家口水库的两个重要入库控制站，流域控制面积31 150 km^2，占流域总面积的70%以上。为了避免潘家口水库蓄水影响，特选用以上两站的降雨径流关系，点绘两个站的降雨径流关系可以看出：1997 年以前两个站的径流量与面平均雨量的点据呈明显带状分布，具有较好的一致性，1997 年以后一致性较差，点据分布集中在关系线的左上方并且呈明显衰减状态（见图 1）。

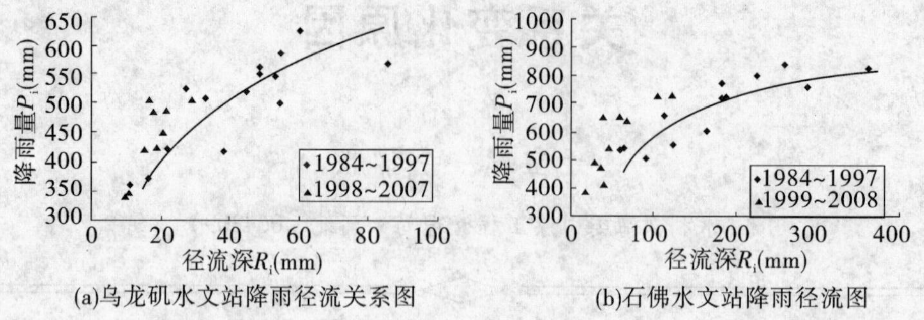

(a)乌龙矶水文站降雨径流关系图　(b)石佛水文站降雨径流图

图 1　降雨径流关系

再分析两个水文站有资料记载以来的年径流量过程线同样可以得出：近些年，滦河流域径流量呈明显的下降趋势，特别是 1999 年以后滦河流域进入了枯水期，乌龙矶站 2000 年的年径流量不及相对丰水的 1994 年的 13%，石佛水文站 2002 年的年径流量不及 1994 年的 5%（见图 2）。

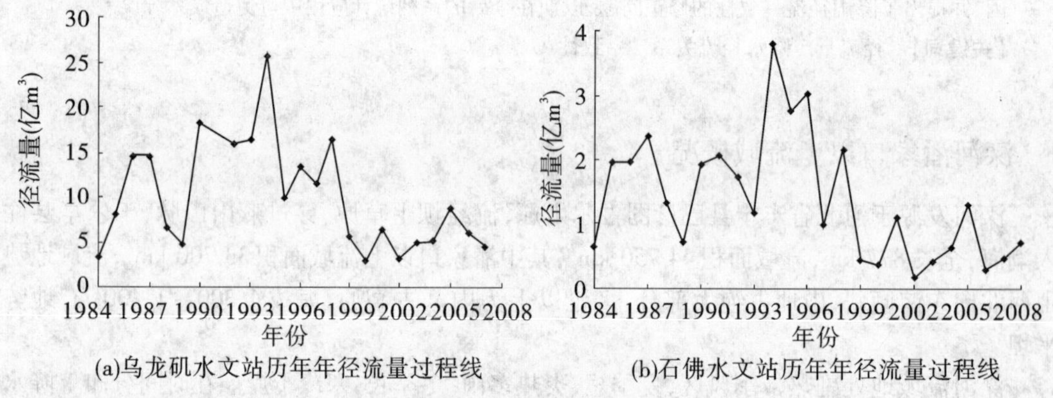

(a)乌龙矶水文站历年年径流量过程线　(b)石佛水文站历年年径流量过程线

图 2　年降水过程线

3　引起降雨径流关系变化的主要因素

在水文资料的观测期内，影响资料一致性的因素很多，如气象条件的变化，流域上修建了蓄水、引水、分洪、滞洪等工程，或发生决口、溃坝、改道等事件，特别是近些年滦河流域社会经济的高速发展，人类活动对流域下垫面的影响日益加剧。这些人工的或天然的原因使流域的径流形成条件有所改变，因而径流的概率分布规律也会有改变。

3.1　气象条件的影响

降雨是产生径流的重要因素，滦河流域降雨成因主要是大气环流和地形的共同作用下形成的。在副热带高压的作用下，夏季东南季风从渤海吹向陆地，带来大量水气与大陆上冷空气相遇，形成锋面降雨。当东南季风越过沿海平原，遇到燕山山脉的阻挡，迎风坡气流被抬升，形成地形雨。气流爬升越高，降雨量越大，到一定高度，因锋面被山阻挡而停滞，增加雨时，加大雨势，降雨量最大。

据气象部门统计,近些年夏季影响本地区的主要天气现象副热带高压较弱,大气环流水汽输送量减少,造成锋面降雨量减少;另外副热带高压位置较常年偏东偏南出海,较少的水汽量很难越过华北平原到达燕山山脉形成地形雨。上述气象要素的变化是造成滦河上游地区年降雨量逐年减少的主要原因(见图3),同时年内降雨不集中,次降雨历时缩短,雨强减弱,无法形成蓄满产流和超渗产流。

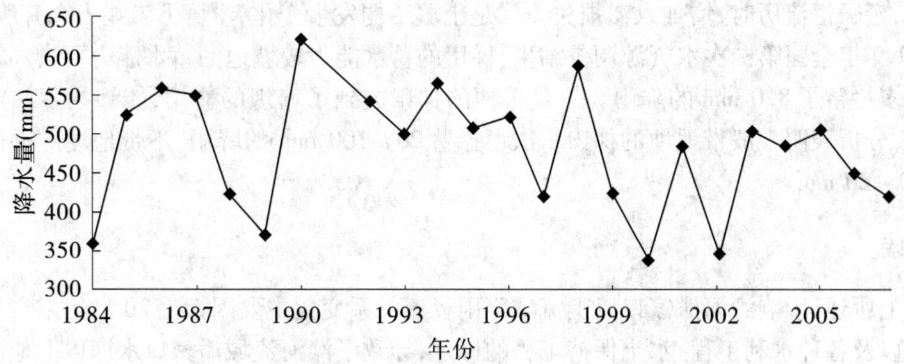

图3 滦河潘家口水库以上流域年平均降水量过程线

3.2 上游水利工程拦蓄

截至2007年,潘家口水库以上流域已建成各型水库50座,总库容3.93亿 m^3,其中大、中型水库7座,总库容3.65亿 m^3,各种塘坝25座,引提水工程239座,引提水量5.06亿 m^3,目前,滦河流域已经建成了由大、中型水库与河道堤防所构成的防御50年一遇的防洪体系,大大减小了洪水威胁,对于这一地区的国民经济发展起到了重大保障作用。但是这些水工建筑物拦蓄水量接近于近些年年平均径流量的70%。这些被拦蓄的大型水体较河川基流增加了水面蒸发量和入渗量。另外,2007年流域上游地区机井数89 488眼,地下供水量6.04亿 m^3,流域地下水已处于超采状态。在山地丘陵地区,径流量主要由地表水和地下水组成,地表水的拦蓄和地下水的超量开采势必会造成径流量的衰减。

3.3 工农业发展的影响

随着社会经济的快速发展,流域内工业生产取得了巨大的成就,经济总量快速增加。采矿业、选矿业、矿山机械业非常发达,经济总量快速增加。2007年上游承德市实现工业生产总值240.00亿元,年平均增长率16.0%。按照国家发展和改革委员会、水利部、国家统计局联合公布的2005年各地区万元工业增加值用水指标中的数据计算,2007年流域内工业用水量为1.30亿 m^3,相当于当年径流量的25%。此外,每年还有迅速增加的趋势。这些企业用水大多取自天然河道,直接导致了流域内径流量的减少。

流域内耕地总资源494.57万亩(1亩=1/15 hm^2),常用耕地面积392.28万亩,其中:水浇地面积176.93万亩,水田面积31.18万亩。中国水利部的数据显示,2007年中国农业用水量占总供水量的61.9%,农业耗水量占总耗水量的74.6%,其中干旱地区耗水率普遍大于湿润地区,农业耗水率高达62%,目前中国每生产1 kg粮食平均需要消耗1 300 kg水。大量的农业用水是流域内径流量减少的另一个重要原因。

3.4 水土保持措施的影响

各种水土保持措施的应用也是对径流产生影响的重要原因,退耕还林、还草,围山转工程,使得流域植被、地形、土壤等下垫面结构有了较大改变。林业对径流的影响主要体现在:

一是对降水有一定的截留作用,这部分的降水大部分要蒸散发掉;二是枯枝落叶层和发达的根系具有调蓄水量的能力,其入渗能力比荒地大,从而增加了降水过程的入渗损失量。截止到2007年,流域内林地面积2 769.08万亩,林地覆盖率46.72%,较20世纪80年代中期增加了近1倍多。

梯田、围山转工程属坡地治理措施,对于中、小洪水其拦蓄作用主要表现在:一是减缓水流速度,延长汇流历时,增加入渗损失。二是土壤结构发生变化,土壤下渗能力会有所增加。根据1999年全国第二次水资源调查结果,梯田的蓄水能力较坡地的蓄水能力要增加2.5%,相当于多拦蓄了8.0 mm的降雨。三是带梗的梯田起到了塘坝的作用,能够拦蓄一定的地表径流,分析表明一般带地埂的梯田一次可拦蓄20~100 mm的降雨,不带地埂的梯田只能拦蓄10~20 mm。

4 结语

综上所述,影响滦河潘家口以上流域降雨径流关系变化主要因素是20世纪末自然环境的变化以及各种水利工程、水土保持工程的修建,导致了滦河流域潘家口水库以上地区径流量逐年衰减。把这些资料混杂在一起作为一个样本进行频率分析,势必会影响到成果的可靠性。降雨径流系列的一致性的改变,已经不适应水文分析计算和水文预报要求。因此,在径流量长系列频率计算分析时,按上述影响径流量因素对整个系列进行还原和修正,力求使样本系列具有同一的分布规律,这样才能减小水文计算和预报的误差。

参考文献

[1] 刘光文.水文分析与计算[M].北京:水利水电出版社,1989.

[2] 于维忠.水文学原理[M].北京:水利水电出版社,1988.

[3] 何晓群.应用回归分析[M].2版.中国人民大学,2007.

【作者简介】 于宝军,男,1968年8月生,河海大学,工程硕士,高级工程师,主要从事水文测验与预报管理工作。

黄河寺沟峡水电站发电取水可靠性分析

安乐平[1] 王 宏[1] 彭建军[2] 李茹虹[1] 李儒泉[1]

(1. 黄河水利委员会天水水土保持科学试验站 甘肃 天水 741000;
2. 浙江省湖州工程疏浚处 浙江 湖州 313000)

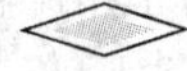

【摘要】 寺沟峡水电站是黄河上游干流梯级开发的第13座水电站,河道来水量能否满足电站设计发电用水要求,取决于上游来水量及水库调度。本文在分析电站上游来水情况的基础上,选择采用"典型年"法对寺沟峡水电站发电用水与来水情况进行了对比分析,得出了取水可靠性与可行性分析结论。

【关键词】 取水 可靠性 寺沟峡水电站

1 概况

黄河寺沟峡水电站(见图1)位于甘肃省永靖县与积石山县交界处黄河干流上,是黄河上游龙羊峡—青铜峡段梯级开发的第13座电站。电站坝址上距规划的大河家水电站29.5 km,下距刘家峡水电站44.5 km。地理位置(全库)东经102°45′~102°05′,北纬35°45′~35°55′,电站装机容量为240 MW,多年平均年发电量9.74亿kWh,水库总库容4 794万m^3,正常蓄水位1 748 m,工程规模为三等中型。

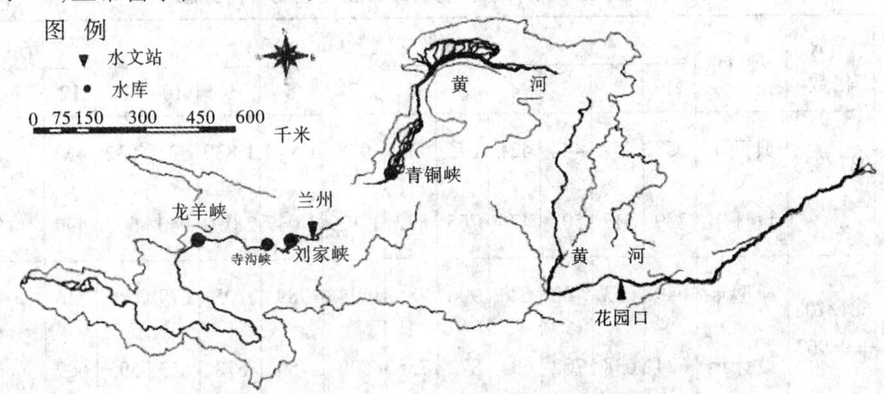

图1 寺沟峡水电站工程地理位置示意图

2 电站发电取水

2.1 循化站天然径流量

电站上游来水量分析采用循化站1956~2000年还原后的天然径流成果,计算天然流量系列的经验频率,采用P-Ⅲ型频率曲线适线,得出多年平均流量为704 m^3/s,年径流量

222.0亿 m^3，循化站 1956～2000 年（天然情况）不同频率平径流量见表 1。

表 1　循化站 1956～2000 年（天然情况）不同频率年径流量

保证率 P(%)	10	20	50	75	90	95
k_p	1.32	1.19	0.98	0.83	0.72	0.65
流量(m^3/s)	931	835	681	581	508	472
径流量($\times10^8$ m^3)	293.6	263.3	214.8	183.2	160.2	148.8

2.2　受水库调节影响的径流年内分配

考虑到寺沟峡水电站发电取水受上游梯级水库调节影响，采用 1986 年上游龙羊峡水库建成后循化站实测资料进行丰、平、枯分析计算比较接近来水现状，1986～2004 年循化站受水库调节影响多年平均流量年内分配见表 2。

表 2　循化站受水库调节影响多年平均流量年内分配（1986～2004 年）

月份	1	2	3	4	5	6	7	8	9	10	11	12	全年
1986～2004 年循化站实测流量(m^3/s)	466	471	491	463	596	677	719	677	639	536	520	480	562
年内分配系数(%)	6.9	7.1	7.3	6.9	8.8	10.1	10.7	10.0	9.5	7.9	7.7	7.1	100
受水库调节影响后流量(m^3/s)	583	600	617	583	743	853	904	845	803	667	650	600	704

根据对 1956～2000 年未受上游水库调节影响的不同保证率的循化站丰、平、枯水年平均流量和年径流量的计算分析，在上游龙羊峡水库建成运行后 1986～2004 年循化站实测径流系列中选择 1989 年、1994 年和 1992 年分别作为丰水年、平水年、枯水年进行调节计算。循化站受水库调节影响多年平均年内分配和丰、平、枯水年径流年内分配见表 3 和图 2。

表 3　循化站受水库调节影响设计典型年径流年内分配

典型年	缩放倍数	项 目	月平均流量(m^3/s)												年平均流量(m^3/s)
			1	2	3	4	5	6	7	8	9	10	11	12	
丰水年(P=10%) 1989 年	931/823 =1.131	典型年	653	457	444	421	676	771	998	1 810	1 830	831	532	433	823
		设计年	739	517	502	476	765	872	1 129	2 047	2 070	940	602	490	931
平水年(P=50%) 1994 年	681/708 =0.962	典型年	649	663	730	628	906	785	624	789	705	679	619	716	708
		设计年	624	638	702	604	872	755	600	759	678	653	595	689	681
枯水年(P=90%) 1992 年	508/511 =0.994	典型年	676	410	364	389	554	689	659	629	393	229	499	631	511
		设计年	672	408	362	387	551	685	655	625	391	228	496	627	508

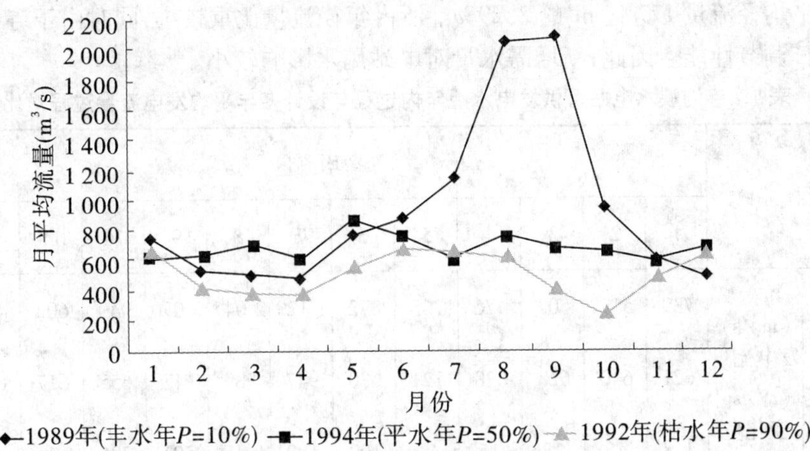

图2　受水库调节影响后设计典型年月平均流量过程

3　可供发电水量分析

寺沟峡水电站可供发电水量就是坝址处上游来水量。坝址处实际来水量为循化水文站实际来水加上循化站—寺沟峡坝址区间来水和区间支流来水,再减去区间工农业用水。考虑到循化站—寺沟峡坝址区间面积小,仅占坝址以上流域面积的1.1%,径流模数相对较小,区间亦无较大支流汇入,加入水量甚微;同时,通过对该河段的现场调查,区间无较大的工农业取水口。所以,循化水文站的实际来水量就是坝址处的实际来水,可认为是可供发电水量。

3.1　来水与用水过程比较分析

来水过程决定用水过程,电站来水量和来水过程受上游龙羊峡等水库年调节影响,加之上游水库相继建成运行,很难通过上游水库调度方式进行寺沟峡水电站可供发电水量计算,认为受梯级水库影响下,循化水文站的实测水文资料已经包含和反映了上游梯级水库的调节影响。同时,今后水电站运行是在受梯级水库调节影响下运行的。因此,可选用受梯级水库调节影响后实测水文资料分析不同保证率来水量,受梯级水库调节后循化水文站来水量就是可供发电水量。将寺沟峡水电站可供发电水量丰、平、枯设计典型年的年内过程与电站设计多年平均发电流量进行比较,分析电站发电用水可靠性。

从表4计算看出,电站上游多年平均来水量为704 m^3/s,设计多年平均发电用水量为664 m^3/s,多年平均来水量满足电站发电用水量要求,从各月多年平均来水量与发电用水量比较看,仅有1月、2月、3月、4月、9月、12月来水量均小于发电用水量,但除9月相差146 m^3/s外,其他各月差值相对较小,说明电站设计发电流量过程基本合理。

对典型年调节计算后丰、平、枯流量与电站发电流量比较,平水年($P=50\%$)流量为681 m^3/s,相应径流量为202亿m^3,与电站设计发电流量658 m^3/s和207.5亿m^3的径流量相差不大,能够满足发电用水量要求。

3.2　规划需水量对电站用水影响分析

根据坝址所在区间现状用水量及规划需水量分析,2010年龙羊峡—兰州区间规划需水量按照2000年基准水平计算为4.06亿m^3,较现状(2000年)用水量0.53亿m^3增加6.66倍。规划需水量约占循化站多年平均天然径流量224.6亿m^3的1.83%,占循化站受水库调节后多年平均径流量 177 亿 m^3 的 2.29%，所占年径流量比重较小，同时由于寺沟峡水电站水库具备日调节性能。因此，新增需水量对电站用水影响较小。

表 4　寺沟峡水电站可供发电水量年内过程与设计多年平均发电流量过程对比

项目		月平均流量												年平均流量（m^3/s）
		1	2	3	4	5	6	7	8	9	10	11	12	
设计年（P=10%）	来水量（m^3/s）	739	517	502	476	765	872	1 129	2 047	2 070	940	602	490	931
	发电流量（m^3/s）	621	640	628	628	721	680	947	989	1 479	655	617	625	769
	变化量（m^3/s）	118	-123	-126	-152	44	192	182	1 058	591	285	-15	-135	162
设计年（P=50%）	来水量（m^3/s）	624	638	702	604	872	755	600	759	678	653	595	689	681
	发电流量（m^3/s）	665	676	646	628	419	463	531	962	806	743	689	668	658
	变化量（m^3/s）	-41	-38	56	-24	453	292	69	-203	-128	-90	-94	21	23
设计年（P=90%）	来水量（m^3/s）	672	408	362	387	551	685	655	625	391	228	496	627	508
	发电流量（m^3/s）	601	612	655	584	517	537	556	561	562	575	424	593	565
	变化量（m^3/s）	71	-204	-293	-197	34	148	99	64	-171	-347	72	34	-57
多年平均	来水量（m^3/s）	583	600	617	583	743	853	904	845	803	667	650	600	704
	设计多年平均发电流量（m^3/s）	629	643	643	613	552	560	678	837	949	658	577	629	664
	变化量（m^3/s）	-46	-43	-26	-30	191	293	226	8	-146	9	73	-29	40

4　取水可靠性与可行性分析结论

4.1　上游天然来水量

通过对黄河循化水文站 1956~2000 年天然径流系列的分析可知，循化站多年平均流量 704 m^3/s，年径流量 222.0 亿 m^3；保证率 P=10% 的流量 931 m^3/s，径流量 293.6 亿 m^3；保证率 P=50% 的流量 681 m^3/s，径流量 214.8 亿 m^3；保证率 P=90% 的流量 508 m^3/s，径流量 160.2 亿 m^3。

4.2　受上游水库调节影响的来水量

通过对循化站 1986~2004 年实测资料的分析，受上游水库调节影响的循化站年平均流量 562 m^3/s,年径流量 177 亿 m^3。循化站年径流量通过上游龙羊峡等水库调节,年内分配趋于均匀化,有利于枯水期电站出力。

由于黄河上游来水受龙羊峡等水库调节,对寺沟峡水电站的发电用水是有利的。1956 ~2000 年未受水库调节影响的流量与电站发电流量比较,年平均流量满足发电流量要求,但 1 ~5 月,11 月、12 月流量不能满足发量流量要求;对典型年调节计算后丰、平、枯流量与电站发电流量比较,平水年($P=50\%$)流量为 640 m^3/s,相应径流量为 202 亿 m^3,与电站设计发电流量 658 m^3/s 和 207.5 亿 m^3 的径流量相差不大,基本能满足电站设计流量。

4.3 循化站径流变化分析

自 1986 年龙羊峡等水库相继建成蓄水后,循化站径流年内变化趋于均匀化,有利于水能资源开发利用。循化站汛期(7 ~10 月)22.4% 的水量经水库调节转移到 11 月、12 月和次年 1 ~5 月。

参考文献

[1] 汪岗,范昭. 黄河水沙变化研究[M]. 郑州:黄河水利出版社,2002.
[2] 冉大川,刘斌,王宏,等. 黄河中游典型支流水土保持措施减洪减沙作用研究[M]. 郑州:黄河水利出版社,2006.
[3] 赵有恩,等. 水土保持坡面措施减洪指标体系的研究[J]. 土壤侵蚀与水土保持学报,1996,2(2):50-56.
[4] 水利部水资源管理中心. 建设项目水资源论证导则[M]. 北京:中国水利水电出版社,2003.

【作者简介】 安乐平,1977 年 11 月生,本科,现任黄河水土保持生态环境监测中心天水分中心副主任,工程师,主要从事建设项目水资源论证、小流域水土流失监测、水土保持科学研究等工作。

考虑多种致灾因子的排涝标准研究*

谢　华　李铁光[2]　黄介生[1]

(1. 武汉大学水利水电学院　武汉　430072；
2. 中国灌溉排水发展中心　北京　100054)

【摘要】 以平原河网地区的排涝问题为例，引入涝灾风险率评估多致灾因子条件下的涝灾发生概率，应用Copula函数建立了暴雨和外江水位的联合概率分布模型，研究了在暴雨和外江水位共同作用下的涝灾概率，并用于排涝工程规模的调算。结果表明，考虑多种致灾因子共同作用下的涝灾概率更能真实地反映涝灾实际发生的概率；基于Copula函数构建的多变量联合概率模型，可以很方便地计算多种致灾因子的各种量级组合下灾害发生概率；对于相同的涝灾风险率，致灾因子有多种不同的组合，不同的荷载组合得到的工程规模不同，排涝工程的设计应按照对排水最不利组合确定工程规模。

【关键词】 涝灾　概率分析　多种致灾因子　Copula函数

我国的沿江平原地区和三角洲河网地区大多地势低平，暴雨涝水能否顺利排除受到外江水位的极大制约。比如一次降雨可能超过了设计暴雨标准，但如果此时遭遇的外江水位很低，这种情况下虽然降雨超出了设计标准，但不一定会造成破坏，涝灾并不一定会发生。在某些情况下虽然降雨没有超过设计标准，但如果此时外江水位很高，暴雨涝水不能顺利排出，涝灾仍然会发生。显然，在这样的地区，涝灾的发生不单纯取决于暴雨这一单一的水文事件，往往取决于暴雨和外江水位等水文荷载整体达到或超过某一极限水平。

当前我国相关规范大都规定"治涝设计标准一般应以涝区发生一定重现期的暴雨不受涝为准"。这种以超过某一量级的暴雨出现的重现期作为排涝标准的方法，实质上只考虑了涝灾的一个致灾因子——暴雨。现状调算排涝工程规模时，通常设定一定频率的外江水位(常采用多年平均最高水位)结合设计暴雨标准作为排涝计算的依据，其本质是假定了涝区设计最大暴雨与外江多年平均最高水位相遭遇，这种人为的假定虽然考虑了外江水位因素，但并没有真实地反映出各致灾因子共同作用下涝灾发生的概率，不符合涝灾发生真实状况。在这样的地区，排涝标准的确定应该综合考虑暴雨和外江水位的共同作用，在科学评估暴雨和外江水位遭遇概率的基础上制定适宜的排涝标准。本文以珠江三角洲河网地区的排涝问题为例，研究多种致灾因子共同影响下的涝灾概率，为确定科学合理的排涝标准提供依据。

* 基金项目：国家自然科学基金资助项目(50909074)；"十二五"国家科技支撑计划项目(2012BAD08B03)。

1 涝灾风险率

1.1 涝灾风险率定义

排涝设计标准的高低影响到治涝工程运行期内涝灾发生概率的大小,排涝标准越高,抗御涝灾的能力就越强,发生超标准涝灾的概率就越小。排涝工程大都由排水管网、沟渠、水闸、泵站、承泄河湖等组成,这些工程按照一定的排涝标准规划设计建设。当出现超过工程排涝能力的暴雨时,涝灾将会发生。因此,涝灾发生的概率就是排涝系统失效的概率。系统失效的风险率定义为"系统在其规定的工作年限内,不能完成预定功能的概率",一般可以概化为系统的荷载效应 L 和系统承载能力 R 之间的关系。当 $L<R$ 时,系统正常工作,灾害不会发生;反之,当 $L>R$ 时,系统将无法完成其功能,此时系统失效,灾害发生。由于存在各种不确定的随机因素,荷载效应 L 和承载能力 R 都是随机变量,因此系统失效 $\{R < L\}$ 是随机事件,其出现概率即为系统的风险率 $Risk$

$$Risk = P(R < L) = \int_r^\infty \int_0^r f_{RL}(r,l)\,\mathrm{d}r\mathrm{d}l \tag{1}$$

式中:$f_{RL}(r,l)$ 为系统荷载 L 和承载能力 R 的联合概率密度函数。

对于排涝系统而言,荷载通常指暴雨、外江水位等致灾因子;承载能力通常指泵站的流量、扬程、水闸的过流能力、河湖滞蓄能力等。由涝灾成因分析可知,荷载效应和承载能力一般是相互独立的,因此风险率的计算公式可改写为下式

$$Risk = P(R < L) = \int_r^\infty \int_0^r f_L(l) \cdot f_R(r)\,\mathrm{d}r\mathrm{d}l \tag{2}$$

实际上,对于一个确定的排水区域而言,通常不考虑承载能力的不确定性,则涝灾风险率与荷载效应的概率是相等的,即

$$Risk = P(L) = F_L(l) = \int_r^\infty f_L(l)\,\mathrm{d}l \tag{3}$$

沿江平原及河网地区涝灾的荷载效应由暴雨 X 和外江水位 Y 两个致灾因子共同决定,其联合概率分布函数表示如下

$$Risk = P(L) = F_L(l) = P(X > x, Y > y) = \iint f(x,y)\,\mathrm{d}x\mathrm{d}y \tag{4}$$

式中:X 为暴雨;Y 为外江水位;$f(x,y)$ 为暴雨和外江水位联合概率密度函数。

沿江平原及河网地区涝灾往往是暴雨和外江水位两个致灾因子共同超过某一量级的结果,通过式(4),可以将风险率转换为暴雨和外江水位的联合分布概率。

1.2 涝灾概率计算方法

显然,式(4)所表示暴雨和外江水位共同作用下的涝灾发生概率在数学上是一个两变量联合概率分布问题。两变量概率分布模型是解决两水文变量联合概率分布的有效工具。传统的两变量模型如两变量正态分布、两变量对数正态分布、混合 Gumbel 分布、Gumbel-Logistic 分布、两变量指数分布、FGM 模型等大多对变量的边际分布、相关关系等有较多的限制而受到局限。近几年来,Copula 函数作为一种优良的构建多变量概率分布模型的方法在水科学领域有较多的应用,其主要特点在于各单因子变量的边缘分布可以采用任何形式,变量之间可以具有各种相关关系。本文选用阿基米德族 Copula 函数中的 Gumbel-Hougard

Copula 函数构建两变量概率分布模型。

$$F(X \leqslant x, Y \leqslant y) = C(u,v) = C\{F_X(x), F_Y(y)\} = \exp\{-[(-\ln u)^{\theta} + (-\ln v)^{\theta}]^{\frac{1}{\theta}}\}$$

$$\theta \in [1,\infty) \tag{5}$$

式中：$C(u,v)$ 为联合概率分布函数（不超过概率）；$u = F_X(x)$，$v = F_Y(y)$ 分别为暴雨 X 和水位 Y 的边际分布函数；参数 θ 为 Copula 函数的参数，该参数与 Kendall's 秩相关系数 τ 有关，$\tau = 1 + 4\int_0^1 \frac{\phi(t)}{\phi'(t)}\mathrm{d}t$，其中 $\phi(t) = (-\ln t)^{\theta}$ 为 Gumbel-Hougard Copula 函数的生成函数。

实际工程中往往关心的是暴雨和外江水位大于某一量级的联合概率，即

$$P(X > x, Y > y) = 1 - F_X(x) - F_Y(y) + C(F_X(x), F_Y(y)) \tag{6}$$

通过式(5)和式(6)，可以很方便地分析不同量级的暴雨和外江水位共同作用下的涝灾发生概率。

2 实例分析

2.1 研究区基本情况

广州市番禺区地处珠江水系下游出海口，区内河网密布，地势低平，全区由大大小小的围区组成，围区四周水系环绕，各围区自成相对独立的排涝区。汛期暴雨期间，围外河网上有西、北江洪水下泄，下有海潮上涌，水位常常居高不下，围内低地涝水难以自排，往往积涝成灾。暴雨涝水只能在退潮时自排或由泵站抽排至外江，该地区的暴雨涝水的排除受到外江水位的极大制约。因此，暴雨和外江水位是该地区涝灾的两个最主要的致灾因子。根据番禺区内市桥气象站 1961～2000 年最大降雨资料和三沙口水文站年最高水位统计资料进行排频分析，不同设计频率的设计暴雨和设计潮位见表 1。

表 1　市桥气象站设计暴雨和三沙口设计潮位成果

类别	均值	特征频率					
		1%	2%	5%	10%	20%	50%
市桥气象站设计暴雨(mm)	142.7	355.6	315.5	262.1	221.3	180.6	125.4
三沙口水文站设计潮位(m)	1.86	2.57	2.45	2.28	2.16	2.02	1.81

2.2 暴雨和外江水位共同作用下的涝灾概率计算

番禺区现状排涝标准大都以 10～20 年重现期的暴雨为设计标准，并设定多年平均洪峰水位或 5 年一遇的最高水位为设计外江水位。究竟这种设计暴雨和设计外江水位遭遇的概率有多大，可以采用式(6)计算。Copula 函数方法不限定变量的边际分布型式，针对本文涉及的暴雨和外江水位两个水文变量，选用皮尔逊Ⅲ型作为边际分布模型。边际分布的参数估计结果见表 2。

表 2　皮尔逊Ⅲ型参数估计结果

水文系列	均值 μ	C_v	C_s/C_v
市桥气象站年最大降雨量	142.77 mm	0.42	4.441
相应三沙口水文站水位	1.136 m	0.31	0.617

根据 1961 ~ 2000 年最大降雨资料和相应三沙口水文站水位统计资料,计算得到 Gumbel-Hougard Copula 函数的参数,得到年最大 24 h 暴雨和相应外江水位联合概率分布函数如下

$$P(X \leqslant x, Y \leqslant y) = C(u,v) = \exp\{-[(-\ln u)^{1.15} + (-\ln v)^{1.15}]^{\frac{1}{1.15}}\} \tag{7}$$

根据式(7)得到的暴雨和外江水位联合概率分布函数,将其代入式(6),可以得到暴雨和外江水位的各种量级组合下涝灾发生的概率和条件概率。图 1 为根据式(6)所得的暴雨和外江水位大于某一量级的联合超过概率。

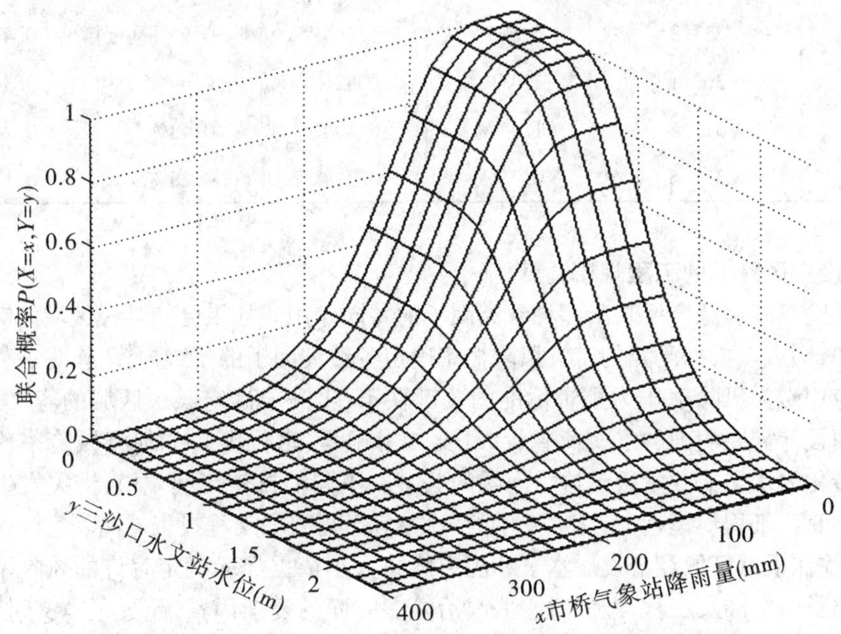

图 1　市桥气象站年 24 h 暴雨与三沙口水文站水位联合分布概率 $P(X > x, Y > y)$

2.3　现状排涝标准下研究区的涝灾概率分析

番禺区现状排涝标准多以 10 ~ 20 年重现期的暴雨为设计标准,并设定多年平均洪峰水位为设计外江水位。根据市桥气象站年最大降雨资料和相应三沙口水文站水位资料进行排频分析,三沙口水文站年最高水位多年平均值 1.86 m,5 年一遇外江水位 2.02 m,2 年一遇外江水位 1.81 m,年平均高潮位多年平均值 0.69 m,发生年最大 24 h 降雨时相应的外江水位多年平均值 1.136 m。10 年、20 年一遇的降雨量分别为 221.3 mm、262.1 mm。将以上各特征水位及频率列入表 3。采用式(6)和式(7)计算暴雨和外江水位共同超过某一量级的概率 $P(X > x, Y > y)$,结果见表 3。

由表 3 的计算结果可知,10 年、20 年一遇的降雨量与多年平均最高外江水位遭遇的概率分别为 $P(R>221.3\ \text{mm}, WL>1.86\ \text{m}) = 0.95\%$ 和 $P(R>262.1\ \text{mm}, WL>1.86\ \text{m}) = 0.69\%$ 。结果表明,大于 10 年一遇 24 h 降雨量与大于多年平均最高外江水位同时发生概率为 0.95%,联合重现期为 105 年;大于 20 年一遇 24 h 降雨量与大于多年平均最高外江水位遭遇概率为 0.69%,联合重现期为 145 年。对于番禺区的排涝工程而言,如果按 10 年一遇暴雨与多年平均最高外江水位遭遇的标准调算排涝工程规模,当发生降雨量和外江水位均大于此设计标准的情况时,将超过工程的排涝能力,涝灾将发生,涝灾发生概率为 0.95%。

表 3　市桥气象站设计暴雨与三沙口水文站各特征水位联合概率 $P(X > x, Y > y)$

频率	降雨量 X(mm)	频率			
		99.99%	50%	41.6%	20%
		水位(mm)			
		1.136	1.81	1.86	2.02
50%	125.4	27.60%	2.40%	1.88%	0.72%
20%	180.6	12.35%	1.58%	1.27%	0.53%
10%	221.3	6.59%	1.16%	0.95%	0.42%
5%	262.1	3.47%	0.82%	0.69%	0.34%
2%	315.5	1.47%	0.49%	0.43%	0.24%
1%	355.6	0.76%	0.31%	0.28%	0.17%

2.4　基于涝灾风险率的工程规模调算

涝灾风险率能够综合反映出多种致灾因子的共同作用。如果已知某一区域能够承受的涝灾风险率,则可以根据已知的涝灾风险概率确定各致灾因子的量级规模,作为工程设计的标准。但应用涝灾风险率作为排涝标准的关键在于:①需要确定某一区域的容许风险率;②工程规模设计需要明确的水文荷载量级(比如暴雨量、水位等),如何将风险率这一概率表述方式转化为水文荷载量级是这一问题的关键。不同的排涝工程规模,涝灾发生的概率不同,通常而言,工程规模越大,涝灾发生的风险越低,需要的工程投资越高。实际确定灾害风险率时,往往根据区域经济发展水平和涝灾威胁大小依据经验确定容许的风险率。由于容许风险率是一个涉及工程、经济、社会、政治等多方面的复杂因素,超过了本文讨论范围,在此主要讨论如何应用风险率标准确定排涝工程规模。

对于相同的涝灾风险率,致灾因子有多种不同的组合,而不同的荷载组合得到的工程规模不同。以番禺区为例,为便于与该区域现状排涝标准作比较,此处假定该区域能够承受的涝灾风险率为0.69%(该值为超过现状设计暴雨和设计外江水位的联合概率),在此风险率标准下,设计暴雨和外江水位有多种不同的组合。

表4为设计联合概率0.69%(重现期145年)条件下暴雨和外江水位的各种不同组合。不同的致灾因子组合虽然具有相同的出现概率(0.69%),但由于荷载量级不同,不同组合调算得到不同的工程规模,究竟选用哪一对组合作为设计暴雨和设计外江水位,需要进行工程规模调算比较,通常按照最不利原则确定工程规模。为此,选取某一具体排涝片,进一步计算在涝灾风险率为0.69%条件下,不同致灾因子组合下所需的排涝工程规模,见表4。显然,在相同的涝灾风险概率条件下(0.69%),当出现20年一遇的暴雨(5%)与多年平均最高水位相遭遇时,对排涝泵站的要求最高,此即为最不利致灾因子组合。当前番禺区采用排涝标准为20年一遇设计暴雨与多年平均最高水位相遭遇,正是在这一标准下的最不利荷载组合。

表4 设计联合概率0.69%时暴雨、外江水位不同遭遇组合

项目	组合1	组合2	组合3	组合4	组合5	组合6	最不利组合
降雨频率	50%	20%	10%	5%	2%	1%	
降雨量(mm)	125.4	180.6	221.3	262.1	315.5	355.6	组合4
水位频率	20.0%	25.0%	32.0%	41.6%	82.5%	99.9%	
水位(m)	2.02	1.97	1.92	1.86	1.66	1.24	
泵站功率(kW)	420	683	905	1 032	976	825	

3 结语

自然灾害的发生往往是多种致灾因子共同作用的结果,涝灾也不例外。涝灾的发生不单纯取决于某一项水文极值,往往取决于同时作用的几种水文荷载整体达到或超过某一极限水平。沿江平原及三角洲河网地区大多地势低平,汛期暴雨涝水的排除受到外江水位的极大制约,暴雨和外江水位是该地区涝灾的两个最主要的致灾因子。因此,排涝标准的确定必须综合考虑暴雨和外江水位共同作用的概率,对涝灾发生的风险概率做出正确的估计。通过采用Gumbel-Hougard Copula函数建立暴雨和外江水位的联合概率分布模型,可以很方便地计算多致灾因子共同作用下涝灾发生的概率。涝灾风险与排涝标准关系密切,排涝标准越高,抗御涝灾的能力就越强,发生超标准涝灾的风险就越小,涝灾风险率与排涝标准和工程规模具有紧密关系。基于风险率的排涝标准能够反映多种致灾因子对涝灾的共同影响,克服了传统排涝标准只考虑单因素致灾因子的不足。对于相同的涝灾风险率,致灾因子有多种不同的组合,而不同的荷载组合得到的工程规模不同,排涝工程的设计必须按照对排水最不利组合确定工程规模。

参考文献

[1] 国家质量技术监督局,中华人民共和国建设部. GB 50288—1999 灌溉与排水工程设计规范[S]. 北京:中国计划出版社,1999.

[2] 上海市建设和交通委员会. 50014—2006 室外排水设计规范[S]. 北京:中国计划出版社,2006.

[3] 刘光文. 水文分析与计算[M]. 北京:水利电力出版社,1989.

[4] Chowdhary H, Escobar L A, Singh V P. Identification of suitable Copulas for bivariate frequency analysis of flood peak and flood volume data[J]. Hydrology Research, 2011, 42(2-3): 193-215.

[5] Roger B Nelson. An Introduction to Copulas[M]. New York: Springer, 2006.

【作者简介】 谢华,男,1975年9月生,工学博士,2007年毕业于武汉大学,现为武汉大学水利水电学院教师,副教授。

四、农业节水与农村供水

PAM与保水剂(施用方法)对土壤水分及春小麦生长的影响

任志宏　于　健　史吉刚　宋日权

(内蒙古自治区水利科学研究院　呼和浩特　010020)

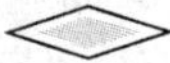

【摘要】 施用PAM和保水剂可减少土壤水分损失,提高土壤含水率,满足作物生长的水分需求。为了研究PAM和保水剂的施用对土壤水分及春小麦生长的影响,本文在河套灌区对PAM(1.5 kg/亩,分子量500万)、PAM(1.5 kg/亩,分子量1 200万)、保水剂(3 kg/亩)和保水剂(5 kg/亩)的4种施用方法进行了试验。结果表明:PAM和保水剂都提高了春小麦整个生育期内0~80 cm土层的土壤含水率,其中处理1的保水效果最显著。PAM和保水剂对各土层土壤水分的影响有差异,其中对0~10 cm土层的影响最为显著,相比对照,各处理0~10 cm土层的土壤含水率平均提高了21.41%;对30~40 cm土层的影响较小,各处理30~40 cm土层的土壤含水率平均提高了8.95%。PAM和保水剂对春小麦不同生育时期内0~80 cm土层的土壤水分影响不同,在春小麦生育期内的耗水高峰时期(五叶期、拔节期和孕穗期)各处理的保水效果比较显著。PAM和保水剂的施用促进了春小麦分蘖和根系向深层土壤生长,提高了春小麦的产量、千粒重,其中处理1的效果最显著,产量增幅29.18%。本文的研究结果可为PAM和保水剂在春小麦生产中的应用提供科学依据。

【关键词】 河套灌区　保水剂　PAM　土壤含水率　春小麦

0　引言

水资源的匮乏已经逐渐成为当今社会制约国家经济持续发展的最主要的因素之一。而农业作为国家经济的命脉,发展节水农业,已经成为当前农业生产中迫切需要解决的问题。而河套灌区作为我国最大的一首制自流灌区,是全国重要的商品粮生产基地,所以对于河套灌区农业的节水增产具有重要的研究意义。

保水剂(Super Absorbent 或 Super Absorbent Polymer, SAP)又称土壤保水剂、高吸水剂、保湿剂、高吸水性树脂、高分子吸水剂,是利用强吸水树脂制成的一种超高吸水保水能力的高分子聚合物。它能迅速吸收和保持自身重量几百倍甚至上千倍的水分,具有反复吸水功能,吸水后膨胀为水凝胶,可缓慢释放水分供作物吸收利用。20世纪60年代,美国农业部首先利用玉米制成淀粉接枝聚丙烯脂类保水剂,作为改善水分状况的重要工具在西部干旱地区推广应用,并取得了良好的效果。20世纪70年代以来,保水剂的研究与应用日益普及,日本在沙漠绿化、英国在水土保持、法国在土壤改良、俄罗斯在节水农业等方面保水剂的应用都取得了明显效果。中国从20世纪80年代开始研制和应用保水剂,现已在60多种作物上试验示范,应用面积超过7万 hm^2。

PAM（聚丙烯酰胺，Polyacrylamide）是一种水溶性线型高分子物质，是由多个单体经聚合而形成的。PAM 作为土壤改良剂，能维系良好的土壤结构，抑制土壤水分蒸发，提高土壤水分含量、孔隙度及阳离子交换量，降低土壤密度，减少土壤侵蚀，提高作物产量和水分利用效率，促进作物生长。20 世纪 50 年代初，英国的蒙萨特公司人工合成一种可瑞利母（Krilium）土壤结构改良剂，几年之后，由于成本太高而中断。80 年代，对土壤结构研究发现：土壤板结主要产生于地表 1～2 mm。这一发现对土壤结构改良剂再次应用奠定了重要基础。90 年代初，美国在地面灌溉中广泛采用 10 mg/kg 浓度 PAM 以增加灌溉水入渗量。以色列等国将 PAM 大量地应用于喷灌与旱作农业。

近年来，随着国民经济的发展，河套灌区灌溉用水愈趋短缺，干旱缺水与土壤结皮已成为河套灌区农业持续发展面临的主要问题。而利用保水剂、PAM 节水技术达到节水增产的目的是目前节水研究的一种新途径和新方法，但有关保水剂、PAM 在河套灌区的机理性研究、应用较少。春小麦是河套灌区的主要农作物，针对河套灌区灌溉水资源不足及土壤易结皮等问题，以春小麦为试材，探讨保水剂、PAM 在河套灌区对土壤水分和春小麦生长的影响，为今后在河套灌区大面积的应用、推广提供技术指导。

1 材料与方法

1.1 试验地概况

试验地位于河套平原西部的磴口县坝楞村，海拔为 1 048.7 m，为中温带大陆性季风气候，年均气温为 7.6 ℃，年日照时数为 3 209.5 h，植物生长期 5～9 月光合有效辐射为 1.68×10^5 J/cm^2，年降水量为 142.7 mm，年均蒸发量为 2 381.8 mm，干燥度为 4.08，无霜期为 136～144 d，年均风速为 3.0 m/s。该区域土壤为棕钙土和漠钙土。试验地土壤为灌淤土，灌淤层达 1.0 m 以上，土壤质地为壤土。试验地耕层土壤有机质含量约 10.0 g/kg，土壤全盐含量在 1.0 g/kg 左右，0～80 cm 土层土壤容重较为一致，平均为 1.48 g/cm^3，耕层田间持水量为 22.0%。供试小麦品种为永良 4 号，行距为 11.0 cm，播种量为 375.0 kg/hm^2。播种时施磷酸二铵 375.0 kg/hm^2，氯化钾 37.5 kg/hm^2。在小麦分蘖期、拔节期、孕穗期和灌浆期进行了灌溉，小麦整个生育期内不同处理的灌水量、灌水次数、施肥、追肥、除草等管理均相同。

1.2 试验设计

试验材料为 PAM 和保水剂两种，均由北京汉力淼新技术有限公司提供。试验共设计了 5 个处理，对照（CK，不施保水剂和 PAM）、处理 1（PAM，分子量 500 万，1.5 kg/亩）、处理 2（PAM，分子量 1 200 万，1.5 kg/亩）、处理 3（保水剂，3.0 kg/亩）、处理 4（保水剂，5.0 kg/亩），重复 3 次，小区面积 12 m× 8 m，总共 5×3＝15 个小区。

PAM 的施用方法采用干撒，具体步骤如下：

（1）对各试验小区进行种植前的平整。

（2）按试验设计进行作物的种植。

（3）将 PAM 按试验处理用量，按 1∶5 比例与过 2 mm 筛标准筛土壤均匀混合后，均匀地撒于试验小区表面。

（4）对小区进行浇水即可。

保水剂的施用方法是采用与土壤混施，具体步骤如下：

(1)将试验材料按试验处理用量,按1∶5比例与过2 mm筛标准筛土壤均匀混合后,均匀的撒于试验小区表面。

(2)人工将试验小区土壤翻至深度为10 cm左右,使试验材料与土壤均匀混合。

(3)对各试验小区进行种植前的平整。

(4)按试验设计进行作物的种植。

(5)对小区进行浇水即可。

1.3 测定项目

测定0~80 cm土层土壤水分。每隔10 d,用土钻每间隔10 cm土层采样1次,烘干法测定不同小区小麦行间0~80 cm土层土壤含水率(质量%)。用根钻法调查根系分布深度和根系生物量。成熟期以1.0 m^2为单位取样,测定小麦的穗数、无效蘖和株高、穗长、小穗数、不孕小穗数、穗粒数、千粒重和茎、叶、颖、籽粒生物量,计算不同处理小麦的经济系数。小区旁设有农田小气候监测仪,测定小麦生长期间的降水量。可根据不同处理的小麦生物量、籽粒产量和小麦生长期间的有效降水量、灌溉量,计算不同处理的田间耗水量、水分利用效率。

1.4 数据处理

试验结果均为3次重复的算术平均值,且所得的数据应用统计学及相关的数理统计分析软件(SAS 9.0)进行处理。

2 结果与分析

2.1 PAM和保水剂对不同土层土壤含水率的影响

图1为施用保水剂和PAM后,整个生育期内0~80 cm不同土层土壤含水率变化过程。在不同的处理下,对土壤含水率的影响差异较大。整个生育期内PAM和保水剂均能提高土壤的含水率,PAM相比保水剂的效果更显著,而且在土层上有较大差异,表层0~20 cm影响更明显,而且表现也比较稳定,总体土壤含水率在5月7日后效果更明显。

各试验处理整个生育期平均土壤含水率见表1,平均土壤含水率增长比例见表2。可以看出,0~10 cm土层和10~20 cm土层的土壤含水率主要受蒸发控制,施用了PAM和保水剂的处理均可显著提高土壤含水量降低土壤蒸发量。不同处理对土壤含水率的差异较大,均能提高表层的土壤含水率,而PAM的效果要比保水剂效果更好。由于在降雨或灌溉中地表土壤受水滴的冲击,造成地表土壤颗粒的破碎,产生地表结皮从而降低了土壤入渗,加快地表径流与土壤侵蚀的发生。保水剂虽能快速地储存水分,可难以形成有效的地面入渗可供植物根系的吸收。而PAM可增强土壤团聚体结构的稳定性,抑制土壤表面结皮,增加土壤入渗与减少土壤侵蚀。PAM对土壤的改良作用从微光上表现在分子链能在相邻的黏粒之间形成“搭接桥”,容易穿透土壤空隙,也可形成土壤颗粒之间搭接,产生较好黏结效果,从而有效地保持了土层间的水分,减少了水分流失。20~30 cm,30~40 cm的土壤含水率与0~10 cm,10~20 cm的土层差异较大。总体上比对照都有一定程度的提高,但受蒸发与作物吸收影响不如0~20 cm土层明显。20~40 cm土层是小麦根系最活跃的区域,由于小麦植株和根系的不断发达,植物体内的耗水量也就要不断变大,进行的光合作用和蒸腾作用逐渐变强,小麦根系在20~40 cm区域内从土壤中吸水量加大,从而60~80 cm土层土壤水分下降很快。由于施用PAM的土壤表层不仅提高了0~20 cm土壤含水率而且使土壤入渗加快，使得下层的土壤含水率得到了充分的补给，上、下层土壤水势差减小，下层土壤水分损失减小，使 40 ~ 80 cm 土壤含水率较对照出现了明显差异。

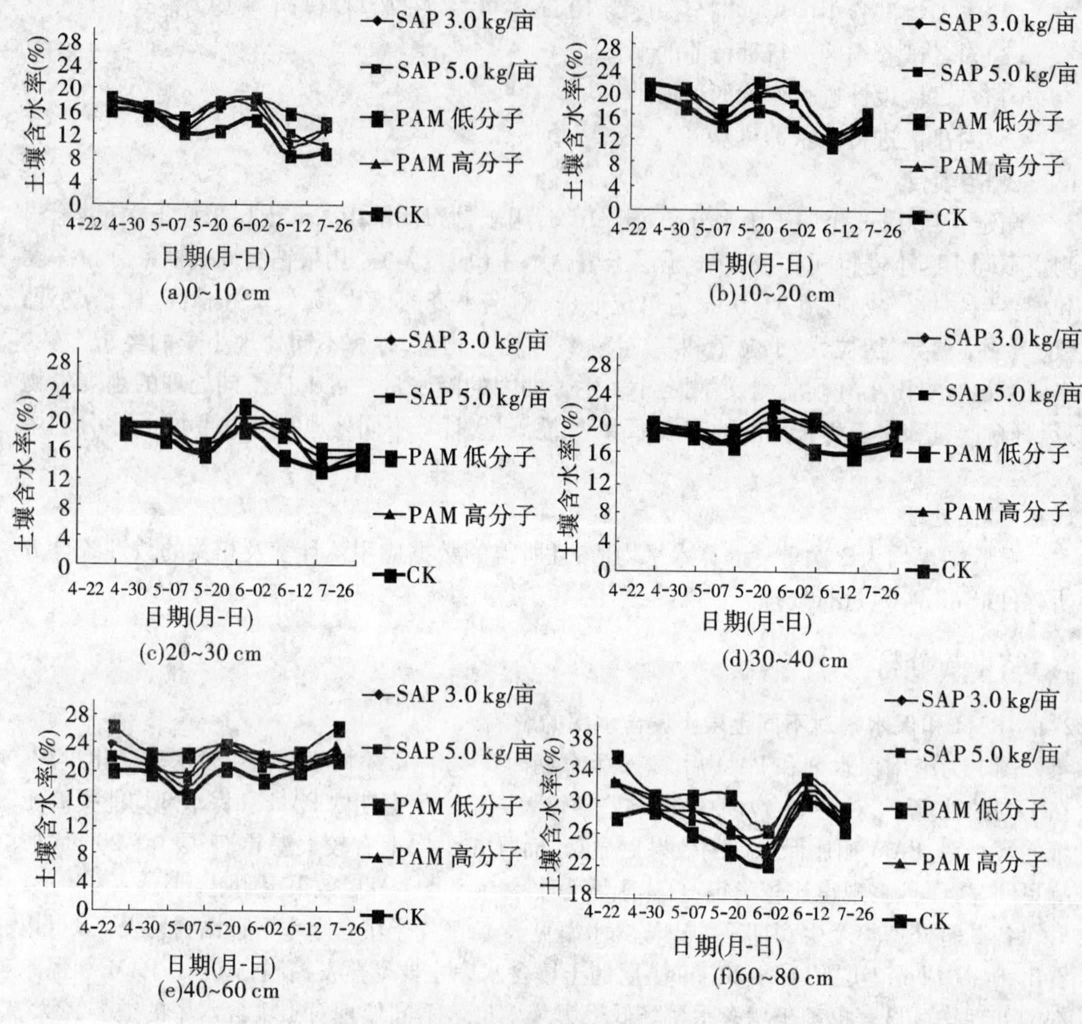

图 1　不同土层土壤含水率

表 1　各试验处理整个生育期平均土壤含水率

处理	不同土层平均土壤含水率(%)					
	0 ~ 10 cm	10 ~ 20 cm	20 ~ 30 cm	30 ~ 40 cm	40 ~ 60 cm	60 ~ 80 cm
处理 1	16.54a，A	18.75a，B	18.68a，B	20.32a，C	23.75a，D	31.00a，E
处理 2	15.03b，A	17.53b，B	17.57b，B	19.08b，C	21.88b，D	28.77b，E
处理 3	15.12b，A	17.81b，B	17.60b，B	18.85b，C	21.37b，D	29.09b，E
处理 4	14.84b，A	17.17b，B	17.51b，B	19.06b，C	21.18b，D	28.31b，E
CK	12.67c，A	15.41c，B	16.24c，B	17.74c，C	19.59c，D	26.55c，E

注：1. 同行中不同小写字母表示处理间存在显著性差异（$P<0.05$）。

2. 同列中不同大写字母表示土层间存在显著性差异（$P<0.05$）。

表 2 各试验处理平均土壤含水率增长比例

处理	不同土层平均土壤含水率增长比例=(处理-CK)/CK					
	0 ~ 10 cm	10 ~ 20 cm	20 ~ 30 cm	30 ~ 40 cm	40 ~ 60 cm	60 ~ 80 cm
处理 1	30.54% a	21.67% a	15.02% a	14.54% a	21.24% a	16.76% a
处理 2	18.63% b	13.76% bc	8.19% b	7.55% b	11.69% b	8.36% bc
处理 3	19.34% b	15.57% b	8.37% b	6.26% b	9.09% b	9.57% b
处理 4	17.13% b	11.42% c	7.82% b	7.44% b	8.12% b	6.63% c
平均值	21.41% A	15.61% B	9.85% D	8.95% D	12.53% C	10.33% CD

注:1. 同行中不同小写字母表示处理间存在显著性差异($P<0.05$)。

2. 同列中不同大写字母表示土层间存在显著性差异($P<0.05$)。

2.2 PAM 分子量和保水剂施用量对土壤含水率的影响

本试验区蒸发强烈,施用低分子量的 PAM 和施用量 3.0 kg/亩保水剂的土壤,提高了土壤的吸水能力和持水能力,效果比较明显。而施用保水剂 5.0 kg/亩与土壤混施后,保水剂均匀的分布于土壤中,在吸水过程中保水剂吸水较为充分,由于保水剂浓度较大,吸水膨胀后使土壤胀裂,甚至暴露于空气中,增大土壤孔隙,这样就会相对的加快土壤水分的蒸发速度,使保水剂的功效降低,对土壤含水率影响较小。PAM 低分子量要比 PAM 高分子量的效果更好,是因为低分子量的 PAM 由于分子链较短,较容易穿透土壤空隙,溶于水之后,有着很强的黏聚作用,也可形成土壤颗粒之间搭接,可有效地固结土壤颗粒,提高了土壤颗粒抗侵蚀能力,减少土壤侵蚀,增强土壤团聚体结构的稳定性,更好的抑制土壤表面结皮。

2.3 PAM 和保水剂对不同时期土壤含水率的影响

各处理在不同时期的不同土层土壤含水率见图 1。播种(4 月 22 日)到三叶期(5 月 7 日)降水偏少,不同处理 0 ~ 80 cm 土层土壤含水量较对照无显著差异。从五叶期(5 月 20 日)开始降水较多,不同处理 0 ~ 80 cm 土层土壤水含量极显著高于对照,且 PAM 低分子量效果更明显。拔节期是小麦快速生长期,消耗的水分相对较多,而此期间降水量相对较少,不同处理 0 ~ 80 cm 土层土壤水含量差异与五叶期相同,均显著高于对照。小麦孕穗期是小麦生长最活跃的时期、耗水的高峰时期,为水分临界期。同时,地表水分蒸发量增大,水分消耗较多,不同处理 0 ~ 80 cm 土层土壤含水量差异与五叶期、拔节期相同;PAM 低分子量极显著高于对照,效果较好。这是由于在此期间地表水分蒸发量加大,在土壤水分吸力的作用下,表层土壤水分的流失得到了有效的缓解,是作物利用的有效水,从而降低了土壤饱和导水率,其原因是在 PAM 溶解过程中低分子量的 PAM 分子链较短,可穿透进入土壤空隙并且能形成土壤颗粒之间的搭接,产生较好的黏结效果,使土壤中大孔隙不断减小而使土壤饱和导水率逐渐降低,从而提高了作物利用的有效水,减少了土壤中水分的流失。成熟期由于小麦的耗水量减少,较对照均无较明显的差异。

2.4 PAM 和保水剂对小麦生长的影响

不同处理均能提高小麦的出苗率、株高、穗长、千粒重、产量等,而施用 PAM 低分子的小麦产量较对照增幅较高,达到 29.18%。根系深度从表 3 知,施用 PAM 小麦的根系量较多,使小麦根系向深层土壤生长。

表 3　不同处理的小麦生物量及产量

处理	总株数（株/m^2）	株高（mm）	穗长（mm）	根系（g）	根系深度（mm）	千粒重（g）	亩产（斤）	增产（%）
处理 1	851b	66.59b	8.87b	2.98	38.56b	33.71c	606.45c	29.18
处理 2	802c	64.03ab	7.66a	2.76	34.12b	30.70ac	585.15ac	24.65
处理 3	764a	62.49a	7.46a	2.81	27.54a	30.14a	568.32a	21.06
处理 4	757a	62.87a	7.55a	2.66	26.18a	28.57b	517.27b	10.19
CK	747a	58.30c	6.75c	2.58	26.11a	24.59d	469.45d	

施用 PAM 的根系深度要比施用保水剂的要深。这是由于虽然保水剂有超强的吸水能力，但由于保水剂吸持的水分并不能完全被植物吸收利用，保水剂所吸收的水分的利用效率还取决于保水剂对水分的吸附力和植物的水分生理特征。植物的水分主要是通过根系从土壤中吸收获得。正常条件下，根系吸水容易，体内生理活动活跃，植株生长快。反之，植株的生长将受到抑制。小麦生长与土壤水分、土壤温度密切相关。幼苗期较高的土壤水分和较低的土壤温度利于春小麦根系生长。春小麦拔节前植株较小，土壤水分消耗主要以棵间蒸发为主，拔节后主要以蒸腾为主。而 PAM 可改良土壤结构，减轻土壤容重，改良土壤的通气状况，减少土壤蒸发量，提高土壤水分。春季施用 PAM 后提高了土壤水分，可减缓土壤温度上升速度，促进了小麦分蘖和根系向深层土壤生长。小麦播种及幼苗期为土壤消融阶段，春潮（泛浆）较重，土壤水分含量较高，土壤水分消耗较少，保水剂的蓄水、保水能力不能充分表现，土壤含水率与对照无显著差异。PAM 可调整土壤结构，在春潮严重的地块可改良土壤的通气状况，因此小麦生长，施用 PAM 的茎叶、根系生物量均高于其他处理。

2.5　对水分利用效率的影响

小麦生长期间有效降水为 52.61 mm，分蘖期、拔节期、孕穗期和灌浆期灌水量共计为 225 mm。从播种到收获不同处理的耗水量均低于对照，有较好的节水，减少无效蒸发的作用。但不同处理间无显著差异。不同处理的水分生产效率均极显著高于对照，且 PAM 高于保水剂。水分生产效率 PAM 低分子较 CK 增幅 36.28%，不同处理之间存在着显著差异，PAM 低分子效果更显著（见表 4）。

表 4　不同处理的耗水量及水分利用效率

处理	播前土壤贮水量（mm）	生育期降水（mm）	田间灌水量（mm）	收获后土壤贮水量（mm）	耗水量（mm）	水分生产效率	水分生产效率增幅（%）
处理 1	277.77	52.61	225	168.52	367.78	1.65	36.28
处理 2	277.77	52.61	225	187.6	386.86	1.51	25.01
处理 3	277.77	52.61	225	166.66	377.34	1.51	24.47
处理 4	277.77	52.61	225	193.15	362.23	1.43	18.02
CK	277.77	52.61	225	178.04	388.72	1.21	—

3　结论

整个生育期内 PAM 和保水剂均能提高土壤含水率，且 PAM 和保水剂对各土层土壤水分的影响有差异，其中对 0 ~ 10 cm 土层的影响最为显著，相比对照各处理 0 ~ 10 cm 土层的土壤含水率平均提高了21.41%;对30~40 cm土层的影响较小,各处理30~40 cm土层的土壤含水率平均提高了8.95%。PAM和保水剂对春小麦不同生育时期内0~80 cm土层的土壤水分影响不同,在春小麦生育期内的耗水高峰时期(五叶期、拔节期和孕穗期)各处理的保水效果比较显著,其中PAM的效果更明显。这是由于PAM可增强土壤团聚体结构的稳定性,抑制土壤表面结皮,增加土壤入渗和减少土壤侵蚀,使土壤中大孔隙不断减小而使土壤饱和导水率逐渐降低,有效地保持了土层间的水分,从而提高了作物利用的有效水,减少了水分流失。随着小麦植株和根系的不断发达,小麦根系在20~40 cm区域内从土壤中吸水量加大,使得土壤水分下降很快。施用PAM的土壤不仅提高了0~20 cm土壤含水率而且使土壤入渗加快,使得下层的土壤含水率得到了充分的补给,上、下层土壤水势差减小,下层土壤水分损失减小。

春季施用PAM和保水剂后提高了土壤水分,可减缓土壤温度上升速度,促进了春小麦分蘖和根系向深层土壤生长,提高了春小麦的产量、千粒重及水分生产效率,其中低分子量PAM效果最显著,产量增幅29.18%。

参考文献

[1] 武继承,郑惠玲,史福刚,等.不同水分条件下保水剂对小麦产量和水分利用的影响[J].华北农学报,2007,22(5):40-42.

[2] 吴娜,赵宝平,曾昭海,等.两种灌溉方式下保水剂用量对裸燕麦产量和品质的影响[J].作物学报,2009,35(8):1552-1557.

[3] 叶燕萍,郑枰,吴松,等.保水剂对岩溶石山土壤及金银花生长的影响[J].中国水土保持,2009(11):16-18.

[4] Lentz R D, Shainberg I, Sojka R E, et al. Preventing irrigation furrow erosion with small application of polymers[J]. Soil Sci Soc Am J, 1992, 56: 1926-1932.

[5] 韩玉国,杨培岭,任树梅,等.保水剂对苹果节水及灌溉制度的影响研究[J].农业工程学报,2006,22(9):70-73.

【作者简介】 任志宏,1985年生,毕业于内蒙古农业大学农业水土工程专业,在内蒙古水利科学研究院任职,主要从事节水灌溉方面的研究。

黄土高塬沟壑区果园水土流失治理模式探讨*

闫晓玲[1] 宋 静[2] 刘海燕[1]

(1. 黄河水土保持西峰治理监督局 庆阳 745000;
2. 黄河上中游管理局 西安 710000)

【摘要】 通过对黄土高塬沟壑区果园水土保持综合治理措施的研究,探讨水土流失治理模式,提出了果园覆草、果园生草、果园套种、旱地果园穴贮肥水地膜覆盖、黄土高原旱地果园土壤蓄水保墒技术、生态果园建设、AGRI SC 免深耕土壤调理剂的使用技术等,对黄土高塬沟壑区塬面防治水土流失提供理论依据。

【关键词】 黄土高塬沟壑区 果园 水土流失治理模式

黄土高塬沟壑区位于黄土高原的南部,是水土流失严重地区之一。防治水土流失,改善生态环境、开展水土保持是黄土高原生态农业建设,实现人与自然和谐共处、社会经济持续发展的核心。水土保持综合措施主要有梯田、造林、种草、建设淤地坝等,果园又是人工造林的一种,塬面水土保持措施主要以果园为主,既可保水保土,又是当地农民主要的经济来源。但果园管理不当还会造成大量的水土流失。因此,果园水土流失综合治理成了塬面治理的主要组成部分。果园水土保持综合治理措施包括果园覆草、果园生草、果园套种、旱地果园穴贮肥水地膜覆盖技术、黄土高原旱地果园土壤蓄水保墒技术、抗旱保水剂在果园中的应用技术、AGRI SC 免深耕土壤调理剂的使用等。

1 果园覆草法

黄土高塬沟壑区果园多因春季和晚秋干旱而影响果实发育,又因夏季雨水大而集中形成严重的地表径流,造成大量的水土流失。果园覆草后,不仅减少了土壤水分的蒸发,而且减轻了地表径流,减少了土壤养分的流失,保持了果园水土,提高了果园土壤的含水量。

通过覆草对果树和土壤的影响研究结合生产实践证明:覆草法在春、夏季对土壤具有明显的降温作用,前期低温推迟了果树的萌芽期和花期,这对果树根系生长不利,但延迟花期可以缓解该区花期霜冻的危害;夏季 7 月、8 月覆草比清耕法地温降低 3.4 ℃、5.2 ℃,可以有效延长根系的生长时间,促进果树根系的生长;在秋末 10 月覆草比清耕法提高地温 1.5 ℃,具有一定的保温作用,有利于果树根系的良好生长和养分积累。覆草法在苹果园应用后苹果产量高,株产量达到 53.7 kg,一等果率较高,达到 70% 以上。

果园覆草使土壤中的转化酶与尿酶活性增大,加速了秸秆转化和有机质分解,增加了土

* 基金项目:水利部科技推广计划项目(TG1017)。

壤中速效养分含量。覆草提高了地表温度，而且地温变化缓和有利于根系生长发育，保证对树体地上部分的供应；覆草可以减少地面蒸发，保持土壤水分，减少干旱对果实生长发育影响，特别对干旱果园意义更大；覆草使土壤有机质含量增加，容重降低，孔隙度增加，透气透水性能增强，改良了土壤，使根系密度增加，特别是吸收根数量增加，保证了果树生长发育需要，使新枝生长、坐果率显著提高。果园覆草是一项行之有效的地面管理措施，对于保持肥水优化果树根际环境十分有利，应予以推广应用。

覆草果园相对湿度增加而且杂草难以清扫，易发生病虫危害，因此应加强病虫害预测、预报，适时防治。果园覆草应坚持连年进行，间断覆草会引起树势衰弱。果园覆草只有在深翻熟化改良土壤前提下进行，才有最大效益。而且每年6月覆草效果最好，覆草前应施入少量氮肥（尿素）并浇水，覆草厚度不低于15 cm，草被上零星地压少量土，以防风吹和火灾。

总之，覆草法是一种较为理想的果园土壤管理制度，它能有效地减少土壤水分蒸发，提高土壤含水量，不仅是有灌溉条件果园节约用水，提高水的利用率的一条重要途径，也是旱地果园一项重要的保墒措施；同时能稳定地温，增加了土壤养分，每公斤麦秸腐熟后可转化成有机质8.5 g，同时提高了有效P、速效K的含量。但连续多年覆草可使果树根系上移，覆草量为37.5 t/hm^2，用草量较大，因此制约了覆草法在草源缺乏地区的推广应用。

2 果园生草法

果园生草法是一种果园土壤管理的自然生态模式，在国外已被广泛推广和应用。在我国，随着果园管理逐步规范、生态农业的日益推进，果园种植豆科牧草也越来越得到了重视，国家农业部已从2000年开始大面积推广以种植三叶草为主的果园生草技术。

2.1 果园生草的水土保持作用

果园生草具明显的防止水土流失作用。据资料介绍，生草地比裸露地地表径流减少40%～60%，减少土壤冲刷量30%～90%；种草木樨的地表径流量比裸露地减少43.8%～61.5%；种草较未种草的雨水渗透深度增加3～9 cm，地表径流减少27.5%，土壤冲刷量减少3.68倍。

黄土高原地区采用带状生草法。在果园行间或梯田地埂实行生草，留足树盘进行耕作施肥等。

2.2 果园生草的方法

果园生草以人工种植生草为主。主要掌握草种的选择和生草栽培技术。

2.2.1 草种选择

草种主要选择紫花苜蓿、草木樨、三叶草、小冠花、多年生香豌豆等，播种时可单播，也可几种草籽混播。

2.2.2 果园种草栽培技术

2.2.2.1 播种量

播种量的确定与种子的真实性、纯净度、千粒重、发芽率、密度、土壤厚度、墒情、整地质量、播种方法、幼苗保存率、地温、利用目的等许多因素有关，一般原则是：大粒种子应少播；种子品质好的要少播，品质差的要加大播种量；条播比撒播要节省种子20%～30%，穴播比条播更省种子；整地质量好可少播，质量差的要加大播种量；早春播种温度较低，不利于发芽，要适当加大播种量。几种适宜果园种植牧草及播种量见表1。

表1 几种适宜果园种植牧草及播种量

名称	播种量(kg/hm^2)	名称	播种量(kg/hm^2)
紫花苜蓿	11.25~15.00	毛苕子	45.0~60.0
红三叶	11.25~15.00	山黧豆	150.0~187.0
白三叶	7.5~11.25	羽扇豆	172.5~195
草木樨	15.0~22.50	豌豆	60.0~150.0
红豆草	60.0~75.00	大豆	60.0~75.0
紫云英	11.25~15.00	秣食豆	37.5~60.0
沙打旺	11.25~15.00	小冠花	0.5~1.0
箭舌豌豆	60.0~90.00	多年生香豌豆	75.0~90.0

2.2.2.2 播种期

果园种植牧草的最佳时期是春、秋两季,即4月中旬至5月上旬、8月中旬至9月中旬。播种方法是在果树行间翻地20~25 cm深,将地整平,灌水湿润后即可播种,撒播或条播均可,条播效果比撒播好。播种后覆盖上浅土,或覆盖地膜,不同牧草播种后出苗时间不同,一般需要10~15 d才能苗齐,有些牧草甚至要一个月才能苗齐,多年生香豌豆就要一个月。

自然生草是根据果园中自然生长的各种草类,对于有害的扯皮草、蓬草、蒿草等要及时拔除,再通过刈割留用。

2.2.2.3 果园生草的管理措施

果园种草后主要是水肥管理、刈割和草种更新。有条件的果园可采用喷灌或滴灌,也可用漫灌,雨量不足则需增加灌溉次数。生草果园应撒施肥料,特别是在春、夏季,需增加N、P、K的用量,苗期以N肥为主,后以P肥为主。每公顷增加过磷酸钙37.5 kg即可增加鲜草绿肥2 250 kg。果园生草5~7年后,草种老化,同时土壤表层形成一个盘根错节的"板结层",对果树根系生长和水肥吸收有影响,应进行草的更新。更新方法有应用除草剂、翻耕灭草、覆盖生草等。

黄土高塬沟壑地区降水量一般小于600 mm,应当采取行间生草、株间清耕覆盖的模式,选择生长量小或蔓生的牧草品种(如白三叶、多年生香豌豆);山旱地梯田果园应当选择耐旱生长量大的牧草品种(如小冠花种)在田埂边,定期刈割树盘覆盖;施肥可以采取穴施或肥水一体化的施肥方式,注意补充P肥;果树整形可采取高干(苹果、桃等)或棚架(葡萄)方式,便于割草机操作,避免草与果实接触而引起果面污染及果实病害。

2.2.2.4 果园可持续发展

果园生草是生产无公害果品,实现可持续发展的基本要求,传统的土壤清耕管理模式已严重制约这一目标的实现。长期单一施用化肥的清耕模式降低土壤肥力,在多雨地区还可引起土壤板结,引起水土流失,有机肥与化肥的配施对提高肥力有较好的作用,果园生草是增加果园有机肥源的重要措施。

果园生草可结合养畜、沼气建立可持续发展的生态果园。草被刈割后直接喂养牲畜,也可晒干做饲料,畜粪可发酵成沼气,沼液、沼渣可直接用做果园肥料。沼气是实现农村节能,增加有机肥源,有效控制农村环境污染,增加农民收入的有效途径。用沼气池内经过发酵的禽畜粪便沼液作为肥料，其N、P、K的含量均高于堆沤肥料，腐殖酸的含量比堆沤肥高出2.6%；腐熟的沼气发酵液含有植物所需的多种养分，如N、P、K和Cu、Fe、Mg、Zn等微量元素以及一些氨基酸（赖氨酸、色氨酸），还含有生长刺激调控物质如维生素、生长激素等。以沼液施肥，既能增加土壤的有机质含量，改善土壤结构，还能极大地提高果树的抗病能力，减少农药残留造成的污染，达到无公害产品的标准。果园生草还可以增加地面生物覆盖指数，减少裸露土壤，减少空气中沙尘含量，改善人居环境。在社会主义新农村建设中具有重要的作用。

3 抗旱保水剂在果园中的应用

KD 1型抗旱保水剂为含有强亲水性基团的高分子吸水树脂，通过其三维网状结构及分子内外侧电解质离子浓度差所产生的渗透压，对水有强烈的亲合作用，因而增强了土壤的蓄水能力，延长了土壤有效水的供应时间。

果树的生长发育受众多条件影响，使用保水剂能改善土壤的供水、通气环境条件，因而促进了果树的生长发育，从表2与果树生长及果实发育有关指标看，使用保水剂促进了新梢的生长和横径的加粗，叶面积和叶绿素相对含量也有一定的增加。营养生长的改善必然促进果实的生长发育，因而幼果的横径和单果重分别增加3.40%～18.02%和5.58%～24.87%，差异达显著性水平。而对总糖的影响变化不大，无规律性，这可能与单果重和产量提高的糖稀释效应有关。

表2 抗旱保水剂对果树生长及果实有关指标的影响

吸水树脂用量（g/株）	新梢长度（cm）	新梢粗度（cm）	叶面积（cm^2）	叶绿素指数	幼果横径（cm）	单果重（g/个）	总糖（%）
0	24.13	0.39	24.51	30.51c	2.94	197c	18.2
25	25.16	0.41	24.87	31.36c	3.08	204bc	17.9
50	27.18	0.42	26.26	34.37b	3.21	214b	18.4
75	29.32	0.44	27.35	36.27a	3.35	232a	17.9
100	29.5	0.45	29.01	36.28a	3.47	235a	18.1

干旱少雨造成的水资源危机对世界农业生产的影响日渐严重。随着黄土高原地区农业产业结构的调整，果树栽培的面积及比例都有大幅度提高，而新发展的相当面积无灌溉条件或灌溉较为困难，因而充分利用当地的有效降水，发展旱作农业或雨养农业将成为这些林果栽培的发展方向。使用抗旱保水剂，利用不断吸胀和释放的缓冲作用，改善根系生长的微域环境条件，促进果树的生长发育，使之产量提高。

4 黄土塬区果园套种技术

随着农业产业结构的调整，高塬沟壑区农村经济已由过去的二元结构逐步转向三元结构，果业已发展成为农村经济中的主导产业。近年来，果树种植面积不断扩大，未挂果的幼树果园在果树总面积中已占相当比例。在由农田转变为果园的过程中，土壤环境必然发生相应的变化。在果树从幼树向成龄生长的1～4年间，为了充分利用土地资源，增加收入，常采用果树间套种其他作物的方式，因而，土壤的理化性状及生物学特征也会因作物的条带性种植而受到影响。

在果树行间进行套种是保水、保土、保肥的有效手段。每年4月,在果树行间套种黄豆、豇豆、红豆、绿豆等豆类作物和绿肥,不仅可以适当增加收入,而且可以利用豆科植物的根瘤菌固氮作用,增加土壤有机质含量,有效改善土壤肥力。在每年的7~9月,当地雨量集中三个月的降水量平均占到全年水量的60%以上,覆盖夏季作物和绿肥,可以阻截雨水的击溅,有效防止土壤冲刷,减少土壤侵蚀量。在旱季,作物的覆盖又可以减少土壤的水分蒸发,有利于果园土壤温度的降低,营造良好的果树生存小环境。在10月份雨季结束时正好进行收割,豆科植物茎叶可埋入树下作绿肥使用。从生态环境保护与建设的角度出发,加强果园的水土保持工作,实现农业经济与环境的全面、协调和可持续发展。水土保持技术在生态果园建设中的合理应用,对于解决果园经营中的水土流失问题,实现农业的持续发展具有重要的意义。

5 旱地果园穴贮肥水地膜覆盖技术

穴贮肥水地膜覆盖技术在果园应用后,其作用主要有:

(1)提高土壤含水量。据试验测定,地膜覆盖地盘,是解决土壤干旱的有效措施,特别是在干旱年份,效果更为显著。它不但降低了年周期中水分的变化梯度,还保持了土壤的疏松和促进水、肥、气,热的协调平衡。

(2)提高春季地温,促进果实早熟,增加经济收入。地膜覆盖在4~5月可使5 cm深处土层的地温提高2.2~4.5 ℃,使根系生长、花芽分化提前。据资料报道,经地膜覆盖的果树,花期和果实成熟期:山楂提前7~13 d、桃提前9~16 d、苹果提前7~15 d。这样,果品上市早,市场竞争能力强,销售价格高。

(3)减少水土流失。一般果树栽植,株行距3 m×4 m,每亩按55株计算,一棵树冠下面挖3~5穴,每个穴上面用长2 m、宽0.8 m的地膜覆盖。这样,在一块地上,地膜覆盖面积占果园土地总面积的20%~33%,降落在地膜上面的雨水,一般都能通过中间洼、四周高、中间留有小孔的地膜进入插有草把的穴内,所以经穴贮肥水、地膜覆盖的果园,有20%~33%面积上的水土基本得到保护,集流就地蓄贮入渗,既为果树生长增加了水分,又减少了水土流失。

穴贮肥水地膜覆盖技术在旱地果园中经过应用,效果明显。它不仅使有限的肥水得到最大程度的利用,促进果树的生长并增加产量,而且适用多种果树,尤其对幼龄果树效果更好,是一项投资少经济效益高的技术措施;它还可改善土壤温、湿条件,提高地力,减少水土流失。该措施简便易行、收益大,很受群众欢迎。

6 果树株间覆盖地膜技术

利用幅宽1.4 m、厚0.06 mm的农用地膜,在果树两侧顺行覆盖两道地膜,即株间覆盖,行间清耕。覆膜宽度2.5 m,覆膜时间为4月上、中旬灌水后。覆膜前使用除草剂西玛津除草,后期长草用草甘磷防治,地膜每年覆盖1次。为防夏季高温,结合行间锄草将草撒在膜面,秋季施基肥时取掉地膜将杂草埋入树下作绿肥使用。

经试验测定,覆膜法可明显提高土壤温度,比清耕法平均提高3.4 ℃,地温越高,覆膜与清耕地温温差越大。覆膜果树比对照可提早发芽4~6 d;7~9月覆膜使地温增高,超过苹果树根系生长的最适温度 14 ~21 ℃，接近生长上限温度 30 ℃，使果树根系进入缓慢生长期。此期地膜上部覆草可明显降低地温，促进果树根系的生长；10 月以后增高地温有利于果树根系后期生长与吸收。覆膜法能提高苹果枝条的生长量，比清耕法枝条长 13.2 cm。据观测，连续多年覆膜，覆膜效果不如初次覆膜的效果好。通过对土壤管理方式中苹果产量和质量有关指标的观测统计以覆膜法苹果产量高，株产达到 52.5 kg，一等果率较高，在 70% 以上，增产效果显著。

覆膜法是幼龄果园一项较好的土壤管理制度，减少了土壤蒸发，提高了土壤含水率，可提高幼树栽植成活率 10% ~20%。提高有效养分含量，促进树体生长发育，可提早萌芽期 3 ~5 d。但土壤有机质矿化率高，有效养分含量降低快，应及时补施有机肥。综合分析，覆膜法仍不失为一种较好的土壤管理制度，特别适宜低温干旱地区的果园应用。

7 结论

果园地类的水土保持措施设置是黄土高塬沟壑区水土流失防治的重要内容。通过科学试验和生产实践，提出果园生草法、果园覆草法、果园套种法、旱地果园穴贮肥水地膜覆盖技术、黄土高原旱地果园土壤蓄水保墒技术、抗旱保水剂在果园中的应用技术等 6 种果园种植及水土流失防治技术模式，为治理黄土高塬沟壑区水土流失和果园发展提供了一定技术支撑。

参考文献

[1] 李怀有. 高塬沟壑区果园土壤管理制度试验研究[J]. 干旱地区农业研究，2001，19(4)：32-37.

[2] 王洪刚. 果园覆草技术综合效应研究[J]. 水土保持研究，2001，8(3)：55-57.

[3] 王守春，余新爱. 果园生草法和 AGRI SC 免深耕土壤调理剂的介绍[J]. 福建水土保持 2004，16(4)：31-32.

[4] 王贤，等. 牧草栽培学[M]. 北京：中国环境科学出版社，2006.

[5] 闫晓玲. 优良牧草多年生香豌豆在水土保持中的作用[C]//第三届黄河国际论坛论文集. 郑州：黄河水利出版社，2007.

【作者简介】 闫晓玲，女，1967 年生，甘肃镇原人，高级工程师，主要从事水土保持生物措施研究工作。

沙棘造林试验研究

张绒君[1]　脱忠平[1]　王志雄[1]　宋　静[2]　潘雪燕[1]　张风霞[1]

（1.黄河水利委员会西峰水土保持科学试验站　庆阳　745000；
2.黄河上中游管理局　西安　710000）

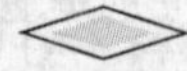

【摘要】　本文进行了不同栽植方法、不同整地方式、苗木定植初始土壤含水量及微地形等对沙棘造林成活率及生长量影响的试验研究。结果表明：在干旱、少雨、多风的黄土高塬沟壑区，沙棘定植初始的土壤含水量对沙棘造林成活率有着显著的影响，只有在土壤含水量大于13.8%时造林，或在造林时采取抗旱或灌溉措施，才能达到造林的目的，避免造林失败；在黄土高塬沟壑区采用鱼鳞坑整地就能满足沙棘造林要求；采用截杆造林能有效提高沙棘造林的成活率；沙棘的生长量与整地方式和微地形有关，其生长量随着山坡上部至下部的土壤水分含量的递增而依次提高。

【关键词】　沙棘　造林　效果

沙棘是我国北方干旱、半干旱地区优良的多用途树种，具有较强的耐旱和固氮能力，其果实、叶片富含多种活性物质与营养成分，广泛用于水土保持林、防风固沙林、饲料林、薪炭林和经济林营造，在生态环境建设与林业产业发展中具有不可替代的地位和作用。沙棘又是典型的克隆植物，其侧根在水平延伸过程中可产生大量的根蘖苗。但是，近年来由于黄土高塬沟壑区持续的干旱，造成沙棘的造林成活率偏低，究其原因主要是近年来黄土高塬沟壑区的人工造林季节常是干旱、少雨、多风的环境，土壤和大气同时干旱导致苗木栽植后失水大于吸水，苗木体内的水分平衡遭到破坏，最终导致苗木干枯死亡。针对这种情况，人们对沙棘纯林、沙棘乔木混交林、沙棘灌木混交林等造林类型进行了较多的研究，提高了沙棘造林的成活率和生长量，而对其他方面特别是苗木定植初始土壤含水量及微地形因素的影响研究相对薄弱，为此我们进行了不同栽植方法、不同整地方式、苗木定植初始土壤含水量及微地形对沙棘造林成活率及生长量影响的试验研究，为提高沙棘造林成活率及探求新的抗旱造林措施提供依据，加快黄土高塬沟壑区人工造林的步伐。

1　试验区概况

沙棘造林试验地选在砚川瓦项目区砚川瓦小流域老山村杨山的半阴坡和阳坡，坡度>25°，属黄土高塬沟壑区，处在砚川瓦流域的上游，地理坐标为北纬35°41′，东径107°30′，该流域植被稀疏，气候类型属于大陆性季风气候，平均海拔高度1 153 m，年平均日照时数为2 423 h，年均气温8.1 ℃，≥10 ℃年积温3 394 ℃，年平均降水量579 mm，季节分布极不均匀，主要集中在7、8、9月三个月，昼夜温差大，空气相对湿度低。土壤为黄绵土。无霜期160 d。

2 试验材料和方法

试验材料采用中国沙棘1年生实生苗、大小基本一致、生长健壮、根系良好的苗木。

2.1 苗木定植初始土壤含水量对沙棘造林成活率的影响

试验区选在砚川瓦小流域老山村杨山的半阴坡和阳坡,2008年春季在阳坡采用鱼鳞坑整地,半阴坡采用鱼鳞坑或水平沟方式整地,每种处理面积2 100 m^2,秋季造林时测定其土壤含水量,2009年10月调查其造林成活率。

2.2 不同整地方式对沙棘造林成活率的影响

试验区选在砚川瓦小流域老山村杨山的半阴坡,2008年春季采用鱼鳞坑、水平沟和不整地三种不同整地方式,每种处理面积2 100 m^2。2008年秋季造林,2009年10月调查其造林成活率。

2.3 栽植季节和栽植方法对沙棘造林成活率的影响

试验区选在砚川瓦小流域杨山的半阴坡,栽植前经过鱼鳞坑整地,2008年秋季或2009年春季造林,栽植前苗木经过截杆和不截杆措施处理,栽植密度为2 m×2 m,每种处理栽植50株左右,重复3次。2009年10月调查其造林成活率。

2.4 微地形对沙棘生长量的影响

试验区选在砚川瓦小流域老山村杨山的半阴坡,2008年春季采用鱼鳞坑或水平沟方式整地,每种处理面积2 100 m^2。2008年秋季造林, 2009年、2011年测定土壤含水量,10月调查其生长量。

3 结果与分析

3.1 不同整地时间的土壤含水量

2008年4月25日和11月23日,分别在砚川瓦小流域杨山的半阴坡进行了鱼鳞坑和水平沟整地,11月26日造林时进行了土壤含水量的测定,其结果见表1。

表1 不同整地时间的土壤含水量

取土日期	2008年11月26日	
整地时间	2008年4月25日	2008年11月23日
平均土壤含水量(%)	16.6	13.8
比较	120	100

由表1可以看出,整地后随即造林和整地后经过休闲的土壤含水量有着极大的不同,当造林前整地为100%时,提前半年整地则为120%,经过长期休闲后土壤含水量有了显著提高。整地后由于土层的翻动,土壤水分大量蒸发,这时土壤含水量降低,造林成活率低。因此在干旱半干旱地区进行造林,最好能提前一个生长季度整地,这样可使土壤有一定的休闲期,促使土壤熟化和增加土壤含水量,从而提高造林成活率。也就是说在春季和雨季造林,最迟应在前一年秋季整地;秋季造林,最迟应在当年春季或雨季整地。

3.2 苗木定植初始土壤含水量对沙棘造林成活率的影响

2008年春季在砚川瓦小流域老山村杨山的半阴坡和阳坡采用鱼鳞坑或水平沟整地,每种处理面积2 100 m^2,秋季造林时测定其根层土壤含水量。2009年1月调查其造林成活率,其结果见表2。

表2 苗木定植初始土壤含水量对沙棘造林成活率的影响

土壤含水量(%)	9.0	13.8	16.6	17.6
成活率(%)	8.0	83.5	89.3	96.1

注:土壤含水量为0~40 cm深土壤含水量的平均值。

由表2可以看出,沙棘定植初始的土壤含水量对沙棘造林成活率有着显著的影响,当定植初始土壤含水量为9.0%以下时,沙棘造林成活率只有8.0%,造林失败;当苗木定植初始的土壤含水量为13.8时,沙棘造林成活率为83.5%。当苗木定植初始的土壤含水量为16.6%时,沙棘造林成活率可达到89.3%。因此,我们在沙棘造林前应先测定土壤含水量,通过土壤含水量的大小来决定是否造林或在造林时进行灌溉,以提高造林的成活率。

3.3 微地形对沙棘生长量影响分析

微地形对沙棘生长量影响分析见表3。由表3可以看出,沙棘的生长量与整地方式和微地形有关。不同整地方式、同一坡面不同部位(微地形)的年平均土壤含水率不同,山坡上部至下部的土壤水分含量依次递增,水平沟整地较鱼鳞坑整地对土壤水分含量的提高显著。沙棘的生长量随着山坡上部至下部土壤水分含量的递增而依次提高,其主要原因是中国沙棘虽然具有较强的耐旱能力,但从起源上讲它属于湿生、中生类型,其生长对降水状况比较敏感。而地形对降水具有再分配作用,因此对沙棘的生长产生明显影响。

表3 微地形对沙棘生长量影响分析

整地方式	坡位	2009年			2010年		
		土壤含水率(%)	树高(cm)	地径(mm)	土壤含水率(%)	树高(cm)	地径(mm)
水平沟	坡上	14.46	64.9	7.85	14.14	126.1	16.06
	坡中	15.27	73.1	8.71	14.93	137.2	17.57
	坡下	15.97	79.4	9.26	15.76	148.2	18.44
鱼鳞坑	坡上	13.62	62.9	7.36	13.15	119.3	15.05
	坡中	14.62	68.4	7.94	14.08	127.3	16.09
	坡下	15.18	71.2	8.81	15.14	135.9	17.45

注:土壤含水率为4~10月0~80 cm深的平均数。

3.4 不同整地方式对沙棘造林成活率及生长量的影响

不同整地方式对沙棘造林成活率及生长量的影响见表4。由表4可以看出,鱼鳞坑、水平沟整地不但可以显著提高沙棘造林成活率,树高、地径、冠幅等年生长量也明显高于不整地,所以说整地是造林的一个重要措施,它具有蓄水保墒、将坡地变为小平地,减少水分蒸发,提高土壤含水率和造林成活率,促进幼林生长的作用。由表4还可看出,水平沟整地的造林成活率,树高、地径、冠幅等年生长量均高于鱼鳞坑整地,单从效果看,水平沟整地要好于鱼鳞坑整地,但其费用较高,从近年来水土保持投资情况和沙棘的造林情况来看,鱼鳞坑也能满足造林的需要。综上所述,在黄土高原进行沙棘造林采用鱼鳞坑整地即可。

表4 不同整地方式对沙棘造林成活率及生长量的影响

整地方式	成活率	费用(元/hm²)	2年高(cm)	3年高(cm)	地径(mm)	冠幅(cm×cm)
鱼鳞坑整地	89.3	5 500	67.5	127.5	16.20	86.5×89.8
水平沟整地	96.1	7 781	72.5	137.2	18.75	94.3×95.6
不整地	38.5	609	51.0	82.6	11.36	62.3×65.6

注：费用按每天50元计。

3.5 不同栽植时间和栽植方法对沙棘造林成活率的影响

截杆技术是截去杆的顶部，只留20 cm左右带根的基部，栽植方法同直栽，只留出顶部2 cm左右。2008年秋季和2009年春季栽植，2009年10月调查其造林成活率，结果见表5。

表5 截杆和不截杆对沙棘造林成活率的影响

栽植时间	栽植方法	调查株数(株)	成活率(%)	株高(cm)
2008年秋季	截杆栽植	100	96.6	70.7
	直栽(不截杆)	100	89.3	68.4
2009年春季	截杆栽植	100	94.8	66.7
	直栽(不截杆)	100	90.1	65.1

由表5可以看出，在干旱少雨多风的黄土高原区，春季造林和秋季造林的成活率变化不大，采取截杆措施可以有效地提高沙棘的造林成活率，其原因一是沙棘有较强的萌芽和再生能力，采取截杆造林更为便利和有效，二是黄土的透气性好，截杆造林后能有效预防多风引起的苗木失水，防止水分过快散失。调查中我们还发现，沙棘秋季直栽后第二年春季，地上部分大多枯干，多数从基部开始发芽，为了促进其生长，须将其地上部分剪掉，这样费工，而采取截杆造林的沙棘第二年春季发芽较直栽的要旺。因此，沙棘造林采用截杆造林较直栽要好，可在干旱少雨多风的黄土高原区及类似地区进行推广。

4 结论

(1)在干旱、少雨、多风的黄土高原区最迟应在栽植前一个生长季进行整地，这样可使造林地的土壤含水量提高20%，从而提高造林的成活率。因此，春季和雨季造林，最迟应在前一年秋季整地。秋季造林，最迟应在当年春季整地。

(2)沙棘定植初始的土壤含水量对沙棘造林成活率有着明显的影响，在沙棘造林前应先测定土壤含水量，只有在土壤含水量大于13.8%时造林，或在造林时采取抗旱或灌溉措施，才能达到造林的目的，避免造林失败。

(3)在黄土高塬沟壑区进行沙棘造林采用鱼鳞坑整地就能满足造林要求。

(4)在黄土高原干旱少雨多风的环境，采用截杆造林，能有效预防多风引起的苗木失水，防止水分过快散失，从而提高沙棘造林的成活率。且对于沙棘而言，由于其有较强的萌芽和再生能力，采取截杆造林更为便利和有效。沙棘造林时应根据当年的气象条件决定是春季造林还是秋季造林。

(5)沙棘的生长量与整地方式和微地形有关。沙棘的生长量随着山坡上部至下部的土壤水分含量的递增而依次提高。

参考文献

[1] 梁宗锁,孙群,王俊峰,等.干旱多风环境下沙棘移栽苗木致死机理与成活率的提高[J].沙棘,2003,16(1):14-19.

【作者简介】 张绒君,女,1962年12月生,1987年6月毕业于西北林学院森林资源保护系,本科,农学学士,高级工程师,主要从事水土保持规划、研究等工作。

不同灌溉条件下棉花水分生产函数的试验研究

杨凤梅

（塔里木河流域阿克苏管理局　阿克苏　843000）

【摘要】 通过田间小区试验,设置不同灌溉处理下棉花水分生产函数的试验研究。结果表明:①棉花产量和水分生产率均与棉花耗水量呈二次抛物线关系;②棉花边际产量和边际水分生产率均随着耗水量的增大而减小,且产量最大时的耗水量与水分生产率最大时的耗水量不完全一致;③提出棉花生育期适宜灌溉处理和土壤水分控制指标,对水资源紧缺的阿克苏河流域实现棉花节水与高产的统一具有重要意义。

【关键词】 棉花　水分生产函数　产量　耗水量　水分生产率

我国水资源严重不足,是国民经济发展,特别是农业持续发展的重要制约因素。灌溉是保证农业稳产和高产的主要手段,提高灌溉水的利用效率,发展节水灌溉是灌溉农业发展的方向。节水高效灌溉制度是根据作物的需水规律,把有限的灌溉水量在作物生育期内进行最优分配,达到节水的目的。而作物水分生产函数的研究为节水高效灌溉制度的制定提供了科学的依据。

水分生产函数反映作物产量随水量的变化规律,是进行科学的节水灌溉最基本、最重要的函数。在水源不充足的条件下,灌溉工程的评估、规划、设计、用水管理以及水资源开发利用规划、水利规划和灌溉经济效益分析计算,都要以它为依据。因此,发达国家从 20 世纪 60 年代起就进行专门研究,建立了很多作物水分生产函数模型,如 Wit(1958)建立了全生育期的初级模型,Stewatr 等人对该模型进行了改进。至 20 世纪 70 年代,已成为农田灌溉试验中最主要的课题,如 Jensen(1968)等建立了阶段缺水的乘法模型,Blank 等建立了加法模型和 Feddes 等建立了动态模型。我国从 20 世纪 80 年代起陆续有数省开展此项试验研究,取得了一系列研究成果与进展,不同学者亦对其进行了不同角度的研究和总结,例如作物水分利用效率分为叶片水平水分利用效率、作物群体水平水分利用效率和作物产量水平水分利用效率三种。其中,作物产量水平水分利用效率是优化灌溉策略和制定灌溉制度的理论基础,因影响作物产量的因素分为可控因素(如水量、品种、肥料等)和不可控因素(如气候等),在研究作物水分生产函数时,为避免诸多因素的影响,国内外学者假设作物的土、肥、气、热等处于相同条件,单独研究水和产量的关系,也就是主要探求可控制且数量有限的水分施加量与产量之间的函数关系,对于不能控制的和供应量不限的另外一些因素,一般视为特定条件下保持一致或者固定不变的因素,构成单因子的水分生产函数。

棉花作为支撑新疆农业发展的主要经济作物,其生产效益如何直接关系到农民脱贫致富、经济发展和社会稳定。而灌溉是新疆农业经济的基础和命脉，是用水大户，面临的形势必然是农业用水要减少，农业产量和产值要提高，出路只有一条，即提高农业用水效率和效益，具体而言就是提高作物水分生产率，即提高单方水消耗的作物产量。为明确提高的方向、途径、措施与办法，在新疆阿克苏灌区特设立了不同灌溉条件下棉花水分生产函数的专题试验研究，对于发展节水农业具有重要意义。由于受监测年数和监测方法的限制，我们这里主要是研究棉花产量水平上的水分生产效率，即棉花耗水量与产量的关系研究，为衡量新疆塔里木河流域阿克苏源流水资源利用状况和灌区的用水管理水平提供依据。

1 材料与方法

2005 ~ 2007 年，在新疆阿克苏地区灌溉实验站进行了为期三年的棉花水分生产函数试验。试区位于新疆阿克苏河流域中游阿克苏市西南方向 6 km 处，为天山南麓中西部、塔里木盆地西北缘的冲积平原，属极典型的干旱大陆气候特征，区内降水稀少而蒸发强烈，年均气温 10.16 ℃，年降水量 80.5 mm，年蒸发量 1 926 mm，是平原区作物和自然生态耗水的驱动因素。试区土壤质地为砂壤土，1 m 土体平均干容重为 1.48 g/cm^3，地下水埋深为 3 m 左右。试验区面积为 0.23 亩（1 亩 = 1/15 hm^2），布置两组（对照）共 2×5 个试验小区，每个试验小区约为 0.001 5 hm^2，1#（10#）为 3.5 m×3.5 m；2#（9#）为 3.5 m×4.5 m；3#（8#）为 3.5 m×5 m；4#（7#）为 3.5 m×4.4 m；5#（6#）为 3.5 m×4.4 m。布置如图 1 所示，1# ~ 5# 地为壤土，6# ~ 10# 为砂壤土，其中 9# 和 10# 砂层严重。每个小区四周均进行了 1.8 m 深的垂直防渗处理（土工膜隔离），以防止水的测渗，底部未作处理。

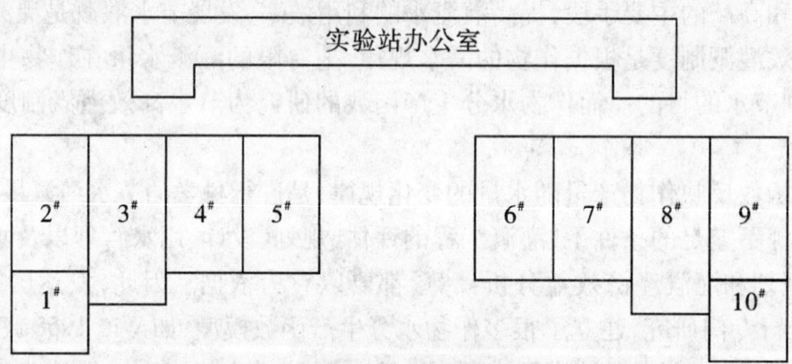

图 1 实验站小区布置示意图

针对棉花试验田进行固定每次灌水日期和灌水定额、采用不同灌水次数条件下作物耗水水分生产函数试验，两组均有 4 ~ 5 个不同的灌溉处理（次灌水量相同，灌水次数分别为 1 次、2 次、3 次、4 次、5 次），同一灌溉处理均有一个重复（对照）。每个小区中部均布设有一根 2 m 长的中子仪测管，主要采用中子仪法，作物生长期（3 ~ 10 月）每 5 d 测定土壤含水率一次，灌水前后和降水后应加测一次。在越冬期（11 月至翌年 2 月）每月观测一次。含水率测定深度：0.1 m、0.2 m、0.4 m、0.6 m、0.8 m、1.0 m、1.25 m、1.5 m、1.75 m、2.0 m。应用中子仪监测每个试验小区不同深度的土壤含水量变化过程，由水量平衡计算出作物的耗水量，收获时准确测定各试验小区的作物产量，较详细地了解农田供水和耗水过程，由此得出较科学和可靠的水分生产函数试验数据，综合分析作物的产量与耗水量的关系。灌溉水量由水表计算得出。棉花供试品种为中棉 35。

作物水分生产函数,是作物水分生产率 *WUE* 与耗水量 *ET* 的关系,或作物产量 *Y* 与耗水量 *ET* 的关系(见图2)。而传统的作物水分生产率定义为每单方水的消耗所生产的农作物(一般指粮食)产量。一般而言,作物产量 *Y* 随着耗水量的加大而增加,但不是直线增加。常用二次抛物线的经验关系来表示作物产量与耗水量的关系。因此,并不是作物的耗水量越大越好,从提高水资源的利用效益考虑,耗水量 *ET* 应控制在一定的范围内。

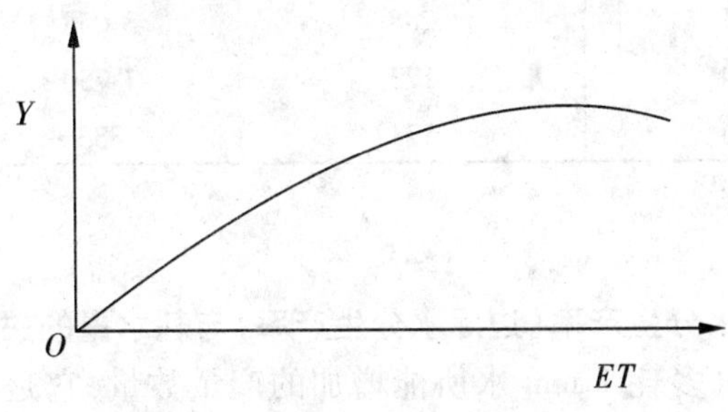

图2 作物产量 *Y* 与耗水量 *ET* 的关系

计算时可以用下式

$$WUE = Y/ET$$

根据农田水量平衡原理

$$ET = SW_1 - SW_2 + I + P \pm BW \tag{1}$$

则有

$$WUE = Y/ET = Y/(SW_1 - SW_2 + I + P \pm BW) \tag{2}$$

式中:*ET* 为作物的耗水量;SW_1 为作物播种前一米土层储水量,mm;SW_2 为作物收获时一米土层储水量,mm;*I*、*P* 分别为生育期灌溉水量和降水量,mm;*BW* 为计算土层与下层的水量交换。

由式(2)可见,要提高棉花水分生产效率,在实践中有以下几种途径:一是在棉花耗水量不变的条件下,提高作物产量;二是在产量不变的条件下,减少棉花耗水量;三是增加棉花耗水量的同时提高作物产量,但要使 *WUE* 提高;四是减少棉花耗水量的同时作物产量有所下降,但仍要使 WUE 提高。由式(1)和式(2)计算得出的棉花水分生产效率 *WUE* 如表1和表2所示。

试验小区棉花生育期不同灌溉处理、灌溉定额(包括播前水)以及作物产量(实收)见表1、表2。

表1 试验小区1#~5#灌水量、耗水量、产量及水分生产率

试验小区编号	灌水量 *I* (m^3/hm^2)	耗水量 *ET* (mm)	产量 *Y* (kg/hm^2)	水分生产率 *WUE* (kg/m^3)
1#	6 000	570	4 337	0.76
2#	4 950	455	4 663	1.03
3#	3 900	350	4 240	1.21
4#	2 850	255	3 475	1.19
5#	1 800	140	1 208	0.87

表 2 试验小区 6# ~10#灌水量、耗水量、产量及水分生产率

试验小区编号	灌水量 I (m^3/hm^2)	耗水量 ET (mm)	产量 Y (kg/hm^2)	水分生产率 WUE (kg/m^3)
6#	6 225	581	3 652	0.63
7#	5 100	411	3 391	0.83
8#	3 975	288	3 058	1.06
9#	2 850	173	1 750	1.02
10#	2 850	173	833	0.48

2 结果与分析

2.1 产量(边际产量)、水分生产率(边际水分生产率)与耗水量的关系分析

边际产量是指棉花每多耗 1 mm 水所能增加的产量数量,它是 $Y \sim ET$ 关系的一阶导数,即

$$M = \mathrm{d}Y/\mathrm{d}ET$$

边际产量说明了当水分投入增加时所引起的产量变动率。

根据试验小区不同灌溉处理下棉花水分生产函数实测资料分析,棉花产量(边际产量)与全生育期总耗水量、棉花水分生产率(边际水分生产率)与全生育期总耗水量的相关关系如图 3、图 4 所示。

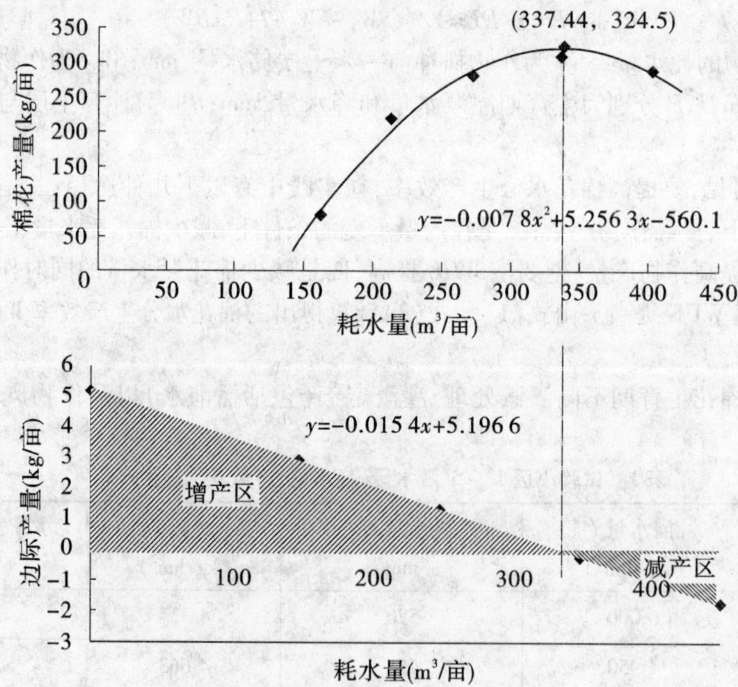

图 3 棉花产量和边际产量与耗水量的关系

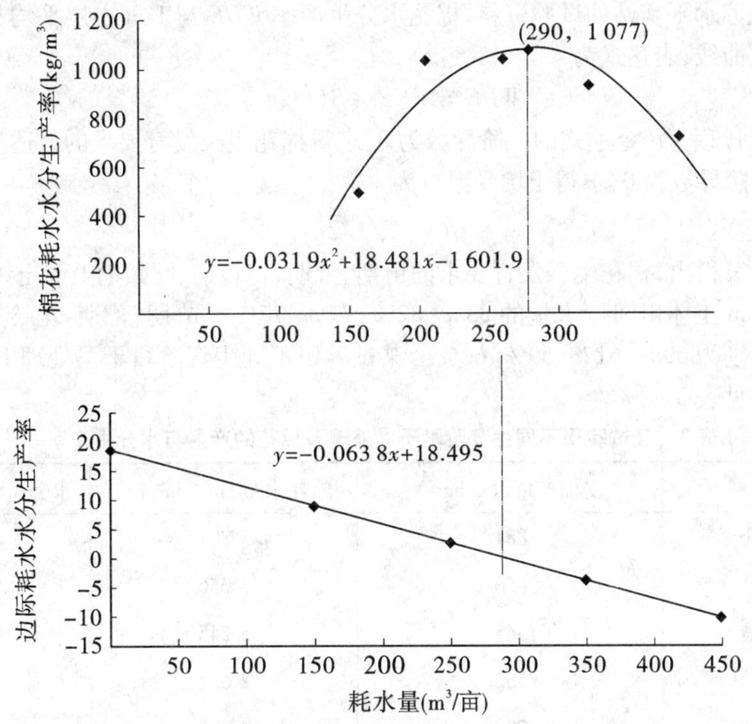

图4　棉花水分生产率和边际水分生产率与耗水量的关系

可见,棉花产量与耗水量之间呈现良好的二次抛物线关系,相关程度较高,其回归方程为

$$Y1 = -0.002\,5ET^2 + 3.187ET - 685(R^2 = 0.976)$$

由 $Y \sim ET$ 关系图可以看出,产量和耗水水分生产率与耗水量呈抛物线关系,随着耗水量的增加而增加,但不是直线增加,开始增加率大,随后增加缓慢,开始呈现"报酬递减"现象,到顶点时,棉花产量最高,此后再增大,不但不能增产,反而引起产量下降。

棉花水分生产效率 *WUE* 与耗水量 *ET* 的关系经回归得

$$MUE1 = -8\times10^{-6}ET^2 + 0.005\,6ET + 0.273\,5(R^2 = 0.967\,6)$$

由 $MUE \sim ET$ 关系图可以看出,*WUE* 随着耗水量的增加,也有一个渐增至渐减的变化过程,但 *WUE* 的最高点与产量的最高点并不一致,相应于 *WUE* 的最高点要低于相应于 *Y* 的最高点值。

可见,当边际产量为零时,耗水量为最高点。边际产量随着耗水量的增大而减小,直至耗水量为506 mm时,增产值为零;边际水分生产率也随着耗水的增大而减小,直至耗水量为435 mm时,边际耗水水分生产率为零,即棉花产量最大时的耗水量约为506 mm左右范围,棉花耗水水分生产率最大时的耗水量为435 mm左右范围,与产量最大的耗水量不完全一致。

2.2　棉花水分生产率与土壤水分状况的关系

大气降水、灌溉水和地下水均须转化为根系层土壤才能被作物有效利用。因此,提高作物水分生产率的几种途径都可能通过科学调控土壤水分得以实现。土壤水分不足或过多时均可影响棉花的正常生长发育,导致水分生产率降低。

根据阿克苏水平衡站的试验资料，棉花水分生产率 WUE 与 1 m 土层平均相对含水率 θ 符合指数二次曲线，表达式为

$$WUE = A\mathrm{EXP}[-B(\theta C)^2]$$

式中：A、B 为回归系数，令上式的一阶导数为 0，求得棉花生长发育要求的最适土壤含水率 $\theta = C$，令上式二阶导数为 0，求得土壤受旱点为

$$\theta = C - \sqrt{1/2B}$$

经拟合计算得出，棉花生长发育要求的苗期、蕾期、花铃期、吐絮期 1 m 土体最适土壤含水率分别为 1 m 土体田间持水量的 83%、82%、77%、77%。苗期、蕾期、花铃期、吐絮期对应的受旱点分别为 69%、61%、59%、62%。某试验田不同生育阶段干旱处理下棉花的产量与水分生产率见表 3。

表 3　某试验田不同生育阶段干旱处理下棉花的产量与水分生产率

处理	皮棉产量(kg/hm^2)	耗水量(m^3/hm^2)	水分生产率(kg/m^3)
适宜水分处理	881	5 801	0.152
苗期干旱	850	5 056	0.168
蕾期干旱	545	5 245	0.103
花铃期干旱	552	2 657	0.20
吐絮期干旱	708	4 985	0.142
蕾后花前干旱	410	4 289	0.095

可见，棉花生育期蕾后花前期干旱处理的产量和水分生产率最低，其次是花铃期和蕾期干旱处理。除去适宜水分处理，苗期干旱处理的产量相对较高，且耗水量明显低于适宜水分处理的耗水量，并且可获得较高的水分生产率。因此，在棉花苗期控制灌水，施加适当的水分胁迫，可促进根系下扎，有利于利用深层的水分和养分，形成合理的群体结构，减少蕾铃脱落，提高水分生产效率。从而得出，不同生育期阶段水分胁迫对棉花产量的影响为：蕾期>花铃期>吐絮期>苗期。

3　结论与讨论

棉花产量随作物耗水量的加大而增加，但不是直线增加，开始较大，随后逐渐减少，到一定程度后作物的产量达最大值，其后耗水量的增加产量不仅不增加，反而降低。所以，并不是作物的耗水量越大越好，从提高水资源的利用效益考虑，通过此次试验，得出砂壤土种植棉花的耗水量控制在 3 945 ~ 4 545 m^3/hm^2 为宜，灌溉定额控制在 445 ~ 454 mm 范围，为当地棉花高效高产节水提供理论依据。

棉花边际产量和边际耗水水分生产率均随着耗水量的增大而减小，且棉花产量最大的耗水量与棉花耗水水分生产率最大的耗水量不完全一致；棉花水分生产率是衡量棉花用水效率的指标之一，根据本流域水资源紧缺现状，在可利用水资源量一定的情况下，研究节水型农田灌溉技术，必须从已往充分满足作物对水分需要的研究中解放出来，从作物在某一时段缺水条件下，获得总的纯收益最大出发，实现棉花高产与节水的统一。

参考文献

[1] 沈荣开,张瑜芳,黄冠华. 作物水分生产函数与农田非充分灌溉研究述评[J]. 水科学进展,1995(3):248-253.

[2] 杨路华,夏辉. 非充分灌溉制度制定过程中 Jensen 模型的求解与应用[J]. 灌溉排水,2002,2(4):13-16.

[3] 夏辉,杨路华. 作物水分生产函数的研究进展[J]. 河北工程技术高等专科学校学报,2003(2):6-8.

[4] 彭世彰,边立明,朱成立. 作物水分生产函数的研究与发展[J]. 水利水电科技进展,2000(1):17-20.

[5] 中华人民共和国水利部. SL 13—2004 灌溉试验规范[S]. 北京:中国水利水电出版社,2005.

[6] 胡顺军,宋郁东,周宏飞,等. 塔里木盆地棉花水分利用效率试验研究[J]. 干旱地区农业研究,2002(3):66-70.

[7] 肖俊夫,刘祖贵,孙景生,等. 不同生育期干旱对棉花生长发育及产量的影响[J]. 灌溉排水,1999(1):23-27.

[8] 韩会玲,康凤君. 水分胁迫对棉花生产影响的试验研究[J]. 农业工程学报,2001(3):37-40.

【作者简介】 杨凤梅,女, 1975 年 7 月生,本科学历,2003 年 7 月毕业于塔里木大学水利工程系,学士学位。塔里木河流域阿克苏管理局工作,工程师。

混凝土节水保湿养护膜在北疆渠道衬砌工程中的应用

魏军霞　卿光辉

（新疆额尔齐斯河流域开发工程建设管理局　乌鲁木齐　830000）

【摘要】 对北疆某工程渠道衬砌中混凝土裂缝产生的原因进行分析并采取控制措施的基础上，结合养护试验中应用节水保湿养护膜对混凝土养护方面起到一定参考作用。

【关键词】 气候干燥　渠道衬砌　混凝土裂缝　节水保湿养护膜

1　引言

新疆地区气候干燥，风沙大，温差大，白天气温高达30～40 ℃，水资源匮乏。北疆某渠道工程，至2010年8月开始进行衬砌试验以来，裂缝的产生问题一直未得到有效控制。通过统计及分析，裂缝的产生主要是养护不到位造成的，采取措施是可以得到有效控制的。通过有效的养护方法可以有效地解决因养护不到位产生的裂缝问题。

2　衬砌混凝土裂缝分析及防治措施

随着新技术、新设备、施工工艺的不断改进，工程质量总体水平不断上升。混凝土在工程建设中占有重要地位，对混凝土的各种性能要求越来越高，不仅要求混凝土工作性能好、强度指标高、耐久性好等，而且还要求混凝土结构有光洁、平滑的外观。但在混凝土工程施工过程中，经常发生混凝土裂缝及蜂窝、麻面、等质量通病，这些通病，有的会缩短了混凝土的使用年限，有的直接影响使用安全和使用功能。最大限度地消除混凝土裂缝和质量通病，保证工程结构安全，是工程管理人员最为关注的问题，现结合本工程实际，对本渠道衬砌工程混凝土裂缝产生原因进行分析。

（1）当结构的基础出现不均匀沉陷，就有可能会产生裂缝，随着沉陷的进一步发展，裂缝会进一步扩大。可通过提前进行土方施工，沉降稳定后进行衬砌施工。

（2）混凝土坍落度太大产生的裂缝。可以通过现场加强质量控制避免此类裂缝产生。通过反复试验，渠坡为1∶2.5时坍落度宜为7～9。

（3）当有约束时，混凝土热涨冷缩所产生的体积涨缩，因为受到约束力的限制，在内部产生了温度应力，由于混凝土抗拉强度较低，容易被温度引起的拉应力拉裂，从而产生温度裂缝。由于太阳暴晒产生裂缝也是工程中最常见的现象。可以通过混凝土分缝分块释放应力得到控制。

（4）切缝时间过晚产生的裂缝。可以加强管理，在350～400 h·℃进行切缝达到有效控制。

(5)在大风天气,混凝土表面水分蒸发较过快,造成混凝土内部水化热过高,在混凝土浇筑数小时后仍处于塑性状态,易产生塑性收缩裂缝。可以通过收光后及时覆盖一布一膜,避免混凝土表面水分散失得到有效控制。

(6)气温高产生的裂缝。采取早、晚衬砌,中午停工,避开高温时段来达到控制此类裂缝目的。

(7)人为因素产生的裂缝。如振捣不密实、抹面时机不当等产生的裂缝,可以通过加强施工人员培训及管理来解决。

(8)养护不到位产生的裂缝。目前,养护方法方式多样,现场养护不当是造成混凝土收缩开裂最主要的原因。混凝土浇筑面不及时覆盖、浇水养护,表面水分迅速蒸发,很容易产生收缩裂缝。特别是在气温高、相对湿度低、风速大的情况下,干缩更容易发生。目前,许多施工工地在浇筑混凝土时,都不能做到及时覆盖保温养护。一般总要等到最后一遍抹光结束后才覆盖,还有好多工地根本不盖。特别是夏天,气温很高,混凝土的水分蒸发很快,养护不到位,浇筑好的混凝土在烈日下曝晒,结果混凝土是前浇后裂。裂缝不可避免地就会产生。养护材料的选择和养护方式是对防治裂缝保证混凝土质量的关键。

3 混凝土节水保湿养护膜在北疆渠道衬砌中的应用

混凝土节水保湿养护膜是以新型可控高分子材料为核心,以塑料薄膜为载体,黏结复合而成,高分子材料可吸收自身重量200倍的水分,吸水膨胀后变成透明的晶状体,把液态水变为固态水,然后通过毛细管作用源源不断地向养护面渗透,同时又不断吸收养护体在混凝土水化热过程中的蒸发水。在一个养生期内养护膜能保证养护体表面保持湿润,达到养护目的。

由于此养护材料已在南水北调渠道工程中应用,根据以前经验,施工中主要防止养护膜铺设施工及养护过程中的破损;二是保证养护膜与混凝土面贴紧,做好防风措施,即铺好后膜四周要压紧,防止风的灌入。在此经验基础上,2011 年 8 月渠道二标对节水保湿养护膜进行试验,根据边坡伸缩缝为 6 m 宽坡面 13.9 m 长尺寸,所用养护膜尺寸为幅宽 6.5 m,卷长 500 m。施工时,先对混凝土坡面清扫湿润,铺设养护膜后,在伸缩缝用柔性棉布压入缝中,坡顶坡底用胶带粘好,覆盖 40 cm 棉布在其上用沙袋压紧。共铺设 30 m 长进行试验。期间风力多为 2 ~4 级,气温 18 ~37 ℃。

经过试验观察,能持久保湿,在 28 d 养护期内,没有补一次水,板面一直保持湿润,相对湿度95%左右;并且能有效平缓温差。通过两次实测,一次中午气温在 35 ℃时,膜内温度 26 ℃;一次早上气温 21 ℃时,膜内温度 23 ℃;有效抑制微裂缝的产生;抗压强度能适当提高。

节水保湿养护膜与其他养护材料及方法对比:用棉线毯养护,由于当地长期刮风、干燥、高热,需 2 h 左右补水一次,可周转使用 1 ~2 次,用水量及人工成本高;土工布养护,保湿效果比养护棉毯好,4 ~6 h 补水一次周转次数可达 3 ~4 次,成本为保湿养护膜的 2 ~3 倍。此外,节水保湿养护膜的核心材料可降解,为环境友好产品。

4 结束语

以上对混凝土衬砌施工中的裂缝问题进行了初步分析,并对混凝土节水保湿养护膜在试验中取得成功进行介绍，保湿养护膜的优点是施工简单、养护效果好、节水，能有效控制干缩裂缝，值得应用，但由于不能重复使用，要大面积推广还需要进一步降低成本。

【作者简介】 魏军霞，1978 年 11 月生，2002 年 7 月毕业于新疆农业大学，本科学历，新疆额尔齐斯河流域开发工程建设管理局，工程师。

灌溉渠系精细化配水模型及其应用研究*

张国华　谢崇宝

（中国灌溉排水发展中心　北京　100054）

【摘要】 国内外现有的渠道配水模型是建立在下级渠道流量相等、上级渠道断面均匀的基础上，针对该种模型的不足，本文建立了精细化配水模型，可以考虑下级渠道流量不等和上级渠道断面变化等情况；在求解算法方面，采用实数编码，通过不断调整优秀个体的变化区间快速逼近非劣解，减少了遗传算法求解该类模型所需要耗去的时间；同时通过构建合理的适应度函数和对约束条件的高效处理，保证了求解过程的始终有效性。应用结果表明，采用本模型和算法得到的配水方案与原配水方案相比，输水渗漏损失减少了15.32%；配水过程也更加均匀，有效减少了闸门调节次数。

【关键词】 灌溉渠道　精细化配水　模型　算法

0　引言

针对配水渠道的轮灌组合优化，前人提出了较多的模型，尽管在算法上也有一些突破，但多是在目标函数上的相互改进，提出了基于遗传算法、自适应遗传算法、粒子群优化算法等两级渠系条件下的优化配水模型，有学者开发了相应计算软件，也有学者基于分解协调思想建立了多级渠系配水优化模型。研究表明，这些模型在解决特定配水问题具有一定的适用性。但这些研究大多局限在配水渠道设计流量单一的情况，未能解决实际渠道设计流量和实际配水流量不等情况下的配水问题，或仅仅解决了下级渠道流量不等时的配水问题，事实上经过调查发现绝大多数较长的渠道都是上游渠道断面比较宽，下游断面比较窄，也就是渠道断面呈现阶梯式变化。在模型求解算法方面，基本遗传算法和自适应遗传算法涉及的参数较多，均需经验确定，可能出现局部收敛，使得它不一定总能获得全局最优解；粒子群优化算法涉及的参数虽少，但变量较多情况下的粒子编码较为复杂，应用基本粒子群优化算法求解多变量模型时也可能收敛到局部最优解。在算法编码方面：一方面大多数研究成果采用二进制方法对变量进行编码，需要在二进制数和实数之间进行频繁的解码转换，耗费大量时间；另一方面编码方法与约束函数处理紧密联系，编码优劣直接关系到求解的速度快慢和求解的质量好坏，而在多约束情况下的渠道配水问题，现有的编码方法将产生许多的无用个体，且计算结果不稳定、程序运行时间较长等。

为此，本文建立了下级渠道流量不等、上级渠道断面变化时的精细化配水模型；在求解

* 基金项目："十二五"国家科技支撑课题"灌溉用水实时调控技术与方法"。

算法方面,本文基于遗传算法的基本理论,采用实数制编码方法,建立合理的目标函数,将约束条件处理与群体编码有机结合,保证求解过程中群体的有效性和合理性,并通过保留优秀个体和收缩变量空间等手段,从而快速有效地获得全局最优解。

1 精细化配水模型的建立

精细化配水模型需要解决的主要问题是:在灌溉周期一定的条件下,确定上下级每条渠道的最优配水流量和配水时间,使得每一时刻配水流量尽可能接近设计流量,并保持上级渠道配水流量均匀,以尽可能减少闸门调节次数和水量损失。

1.1 目标函数

设有 K 段上级渠道的设计流量和实际配水流量分别为 Q_{sk}、Q_{auik}, $i=1,2,\cdots,T$, $k=1,2,\cdots,K$;N 条下级渠道的设计流量和实际配水流量分别为 q_j、q_{aij}, $i=1,2,\cdots,T$, $j=1,2,\cdots,N$,假定下级渠道上下游断面和流量一致。以轮期 T 内所有配水时段上下两级渠道的输水渗漏损失量最小为目标建模,即

$$\text{Min}VW = VW_u + VW_d \tag{1}$$

$$VW_{\text{u}} = \sum_{i=1}^{T}\sum_{k=1}^{K} f(A_{\text{uk}}, m_{\text{uk}}, Q_{\text{au}ik}, L_{\text{uk}}, t_{\text{au}ik}) \tag{2}$$

$$VW_{\text{d}} = \sum_{i=1}^{T}\sum_{j=1}^{N} f(A_j, m_j, q_{\text{a}ij}, L_j, t_{\text{a}ij}) \tag{3}$$

式中:VW_{u}、VW_{d} 分别为轮期内上、下两级渠道的输水渗漏损失量,m^3;$Q_{\text{au}ik}$、$q_{\text{a}ij}$ 分别为上、下级渠道在 i 时段的实际配水流量,m^3/s;A_{uk}、A_j、m_{uk}、m_j 分别为上、下级渠道的渠床透水系数和指数;L_{uk}、L_j 分别为上下级渠道长度,km;$t_{\text{au}ik}$、$t_{\text{a}ij}$ 分别为上下级渠道在 i 时段的实际配水时间,s。

渠道渗漏损失水量的经验公式为

$$VW = f(A,m,Q,L,t) = [A \cdot L \cdot Q^{(1-m)} \cdot t]/100 \tag{4}$$

参数同上。

1.2 约束条件

(1)时间约束:各下级渠道的配水时间应在轮期 T 内,即

$$0 \leqslant t_{1ij}(i=1,2,\cdots,T;j=1,2,\cdots,N) \tag{5}$$

$$t_{2ij} \leqslant T(i=1,2,\cdots,T;j=1,2,\cdots,N) \tag{6}$$

$$t_{\text{a}ij} = t_{2ij} - t_{1ij}(i=1,2,\cdots,T;j=1,2,\cdots,N) \tag{7}$$

式中:t_{1ij}、t_{2ij} 分别为第 j 条下级渠道配水开始和结束时间;其他符号意义同前。

(2)流量约束:上下级渠道的配水流量应符合渠道设计流量和最大允许流量的要求,即

$$Q_{\text{au}ik} \leqslant Q_{k\max}(i=1,2,\cdots,T;k=1,2,\cdots,K) \tag{8}$$

$$Q_{\text{au}ik} \approx Q_{sk}(i=1,2,\cdots,T;k=1,2,\cdots,K) \tag{9}$$

$$Q_{\text{au}ik} = \sum_{j=1}^{N} q_{\text{a}ij} \cdot x_{ij}(i=1,2,\cdots,T;j=1,2,\cdots,N;k=1,2,\cdots,K) \tag{10}$$

$$q_{\text{a}ij} = \alpha_{ij} q_j(i=1,2,\cdots,T;j=1,2,\cdots,N) \tag{11}$$

式中:$Q_{k\max}$ 为第 k 段上级渠道最大允许流量,一般取 $1.2Q_{sk}$;x_{ij} 为开关变量,当第 j 条下级

渠道第 i 时段有配水时取 1，否则取 0；α_{ij} 为第 i 时段、第 j 条下级渠道实际配水流量与设计配水流量的比值，$0.6 \leqslant \alpha_j \leqslant 1.0$；其他符号意义同前。

（3）水量约束：各下级渠道的实际需要配水量应等于其实际配水流量与实际配水时间的乘积，即

$$W_j = \sum_{i=1}^{N} q_{aij} \cdot t_{aij} \cdot x_{ij} (i = 1,2,\cdots,T; j = 1,2,\cdots,N) \tag{12}$$

式中：W_j 为第 j 条下级渠道实际需要的配水量；其他符合意义同前。

2 模型求解

本文基于遗传算法的基本理论，采用实数制编码方法，将约束条件处理与群体编码有机结合，保证求解过程中群体的有效性和合理性，并通过保留优秀个体和收缩变量空间等手段，快速有效地获得全局最优解。

2.1 编码设计

采用实数制编码，避免使用二进制编码需要在实数与二进制数之间的频繁转换。同时，为减少模型约束条件数，本文将第 i 时段、第 j 条下级渠道实际配水流量与设计配水流量的比值 α_{ij} 和第 j 条下级渠道配水结束时间 t_{2ij} 为基本变量，则约束条件（7）、（11）、（12）自动满足要求。

2.2 适应度函数构造

基于模型的目标函数（1）、（2）、（3），将约束条件（8）、（9）、（10）作为罚函数处理。

2.3 约束条件处理

除约束条件（5）和（6）外，其他约束条件均已在编码设计和适应度函数中进行了处理。约束条件（5）和（6）的处理方法如下：

（1）当第 j 条下级渠道的配水结束时间 $t_{2ij} > i$ 和 $t_{1ij} < i$ 时：如果 $i - t_{1ij} \geqslant 1$，第 j 条下级渠道第 i 时段的实际配水时间 t_{aij} 为一个时段，$x_{ij} = 1$；否则为 $(i - t_{1ij})$ 个时段，$x_{ij} = 1$。

（2）当第 j 条下级渠道的配水结束时间 $t_{1ij} > i - 1$ 和 $t_{2ij} < i$ 时：第 j 条下级渠道第 i 时段的实际配水时间 t_{aij} 为 $t_{aij} = (t_{2ij} - t_{1ij})$ 个时段，$x_{ij} = 1$。

（3）当第 j 条下级渠道的配水结束时间 $i - 1 < t_{2ij} < i$ 和 $t_{1ij} < i - 1$ 时：第 j 条下级渠道第 i 时段的实际配水时间 t_{aij} 为 $[t_{2ij} - (i - 1)]$ 个时段，$x_{ij} = 1$。

（4）如果（1）、（2）、（3）都不满足时：第 j 条下级渠道第 i 时段的实际配水时间为 0，$x_{ij}=0$。

2.4 遗传算法的改进

许多学者对遗传算法进行了有效改进并将其应用到灌溉渠道优化配水研究中，取得了较好的效果，本文借鉴前人的研究成果，对算法进行了一些改进并将其应用到求解本模型中。

（1）选择操作阶段，首先通过构造的罚函数计算各个体的适应度函数，然后通过轮盘选择方式选择子代群体，其中将适应度值位于前 n 个的个体直接保留进入子代群体。

（2）接着依次进行交叉和变异操作，再分别存储经过选择、交叉和变异操作后得到的新的群体，此时的新群体规模为 $3M$，其中需要注意的是选择、交叉和变异操作都是针对原父代群体进行操作的。

（3）对新群体按照适应度值大小进行排序，取排在最前面的 M 个个体作为新的父代群体。

（4）将上述过程反复演化（一般取 2 ~ 3 次）后得到群体所对应的变量变化区间，作为变

量新的初始变化区间，再进行选择、交叉、变异操作，如此加速循环，优秀个体的变化区间将逐步调整和收缩，直至达到预设的收敛目标。

(5)输出最后群体中最优秀的个体及其变化区间，整个算法结束。

3　实例应用

本文选用西北某灌区南支渠共33条斗渠冬灌的配水过程资料，渠体土质疏松，支斗渠均采用混凝土衬砌，坡降为1/3 000～1/1 000，灌溉期为21 d，详见表1。

表1　西北某灌区南支渠基本参数

斗渠序号	下级渠道(斗渠) 设计流量(m^3/s)	下级渠道(斗渠) 长度(km)	上级渠道(支渠) 不同断面设计流量(m^3/s)	上级渠道(支渠) 长度(km)
1	0.13	0.5	0.9	1.041
2	0.016	0.5		
3	0.016	0.5		
4	0.38	0.216	0.8	0.659
5	0.134	0.286		
6	0.016	0.5	0.7	0.394
7	0.016	0.5		
8	0.016	0.5		
9	0.089	0.485		
10	0.082	0.125		
11	0.085	0.175		
12	0.016	0.5	0.4	0.339
13	0.096	0.565		
14	0.135	0.75		

斗渠序号	下级渠道(斗渠) 设计流量(m^3/s)	下级渠道(斗渠) 长度(km)	上级渠道(支渠) 不同断面设计流量(m^3/s)	上级渠道(支渠) 长度(km)
15	0.112	0.362	0.5	0.592
16	0.016	0.5		
17	0.016	0.5		
18	0.016	0.5		
19	0.082	0.15		
20	0.164	0.306		
21	0.016	0.5	0.2	0.493
22	0.016	0.5		
23	0.016	0.5		
24	0.016	0.5		
25	0.135	0.186		
26	0.016	0.5		
27	0.016	0.5	0.1	0.386
28	0.016	0.5		
29	0.016	0.5		
30	0.016	0.5		
31	0.016	0.5		
32	0.016	0.5		
33	0.016	0.5		

将南支渠的基本资料输入本文建立的配水模型，其中，渠床透水系数A取3.4，透水指数m取0.5，灌水轮期为21 d；采用改进后的遗传算法求解得到的支渠配水过程见图1。计算结果显示，采用经验方法编制的支渠和斗渠配水时间集中、总体流量小，个别斗渠实际配

水流量大于其设计流量或小于最小允许流量;精细化配水方案的斗渠实际配水流量均在其允许的流量范围内,且与设计流量基本一致,配水过程搭配合理,配水时间也减少了 1 d;由图 1 可知,可以看出,精细化支渠实际配水流量均匀,有效减少了支渠闸门的调节次数,便于实际配水。经计算,优化配水方案的输水渗漏损失较分组灌溉配水方案的输水渗漏损失减少了 15.32%,表明精细化配水方案的配水质量高。

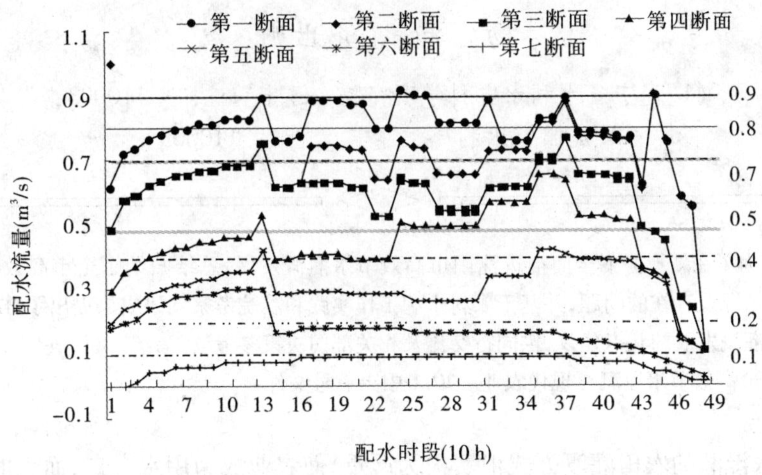

图 1　采用改进后的遗传算法求解得到的支渠配水过程

4　结论

与已有成果相比,本文的模型考虑了渠道上下游断面水力参数的不同问题,拓宽了模型的应用范围,且模型简单,便于实际应用;对约束条件的高效处理和算法的改进,保证了计算过程中群体的始终有效性。另外,通过不断调整优秀个体的变化区间快速缩小算法的收敛范围,从而大大缩短了程序运行时间。

应用结果表明,采用本文的模型和算法得到的精细化配水方案较原配水方案相比,输水渗漏损失减少了 15.32%;配水过程均匀,减少了闸门调节次数,表明模型及其求解算法是可行的,能获得较好的配水质量,可为下级渠道流量不等、上级渠道上下游断面不同情况下的渠系精细化配水提供参考。

参考文献

[1] 吕宏兴,熊运章,汪志农.灌溉渠道支斗渠轮灌配水与引水时间优化模型[J].农业工程学报,2000,16(6):43-46.

[2] 马孝义,刘哲,甘学涛.下级渠道流量不等时渠系优化配水模型与算法研究[J].灌溉排水学报,2006,25(5):17-20.

[3] 赵文举,马孝义,刘哲,等.基于自适应遗传算法的渠系优化配水模型研究[J].系统仿真学报,2007,19(22):5137-5140.

[4] 金菊良,丁晶.遗传算法及其在水科学中的应用[M].成都:四川大学出版社,2000.

[5] 张国华,张展羽,邵光成,等.基于粒子群优化算法的灌溉渠道配水优化模型研究[J].水利学报,2006,37(8):1004-1008.

【作者简介】 张国华,男,1980 年 9 月生,博士,2008 年 3 月毕业于河海大学,获工学博士学位,高级工程师,现工作于中国灌溉排水发展中心。

沈阳市高效节水灌溉发展现状、问题与建议

杨　洵[1]　齐世卿[2]

(1. 辽宁省水利水电科学研究院　辽宁　沈阳　110003;
2. 沈阳市水利局　辽宁　沈阳　110003)

【摘要】 为了在新形势下全面推进沈阳市高效节水灌溉发展,总结归纳了全市高效节水灌溉建设基本情况及存在的问题,并基于农村水利工作实践和苏皖考察调研成果提出了推进沈阳市高效节水灌溉发展的几点建议,供同行交流及相关部门决策参考。

【关键词】 高效节水灌溉　现代农业　2011 中央一号文件

传统大水漫灌的农田灌溉方式浪费极为严重,使农业成为用水大户,而水的有效利用率只有 30% ~40%,因此改变人们千百年来传统的灌溉习惯,推广高效节水灌溉技术,提高用水效率与效益是一项重任,也是缓解我国水资源紧缺的途径之一,更是现代农业发展的必然选择。2011 年 1 月中央一号文件即《中共中央国务院关于加快水利改革发展的决定》明确要求"大力发展节水灌溉,推广渠道防渗、管道输水、喷灌滴灌等技术。"根据 2011 年中央一号文件要求,辽宁省委下发一号文件《中共辽宁省委辽宁省人民政府关于贯彻落实〈中共中央 国务院关于加快水利改革发展的决定〉的实施意见》,辽宁省农委、水利厅联合下发《关于印发辽宁省实施 1 000 万亩滴灌节水农业工程推进工作方案的通知》,计划"十二五"期间全省计划发展高效节水滴灌面积 1 000 万亩,加快高效节水灌溉发展。为落实上述文件要求,沈阳市农田水利工作者深入开展高效节水灌溉专题调研,并赴安徽、江苏考察学习。笔者参加了上述专题调研和考察学习,现将沈阳市高效节水灌溉发展现状、问题与建议归纳如下。

1　沈阳市高效节水灌溉发展现状

截至 2010 年底,沈阳市共有耕地面积 982.5 万亩,有效灌溉面积 408.77 万亩,占耕地面积的 41.6%,各种作物灌溉面积及所占比例见图 1。其中,高效节水灌溉面积为 75.57 万亩,仅占有效灌溉面积的 18.49%,主要高效节水灌溉形式为喷灌、管灌和微灌,见图 2。总体来说,沈阳市高效节水灌溉发展缓慢,在全国及全省处于落后水平。

2　沈阳市高效节水灌溉发展中存在的主要问题

2.1　投资少,高效节水灌溉发展水平和标准较低

到 2010 年底,全市高效节水灌溉面积 75.57 万亩,仅占有效灌溉面积的 18.5%,其余 81.5% 为传统的地面灌溉。在高效节水灌溉工程建设中忽视了区域排水系统建设,很多设

施农业区排水不畅或根本没有排水渠道，造成雨季棚区积水，作物受淹，损失严重。全市非设施菜田灌溉面积66万亩的主要水源工程为农民自打的小井，抗旱能力差，经常出现地下水位低、小井出水不足等问题，无法保证作物用水需求。

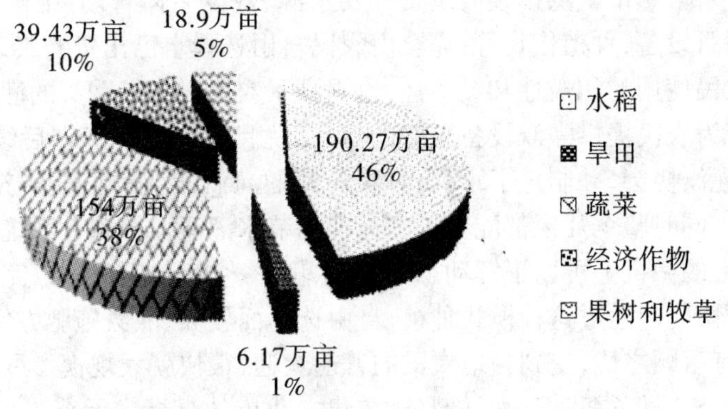

图1　沈阳市主要作物有效灌溉面积分布

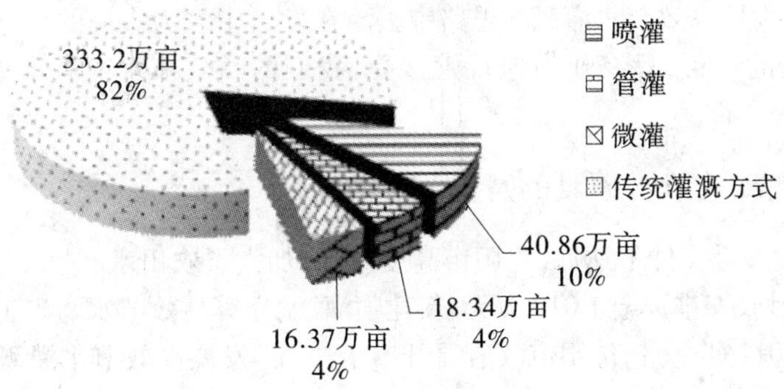

图2　沈阳市高效节水灌溉类型及比例

2.2　高效节水灌溉工程分布零散

从近几年全市各渠道建设的高效节水灌溉工程看，分布十分零散，造成很多地区虽然高效节水灌溉发展面积较大，但真正成规模的大型片区却寥寥无几，这给工程的后续管理带来很大困难。部分农民对高效节水灌溉不接受，导致因为小部分群体的排斥，而使已发展的片区不能继续扩大规模，只能另外选择地区来发展，如此使高效节水灌溉工程零零散散地分布在各地，无法形成大规模集中连片的示范基地，不能充分体现工程高效特点。

2.3　高效节水灌溉后续管理不健全

节水灌溉设备下发给农户，或者直接在项目地配套安装后，管理部门往往放松了对设备后续使用情况的关注，当灌溉设施出现滴灌孔堵塞等问题，不能继续进行灌溉工作，没有技术服务队伍跟进，农户自己又不会维修，因此很多灌溉设施只用了一年即废弃，投入的成本不能回收，无法发挥设备的长期效益，造成资源的严重浪费，并且影响了农户使用这类灌溉设施的积极性，严重制约了沈阳市高效节水灌溉的发展。

3　江苏常熟市发展高效节水灌溉的主要经验

在江苏常熟市考察过程中，一种新的理念给我们启发。常熟市高效节水灌溉发展处于江苏省领先地位,工程集中连片、规模庞大、灌溉技术多样、自动化水平高、后续管理方便有效,为我们提供了很好的参考。常熟市高效节水灌溉发展如此之快,很重要一点在于当地政府一切工作本着为农民服务的思想去做。

在发展过程中,常熟市董浜镇政府负责开发水源、敷设主体管网、维修道路并配套自动化系统,其他如灌溉设备,种植作物等,完全由农民自由选择自己出资购买,给农民很大的发挥空间,提高了农民使用的积极性和灵活性,也提高了农民对自己购买的灌溉设备的重视程度,而政府只负责为农民提供灌溉设备较好的厂家,优良的种子等。在后续管理中,董浜镇政府成立的节水灌溉协会,并制定了章程,对农户家的灌溉设备进行维修养护,真正体现服务性政府的理念。同时整个片区都配套了高标准的节水灌溉监控服务系统,农民用水时,工作人员在服务中心的微机上可以了解动态,计算水量,一个工作人员可以完成整个片区的微机管理工作,方便快捷。通过这种模式种植,自由选择面较大、灌溉施肥方便省力、灌溉设备的维修养护也没有后顾之忧,受到当地农民的普遍欢迎,很快就大规模发展起来。

常熟市的发展模式值得我们学习借鉴。在建立我市的高标准示范工程中,根据我市实际情况,要把常熟市的一些先进经验融入进去,更重要的是要把为农民服务,让农民收益的理念融入到我们发展高效节水灌溉的思路中去。在发展中多为农民的利益考虑,才能让工程真正受到农民的认可,只有让当地百姓接受,我们的高效节水灌溉工程才能有长足的发展。

4　沈阳市发展高效节水灌溉的意见与建议

2011 年中央一号文件下达后,沈阳市高度重视,加大了农田水利投入。2011 年已下达二批农田水利计划,安排资金 1 010 万元,新建喷灌、滴灌等高效节水灌溉工程 6 项,发展面积 2.35 万亩。2011 年 9 月,沈阳市政府召开“十二五”发展高效节水灌溉工作会议,确定“十二五”期间发展滴灌面积 115 万亩,其中 2011 年发展 15 万亩,总投资 2.04 亿元,其中辽宁省水利厅补贴 1.02 亿元。由此将迎来了全市高效节水灌溉发展的新高潮。为适应新的形势和发展要求,针对沈阳市高效节水灌溉发展现状,提出以下建议。

4.1　提高认识,加快发展

目前,我市的农田水利现状,和六七十年代相比没有本质的超越,有的地区甚至不如当年。这是因为改革开放 30 年,全国的发展理念不断创新进步,但水利发展还依然停留在粗放式的农田水利基础设施建设上,还在使用六七十年代时水利建设的剩余成果,对损坏工程进行维修改建,对逐步暴露的诸多问题束手无策。陈雷同志在 2010 年 12 月 24 日召开的全国水利工作会议上指出:“十二五”时期,是加强水利重点薄弱环节建设、加快民生水利发展的关键时期,是深化水利改革、加强水利管理的攻坚时期,也是推动传统水利向现代水利、可持续发展水利转变的重要时期。沈阳市的农村水利工作者,要深切认识到沈阳市水利改革的紧迫性和必要性,转变落后思想和观念,积极领会中央一号文件精神,以新的理念和思路、长远发展的眼光来研究问题,使沈阳市的水利发展水平在 5 ~ 10 年内有一个质的飞跃。

4.2　科学规划,合理布局

要想做好我市水利工作,少走弯路,就要在项目建设前就对全市的高效节水灌溉发展进行合理布局和科学规划,而各县、区(市)根据市里的布局规划内容,形成本地切实可行的实施方案。规划要从各地的自然条件、种植作物、当地政策、农民积极性等多方面考虑,制定适合当地的高效节水灌溉发展思路。譬如,辽中县水源较为充沛,地势平坦,且西部地区由于水质含铁较多,发展滴灌技术时间一长,容易发生滴口堵塞,因此更加适合发展喷灌、管灌等灌溉技术;法库县水源条件较差,但水质较好,十分需要发展高效节水技术,是发展滴灌技术的理想地点;康平县是沈阳市水源条件最困难的地区,因此在康平县发展高效节水灌溉,必须立足于抗旱,要在保证水源条件的情况下,发展高效节水灌溉技术。

确定各地适宜发展的灌溉方式后,还要考虑发展高效节水灌溉的地块落实位置、建设时间先后、后续管理制度等相关问题;在施工过程中,提前对片区进行科学合理的设计和规划,也尤为重要。我们在管理中发现,部分地区忽视规划设计的合理性,盲目上项目,没有充分研究论证就开发水源、配套设备,但在施工过程中或竣工后却发现诸多问题(如水源井不出水,输水管道埋藏不了等),再想变更设计已来不及,造成了资金的严重浪费。因此,要全面认识科学规划的意义,具体的工程项目在实施前,要充分考虑各种情况,避免工程变更设计、多次维修、重复建设等情况出现。

4.3 因地制宜,突出重点

高效节水灌溉工程技术含量高,但造价也高,虽然新时期加大了水利资金的投入,但是对于沈阳市近1 000万亩耕地来说,资金也是有限的。这就要求我们在建设项目时要坚持两项发展原则:一是突出重点,突出特色。要重点支持1 000亩以上规模种植的农场、农业企业、农民专业合作社和种植大户等,要优先发展经济效益较高,有投资价值的如寒富苹果、树莓等果树的节水滴灌,重点发展花生、马铃薯、菜籽、红干椒等特色农作物的节水滴灌。便于工程后续的管理和维护,提高了资金投入的效益;二是注重实效,量水而行。在项目确定之前要充分考虑工程建设完成后的实际效益,是否有利于水资源节约、作物增产、农民增收,还要充分考虑当地现有的水源条件和可开发利用的水资源潜力,没有投资价值、不能提供水源的地区决不能盲目建设。

4.4 提高标准,大力发展高效节水现代农业

目前,沈阳市的高效节水灌溉工程,大多还只限于开发水源、铺设管路、安装设备等最基本的内容,这也是高效节水灌溉的最低标准。一些高效节水灌溉片区内部的道路泥泞不堪、排水沟道不畅、杂草丛生,严重影响了工程的形象。我们在发展灌溉技术的同时,忽视了水利工程的减灾能力、生态效益和旅游资源,而仅仅把“高效”二字局限于节水效益这一方面,没有尝试开发工程的其他功能,给农民、给社会创造更多的经济效益。建立真正的高效节水农业,要将节水、节肥、省时、省力、高产、减灾、生态建设、旅游开发等多方面效益相结合,联合农委、财政、林业等多个部门,共同打造沈阳市的多功能现代农业园区。水利部门要积极主动地向市领导谏言,向相关部门宣传,整合资金,联合推动沈阳市的现代农业向前发展。

4.5 创新体制,提高社会化服务水平

在沈阳市高效节水灌溉发展存在问题中我们已经提到,工程分散,农民积极性不高、设备损坏无人维修是制约沈阳市灌溉发展的一个问题。针对这些问题,我们借鉴常熟市发展高效节水灌溉的主要经验,可以从以下几个方面着手:首先要建立高效节水灌溉设备后期维护服务队伍,这支队伍应由科研人员、技术人员和管理人员组成,并由当地乡镇政府管理,为设备有损坏的农户提供上门服务,定期下乡为农民做设备使用、保养和维修的讲解;此外,从可持续发展的角度在广大农村加强高效节水灌溉工程的宣传,评选节水灌溉先进农户,并给予一定物质奖励;创建高效节水发展信息交流的平台,定期召开高效节水灌溉技术研讨会,

带领与会人员赴先进地区学习考察,开阔水利工作人员的视野和思路,学习先进的节水灌溉技术和管理体制等。现代农业、现代水利是多因素的系统工程,涉及当地的经济发展水平、科技水平、生产关系水平、群众思想、习惯、素质与文化水平、市场经济规律、发展的机遇、体制、机制等因素,涉及种植业、土地、农机具等方方面面的问题,没有政府的主导作用是难以有效推动的。因此,政府要统筹考虑、多措并举、攻坚突破,提高社会化服务水平,使高效节水灌溉走上健康有序的发展轨道。

参考文献

[1] 杨运革.高效节水灌溉是现代农业持续发展的必然选择[J].山西水利,2011(2):55-56.
[2] 位铁强.因地制宜发展高效节水灌溉为农村经济持续发展提供有力支撑[J].节水灌溉,2011(9):11-13.

【作者简介】 杨洵,女,1986年4月生;硕士,毕业于大连理工大学,现工作在辽宁省水利水电科学研究院,助理工程师。

基于 MODIS 荒漠草原土壤湿度遥感反演

邬佳宾　魏永富　李锦荣　王明新

（水利部牧区水利科学研究所　呼和浩特　010010）

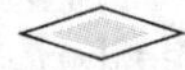

【摘要】 土壤湿度是决定荒漠草原植被覆盖的重要因子，研究典型荒漠草原植被生长期土壤湿度及其变化规律对于草原生态科学研究具有重大的科学意义和现实意义。本文在概述研究区情况和土壤湿度的主要获取方法后，介绍了以 MODIS 数据为基础，利用 RS 和 GIS 技术手段提取土壤湿度方法的原理与方法，并得到 2011 年达茂旗荒漠草原植被生育期的土壤湿度反演成果，最后对成果进行了分析和讨论。

【关键词】 遥感　荒漠草原　土壤湿度　MODIS

1　研究区概况

达尔罕茂名联合旗（简称达茂旗）荒漠草原区有天然草场 103.88 万 hm^2，其中可利用面积为 81.27 万 hm^2。气候干旱多风，属明显的大陆性气候，年均降雨量为 175～250 mm，且主要集中于 7、8、9 月，这三月降雨量占全年降雨量的 60% 以上。年平均气温 3.6 ℃，地形为高平原台地和起伏丘陵。受达茂旗自然条件的制约，该草地类型结构简单，层次分明，以石生针茅、短花针茅建群的植物群落分布广，面积大，占本草地类型的 45% 以上，伴生种有多年生小半灌木冷蒿和旱生丛生禾草。土壤湿度是决定荒漠草原植被覆盖的重要生态环境因子，研究典型荒漠草原植被生长期土壤湿度及其变化规律对于草原生态科学研究具有重大的科学意义和现实意义。

目前，获取土壤湿度的方法有三大类，分别为实测法、模型方法和遥感方法。实测法作为常规方法可以准确估测土体剖面的含水量，但是其样点较稀疏，代表范围有限，数据收集时效性差，人力财力消耗较大，尤其在区域土壤含水量监测中，存在以点代面的问题，而且较难给出土壤不同含水量区域之间的分界线。模型法根据能量平衡原理，寻找影响土壤湿度的驱动因子，建立水分平衡方程求解土壤湿度，这种方法需要有大量气象数据的支持，而且参数复杂、确定困难，估测误差较大。遥感监测土壤水分的方法具有宏观、高时效、经济等特点，且可以划分区域土壤含水量等值线图，便于对区域不同含水量范围的分析和研究。本文以达茂旗荒漠草原为研究区域，以 2011 年草地植被生长期的 MODIS 数据为基础，通过 *TVDI* 方法反演其土壤湿度，并对结果进行了分析和讨论。

2　原理与方法

2.1　温度植被干旱指数法

国内外学者研究了各种空间分辨率和时间分辨率的地表温度与植被指数的关系，发现二者之间存在明显的负相关关系。Price 等研究发现如果研究区的植被覆盖和土壤湿度的变化范围较大,则以遥感资料获得的归一化植被指数(*NDVI*)和地表温度(T_s)为横纵坐标的散点图呈三角形。Moran 等从理论角度分析认为 T_s 与 *NDVI* 的散点图呈梯形关系。Sandholt (2002)等在研究土壤湿度时发现,T_s-*NDVI* 的特征空间中有许多等值线,于是提出来温度植被干旱指数 *TVDI* 的概念。*TVDI* 由植被指数和地表温度计算得到,计算公式如下所示

$$TVDI=\frac{T_s-T_{smin}}{T_{smax}-T_{smin}} \tag{1}$$

式中:T_s为任意像元的地表温度;T_{smin}、T_{smax}分别为某一 *NDVI* 对应的最低、最高地表温度。

TVDI 原理如图 1 所示,干边上 *TVDI*=1,湿边上 *TVDI*=0。*TVDI* 越大,土壤湿度越低;*TVDI* 越小,土壤湿度越高。估计这些参数要求研究区域的范围足够大,地表覆盖从裸土变化到比较稠密的植被覆盖,土壤表层含水量从萎蔫含水量变化到田间持水量。

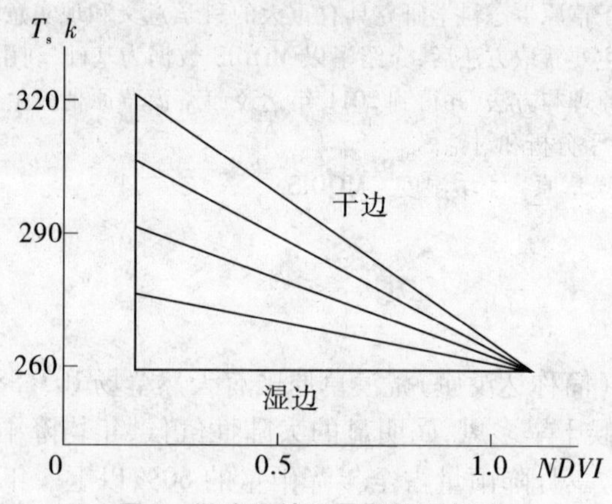

图 1　温度植被干旱指数原理图

Moran 等在假设 T_s~*NDVI* 特征空间呈梯形的基础上,从理论上计算梯形四个顶点坐标的研究结果表明,在不同的植被覆盖条件下,T_s~*NDVI* 特征空间中最低温度随植被盖度不同而变化。因此,可以利用简化的特征空间对 T_{smin} 和 T_{smax} 同时进行线性拟合,拟合方程如下所示

$$T_{smin}=a_1+b_1NDVI \tag{2}$$

$$T_{smax}=a_2+b_2NDVI \tag{3}$$

式中:a_1、b_1分别为干边拟合方程系数;a_2、b_2分别为湿边拟合方程系数。

这样将式(2)、式(3)代入式(1)即可计算温度植被干旱指数。

2.2　计算过程

根据上述原理,即可以通过 *TVDI* 方法计算土壤湿度。

首先完成遥感数据的预处理,包括几何校正、辐射校正、影像拼接裁剪等。这些都可以通过 ENVI 遥感图像处理软件实现,这一软件提供了既定的地理信息校正 MODIS 数据的功能,可利用 MODIS 数据中的地理信息进行几何校正,在进行几何校正的同时还可以进行影像的投影。ENVI 软件在对影像赋予地理参考的同时,还通过影像的自带的描述信息进行辐射定标,即自动把影像的 *DN* 值转化为实际的辐射强度和反射率。经过这些操作,MODIS 图像具有了地理参考,而且每个像元值代表了实际的反射率和辐射强度。然后利用与影像具有相同的地理参考的达茂旗的矢量边界图对影像进行裁剪,得到研究区域的遥感基础资料,

有时候如果一景图像不能覆盖研究区域,还要进行影像的拼接处理。

完成影像预处理后通过遥感软件实现 *NDVI* 及 T_s 的计算。*NDVI* 通过 MODIS 影像 1、2 波段进行计算;T_s 采用劈窗算法获得,劈窗算法的推导是一个复杂的过程,限于篇幅,本文不再赘述。

得到研究区域的 *NDVI* 和 T_s 后,对二者的空间特征进行分析,并得到特征空间干边、湿边的拟合方程。利用方程(1)、方程(2)、方程(3)进行计算即可得到温度植被干旱指数来表征土壤湿度。MODIS 数据具有时间分辨率较高的特点,可以天为单位进行数据分析计算,研究中如果不需要这么密集的数据采集,同时为了避免云对数据质量的影响,可采用最大值合成法(MVC)合成所需时间段的数据以达到研究的要求。

3 土壤湿度反演成果

按照上述原理和方法,本文以覆盖达茂旗全境的 MODIS 数据为基础,利用 *TVDI* 方法,对达茂旗荒漠草原 2011 年植被生长期(5 月 20 日至 9 月 10 日,共 110 d)每日的 *NDVI* 和 T_s 进行了信息提取。由于每日的相关指数变化不大,同时为消除大气云层对反演成果的影响,因此本文采用 MVC 方法每 10 d 合成 1 期数据,共计得到按照时间序排列的 11 期 *TVDI* 栅格数据。限于篇幅,本文仅将 2011 年 8 月中旬(8 月 11 ~20 日)T_s ~ *TVDI* 空间特征干、湿边方程拟合图及 *TVDI* 分带图进行展示,如图 2、图 3 所示,其余日期数据不再一一展示。

由图 2 可知,干边、湿边拟合方程相关系数分别达到 0.81、0.89,可见其拟合效果较好,成果较为可靠。从图 3 可以看到,在 8 月中旬植被生育盛期,植被覆盖高的地区 *TVDI* 指数较低,说明土壤湿度相对较高,说明植被有保持土壤水分的作用。

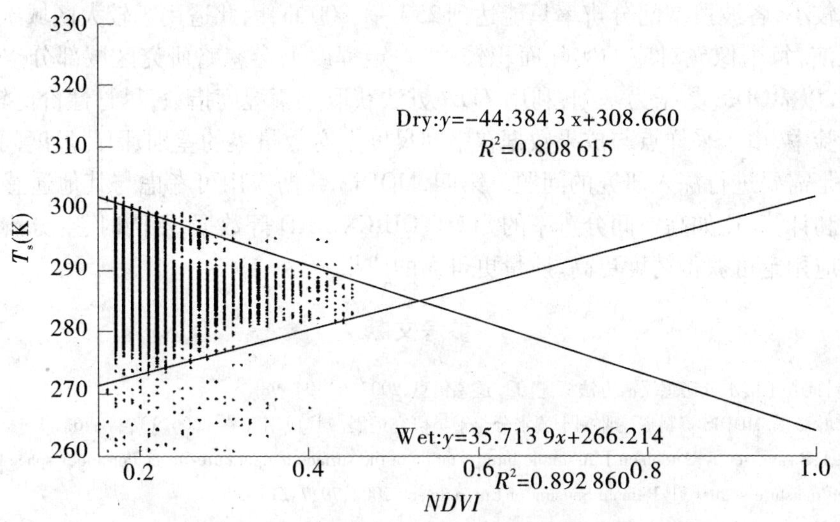

图 2 T_s ~ *NDVI* 空间特征干边、湿边拟合图

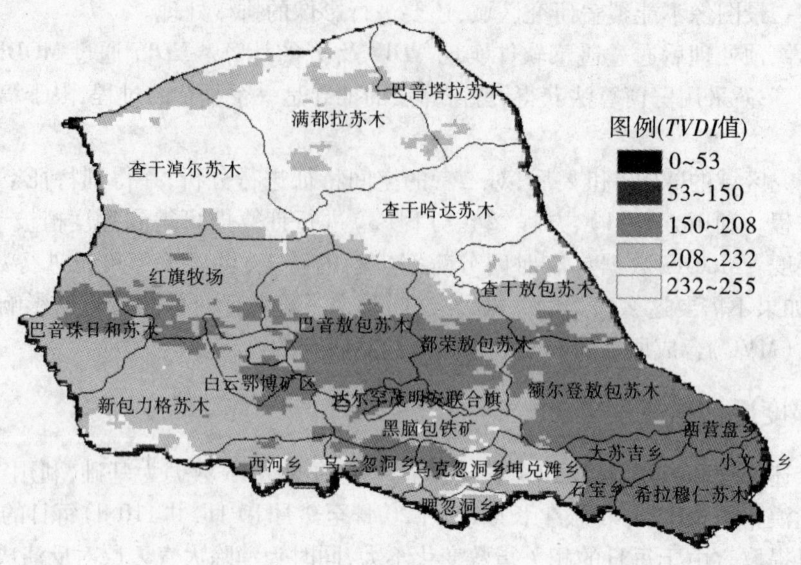

图3 2011 年 8 月中旬达茂旗荒漠草原 *TVDI* 分带图

4 分析和讨论

通过上述研究,笔者得到如下几个结论:第一,利用 MODIS 数据获取区域土壤湿度信息迅捷、可靠,同时时间分辨率明显优于其他遥感数据,卫星每天过境,可以及时准确地掌握研究区土壤湿度情况,为草原旱情预警和灾害评估提供了良好的手段。第二,MODIS 数据空间分辨率较小,各波段空间分辨率只能达到 250 ~ 1 000 m,因此适用于较大区域土壤湿度的分析和反演,每个像元对应的实际面积较大,一定程度上会忽略研究区域部分空间变异特性。第三,以 MODIS 数据为基础,利用 *TVDI* 方法获取土壤湿度指数,尽管性价比较高,但是在成果检验中,由于采样点与成果数据在空间尺度上有数量级的差别,因此尺度问题如何准确转换将是需要进行深入研究的问题。第四,MODIS 数据应用可考虑与其他遥感影像结合进行研究和计算,比如高空间分辨率的 SPOT、QUICKBIRD 等数据,不同时、空分辨率的遥感数据联合应用定可获得精度更高、质量更可靠的成果。

参考文献

[1] 肖斌,沙晋明. 土壤水分遥感反演方法概述[J]. 遥感信息,2007(6):94-98.

[2] 佘万明,叶彩华. MODIS 资料遥感监测土壤水分与干旱研究进展[J]. 应用气象,2006(1):44-46.

[3] Sandholt I, Rasmussen K, Andersen J. A simple interpretation of the surface temperature/vegetation index space for assessment of surface moisture status[J]. Remote Sensing of Environment, 2002, 79, 7:213-224.

[4] Goward S N, Xue Y, Czajkowski K P. Evaluating land surface moisture conditions from the remotely sensed temperature/vegetation index measurements an exploration with the simplified simple biosphere model [J]. Remote Sensing of Environment, 2002, 79(2):225-242.

[5] 齐述华,等. 利用温度植被旱情指数(TVDI)进行全国旱情监测研究[J]. 遥感学报,2003,7(5):420-427.

【作者简介】 郭佳宾,1981 年 12 月生,硕士,2006 年 7 月毕业于内蒙古农业大学,现工作于水利部牧区水利科学研究所,工程师。

连续枯水状态下的引黄灌区小型农村水利工程建设

曹 麟

（宁夏水电技师学院 银川 750006）

【摘要】 本文针对近年来水量锐减的现状，从小农水工程的重要作用出发，探讨了连续枯水状态下引黄灌区小农水的建设方向，并以宁夏青铜峡灌区为例提出了小农水的具体改革措施：①分区滞水，开源节流；②降低工程的设计标准；③全面推进农民用水协会建设，加强枯水状态下的工程监护。

【关键词】 连续枯水 小型农村水利工程 建设

随着全球气候的不断恶化，降水量日渐减少，我国多处河流进入连续枯水状态。以黄河为例，由于上游生态退化加剧，干旱少雨，天然来水量持续减少，自2000年以来，黄河上中游连续12年出现持续干旱，上游主流和支流来水量比多年平均值减少最多达到50%。伴随着历史上罕见的枯水形势，龙羊峡和刘家峡两大水库蓄水量多次降到建库以来的最低水位，以宁夏和内蒙古为主的两大引黄灌区有数百万亩农田无水可灌，农业生产受到巨大影响。连续枯水状态下，当前农田水利工程的建设呈现出与种植结构调整相适应的特性，在设计标准、取水点布局及水质处理上都有了和以往充分灌溉不同的特点，因此全面研究连续枯水状态下引黄灌区小农水的建设十分必要。

1 连续枯水年的外在表象

1.1 河道断流，需工程联合调配

连续枯水年的最大特点是河川径流量相比较于多年平均值而言大幅衰减。水工建筑物较少或无法实现水资源的综合调配时，多以河道断流为外在表象。如表1所示，在对1 550条典型河道的调查中，共有60条河流发生断流，断流河长约占总河长的51.5%。

表1 典型调查河段断流情况一览

	河流数（条）	总长度（km）	断流河长（km）
松花江	3	1 376	254
辽河	8	3 916	1 531
海河	30	6 906	4 106
黄河	7	2 465	327
淮河	5	2 542	3 934
西北诸河	7	5 188	1 384
合计	60	22 393	11 536

近些年,在水库联合调度的作用下,连续枯水年的影响以工农业生产遭受较大影响、农业灌溉面积大幅压减为主要外在表象。以黄河为例,黄河曾在1922~1932年出现过连续11年的枯水期,进入20世纪90年代,黄河几乎年年断流,如图1所示,1997年黄河的断流时间长达226 d,仅山东省直接经济损失就达135亿元。从1999年起,国家授权黄河水利委员会对黄河水资源实行统一调度、合理配置,这才实现了自该年起至今连续13个枯水年无断流的奇迹。尤其是在水量位列黄河历史第2、第3极枯水年的2000年和2001年,如果没有以小浪底水库为主的水库联合调度,有限的黄河水资源根本无法保证大旱之年不断流。

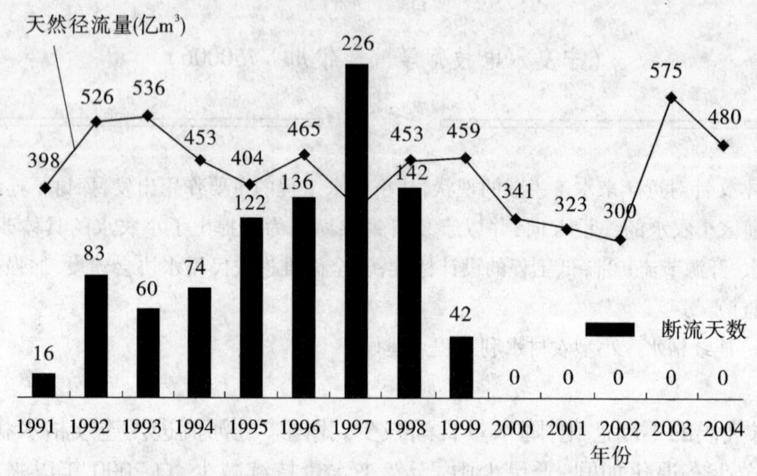

图1 黄河天然径流量与断流天数示意图

1.2 水质恶化,污染治理压力重大

连续枯水状态下,用于河道生态维护的水量会相应减少,因此满足水体自净能力要求的水体总量也多难以保证,由此造成了更大面积的水体富营养化和湿地退化,这些因素与由于工业快速发展而导致的水污染相互叠加,使得水质更趋恶化,污染治理压力不断加大。如图2所示,研究表明,常态下最小生态用水要占到地表水资源量的32%,而遭遇枯水年时,这一比例不足10%,甚至出现了生态用水全部被工业用水挤占,随后又以劣质水的形式返回到河道的现象。以2009年为例,由于遭受历史罕见的旱情,全年生态环境补水量仅占总用水量的1.7%,全国16.1万km的河流中Ⅳ类以上的河长超过了总评价河长的40%,64.8%的湖泊和28.2%的水库呈现富营养化状态,3 411个水功能区中全年达标率仅为47.4%。在黄河流域,一些支流污染已非常严重,以宁夏为例,2009年的监测结果显示,全境劣Ⅴ类水质河长超过50%;清水河作为黄河的主要支流,实际上已经成为一条排污沟,全程都是劣Ⅴ类水。

2 连续枯水对引黄灌区小农水工程的影响

2.1 可取水量锐减对小农水的影响

天然来水的减少必然导致用水指标减小。为保证GDP的发展,不少省份都采用了削减农业灌溉用水的办法,由此导致小农水工程可取水量锐减,工程长期低水位运行,原设计方案中维持渠道边坡稳定的水体压力大幅缩减,原设计标准无法实现,内侵、底部冻胀破坏概率增大。因此,在进行工程设计时,连续枯水状态下的渠系类小农水工程底宽应较常规偏小。此外,由于灌溉用水保证率降低,水资源调配压力增大,小农水工程运营管理中的大幅削减农业灌溉面积、大力发展节水灌溉技术,有效提高灌溉水利用效率需求增高。

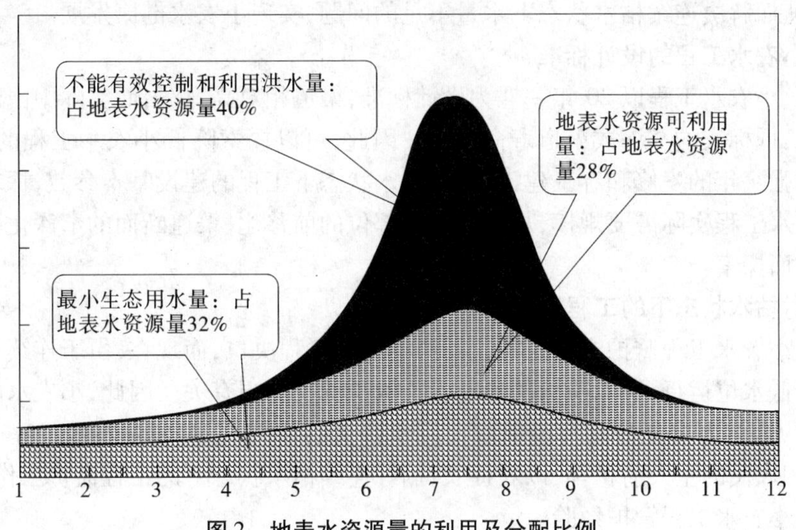

图2 地表水资源量的利用及分配比例

2.2 可取水点锐减对小农水的影响

连续枯水状态下河道水位较正常运行水位偏低,导致河道流速相应减小,河心洲发展迅速,边滩发育、河道取水口受泥沙淤积影响严重,甚至出现取不上水的现象。加上小农水工程所在渠道多未经砌护,因此连续枯水状态下的小农水设计必须考虑通过增大渠道比降、增设节制闸、冲沙闸的办法减小因可取水点锐减对小农水的影响。

2.3 可取水质恶化对小农水的影响

由于监测手段不完善,农村供用水工程中的水源地保护强度远不如城市,水质污染对引黄灌区人饮类的小农水工程建设成效影响巨大。国家发展和改革委员会、水利部、卫生部组织的全国农村饮水安全现状调查评估结果显示,在对农村饮水不安全因素的人口调查中,水质不达标的人口占据70%,而水量、方便程度和保证率不达标仅占30%。因枯水状态下水流流量小、流速慢,不仅更易造成单位水体氨氮超标,也容易将蕴含在土壤中的有毒物质溶洗而出,因此枯水状态下的氟化物超标、砷超标和苦咸水等水污染事件更易发生。

最后,受传统发展观念的影响,以高投入、高消耗、低产出、高排放为特征的粗放型经济发展局面在引黄灌区相关省份仍未得到根本性扭转。资料显示,近年来高污染行业逐步向西部地区转移和聚集,1990~2009年,西部地区工业COD和SO_2排放量占全国的比重分别由14%和18%上升至35%和36%。以内蒙古、甘肃、宁夏为代表的引黄灌区也正是我国高污染行业比重较高的地区,因此必须通过增设处理装置或建筑物等办法对不达标水进行治理后再进行灌溉。

3 连续枯水下引黄灌区小农水建设发展方向

3.1 分区滞水,开源节流

因过去取水便利,引黄灌区多年来一直采用的是自流排灌的灌溉方式,一些地区甚至还在采用大水漫灌的方式种植作物,水稻亩均供水量多达2 500 m^3。因此,在连续枯水状态下,除了调整种植结构、水改旱外,还可以采用窑窖结合、井渠结合的办法,分区滞水,高效用水。通过有效截留短期高强度暴雨,加强对工业污染废水的治理与回用,开发、节约一切可用水源,从而解决连续枯水状态下水量不足的问题,实现小农水健康发展。

3.2 降低小农水工程的设计标准

常态下,小农水工程以20年一遇为设计标准,渠道相对高宽浅,而实际中由于工程长期低水位运行,低标准的窄深式渠道寿命更长。因此,可以探索降低小农水工程的设计标准。通过不断加强每年的岁修和春工建设,积累枯水状态下工程的建设基本参数,探求低水位运行下的小农水工程实际需要规模,从工程布置部位的前移化、渠道断面的窄深化和水量调配的综合化方面探索。

3.3 加强对枯水状态下的工程监护

过去的小农水工程监护主要是防止暴雨导致的工程决口,而现在,由于连续枯水年下工程长期处于低水位运行,所以险情的出现多与资源调度不当有关。因此,小农水的管护比建设更加重要,这就需要充分发挥当前农民用水协会的作用,在明晰小农水工程产权的基础上,充分发挥农民的主人翁精神,以农促农,搞好连续枯水状态下的工程监护工作,积累低水位运行下的小农水工程管护经验。

4 实例研究——青铜峡灌区小农水的建设与发展

4.1 灌区小农水概况

青铜峡灌区位于宁夏回族自治区境内,是一个以灌溉478万亩农田为主的大型灌溉供水工程,也是宁夏回族自治区农业和国民经济发展的重要基础设施。青铜峡灌区以黄河为界分为河东、河西两部分,灌溉面积分别为101万亩和377万亩。有总干渠、干渠及支干渠18条,总长1 084.3 km,总引水流量603 m^3/s,年引水总量66.4亿m^3。骨干排水沟道24条,总长660.1 km,控制排水面积628.7万亩,排水能力558.6 m^3/s,总排水量41.0亿m^3,按渠系设置6个渠道管理处,建有各类渠道建筑物3 968座。

4.2 灌区小农水存在的问题

4.2.1 配水指标锐减,工程管护资金紧缺

以往青铜峡灌区的农民从未因农田缺水而犯愁,但随着人口的增加和经济的迅速发展,青铜峡灌区的水资源供需矛盾日益显现。自2003年以来,水利部下达的《黄河水量调度方案》分配给宁夏的总引黄水量指标常有削减,最多时比正常年份压减了33%,致使宁夏河套灌溉区春季溉用水缺口达12.3亿m^3,有200万亩农田无法正常灌溉,占全灌区农田总面积的1/3。由于各渠道管理单位多为自收自支单位,水费是后期工程管护的主要资金来源,所以来水的锐减直接导致水费收入大幅下降。

4.2.2 泥沙淤积严重,岁修劳动强度大

青铜峡灌区的多数渠道渠线较长,且多居农田荒野,无人值守。全灌区除西干渠实现了全断面砌护外,均为多种断面联合供水。由于支渠以下基本上都是土渠,风吹日晒雨淋之下,极易老化失修。加上近几年配水量日渐减少,负责渠道巡护的专业技术人员编制有限,个别渠段尤其是末梢段常因无水可灌出现淤废,因此岁修工程强度大。

4.2.3 节水力度低,农民用水协会方兴未艾

一直以来青铜峡灌区的节水力度都比较低,虽然有渠道砌护及田间工程改造,但灌区的

渠系水利用系数仅有0.43。组建农民用水协会后,青铜峡灌区实行了支斗渠承包管理,让用水农户对支渠以下的工程管理、灌溉用水、水费收缴进行自主管理,自我经营,初步形成了以"水管站+农民用水户协会+一把锹淌水"模式为主的支斗渠管理体制。但因为协会的发展程度不一,整个灌区的水资源利用效率还有待提高。

4.3 连续枯水状态下引黄灌区小农水改革的具体措施

目前,连续枯水状态下青铜峡灌区小农水改革的归因集中在灌溉面积与种植结构基底不清、工程基本建投资有限、农民用水协会管理有待完善三个方向,因此未来引黄灌区的小农水改革也应由此入手。

首先,应着重做好青铜峡灌区的灌溉面积与种植结构的调查工作。完善的调查与分析有助于根据连枯状态下的均值确定适当的灌溉面积和最优的种植结构,从而实现最严格水资源管理制度下满足三条红线要求的小农水建设。

其次,应通过多种渠道设法增加枯水期可利用的资源量。通过实施开源节流、分区滞水,加大污水回用等措施,全面协调,统筹发展,在培养农民水管技术带头人的基础上,大力发展节水型灌区建设,不断积累连续枯水状态下小农水工程设计、施工和运行管护的经验。

最后,要全面推进农民用水协会的建设,并想方设法引进资金完成小农水工程的建设,建议从种粮补贴中提取固定份额作为小农水的基本建设基金,并将水的零售价确定主动权下放给农民用水协会,从而将"一票到户、一把锹淌水"真正落到实处。

5 结语

当前,高端科技及模型模拟技术的研发与应用已经使水文预报的精度逐步提高,预报期也已经日渐缩短,随着全球极端气候的日益频发,为三农建设服务的小农水将日趋重要,而基于特殊状态下的连续枯水或连续丰水研究,将有效降低因极端事件常态化而导致的工程老大笨粗,使小农水向精细化、适应化方向发展,并在摸清本底条件、加大资金投入力度、完善设计方法与标准的基础上实现合理高效利用水资源,全面促进农村水利事业的发展。

参考文献

[1] 王浩.实行最严格的水资源管理制度理论基础与科技支撑[R].北京:中国水利水电科学研究院,2011.

[2] 水利部.2009年中国水资源公报[M].北京:中国水利水电出版社,2010.

[3] 郑梅.蚌埠市农村饮用水安全现状及保障措施[J].现代农业科技,2010(9):286-287.

[4] 卢伟,高国力.出台生态补偿条例是解决西部问题的关键[N].中国经营报,2012-01-07.

[5] 田成龙,孙学平,张允,等.宁夏引黄灌区灌溉面积及作物种植结构遥感调查成果报告[R].宁夏水利厅灌溉管理局,宁夏遥感测绘勘察院,2011.

【作者简介】 曹麟,女,1975年3月生,2009年毕业于河海大学,硕士,宁夏水电技师学院水利系主任,高级讲师,学科带头人。

赤泥对水稻植株各部位镉含量及水稻生物量的影响

刘慧莹　匡向阳　罗　琳

（湖南省永州水文水资源勘测局）

【摘要】 稻田镉污染直接影响水稻生长和稻米品质。本文主要通过测定水稻各部位镉的含量以及成熟期水稻株高、穗长、千粒重等指标，分析施用赤泥对水稻植株各部位镉富集以及水稻生物量的影响。结果表明，镉在水稻植株不同部位的浓度大小依次为根>茎≫叶>壳>糙米，表明施用赤泥可减少镉在水稻各部位的累积，并能起到促进水稻生长的作用；按 7 500 kg/hm^2 施用赤泥可以较好地减少水稻植株对镉的吸收。

【关键词】 赤泥　水稻生物量　镉

1　前言

水稻是我国主要的粮食作物之一，主要分布于我国南方广大地区。稻米的质量安全和无公害生产涉及大多数人的健康和安全，在国内外已经引起高度重视。由于农业生产的发展和长期以来对有色金属矿产的开采和冶炼，使我国超过 2×10^7 hm ^{2}的良田，约占农业土壤的 20%，已受到重金属的污染。

镉是广泛存在于自然界的一种重金属元素，是一种对动植物都具有毒害作用的痕量重金属元素，具有高的移动性和低的中毒浓度，对各种生物有机体具有潜在的毒性。在一定浓度范围内，镉对植物显示出比其他重金属更大的毒性。镉在土壤中的存在形态有水溶态、可交换态、碳酸盐态、有机结合态、铁锰氧化态及硅酸态等 6 种形态。水溶态和可交换态为植物有效态，易被植物吸收，其余的几种形态均为难溶态，不易被植物吸收。当作物组织中镉浓度积累到一定水平时，会出现毒害症状，包括生长迟缓、褪绿、矮化、产量下降等，严重时甚至会导致作物死亡。

近年来，由于工业"三废"的不合理排放、固体废弃物（尤其是城市垃圾）处理不善、污水灌溉、污泥农用以及施用含重金属元素的肥料等，都导致了土壤中重金属镉含量急剧增加。水稻——食物链中的初级生产者，当生长于有毒重金属污染的土壤中时，过量的有毒重金属在其根、茎、叶以及籽粒中大量积累，不仅严重影响水稻的生长发育，而且还严重影响稻米品质，经食物链危及人类和动物等。如 20 世纪 60 年代，在日本就曾发生了因食用大量镉污染粮食，主要是大米而引发骨痛病。为此，世界各国都对粮食、食品中的镉含量制定了最高的限制标准，限制标准的制定对水稻等粮食作物生产提出了更高的要求和挑战。

随着工农业生产的发展，重金属对土壤和农作物的污染问题越来越突出。重金属具有很强的蓄积性、隐蔽性、不可逆性和长期性。因此,寻找一种解决方法势在必行。根据国内外文献报道,赤泥——氧化铝工业生产的废料,它的堆放不仅占用大量耕地,还对周围环境产生污染。但同时赤泥又具有高碱度,较稳定的化学成分、非常细的分散度、高的比表面积、较好的吸附性能。如果利用其环境修复特性,将它用做环境修复材料处理土壤环境中的重金属污染,具有成本低、工艺简单等优点,可达到以废治废的效果。

目前,开展了大量将赤泥应用于重金属污染土壤的研究,而将其应用在南方水稻的报道不多,对赤泥应用于南方稻田的最佳施用量报道更少。

本试验在大田试验下,采用不同赤泥施用量,结合适当施肥配方,通过测定成熟期水稻根、茎、叶、壳和籽实中的镉含量,分析赤泥施用对水稻不同部位镉含量分布的影响。同时结合在水稻成熟期测量其植株株高、穗长、千粒重等指标,分析赤泥施用对水稻生长量的影响。试验所得数据可以为赤泥应用于湖南农田生态系统作重金属稳定剂提供科学依据。

2 材料与方法

2.1 试验方法

本试验共设6个不同的处理,每个处理20 m^2,每个处理做3个重复,每个处理除了赤泥施用量不同,均添加20 kg有机肥。处理1~6赤泥施用量依次为0 kg、5 kg、10 kg、15 kg、20 kg、25 kg。

2.2 仪器与试剂

2.2.1 仪器

原子吸收分光光度仪(TAS-990 super F型 北京普析通用仪器有限责任公司),30—404型平板电热加热板(编号555,沪南试验仪器厂),电子分析天平(岛津 AUY220型),50 mL具塞容量瓶、150 mL锥形瓶、0.5 mL、2 mL、5 mL、10 mL移液管,电热恒温鼓风干燥箱(DHG-9246A型,上海精宏实验设备有限公司)。

2.2.2 试剂

浓硝酸(优级纯);高氯酸(优级纯);硝酸(0.1%)吸取1 mL浓硝酸至1 000 mL容量瓶中,用超纯水定容至刻度线;镉标准储备液1 000 mg/L,由国家标准物质研究中心提供;镉标准工作液10 mg/L,吸取镉标准储备液1 mL,用0.1%硝酸稀释至100 mL。

试验所用水均为超纯水

2.2.3 镉标准溶液的配制

用移液管分别吸取0 mL、0.50 mL、1.00 mL、1.50 mL、2.00 mL、2.50 mL镉标准工作液(10 mg/L)于50 ml的容量瓶中,用0.1%的稀硝酸定容至刻度线,分别配制成0 mg/L、0.10 mg/L、0.20 mg/L、0.30 mg/L、0.40 mg/L、0.50 mg/L的镉标准系列溶液,将其摇匀置于冰箱中储存、备用。

2.3 试验材料

2.3.1 供试作物

供试水稻品种:湘早籼31号。

2.3.2 供试土壤

供试土壤为河流冲积物发育的酸性潮泥田。供试土壤的理化性质如下:碱性氮336.0 mg/kg、有效磷5.8 mg/kg、速效钾92.0 mg/kg、全氮1.83 g/kg、有机质30.2 g/kg、全镉 4.51 mg/kg、pH 值 5.90。

2.3.3 供试赤泥

供试赤泥由中国长城铝业集团提供，为拜耳-烧结联合法赤泥，其基本理化性质如下：pH 值 12.7 、SiO_2 20.55% 、Al_2O_3 5.98% 、Fe_2O_3 8.5% 、CaO39.9%、K_2O 0.88% 、全镉 0.092 mg/kg。

2.4 样品的采集与处理

2.4.1 水稻样品的采集及预处理

在水稻成熟期，每个处理中按照梅花采样法取长势较好的 5 株水稻，将水稻做好标记带回实验室，先用自来水洗净，再用蒸馏水冲洗，然后在烘箱中 105 ℃杀青 0.5 h，接着在 75 ℃恒温烘干至恒重，将水稻的根、茎、叶、稻壳和糙米分开，粉碎，研细，并全部过 2 mm 筛，然后将样品分别收集在自封塑料袋中待用。

2.4.2 水稻各部位样品的消解

准确称取磨碎过筛混和均匀的样品 0.500 0 g，置于 150 mL 的锥形瓶中，加 15 mL 的混合酸（硝酸+高氯酸，4+1），放数粒玻璃珠，加盖浸泡过夜，锥形瓶上放一小漏斗，置于电热板上消解，若液体变棕黑色，再加少量混合酸，直至冒白烟、消化液呈无色透明或略带黄色，取下自然冷却，再加 10 mL 水继续加热，直至冒白烟后取下放冷，同时做试剂空白试验。将试样消化液移入 50 mL 的容量瓶中，用超纯水少量多次洗涤锥形瓶，洗液合并于容量瓶中并定容至刻度，混匀备用。

2.4.3 测量水稻植株的株高

在水稻成熟期，按梅花形布点选取其中长势较好的 5 株水稻，测量从土壤表面到水稻剑叶叶端的高度，计数并求其平均值，得到水稻植株的株高。

2.4.4 测定水稻谷粒千粒重

水稻收割后，将谷粒脱下，均匀混合后数 1 000 粒谷粒，称重，求其平均值，得到水稻谷粒千粒重。

2.4.5 测定水稻穗长

水稻收割后，随即挑选 5 株水稻，测量稻穗末端到稻穗尖端的长度，计数并求其平均值，得出这一处理的水稻植株的穗长。

2.4.6 镉标准曲线绘制

将先前已经备好的系列浓度（0.1 mg/L、0.2 mg/L、0.3 mg/L、0.4 mg/L、0.5 mg/L）的镉标准溶液用原子吸收分光光度仪测定其不同浓度下的吸光度，然后根据其浓度和吸光值绘制标准曲线。得镉标准溶液的标准曲线为 $Abs = 0.356C + 0.0023$，线性相关系数 $R_2 = 0.9994$。

2.4.7 计算

水稻样品中的镉含量按下式计算

$$X = (A_1 - A_2) \times V/M$$

式中：X 为试样中的镉含量，mg/kg；A_1 为测定用试样液中镉含量，mg/L；A_2 为试剂空白液中镉含量，mg/L；M 为试样质量或体积，g；V 为试样处理液的总体积，mL。

3 结果分析

3.1 水稻不同部位镉的分布

重金属镉一旦进入环境,尤其是进入土壤-水稻系统中就很难去除。过量的镉在水稻的根、茎、叶以及籽粒中大量积累,不仅影响水稻产量和品质以及整个农田生态系统,并可通过食物链危及动物和人类的健康。

水稻不同部位镉含量见图1。由图1可知,施用赤泥的处理当中根、茎、叶、壳、糙米吸收的镉含量比没有施用赤泥的要低。随着赤泥添加量的上升,水稻不同部位镉含量都呈现递减趋势,尤以水稻根系中镉递减趋势最为明显,说明施用赤泥有利于减少水稻中镉含量,改善水稻品质。

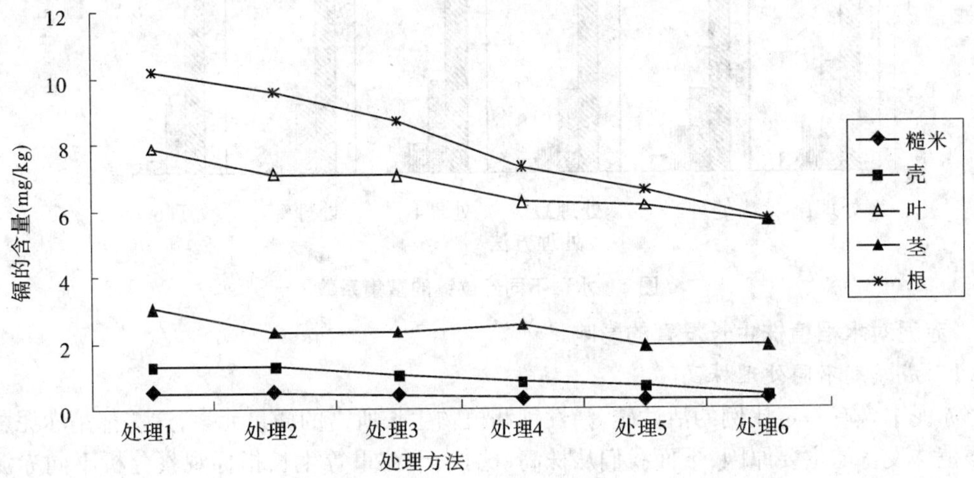

图1 水稻不同部位镉含量

3.2 镉在水稻不同部位富集能力的比较

植物从沉积物中吸收、富集的重金属,可用富集系数来反映植物对重金属富集程度的高低或富集能力的强弱。重金属富集系数是指植物某一部位的元素含量与土壤中相应元素含量之比,它在一定程度上反映着沉积物-植物系统中元素迁移的难易程度,说明重金属在植物体内的富集情况。重金属镉在水稻植株不同部位的富集系数如表1所示。

表1 镉在水稻不同部位富集系数

部位	处理1	处理2	处理3	处理4	处理5	处理6
根	2.26	2.12	1.93	1.62	1.46	1.27
茎	1.75	1.58	1.56	1.39	1.35	1.26
叶	0.67	0.51	0.51	0.56	0.42	0.42
壳	0.28	0.28	0.22	0.18	0.14	0.09
糙米	0.11	0.1	0.08	0.07	0.07	0.07

下面用图2进一步阐述镉在水稻体内的迁移规律。

外源镉进入土壤后,镉与土壤中的 OH^- 和 Cl^- 形成络合物则易于移动。由图2可知,镉的迁移能力较强。镉在植物体内各部分的分布为:根>茎>叶≫壳>糙米。同时,随着赤泥施用量的增加，富集系数逐渐降低。由此可认为，随着赤泥施用量的增加，通过吸附等作用可以改变水稻根际环境中重金属镉含量且降低了镉在土壤中的迁移能力，从而根系从环境中吸收镉总量也减少，这说明施用赤泥有利于控制农田生态系统重金属污染的风险，是建立可持续农业的有效途径之一。

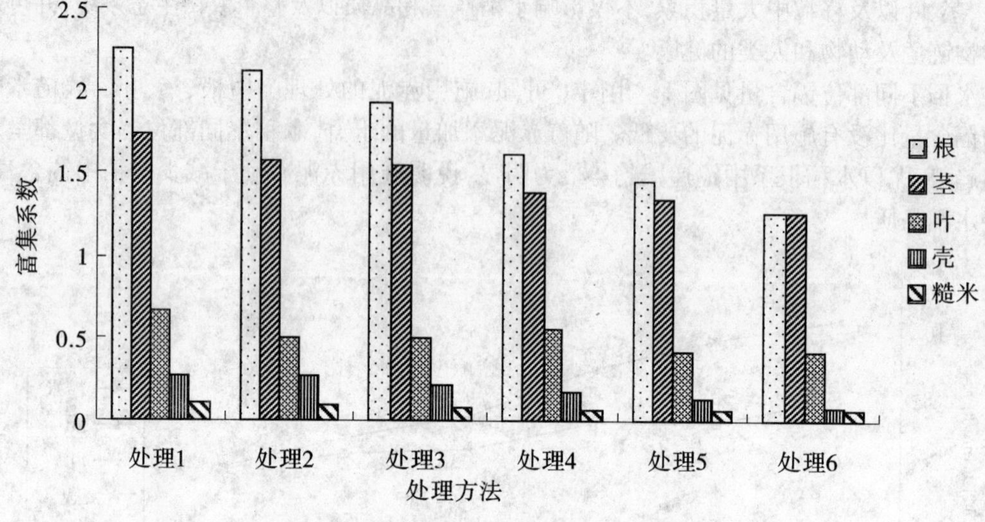

图2　水稻不同部位镉的富集系数

3.3　赤泥对水稻植株生长发育的影响

3.3.1　成熟期不同处理株高

赤泥中含有一定比例的活性硅、活性钙和植物生长所需的微量元素，这些都是赤泥影响水稻植株发育的主要原因，下面我们从株高、穗长、千粒重等生长指标观察分析不同赤泥施用量对水稻植株生长发育的影响。

由图3可知，水稻成熟期处理3和处理4的长势较其他处理要好，株高分别为100.92 cm和101.42 cm，相比处理1高出了约3 cm，增幅约为3%。试验中赤泥添加量的多少是最主要因素，由此可以认为水稻长势之所以有一定的差距，可能与赤泥的添加量有关。为了更好地证明赤泥的改良效用，下面从水稻千粒重和穗长进一步分析其作用。

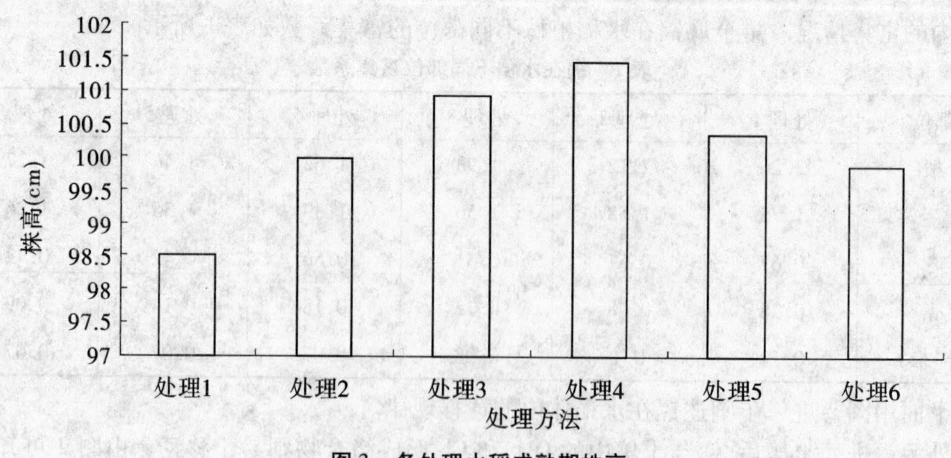

图3　各处理水稻成熟期株高

3.3.2 成熟期不同处理穗长

在水稻成熟后我们对不同处理的水稻穗长进行测量，从处理 1 到处理 6 依次为 20.68 cm、22.26 cm、22.32 cm、22.72 cm、22.72 cm、22.14 cm。一般来说穗长越长，同一稻穗上的出籽率越高，有可能增加水稻的结实率，有利于提高水稻产量。

各处理水稻成熟期穗长见图 4。由图 4 可知，处理 4 的穗长相比处理 1 增加了约 2 cm，增幅约为 10%。可知，施用赤泥的处理当中穗长明显要比没有施用赤泥的长，说明赤泥添加对水稻生长确实有一定的促进作用。但同时我们发现，处理 6 添加的赤泥量虽然比处理 5 的多，但穗长却比处理 5 短，赤泥可以用于改良土壤性质，应用于南方稻田酸性土壤当中，降低土壤中重金属镉含量，提高水稻品质。但是并不是赤泥施用量越多越好，赤泥呈碱性，而南方的水稻适宜在酸性环境下成长，添加过量的赤泥反而会抑制水稻的生长。

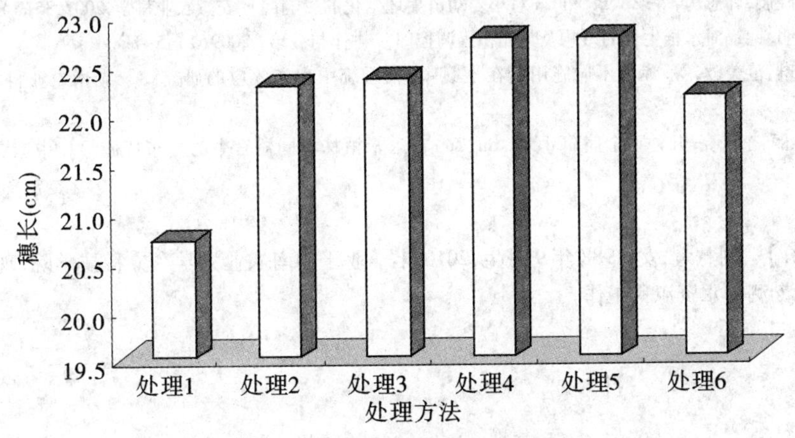

图 4 各处理水稻成熟期穗长

3.3.3 成熟期不同处理千粒重

在水稻成熟收割后我们对不同处理的水稻千粒重进行称重，从处理 1 到处理 6 依次为 23.57 g、24.16 g、24.73 g、25.32 g、24.54 g、24.38 g。

从上述数据可知，处理 1 千粒重比较低，处理 4 的千粒重最大，处理 3 其次。由水稻株高、穗长和千粒重三个指标综合得出添加了赤泥的处理都要比没加赤泥的处理 1 长势好。同时，随赤泥的添加，水稻株高、穗长、千粒重都呈现先上升后下降的趋势。对于这个现象可以推断：在施加适量赤泥的条件下水稻产量较好，因为赤泥具备有一定的吸附重金属离子的能力，适量的赤泥可以降低土壤中重金属离子的含量，从而减少水稻植株对重金属离子的吸附，而且适量的赤泥可以补充土壤的微量元素。处理 5、6 可能是由于加入了过量的赤泥导致产量稍有下降。赤泥呈碱性，加入过量的赤泥到稻田当中改变土壤的理化性质，使土壤呈碱性。南方水稻本来是适应于生长在酸性土壤，生长环境的改变造成水稻产量的减少。

4 结语

通过试验可以得到以下结论：

(1)在试验中，综合考虑水稻产量、经济效益和品质，对比可知赤泥的最佳投入量为每 20 m^2 中添加 15 kg，即 7 500 kg/hm^2。

(2)与不施用赤泥的处理相比，添加适量赤泥可以一定程度增加水稻产量。赤泥可以用于改良南方稻田酸性土壤，起到重金属稳定剂的作用，降低土壤中生物有效态重金属镉含量，从而降低水稻中镉含量。

(3)在其他生长环境相同的条件下，不同量的赤泥施用后，重金属镉在植物体内各部分的分布为根>茎≫叶>壳>糙米。比较 6 个处理的不同部位镉含量，可知施用赤泥的处理当中，根、茎、叶、壳、糙米吸收的镉含量比没有施用赤泥的要低。

(4)适量的赤泥可补充土壤中微量元素，但赤泥呈碱性，而南方的水稻适宜酸性，过量的添加反而抑制水稻的生长。

参考文献

[1] 赵中秋，朱永官，蔡运龙. 镉在土壤-植物系统中的迁移转化及其影响因素[J]. 生态环境, 2005,14(2):282-286.

[2] 马海涛，李晓晨，郭志勇，等. Zn、Pb 和 Cd 对小麦幼苗生理生化的影响[J]. 安徽农业科学，2007,35(3):647-648.

[3] 覃都，王光明. 锰镉对水稻生长的影响及其镉积累调控[J]. 耕作与栽培, 2009(1):5-7.

[4] 魏建宏，罗琳，范美蓉，等. 赤泥不同施用量在土壤-水稻系统中生态效应的研究[J]. 湖南农业科学，2009(10):39-42.

[5] 李静，依艳丽，李亮亮. 几种重金属(Cd、Pb、Cu、Zn)在玉米植株不同器官中的分布特征[J]. 中国农学通报，2006(4):12.

【作者简介】 刘慧莹，女，1988 年 9 月生，2010 年毕业于湖南农业大学东方科技学院，现在湖南省永州水文水资源勘测局水资源科工作。

五、水力学与泥沙

齐家川示范区水土保持综合治理研究

张绒君　赵安成　李怀有　白文瑷　陈昊潭

（黄河水土保持西峰治理监督局　庆阳　745000）

【摘要】 齐家川示范区以小流域为单元进行综合治理，治理措施包括工程措施、植物措施、小型拦蓄工程措施、节水灌溉措施和技术推广措施。示范区以径流高效利用和水资源合理配置为重点和特色，建立了集塬面、沟坡、沟道径流集蓄利用为一体，节灌、补灌相结合的崔沟径流高效利用示范村，搭建了传统水土保持治理措施与节水灌溉、集水造林等新技术措施衔接、组装与集成的平台，提高了地表径流平均利用率，实现了既定目标。

【关键词】 水土保持　综合治理　技术体系

依据黄土高塬沟壑区水土流失规律研究成果，径流来自于塬面，泥沙来自于沟谷，塬水下沟后，流域产沙量增加77%左右，侵蚀模数增加1.26～1.40倍。塬面径流占流域总量的67.4%，泥沙占流域总量的12.3%。其中，村庄道路（多为胡同）是主要产区，径流量占塬面部位的87%、泥沙量占塬面部位的92%；坡面径流占流域总量的8.6%，泥沙占1.4%；沟谷部位径流占流域总量的24%，泥沙占86.3%。在沟谷中沟床和红土泻溜是主要产沙区，占沟谷泥沙的96%，占流域的83%。

因此，要从根本上解决高塬沟壑区的水土流失问题，就必须控制塬面径流，一般治理的模式是：在塬面建设基本农田、改道路、堵胡同、修建小型拦蓄工程，水窖、涝池、沟头防护等措施，沟道修建骨干坝，坡面修水平梯田和水平阶。但对拦蓄工程所蓄集的径流如何利用，则较少注重，因此开展集塬面、沟坡、沟道径流集蓄利用为一体，节灌、补灌相结合的技术研究，集成传统的水土保持治理措施与节水灌溉、集水造林等新技术措施，提高地表径流平均利用率等研究，对水土保持综合治理具有重要指导意义。

1　齐家川示范区概况

黄河水土保持生态工程齐家川示范区地处甘肃省庆阳市西峰区境内，由齐家川、清水沟、老庄沟和南小河沟4条小流域组成，总面积166.57 km^2，其中齐家川小流域面积为48.35 km^2，清水沟小流域面积为43.21 km^2，老庄沟小流域面积为38.71 km^2，南小河沟小流域面积为36.30 km^2。项目区涉及西峰区的什社、温泉、董志、陈户、后官寨5个乡（镇）31个行政村及黄河水利委员会西峰水土保持科学试验站的南小河沟试验监测站。

项目区属黄土高塬沟壑区，主要由塬、坡、沟3大地貌单元组成，1 km以上支毛沟70条，塬面坡度多小于5°，沟坡多大于25°；多年平均降水量579 mm，降水季节分布不均，主要集中在7～9月，约占全年降水量的68%，且以暴雨为主；11月底至翌年2月底为大地封冻期，冻土时间长达 3 个多月，冻土深 60 ~ 106 cm；年平均日照时数为 2 423 h，无霜期 160 d；气象灾害以干旱威胁最重、影响最广、发生最多，且常伴有大风、沙尘暴、干热风、霜冻等。

2 治理措施总体布局与技术体系

治理措施总体布局与技术体系见图 1。

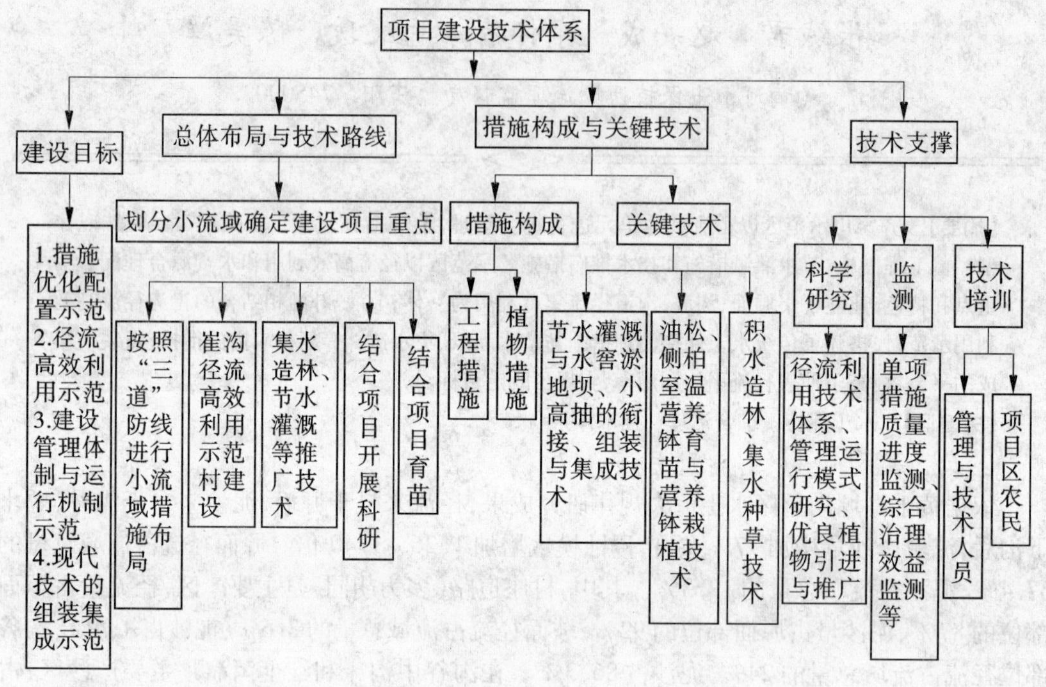

图 1 治理措施总体布局与技术体系

2.1 治理措施总体布局

示范区综合防治措施总体布局是塬面以完善水平梯田、胡同堰、道路涝池、庄院水窖为主体，建设径流集蓄设施，以上水、输水、大田喷灌、果园微灌为主体，建设径流高效利用设施，控制塬面径流下沟，使水土保持措施与主导产业开发紧密结合；沟坡以集水造林、种草和坡改梯为主要措施进行坡面防护体系建设，提高植被覆盖率，减轻坡面侵蚀，改善生态环境；沟谷以沟头防护、沟边埂、支毛沟谷坊、骨干工程为主体配置沟道治理措施体系，使侵蚀活跃的沟头及沟道得到有效防治。

2.2 技术体系

(1)以小流域为单元进行综合治理。在“保塬固沟”的要求下，按照黄土高塬沟壑区“三道”防线综合治理模式进行措施布设，单项工程依据不同立地条件，因地制宜地进行工程设计。在塬面治理中，重点实施梯田工程(该项目区重点为塬边、塬嘴)，着力进行塬面农田整治，改善农业生产条件；大力修建沟头防护、涝池、水窖等塬面小型蓄水工程，集蓄塬面道路、村庄径流，为塬面径流高效利用创造水源条件。

在坡面治理中，除实施必要的坡改梯工程外，重点实施坡面林草措施，调整土地利用结构，改善生态景观，为舍饲养殖提供保障，建设林果产品生产基地。乔木林的栽植树种逐步由以刺槐、白榆为主的造林先锋树种向以油松、侧柏为主的常绿树种转变；造林整地工程技术标准逐步由简单的造林预整地方式向以集水造林技术为支撑的坡面集水整地方式转变；林种结构逐步由纯林向乔灌混交林转变；栽植技术逐步由裸根苗栽植向营养钵带土栽植方向转变。

在沟道治理中，支毛沟治理按照沟头防护工程、沟边埂、沟坡乔灌草护坡林、沟底防冲林、小型淤地坝或谷坊工程模式进行布设；主沟或支沟治理按照沟头防护工程、沟边埂、沟坡乔灌草护坡林、沟底防冲林、治沟骨干工程、坝地或沟川台地开发利用模式进行布设。

(2)小流域树立示范实体样板。在清水沟小流域建设崔沟径流高效利用示范村，树立技术示范实体样板。在引进节水灌溉、集水造林、营养钵栽植等新技术的基础上，与传统梯林草坝等常规治理技术结合，建立有黄土高塬沟壑区特色的径流利用工程体系与技术体系。塬面建立以果园、蔬菜补灌人畜饮水为主，大田补灌为辅的就地拦蓄就地利用的塬面径流高效利用体系；坡面建立以满足林草植被正常生长水分需求，就地拦蓄就地利用坡面径流高效利用体系；沟道建立以满足塬面、坡面、沟道两岸或下游川台坝地生产开发用水需求的异地拦蓄、异地利用沟道径流高效利用体系。

(3)重点引进节水灌溉、集水造林、营养钵育苗、营养钵栽植等新技术，大力推广苹果丰产栽培、地膜覆盖、温棚养殖等生产技术。

节水灌溉、集水造林等新技术如图2所示。

(a)果园节水灌溉

(b)沟道径流拦蓄

(c)集水造林

(d)喷灌

图2　节水灌溉、集水造林等新技术

3 治理措施与关键技术

齐家川示范区所实施的治理措施包括工程措施、植物措施、小型拦蓄工程措施、节水灌溉措施和技术推广措施5个方面,工程措施主要有治沟骨干工程、小高抽上水工程、谷坊、梯田等,植物措施主要有乔木林、灌木林、经济林、果园、农田防护林、人工种草等,小型拦蓄工程措施主要有水窖、涝池、沟头防护等,节水灌溉措施主要有微灌、喷灌、低压管灌、施药微灌复合系统、温室育苗微喷系统等,技术推广措施主要有苹果丰产栽培技术、地膜覆盖种植技术、温棚养殖技术、美国香豌豆和牧场草种植推广等。

5个方面治理措施的关键技术为:①节水灌溉技术与水窖、涝池、沟道骨干工程、小高抽上水工程、温室生产设施的衔接、组装和集成;②油松、侧柏温室营养钵育苗技术与营养钵带土栽植技术;③与造林立地条件相适应的集水造林与集水种草技术。

4 技术支撑

开展"齐家川示范区综合防治措施体系与管理机制研究"、"齐家川示范区径流高效利用技术体系研究"和"齐家川示范区水土保持优良植物推广"3项科研课题的研究,解决了项目建设、项目管理、技术推广过程中的技术难题,项目研究人员主持或直接参与了有关工程项目的施工设计工作,为项目建设提供关键技术支撑,为项目科学管理提供咨询服务。

开展项目效益、质量、进度等综合监测,实时掌握项目各方面的进展动态,为项目管理和后期评估服务。结合技术引进、新技术消化吸收以及新技术推广,大力培训从事实际生产的农民技术人员,帮助他们掌握关键技术,提高生产经营水平,同时针对项目管理工作的难点和要求,培训项目管理技术人员,促进项目的规范化和科学化管理。

5 水资源优化配置和高效利用

据有关初步研究成果,黄土高塬沟壑区水土流失的基本规律是径流来自于塬面,泥沙来自于沟谷,塬水下沟后,流域产沙量增加77%左右,侵蚀模数增加1.26~1.4倍。塬面径流占流域总量的67.4%,泥沙占流域总量的12.3%,其中村庄、道路是塬面径流的主要产区,其径流量占塬面总量的87%、泥沙量占塬面总量的92%;坡面径流占流域总量的8.6%,泥沙占1.4%;沟谷部位径流占流域总量的24%,泥沙占86.3%,沟谷中沟床和红土泻溜是流域泥沙的主产区,其产沙量占沟谷泥沙量的96%,占流域总量的83%。要从根本上解决黄土高塬沟壑区的水土流失问题,就必须控制塬面径流,传统治理模式虽然在塬面修建了水窖、涝池等小型拦蓄工程,但是对拦蓄工程所蓄积的径流如何利用,能否利用,怎样利用以及利用到什么地方,为谁所用则在设计时较少注重,因此注重对拦蓄径流的高效利用,实际上是对传统水土保持技术措施内涵的延伸与扩充;注重径流高效利用,不但能减轻流域水土流失,同时又使稀缺的降水资源化,解决了本区干旱缺水危害,达到了双赢目标。

在初步设计阶段,齐家川示范区依据黄土高塬沟壑区水土流失的基本规律,把径流高效利用和水资源合理配置作为建设的重点和特色,坚持把径流利用作为一项主要技术支撑措施予以实施,通过建立集塬面、沟坡、沟道径流集蓄利用为一体,节灌、补灌相结合的崔沟径流高效利用示范村这一特色样板工程,搭建了传统水土保持治理措施与节水灌溉、集水造林等新技术措施衔接、组装与集成的平台,使崔沟径流高效利用示范村的地表径流平均利用率由示范前的2.27%提高到示范后的18.9%;地下浅水由示范前的5.39%,提高到示范后的24.8%。效益增长迅猛,年增加产值64万元,年均净收益42.4万元,人均增加粮食10 kg,增加果品151 kg,年人均增加纯收入231.2元,实现了既定目标,使其成为示范区建设的一个亮点工程,彰显了黄土高塬沟壑区水保生态工程的特色与巨大的效益空间。

6 存在的问题与建议

(1)对威胁城镇安全且危害较大的沟头,治理力度小,投入不足。按照以往小流域综合治理的惯例,沟头防护工程按小型蓄排水工程对待,投资包含在综合治理每平方公里的造价之中,但是实际上许多伸入董志塬腹地的大型沟头的集水面积并不亚于一个中小型淤地坝的控制面积,已实施的该类大型沟头防护投资约16万。受规定和投资规模限制,设计时以一些小型沟头建设防护工程代替大型沟头防护治理,造成"抓小放大",使真正危害城市村镇安全、威胁道路交通、破坏性大的大型活动沟头长期得不到有效治理。当遇到大暴雨或特大暴雨时,综合防治体系的潜在漏洞表露无遗,塬水下沟,沟头前进,吞食塬面,破坏之大触目惊心。建议在以后的综合治理项目中,应注重对危害城镇安全和乡村主干交通道路的大型活动沟头治理,加大沟头防护工程的投资力度,把大型活动沟头的防护工程提高到与治理骨干工程或中型淤地坝同等重要的位置,提升沟头防护工程在黄土高塬沟壑区治理中的地位和作用。

(2)进一步加强治理成果的管护力度。坡面林草措施量大、面广、战线长,是管护的难点和关键;林牧矛盾问题突出,个别经济条件差的边远山区乱牧毁林(草)现象时有发生,需进一步研究和解决;治理过程中管护费用较低、竣工验收后管护费用无保障,往往造成大片林草植被毁。在今后的项目建设中,应加强有关法律法规的制定和宣传,从思想上解决林木的管护问题。依靠项目区所在地的水土流失预防监督执法队伍和乡村两级政府,加强巡查和督导,加大对乱牧毁林案件和破坏水土保持设施案件的查处力度,为治理成果的管护提供保障。淤地坝的管护要从产权、使用权、收益权、防汛服从权方面入手,建立责权利统一的维修管护体制和运行机制。

(3)慎重采用个别治理技术。齐家川示范区建设项目设计采用了沟道土谷坊治理技术,由于谷坊的设计标准较低(土谷坊防御标准为10年一遇3 h最大暴雨),工程规模较小,且为全拦全蓄,无任何排泄措施,致使遇到较大暴雨洪水时,谷坊群极易被连级冲毁,水土保持效益不显著。鉴于此经验教训,在今后该类型区的水土保持治理设计中,应慎重采用沟道土谷坊技术。

(4)水土保持建设项目投资概算调整。水土保持基本建设项目多年来的建设实践证明:水土保持生态建设项目跨年度项目相对较多(一般为5年)、工期相对较长,受物价和国家政策变化等诸多因素影响,水土保持建设项目投资成增加趋势。目前,水土保持项目在实施过程中未见到调整概算,增加投资的情况,致使项目设计初期的跨年度概算不完全符合多年的物价实际情况。从该项目建设实践看,急需建立水土保持建设项目投资概算调整体制和畅通渠道,以保障基本建设项目的质量和建设各方的正当利益。

(5)注重节水灌溉、集水造林等新技术应用,可提高干旱地区的水资源利用率和经济效益。通过建立集塬面、沟坡、沟道径流集蓄利用为一体,节灌、补灌相结合的崔沟径流高效利用示范村这一特色样板工程,搭建传统水土保持治理措施与节水灌溉、集水造林等新技术措施衔接、组装与集成的平台，使示范村的地表径流平均利用率和经济效益显著增加，人均增加粮食 10 kg，增加果品 151 kg，年人均增加纯收入 231.2 元，彰显了黄土高塬沟壑区水保生态工程的特色与巨大的效益空间，值得在黄土高原水资源匮乏地区推广，但是新技术投资超出目前的国家投资定额(8.0 万元/km^2)，完成比较困难，建议今后水土保持生态项目建设投资审核时，注意其特殊性，可将其投资单列，不按照每平方公里投资定额核减。

参考文献

[1] 李倬. 黄河中游黄土高塬沟壑区的水土流失特点和治理探讨[C]//黄土高原水土流失及其综合治理研究——西峰水保站试验研究成果及论文汇编(1989~2003). 郑州：黄河水利出版社，2005.

[2] 赵安成，李怀有. 黄土高塬沟壑区水资源调控利用技术研究[M]. 郑州：黄河水利出版社，2006.

【作者简介】 张绒君，女，1962 年 12 月生，1987 年 6 月毕业于西北林学院森林资源保护系，本科，农学学士，高级工程师，主要从事水土保持规划、研究等工作。

青藏高原开发建设项目水土保持治理措施的新探索

——新建柴达尔至木里地方铁路工程水土保持措施的实践与经验

安润连

（黄河水土保持西峰治理监督局　庆阳　745000）

【摘要】 我国青藏高原属大陆性高原气候区，海拔高，气候寒冷，人烟稀少，多为牧区，很少有大型开发建设项目，因此也很少有人工水土保持的先例。青海省新建柴达尔至木里地方铁路工程在建设中率先进行植物种植试验研究并取得成功，实施了混凝土网格种草护坡、纯种草护坡、草皮护坡、草皮排水沟及临时用地种草等措施，开创了青藏高原地区开发建设项目水土保持植物种植管护的先列，为青藏高原开发建设项目水土保持措施的多元化提供了科学依据，标志着该地区水土保持事业进入了一个新时期。

【关键词】 青藏高原　水土保持　措施

1　项目区自然概况

新建柴达尔至木里地方铁路工程项目区地处青藏高原的祁边-昆仑断陷盆地内，海拔高程3 000～4 300 m，线路经过地段主要为哈尔盖河多级河流冲积型狭长阶地地貌和大通河河谷阶地及滩地。哈尔盖河谷阶地区河谷宽浅，地形平坦，高寒草甸发育良好，零星分布沼泽地。大通河河谷阶地为连续分布的高寒草甸，并有较大面积分布的斑块状高寒沼泽草甸。

项目区属大陆性高原气候区，日照时间长，气温日差较大。多年平均气温-0.3～-0.5 ℃，多年平均降水量341.6～379.4 mm，多年平均蒸发量1 464～1 791 mm；年平均8级以上大风日数47～57 d，年平均风速3.2～3.4 m/s，主导风向西北；最大冻土深度350 cm。

由于地处高寒人烟稀少地区，历史上无人工种植植被的先例。

2　工程建设概况

2.1　工程性质及位置

新建柴达尔至木里地方铁路是由青海省地方铁路管理局建设的一条运煤专用铁路，是连接海西蒙古藏族自治州和海北藏族自治州煤炭资源分布区的重要运输通道。铁路起点与青藏铁路哈尔盖至柴达尔支线的柴达尔车站相接，终点为天峻县木里煤矿，沿途经过刚察县、祁连县和天峻县。

2.2 工程水土保持措施原始设计

2005 年 7 月,铁道第一勘察设计院编制的《新建铁路柴达尔至木里可行性研究报告》,为路基边坡及附属工程设计浆砌石护坡 68 543 m,浆砌石挡土墙 960 m。

2.3 工程水土保持设计变更及建设情况

由于项目区特殊的自然条件,浆砌石工程在热胀冷缩和冻融侵蚀的作用下变形损坏严重。铁路建设单位邀请青海省高原生物科学研究院的工程技术人员现场进行植物品种的种植试验。冷地早熟禾、垂穗披碱草和多叶老芒麦种植试验成功。据此,设计单位对原设计的浆砌石护坡进行设计变更,根据路基边坡土壤、边坡高度及所在区域的自然条件,分别变更为浆砌石拱形骨架种草护坡(见图 1(b))、混凝土网格骨架种草护坡(见图 1(a))、纯种草护坡(见图 1(c))和草皮护坡,并将所在区域的浆砌石排水沟变更为草皮排水沟(见图 1(d))。设计变更后,共建成浆砌石拱形骨架种草护坡 19 100 m,混凝土网格骨架种草护坡 120 595 m,防护面积 152.07 hm^2;种草护坡 54 959 m,防护面积 48.38 hm^2,草皮移植护坡 33 684 m,防护面积 30 hm^2;草皮排水沟 55 704 延米;取土场、弃渣场、施工营地等临时用地种草 132.67 hm^2,路基边坡防护及临时用地地表植被恢复顺利完成。

(a)混凝土网格骨架种草护坡

(b)浆砌石拱形骨架种草护坡

(c)纯种草护坡

(d)草皮排水沟

图 1 水土保持设计变更

新建柴达尔至木里地方铁路工程于 2006 年 4 月开工建设, 2010 年 7 月 3 日全线竣工验收,边坡防护工程建于 2008 ~ 2009 年。经过 2 ~ 3 年的运行,所种植的植物措施生长良好,说明采用的治理措施合理,选择的植物种类准确。

3 工程采用的植物品种

3.1 植物品种及其特性

3.1.1 垂穗披碱草

垂穗披碱草,拉丁名 Elymus nutans Griseb,别名钩头草,湾穗草。禾本科,多年生。高 50 ~ 70 cm,根茎疏丛状,须根发达。秆直立,具3节,基部节稍膝曲。叶扁平,长6 ~ 8 cm,宽 3 ~ 5 mm,两边微蹙草或下部平滑,上面疏生柔毛。在我国主要分布于我国内蒙古 、河北、陕西、甘肃、宁夏、青海、新疆,四川、西藏等省(区)。4 月下旬至 5 月上旬返青,6 月中旬至 7 月下旬抽穗开花,8 月中下旬种子成熟,全生育期 102 ~ 120 d。抗寒耐低温,喜生长在平原、高原平滩以及山地阳坡、沟谷、半阴坡等。常与芨芨草(Achnathe-rumsplendens)、紫花芨芨草(A. Purpurascens)等组成芨芨草-垂穗披碱草草场;可与冷地早熟禾(Poacrymophila)、草地早熟禾(P. Pratensis)混播。

3.1.2 冷地早熟禾

冷地早熟禾,拉丁名 Poacrymophila,禾本科,多年生。根须状,秆从生、直立、稍压扁,高 30 ~ 65 cm,具2 ~ 3 节。叶鞘平滑,基部略带红色,长3 ~ 9.5 cm,宽0.7 ~ 1.3 cm。圆锥花序狭窄而短小,花序长4.5 ~ 8.0 cm。小穗灰绿带紫色,长3 ~ 4 毫米,含1 ~ 2 朵小花。在我国主要分布于青海、甘肃、西藏、四川、新疆等地区。在青海省播种后第二年 4 月下旬至 5 月上旬返青,5 月中旬至 6 月上旬孕穗,6 月上旬至 7 月上旬抽穗开花,8 月下旬种子成熟,生育期 105 ~ 115 d。冷地早熟禾根茎发达,分蘖能力强,常见于海拔 3 200 ~ 3 600 m 的河谷、缓坡、湿润平滩、沼泽化草甸周围。

3.1.3 老芒麦

老芒麦 ,拉丁名 Elymus sibiricus Linn,别名西伯利亚披碱草、多叶老芒麦,禾本科 ,披碱草属 ,多年生,疏丛型,须根密集而发育。秆直立或基部稍倾斜,粉绿色,具3 ~ 4 节,3 ~ 4 个叶片,各节略膝曲。叶鞘光滑,下部叶鞘长于节间,叶舌短,膜质,长0.5 ~ 1 mm。叶片扁平,内卷,长 10 ~ 20 cm,宽 5 ~ 10 mm。穗状花序疏松下垂,长 15 ~ 25 cm,具 34 ~ 38 穗节,每节 2 小穗。

老芒麦根系发达,入土较深。春播第一年根系的分布以土层 3 ~ 18 cm 处为最密,第二年根系入土可达 125 cm,是优良的水土保持草种。老芒麦属旱中生耐严寒植物,在年降水量 400 ~ 500 mm 的地区,可旱地栽培,在 -3 ℃的低温下幼苗不受冻害,冬季气温下降至 -38 ~ -36 ℃时,能安全越冬,越冬率为 96% 左右。在青藏高原秋季重霜或气温下降到 -8 ℃时,仍能保持青绿。老芒麦对土壤的要求不严,在瘠薄、弱酸、微碱或含腐殖质较高的土壤中均生长良好。在 pH 值为 7 ~ 8,微盐渍化土壤中亦能生长。老芒麦具有广泛的可塑性,能适应较为复杂的地理、地形、气候条件。

3.2 种植技术

3.2.1 种草技术

(1)种植土配置:将有机质、肥料、保水剂、消毒剂等,按一定比例加入腐殖土(剥离表土)中,混合成种植土。种植土中有机质≥33.0%;腐殖质≥12.0,氮、磷、钾不低于 5.0%,水分(游离水)约 20.0%。

(2)边坡处理:在种草前,先在路基边坡覆 10 ~ 20 cm 种植土,用六齿耗清理草根和杂物,并沿等高方向起垄,垄高约5 cm,行距5~8 cm,保护种子不受水流冲刷影响。

(3)种子选配:采用垂穗披碱草与冷地早熟禾混播,或多叶老芒麦与冷地早熟禾混播。播种量及种子比例见表1。

表1 播种量及种子比例

种子配置	播种量(kg/hm²)	百分比(%)
穗披碱草或多叶老芒麦	50~60	70
冷地早熟禾		30

(4)种子质量要求及处理:采用一级种子,播前用45 ℃的温水加入适量生根粉浸泡12~24 h。

(5)播种时间及方式:5月底6月初播种,使草籽有充分时间生长出一心两叶,以便越冬,人工播种或喷播。

(6)养护:播后覆膜,洒水养护。20 d左右出苗,出苗后视长势揭膜,避免膜内温度过高将幼苗烧死,养护期不少于30 d。

(7)注意事项:播种前应先在坡面洒水浸润边坡土体,播种初期定期浇水或根据天气情况洒水,确保种子发芽所需的水分及温度。幼苗长出后期定期浇水施肥,使其快速成坪。成坪后注意防止病虫害,并加强看护,避免牲畜及人为踩踏和弃土弃碴压盖。

3.2.2 草皮排水沟铺设技术

(1)草皮质量要求及采取时间:挖取草皮要选在草本植物的分蘖期—结实期,即5~8月。草皮挖取后,若不能及时移植,应加强养护,每天洒水不少于3次,水温控制在10~20 ℃。

(2)草皮挖取技术:为保证草皮根系的完整性,草皮厚度要大于根系埋入地下的深度,本工程铺设草皮水沟所用草皮厚为15 cm。草皮大小考虑方便人工搬运,切块为20 cm×20 cm~50 cm×50 cm。

(3)草皮铺砌:铺草皮是一种较为传统的施工工艺。在路基边坡进行草皮铺装,是根据坡面冲刷情况、边坡坡度、坡面水流速度等具体条件,分别采用平铺(平行于铺面铺装)、水平叠铺(平行于水平地面)、垂直叠铺(垂直于坡面)、斜交叠铺(与坡面垂直成一个小于90°的角)等形式。草皮排水沟断面尺寸设计为底宽1 m,深40 cm,边坡1∶1的梯形断面。铺设前将底部夯实,草皮铺设时块与块之间留3~5 cm的缝,铺设完成后将缝间用拌有植物种子的腐殖土填实。

(4)养护与管理:草皮移植过程中根系损伤较大,在新的环境中重新扎根之前,必须渡过一个休克期。在此期间,要求始终保持湿润状态,若水分供给不足,会导致草皮死亡。同时要加强看管,或用铁丝网隔离,保证在初期生长阶段不受啃食和踩踏损坏。

3.2.3 结论与建议

柴达尔至木里铁路工程水土保持的成功经验,填补了区域空白,值得同类地区推广。建议有关科学研究单位进一步加强青藏高原地区水土保持科学研究,推进该地区水土保持事业的深入发展和水土保持技术的更加成熟。

【作者简介】 安润连,女,1960年10月生,毕业于西北农林科技大学,大学本科学历,就职于黄河水土保持西峰治理监督局,工程师。

淮南煤化工基地取水口河段河道演变分析

余彦群　王再明　洪　成　丁　翔

（中水淮河规划设计研究有限公司　蚌埠　233001）

【摘要】 六坊堤河段为淮河干流典型的分汊河段，本文通过对该河段来水来沙条件、河道近期演变的分析，得出本段河道左、右汊的冲淤特征将维持较长时期，河段分汊格局不易改变，本段河道河势变化趋势基本稳定，规划工程实施后对河道的演变不会产生大的影响，取水工程的布置是合适的。建议对取水头上、下游一定范围内的岸坡进行防护处理，对管道穿越淮北大堤段进行截渗处理，确保淮北大堤的防洪安全和六坊堤河段的河势稳定。

【关键词】 淮河　六坊堤　河道演变　取水工程

1　基本情况

1.1　取水工程

淮南煤化工基地位于安徽省淮南市西北部，拟建成涵盖煤化工、新型建材业、先进制造业、仓储物流产业、创新型产业、综合服务业等多种产业类型的产业新城。该基地的建设和运行需消耗大量的水资源，为此需配备完善的输配水系统。输配水系统的主要任务是抽引淮河水进入煤化工基地的水厂，解决化工生产及生活用水需要。

淮南煤化工基地取水工程位于淮河干流六坊堤河段，该段上起灯草窝生产圩上口（凤台大桥下游 1.06 km），下至下六坊堤下口（平圩大桥上游 1.60 km）。为了取水工程能够长期稳定地取到所需水量，需研究取水口处淮河所在河段河道近期演变情况，从河势角度分析拟建取水工程的可行性，并定性分析河势变化对工程建设的影响以及工程建设后对河势的影响，为工程建设和运行期有关决策、实施和运营管理提供必要的依据或参考。

1.2　淮河六坊堤河段河道

淮河六坊堤河段分为左、右两汊，左汊经灯草窝生产圩及上、下六坊堤行洪区的左侧通过，长约 24.06 km；右汊经超河及上、下六坊堤行洪区右侧通过，长约 24.08 km，两汊在平圩处汇合。在左、右两汊之间又有两道横汊，分别是灯草窝生产圩与上六坊堤之间的横汊，以及上六坊堤与下六坊堤之间的横汊。原六坊堤行洪区为一个整体，1958 年实施了“二道河改道工程”，在六坊堤之间开挖一条长 3.00 km，宽 0.15 km 的引河，以引右汊河水向左汊汇流，引河又称二道新河，两岸筑格堤，与原行洪区堤防相连，将六坊堤行洪区分为上、下两部分，形成上六坊堤行洪区和下六坊堤行洪区，六坊堤河段河道形势见图 1。

1.3　来水来沙特性

1.3.1　来水特性

该河段上游最近的水文站为鲁台子站，距该段距离约 30 km，区间无大的支流汇入，因此鲁台子站的水沙资料可代表该段的来水来沙条件。鲁台子站集水面积 88 630 km^2,是淮河中游主要控制站,根据 1950 ~ 2007 年水文资料,统计出鲁台子站来水特征值(见表 1)。

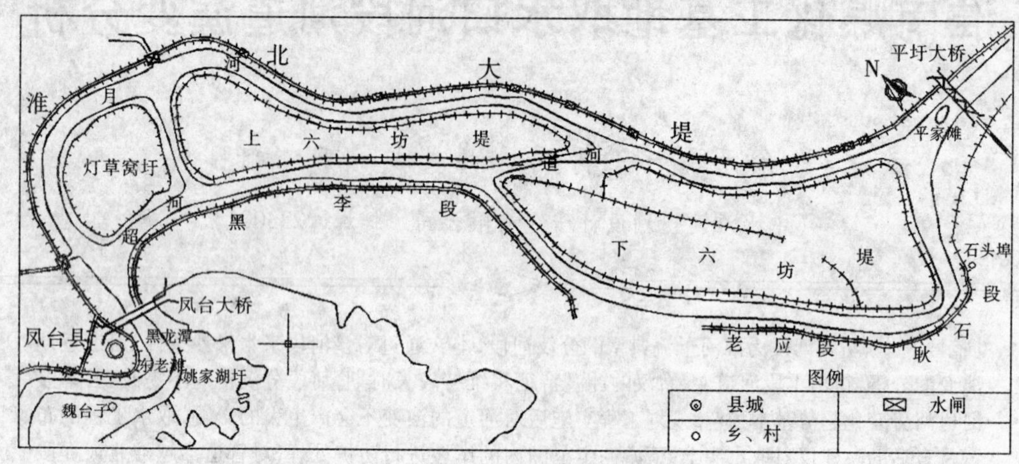

图 1 淮河六坊堤河段河道形势

表 1 鲁台子站来水特征值

统计年限(年)	多年平均径流量(亿 m^3)	最大年径流量(亿 m^3)	最小年径流量(亿 m^3)	多年平均汛期径流量(亿 m^3)	最大年汛期径流量(m/s)	最小年汛期径流量(m/s)	多年平均流量(m/s)	单日最大流量(m/s)
1950 ~ 2007	220.56	524.85	34.93	137.33	421.81	12.65	698.83	12 500
出现时间	—	1956 年	1966 年	—	1956 年	2001 年	—	1954 年 7 月 25 日

从表 1 可以看出,鲁台子站多年径流量 220.56 亿 m^3,多年平均流量 698.83 m^3/s;鲁台子站径流量的年际变化较大,1956 年实测年径流量 524.85 亿 m^3,是最小年份 1966 年 34.93 亿 m^{3}的 15 倍,变差系数 C_v = 0.55;不同年份的汛期径流量变化更大,1954 年实测汛期径流量 421.81 亿 m^3,是最小年份 2001 年径流量 12.65 亿 m^{3}的 33 倍。

1.3.2 来沙特性

通过对鲁台子站含沙量和输沙量的各年代平均值统计分析(见表 2),鲁台子站含沙量逐年代减少,20 世纪 50 年代含沙量最大,为 0.651 kg/m^3,2000 年后含沙量最小,为 0.106 kg/m^3;输沙量从 1950 ~ 1990 年逐年代减少,20 世纪 50 年代年均输沙量最大,为 1 792 万 t,90 年代年均输沙量最小,为 243 万 t,而 2000 年后由于水量较丰,输沙量较 20 世纪 90 年代有所增加,年均输沙量为 266 万 t。

表 2 鲁台子站泥沙及径流量各年代平均值

统计年限(年)	含沙量(kg/m^3)	输沙量(万 t)	径流量(亿 m^3)
1950 ~ 1959	0.651	1 791.58	275.08
1960 ~ 1969	0.582	1 300.13	223.31
1970 ~ 1979	0.503	864.81	171.91
1980 ~ 1989	0.307	752.22	244.83
1990 ~ 1999	0.144	242.51	168.67
2000 ~ 2007	0.106	265.87	251.12

2 河道演变分析

2.1 河道近期演变

本河段计有1917年、1950年、1954年、1971年、1983年、1992年和2005年7次河道测量资料,林怡然对1950年、1954年、1971年和1983年4次河道测量资料进行了对比分析,六坊堤河段河道主槽容积(高程20.5 m以下)变化情况见表3。六坊堤河段左汊上段淤积,中段冲淤基本平衡,下段1950~1971年冲刷,1971~1983年冲淤基本平衡;右汊上、中段冲刷,下段1950~1954年淤积,1954~1983年冲淤基本平衡。

表3 六坊堤河段河道主槽容积变化汇总 (单位:万 m^3)

河段		1950年	1954年	1971年	1983年
左汊	灯上分汊以上(5.75 km)	1 058	1 014	898	896
	灯上分汊—二道河出口(11.2 km)	1 347	1 291	1 169	1 226
	二道河出口以下(7.5 km)	860	970	1 266	1 210
右汊	超河(2.95 km)	268	325	492	493
	灯上分汊—二道河进口(6.3 km)	918	1 130	1 281	1 357
	二道河进口以下(15.2 km)	3 078	2 855	2 938	2 895

本文选用1992年和2005年测量河道资料,对该段左、右汊河道深泓纵剖面和主槽面积进行了对比(2010年测量了局部河段水下地形资料,本文不进行对比)。图2为六坊堤河段河道深泓纵剖面图。

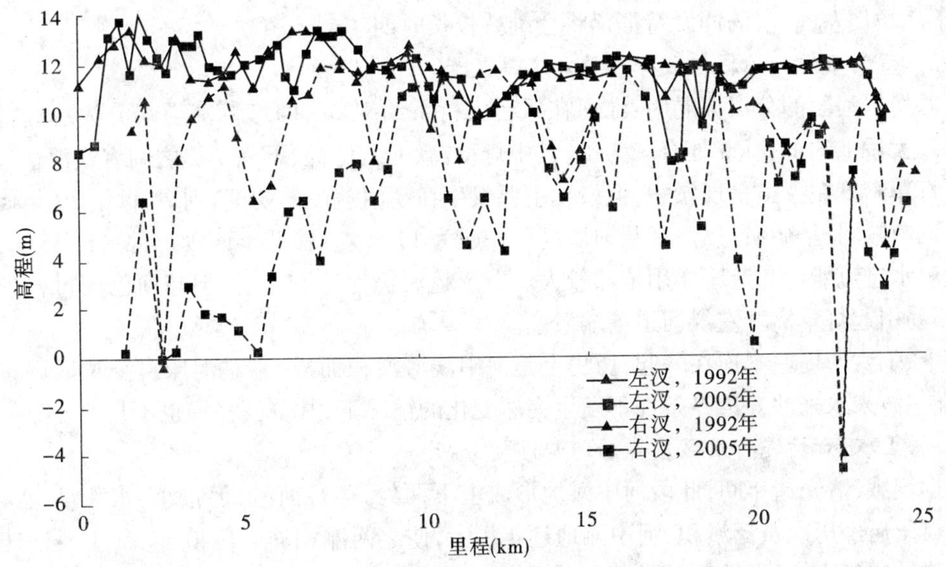

图2 六坊堤河段河道深泓纵剖面

六坊堤河段河道主槽(高程19.5 m以下)横断面变化情况详见表4。

表 4　六坊堤河段河道横断面变化汇总

河段		2005 年测量		1992 年测量		差值(1992-2005 年)	
		主槽平均面积(m^2)	主槽容积(万 m^3)	主槽平均面积(m^2)	主槽容积(万 m^3)	主槽平均面积(m^2)	主槽容积(万 m^3)
左汊	二道河以上	1 157	1 839	1 234	2 012	77	173
	二道河以下	1 520	1 202	1 483	1 173	-37	-29
右汊	二道河以上	2 880	2 448	1 968	1 615	-912	-833
	二道河以下	2 275	3 418	1 925	2 893	-350	-525
合计			8 907		7 693		-1 214

由上可见,1992 ~ 2005 年期间,六坊堤河段左汊深泓高程变化幅度较小,右汊深泓冲深;左汊上段微淤,左汊下段冲淤基本平衡,右汊全段均冲刷。

2.2　河道演变趋势分析

2.2.1　六坊堤河段左、右汊的冲淤特征将维持较长时期

通过工程所在河段近期演变分析:六坊堤河段左汊上段微淤,左汊下段冲淤基本平衡,右汊呈冲刷。从取水口处断面看,断面虽有冲淤,但基本保持良好形态,横向、纵向未发生剧烈变形。

六坊堤左汊上段的淤积、右汊的冲刷与超河的冲刷发展关系密切。近期超河河道仍呈冲刷趋势,断面进一步扩大,冲刷部位主要以进、出口刷深为主。随着超河段冲淤基本达到相对平衡状态,六坊堤左汊上段的淤积和右汊的冲刷才有可能减弱或停止。从目前来看,上、下六坊堤左、右汊的冲淤特征仍将会维持较长时期。

2.2.2　淮河干流六坊堤河段分汊格局不易改变

六坊堤河段的分汊格局为:淮河在凤台大桥下约 1 km,即灯草窝上端分为左、右两汊,在平圩大桥上约 1.6 km 处合二为一,其中灯草窝、上六坊堤、下六坊堤之间有横汊连通。

灯草窝圩左汊虽呈现淤积,但淤积主要发生在超河疏浚的初期,到 20 世纪 70 年代后淤积已逐渐减弱,呈微淤态势;现状河底高程一般为 14 m 左右,低于该段正常水位 18 m,中等以上洪水过流能力和造床作用依然较大。右汊超河仍处冲刷状态,但超河已度过剧烈冲刷期,不易引起灯草窝圩左汊河道萎缩淤塞。

上、下六坊堤原本连在一起,历史上是河中浅滩淤积而成,它是河床自身长期调整的结果,在上游来水来沙等边界条件不发生突然变化的情况下,其分汊格局也不易改变。

2.2.3　上、下六坊堤行洪区废弃后河势演变分析

据记载,清光绪年间,此段河中滩地形成上下两处,称月河滩,但很小,水涨没,水落出,不可耕。后经历次黄泛淤积,河中滩地逐年扩大,使之两滩相连。1940 年,河中滩地四周筑堤围耕。1946 年河中滩地定名为六坊堤。1950 年淮河大水后,组织群众复堤,形成长 38.56 km的圩堤,堤顶高程 23.3 ~ 22.56 m,堤顶宽 3 m。1954 年大水,六坊堤漫决,汛后复堤,并确定为行洪区。1956 年大水,六坊堤行洪,汛后堤防进行加高培厚,建成后堤顶高程 23.7 ~ 23.5 m,堤顶宽 4 m。

在《淮河干流行蓄洪区调整规划》中,六坊堤河段的行洪区调整工程的方案为废弃上、

下六坊堤,工程措施为铲除灯草窝生产圩圩堤,铲除上、下六坊堤行洪区堤防。从上、下六坊堤行洪区的形成过程看,有无行洪区堤防的情况下,上、下六坊堤一直四周临水。因此,上、下六坊堤行洪区调整工程对该段河道的演变不会产生大的影响,其影响的程度还需物理模型来进一步来验证。

3 结论及建议

综上所述,六坊堤河段为淮河干流典型的分汊河段,本段来沙量在径流量未出现系统增加或减少的情况下自20世纪50年代以来明显减少;经1992年和2005年河道断面资料分析,本段左汊上段微淤,下段冲淤基本平衡;右汊全段冲刷。淮南煤化工基地取水工程布置在左汊下段起始处,根据本段的来水来沙条件和冲淤分析,取水工程处的河道冲淤特征将维持较长时期,河段分汊格局不易改变,规划工程实施后对河道的演变不会产生大的影响,本段河道河势变化趋势基本稳定,取水工程的布置是合适的。

由于取水工程的取水头和管道的布设,对河道岸坡和淮北大堤有一定影响,建议对取水头上、下游一定范围内的岸坡防护处理,对管道穿越淮北大堤段进行截渗处理,确保淮北大堤的防洪安全和六坊堤河段的河势稳定。

参考文献

[1] 刘玉年,何华松,等.淮河中游河道特性与整治研究,2010.

[2] 林怡然.淮河凤台以下两汊河道河床的变化[J].治淮科技,1994(2).

[3] 刘玲,余彦群,等.淮河干流行蓄洪区调整规划,2008.

【作者简介】 余彦群,安徽宿州人,1972年2月生,1995年6月毕业于武汉水利电力大学(现武汉大学),2008年7月毕业于河海大学,获工程硕士学位,现在中水淮河规划设计研究有限公司任高级工程师,为河海大学研究生培养基地(淮河水利委员会)硕士生指导教师。

气候变化情景下未来坡面土壤侵蚀的预估*

马 良[1] 左长清[2] 邱国玉[3]

(1. 山东省水利科学研究院 济南 250013;
2. 中国水利水电科学研究院水利部水土保持生态工程技术研究中心 北京 100038;
3. 北京大学深圳研究生院环境与能源学院 深圳 518055)

【摘要】 利用IPCC AR4中17个全球大气环流模式在SRES A1B、A2和B1三种典型排放情景下的未来气温和降水预测结果,结合坡面土壤侵蚀WEPP模型,在对模型验证效果良好的基础上,参照集合预报方法,对未来89年(2011~2099年)气候变化下典型红壤坡面的土壤侵蚀进行预估。研究结果表明,各模式均倾向于未来降水量比现状增加,径流量很可能增加,坡面侵蚀可能增加。未来三种情景下的坡面土壤侵蚀水平均比现状年有所增加,其中温室气体排放浓度较高的情景(A2)下发生的降雨产流及侵蚀高于其他情景(A1B、B1)。至21世纪末降雨、径流及土壤侵蚀可能呈现出持续增加的趋势,而中后期(2051~2099年)典型红壤坡面的土壤侵蚀最为严重。

【关键词】 气候变化 土壤侵蚀 坡面 WEPP模型 红壤

1 引言

当前全球气候变化特别是降水的变化对土壤水蚀带来极为复杂的影响。IPCC第4次评估报告(AR4)也认为,多数地区强降雨事件出现频次的增加很可能加重土壤侵蚀率。同时,土壤侵蚀也是CO_2、CH_4等温室气体的主要排放源,加剧了温室效应和气候变化。深刻认识气候变化—土壤侵蚀的相互关系及响应,尤其加强对水土流失严重及环境敏感性地区气候变化的适应研究,探索变化环境下的区域水土保持和生态恢复的有效管理途径,是在新形势下对水土保持科学提出新的挑战。

土壤侵蚀的预估是建立在未来气候情景预测和气候模式模拟的基础之上的,从20世纪90年代研究之初采用人为递增或倍增降水量作为边界条件下的侵蚀量经验预估,发展到如今利用全球耦合气候系统模式的区域化或降尺度后,再结合物理过程的土壤侵蚀模型进行模拟预测,精度极大提高。但当前的预测方法多采用单个或少数几个(最多6个)的气候模式,尚未有大量气候模式集合预报的研究报道。由于不同气候模式的模拟表现各不相同,甚至结果相悖,少数模式并不足以反映未来某地区的气候变化,亟须开展多模式的集合预报研究。

* 资助项目:水利部"948"项目(201029),水利部公益性行业科研专项经费项目(201101057)。

本文利用 IPCC AR4 可获得数据的 17 个大气环流模式(General Circulation Model, GCM),在 SRES B1、A1B、A2 三种典型排放情景下研究区未来的降水、温度等气象参数;经天气发生器法降尺度,由月气象数据生成 2011 ~ 2099 年间日连续数据;驱动基于 2001 ~ 2006 年 229 场降雨侵蚀观测试验率定验证的坡面土壤侵蚀 WEPP 模型,预估未来不同情景、不同模式、不同时期的坡面侵蚀量,探寻未来坡面土壤侵蚀的发生发展特征。

2 研究区概况

本研究选择地处全球环境变化速率最大的东亚季风区的中心——我国南方红壤区为研究对象,该区水土流失对气候变化的响应具有敏感脆弱的特点,具有典型代表性。

本研究在该区布设典型坡面侵蚀的标准径流小区,地理位置为德安国家水土保持科技示范园内,位于江西省北部鄱阳湖水系的德安县郊燕沟小流域。当地为亚热带季风气候区,多年平均降水量 1 300 ~ 1 400 mm,受东亚季风气候影响年内存在明显的干、湿两季,降水呈双峰分布。土壤类型为红壤,成土母质是第四纪红黏土,泥质岩类风化物。土质为中壤土、重壤土和轻黏土,并具有酸、黏、板、瘦等特性。

本研究自 2000 年始建设的标准径流小区长 20 m、宽 5 m、投影面积 100 m^2,坡度为 12°。小区坡面裸露,降雨产流和产沙通过集流槽进入径流池,取样观测。小区周边还设置了一处气象观测站,用于降水、气温、蒸发等气象要素的观测。

3 研究方法

3.1 未来气候情景及气候模式的选择

目前普遍采用的 SRES 情景是在 2000 年的 IPCC 第 3 次评估报告中(TAR),为替代之前的 IS92 情景提出,并在多年的气候变化预估研究中延续采用。在 IPCC 第 5 次评估报告(AR5,预期 2014 年出版)新情景之前,SRES 仍是进行未来气候变化评估的最佳选择。本文中,选择普遍认为最具代表性的 SRES B1、A1B、A2 三个情景作为描述预估气候情景,分别对应未来温室气体的低、中、高排放。

IPCC AR4 收录了全球不同研究中心 20 余个具有典型代表性的大气环流模式。由于个别模式在研究区处数据有缺失,经筛选本研究确定了 IPCC 数据分发中心 17 个可获得数据的公开模式(http://www.ipcc-data.org/),如表 1 标"●"所示。

3.2 气候模式的天气发生器法降尺度

尽管现代大气环流模式开展复杂的环流和交互作用的模拟能力获得飞速发展,但仍是较低解析度的计算,在高分辨率的区域气候模拟时,应进行时间或空间的降尺度。本文为获得研究区未来完整日气象序列,必须对获得的 GCM 月数据进行降尺度计算。

降尺度采用天气发生器 Cligen 法,其本质是通过对未来月气象数据的分布特征的统计,采用一阶马尔可夫链或者干湿天延续天数计算日尺度的降雨概率。通过修改天气发生器 Cligen 参数,包括月雨量均值、雨量标准差、雨量偏差 skew、转移概率(降雨-降雨概率 $P_{w/w}$、降雨-不降雨概率 $P_{w/d}$)、月最高温度均值及标准差、月最低温度均值及标准差等,实现对各 GCMs 未来月气象数据降尺度的目的,得到未来连续的日降水量(P)、日最高温(T_{max})、日最低温(T_{min})序列。

表 1　IPCC AR4 采用的大气环流模式的一览

国别	机构	机构缩写	模式名称	SRES 情景设置		
				B1	A1B	A2
China	Beijing Climate Center	BCC	CM1	○	○	○
Norway	Bjerknes Centre for Climate Research	BCCR	BCM2.0	●	●	●
Canada	Canadian Center for Climate Modelling and Analysis	CCCma	CGCM3 (T47 resolution)	○	○	○
			CGCM3 (T63 resolution)	○	○	○
France	Centre National de Recherches Meteorologiques	CNRM	CM3	●	●	●
Australia	Australia's Commonwealth Scientific and Industrial Research Organisation	CSIRO	Mk3.5	●	●	●
Germany	Max-Planck-Institut for Meteorology	MPI-M	ECHAM5-OM	○	●	●
Germany	Meteorological Institute, University of Bonn	MIUB	ECHO-G	●	●	●
Korea	Meteorological Research Institute of KMA	NIMR				
Germany	Model and Data Groupe at MPI-M	METRI				
		M&D				
China	Institude of Atmospheric Physics	LASG	FGOALS-g1.0	○	○	○
USA	Geophysical Fluid Dynamics Laboratory	GFDL	CM2.0	●	●	●
			CM2.1	●	●	●
USA	Goddard Institute for Space Studies	GISS	AOM(C4×3)	●	●	○
			E-H(MODELE20/HYCOM)	○	●	○
			E-R(MODELE21/Russel)	●	●	●
Russia	Institute for Numerical Mathematics	INM	INMCM3	●	●	●
France	Institut Pierre Simon Laplace	IPSL	CM4	●	●	●
Japan	National Institute for Environmental Studies	NIES	MIROC3.2 hires	○	○	○
			MIROC3.2 medres	○	○	○
Japan	Meteorological Research Institute	MRI	CGCM2.3.2	●	●	●
USA	National Centre for Atmospheric Research	NCAR	PCM1	●	●	●
			CCSM3	●	●	●
UK	UK Met. Office	UKMO	HadCM3	○	○	○
			HadGEM1	○	●	●
Italy	NationalInstitute of Geophysics and Volcanology	INGV	ECHAM4.6	○	●	●

注:"●"表示研究中选择的模式,"○"表示月数据缺失;试验区坐标:28°46′21.61″N,115°24′10.31″E。

3.3 坡面土壤侵蚀 WEPP 模型的率定及验证

研究中选取试验区 2001 ~2003 年 116 场次降雨产流、侵蚀实测数据进行坡面土壤侵蚀 WEPP 模型的率定。率定的参数主要集中在气象、管理、坡面及土壤四类。其中,根据观测所得的日降水、温度极值修改气象参数;管理参数参考休耕状态;以试验设计调整坡面参数;以实测的土壤肥力、抗剪切力、抗蚀性等土壤理化性质率定土壤参数。

模型率定后,采用 2004 ~2006 年 113 次降雨产流和侵蚀数据进行模型的验证。经验证,年侵蚀量模拟值与观测值之间相关系数为 0.92,Nash-Suttclife 效率系数 E_{ns} 为 0.89;次侵蚀量模拟值精度也较好,如图 1 所示,模拟值与观测值之间相关系数为 0.83,E_{ns} 为 0.67。

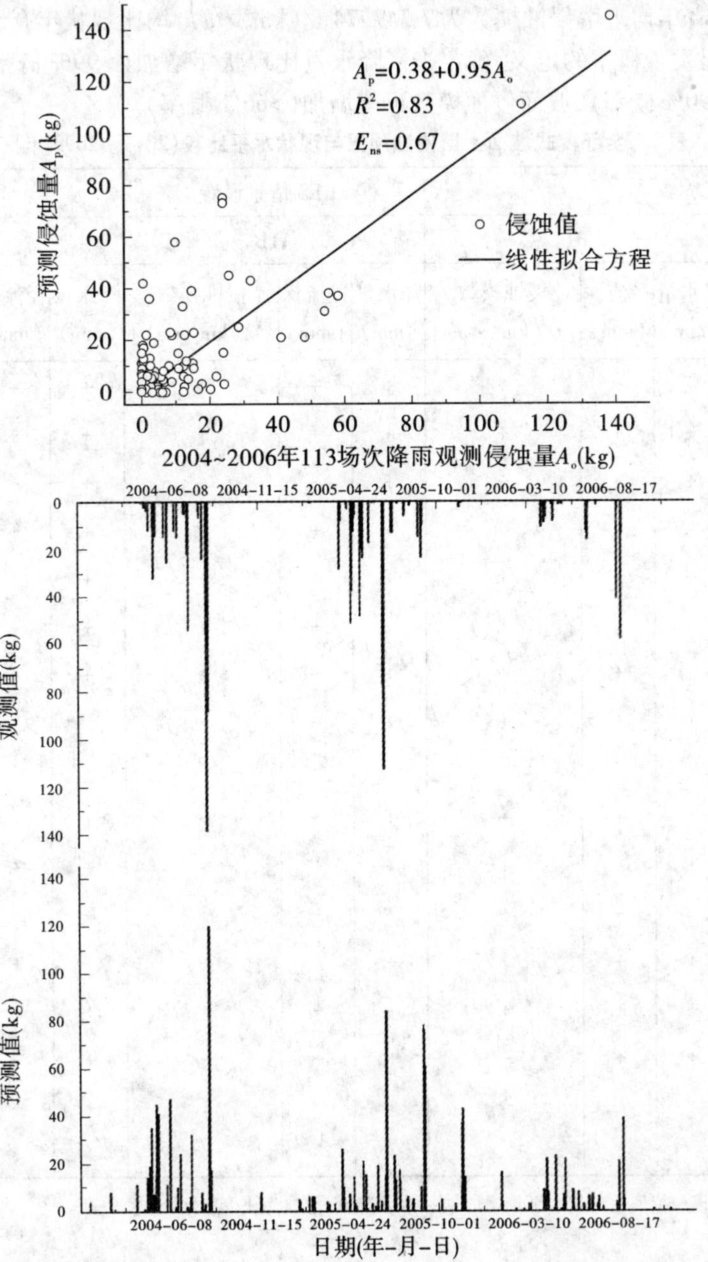

图 1 WEPP 模型次侵蚀量验证结果(113 个样本)

4　结果与分析

通过17个大气环流模式在SRES A1B、A2和B1三种典型排放情景下气温和降水的降尺度结果，驱动坡面土壤侵蚀WEPP模型，获得未来至21世纪末的典型红壤坡面土壤侵蚀变化特征。

表2给出了17个模式模拟未来降雨量、径流深及侵蚀模数与现状年的对比。现状年水平下全年平均降水量为1 325.00 mm、典型红壤坡面年均径流深799.87 mm、年均土壤侵蚀模数为4 956.51 t/(km^2·a)。各模式预估未来多年平均降水量为2 011.35 mm，年均径流深1 094.53 mm，年均土壤侵蚀模数为7 347.74 t/(km^2·a)，均比现状年有了大幅增加。参考IPCC AR4对发生概率的定义，确定未来降水量比现状年增加(>99%概率)，径流深很可能随之增加(>90%概率)，坡面侵蚀模数可能增加(>66%概率)。

表2　全部模式坡面土壤侵蚀预报与现状水平比较(2011～2099年)

模式名称	SRES情景设置								
	B1			A1B			A2		
	降水量(mm)	径流深(mm)	侵蚀模数[t/(km^2·a)]	降水量(mm)	径流深(mm)	侵蚀模数[t/(km^2·a)]	降水量(mm)	径流深(mm)	侵蚀模数[t/(km^2·a)]
BCM2.0	+	+	+	+	+	+	+	+	+
CM3	+	+	+	+	+	+	+	+	+
Mk3.5	+	−	−	+	−	−	+	+	+
ECHAM5-OM				+	+	+	+	+	+
ECHO-G	+	+	+	+	+	+	+	+	+
CM2.0	+	+	+	+	+	+	+	+	+
CM2.1	+	+	+	+	+	+	+	+	+
GISS_AOM	+	+	+	+	−	−			
GISS_EH				+	−	−			
GISS-ER	+	+	+	+	+	+	+	+	+
INMCM3	+	+	+	+	+	+	+	+	+
CM4	+	+	+	+	+	+	+	+	+
CGCM2.3.2	+	+	+	+	+	+	+	+	+
PCM1	+	+	+	+	+	+	+	+	+
CCSM3	+	+	−	+	+	−	+	+	+
HadGEM1				+	+	+	+	+	+
ECHAM4.6				+	+	−	+	+	+

注：表中"+"表示该模拟值比现状水平年增加；"−"表示降低，空白为缺少数据。

4.1 不同情景红壤坡面土壤侵蚀特征的预估与分析

4.1.1 B1 情景下红壤坡面土壤侵蚀的预估

图 2 给出了预估未来 89 年 B1 情景下 13 个气候模式的土壤侵蚀模数年序列和与现状年对比箱图。经统计,不同模式的表现各不相同,多年平均侵蚀模数以 GISS_ER 为最高,达 11 861.24 t/(km^2 · a),比现状年增加一倍以上(139.31%);最低为 CCSM3 模式,仅有 3 751.62 t/(km^2 · a),为现状年侵蚀水平的 3/4。全部 13 个模式中仅有 CCSM3、MK3.5 两种模式的侵蚀模数低于现状年,其余均超出现状年水平,幅度在 14.73%(CM4)~139.31%(GISS_ER)。

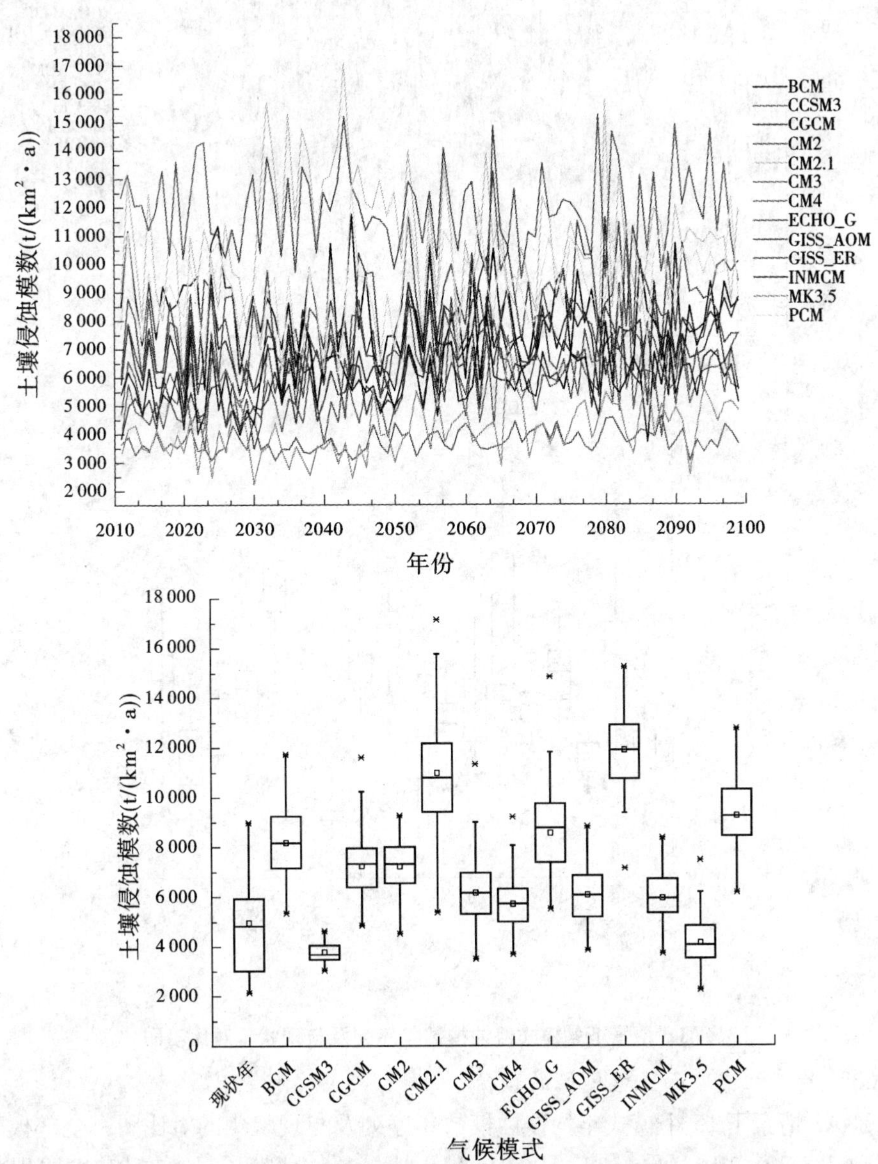

图 2 B1 情景下各模式侵蚀模数年序列及与现状年对比箱图

4.1.2 A1B 情景下红壤坡面土壤侵蚀的预估

A1B 情景下 17 个模式的多年平均侵蚀模数同以 GISS_ER 为最高,达11 811.69 t/(km^2 · a),

比现状年增加 138.31%;最低为 GISS_AOM 模式,仅有 2 807.53 t/(km^2 · a),为现状年的 56.64%,如图 3 所示。全部模式平均值比现状年增加了 43.78%。17 个模式中有 12 个模式比现状年增加,其余 5 个模式低于现状年水平,分别是 MK3.5、GISS_AOM、GISS_EH、CCSM3 和 ECHAM4.6,个数比 B1 情景稍多。

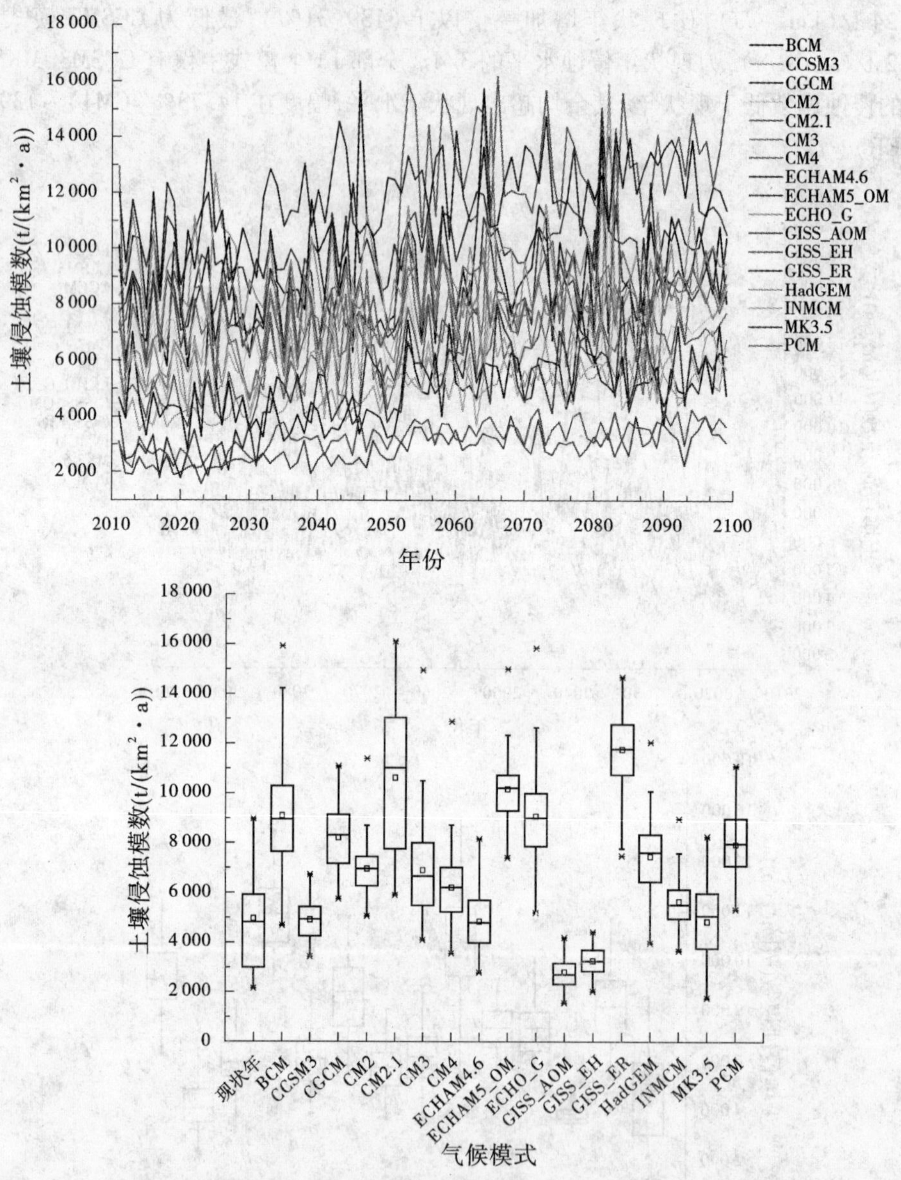

图 3 A1B 情景下各模式侵蚀模数年序列及与现状年对比箱图

4.1.3 A2 情景下红壤坡面土壤侵蚀的预估

图 4 是 A2 情景下 15 个模式年均侵蚀模数年序列及与现状年的对比图。全部模式平均年侵蚀模数达 8 201.19 t/(km^2 · a),比 B1、A1B 情景预估值有了显著增加,也比现状年增加一半以上。A2 情景下同样以 GISS_ER 模式的侵蚀模数为最高,达12 874.87 t/(km^2 · a),比现状年增加 159.76%;最低为 MK3.5 模式,但也有 5 330.10 t/(km^2 · a),比现状年增加

7.54%。全部 15 个模式预估值全部比现状年有所增加,这是在 B1、A1B 两个情景中所不曾出现的。

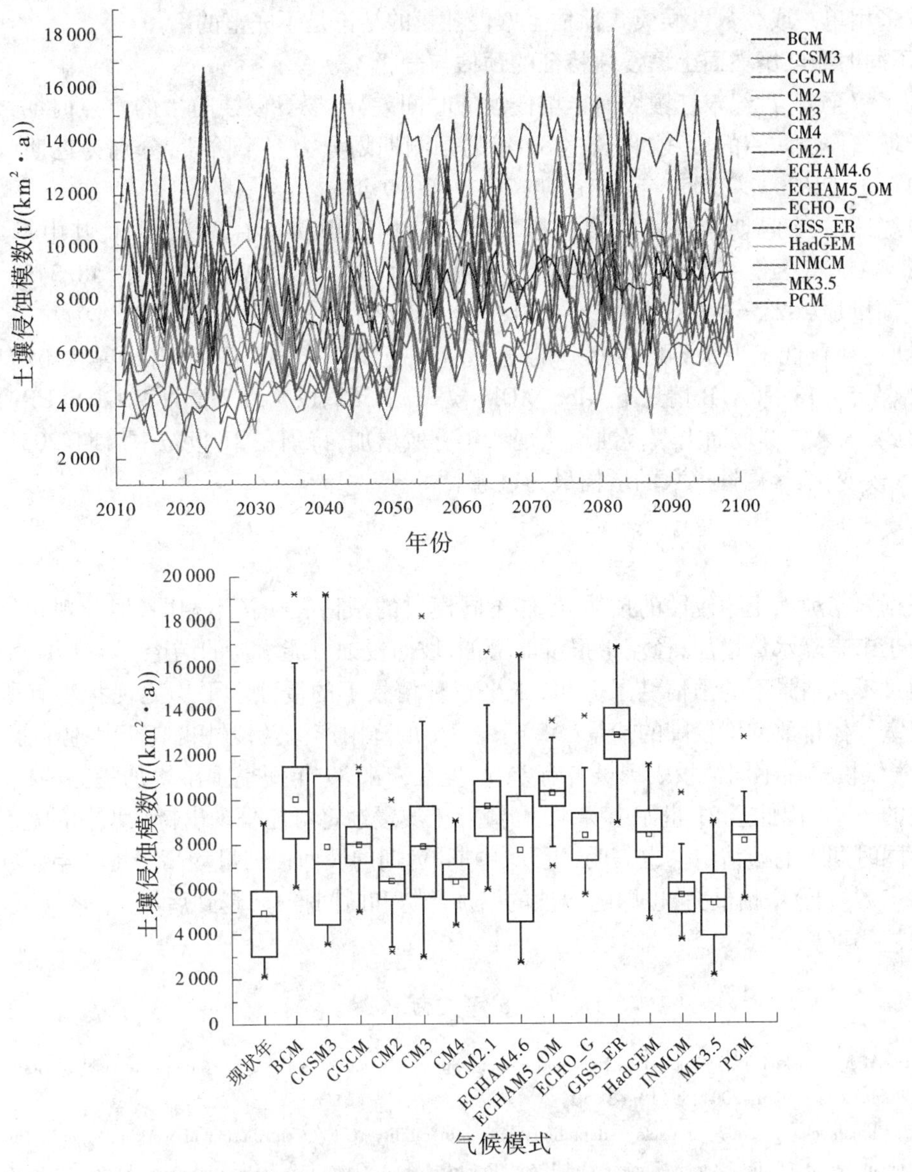

图 4 A2 情景下各模式侵蚀模数年序列及与现状年对比箱图

4.1.4 不同情景红壤坡面土壤侵蚀预估的对比分析

通过以上不同情景下各模式对 2011 ~ 2099 年侵蚀模数预估的对比分析,A2 情景比 B1、A1B 情景有更高的坡面土壤侵蚀量,各模式平均值比现状年增加一半以上(65.46%),而后两个情景仅分别增加 47.18% 和 43.78%,也即未来温室气体排放浓度水平越高,越易发生更严重的坡面侵蚀。

SRES B1、A1B、A2 三种情景下,均以 GISS_ER 模式模拟的年侵蚀模数最高,该模式也在 A2 情景下达到峰值,为 12 874.87 t/(km^2 · a),比现状年增加 159.76%。经时间序列分析,无论在何种情景下,未来 89 年间呈现增加趋势序列的模式在总体上占多数,B1 情景下增加序列的模式占总数的77%、A1B 情景下为88%、A2 情景下为73%。

通过上述结果表明,联合采用多种模式在不同情景下开展预估的方法更加科学全面,证实了仅采用单一或少数气候模式预估土壤侵蚀量的结论是不完整的。

4.2 不同时期红壤坡面土壤侵蚀特征的预估与分析

为了解至21世纪末红壤坡面土壤侵蚀的时间差异,探寻侵蚀防治的重点时期,也更好地指导坡面侵蚀防治的实践,将长达89年的预测期划分为三个阶段,分别为近期(2011~2030年)、中期(2031~2050年)和远期(2051~2099年)。

经统计,三个时期各模式年侵蚀模数平均值均比现状年有显著增加,其中近期增加23.42%(A1B 情景)~41.86%(A2 情景);中期增加33.50%(A1B 情景)~48.34%(A2 情景);远期增加262.56%(B1 情景)~331.82%(A2 情景),尤其远期增加最为显著。在近期、中期还存在低于现状年侵蚀水平的模式,但在远期各模式均呈现出比现状年增加的特征,增幅在53.18%(A1B 情景下 GISS_AOM 模式)~552.17%(A2 情景下 GISS_ER)之间。

因此,未来红壤坡面土壤侵蚀均比现状年有所增加,特别在21世纪中后期(2051~2099年)受气候变化,年侵蚀模数的增幅最为显著。

5 结论

根据未来坡面土壤侵蚀的预估,虽然不同模式的预估表现各不相同,但与现状年相比,均倾向于未来降水量增加,径流量很可能增加,坡面侵蚀可能增加的结论。

通过不同情景下的预估结果分析,未来三种情景下的侵蚀水平均比现状年有所增加。未来温室气体排放浓度较高的情景(A2)下,发生的降雨产流及侵蚀也高于其他情景(A1B、B1)。气候模式在不同情景表现虽各有差异,但在未来89年间呈现出增加趋势的模式均超过总数的70%,因此至21世纪末降雨、径流及土壤侵蚀将可能呈现出持续增加的趋势。

不同时期下的预估结果也表明,近期、中期、远期的坡面侵蚀量均高于现状年,其中远期(2051~2099年)增幅最大,也即随侵蚀的增加趋势和时间推移,越到后期,未来红壤坡面土壤侵蚀越严重。

参考文献

[1] Nearing M A, Pruski F F, O'Neal M R. Expected climate change impacts on soil erosion rates: a review[J]. Journal of Soil and Water Conservation, 2004,59(1):43-50.

[2] IPCC. Climate change 2007: impacts, adaptation, and vulnerability[C]//Contribution of working group Ⅱ to the forth assessment report of the intergovernment panel on climate change. Cambridge, UK and New York, USA: Cambridge University Press, 2007.

[3] Rattan L. Soil Carbon Sequestration Impacts on Global Climate Change and Food Security[J]. Science, 2004, 304(5677): 1623-1627.

[4] Zhang X C, Nearing M A, Garbrecht J D, et al. Downscaling monthly forecasts to simulate impacts of climate change on soil erosion and wheat production[J]. Soil Science Society ofAmerica Journal, 2004, 68(4): 1376-1385.

[5] Zhang X C. A comparison of explicit and implicit spatial downscaling of GCM output for soil erosion and crop production assessments[J]. Climatic Change, 2007, 84(3-4): 337-363.

【作者简介】 马良,1980年8月生,博士,2011年毕业于北京师范大学资源学院,2004年始在山东省水利科学研究院工作至今,工程师。主要研究方向为土壤侵蚀、生态水文及对气候变化响应。

曲条面引水渠首在阿韦滩渠首改造工程中的应用

李虹瑾[1] 陈 晖[2] 陈顺礼[3]
(1. 新疆水利水电规划设计管理局 新疆 830000;
2. 新疆水利厅造价管理总站乌鲁木齐 830000;
3. 新疆水利水电勘测设计研究院 乌鲁木齐 830000)

【摘要】 本文介绍了针对阿韦滩渠首改造过程中提出的一种新型引水渠首——曲条面引水渠首。该渠首利用弯道凸岸沉沙、凹岸引水的原理,对原底栏栅式引水渠首进行改扩建,在条形冲沙槽进水口前修建曲线形挡沙坎,采用曲面引水实现第一级水沙分离,凹岸表层较清的水流进入条形冲沙槽后,更清的水流溢入两侧的溢流槽,最终实现第二级水沙分离。工程建成运行后,发现通过这两级的水沙分离过程,最终实现了曲条面引水渠首引水、排沙、水沙分离效果,值得推广应用。本工程已获国家专利,专利号:ZL200620173129.3。

【关键词】 曲条面 引水渠首 水沙分离

1 问题的提出

新疆河流多系山溪性多沙内陆河,河道流程短,水量小,河床多为沙砾石阶地,厚度可深达数十米。汛期汹涌的洪水挟带大量泥沙而下,因此如何解决好引水排沙,是新疆引水工程的一个关键问题。

位于新疆呼图壁县石梯子乡的阿苇滩渠首为拦河渠首,重点解决河道右岸石梯子乡阿苇滩灌区的农业用水问题,控制面积为 3.8 万亩(1 亩 = 1/15 hm^2)。出于引水和防沙排沙的考虑,1994 年修建有一孔尺寸为 1.5 m×1.6 m 的引水闸和一座长 40 m 的底栏栅工程,由于底栏栅式引水渠首栏栅空隙易被推移质或漂浮物堵塞,经常需要清理,不便管理。随着工程运行时间的推移,大量粒径小于栅隙的泥沙进入引水廊道造成廊道内泥沙淤积严重,无法发挥其应有的作用,渠首最大引水流量 1.7 m^3/s,小于设计引水流量 2.2 m^3/s,因此渠首泥沙问题亟待解决。针对存在的问题,在阿韦滩原底栏栅式引水渠首改造过程中,提出一种新型的引水渠首——曲条面引水渠首,大大解决了原有引水渠首存在的泥沙问题,发挥了引水渠首引水防沙的功能。

2 曲条面引水渠首简介

新疆呼图壁河灌区阿苇滩渠首的改造主要是针对渠首引水廊道内泥沙淤积问题提出的,改建和扩建工程应遵循合理、经济、实用的原则。

曲条面引水渠首，是一种新型的引水渠首，采用曲面引水、条形冲沙槽冲沙、溢流槽双向溢流面进水。

该渠首应用了曲面引水的原理，在原引水廊道前修建曲线型挡沙坎，使整个水流沿曲线型挡沙坎呈弧线前进，将河道推移质带到原冲沙闸前，冲入下游河道，表层较清水流进入条形冲沙槽，实现第一级水沙分离，使进渠泥沙大大减少。

在溢流槽的设计中，充分利用原有廊道，在廊道前采用条形冲沙槽和溢流槽相间布置，条形冲沙槽末端设冲沙闸，闸后设置渡槽。水流进入条形冲沙槽后，由闸门控制下形成壅水，使水流流速降低，致使泥沙在重力作用下形成沉积。当冲沙渠内水位超过渠堤时，水流自然溢入相邻的矩形溢流槽内，然后进入引水廊道，引入干渠，实现第二级水沙分离。当冲沙渠内泥沙淤积到一定厚度之后，开启末端冲沙闸，水流挟带泥沙经渡槽流入下游，由于冲沙渠末端高程较泄洪冲沙闸底板高程高出 1.25 m，冲沙渠出水口与泄洪冲沙闸之间形成自然陡坡，冲沙时，挟沙水流可沿陡坡流入泄洪冲沙闸后主河道。

曲条面引水渠首由浆砌石护底整治段、进口挡沙坎、条形冲沙槽、溢流槽、冲沙闸、引水廊道、原泄洪冲沙闸和左岸溢流堰组成。本次改建工程需在原有底拦栅渠首基础上新建进口挡沙坎、条形冲沙槽、溢流槽、冲沙闸，改建引水廊道。主要建筑物平面布置见图 1。

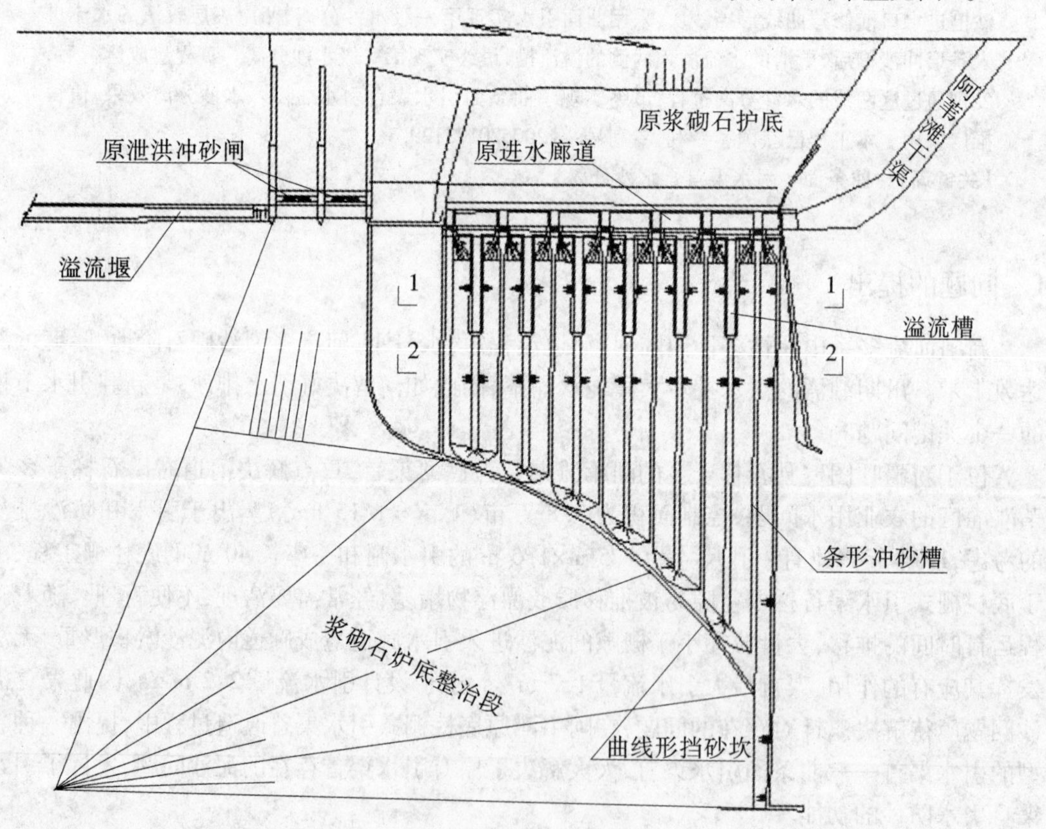

图 1　曲条面引水渠首平面布置及剖面

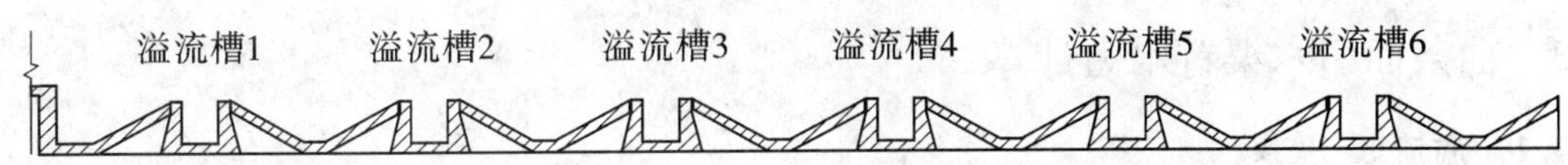

2—2剖面

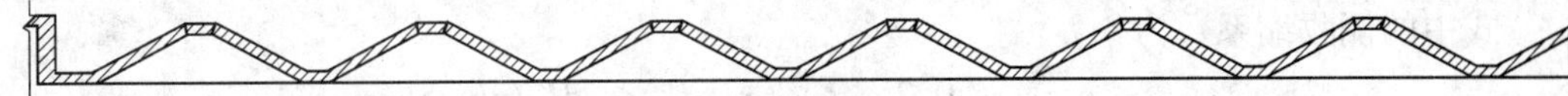

续图 1

曲条面引水渠首各组成部分的作用及功能如下。

2.1 浆砌石护底整治段和进口挡沙坎

为了使取水口上游河段具有稳定的河床宽度，降低河床糙率，利于冲沙，且保证溢流槽流量，需要对上游河床、河滩及河岸采取综合整治措施。本次设计中，在冲沙闸上游70 m内及取水口前端河底，采用30 cm厚C20F200W4细石混凝土砌石护底，在挡沙坎前端右岸修建护堤。

在条形冲沙槽进水口前修建曲线形挡沙坎，利用弯道凸岸沉沙、凹岸引水的原理实现第一级水沙分离。为保证曲线型挡沙坎具有良好的横向环流及水流稳定条件，浆砌石护底整治段后的曲线型挡沙坎总长50 m。

2.2 冲沙槽和溢流槽

进口挡沙坎后采用冲沙槽、溢流槽垂直于廊道相间布置的方式，通过条形冲沙槽末端的冲沙闸门关闭，水流流速降低，形成壅水，致使泥沙在重力作用下下沉，表层清水溢入相邻的溢流槽内，实现第二级水沙分离。经计算引水廊道前布置了7条条形冲沙槽、6条溢流槽。

由于7条条形冲沙槽的起始端为曲线形挡沙坎，末端为冲沙闸，所以长度不等，均为梯形断面。6条溢流槽的末端和条形冲沙槽一致，起始端为末端向前12 m的位置，矩形断面。

2.3 冲沙闸和渡槽

7条条形冲沙槽末端各设1孔冲沙闸，用于第二级水沙分离。当条形冲沙槽内泥沙淤积到一定厚度时，开启闸门，水流挟带泥沙经渡槽流入河道下游。闸墩上放置手电两动启闭机。在7条条形冲沙槽末端的冲沙闸后接渡槽，为矩形无横杆渡槽。

2.4 引水廊道

经过二级水沙分离的清水进入引水廊道。原有廊道经改建，把廊道上的栅条去除，在原来的基础上增加0.2 m即可实现引加大流量3.0 m^3/s的水至阿韦滩干渠。

2.5 原泄洪冲沙闸

将上游泥沙冲向下游河道。渠首冲沙闸目前运行良好，仍可正常使用。

2.6 左岸溢流堰

由于二级水沙分离造成引水面抬高，高出原溢流堰堰顶高程。通过计算复核，此次改造将对其加高1.25 m，溢流堰原顶部高程从849.7 m变为850.95 m。

3 曲条面引水渠首水力计算

3.1 溢流堰

由于 $0.67<\delta/H<2.5$（δ 为堰的厚度；H 为堰上作用水头），溢流堰的过流能力可按实用堰流量公式进行计算，即

$$Q=\sigma\varepsilon mB\sqrt{2g}H_0^{3/2}$$

式中的侧收缩系数为

$$\varepsilon=1-0.2[\xi_k+(n-1)\xi_0]H_0/nb$$

3.2 原泄洪冲沙闸

分析表明，当 $2.5<\delta/H<10$ 时，泄洪冲沙闸可按宽顶堰处理。为减小闸后冲刷坑深度，本次设计将闸门开度控制在 2 m 以内。由于 $e/H\leqslant0.75$（e 为闸孔开度），泄洪闸为闸孔初流，其流量计算公式为

$$Q=\left[0.4\left(\frac{h_\mu-e}{R}\right)^2+0.5\right]be\sqrt{2g(H-0.7e)}$$

3.3 条形冲沙槽

条形冲沙槽闸门的最大开度为 1.4 m，由于 $e/H\leqslant0.75$，因此冲沙闸的过流公式可按闸孔出流（平板闸门）公式进行计算

$$Q=\mu_0be\sqrt{2gH_0}$$

$$\mu_0=0.60-0.18\frac{e}{H}$$

3.4 溢流槽

在设计洪水和校核洪水流量时，溢流槽均为淹没出流。由于 $2.5<\delta/H<10$，因此可按宽顶堰处理。

流量公式

$$Q=\sigma\varepsilon mN\sqrt{2g}H_0^{3/2}$$

侧收缩系数

$$\varepsilon=1-0.2[\xi_k+(n-1)\xi_0]H_0/nb$$

流量系数（直坎）

$$m=0.33+0.01\frac{3-P/H}{0.46+0.75P/H}$$

以上计算公式中：

Q 为泄洪流量，m^3/s；H、H_0 为堰前水头，m；B、b 为闸孔（堰）净宽，m；e 为闸孔开启度，m；g 为重力加速度，9.8 m/s^2；R 为弧形闸门半径，m；h_μ 为弧形闸门转运轴距闸床高度，m；n 为闸孔数；m 为流量系数；μ_0 为流量系数；σ 为淹没系数；ε 为侧收缩系数。

各建筑物过流能力计算结果见表 1。

表1　渠首各建筑物泄洪组合

运行方式	水位(m)	泄流量(m^3/s)			
		溢流堰	冲沙闸	冲沙渠	溢流槽
设计	851.48	192.02	59.62	29.96	130.24
校核	851.78	312.72	64.72	33.07	158.49

经计算,改建后渠首各建筑物总的泄洪流量均能满足设计要求。

4　曲条面引水渠首冲沙效果

工程建成后,通过第一、第二级的水沙分离,最终实现了新型渠首——曲条面引水渠首“引水排沙、水沙分离”的目标,参见图2。

4.1　第一级水沙分离——曲线形导沙坎

曲条面引水渠首利用弯道凸岸沉沙、凹岸引水的原理,通过曲线形挡沙坎使凹岸表层较清的水流进入条形冲沙槽,将河道流沙等推移质带到河道的原泄洪冲沙闸前,冲入下游河道,实现第一级水沙分离,运行效果见图3。

图2　两级水沙分离图

图3　第一级水沙分离

4.2　第二级水沙分离——条形冲沙槽和溢流槽

由于条形冲沙槽末端的冲沙闸门关闭,形成壅水,水流流速降低,致使泥沙在重力作用下下沉,形成淤积。表层清水自然溢入相邻的溢流槽内,上层清水顺着引水明渠最终进入引水廊道,实现第二步水沙分离,详见图4。当条形冲沙槽内泥沙淤积到一定厚度时,开启闸门,水流携带泥沙经渡槽至导流渠流入河道下游。

第二级水沙分离后,清水由引水廊道引入下游阿韦滩干渠,泥沙在冲沙闸前淤积。闸门打开的时候,泥沙经冲沙闸下泄,冲沙后的水携沙经冲沙泄槽带向下游,冲沙过程见图5。

图5为洪水期水流溢出来和泄洪冲沙闸门打开时泄槽后水流的对比照片。从照片中可以看出:经过两级水沙分离,引水廊道中溢出的水明显清于从泄槽出来的浑水,所以充分体现两级水沙分离的良好效果。

图4 第二级水沙分离

图5 泄槽和引水廊道水流对比图

5 曲条面引水渠首运行管理

本渠首是在原有的底栏栅渠首基础上改建的新型渠首。原有的底栏栅渠首不能拦阻粒径小于栅条间隙的泥沙进入引水廊道,杂物、大石头会卡在栅条上后致使过栅水流流量降低。虽然栏栅设计的恰好能过小石头,但是在通过介于中间的石头、还有流速低时,卡在上面的石头,特别是发洪水时很多无法预测的大石头砸在栏栅上,容易造成栅条破裂。底栏栅栅条固定在嵌槽内,清淤及检修不方便。每次人工清理石头等杂物,上游等都要停水清沙,给运行管理带来极大不便以致无法满足下游灌溉需要。

本曲条面引水渠首采用7条条形冲沙槽与6条溢流槽相间布置,7条条形冲沙槽后布置7孔泄洪冲沙闸。在常规工况下,7条条形冲沙槽中,有2孔泄洪冲沙闸门打开,即2条条形冲沙槽用于平时冲沙,5条条形冲沙槽引水。经计算,如按上述工况运行,冲沙间隔时间为3 d,即闸门关闭3 d闸前淤积的泥沙平均需要12 min即可全冲干净。工程建成后,经实际观测,冲沙间隔、时间和计算结果基本一致。

本曲条面引水渠首相对传统渠首最大的特点是两级水沙分离、运行管理方便。相对传统底栏栅无法预测的大石头砸在栏栅上,容易造成栅条破裂,而新型曲条面引水渠首在第一级分沙时就将河道流沙等推移质带到河道的原泄洪冲沙闸前,冲入下游河道,将大石头等排除在外。第二级分沙除上述常规工况下引水冲沙外,还可根据实际情况进行相应的调节,将泄洪冲沙闸前泥沙全部带到下游,避免了传统底栏栅为了清沙还要上游等都要停水带来的不必要的麻烦。

6 结语

新疆境内河流为多泥沙河流,目前国内已经研制出了不少水沙分离系统,如排沙漏斗、浑水水力分离清水装置等,并在工程中得到了成功应用。曲条面引水渠首作为一种新型水沙分离系统,是在阿韦滩原底栏栅式引水渠首改造过程中提出的一种新型引水渠首。该实用新型渠首利用弯道凸岸沉沙、凹岸引水的原理,通过两级水沙分离"引水排沙、水沙分离",再取上层清水,实现了阿苇滩渠首改造的成功。该新型渠首在运行期间冲沙效果较明显,充分使理论和实践相结合。通过这两级的水沙分离过程,实现了方便顺畅地引水、排沙,水沙分离效果,明显改善了阿韦滩渠首工程运行中存在的问题,为今后该新型渠首推广应用奠定基础。该新型渠首通过运行管理发现一些问题,因此在运行管理中如何进行科学调度需要不断深入研究。建议随着运行时发现的问题逐渐改良该渠首，相信通过进一步完善和发展后，该渠首应该是一种很有发展前景的新型引水排沙渠首。

参考文献

[1] 张立德. 新疆引水渠首[M]. 新疆:新疆人民出版社,1994.

[2] 高亚平,何晓宁,等. 新疆呼图壁河灌区续建配套与节水改造工程 2005 年实施方案[R]. 新疆:新疆兴利水利水电勘查设计所,2007.

【作者简介】 李虹瑾,女,1980 年 6 月生,2006 年毕业于新疆农业大学,硕士研究生学位,新疆水利水电规划设计管理局,中级职称。

玻璃钢管道糙率值实证分析

蒲振旗　徐元禄　周　骞

（新疆额尔齐斯河流域开发工程建设管理局　乌鲁木齐　830000）

【摘要】 管道水力学计算的主要内容是确定水头损失，而管道断面粗糙程度的界定、即糙率系数 n 值的取值对计算结果有很大的影响。本文以已建成投入运行的工程为研究对象，通过实测相关参数，就玻璃钢管道糙率值进行了实证分析，得出了较目前玻璃钢管道糙率推荐值或设计取值不同的结论。

【关键词】 玻璃钢管道　糙率值　实证分析

管道水力学计算的主要内容是确定水头损失。水头损失包括沿程水头损失和局部水头损失两部分，通常根据这两种水头损失在总水头损失中所占比重的大小，将管道分为长管及短管，长管是指水头损失以沿程水头损失为主，其局部损失和流速水头在总损失中所占比重很小，计算时可以忽略不计且不影响计算精度的管道；短管是局部损失及流速水头在总损失中占有相当的比重（一般认为局部损失及流速水头大于沿程损失的5%），计算时不能忽略的管道。影响管道过水能力的主要因素有管道断面形状、尺寸、断面粗糙程度以及进出口建筑物和管线纵向布置型式等。不论是按长管抑或按短管计算，管道断面粗糙程度的界定，即糙率系数 n 值的取值对计算结果有很大的影响。目前，有些专著和玻璃钢生产厂家对玻璃钢管糙率值，即 n 值的推荐值为 0.008 4，大多数设计单位在设计中取值为 0.009。本文以已建成投入运行的新疆北疆供水工程之组成部分——小洼槽倒虹吸工程为研究对象，通过实测相关参数，经分析计算，得出了与上述 n 值有较大差距的结论，供同行讨论、参考。

1　工程简介

1.1　主要设计参数

新疆北疆供水工程中的小洼槽倒虹吸全长 5 765 m，采用双管线布置型式，地埋式敷设，管内经 3 100 mm，单管设计过水流量为 15.25 m^3/s（双管为 30.5 m^3/s ），加大流量为 17.5 m^3/s（双管为 35 m^3/s），工作压力 0.46 MPa，管材为玻璃纤维缠绕增强热固性树脂夹砂压力管（简称 FRP 管或玻璃钢夹砂管），管内径 3.1 m，为我国同类管材中管径最大的工程。本工程于2005 年完工并试运行1 个月，2006 年至原型观测时已正常运行8 个月。工程设计工作压力 0.46 MPa，试验压力 0.69 MPa，管内真空压力 0.1 MPa。直管段管壁厚度 58.5 mm，其中内衬厚度 2.5 mm，结构层厚度 55.5 mm，外保护层厚 0.5 mm，标准管有效长度 12 000 mm。标准管结构见图 1。

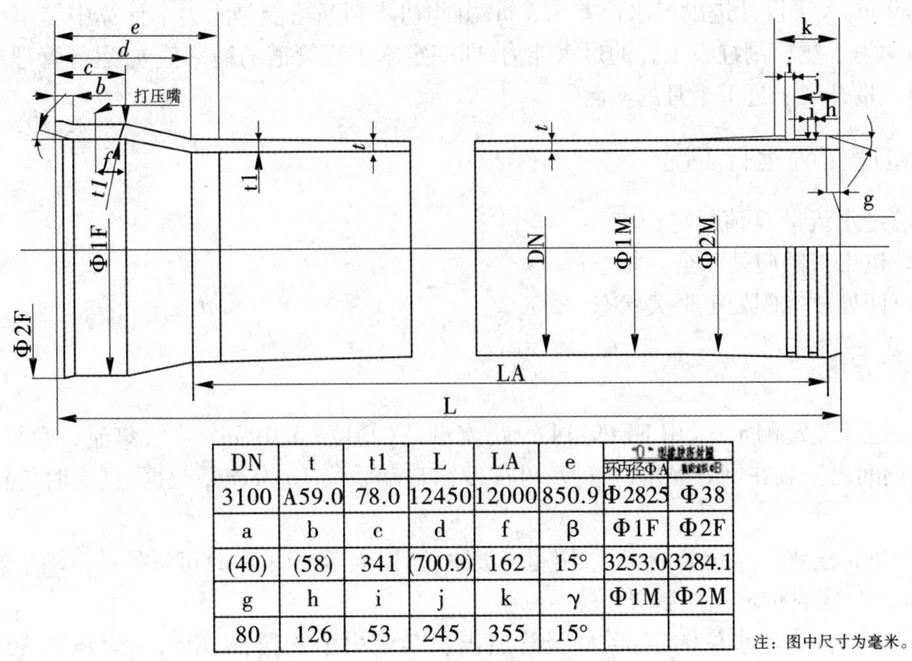

DN	t	t1	L	LA	e	“O”[illegible] 环内径ΦA	[illegible]
3100	A59.0	78.0	12450	12000	850.9	Φ2825	Φ38
a	b	c	d	f	β	Φ1F	Φ2F
(40)	(58)	341	(700.9)	162	15°	3253.0	3284.1
g	h	i	j	k	γ	Φ1M	Φ2M
80	126	53	245	355	15°		

注：图中尺寸为毫米。

图1　标准管结构

1.2　主要工程量

倒虹吸管道总长2×5 346 m，其中FRP管长2×5 187 m（标准管826根，非标准管60根）、钢制管件总长2×159 m，沿管线有2×17座镇墩，2×8座进、排气阀井，2×14座人孔井，2×2座放空阀井等构筑物；单线与管道水力学计算有关的管道连接接头共462个、折坡点（弯管）14处、三通管24个、出口平板钢闸门1扇。管道连接型式见图2。

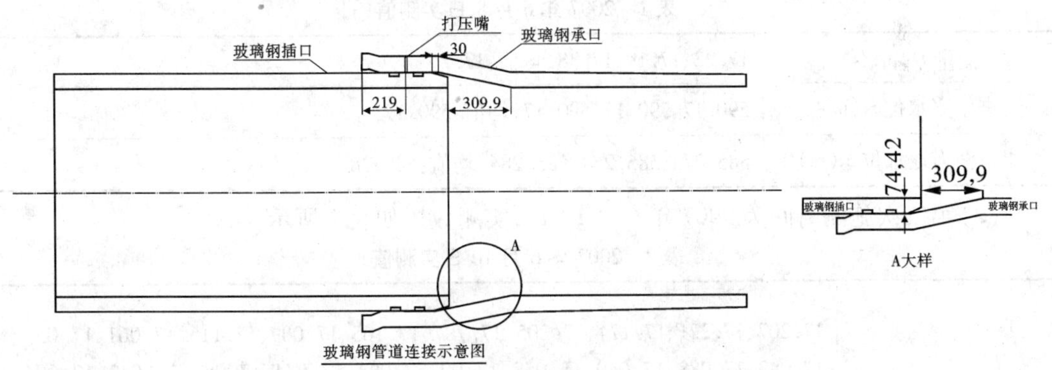

图2　管道连接示意图

2　问题的提出

2007年6月上旬，北疆供水工程输水稳定，输水流量接近倒虹吸设计加大流量17.5 m^3/s，管道以单线方式运行，运行管理人员在巡视检查过程中发现，倒虹吸工程进口水位明显高于设计水位值，经组织技术人员对可能影响上游水位高原因进行检查，发现倒虹吸进出口闸门均处于全开状态，进口拦污栅前亦无杂物拥堵，经初步实测，倒虹吸进出口水位差为4.9 m,大于设计进出口水位差4.3 m,由此初步判断可能是水力学计算中糙率 n 值取值偏小。为了复核倒虹吸工程的过水能力,即率定本工程管道的糙率值 n,运行管理单位组织专门力量开展了近1个月的实测。

3 倒虹吸工程运行工况

(1)过水流量 Q 值稳定;

(2)进出口闸门处于全开状态;

(3)两处放空阀处于全关状态。

4 实测方法

(1)过流流量测定:采用UF-9114声路流量计(精度±0.01 m^3/s)采集流量数据,该设备施工期间已安装在倒虹吸出口管段,可每秒钟自动采集16次流量数据,且实时传输至控制屏。

(2)进口进水室水位值测定:采用超声波水位计(精度±0.01 m)可每秒钟自动采集水位数据16次且实时传输至控制屏。

(3)出口消力池水位测定:倒虹吸管道出口与消力池连接,因出口消力池长30 m、宽23 m、尾端坎高4.2 m,容积较大,经观察,在流量稳定时,消力池水位变幅很小,故出口消力池水位采用测绳人工施测,且两次施测间隔要大一些,每次实测期间采集的水位值要少一些。

5 实测水力学参数

(1)第一次施测时间为2007年6月8日,实测数据如表1所示。

表1 2007年6月8日实测值

流量 Q(m^3/s)	17.22、17.19、17.38,均值:17.26
进水室水位 z_1(m)	590.17、590.17、590.17,均值:590.17
出口消力池水位 z_2(m)	585.274、585.284、585.284,均值:585.28

(2)第二次施测时间为2007年6月10日,实测数据如表2所示。

表2 2007年6月10日实测值

流量 Q(m^3/s)	17.207、17.221、17.174、17.205、17.187、17.165、17.009、17.115、17.061、17.035、17.083、17.088、17.040、17.058、17.00、17.094、17.074、17.082、17.042、17.004、17.007、17.092、17.116、17.153、17.092、17.088、17.143、17.174、17.115、17.130、17.137、17.126、17.140、17.105、17.068、17.036、17.023、17.00、17.058、17.109、17.026、17.048、17.009、17.014、17.039、17.10、17.077、17.073、17.106、17.10、17.065、17.098、17.095、17.087、17.088、17.019、17.004,均值:17.086

续表 2

进水室水位 z_1(m)	590.031、590.020、590.005、590.022、590.020、590.020、590.012、590.022、590.020、590.025、590.036、590.014、590.018、590.001、590.053、590.011、590.005、590.020、590.100、590.041、590.011、590.070、590.002、590.040、590.002、590.013、590.010、590.003、590.007、590.010、590.001、590.018、590.016、590.015、590.001、590.015、590.019、590.024、590.000、590.012、590.011、590.002、590.031、590.060、590.030、590.021、590.014、590.012、590.011、590.024、590.021、590.022、590.021、590.024、590.020、590.021、590.010,均值:590.02
消力池水位 z_2(m)	585.274、585.261、585.258、585.268、585.298、585.292、585.300、585.278、585.273、585.320、585.290、585.263、585.298、585.279、585.289、585.304、585.291、585.276、585.295、585.273,均值:585.284

(3)第三次施测时间为 2007 年 6 月 11 日,实测数据如表 3 所示。

表 3 2007 年 6 月 11 日实测值

流量 Q(m^3/s)	16.789、16.780、16.745、16.740、16.660、16.672、16.664、16.650、16.633、16.662、16.615、16.619、16.704、16.696、16.715、16.758、16.781、16.828、16.860、16.724、16.719、16.757、16.741、16.768、16.829、16.850、16.851、16.861、16.910、16.916、16.872、16.922、16.920、16.935、16.896、16.878、16.920、16.916、16.813、16.747、16.759、16.785、16.809、16.802、16.856、16.798、16.873、16.924、16.862、16.796、16.752、16.778、16.822、16.845、16.826、16.855、16.854,均值:16.795
进水室水位 z_1(m)	589.90、589.92、589.89、589.93、589.91、589.89、589.90、589.87、589.86、589.87、589.88、589.86、589.87、589.89、589.90、589.87、589.86、589.91、589.89、589.88,均值:589.887
消力池水位 z_2(m)	585.264、585.264、585.264,均值:585.264

(4)第四次施测时间为 2007 年 6 月 14 日,实测数据如表 4 所示。

表 4 2007 年 6 月 14 日实测值

流量 Q(m^3/s)	16.7、16.6、16.7、16.6、16.5、16.595、16.6、16.9、16.8、16.6、16.7、16.6、16.6、16.7、16.6、16.6、16.7、16.7、16.8、16.6、16.6、16.5,均值:16.65
进水室水位 z_1(m)	589.74、589.75、589.74、589.74、589.75、589.76、589.76、589.73、589.76、589.75、589.75、589.75、589.76、589.70、589.73、589.77、589.76、589.77、589.73、589.73、589.74、589.74,均值:589.75
消力池水位 z_2(m)	585.194、585.194、585.194、585.204、585.204、585.204、585.194、585.204,均值:585.199

(5)第五次施测时间为 2007 年 7 月 9 日,实测数据如表 5 所示。

表5 2007年7月9日实测值

流量 $Q(m^3/s)$	16.73、16.76、16.86、16.88、16.76、16.73、16.57、16.6、16.61、16.57、16.57、16.69、16.6、16.57、16.67、16.67、16.67、16.65、16.67，均值:16.675
进水室水位 z_1(m)	589.9、589.91、589.9、589.92、589.9、589.91、589.94、589.93、589.92、589.93、589.91、589.93、589.93、589.93、589.92、589.91、589.91、589.93、589.91，均值:589.92
消力池水位 z_2(m)	585.244、585.244、585.244、585.244、585.244，均值:585.244

6 实测值合理性分析

（1）进水室水位：第一次实测水位变幅值为0；第二次实测水位变幅为590.00～590.07 m，变幅值为0.07 m；第三次实测水位变幅为589.86～589.93 m，变幅值为0.07 m；第四次实测水位变幅为589.73～589.77 m，变幅值为0.04 m；第五次实测水位变幅为589.90～589.94 m，变幅值为0.04 m。五次实测水位变幅值占均值水位最大仅0.12‰，最小为0。导致水位变幅的主要因素是风力作用使进水室水面产生轻微涌浪。

（2）消力池水位：第一次实测水位变幅为585.274～585.284 m，变幅值为0.01 m；第二次实测水位变幅为585.258～585.320 m，变幅值为0.062 m；第三次实测水位变幅值为0；第四次实测水位变幅为585.194～585.204 m，变幅值为0.01 m；第五次实测水位变幅值为0。五次实测水位变幅值占均值水位最大仅0.105‰，最小为0。导致水位变幅的主要因素是风力作用使消力池水面产生轻微涌浪。

（3）流量值。第一次实测流量变幅为17.19～17.38 m^3/s，变幅值为0.19 m^3/s；第二次实测流量变幅为17.0～17.221 m^3/s，变幅值为0.221 m^3/s；第三次实测流量变幅为16.615～16.935 m^3/s，变幅值为0.320 m^3/s；第四次实测流量变幅为16.5～16.9 m^3/s，变幅值为0.4 m^3/s；第五次实测流量变幅为16.57～16.88 m^3/s，变幅值为0.31 m^3/s。五次实测流量变幅值占均值流量最大为2.4%，最小为1.10%，小于UF-9114流量计所标识计量误差小于5%的误差范围。

7 水力学计算

7.1 基本公式

为了方便计算，取上游进水室为1—1断面、下游消力池为2—2断面（计算简图如图3所示），此区间的边界条件比较清楚，简化后的能量方程如下

$$h_{w1-2} = \frac{v^2}{2g}\left(\sum \zeta_j + \lambda \frac{L}{d}\right) = z_1 - z_2$$

式中：z_1 为倒虹吸进口进水室水位，m；z_2 为倒虹吸出口消力池水位，m；z 为进出口水位差，即总水头损失，$z=z_1-z_2$，m；v 为管内平均流速，m/s，$v=\frac{4Q}{\pi d^2}$；$\sum \zeta_j$ 为局部水头损失系数之和；λ 为沿程水头损失系数，$\lambda=\frac{8g}{C^2}$，$C=\frac{1}{n}R^{1/6}$；g 为重力加速度，m/s^2；C 为谢才系数；n 为管

道糙率；R 为断面水力半径，m，对于圆管，$R=\frac{d}{4}$；d 为管道内径，m；L 为管道长度，m；Q 为流量，m^3/s。

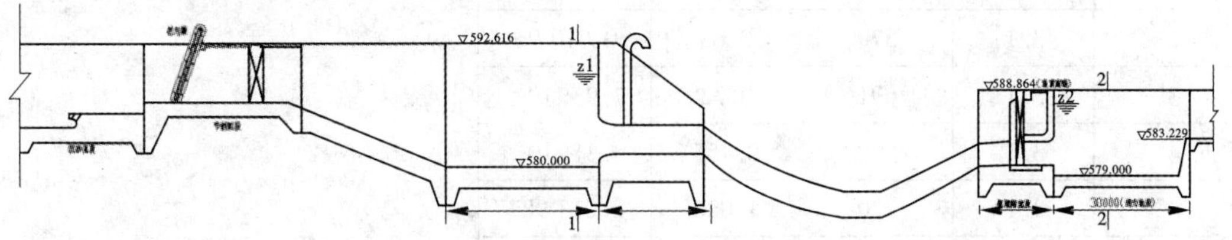

图3 小洼槽倒虹吸结构示意图

7.2 局部水头损失计算

$$\sum\zeta_{j}=\zeta_{进口}+\sum\zeta_{弯管}+\zeta_{门槽}+\zeta_{出口}$$

(1)弯管局部水头损失 $\sum\zeta_{弯管}$。本工程单管线总计设置弯管13个，各弯管几何参数、局部水头损失计算详见表6。

(2)进口局损 $\zeta_{进口}$。进口为圆角，取 $\zeta_{进口}=0.1$。

(3)末端闸门槽局损 $\zeta_{门槽}$。出口平板闸门全开，取 $\zeta_{门槽}=0.4$。

(4)出口局损 $\zeta_{出口}$。倒虹吸管线出口接消力池，属淹没出流，取 $\zeta_{出口}=1.0$。

由此得出 $\sum\zeta_{j}=1.669\ 775$。

7.3 管道糙率值计算

(1)按第一次实测值计算糙率值 n，流量、上下游水位均取均值。

①将钢制管件、玻璃钢管沿程损失分别计算。

$$h_{w1-2}=\frac{v^2}{2g}\left(\sum\zeta_{j}+\lambda_1\frac{L_1}{d_1}+\lambda_2\frac{L_2}{d_2}\right)=z$$

式中：λ_1、λ_2 分别为玻璃钢管段及钢管段沿程水头损失系数；d_1、d_2 分别为玻璃钢管和钢管内径，m，本工程 $d_1=d_2=3.1$ m；L_1、L_2 分别为玻璃钢及钢管段长度，m。

取钢制管件段糙率 $n=0.011$（属新管），将已知数值 $Q=17.26\ m^3/s$、$z=4.89$ m、钢制管件总长159 m、玻璃钢管长5 187 m等代入以上公式求得玻璃钢管道糙率值 $n=0.010\ 62$。

②将整个管线视为玻璃钢管道，将管线总长5 346 m等已知数据代入以上公式求得管线综合糙率值 $n=0.010\ 63$。

(2)按第二次实测值计算糙率值 n。

①将钢制管件、玻璃钢管沿程损失分别计算：将已知数值 $Q=17.086\ m^3/s$、$z=4.736$ m 等已知数据代入以上公式求得玻璃钢管道糙率值 $n=0.010\ 59$。

②将整个管线视为玻璃钢管道，将已知数据代入以上公式求得管线综合糙率值 $n=0.010\ 60$。

(3)按第三次实测值计算糙率值 n。

①将钢制管件、玻璃钢管沿程损失分别计算：将已知数值 $Q=16.795\ m^3/s$、$z=4.623$ m 等已知数据代入以上公式求得玻璃钢管道糙率值 $n=0.010\ 61$。

②将整个管线视为玻璃钢管道，将已知数据代入以上公式求得管线综合糙率值 $n=0.010\ 62$。

表 6　弯管局部水头损失计算表

弯管编号	d(m)	ρ(m)	θ(°)	ζ	计算公式
1#	3.1	446	0.51	9.86×10^{-3}	$\zeta=[0.131+0.163\,2(\frac{d}{\rho})^{7/2}](\frac{\theta}{90})^{1/2}$ 式中：ζ 为局部水头损失系数；d 为弯管内径，m；ρ 为弯管转弯半径，m；θ 为弯管圆心角(°)。
2#	3.1	379	0.63	0.010 96	
3#	3.1	691	0.33	7.93×10^{-3}	
4#	3.1	500	0.455	9.31×10^{-3}	
5#	3.1	220	1.026	0.013 987	
6#	3.1	258	0.924	0.013 27	
7#	3.1	139	1.714	0.018 08	
8#	3.1	277	0.863	0.012 828	
9#	3.1	207	1.10	0.014 48	
10#	3.1	656	0.366	8.354×10^{-3}	
11#	3.1	436	0.524	9.996×10^{-3}	
12#	3.1	228	1.047	0.014 13	
13#	3.1	60	3.707	0.026 59	
合计				0.169 775	

(4)按第四次实测值计算糙率值 n。

①将钢制管件、玻璃钢管沿程损失分别计算：将已知数值 $Q=16.65\ m^3/s$、$z=4.551$ m 等已知数据代入以上公式求得玻璃钢管道糙率值 $n=0.010\ 62$。

②将整个管线视为玻璃钢管道，将已知数据代入以上公式求得管线综合糙率值 $n=0.010\ 63$。

(5)按第五次实测值计算糙率值 n。

①将钢制管件、玻璃钢管沿程损失分别计算：将已知数值 $Q=16.675\ m^3/s$、$z=4.676$ m 等已知数据代入以上公式求得玻璃钢管道糙率值 $n=0.010\ 768$。

②将整个管线视为玻璃钢管道，将已知数据代入以上公式求得管线综合糙率值 $n=0.010\ 775$。

8　结论

(1)所选倒虹吸进出口计算断面边界条件清楚，进出口局部水头损失系数选取准确。

(2)管线弯管边界条件详尽，弯管局部水头损失系数计算准确。

(3)管道连接为承插连接型式，形成多达 462 个错台(见图 2)，肯定会产生局部水头损失，因目前尚无成熟的经验公式进行计算，故未予计入。

(4)管道沿线设置有 24 个三通件(其中检修人孔 14 个，内径 700 mm；进排气阀 8 个，内径 600 mm；底部放空 2 个，内径 300 mm)，也会产生局部水头损失，本次计算中也未予计入。

(5)五次实测值计算结果：玻璃钢管道糙率值 $n=0.010\ 59\sim0.010\ 768$，绝对差值为 1.78×10^{-4}，均值为 0.010 64，比较差为 1.67%。所得结果较本工程设计取值(0.009)大

17.67%～19.72%,较玻璃钢生产厂家推荐值(0.008 4)大22.5%～28.27%。因计算中未计入管道接头、三通件产生的局部水头损失,因此求得的糙率值应称玻璃钢管道的综合糙率值。

将整个管线视为玻璃钢管道求得的管线综合糙率值 $n=0.010\ 60 \sim 0.010\ 775$,均值为0.010 65,与两种管材分别计算水头损失所得结果相比差别很小,因此在进行玻璃钢管道水力学计算时,当异型管材长度(管径相同情况)占管线总长度比重小于5%时,为使计算变得简单快捷,可将整个管线视为玻璃钢管材,其计算结果完全满足精度要求。

参考文献

[1] 岳红军.玻璃钢夹砂管道[M].北京:科学出版社,1998.
[2] 余际可,魏璟,罗尚生,等.倒虹吸管[M].北京:水利电力出版社,1983.
[3] 成都科技大学水力教研室.水力学[M].2版.北京:高等教育出版社,1982.

【作者简介】 蒲振旗,男,1978年1月生,2002年7月毕业于石河子大学,本科,工程师,研究方向为水利工程建设管理,工作单位:新疆额尔齐斯河流域开发建设管理局。

河段水位预报 RC(阻容)模型

胡兴艺

(湖南省水文水资源勘测局　长沙　410007)

【摘要】 本文利用 RC 电路对理想河流区段进行了模拟,建立了简单的电路模型,推导出模型公式,并通过实例对模型进行了验证,该方法对水利工程建设和水的研究有一定参考价值。

【关键词】 水位预报; 模型; RC 电路; 输入响应

1 前言

洪水预报是一项很复杂的任务,主要是预报对象错综复杂。水文模型通常是指能对径流的产生、发展进行模拟、验证或预测的数学方法。根据侧重点不同,水文预报模型大致可分为确定模型和随机模型、线性模型和非线性模型、时变模型和时不变模型、模拟模型和预报模型、集总模型和分散模型。随着流域面积的增大,流域的特性也随之而发生变化,流域的线性特性也越来越明显,这是线性系统模型得以应用的根本条件。

2 线性系统

某个系统,当输入发生变化后,相应输出的变化过程称为动态过程。如果系统的动态过程在数学上可归结为以下的线性微分方程:

$$\begin{aligned} & a_n \frac{\mathrm{d}^n y}{\mathrm{d}t^n} + a_{n-1} \frac{\mathrm{d}^{n-1} y}{\mathrm{d}t^{n-1}} + \cdots + a_1 \frac{\mathrm{d}y}{\mathrm{d}t} + a_0 y(t) \\ = & b_m \frac{\mathrm{d}^m u}{\mathrm{d}t^m} + b_{m-1} \frac{d^{m-1} u}{\mathrm{d}t^{m-1}} + \cdots + b_1 \frac{\mathrm{d}u}{\mathrm{d}t} + b_0 u(t) \end{aligned} \tag{1}$$

式中: $u(t)$ 为系统输入; $y(t)$ 为系统输出。

那么,该系统就称为线性动态系统。

在水文上通常将上式离散化,令采样时间间隔 t 相同,得到线性系统的差分方程模型。例如对一个单输入、单输出离散线性系统的 n 阶差分方程可表示为

$$\begin{aligned} & y(t) + a_1 y(t-1) + \cdots + a_n y(t-n) \\ = & b_0 u(t) + b_1 u(t-1) + \cdots + b_n u(t-n) \end{aligned} \tag{2}$$

线性系统通常用激励与响应来分析,激励是指系统的输入,响应是指系统的输出。

3 模型建立

设河流区间为 AB,A 点为水位 $V_a(t)$,B 点为水位 $V_b(t)$,如图 1 所示。我们可以把此区

间看做是有一个输入端和输出端的线性系统,如图2所示。其中,$V_a(t)$为随时间t变化的输入函数,$V_b(t)$为随时间t变化的输出函数,即所谓的激励和响应。

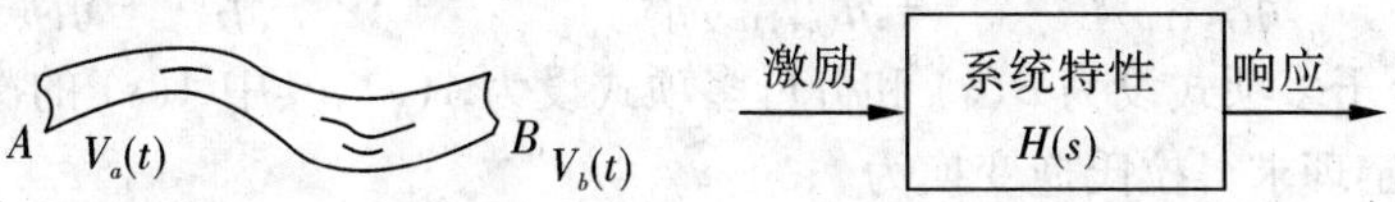

图1 河流区间示意图　　图2 线性系统模型

对于一个线性系统,它可以用一个常系数线性微分方程来描述。在电子线路中,RC串联电路电容器两端的电压$u_c(t)$所满足的关系式为

$$RC\frac{\mathrm{d}u_c}{\mathrm{d}t}+u_c=e(t) \tag{3}$$

这是一个一阶常系数线性微分方程,电路中外加电动势$e(t)$是系统的激励,电容器两端的电压$u_c(t)$是系统的响应,如图3所示,虚线框中的电路结构决定于系统内部的元件参数和连接方式。

根据河流的水力学及RC电路特性,河流区间AB可用一组RC串联电路来近似模拟,如图4所示,它是一个n阶非线性电路。由于河流具有容积特性,它与电路上的电容特性相似,以及河流的蒸发和径流的作用类似电路上的电阻特性。因此,对于一段河流来说,完全可以采用由电阻和电容组成的RC电路来模拟。

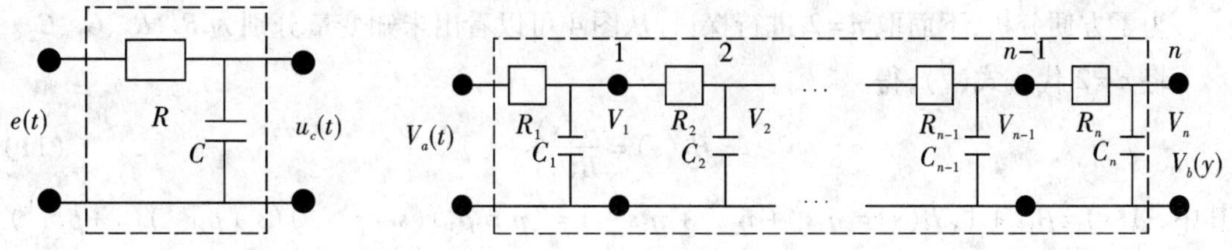

图3 基本RC电路　　图4 串联型RC电路

根据阻抗分压关系写出系统函数$H_n(s)$如下

$$H_n(s)=\frac{V_b(s)}{V_a(s)} \tag{4}$$

根据电路分析和递推法,求得

$$H_n(s)=\prod_{i=1}^{n}\frac{Z_i}{R_i+Z_i} \tag{5}$$

式中:$Z_1,Z_2,\cdots,Z_{n-1},Z_n$为节点$1,2,\cdots,n-1,n$对地分压阻抗。

由式(5)可推导得出

$$H_n(s)=\frac{1}{\prod\limits_{i=1}^{n}\left(\frac{R_i}{Z_i}+1\right)}=\frac{1}{\prod\limits_{i=1}^{n}(\alpha_i+1)}=\frac{1}{\sum\limits_{m=0}^{n-1}\sum\limits_{i=1}^{C_n^{n-m}}\overbrace{\alpha_j\cdots\alpha_k}^{n-m}+1} \tag{6}$$

式中:$\overbrace{\alpha_j\cdots\alpha_k}^{n-m}$为在$\alpha_1,\alpha_2,\cdots,\alpha_{n-1},\alpha_n$中取$n-m$个组合相乘,其中$j,k$取$1,2,\cdots,n-1,n$。

通过推导,可得

$$\alpha_i=\frac{\mu_{n-i+1}s^{n-i+1}+\mu_{n-i}s^{n-i}+\cdots+\mu_2s^2+\mu_1s}{\lambda_{n-i}s^{n-i}+\lambda_{n-i-1}s^{n-i-1}+\cdots+\lambda_2s^2+\lambda_1s+1}$$

将上式带入式(6),可得

$$H_n(s)=\frac{\beta_{(n-1)n/2}s^{(n-1)n/2}+\beta_{(n-1)n/2-1}s^{(n-1)n/2-1}+\cdots+\beta_2s^2+\beta_1s+1}{\eta_{(n+1)n/2}s^{(n+1)n/2}+\eta_{(n+1)n/2-1}s^{(n+1)n/2-1}+\cdots+\eta_2s^2+\eta_1s+1} \tag{7}$$

将 $H_n(s)$ 分子多项式设为 $B(s)$ 和分母多项式设为 $A(s)$,其中 $A(s)$ 的次数为 $B(s)$ 的 n 次倍,根据留数定理求得拉氏逆变换为

$$h_n(t)=\pounds^{-1}[H_n(s)]=\sum_{k=1}^{(n+1)n/2}\underset{s=p_k}{\mathrm{Res}}[H_n(s)e^{st}] \qquad t>0 \tag{8}$$

设 $B(s)$ 有 $(n+1)n/2$ 个单零点 $p_1,p_2,\cdots,p_{(n+1)n/2}$,即这些点都是 $\frac{A(s)}{B(s)}$ 的单极点,代入海维赛(Heaviside)展开式,有

$$h_n(t)=\sum_{k=1}^{(n+1)n/2}\frac{A(p_k)}{B'(p_k)}e^{p_kt} \qquad t>0 \tag{9}$$

显然,系统函数 $H_n(t)$ 的逆变换为 $h_n(t)$,$V_b(t)$ 可以利用 $h_n(t)$ 与 $V_a(t)$ 的卷积积分求得,即

$$V_b(t)=\int_0^t h_n(\tau)V_a(t-\tau)\mathrm{d}\tau \tag{10}$$

4 模型分析

为了方便分析,下面取 $n=2$ 进行探讨,从图 4 可以看出未知变量分别为 R_1、R_2、C_1、C_2。

将 $n=2$ 代入式(7),得

$$H_2(s)=\frac{A(s)}{B(s)} \tag{11}$$

其中,$A(s)=\beta_1s+1$,$B(s)=\eta_3s^3+\eta_2s^2+\eta_1s+1=(\rho_1\rho_2\rho_3)(s+\rho_1{}^{-1})(s+\rho_2{}^{-1})(s+\rho_3{}^{-1})$

令

$$C(s)=(\beta_1s+1)/\rho_1\rho_2\rho_3$$

$$D(s)=(s+\rho_1{}^{-1})(s+\rho_2{}^{-1})(s+\rho_3{}^{-1})$$

由式(9)得

$$h_2(t)=\sum_{k=1}^{3}\frac{C(-\rho_k{}^{-1})}{D'(-\rho_k{}^{-1})}e^{-\rho_k{}^{-1}t}=K_1e^{-\rho_1{}^{-1}t}+K_2e^{-\rho_2{}^{-1}t}+K_3e^{-\rho_3{}^{-1}t} \tag{12}$$

其中

$$K_1=-\frac{\rho_1-\beta_1}{(\rho_1-\rho_2)(\rho_1-\rho_3)} \qquad K_2=-\frac{\rho_2-\beta_1}{(\rho_2-\rho_1)(\rho_2-\rho_3)} \qquad K_3=-\frac{\rho_3-\beta_1}{(\rho_3-\rho_1)(\rho_3-\rho_2)}$$

代入式(10)求得系统输出函数 $V_b(t)$,即

$$V_b(t)=\int_0^t h_2(\tau)V_a(t-\tau)\mathrm{d}\tau \tag{13}$$

设 $V_a(t)$ 为一个离散时间函数,采样时间间隔为 T,则 $t=mT$,按照卷积积分的数值计算方法,上式可变为

$$V_b(mT)=T\sum_{k=1}^{m}h_2(k)V_a(m-k) \tag{14}$$

上式结果可利用计算机编程计算,为了便于分析,这里采用矩形法,还可以利用梯形法、

辛普生法以及缩短采样时间间隔 T 来减小近似误差。

将式(12)代入式(14)有

$$V_b(mT) = T\sum_{k=1}^{m}(K_1 e^{-\rho_1{}^{-1}k} + K_2 e^{-\rho_2{}^{-1}k} + K_3 e^{-\rho_3{}^{-1}k})V_a(m-k)$$

$$= -T\sum_{k=1}^{m}\left[\frac{\rho_1-\beta_1}{(\rho_1-\rho_2)(\rho_1-\rho_3)}e^{-\rho_1{}^{-1}k} + \frac{\rho_2-\beta_1}{(\rho_2-\rho_1)(\rho_2-\rho_3)}e^{-\rho_2{}^{-1}k} + \frac{\rho_3-\beta_1}{(\rho_3-\rho_1)(\rho_3-\rho_2)}e^{-\rho_3{}^{-1}k}\right] V_a(m-k) \tag{15}$$

设有一组实测水位 $V_o(j)$，令 $M = \sum_{j=0}^{w}[V_o(j) - V_b(j)]^2$，根据多元函数的定理，要使得 M 取得最小值，必须满足

$$\begin{cases}
\dfrac{\partial M}{\partial \rho_1} = \sum_{j=0}^{w}[V_o(j)+V_b(j)]\left\{\sum_{k=1}^{jT^{-1}+1}\left[\dfrac{(\beta_1-\rho_2)(\beta_1-\rho_3)-(\beta_1-\rho_1)^2}{(\rho_1-\rho_2)^2(\rho_1-\rho_3)^2}e^{-\rho_1{}^{-1}k} + \dfrac{(jT^{-1}+1)\rho_1^{-2}(\rho_1-\beta_1)}{(\rho_1-\rho_2)(\rho_1-\rho_3)}e^{-\rho_1{}^{-1}k}\right.\right. \\
\qquad \left.\left. + \dfrac{(\rho_2-\beta_1)}{(\rho_2-\rho_1)^2(\rho_2-\rho_3)}e^{-\rho_2{}^{-1}k} + \dfrac{\rho_3-\beta_1}{(\rho_3-\rho_1)^2(\rho_3-\rho_2)}e^{-\rho_3{}^{-1}k}\right] V_a(jT^{-1}+1-k)\right\} = 0 \\
\dfrac{\partial M}{\partial \rho_2} = \sum_{j=0}^{w}[V_o(j)+V_b(j)]\left\{\sum_{k=1}^{jT^{-1}+1}\left[\dfrac{(\rho_1-\beta_1)}{(\rho_1-\rho_2)^2(\rho_1-\rho_3)}e^{-\rho_1{}^{-1}k} + \dfrac{(\beta_1-\rho_1)(\beta_1-\rho_3)-(\beta_1-\rho_2)^2}{(\rho_2-\rho_1)^2(\rho_2-\rho_3)^2}e^{-\rho_2{}^{-1}k}\right.\right. \\
\qquad \left. + \dfrac{(jT^{-1}+1)\rho_2{}^{-2}(\rho_2-\beta_1)}{(\rho_2-\rho_1)(\rho_2-\rho_3)}e^{-\rho_2{}^{-1}k} + \dfrac{\rho_3-\beta_1}{(\rho_3-\rho_1)^2(\rho_3-\rho_2)}e^{-\rho_3{}^{-1}k}\right] \\
\qquad \left. V_a(jT^{-1}+1-k)\right\} = 0 \\
\dfrac{\partial M}{\partial \rho_3} = \sum_{j=0}^{w}[V_o(j)+V_b(j)]\left\{\sum_{k=1}^{jT^{-1}+1}\left[\dfrac{(\rho_1-\beta_1)}{(\rho_1-\rho_2)^2(\rho_1-\rho_3)}e^{-\rho_1{}^{-1}k} + \dfrac{(\rho_2-\beta_1)}{(\rho_2-\rho_1)^2(\rho_2-\rho_3)}e^{-\rho_2{}^{-1}k}\right.\right. \\
\qquad \left. + \dfrac{(\beta_1-\rho_1)(\beta_1-\rho_2)-(\beta_1-\rho_3)^2}{(\rho_3-\rho_1)^2(\rho_3-\rho_2)^2}e^{-\rho_3{}^{-1}k} + \dfrac{(jT^{-1}+1)\rho_3{}^{-2}(\rho_3-\beta_1)}{(\rho_3-\rho_1)(\rho_3-\rho_2)}e^{-\rho_3{}^{-1}k}\right] \\
\qquad \left. V_a(jT^{-1}+1-k)\right\} = 0 \\
\dfrac{\partial M}{\partial \beta_1} = \sum_{j=0}^{w}[V_o(j)+V_b(j)]\left\{\sum_{k=1}^{jT^{-1}+1}\left[\dfrac{1}{(\rho_1-\rho_2)(\rho_3-\rho_1)}e^{-\rho_1{}^{-1}k} + \dfrac{1}{(\rho_1-\rho_2)(\rho_2-\rho_3)}e^{-\rho_2{}^{-1}k} + \right.\right. \\
\qquad \left.\left. \dfrac{1}{(\rho_3-\rho_1)(\rho_2-\rho_3)}e^{-\rho_3{}^{-1}k}\right] V_a(jT^{-1}+1-k)\right\} = 0
\end{cases} \tag{16}$$

其中

$$V_b(j) = T\sum_{k=1}^{jT^{-1}+1}\left[\frac{\rho_1-\beta_1}{(\rho_1-\rho_2)(\rho_1-\rho_3)}e^{-\rho_1{}^{-1}k} + \frac{\rho_2-\beta_1}{(\rho_2-\rho_1)(\rho_2-\rho_3)}e^{-\rho_2{}^{-1}k} + \frac{\rho_3-\beta_1}{(\rho_3-\rho_1)(\rho_3-\rho_2)}e^{-\rho_3{}^{-1}k}\right] V_a(jT^{-1}+1-k)$$

在最小偏差条件下，利用以上的最小二乘法和求偏导数方法可得出 ρ_1、ρ_2、ρ_3、β_1 代数方程组，再利用牛顿迭代法求解。由于这些常数和 RC 电路模型中常量个数相等，通常我们不再去求解 R_1、R_2、C_1、C_2 的实际值，它们在模型中只作为过渡变量。

5 实例验证

在进行模型验证时，对环境复杂的河流区域，其计算结果会产生一定偏差。因此，选择理想的河段进行验证，其效果较好。本文以湖南沅水桃源至常德区段为例进行分析，说明模

型的使用过程和计算方法。随着计算机的广泛应用,可利用 Matlab 或 Mathimatics 等数值计算语言将逆变换的有关运算编程计算出结果,使分析过程简化。由于本文篇幅有限,桃源至常德河段水位 *RC* 预报模型参数演算过程省略。

通过对桃源、常德站水位按 1 h 时段实时监测,即 $T=1$,以及计算机编程计算得出 $\rho_1 = 0.7123007$,$\rho_2 = 0.3386039$,$\rho_3 = 0.9376608$,$\beta_1 = 0.1440911$,$K_1 = 6.747$,$K_2 = -0.86888$,$K_3 = -5.878$。

分别代入式(13)并化简得

$$h_2(t) = \frac{6.747}{e^{t/0.7123007}} + \frac{-0.86888}{e^{t/0.3386039}} + \frac{-5.878}{e^{t/0.9376608}} \tag{17}$$

上式即为桃源至常德河段水位 *RC* 预报模型,由卷积公式(13)可求得预报结果,如表 1 所示。

表 1 湖南省沅水桃源、常德 2011 年 6 月 7 日至 20 日水位资料分析

月	日	时	桃源	常德		
			实测 $V_a(t)$	实测 $V_o(t)$	计算 $V_b(t)$	误差 $v(t)$
6	7	8:00	35.7	32.37		
6	7	20:00	36.55	33.23		
6	8	8:00	36.7	33.6		
6	8	20:00	37.74	33.99		
6	9	8:00	39.04	35.15	35.12	0.03
6	9	20:00	38.18	35.26	35.61	-0.35
6	10	8:00	37.06	34.5	35.26	-0.76
6	10	20:00	36.98	34.32	34.92	-0.60
6	11	8:00	36.84	34.18	34.69	-0.51
6	11	20:00	36.72	34.04	34.52	-0.48
6	12	8:00	36.19	33.87	34.22	-0.35
6	12	20:00	35.81	33.48	33.87	-0.39
6	13	8:00	35.73	33.32	33.64	-0.32
6	13	20:00	35.72	33.25	33.52	-0.27
6	14	8:00	35.76	33.23	33.49	-0.26
6	14	20:00	36.24	33.49	33.68	-0.19
6	15	8:00	35.87	33.53	33.67	-0.14
6	15	20:00	35.51	33.25	33.48	-0.23
6	16	8:00	35.34	33.09	33.28	-0.19
6	16	20:00	35.22	32.95	33.12	-0.17
6	17	8:00	35.09	32.85	32.98	-0.13
6	17	20:00	35.09	32.74	32.91	-0.17
6	18	8:00	35.06	32.76	32.86	-0.10
6	18	20:00	35.17	32.76	32.88	-0.12
6	19	8:00	35.93	33.18	33.22	-0.04
6	19	20:00	35.58	33.34	33.31	0.03
6	20	8:00	35.36	33.16	33.23	-0.07
6	20	20:00	35.21	33.07	33.10	-0.03

由于该模型 $n=2$,选取的参数个数有限,以及外界环境的影响,如常德站位处洞庭湖区,水位受湖水水位影响较大,预报误差是难免的。从表中的结果来看,计算水位与实际水位基本吻合,预报误差在允许范围内。

6 结论

水文预报是利用各种预报模型和现代计算机技术对河流流域上发生的水文过程进行模拟,本文建立的河段水位预报模型则是对特定的河流区间进行模拟,根据上游水位和当前下游水位,预报在一定的洪峰传播时间后下游将达到的水位。该预报方法非常方便,它可作为其他洪水预报模型的一种辅助手段,在实际应用中具有一定意义。从本文的结论性公式中可以看出,为了提高预报精度,特别是对于流域复杂的河流,我们可以选取尽可能大的 n 值,参数个数将为 $2n$,同时参数率定要选取尽可能多和采集间隔短的实测资料,并且利用现在计算机数值分析软件技术,求出最佳的模型参数。

参考文献

[1] 胡兴艺. 广义线性水库模型及其 RC(阻容)网络模拟[J]. 长江科学院院报,2009(5).
[2] 芮孝芳. 洪水预报理论的新进展及现行方法的适用性[J]. 水利水电科技进展,2001(5).
[3] 王厥谋,等. 综合约束线性系统模型[J]. 水利学报,1986(7).

【作者简介】 胡兴艺,1973 年 7 月生,河海大学研究生,工程硕士,湖南省水文水资源勘测局,高级工程师,湖南省评标专家。

潼关高程变异特征及其影响因素分析

王　亮　周怀东　王世岩　毛战坡　易伟雄

（中国水利水电科学研究院　北京　100038）

【摘要】 本文分析了三门峡水库潼关高程的历史变化情况及其主要影响因素，结果表明：①潼关流量基本呈现逐年下降的趋势，相关分析表明潼关年平均流量与年份之间呈显著负相关；②潼关输沙量的大小与径流量的大小密切相关，且潼关逐年输沙量呈下降趋势，相关分析表明输沙量与年份显著之间呈负相关，这说明黄河潼关段输沙能力呈逐年下降趋势；③利用潼关监测站降雨量数据，发现库区降雨量呈逐年下降趋势。此外，本文还探讨了不同时期潼关高程的主要影响因素的变化情况及其成因，得出随着流量的逐年降低三门峡水库的调节能力呈逐渐下降趋势，在现有情况下来水来沙条件已经成为影响潼关高程的主要因素。

【关键词】 三门峡库区　潼关高程　黄河　输沙率

1　引言

黄河是我国乃至世界上最为典型的多沙河流，它发源于青海省巴颜喀拉山北麓，流程 5 464 km，注入渤海。黄河实测多年平均水量 580 亿 m^3，沙量 15.6 亿 t，干流最高含沙量达 920 kg/m^3，平均含沙量 33.6 kg/m^3。黄河是世界第六长河，输沙量、含沙量均为世界之最，是一条举世闻名的高含沙量河流。1960 年 9 月，黄河上第一座综合性水利枢纽——三门峡水库建成，三门峡水库位于黄河中游河南省三门峡市和山西省平陆县交界处，是一座以防洪、防凌为主，兼有灌溉、发电、供水等综合大型水库。

然而由于种种原因，原规划设计对黄河泥沙问题的复杂程度和治理形势估计不足，致使水库运用初期发生了严重的淤积，先后进行了两次改建和三次水库运用方式的调整。其中，潼关高程变化是揭示渭河下游河道淤积状况的重要指标。潼关高程指黄河潼关水文站 6 号断面洪水流量在 1 000 m^3/s 时的相应水位。在潼关处，黄河受秦岭阻挡，转向东流，中条山和华山将该处的河谷压缩到 850 m，形成一个卡口。从潼关到三门峡坝址，黄河穿行在秦岭和中条山的阶地之间，黄土台地大多位于高程 380 ~ 420 m 处，河谷变窄，两岸地面沟壑冲刷，高低起伏。河道上宽下窄，滩高漕深，主流被束缚在狭窄的河槽内，蜿蜒曲折，流至三门峡坝址处的河滩宽度约为 300 m，呈带状河道型库区。影响潼关高程的主要因素有：来水量，来沙量，洪水来源与组成，洪峰流量过程与出现时间，洪峰级别，洪次，悬移质含沙量及床沙组成（颗粒级配），潼土古段河道比降、河势，三门峡水库运用水位，渭河入黄位置，黄河、渭河、北洛河洪峰先后次序等。自 1960 年三门峡水库（见图 1）开始蓄水运用以来，潼关 1 000 m^3/s流量的水位已经上升 5 m 左右，致使渭河下游河床不断淤积抬高，防洪任务加重，同时也造成关中广大地区地下水位上升，土地盐碱化，影响农作物的生长和产量。关于降低憧关高程、减轻渭河下游防洪压力一直是水利界关注的热点之一。

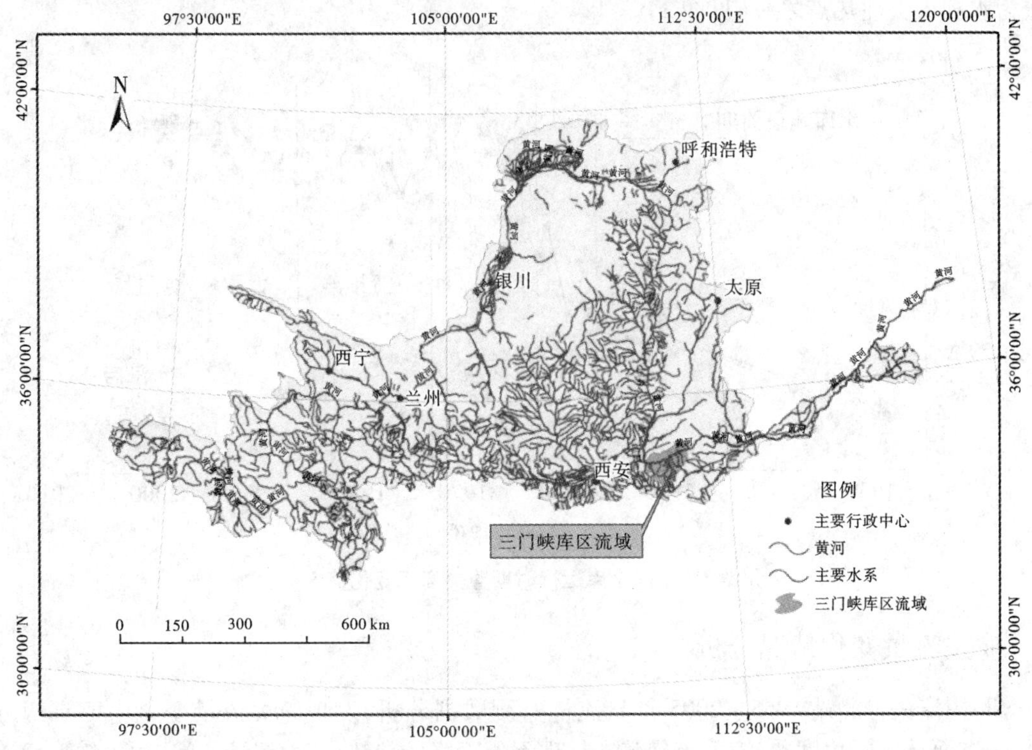

图1 三门峡库区位置示意图

潼关高程的抬升是多种因素的合力造成的,其中水沙条件及三门峡水库运行方式是影响潼关高程的重要因素。一方面,水库运用方式对潼关高程产生影响的实质是,通过抬高水流侵蚀基准面,降低水流比降和水流冲刷力,使潼关高程在年内得不到有效的冲刷而使潼关高程保持持续上升状态;另一方面,大量研究表明潼关高程的变化主要是泥沙淤积导致的,因此水沙条件对潼关高程的影响也相当显著,但是泥沙的淤积与水量及水体含沙量之间存在密切关系,而水库运行方式对库区流量及水体含沙量又存在显著影响,因此导致潼关高程抬高的多因素间相互关系极其复杂。所以,如何通过分析较长历史时段库区水体含沙量、水量及水库运行方式的变化情况对潼关高程的影响情况,并给出科学合理的解释,具有显著的现实意义及科学意义。

2 潼关水位的历史变化情况

三门峡水利枢纽于1957年4月正式开工,同年11月25日开始截流,1960年9月15日蓄水运用,1962年3月水库淤积严重,改为滞洪排沙运用,1973年采用蓄清排浑的运行方式。水库正式开工以前潼关高程最高水位为323.64 m,平均水位为322.33 m;1960年水库开始运行至1962年潼关高程最高达到325.93 m,较水库开工前升高2 m以上,1963~1973年滞洪排沙期间,潼关高程最高水位为328.71 m(1968年汛前),潼关平均水位为327.60 m;1973~1986年间采用"蓄清排浑"的运用方式,库区基本保持冲淤平衡,潼关高程相对稳定;1986年以后由于龙羊峡水库投运、工农业用水增加及降雨偏少等因素影响,黄河水量特别是汛期水量大幅减少,年内来水过程趋于均匀,洪水冲刷能力降低,潼关高程有所上升;2002 年随着小浪底水利枢纽的投入运用,关于降低潼关高程、减轻渭河下游防洪压力成为水利界关注热点之一(见图 2)。

图 2　不同水库运行时期潼关高程变化情况

3　潼关流量变化情况分析

利用潼关检测站 1965 ~2005 年日流量监测数据分析,发现潼关流量变化主要有两方面:一方面从年际角度看:潼关流量基本呈现逐年下降的趋势,其中统计期间 1967 年平均流量最高,为 1 990.75 m^3/s,最小流量出现在 1997 年(平均流量为 473.69 m^3/s),相关分析表明潼关年平均流量与年份之间显著负相关($P<0.01$,说明:本文中所有数理统计均采用 spss 软件),流量呈逐年下降趋势;另一方面从年内角度看:潼关流量年内变异逐渐减少,即年内流量趋于平稳,年内流量波动呈逐渐减小趋势(见图 3)。

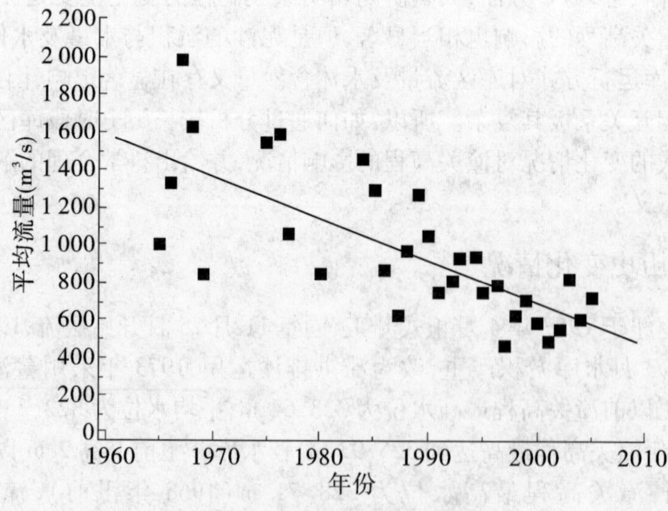

图 3　潼关流量年平均值变化情况

4 潼关输沙率变化分析

年内输沙量变化情况具有明显的周期性,潼关多年平均输沙量为10.43亿t,一年中最大输沙量在汛期(7~10月),最小输沙量在枯水期(11月至次年6月),汛期输沙量约占全年输沙量的78%(多年平均值),部分年份甚至可以达到90%(见图4)。以1993年潼关流量、水位及输沙率逐日时间变化曲线为例,流量最大值出现在汛期,输沙率最大值也出现在汛期且变化规律基本与流量同步,相关分析表明输沙率与流量呈显著正相关(相关系数为0.810,$P<0.01$)。

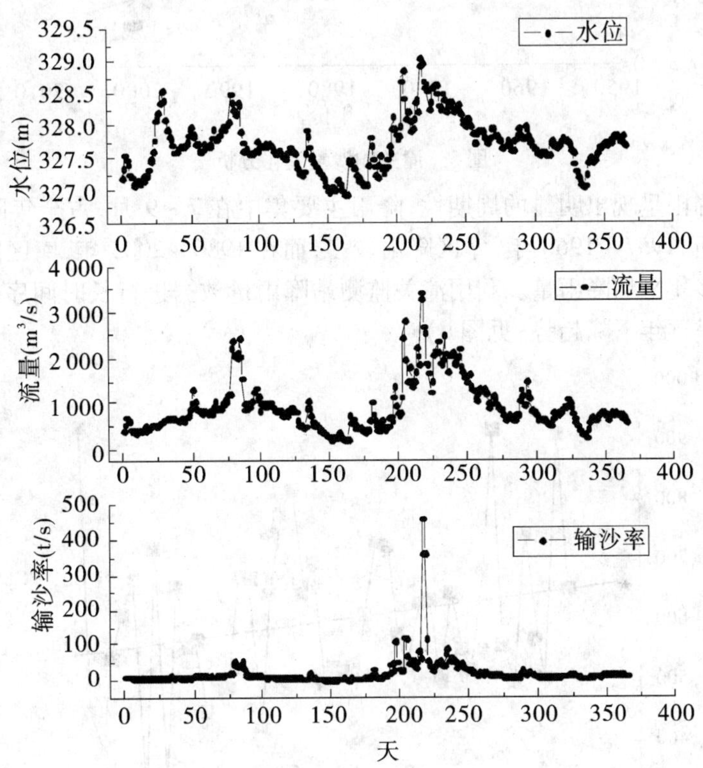

图4 1993年潼关水位、流量、输沙率变化曲线

潼关处多年输沙量年际差别较大,逐年变化趋势明显,潼关多年输沙量变化图(见图5)表明潼关逐年输沙量呈下降趋势,输沙量的大小与当年径流量密切相关。相关分析表明,输沙量与年份呈显著负相关(相关系数为-0.382,$P<0.01$)。这说明黄河潼关段输沙能力呈逐年下降趋势。

5 库区降雨量的年际变化情况

三门峡库区属半干旱的大陆性季风气候,降雨时空分布不均,冬、春两季少雨,夏、秋两季暴雨集中。根据三门峡、华县、潼关、洑头、河津等多年降雨、蒸发资料,库区多年平均降雨量为587 mm,降雨主要集中在7~9月,占全年降雨量的50%。多年平均蒸发量为1 430 mm,最高年达2 260 mm。由于干燥度高,且降雨过于集中,库区素有“十年九旱”之说。

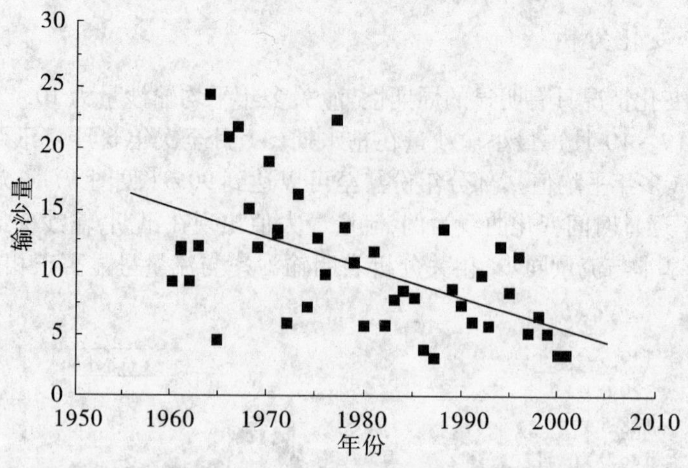

图 5 潼关输沙量变化分析

库区年内降雨呈现出明显的周期性,降雨主要集中在 7 ~ 9 月,占全年降雨量的 50%。从年际角度分析,1957 ~ 1966 年,库区降雨较多,而在 1986 ~ 2003 年,库区降雨较少,多数年份降雨低于多年平均降雨量。利用潼关监测站降雨量数据进行长时间序列分析表明,潼关区域降雨量呈逐年下降趋势(见图 6)。

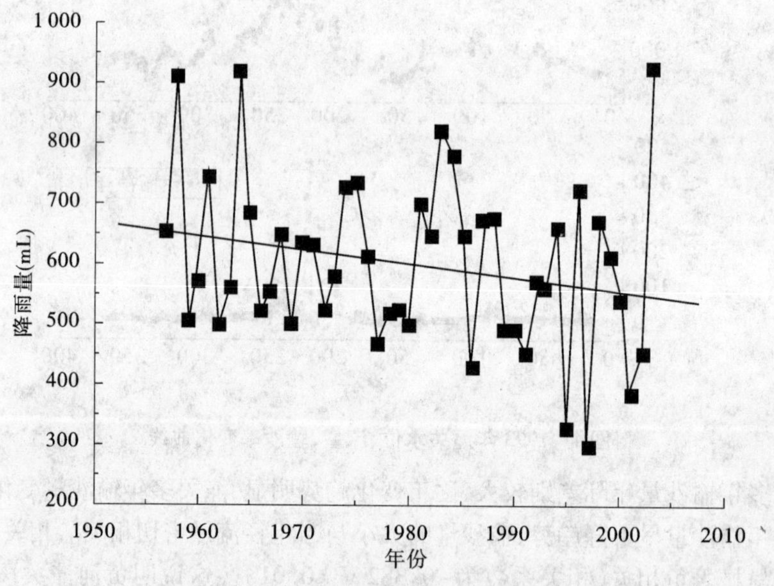

图 6 潼关地区降雨量变化情况

6 潼关高程变化趋势分析

上述分析结果表明,潼关高程 1986 年后稳定呈逐年缓慢上升趋势,总体稳定在 328 m 附近,而潼关处输沙能力却呈逐年下降趋势。大量研究表明,影响潼关高程的主要因素较多且非常复杂,其中包括自然因素及社会因素,如天然河道条件、水沙变化和三门峡水库运行方式等。从研究内容角度,大量关于潼关高程的研究主要针对潼关水沙条件及三门峡水库的运行方式对潼关高程的影响情况。三门峡水库建库前,潼关河段属于天然河道,主要受上游来水来沙条件影响，河道处于自然冲淤状态，并且从一个较长的历史时期来看潼关附近是处于微淤状态的。三门峡建库后，在短短几年的时间里，泥沙淤积导致潼关高程大量抬高，潼关高程抬高到 328.64 m，较建库前抬高了 5 m。1969 ~ 1973 年间水库除有防凌等任务外几乎敞泄，使潼关高程在水量严重偏枯的条件下降低了 2 m。1974 ~ 1979 年三门峡水库发电后，坝前水位抬高，潼关高程抬高。1979 ~ 1986 年潼关流量较大，潼关高程有所下降。1986 年以后至数据统计期间潼关高程又呈逐步增高趋势。

通过分析不同的文献研究内容及历史数据可以看出，影响库区泥沙淤积和潼关高程变化的主要因素为来水来沙条件、坝前运用水位。但是随着时间的发展，本文认为不同因素重要性已经发生改变，且大体上可以分为三个时期：水库建设前，水库建设后至 1986 年，1986 年后至今。第一个时期（水库建设前），大量研究表明水库淤积主要处于自然冲淤状态，且淤积量较小。叶青超等采用地质沉积结构原理，估算出三国时期至 1960 年小北干流的分段淤积厚度得出：夹马口—潼关三段的淤积厚度为 37.6 m，并由此计算出三段的年均淤积厚度分别为 0.021 m。邢大韦等根据地质剖面分析，认为全新世 10 000 年间潼关隆起沉积厚度为 16 m，年淤积速率为 0.001 6 m。焦恩泽等估算出 1583 ~ 1960 年的 377 年间滩面净淤厚为 15.47 m，相应潼关高程年均淤厚为 0.041 m，因此第一时期的主导因素为自然来水来沙条件，且呈微淤积状态。第二个时期（水库建成后至 1986 年）：水库建成后潼关高程的抬升已经是不争的事实；但是图 2 表明除蓄水拦沙期外，在滞洪排沙期及蓄清排浊期，潼关高程均具有抬高及下降的双重过程，这种过程必然受到水库来水来沙条件以及坝前运用水位的影响，这说明在这个时期内通过水库运行方式的调整及在适合的水沙条件下，三门峡水库仍然具有调节潼关高程的能力。但是随着区域流量逐渐下降下，黄河潼关段输沙能力呈逐年下降，三门峡水库的调节能力已经呈现下降的趋势；而且已经有研究指出，1974 年以后水库运用方式对潼关高程的影响不大：吴保生等通过对比坝前水位与潼关高程及入库水量的年际变化规律，认为水库“蓄清排浑”运用以来，潼关高程升降变化主要受入库水沙条件的影响，潼关高程与入库水量之间具有较好的关系，而与水库运用水位的关系不大。郭庆超等同样认为 1974 年以后采用“蓄清排浑”运用，潼关高程的变化主要受来水来沙影响，而与水库运用关系不明显。这些研究说明三门峡水库曾经具有的调节能力已经受到库区来水来沙影响呈逐渐下降趋势。

由于 1986 年以后潼关高程持续稳定在 328 m 附近，且没有大的变化波动基本呈缓慢上升趋势，而潼关多年平均流量却呈逐年下降趋势，因此第三个时期以 1986 年为时间节点的原因。影响潼关高程的自然因素主要为潼关水沙条件，其中包括洪水过程特征、洪峰及洪水历时等，大量研究表明潼关输沙能力下降与潼关流量的逐年下降有显著关系，特别是高含沙小洪水易造成淤积，对潼关高程往往不利；姜乃迁等认为三门峡水库“蓄清排浑”运用以来潼关高程近年来不断上升的主要原因是汛期来水量大幅度减少，而且本文也发现输沙率与流量呈显著正相关，因此近期入库水量减少是潼关高程居高不下的重要原因。通过分析潼关地区多年年平均降雨量发现，该区域降雨量呈逐年减少趋势，而且相关分析表明潼关流量与区域降雨量显著相关，因此如果三门峡库区的流量仍呈逐渐下降趋势，潼关高程在长时期内很可能依旧处于淤积状态，而这种状态对关中平原甚至西安都是不利的。

7 结论

通过对潼关高程演变过程变异特征及其影响因素分析,得到如下初步认识:

(1)潼关高程的变化受到三门峡水库运用方式和来水条件的共同影响,然而随着水库流量的变化,影响因素重要性已经发生改变。研究表明,通过改变三门峡水库的运用方式,在适当的来水来沙条件下潼关高程是可以下降的,但是随着流量的降低三门峡水库的调节能力呈逐渐下降趋势,来水来沙条件已经成为影响潼关高程的主要因素。

(2)通过分析潼关地区多年年平均降雨量发现,该区域降雨量呈逐年减少趋势,而且相关分析表明潼关流量与区域降雨量呈显著相关,这说明潼关流量的下降一部分是自然因素造成的,但是我们还应考虑到库区上游社会经济的高速发展对库区来水量的影响;因此如何通过合理的水资源管理措施,使三门峡水库在适合的水沙条件下发挥更大的调节能力,是现在必须解决的一个科学问题同样也是一个社会问题。

参考文献

[1] 焦恩泽,侯素珍,林秀芝,等.潼关高程演变规律及其成因分析[J].泥沙研究,2001(2):8-11.

[2] 张根广,王新宏,赵克玉,等.潼关高程抬升成因相关分析[J].泥沙研究,2004.56-62.

[3] 邢大韦,等.影响三门峡库区潼关高程的主要因素和控制措施[C]//三门峡水利枢纽运用四十周年论文集.黄河水利出版社,2001 .

[4] 吴保生,夏军强,王兆印.三门峡水库淤积及潼关高程的滞后响应[J].泥沙研究,2006:9-16.

[5] 姜乃迁,张翠萍,侯素珍,等.潼关高程及三门峡水库运用方式问题探讨[J].泥沙研究,2004.23-28.

【作者简介】 王亮,1980 年 1 月生,博士,毕业于中国科学院生态环境研究中心,现工作于中国水利水电科学研究院。

黄河乌兰布和沙漠段入黄风积沙监测研究

郭建英　邢恩德　李锦荣　荣　浩　崔　崴　刘铁军

（水利部牧区水利科学研究所　呼和浩特　010010）

【摘要】 黄河内蒙古段途经乌兰布和沙漠，沙丘直接侵入河道，入黄沙量连年不断，使河床淤积，河床的行洪、行凌能力降低，局部河段发展为“地上悬河”，洪灾、凌灾频繁发生，严重威胁下游水利工程安全运行和河道大堤两岸居民的安危。本文对乌兰布和沙漠沿黄河段风沙入黄的初步监测结果进行了分析与探讨，预期揭示沿黄段入黄风沙运移规律并精确计算出风积沙入黄沙量，以望引起国家及有关部门和学者就乌兰布和沙漠对黄河危害的高度重视，形成多渠道、多学科联合攻关研究的平台，使研究结果更好地为科学治理黄河泥沙问题及完善黄河沿岸综合防护体系提供科学依据。

【关键词】 黄河　乌兰布和沙漠　风积沙　入黄沙量

1　基本情况

1.1　区域概况

黄河内蒙古段穿行于我国北方干旱向半干旱地区的过渡地带。区域内干旱少雨，且大风频繁，两岸的风沙活动强烈，直接进入黄河，严重影响黄河的泥沙量。其中，乌兰布和沙漠位于黄河内蒙古段（乌海—磴口）西岸，此处沙漠化发展强烈，沙丘密集高大，沙丘链高 7 ~ 20 m，最高达 65 m，植被稀少，目前已有 50 km 左右的流动沙丘直接侵入黄河河道。

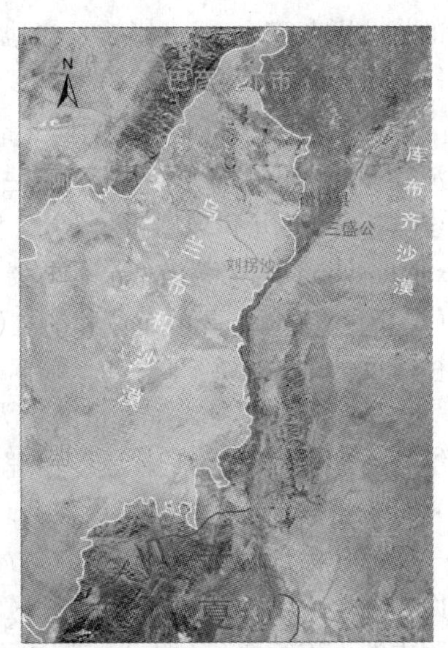

图 1　黄河内蒙古段乌兰布和沙漠沿岸遥感影像图

乌兰布和沙漠沿黄段位于内蒙古乌海市乌达区与巴彦淖尔市磴口县之间，右与鄂尔多斯杭锦旗相邻（见图1）。该区气候干燥，雨量稀少，年平均降水量 142.7 mm，年平均气温 8.0 ℃，风沙频繁，地表风蚀强烈，年平均风速 3.7 m/s，大风和风沙一年四季均有出现，以 3 ~5 月最多，风向多为西风及西南风，多年平均大风日数 10 ~32 d，多年平均扬沙日数 75 ~79 d，沙尘暴日数 19 ~22 d，属于典型的中温带大陆性干旱季风气候。

1.2　风成沙的危害

黄河沿岸乌兰布和沙漠流动沙丘延绵不断，大部分为新月形流动沙丘,西北高东南低,整个地势倾向黄河,且地势倾向与主风方向相同,致使大量风成沙直接向黄河倾泻,风成沙入黄以粗沙(>0.1 mm)为主,是黄河粗沙的重要策源地。据当地有关人士推测,乌兰布和沙漠每年直接输入黄河的泥沙为2 800万~6 000万t,是单位长度内使黄河水含沙量增幅最大的地区之一。如2010年3月19日磴口县的一次特大沙尘暴天气,平均风速11.3 m/s,最大风速23.5 m/s,持续20多h,仅黄河刘拐子沙头段(8.5 km)处,沙漠一次性向黄河输沙约为10万t。

图2　黄河沿岸沙丘的水流侵蚀

黄河泥沙含量的剧增,使黄河河床淤积抬高速度加快,造成乌兰布和沙漠沿黄刘拐子沙头段20 km黄河主流摆动,河岸崩塌,冲毁农田和林地,对下游三盛公黄河枢纽工程和其他水利工程构成严重威胁;河道淤积造成同水位下过流量不断减少,形成"地上悬河",给当地防凌防汛任务形成巨大的压力,使内蒙古黄河段成为防凌防汛最为严重的河段;乌兰布和沙漠使河套引水总干渠也饱受流沙侵袭的危害,每年用于干、支、斗、农、毛渠的清淤费高达500万元;河道淤积还使每年护岸防洪工程耗费大量人力、物力和财力,严重威胁人民群众的生命财产安全,极大影响了当地和下游地区社会经济的可持续发展。

1.3　生态现状及治理经验

乌兰布和沙漠地处我国西北地区荒漠半荒漠的前沿地带,总面积为1.15万km^2。其中流动沙丘占36.9%,半固定沙丘占33.3%,固定沙丘占29.8%。黄河沿岸主要是流动沙丘。土壤以灰漠土和风沙土为主。植被以白刺(Nitraria tang utorum)和霸王(Zygophyllum xanthoxylum)构成荒漠植被的重要建群种。另外,还有梭梭(Haloxylon amm odendron)、沙米(Agriophyllum squarrosum)和沙竹(Phyllostachys propinqua)、油蒿(Arte misia ordosica)等,植被盖度为5%~15%。

多年来国家投入了大量的物力、财力,在黄河沿岸先后实施了生态建设工程,水土保持综合治理工程、日元贷款风沙治理等项目,各工程项目均以治理流动沙丘、减少入黄泥沙,改善沿黄生态环境为主要建设目标。经过多年的治理,项目治理区取得了一定的效果,积累了一些成功治理沙漠的经验,如草方格沙障、红土压沙、营造草灌植被带等方法。但由于治理面积远小于危害面积,生态治理区相互独立,没有连成一片,形成许多治理空缺,加之治理过程中的不连续性和对防治措施的科技投入十分有限,使得黄河沿岸的流动沙丘没有得到有效控制,严重威胁黄河河道的安全运行。

2　研究的重要性和必要性

国家中长期科学和技术发展规划纲要(2006~2020年)已把黄河中上游、荒漠地区典型生态脆弱区生态系统的动态监测技术,黄河综合治理纳入为我国今后发展和研究的重点领域;全国水土保持科技发展规划纲要(2008~2020年)已明确把晋陕蒙粗沙多沙区不同尺度区域水土流失分布及其水沙调控纳入为我国主要土壤侵蚀区水土保持研究重点;水利部有关科技发展规划将黄河防洪防凌减灾问题、黄河水沙调控问题、水土流失问题等作为近年研究的重点内容;当地各级政府非常重视黄河沿岸的水土保持工作,将该区水土流失治理作为环境建设和改善人民生活的主要内容,列入重要议事日程。因此,黄河内蒙古段乌兰布和沙漠入黄沙量监测研究,符合国家和水利部有关科技发展规划需求。

开展乌兰布和沙漠沿黄段水土流失监测是综合治理工作的基础,能及时、准确、全面地反映水土保持生态建设情况、水土流失动态及其发展趋势,为水土流失防治、监督和管理决策服务,为国家生态建设提供科学依据。长期以来,由于缺乏先进的监测设备与科学技术的支撑,致使乌兰布和沙漠黄河沿岸段风成沙入黄量的估算结果说法不一,从0.28亿~0.60亿m^3/a,数据相差太大,严重影响了各级部门的防治决策。因此,开展黄河内蒙古段乌兰布和沙漠入黄沙量监测研究,可全面了解和掌握乌兰布和沙漠黄河沿岸风沙流的运移特征、沙丘移动规律和精确估算入黄风成沙量,将为研究黄河泥沙问题、维持黄河健康生命、不断完善黄河沿岸综合防护体系提供科学的决策依据。

3 监测研究内容及初步结果分析

3.1 监测研究内容

本项监测研究在广泛收集、整理、分析该区基础资料和相关研究成果的基础上,针对乌兰布和沙漠黄河(乌海—磴口段)沿岸风成沙入黄淤积河道的现象,对黄河沿岸流动沙丘的风沙流结构、输沙率与风速及高度的相关关系、风沙运移规律、风成沙入黄量、沙丘移动速度、河岸坍塌速度、坍塌量等,进行量化的监测研究,探讨乌兰布和沙漠风成沙入黄的危害,以期为黄河(乌海—磴口段)及下游库区淤积防治提供科学的基础数据。

本项目采用的监测设备主要有自动气象站用于统计不同时间段的风速、风向、沙丘表层土壤含水量;全方位定点集沙仪、全方位沙粒跃移梯度集沙仪测定不同方位和不同高度的积沙量;观测桩布设于流沙侵入河道的典型地区,每隔20~30 m布设一对,总长2 650 m,依据汛期和风沙活动的强度,定期测量沙丘的向前推移距离和河道水位的变化;通过三维激光地貌扫描仪精准扫描,生成项目区地形的点云图及DEM图,监测该区域不同时期地形地貌的变化情况,并可精确计算不同时期监测区域的土方量变化以及流动沙丘向河道推动距离;同时结合遥感影像资料分析多年黄河沿岸流动沙丘直接接触河道的有效长度及河岸坍塌的变化情况。

3.2 初步监测结果分析

乌兰布和沙漠黄河沿岸风成沙入黄主要表现为沙丘前移、风沙流及岸边坍塌三种形式。根据2010年5~6月对黄河刘拐子沙头段风沙观测场的野外观测,当气象站高度(2 m)风速达到5.3 m/s时,流动沙丘表面即可起沙。

3.2.1 测桩法监测结果分析

通过对沙漠沿黄段2 650 m断面布设观测桩监测,每隔2 m测一个沙丘垂直断面的水平向前移动距离,经对2010年5~6月11 503个观测数据的分析,沙丘向黄河沿岸推进的最大水平距离为9.66 m,平均推进3.78 m,如图3、图4所示。

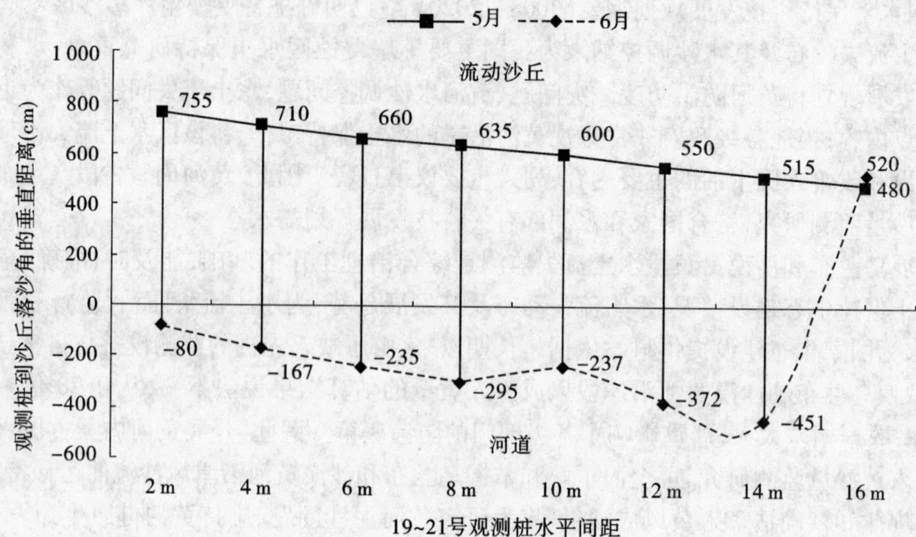

图3 典型沙丘向前移动断面图(一)

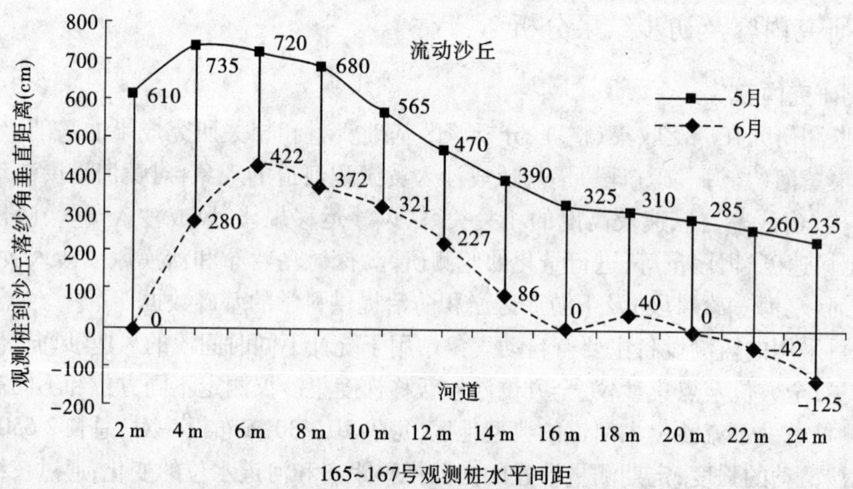

图4 典型沙丘向前移动断面图(二)

3.2.2 三维激光扫描仪监测结果分析

与传统测量技术的单点获取方式不同,地面三维激光扫描是从水平到垂直的全自动高精度步进式扫描测量,通过不同站点的连续扫描,可以得到完整的、全面的三维空间信息,满足空间信息数据库的数据源和不同应用需要。本文利用三维激光扫描仪对黄河沿岸239 621.32 m^2的地形于2010年5月和6月两次进行了扫描,由后处理软件对采集的点云和影像数据进行处理,将数据转换成绝对坐标系中的模型,生成DEM图(见图5)。经对黄河沿岸1 244.56 m流动沙丘的分析,风成沙5~6月进入黄河河道的沙量为36.29万t,沙丘平均向前移动3.94 m,与用测桩法监测的结果3.78 m基本相近。

上述监测结果仅为短期的初步监测,年均入黄沙量尚需对大于起沙风速的持续时间、有效风向、风蚀量、输沙率、河岸坍塌沙量等进行长期的动态监测,分析总结其特征和规律的基础上计算出,有待于进一步的监测研究。

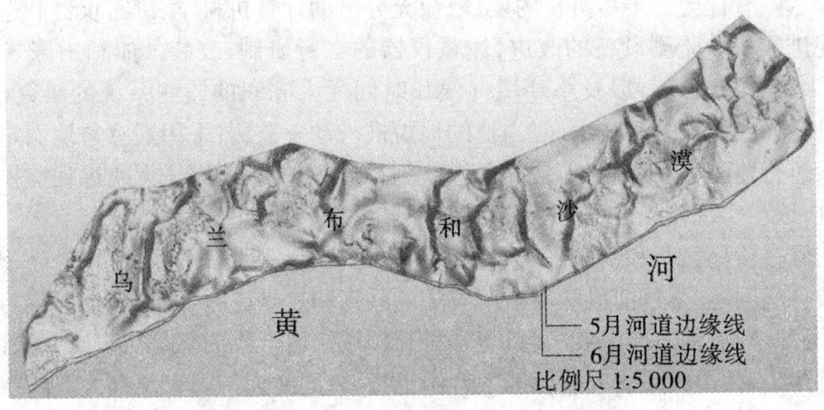

图5 监测区不同时期地形 DEM 图

4 研究展望

研究乌兰布和沙漠黄河沿岸入黄风积沙运移规律及防控机制是一项涉及面广、复杂多样、跨多学科的科研工作。还需从以下几个方面开展综合、系统、连续、动态、深入的监测研究。

4.1 沿黄段风积沙运移规律研究

乌兰布和沙漠沿黄段风沙运动具有不确定性和复杂性的特点,需长期对该地区不同风力及沙物质条件下,风沙流固体流量结构、梯度分布、风沙流输沙量与各种影响因子之间的多元数量关系进行研究,揭示黄河内蒙古段风沙流活动的时空分布特性。建立乌兰布和沙漠沿黄段风沙流输沙率函数模型,并进行黄河沿岸风沙运动过程的模拟与预测。

4.2 乌兰布和沙漠入黄风积沙与河道泥沙耦合关系及运移机制研究

由于受乌兰布和沙漠风积沙入黄的影响,使该段河道水沙过程变异大、河道萎缩、主槽摇摆不定等新情况、新问题不断出现,对其进行深入研究的科学意义和社会意义重大。因此,需对该地区风积沙入黄后与河道泥沙的耦合关系、风积沙入黄后的运移与扩散、泥沙分配和再分配、输水河段含沙量的时空变化、上下游监测站含沙量的关系分析、上下游监测断面淤积变化、水位变化对沿岸地形冲淤及其速率等方面开展深入研究。

4.3 沿黄段风积沙防控机制与水土保持关键技术研究

乌兰布和沙漠沿黄段气候、地形复杂多变,治理难度特别大,其治理成果很难发挥长期效益,急需加强对沿黄段风积沙防控机制与水土保持关键技术的研发,对防治风沙对黄河的危害和改善黄河两岸生态环境,促进区域经济可持续发展都具有十分重要的意义。需采用高新技术及多种方式开展乌兰布和沙漠沿黄段水土流失不同防治措施及其组合措施的防控机制和防护效益试验研究,提出合理的水土保持措施配置,为该区域黄河两岸水土流失治理工程提供科学依据,使乌兰布和沙漠沿黄段风沙危害降到最低程度。

4.4 风积沙监测技术研究

乌兰布和沙漠沿黄段流动沙丘远离城镇和村庄,交通不便,无常规电源等基础设施可用,且现有的监测设备多为自行研制的观测设备,可移动性差,智能化和精度低,不具备连续观测的能力,加之该地区风积沙入黄后与河道泥沙的耦合关系、输水河段含沙量的时空变化、气候变化等环境因子复杂多变,对该区域进行入黄风积沙的长期、连续、动态监测造成了极大的限制。随着社会进步和科技的发展，应充分借助计算机科学、机密仪器制造等其他行业的发展促进风积沙监测设备的改进，注重仪器的自行研制、改装。如利用嵌入式技术与GPRS技术对野外风蚀风沙以及小环境气候长时间无人守职的监测系统的研究；引进测量领域的三维激光扫描仪测量技术，对当地的地貌进行动态监测；采用遥感与地面定位观测相结合的手段进行监测；应用高清晰自动录像监测；实现野外环境因子无线远程监控系统等的研发与引进，逐步实现监测设备的标准化和监测手段的智能化。

黄河内蒙古段乌兰布和沙漠入黄风积沙监测研究是一项多尺度、多场耦合、随机性、非线性的复杂科学问题，是目前科学前沿所关注的共性和热点课题，涉及水土保持与荒漠化防治、水文水资源、河流动力与泥沙、气象等学科和领域。由于我们只是初步研究探讨，研究方法和思路有待完善，望相关研究领域的学者给予指教，同时期待国家及有关部门就乌兰布和沙漠沿黄段风积沙监测研究引起高度重视，形成多渠道、多学科联合攻关研究的平台。使研究结果更好地为科学治理黄河泥沙问题及完善黄河沿岸综合防护体系提供科学依据。

参考文献

[1] 中华人民共和国水利部，中国科学院，中国工程院. 中国水土流失防治与生态安全·北方农牧交错区卷[M]. 北京：科学出版社，2010.

[2] 马玉明，等. 沙漠学[M]. 呼和浩特：内蒙古人民出版社，1998.

[3] 春喜，陈发虎，范育新. 乌兰布和沙漠的形成与环境变化[J]. 中国沙漠，2007，27(6)：927-931.

[4] 贾铁飞，石蕴棕，银山. 乌兰布和沙漠形成时代的初步判定及意义[J]. 内蒙古师范大学学报：自然科学版，1997，3：46-49.

[5] 李清河，包耀贤，王志刚，等. 乌兰布和沙漠风沙运动规律研究[J]. 水土保持学报，2003，17(4)：86-89.

【作者简介】 郭建英，男，1979年9月生，博士，2010年毕业于北京林业大学水土保持学院，现工作于水利部牧区水利科学研究所、工程师。

基于 FEPG 计算的微小流道结构的多相模型新算法研究

高 林

（中国水利水电科学研究院 北京 100038）

【摘要】 本文通过对多种流道结构形式的滴头进行 FEPG 模拟并对试验研究结果进行了对比，开发了基于 FEPG 和 FLUENT 的多相流数值模拟软件，利用最新的 CBS 算法研发出一套能够应用于微小流道计算的程序并提出了较优的流道结构，同时也建立了描述滴头流量与流道长度及工作压力关系的数学模型，并对滴头内流场进行了模拟与分析，结果说明该双流体模型有限元程序是有效的，可用于微小流道结构的流场模拟。

【关键词】 FEPG CBS 滴头 流道 数值模拟 流态指数

滴灌是一种省水、省力的高效灌水模式，在我国快速地发展。滴头是滴灌系统的关键部件，其结构形式和水力性能对滴灌系统有着重要的影响。

由于滴头内部的流动状态对滴头的水力性能和抗堵塞性能影响很大，但其尺寸小，无法采用传统的流场测量方法。FEPG(finite element program generator)是一款国产有限元程序自动生成系统，其主要设计思想是采用元件化的程序设计方法和有限元语言，根据有限元方法的数学原理及其内在规律，以类似于数学公式推理的方式，由有限元问题的微分方程表达式及其求解算法自动产生有限元程序，目前在流体界应用较为广泛，例如河海大学的娄一清、陈睿等人研究了 FEPG 在流体渗流中的应用；华南理工大学的黄鹏、魏兴钊等研究了 FEPG 在高压气体数值模拟应用的情况等，模拟结果比较理想。目前，CFD(Computational Fluid Dynamics)数值计算方法对水流的流场模拟已较为成熟，曾经有文献都对此进行了研究，李永欣等更结合实物对比试验，模拟所得出口流量和实测结果值吻合较好，验证了该方法的有效性，为滴头设计提供了数字试验研究依据；扬州大学的成立、刘超等对于进水流道内水流应用 SMPLEC 方法进行了定长计算，但鉴于目前的 CFD 流体软件计算一般为二阶精度，而 FEPG 计算则可以达到将近三阶精度，并且针对于本文的多场耦合和强变形等高性能计算具有精度高、计算收敛快等特点。本文利用算子分裂法将 N-S 方程表述的物理过程分解为扩散和对流两个过程，导出了扩散方程的弱形式，进而建立起它的有限元方程，再利用最小二乘法建立起对流方程的有限元方程，所产生的程序对二维的黏性不可压流体的流动问题进行了数值模拟。通过利用 FEPG 数值模拟，研究不同结构形式滴头的水力性能，同时进行了算例分析和对比。

1 滴头流道水力性能研究方法

1.1 滴头内部流道建模

滴头内部水流运动可视为不可压缩流体(密度 ρ 为常数)。基本控制方程为连续性方程和 Navier–Stokes 方程。大多数情况下,滴头内部流动属于湍流范围,需对瞬态 Navier–Stokes 方程作时均处理,同时补充湍流方程。滴头的边界条件为压力进、出口。最后通过数值模拟所得流场分布,进而计算滴头出口流量。

滴头内部流道的壁面对紊流有明显的影响,壁面的表面粗糙度根据目前塑料滴头的成型工艺水平设置为 0.001 mm。标准壁面函数的基本思想是在壁面区直接使用半经验公式将壁面上的物理量与湍流核心区内的求解变量联系起来。

1.2 滴头结构形式的选取

市场上滴头流道形式多样,齿形各异。第一类(E_1)流道是最常见的锯齿形,流道齿为等腰三角形构建;第二类(E_2)流道的流道齿为直角三角形构建;第三类(E_3)流道的流道齿形式特殊且形式不同,流道整体圆弧处理;第四类(E_4)流道是圆柱形滴头的流道,流道齿形式整齐统一。

1.3 滴头结构尺寸

各滴头的流道齿主体是以三角形为基础构建的,具体差异主要表现在流道结构控制参数(齿宽、齿底距、齿高、齿尖角度、流道宽和流道长等)的变化,如图 1 所示。分析各参数关系可知,齿宽、齿底距固定齿间距;齿宽、齿高固定齿尖角度;齿宽、齿底距、齿高固定流道宽;流道深的变化不影响流道齿平面结构形式,所以将齿宽、齿底距、齿高作为流道齿的构建三要素。

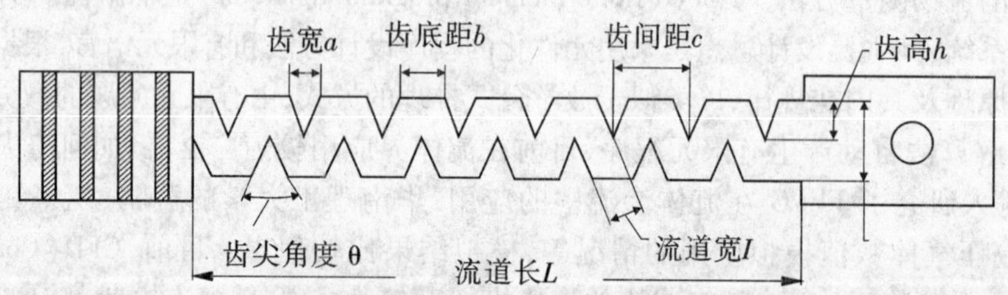

图 1 流道结构形式

本文仅研究过流面积 A 相近的情况下,各流道类型的水力性能。考虑到滴头流道的敏感尺寸为 0.7 mm,根据上述尺寸范围,各种不同类型的滴头结构参数取值为 $a = 0.8$ mm, $b=1.2$ mm, $h=1.2$ mm, $d = 0.8$ mm, $H = 2.4$ mm, $L= 20$ mm。圆角半径 r 参考现有滴头样本的取值。

2 滴头两相流理论模型的构建

对于滴头流道流体流动,有以下基本假设:第一,由于单流体模型中,泥沙颗粒运动状况被假设为与流体完全一致。而在双流体模型中,泥沙和流体可以具有不同的运动状况。而在本模型中,泥沙颗粒被假设成连续介质。第二,假设水流充满整个流道,并且颗粒在流道内是悬浮状。第三,在组分数量上没有考虑级配。第四,阻力交换系数的选择是采用经验系数。有以上的基本假设,我们可以把含沙水流的流动情况被处理成两相流体流动情况。其

泥沙输运的双流体模型如下

$$\frac{\partial \alpha_1 \rho_l}{\partial t} + \nabla \cdot (\alpha_1 \rho_1 u_1) = 0 \tag{2-1}$$

$$\frac{\partial \alpha_1 \rho_1 u_1}{\partial t} + \nabla \cdot \alpha_1 \rho_1 u_1 u_1 = -\nabla p + \nabla \cdot (\alpha_1 \mu \nabla u_1) + K(u_p - u_1) + \alpha_1 \rho_1 g - F_{\text{lift}} \tag{2-2}$$

$$\frac{\partial \alpha_p \rho_p}{\partial t} + \nabla \cdot (\alpha_p \rho_p u_p) = 0 \tag{2-3}$$

$$\frac{\partial \alpha_p \rho_p u_p}{\partial t} + \nabla \cdot \alpha_p \rho_p u_p u_p = \nabla \cdot (\alpha_p \mu \nabla u_p) + K(u_1 - u_p) + \alpha_p \rho_p g + F_{\text{lift}} \tag{2-4}$$

颗粒被假设成圆球状颗粒，因此阻力项系数 K 表达式如下

$$K = \frac{3}{4} C_d \frac{\alpha_s \alpha_1 \rho_1 | u_p - u_1 |}{d} \tag{2-5}$$

其中，式(2-1)、式(2-2)和式(2-3)、式(2-4)分别为液相方程和固相方程，参数 $C_d = \frac{24}{\alpha_1 Re_p}[1 + 0.15 (\alpha_1 Re_p)^{0.687}]$，升力 F_{lift} 计算如下

$$F_{\text{lift}} = -0.5 \alpha_p \rho_p (u_p - u_1) \times \nabla \times u_p$$

式中　α_p 为迷宫流道中泥沙的浓度；ρ_p 为迷宫流道中悬浮泥沙颗粒的干沙密度；u_p 和 u_1 分别为固相和液相的速度。

采用有限元方法求解上述双流体模型，需要写出方程(2-4)的虚功形式，为了书写方便，考虑如下无量纲通用形式的连续性和 N-S 方程，以下是方程(2-1)至方程(2-4)的虚功形式

$$\frac{\partial \alpha}{\partial t} + \nabla \cdot \alpha u = 0 \tag{2-6}$$

$$\frac{\partial \alpha \phi}{\partial t} + \nabla \cdot \alpha u \phi = -\nabla p + \nabla \cdot \frac{\alpha}{Re} \nabla \phi + S_c \tag{2-7}$$

式中　ϕ 为通用变量；α 为通用体积分数；S_c 为源项，包括体力、阻力和升力等。采用算子分裂法和 CBS 方法离散并求解方程(2-6)、方程(2-7)。对于对流扩散方程，采用算子分裂法将其分离成扩散项和对流项两部分计算如下

$$\frac{\partial \alpha \phi}{\partial t} + \nabla p - \nabla \cdot \frac{\alpha}{Re} \nabla \phi = S_c \tag{2-8}$$

以及

$$\frac{\partial \alpha \phi}{\partial t} + \nabla \cdot \alpha u \phi = 0 \tag{2-9}$$

需要注意的是，粒/泥沙运动方程中不存在形如方程中压力梯度项 ∇p，它仅存在于水(相)中。方程中虚功形式如下(在求解水的扩散方程时，水被简化成不可压流体 $\nabla \cdot u = 0$)：

$$\left(\frac{\partial \alpha \phi}{\partial t}, \delta \phi\right)_\Omega - (p, \nabla \delta u)_\Omega - (\nabla \cdot u, \delta \text{p})_\Omega + \int_{\partial \Omega} pn \cdot \delta u \text{d}S$$
$$+ \frac{\alpha}{Re}(\nabla \phi, \nabla \delta \phi)_\Omega - \int_{\partial \Omega} \frac{\partial \phi}{\partial n} \delta \phi \text{d}S = (S_c, \delta \phi)_\Omega \tag{2-10}$$

对于二维悬浮颗粒的模型，假设升力和重力平衡，因此源项 S_c 中仅含有阻力项，对此，可以将方程(2-10)中右端表达式具体形式写出如下

$$(S_c,\delta\phi)_\Omega = K(u_p,\delta u)_\Omega - K(u_1,\delta u)_\Omega \tag{2-11}$$

再作部分处理。式(2-8)-式(2-9)×ϕ，有如下对流方程

$$\frac{\partial\phi}{\partial t} + u\cdot\nabla\phi = 0 \tag{2-12}$$

因此，可以写出方程(2-6)与方程(2-12)的虚功形式

$$(\phi,\delta\phi)_\Omega = -\,\mathrm{d}t\,(u^n\cdot\nabla\phi^n,\delta\phi)_\Omega - \mathrm{d}t\,(S_m,\delta\phi)_\Omega - \frac{\mathrm{d}t^2}{2}(u^n\cdot\nabla\phi^n,u^n\cdot\nabla\delta\phi)_\Omega \tag{2-13}$$

式中　$S_m = \alpha\nabla\cdot u^n$，上标 n 表示上一个时刻的 ϕ 值；$\mathrm{d}t$ 为时间步长。

3　计算多维体系的思考及单元结构化网格的划分与程序生成

考虑到液固两相流中，流道内部悬浮颗粒是与水混合在一起共同流动的，其中水流挟带泥沙颗粒能力与水流速度、颗粒大小和浓度都有关系。

颗粒在流道中运动受到的综合作用力包括重力、水流拖拽力、水流升力、颗粒间作用力等，而重力、水流拖拽力、水流升力、颗粒间作用力与悬浮颗粒浓度有很大的关系。由于颗粒在水流动中处于悬浮状态，其重力与水流升力可以考虑相互抵消。从使用流道的中心纵轴界面来观察颗粒的堵塞分布在悬浮颗粒浓度受流场的影响，当水流与颗粒速度差值增加，流场发生紊乱时，颗粒与水流、颗粒之间的摩擦和碰撞也会增加，此时的浓度也就会相应发生变化。

在流道进行网格划分时，网格划分采用的是手工结构化网格划分，网格大小为 0.1 mm，根据具体的结构形式划分，局部进行加密，最终网格数量在两万个左右。

非线性稳态问题求解流程如图 2 所示。

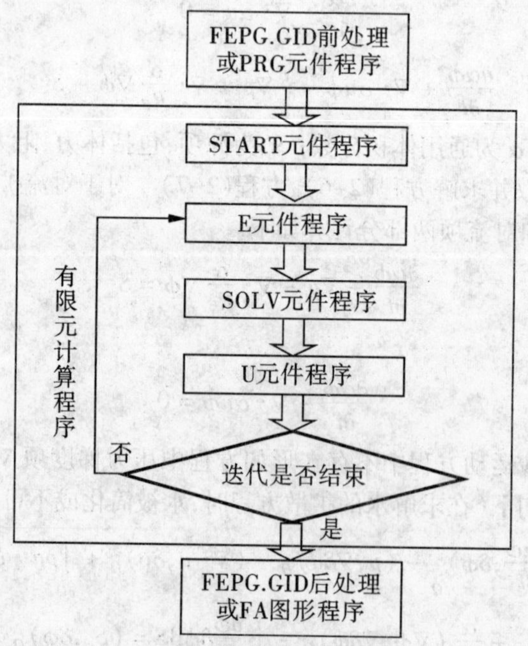

图 2　非线性稳态问题求解流程

对于方程 $\left(\frac{\partial\alpha\phi}{\partial t},\delta\phi\right)_{\Omega} - (p,\nabla\delta u)_{\Omega} - (\times\cdot u,\delta p)_{\Omega} + \int_{\partial\Omega} pn\cdot\delta u\mathrm{d}S + \frac{\alpha}{Re}(\nabla\phi,\nabla\delta\phi)_{\Omega} - \int_{\partial\Omega}\frac{\partial\phi}{\partial n}\delta\phi\mathrm{d}S = (S_c,\delta\phi)_{\Omega}$

采用简单迭代法求解并由其弱形式,可编写生成各个单元刚度程序:分成对流项和扩散项,求解方程难度很高,N-S 方程是一个高度非线性方程,其难度首先体现在对流项,所以采用算子分裂方法把对流项和扩散项分解,联立求解进行结果叠加。下面列出对流相的单元刚度程序,扩散相的单元刚度程序和颗粒浓度计算程序见附件程序:

```
    Field B: CBS for Fluid 1 Convective part:
FUNC
@l grad.xy m gu x y u
@l grad.xy m gv x y v
@l grad.xy m gp x y p
$cv f_i = +gup_i_j * u_j
fvect ghu 2
fvect ghv 2
fvect ghp 2
@l grad.xy f ghu x y hu
@l grad.xy f ghv x y hv
@l grad.xy f ghp x y hp
\fmatrix ghup 3 2
fvect l1 1
fvect l2 1
fvect l3 1
@a l1_i = +[ghu_j] * u_j
@a l2_i = +[ghv_j] * u_j
@a l3_i = +[ghp_j] * u_j
@w lhu l1
@w lhv l2
@w lhp l3
stiff
null =
dist = [hu;hu] * 0.0
load = -[hup_i] * f_i * dt-[lhup_i] * f_i * dt * dt/2
```

4 模拟结果流道入口及其他部位的颗粒浓度分布

采用结构化网格体系,每个网格均为 9 节点四边形单元,由于压力的求解需要满足 LBB 条件,因此在流场计算中实际采用单元类型为混合阶单元。模型的雷诺数为 60,流道入口宽度为 1 mm,长度约为 40 mm,采用准定长假设,给定初始浓度分布及入口流量,时间步长为 0.01 ms。

图 3 中的(a)和(b)分别是流道拐角处的沙粒浓度与沙粒速度矢量图,从图中可以看出,沙粒在流道内齿间处速度较大,齿根处速度较小,并在附近有旋涡产生,该涡结构使得流道齿后的浓度降低,并且齿根处的浓度也没有明显集中。

图 3 E2 流道拐角处沙粒浓度与速度矢量图

从图 3 中看到在流道拐角的齿间部有非常明显的涡结构,使得该处的沙粒浓度分布较为均匀。但流道中部则有较大的浓度集中,流道齿根前也有更大的浓度集中,在具体试验中此处也发现了堵塞现象,表现出与流道其他部位有所不同。

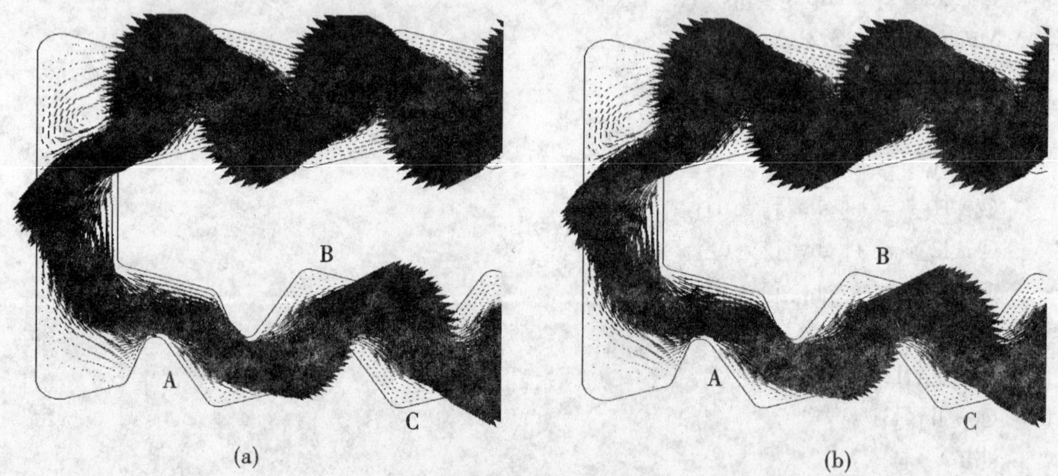

图 4 E2 流道拐角处沙粒速度矢量与水流速度矢量图

图 4 的(a)和(b)分别是流道拐角处沙粒速度矢量图与水流速度矢量图的比较,由于沙粒粒径较小,所以在 A、B 和 C 三处的水沙两相速度值上差异并不大(见表 1)。

表 1 不同部位沙粒和水的速度

两相	A 齿间	B 齿间	C 齿间	C 齿根
沙粒	0.217	0.241	0.234	250
水	0.259	0.302	0.275	500

图 5 的(a)、(b)是利用 Flute 软件模拟技术得到的 E3 和 E2 流道内水流的流场分布图，图中速度矢量颜色由蓝变红，代表流速由小变大；箭头的大小和方向代表了速度的大小和方向。

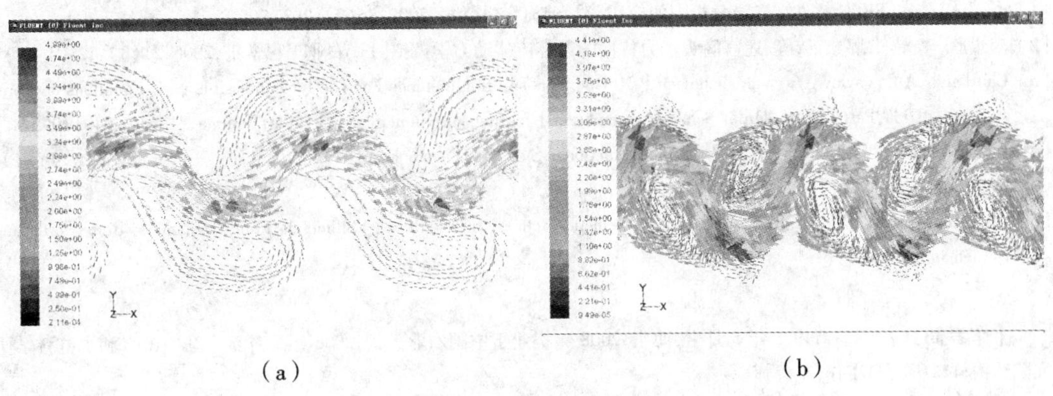

(a) (b)

图 5　E3 和 E2 流道速度矢量图

由于流速在流道各个单元内流动分布较为相似，所以选取四种滴头流道同断面齿间到齿根处沙粒的速度进行比较，如图 6 所示：发现四种流道的同断面速度为 E2> E1> E3 >E4。

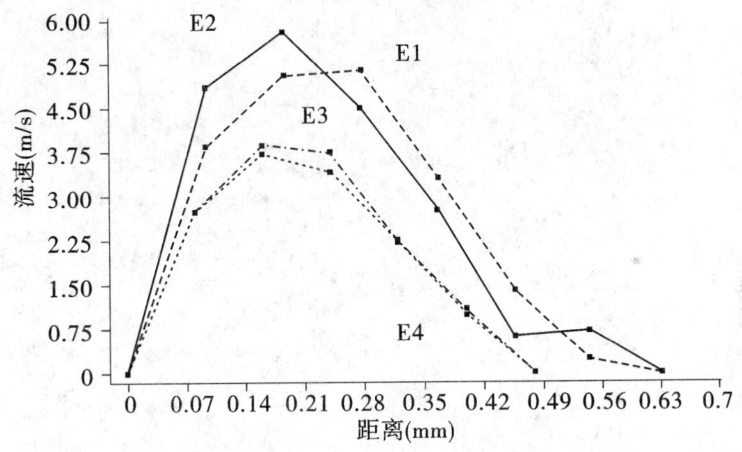

图 6　四种流道断面流速变化图

5　结论

根据迷宫流道水沙流动的特点，建立了滴头流道的两相流模型方程，并采用 FEPG 软件进行了求解。结果表明：

(1)本文提出的 CBS 算法用于实际计算迭代次数少、计算效率高并与常用的有限元方法所求得的解相符合，保证了精度，这证明了本算法是可行的。

(2)在基于 FEPG 软件上开发了两相流应用程序，利用所产生的程序对二维的黏性不可压流体问题进行了数值模拟并和 Flute 软件模拟的结果进行了对比，说明了采用的数值模拟程序是合理的，并可以应用到进一步的计算中。

(3)在 E2 型滴头流道内部，悬浮流动的泥沙颗粒总是在流道内的齿前集中，相对齿后的浓度则较小些，且在拐角处有与其他流道明显的不同速度和泥沙浓度分布，但由于颗粒的粒径小,所以流道内水相和颗粒相的速度差异并不大。

参考文献

[1] 娄一清,陈睿. FEPG 软件在渗流分析中的应用[J]. 河海大学学报,2008,36(2):203-208.

[2] 李永欣,李光永,邱象玉,等. 迷宫滴头水力特性的计算流体动力学模拟[J]. 农业工程学报,2005,21(3):12-16

[3] Churbanov A G, Pavlov A N, Vabishchevich P N. Operator-splitting Methods For The Incompressible N-S Equations on Non-staggered Grids, Part 1: First-Order Schemes. Inernational Journal For Numerical In Fluids, Volume, 21, 617-640(1995).

[4] Kloucek Peter, Rys Franz S. Stability Of The Fractional Step O-scheme For The Nonstationary N-S Equations, SIAM, J. Numer Anal. 31:5-11.

[5] Zienkiewicz O C, Taylor R L. The Finite Element Method, fifth edition, Volume 3: Fluid Dynamics, Butterworth and Heinemann, 2000.

【作者简介】 高林,1978 年 5 月生,博士,2008 年毕业于中国农业大学,2008 年 7 月至今在中国水利水电科学研究院结构材料所(企)工作,工程师。

大型灯泡贯流泵工况调节与水力性能研究

秦钟建　方国材　胡大明

（中水淮河规划设计研究有限公司　安徽　蚌埠　233001）

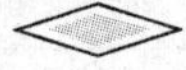

【摘要】 南水北调东线工程多座泵站采用大型灯泡贯流泵机组，为了研究灯泡贯流泵不同工况调节方法及其水力性能的特点，根据某大型调水泵站水泵装置模型试验成果进行了变速调节水力性能的计算，并在满足设计流量前提条件下与变角调节水力特性进行了对比分析，结果表明变角和变速调节水泵运行工况的方法均适用于该调水泵站，且设计、平均和最高扬程相应工况点的效率也基本相等。本文分析了大型灯泡贯流泵装置采用变角、变速及变角变速综合调节方法调节水泵运行工况时水泵装置的水力性能特点，为灯泡贯流泵工况调节方法的选择提供参考。

【关键词】 灯泡贯流泵　工况调节　变角　变速　水力性能

1　引言

大型灯泡贯流泵机组具有优良的水力性能，已被南水北调东线第一期工程的多座泵站所采用。为适应泵站经济运行，泵站根据自身特点采用改变叶片角度或改变水泵转速的方法调节水泵运行工况。目前，灯泡贯流泵机组自身水力性能的研究较多，而其不同工况调节方法及其水力性能的对比研究较少。本文根据某大型调水泵站水泵装置模型试验的成果，对采用变角和变速的调节方法改变水泵运行工况点的水力性能进行对比研究，为类似机组工况调节方式的选择提供参考，更好地满足不同泵站工况调节和经济运行的要求。

2　变角调节水力性能

2.1　装置模型试验

某大型调水泵站设计调水流量75 m^3/s，设计净扬程2.40 m，平均净扬程2.08 m，最高净扬程3.10 m，最低净扬程0.10 m，采用后置式机械全调节灯泡贯流泵机组4台套（备用1台套），水泵叶轮直径为2 850 mm，额定转速为120 r/min，单泵设计流量25.0 m^3/s，配套电动机功率为1 250 kW，水泵与电动机通过行星齿轮箱连接传动，齿轮箱传动比为6.25。

为保证该泵站灯泡贯流泵机组的水力性能满足南水北调工程的设计要求，利用CFD仿真技术模拟分析了流道内的压力场和流速场，优化了贯流泵装置过流部件的型线，并进行了水泵装置模型试验。试验模型泵叶轮直径为315 mm，转速为1 230 r/min，模型与原型泵装置几何尺寸比为1∶9.048。水泵装置模型完全模拟了原型泵装置，并进行了装置−6°～+4° 6个叶片角度的$Q\sim H$、$Q\sim\eta$和$Q\sim P$等水力性能试验，模型泵装置综合性能曲线见图1。

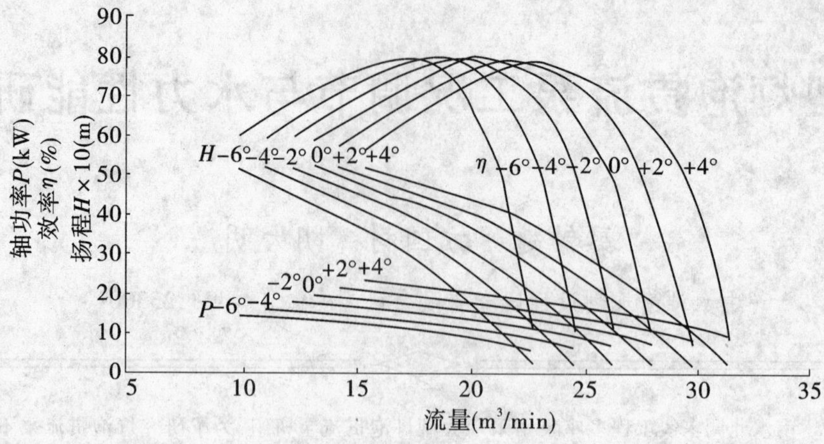

图 1　模型泵装置性能曲线图

2.2　原型泵装置水力性能

装置模型试验的目的是根据模型装置的试验成果换算出原型泵装置的水力性能等参数。由于模型泵装置与原型泵装置的过流部分形状相似、几何尺寸成比例，根据式(1)和式(2)分别换算出原型泵装置 $Q \sim H$ 和 $Q \sim \eta$ 等水力性能并绘成性能曲线，如图 2 所示。设计流量下泵站设计、平均和最高扬程工况点参数如表 1 所示。

$$\frac{Q_P}{Q_M} = \left(\frac{D_P}{D_M}\right)^3 \frac{n_P}{n_M} \tag{1}$$

$$\frac{H_P}{H_M} = \left(\frac{D_P}{D_M}\right)^2 \left(\frac{n_P}{n_M}\right)^2 \tag{2}$$

式中　Q 为流量；H 为扬程；D 为叶轮直径；n 为转速；下标 P 为原型泵；下标 M 为模型泵。

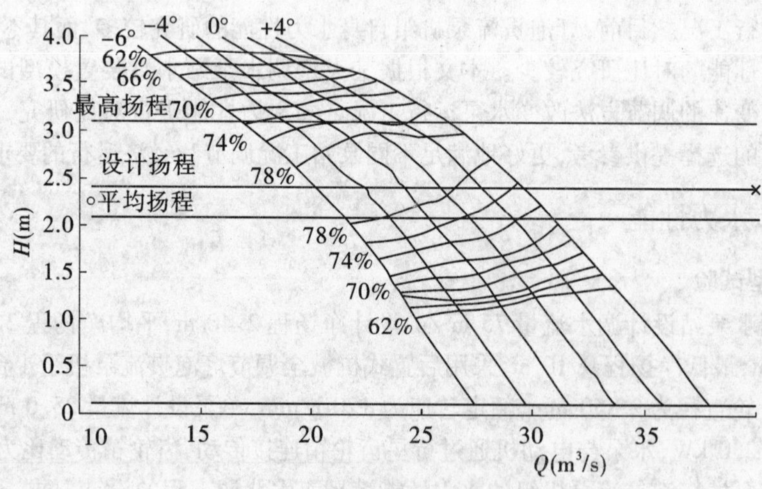

图 2　原型泵装置性能曲线图

2.3　水力性能特点

由图 1 和图 2 知：灯泡贯流泵装置变角调节的水力性能和大中型立式轴流泵装置的水力性能相似，即当调节水泵叶片角度时最高效率点(区间)的变化轨迹大致水平移动，适用于调节变化幅度较大且效率要求较高的泵站。因此，该泵站灯泡贯流机组采用的变角调节方式也是国内大中型立式全调节轴流泵采用最广泛的工况调节方式;同时,由表1知在满足泵站设计流量的前提下,设计、平均和最高扬程等工况均可通过调节水泵叶片角度的方法确保该调水泵站灯泡贯流泵装置在较高效率区间内运行,满足南水北调泵站高效、节能的要求。

表1 变角、变速调节各工况水力性能对比表

工况	设计		平均		最高	
调节方法	变角	变速	变角	变速	变角	变速
角度(°)	-1.0	0	-2.4	0	+2.8	0
额定转速百分比(%)	100	98.2	100	95.2	100	104.5
转速(r/min)	120	117.8	120	114.2	120	132.5
泵站净扬程(m)	2.40	2.40	2.08	2.08	3.10	3.10
流量(m^3/s)	25.0	25.0	25.0	25.0	25.0	25.0
效率(%)	78.0	78.0	77.0	77.8	76.6	78.0

3 变速调节水力性能

3.1 运行工况点的确定

在水泵转速变化不太大的情况下,根据式(1)和式(2)及 $D_P/D_M=1$,即式(3)和式(4),可确定原型泵装置采用变速调节时不同转速的运行工况点参数。

$$\frac{Q_1}{Q_2}=\frac{n_1}{n_2} \tag{3}$$

$$\frac{H_1}{H_2}=\left(\frac{n_1}{n_2}\right)^2 \tag{4}$$

式中:下标1、2分别为水泵不同转速下的工况点。

由表1和图2知:水泵装置设计工况点的叶片角度为-1°,以此角度进行变速调节确定其他工况点是比较合适的,然而为了便于结合试验数据研究水泵调速的水力特性,则选择0°叶片角的水力特性曲线为该机组额定转速时的特性曲线,并以此为基础进行变速调节水力特性分析。根据0°叶片角水泵装置效率为78%、76%和74%工况点的流量、扬程及抛物线常数,绘制的不同转速的水力特性曲线及相似工况等效率抛物线如图3所示,设计流量工况下的泵站设计、平均和最高扬程工况点参数如表1所示。

3.2 水力性能特点

由表1数据对比知:在设计扬程工况下两种工况调节方法均能使原型泵装置效率达到78%;在未计入变速调节装置(变频设备)工作效率时,采用变速调节方法的原型泵装置效率在平均和最高扬程工况下较采用变角调节的稍高,而当计入变速调节装置工作效率98%~99%时,则两种工况调节方式的水泵装置效率基本相等;由图3知:采用降速调节水泵工况时,水泵装置的流量和扬程等参数均同时下降、等效率区间的范围也逐渐变窄,而采用增速调节时装置的流量、扬程参数和效率区间等则呈相反变化的趋势。

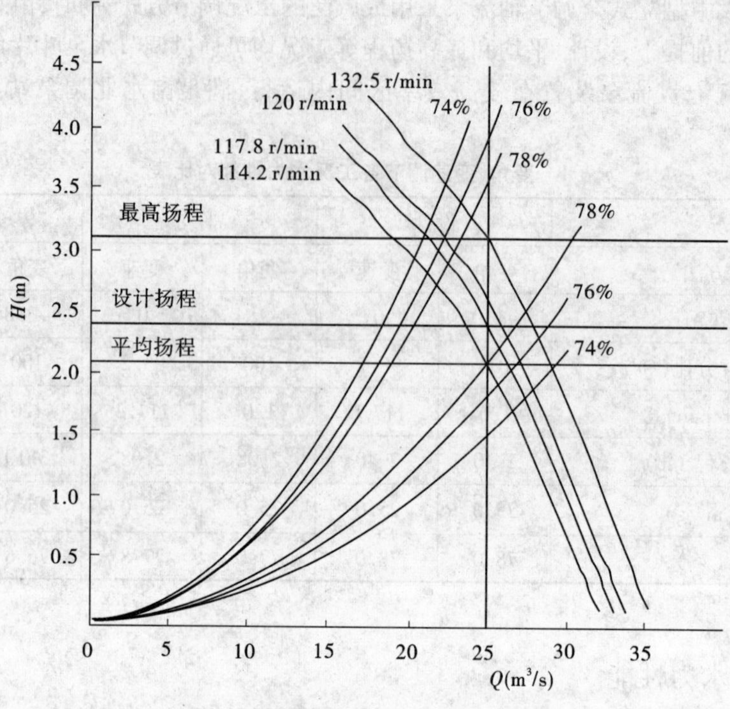

图 3　变速调节工况下水力性能曲线图

4　综合调节水力性能

水泵装置采用综合调节方法,是同时采用变角和变速两种调节方式对水泵装置进行工况调节的,其具有变角和变速调节的共同特性。该大型调水泵站原型泵装置的额定转速和85%额定转速下不同叶片角度的性能曲线分别如图 4 中的实线和虚线绘制的性能曲线。由图 4 知:采用综合调节方式扩大了水泵装置的高效率运行区间,更易满足泵站流量、扬程调节的要求。

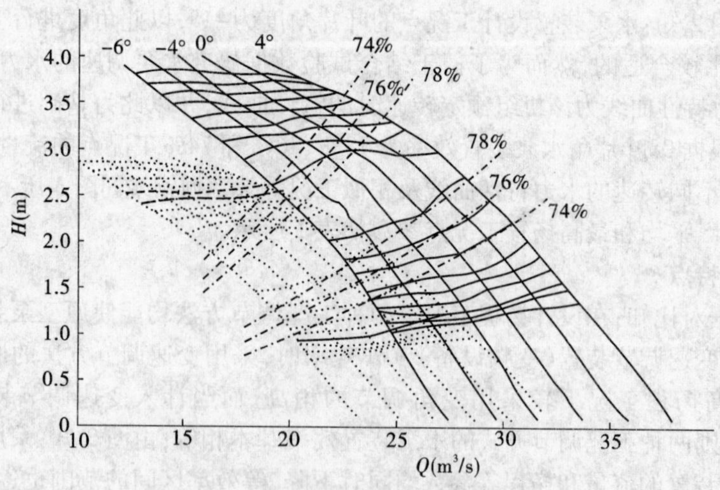

图 4　原型泵装置综合调节性能曲线图

5 结语

(1)根据理论分析可知,采用变角或变速调节水泵运行工况均能使该大型调水泵站的灯泡贯流泵装置在设计扬程、平均扬程和最高扬程运行工况下具有较高的效率,且各相应工况点的效率也基本相等,均满足南水北调东线工程的设计和运行要求。

(2)灯泡贯流泵和大中型立式轴流泵工况调节方法与水力性能具有相似的特性,因此该研究对立式轴流泵的工况调节方法的选择具有参考意义。

(3)大型灯泡贯流泵装置工况调节与水力性能的研究是在某大型调水泵站的灯泡贯流泵装置模型试验的基础上进行的理论研究,而未进行原型泵装置的实测试验,它与原型泵装置的实测试验之间的水力特性差异仍有待进一步研究。

(4)采用变角或变速方法调节水泵装置运行工况的大型灯泡贯流泵在国内已均有采用,而综合调节方式水泵机组在国内则很少采用。由于本研究在进行水力性能研究时未涉及不同工况调节方法的设备制造、运行管理及工程造价等因素,因此不同泵站灯泡贯流泵工况调节方式的选择应结合工程的实际特点、调节要求及必要性等经技术经济比较后确定。

参考文献

[1] 秦钟建,伍杰,张仁田. 蔺家坝灯泡贯流泵机组水力性能及结构分析[J]. 排灌机械,2009,27(3):177-180.

[2] 李万荣. 南水北调东线一期工程蔺家坝泵站水泵模型装置验收试验[J]. 南水北调与水利科技,2009,7(6):278-281,342.

[3] 刘大恺. 水轮机[M]. 北京:中国水利水电出版社,2008.

[4] 刘超. 水泵及水泵站[M]. 北京:科学技术文献出版社,2003.

[5] 日本农业土木事业协会. 泵站工程技术手册[M]. 丘传忻,林中卉,黄建德,等译. 北京:中国农业出版社,1998.

【作者简介】 秦钟建,1979 年 2 月生,本科,2002 年毕业于扬州大学,现工作于中水淮河规划设计研究有限公司,工程师。

东北黑土区土壤风力侵蚀观测及发展趋势研究*

刘铁军 珊 丹 刘艳萍 郭建英 高天明

(水利部牧区水利科学研究所 呼和浩特 010020)

【摘要】 东北黑土区是我国重点粮食产区,其粮食生产能力及可持续性关系到国家的粮食安全战略,目前该区水土流失日趋剧烈,导致黑土功能严重退化,耕作土壤退化、沙化及生产力明显降低,严重制约我国粮食安全可持续发展。本文观测了黑土区旱作耕地风蚀,分析了黑土区风蚀主要因子,并从黑土区所处的地理位置、区域近40年气候变化规律、东北沙地沙漠化对黑土区土壤风蚀的影响方面分析了东北黑土区土壤风蚀的发展趋势,为该区土壤风力侵蚀研究及防治提供依据。

【关键词】 东北黑土区 风力侵蚀 水土流失 风蚀因子

1 引言

东北黑土区是我国重点粮食产区,其粮食生产能力及可持续性关系到国家的粮食安全战略,广义黑土区指有黑色表层土分布的区域,面积101.85万km^2,其中:黑龙江省45.25万km^2,吉林省18.70万km^2,辽宁省12.29万km^2,内蒙古自治区25.61万km^2。气候类型属于中温带大陆性季风气候区,冬季寒冷漫长,春季干旱多风,年平均降水量380~743 mm,全年平均风速2.5~4.5 m/s。区内主要土壤类型有暗棕壤土、黑土、黑钙土,黑土主要分布在松嫩平原的北部和东部,黑钙土集中分布在松嫩平原的草甸草原。黑土土壤疏松、抗蚀能力弱,由于降雨、风力和长期以来人口增加导致的过度垦殖等不合理开发利用,使该区水土流失日趋严重。

黑土区水土流失日趋剧烈,导致黑土功能严重退化,耕作土壤退化、沙化及生产力明显降低,严重制约我国粮食的可持续发展。国内学者针对黑土区土壤侵蚀研究主要集中在水力侵蚀上,风力侵蚀研究成果较水力侵蚀少,深入机制性研究工作有待加强。笔者在主持国家自然基金项目黑土地冻融风蚀机制研究的基础上,通过几年来多次野外调研及实地观测发现东北黑土区风力侵蚀研究及防治亦应引起社会的关注。本文在对黑土区风蚀进行野外观测的基础上,对该区风蚀发展趋势进行了系统分析,提出黑土区风力侵蚀研究及防治的迫切性,为该区风蚀防治保证耕地健康可持续发展提供相应依据。

* 基金项目:国家自然基金资助项目(40901136)。

2 研究区概况及试验测试方法

野外试验项目区位于内蒙古兴安盟乌兰浩特市西北方向义勒力特镇合发村境内旱作黑土耕地，地理坐标为东经 E121°58′42.5″，北纬 N46°11′38.7″，距离乌兰浩特市区距离为 20 km，该区多年平均降雨量 365 mm，年平均气温 6.1 ℃，年平均风速 3.2 m/s，年平均日照时数 2 869.7 h，作物主要以玉米、黄豆等为主。研究区为乌兰浩特市境内具有代表性的旱作黑土耕地，土壤以黑钙土为主，试验项目区周边宽旷，2010 年春季 4 月初进行研究区野外风蚀相关测试。

风蚀气象因子试验测试方法为应用微型气象站连续观测研究区春季风速较大日 0.5 m、1.0 m、1.5 m 高度的平均风速、气温、空气相对湿度，每小时观测 1 次；在试验区野外现场进行了起沙风速的观测，观测方法为测定 2 m 高度风速的同时，观测地表起沙记录 2 m 高度的瞬时风速，将多次观测值进行平均；风蚀过程中输沙通量的测定，采用 SC－I 型集沙仪，集沙仪为铁皮制，共有 18 个 5 cm× 5 cm 的正方形接纳孔，上下孔之间按 5 cm 等间距排列，每一层同一高度安装两个集沙孔（共 9 层），开口均与地面平行。该沙尘采集器可以控制沙尘采集器始终随主风向自由旋转，使得沙尘采集器入口始终正对主风向，集沙口下端距离地面 5 cm，这样集沙仪收集到的跃移沙粒是单位时段内 5～90 cm 高度处的风蚀沙粒物。

对研究区黑土地耕作层深度以内的土样进行理化性质及机械组成化验，结果如表 1 所示。表 1 给出了研究区 0～14 cm 深度土样的理化性质（每 2 cm 深度内取样测定，7 个深度的土样结果），表 2 给出了研究区土壤颗粒的机械组成。

表 1 研究区土壤理化性质特性表

土样	pH 值	有机质（g/kg）	全氮（g/kg）	速效氮（mg/kg）	全磷（g/kg）	速效磷（mg/kg）	全钾（g/kg）	速效钾（mg/kg）
0～2 cm	8.00	28.49	1.45	41.41	0.09	6.40	20.5	144.84
2～4 cm	8.13	29.34	1.41	30.75	0.084	5.36	20.6	133.47
4～6 cm	7.99	27.04	1.36	43.39	0.092	4.68	20.62	130.92
6～8 cm	7.96	28.38	1.39	36.54	0.086	4.58	19.39	126.95
8～10 cm	7.78	29.47	1.46	36.16	0.079	4.86	18.96	126.19
10～12 cm	8.00	32.50	1.54	30.83	0.106	1.79	18.79	118.09
12～14 cm	7.75	30.39	1.55	27.02	0.067	1.97	19.39	120.71
均值	7.94	29.37	1.45	35.16	0.09	4.23	19.75	128.74

表 2 研究区土壤颗粒机械组成表

粒径（μm）	$d>400$	$200<d<400$	$100<d<200$	$60<d<100$	$40<d<60$	$30<d<40$	$20<d<30$	$10<d<20$	$d<10$
含量（%）	0	1.85	9.94	18.05	15.16	11.15	15.33	14.29	14.23

3 试验结果分析

3.1 研究区起沙风速观测

经过多次观测起沙风速取平均值最后得出研究区起沙风速为 6.2 m/s，根据起沙风速结合风速日变化曲线可以统计发生风蚀的时间段。

3.2 风蚀过程输沙通量计算分析

将6套集沙仪按照梅花状排列布置于研究区黑土耕地上，测定挟沙气流垂直断面输沙通量。观测时间为试验区风速较大日即2010年4月7日至4月9日。将6组集沙仪收集沙量进行平均，分析输沙量沿高度的变化并计算单位面积集沙量，计算表如表3所示。

通过对试验区黑土耕地输沙量沿高度的变化规律分析（见图1）可知，输沙量沿高度呈指数函数递减，函数式为

$$Q = 6.3138\,e^{-0.2059Z} \tag{1}$$

式中：Q 为集沙量，g；Z 为距地面高度，cm。

将式(1)进行平面积分，通过计算可知，距地表0～10 cm高度内集沙量占全部集沙量的29%，0～40 cm高度内集沙量占全部集沙量的72%，50 cm高度以内跃移颗粒的百分含量将近占全部集沙量81%，60 cm高度以内跃移颗粒的百分含量将近占全部集沙量的92%，说明该区域输沙主要集中在床面上60 cm高度之内。计算试验区不同高度的输沙通量，最大值产生在0～10 cm高度内，值为2.23 g/(m^2·s)。

表3 土壤风蚀输沙量计算表

集沙口层号	集沙层距地面高度(cm)	集沙口中心距地面高度(cm)	集沙口面积(cm^2)	输沙量(g)	百分含量(%)	输沙量(g/cm^2)
1层	5～10	7.5	25	5.95	25.40	0.238
2层	15～20	17.5	25	3.53	15.05	0.141
3层	25～30	27.5	25	3.03	12.91	0.121
4层	35～40	37.5	25	2.98	12.70	0.119
5层	45～50	47.5	25	2.23	9.50	0.089
6层	55～60	57.5	25	1.98	8.43	0.079
7层	65～70	67.5	25	1.60	6.83	0.064
8层	75～80	77.5	25	1.25	5.34	0.050
9层	85～90	87.5	25	0.90	3.84	0.036

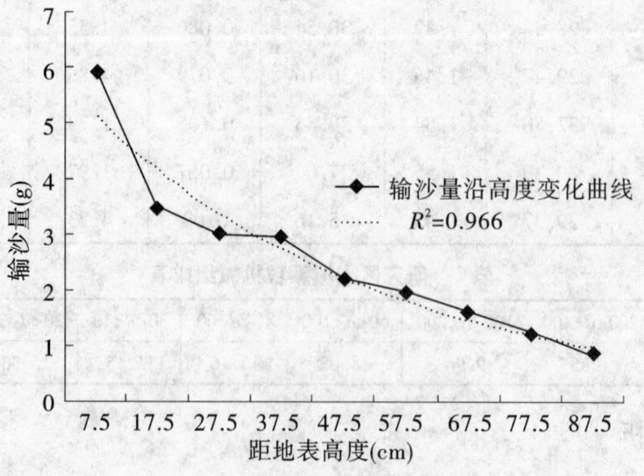

图1 输沙量沿高度变化曲线图

4 黑土区土壤风力侵蚀发展趋势分析

4.1 区域气候变化对风蚀发展趋势影响分析

气候变化对风力侵蚀的影响关键指标为气温、降雨、风速。

东北地区近44年(1959~2002年)气象资料分析得出气温有明显增加趋势,增温率为0.342 ℃/10a,降雨量有明显的减少趋势,气候倾向率为-11.815 mm/10a,说明东北地区总体气候变化具有明显暖干化趋势,并且环境脆弱区的三江平原和科尔沁沙地及周边地区的气候有更明显的暖干化趋势。气候暖干化,导致土壤含水量降低,抵御风力侵蚀能力降低,这将为该区土壤风力侵蚀提供有利条件。风速为风力侵蚀发生的先决条件,也是动力源泉。风速大小的变化趋势亦直接影响风力侵蚀强弱的发展趋势。通过对我国近46年东北地区平均风速变化情况进行分析,风速平均值呈整体下降趋势,气候变率为-0.23 m/(s·10a),但是嫩江、鸡西、长白等风蚀区逐年平均风速变化出现了正增长趋势,其中嫩江最高,气候变率为每10年上升0.08 m/s。

东北黑土区气候暖干化的发展趋势势必导致该区风蚀的进一步发展,虽然该区风速平均值呈总体下降趋势,但是部分区域依然出现正增长趋势,即风蚀区的风速正增长,也将导致风蚀强度增加及风蚀面积扩张的发展趋势。

4.2 东北沙地对黑土区风蚀发展趋势影响分析

东北黑土区及周边有我国的两大沙地,分别为松嫩沙地与科尔沁沙地。其中松嫩沙地位于东北黑土区中西部,东北典型黑土区西部,属于嫩江下游及东辽河中游北部区域。20世纪30~40年代日本殖民者对植被资源进行了空前掠夺,中华人民共和国成立后又有数次大规模毁林开荒造成沙化土地日趋严重。至2000年,沙区沙漠化土地总面积6 767.50 km^2。其中,重度沙漠化土地200.22 km^2,中度沙漠化土地231.04 km^2,轻度沙漠化土地5 613.58 km^2,潜在沙漠化土地722.66 km^2。松嫩沙地沙漠化导致沙地面积逐年扩张,植被覆盖降低,引起区域气候变化,为东北典型黑土区土壤的风力侵蚀创造气候发生条件,使沙地周边典型黑土地土壤风蚀土地退化沙化,黑土厚度及面积逐年减小。

科尔沁沙地是我国东北西部长达400余km的一条大沙带,地处东北平原向内蒙古高原的过渡地带,位于我国东北黑土区西南边缘。至2005年该区沙漠化土地面积为22 422.4 km^2,原沙漠化土地发展程度明显减轻,而非沙漠化土地出现较为严重的沙漠化过程,监测期间(2000~2005年)有113.3 km^2非沙漠化土地变为沙漠化土地,沙漠化速度为22.7 km^2/a。科尔沁沙地紧邻东北黑土区西南部,其沙漠化进程会直接威胁黑土区南部黑土耕地安全,导致农田土壤风蚀进而土地退化、沙化,黑土地面积由南向北逐年减小。另外,从科尔沁沙地近40年(1961~2000年)气候背景分析得知,近40年来年均温度呈现显著的直线上升趋势,平均增温1.51 ℃,增温幅度在0.90~1.79 ℃;年均温增加主要表现在冬季增温和春季增温,而年降水量未发现显著上升或下降的趋势。可见该区区域气候变化规律为暖冬暖春,冬季与春季会逐年干旱即风季干旱,正是土壤风蚀发生的有利条件。

5 结论

(1)国家正在振兴东北老工业基地,同时人口数量与密度的逐年增加,势必会对黑土区自然环境进行扰动、改变原有景观结构,稍有不慎就会诱发人为影响区内水土流失的发生与发展,进而导致周边植被覆盖降低,在东北气候变化趋势暖干化的大背景下,易发生土壤风蚀。

(2)我国东北地区未来气候变化趋势向暖干化发展,而东北风蚀区风速变化趋势却有上升趋势,这将为东北地区风沙活动提供有利先决条件,导致东北沙地沙漠化进程的加快。而东北黑土区中西部为松嫩沙地,西南部又与科尔沁沙地毗邻,这两大沙地沙漠化发展,势必会使东北黑土区土地形成由中部向四周,由南部向北部退化、沙化,黑土地面积逐年减少,土壤风蚀区域逐年增加的局面。

(3)土壤风蚀导致黑土粗化、沙化,土地生产力下降,黑土层厚度及黑土区面积减小,最终将严重威胁我国粮食安全战略,制约区域经济的可持续发展,可见东北黑土区风力侵蚀需要引起关注,加强土壤风蚀过程及机制性科学研究,采取积极有效的防治措施。

参考文献

[1] 孙凤华,杨素英,陈鹏狮. 东北地区近 44 年的气候暖干化趋势分析及可能影响[J]. 生态学杂志,2005,24(5):751-755.

[2] 杨雪艳. 中国东北地区风的气候变化特征及大风的成因研究[C]. 兰州大学,2008.

[3] 姜凤岐,等. 科尔沁沙地生态系统退化与恢复[M]. 北京:中国林业出版社,2002.

[4] 李爱敏,等. 21 世纪初科尔沁沙地沙漠化土地变化趋势[J]. 地理学报,2006,61(9):976-984.

[5] 赵云龙,唐海萍,李新宇. 近 40 年来科尔沁沙地沙漠化过程的气候背景分析[J]. 干旱区资源与环境,2004,18(5):8-14.

【作者简介】 刘铁军,男,1979 年 10 月生,硕士,2006 年毕业于内蒙古农业大学,工学硕士,现在水利部牧区水利科学研究所工作,工程师。

人类活动对泾河流域水沙过程影响的模拟分析

龚家国　王　浩　贾仰文　周祖昊　刘佳嘉

（中国水利水电科学研究院　北京　100038）

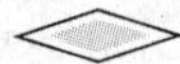

【摘要】 人类活动（下垫面变化）对流域水沙过程的影响具有双重性，综合作用下流域水沙过程变化趋势怎样变化，是人们普遍关心的问题。利用流域分布式水沙耦合模型，从土地利用和水土保持措施角度对不同人类活动（下垫面变化）情景下的泾河流域水沙演变过程进行模拟，尝试回答了1985～2000年人类活动的增洪增沙与减洪减沙综合作用结果，以及2000年水土保持工程措施情景的减水减沙效益。研究成果对支撑流域水土保持措施科学配置，建立科学的水土保持体系，正确评估水土保持措施效益等方面具有重要作用。

【关键词】 分布式水沙耦合模型　人类活动　减水减沙　泾河流域　黄土高原

1　研究背景

人类活动（下垫面变化）对流域水沙过程的影响具有双重性。一方面表现为陡坡开荒，毁林开荒，大规模工程建设等造成的水土流失。另一方面表现为，通过工程措施、植被措施、耕作措施等进行大规模水土保持治理减少水土流失。近年来的研究表明人类活动是影响流域水沙过程的主要因素之一。景可等采用相关分析研究认为黄土高原地区1919～1949年人类活动造成的加速侵蚀占18.4%，1950～1983年间人类活动影响进一步增强了流域加速侵蚀的趋势，达到25%。许炯心研究认为黄河中游多沙粗沙区1998～2006年的水沙演变影响因素中人类活动的影响已经成为支配产沙过程的主导因素，对产沙量变化的贡献率达到65%。高鹏研究认为1950～2008年间降水和人类活动是影响黄河中游水沙变化的主要驱动因素。降水和人类活动的减水贡献率分别为30%、70%；减沙贡献率分别为20%、80%。然而这些结论均是基于资料分析得出的结论。利用分布式水沙耦合模型再现流域水沙演变过程，利用土地利用数据和水土保持数据对不同的人类活动（下垫面变化）情景下的流域水沙变化进行模拟，深入探求其深层次的科学机制，对开展大规模水土保持与生态治理，增强人类应对气候变化的能力具有重要作用。

2　研究区域及方法

2.1　研究区域概况

泾河位于黄土高原中部，处于六盘山和子午岭之间（106°20′～108°48′E，34°24′～37°20′N），属温带半干旱半湿润大陆性季风气候。流域总面积约4.5万km^2，其中水土流失面积 3.3 万 km^2。流域内地貌复杂多样,主要有黄土丘陵沟壑区、黄土高塬沟壑区、土石山区、黄土丘陵林区和黄土阶地区等 5 个地貌类型区,其中以黄土丘陵沟壑区、黄土高塬沟壑区所占面积最大,分别占流域总面积的 41.13% 和 39.17%。全流域中黄绵土是最主要的土壤类型,占地面积达到 29 104 km^2,占到流域面积的 75.10%。其次是黑垆土,占全流域面积的 13.29%。褐土、灰褐土、红黏土、粗骨土、山地草甸土在流域内也有一定的分布。

2.2 研究方法

采用作者博士论文建立的分布式流域水沙耦合模型进行泾河流域水沙过程模拟,在模型率定与验证的基础上通过情景分析研究不同人类活动条件下泾河流域水沙过程。模型以 WEP-L(Water and Energy Transfer Process in Large River Basins)模型为平台,将流域分为坡面和河(沟)道两种典型的地貌类型单元进行流域泥沙过程模拟模型构建。其中坡面水沙过程主要包括雨滴溅蚀过程、薄层水流侵蚀过程、股流侵蚀过程和重力侵蚀过程。河(沟)道输沙过程模拟采用沿水深积分后的一维恒定水流泥沙扩散方程。河道断面的水流挟沙力计算分为沟道和河道两种情况:采用费祥俊公式计算沟道挟沙力,采用张红武公式计算河道水流挟沙力。对水库水沙过程的计算是基于水位—库容曲线和溢洪道参数等水库属性数据,利用水量平衡计算水库出流量过程;对水库泥沙过程模拟方法是:将水库分为建成初期和稳定运行期,采用排沙比计算水库出水过程的输沙率。

2.2.1 输入数据

分布式模型的输入数据较多,包括 DEM,逐日降水、风速、气温、日照、湿度等气象数据,径流、输沙率及泥沙、水土保持、土壤及水文地质、植被、土地利用、水库、水土保持等。其中人类活动主要通过改变流域下垫面来影响流域水沙过程。其中植被措施和增加水土流失的人类活动主要表现为土地利用类型的变化,而工程措施主要表现为修建水平梯田以及建设淤地坝形成的坝地面积变化。

模型以 30 m 分辨率 DEM 数据为基础,采用 ArcGIS9.2 默认设置重采样为 1 000 m 分辨率的 DEM 为平台。以 50 个栅格数为河道阈值进行流域数字化河网和子流域套等高带计算单元数字特征的提取,流域共划分为 361 个子流域共 2 257 个等高带。

选取的降雨情景时段为 1956 ~ 2000 年。用于情景模拟的数据主要为 1985、1995、2000 年三期土地利用数据和 1986 ~ 2000 年梯田和淤地坝分布数据。其中,三期土地利用数据均为中科院遥感所基于 LandsatTM 遥感数据生产,其斑块的地面分辨可达到 30 m。

从表 1 可以看出,泾河流域土地利用类型涵盖 6 种一级分类、22 种二级分类,其中占主导的土地利用类型为丘陵旱地和中覆盖度草地,其次为平原旱地和低覆盖度草地。不同时期的土地利用格局虽然没有较大变化,但其他林地和低覆盖度草地的面积逐年递增明显,同时对流域产沙过程影响较大的山地旱地、丘陵旱地、大于 25°坡旱地、裸土地等土地利用类型变化显著;相对于 1985 年,泾河流域植树造林等减少水土流失的措施增加明显,同时草地退化也较为显著,特别是对水土流失有重要影响的耕地和裸土地均有波动增加的趋势。因此,两方面综合作用下流域水沙过程变化趋势怎样变化,是人们普遍关心的问题。

表1　泾河流域土地利用情况

土地利用类型	面积(km²),比例(%)				
	1985 年	1995 年	相对 1985 年变化率(%)	2000 年	相对 1985 年变化率(%)
平原水田	0.36(0.00)	0.36(0.00)	0.00	0.36(0.00)	0.00
山地旱地	1 376.16(3.24)	1 401.92(3.30)	1.87	1 387.19(3.26)	0.80
丘陵旱地	12 697.38(29.86)	12 959.03(30.48)	2.06	12 982.89(30.54)	2.25
平原旱地	4 284.07(10.08)	4 145.83(9.75)	-3.23	4 123.06(9.70)	-3.76
>25°坡旱地	27.33(0.06)	31.74(0.07)	16.14	29.04(0.07)	6.26
有林地	666.65(1.57)	654.79(1.54)	-1.78	653.09(1.54)	-2.03
灌木林	2 201.67(5.18)	2 018.43(4.75)	-8.32	2 065.25(4.86)	-6.20
疏林地	1 127.76(2.65)	1 067.63(2.51)	-5.33	1 083.48(2.55)	-3.93
其他林地	65.39(0.15)	80.54(0.19)	23.17	92.69(0.22)	41.75
高覆盖度草地	965.07(2.27)	879.66(2.07)	-8.85	898.94(2.11)	-6.85
中覆盖度草地	15 656.79(36.83)	12 146.67(28.57)	-22.42	12 107.76(28.48)	-22.67
低覆盖度草地	2 733.63(6.43)	6 379.07(15.00)	133.36	6 276.53(14.76)	129.60
河渠	20.75(0.05)	38.25(0.09)	84.34	24.24(0.06)	16.82
湖泊	4.01(0.01)	4.34(0.01)	8.23	4.03(0.01)	0.50
水库坑塘	14.09(0.03)	11.42(0.03)	-18.95	12.25(0.03)	-13.06
滩地	137.36(0.32)	92.92(0.22)	-32.35	126.48(0.30)	-7.92
城镇用地	35.55(0.08)	39.06(0.09)	9.87	41.02(0.10)	15.39
农村居民点	492.76(1.16)	482.65(1.14)	-2.05	525.91(1.24)	6.73
其他建设用地	6.84(0.02)	8.51(0.02)	24.42	8.29(0.02)	21.20
沙地	0.36(0.00)	0.36(0.00)	0.00	0.36(0.00)	0.00
裸土地	2.19(0.01)	72.56(0.17)	3 213.24	72.93(0.17)	3 230.14
裸岩石砾地	0.00(0.00)	0.27(0.00)	—	0.27(0.00)	—

此外,水土保持措施包括工程措施、植被措施以及耕作措施等。由于人工林草地建设等信息已包含在不同时期的土地利用数据中,这里的水土保持数据主要指梯田和淤地坝建设形成的水平梯田和坝地的空间分布数据。本研究利用的数据为黄河流域范围内各省1986～2000 年水利统计资料中采集的。从图 1 可以看出,1986～2000 年,泾河流域的水平梯田和坝地面积基本以线性变化的趋势逐年增长。

2.2.2　情景设置

本研究共设定 4 种不同的人类活动情景:

(1)情景 1。历史下垫面情景,即真实下垫面变化情景下模拟 1956～2000 年泾河流域水沙变化过程,用于还原真实下垫面变化情况下流域水沙变化情况。

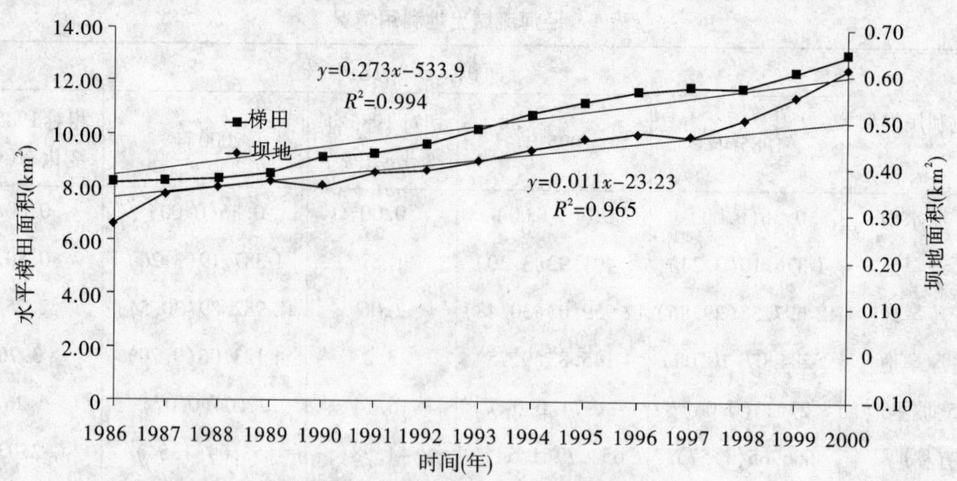

图1　泾河流域水土保持实施面积

(2)情景2。1985年土地利用下垫面情景,即下垫面保持1985年土地利用和水土保持工程措施状态不变条件下1956～2000年模拟泾河流域水沙变化过程。与情景1对比可以说明1985年水土保持措施对流域水沙关系的影响。

(3)情景3。2000年土地利用下垫面情景,即下垫面保持2000年土地利用和水土保持工程措施状态不变条件下1956～2000年模拟泾河流域水沙变化过程。与情景2对比可以说明,1985～2000年间人类活动减水减沙与增水增沙过程的对比关系。

(4)情景4。2000年土地利用-水土保持措施下垫面情景,即下垫面保持2000年土地利用但不含水土保持工程措施状态,模拟泾河流域1956～2000年水沙变化过程。与情景3对比可以说明不同降雨情景下梯田、坝地等工程类水土保持措施的水土保持效益。

3　结果与讨论

3.1　模型率定与验证

用于模型率定和验证的径流泥沙数据均为实测资料,其中逐日径流数据时段为1979～1990年,输沙率数据时段为1979～1990年(缺1988年)。将1979～1985年水沙资料用于模型的参数率定,1986～1990年(缺1988年)作为模型验证。

选取东干流上游黄土丘陵沟壑区的洪德水文站、东干流把口站雨落坪站、西干流把口站杨家坪站和下游张家山站用于模型的率定和验证。验证结果见表2。从验证结果看,泾河流域逐月径流过程Nash效率可以达到0.72以上,相关系数可以达到0.84以上,相对误差小于±17%,逐月输沙率过程Nash效率可以达到0.63以上,相关系数可以达到0.76以上,相对误差不超过±15%。

3.2　人类活动对泾河流域水沙过程演变的影响分析

研究采用代表时段方法,分别对1956～1959年、1960～1969年、1970～1979年、1980～1989年、1990～2000年、1956～1979年、1980～2000年等7个时段的河道水沙量变化进行对比分析。

表2 泾河流域部分水文断面水沙过程率定与验证结果

水文站	水沙要素		Nash 效率	相关系数	相对误差(%)
洪德	率定期	径流量	0.85	0.91	16.80
		输沙率	0.81	0.82	11.85
	验证期	径流量	0.89	0.85	12.3
		输沙率	0.76	0.85	12.2
杨家坪	率定期	径流量	0.87	0.91	7.29
		输沙率	0.66	0.89	7.46
	验证期	径流量	0.72	0.91	6.54
		输沙率	0.79	0.90	1.36
雨落坪	率定期	径流量	0.81	0.86	7.30
		输沙率	0.63	0.76	6.87
	验证期	径流量	0.84	0.96	7.30
		输沙率	0.81	0.91	-7.01
张家山	率定期	径流量	0.84	0.87	-5.43
		输沙率	0.80	0.84	14.13
	验证期	径流量	0.83	0.84	7.40
		输沙率	0.81	0.83	14.13

3.2.1 径流量变化分析

从表3可以看出,情景1下流域径流量与降雨的变化趋势一致,自20世纪60年代起流域径流量开始明显减少。根据径流的变化可以以1980年为界,可以明显地分为丰枯两个时段,其中1956～1979的多年平均径流量达21.33亿m^3,是1956～2000平均年径流量的1.24倍;而1980～2000年的多年平均径流量为12.56亿m^3,仅为1956～2000平均年径流量的72.85%。

与情景1相比,情景2对应情景时段的平均年流域径流量变化幅度均在±2%以内,说明情景2相对于情景1人类活动对流域的产汇流条件影响较小。相对于情景1,情景2的1956～1979年多年平均径流量增加0.27亿m^3,而1980～2000年年均径流量减少0.07亿m^3,在1956～2000年时段总体上表现为径流量增加;情景3在不同时段的径流量均表现为增加趋势,说明1956～2000年以来人类活动造成的径流增量大于水土保持减水量。对比情景3和情景2可以发现,人类活动在扣除1985～2000年之间新增的水土保持措施减水效益的基础上,在不同时段的径流量均增加。

情景3及情景4与情景1相比,在不同时段均表现出径流增加趋势。说明在不同时段,2000年下垫面情景下的人类活动增水量大于水土保持措施的减水量。同时对比情景3和情景4可以看出,工程类水土保持措施减水效益明显:1956～1959年、1960～1969年、1970～1979年、1980～1989年、1990～2000年对应时段分别减水0.76亿m^3、1.09亿m^3、

1.3 亿 m^3、0.74 亿 m^3、0.98 亿 m^3，其中由于 1956～1979 时间段降雨较多，平均减水效益为 1.13 亿 m^3，而在 1980～2000 年时间段为 0.87 亿 m^3，总体上年均减水效益为 1 亿 m^3。

表 3　不同下垫面情景河道径流量与输沙量变化

时段(年)	年均降雨量(mm)	年均径流量($\times10^8 m^3$)				年均输沙量($\times10^8$ t)			
		情景 1	情景 2	情景 3	情景 4	情景 1	情景 2	情景 3	情景 4
1956～1959	526.80	17.91	18.19	18.59	19.35	2.22	2.27	2.34	2.57
1960～1969	555.18	24.07	24.41	25.17	26.26	2.53	2.54	2.67	2.99
1970～1979	496.55	19.96	20.16	20.68	21.98	1.94	1.95	2.03	2.31
1980～1989	492.07	17.35	17.34	17.87	18.61	2.07	2.09	2.19	2.35
1990～2000	464.51	8.20	8.08	8.33	9.31	0.62	0.60	0.64	0.76
1956～1979	526.02	21.33	21.60	22.20	23.33	2.23	2.25	2.35	2.64
1980～2000	477.64	12.56	12.49	12.87	13.74	1.31	1.31	1.38	1.52
1956～2000	503.44	17.24	17.35	17.85	18.85	1.80	1.81	1.89	2.11

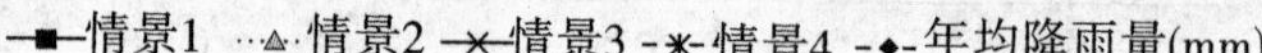

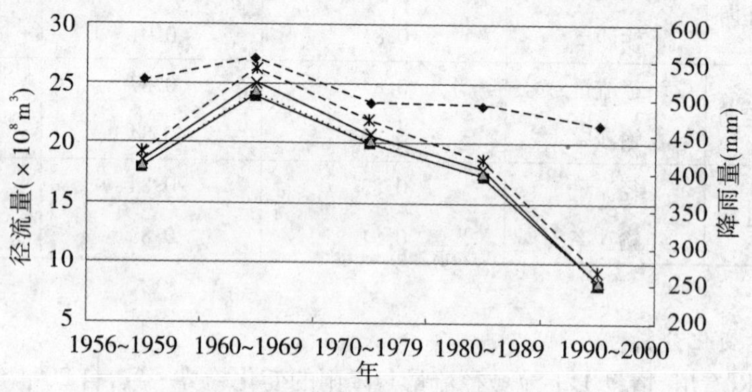

图 2　泾河流域降雨量～径流量年代变化

结合图 2 可以看出，泾河流域出口的径流量与降雨量的年代变化增减变化趋势一致，说明自然条件对泾河流域径流的影响仍占主导地位。但不同情景下的响应关系不同，其中情景 4 在 1956～1959 年到 1960～1969 年期间的径流量增长率远大于降雨增长率，而从 1970–1979 年开始，在降雨减少速率较小的情况下径流量迅速减少，说明梯田、坝地等水土保持工程措施对调节流域降雨—产流关系起到了重要作用。

3.2.2　输沙量变化分析

从表 3 可以看出，情景 1 下自 20 世纪 60 年代起流域输出泥沙开始明显减少，但与流域降雨、径流量不同的是，在 1980～1989 年时段的侵蚀量较 1970～1979 年时段多出 0.13 亿 t，说明该时期的人类活动的增沙量大于减沙量。以 1980 年为界，流域在 1956～1979 年的多年平均输沙量 2.23 亿 t，1980～2000 年多年平均输沙量 1.31 亿 t，分别是 1956～2000 年多年平均的 1.24 倍和 0.73。

与情景 1 相比，情景 2 在 1956～1985 年、1990～2000 年间人类活动的增沙量大于减沙量。这种情况在情景 3 中体现更为直接，分别变化 5%、6%、5%、5%、3%，说明不同时期的人类活动增加的产沙量大于其减少的产沙量。对比情景 3 和情景 2 可以发现，在扣除

1985～2000年之间新增的水土保持措减沙效益的基础上，不同时段人类活动造成的流域增加输沙量分别为0.07亿t、0.13亿t、0.08亿t、0.10亿t、0.04亿t。

对比情景3和情景4可以看出，工程类水土保持措施减沙效益明显：对应时段分别减沙0.23亿t、0.32亿t、0.28亿t、0.16亿t、0.12亿t，其中1956～1979年时段平均减沙效益为0.29亿t，而在1980～2000年时间段为0.14亿t，总体上年均减沙效益为0.22亿t。

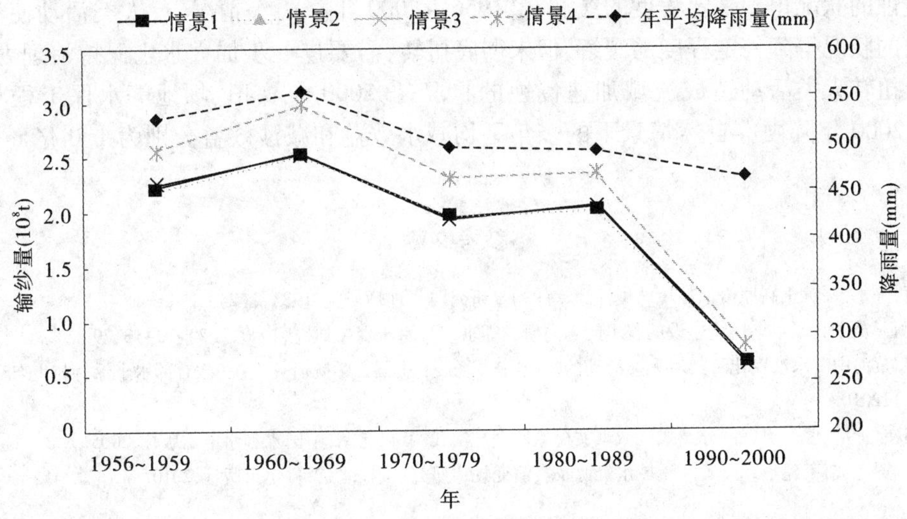

图3 泾河流域降雨—输沙量年代变化

由图3可以看出1956～1959年、1960～1969年和1970～1979年三个时段间的流域产沙量变化趋势与降雨变化相同。而自1980～1989年开始，水土保持措施等人类活动开始对流域减沙产生显著的正面影响。其中情景4的减沙速率明显较大，与情景2对比发现，自1980年以来的非工程水保措施对流域减沙的贡献较大。

从图4可以看出，单位径流输沙量中以情景4最大，而情景3最小。其中情景3在1956～1959年、1960～1969年、1970～1979年三个时段的单位径流输沙量显著低于其他情景，说明1980年以来的水土保持措施使人类活动显著改变了泾河流域的水沙关系，使流域加速侵蚀的状况得到一定程度的缓解。

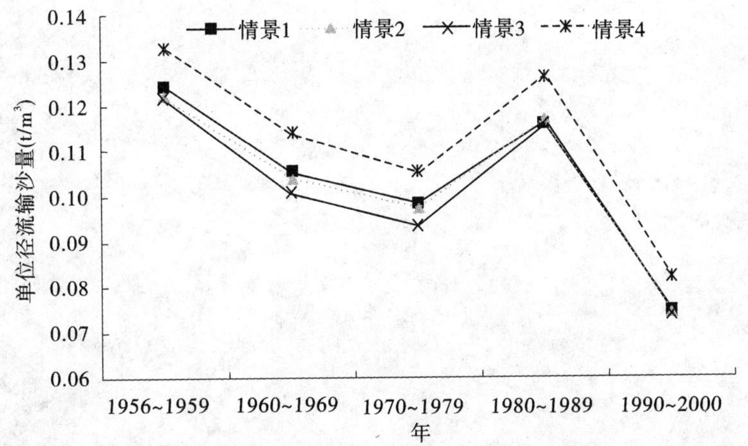

图4 不同情景下流域单位径流输沙量

4　结论

对流域不同时期的水沙变化关系进行了情景设置分析。利用模型分析，再现了自 1980 年以来我国开展大规模水土保持的客观实际。分析表明：①1980 年以来的非工程水保措施对流域减沙的贡献较大，水土保持措施使人类活动显著改变了泾河流域的水沙关系，使流域加速侵蚀的状况得到一定程度的缓解；②1956 ~ 2000 年降雨等情况下，人类活动使流域加速侵蚀的状况得到一定程度的缓解，但未彻底扭转，需要进一步加大水土保持治理力度，从根本上扭转人类活动造成流域加速侵蚀的状况；③2000 年梯田、坝地等水保工程状况在 1956 ~ 2000 年降雨等自然情景下的多年平均减水效益和减沙效益分别为 1.0 亿 m^3、0.22 亿 t。

参考文献

[1] 景可，陈永宗. 黄土高原侵蚀环境与侵蚀速率的初步研究[J]. 地理研究，1983，2(2)：1-11.
[2] 许炯心. 黄土高原生态环境建设的若干问题与研究需求[J]. 水土保持研究，2000，7(2)：10-13，79.
[3] 高鹏. 黄河中游水沙变化及其对人类活动的响应[D]. 杨凌：中国科学院研究生院（教育部水土保持与生态环境研究中心），2010.
[4] 贾仰文，王浩，倪广恒，等. 分布式流域水文模型原理与实践[M]. 北京：中国水利水电出版社，2005.
[5] 王浩，贾仰文，王建华，等. 黄河流域水资源及其演变规律研究[M]. 北京：科学出版社，2010.

【作者简介】　龚家国，男，1977 年 10 月生，博士，2011 年 7 月毕业于中国水利水电科学研究院，现就职于中国水利水电科学研究院，工程师，主要从事流域水沙过程、土壤侵蚀及水文水资源等方向的研究。

黄土丘陵沟壑第三副区小流域水土流失监测技术应用研究

张琳玲　张满良　张海强　袁宝琴

（黄河水利委员会天水水土保持科学试验站　甘肃　天水　741000）

【摘要】 水土保持监测对水土保持工作起着重要作用，黄河水利委员会天水水土保持科学试验站自20世纪40年代开始进行水土流失监测，积累了丰富的监测成果，在监测技术方面不断探索完善，已掌握了比较完善的监测方法。本文以黄土丘陵沟壑第三副区典型小流域罗玉沟流域为研究对象，通过对其水土流失监测站网布设及监测手段、方法等技术的应用，分析小流域水土流失监测技术、方法改进，提出了小流域监测技术方法的应用和发展前景，以期为同类型区水土流失监测提供参考。

【关键词】 黄土丘陵　小流域　水土流失　监测技术

1　前言

水土流失是世界性的环境灾害问题之一，水土流失的发生和发展受降水、土壤、地形、植被覆盖和土地利用类型等多种因素的影响，是一种复杂的人文和自然地理过程。水土流失监测是研究区域水土流失规律的基础，传统的水土流失监测方法耗时多、周期长，往往不能对大范围的水土流失类型、强度、分布和变化等情况实施监控，更无法适时定量地监测水土保持效果。随着水土流失规律研究的不断深入，传统的监测方法已经不能满足需要，监测技术的改进和先进仪器、设备的引进是提高水土流失监测水平的重要措施。

本文以黄土丘陵沟壑第三副区的罗玉沟典型试验流域为对象，对该流域监测站网的布设进行充分论证，根据试验研究内容要求的需要补充完善站网，更新原有监测仪器设施，改善监测手段、方法，提出小流域水土流失监测技术的应用效果。

2　研究区域概况

罗玉沟流域位于甘肃省天水市北郊，属黄土丘陵沟壑第三副区。流域面积72.79 km^2，呈狭长形，羽状沟系，平均宽度3.37 km，主沟长21.63 km。流域内有大小支沟138条，沟壑密度3.54 km/km^2。多年平均降水量548.9 mm，沟道径流以地表径流为主，多年平均径流模数30 690 m^3/km^2。全流域分为土石山区、杂色土区和黄土区三个类型区，共有土壤11种，山地灰褐土是本流域典型的地带性土壤，其分布占全流域的91.7%，另有鸡粪土、黄坂土、黑红土和杂色土等。

3　小流域水土流失监测站网布设及监测方法

3.1　降雨量监测

3.1.1　雨量站布设

降雨是小流域水土流失的主要因子之一，为了系统、全面掌握试验流域降雨分布和降雨量等参数，按照规范要求在罗玉沟流域布设有雨量站24个，其中上游5个、中游10个、下游9个。雨量站均配备有JDZ-1型数字雨量计和普通雨量筒，进行全天候降水量观测，人工和自动化观测仪器相互校核验证。

3.1.2　观测内容和方法

降雨量观测主要包括次降雨过程及降水量、降雨强度、降雨历时。分为汛期和非汛期两个阶段，多年来一直采用普通雨量筒和自记雨量计进行监测。为了提高监测效率、增加观测科技水平，对引进的先进的数字雨量计、遥测雨量计进行对比观测。

3.1.3　监测结果分析

雨量人工观测与自动化仪器观测误差受降雨强度影响，呈正相关关系。观测结果符合规范要求的3%。最大误差4.2%，最小误差0.6%，平均2%左右（见表1）。

表1　罗玉沟流域降雨量对比观测成果分析表

观测站名称	观测日期			降雨量（mm）		误差（%）
	年	月	日	普通雨量筒	JDZ-1型	
唐家河	2009	6	5	14.9	14.5	2.7
唐家河	2009	7	16	12.1	11.6	4.1
唐家河	2009	7	26	14.2	13.6	4.2
廖家岘	2010	5	25	21.8	20.9	4.1
廖家岘	2010	5	31	22.3	23.2	-4.0
廖家岘	2010	7	25	15.9	15.8	0.6

3.2　径流泥沙监测

3.2.1　径流泥沙测站布设现状

径流泥沙输出量的观测是掌握小流域水土流失发生程度的重要内容，为全面、准确监测试验流域水土流失程度，在罗玉沟流域沟口设有左家场把口站1个，安装自动化观测系统。

3.2.2　监测的内容和方法

径流泥沙测站观测内容为水位、流速、流量、含沙量等。

在试验流域罗玉沟沟口布设有长100 m的监测断面，水位观测采用YCH-1D型雷达式水位计进行监测；流速观测采用LA15-1型电波流速仪和浮标法。

通过观测的水位，运用率定的水位—流量关系曲线，推求流量。非汛期常流水流量较小，无法观测水位时，采用人工施测水面宽和深，计算过水断面面积，结合浮标法测得的流速，推求过水断面流量。

3.2.3　泥沙测验

泥沙测验采用水边一点法人工取样，置换法处理沙样得到泥沙含量。按照规范要求，泥沙取样频次与流速测次一致，洪水期根据洪水过程变化情况适当增加测次，以保证能够测到洪峰和洪谷时的泥沙变化过程，满足输沙量、泥沙级配分析的要求。

3.2.4 结果与分析

水位、流速人工监测数据与仪器观测数据平均误差在4%左右，最大误差11.63%，最小误差1.54%，能够满足监测要求（见表2）。泥沙沿用传统的置换法，监测结果见表3。

表2 YCH-1D 水位计、LA15-1 流速仪与人工观测对照表

径流站		YCH-1D 水位计（m）	人工观测（m）	误差（%）	LA15-1 流速仪（m/s）	浮标观测（m/s）	误差（%）
左家场	1	0.77	0.79	2.53	6.9	6.7	-2.99
	2	0.48	0.43	-11.63	5.3	5.0	-6.00
	3	0.73	0.70	-4.29	6.6	6.5	-1.54

表3 不同年份段监测结果表

流域	年份	降水量（mm）		径流量（万 m^3）		输沙量（万 t）		年代均值与多年均值差				
		全年	汛期	全年	汛期	全年	汛期	年降水量（mm）	相差（%）	年径流量（万 m^3）	相差（%）	年输沙量（万 t）
罗玉沟	1986～1989	560.1	439.1	429.8	312.4	140.4	79.0	8.0	1.4	214.2	49.8	63.1
	1990～1999	520.9	423.9	175.0	151.2	89.2	24.2	-31.2	-6.0	-40.6	-23.2	11.9
	2000～2010	519.6	488.6	155.2	151.8	37.9	29.2	-32.5	-6.3	-60.4	-38.9	-39.5
	1985～2010	552.1	454.8	215.6	182.4	77.3	36.8					

4 沟道重力侵蚀监测

根据实地调查，黄土丘陵沟壑区小流域水土流失有60%左右的发生在沟道，试验流域沟道侵蚀是小流域水土流失的主要来源，及时准确地掌握小流域沟道重力侵蚀的发生发展，是小流域水土流失监测的重要内容。

4.1 沟道重力侵蚀监测内容

为了系统、及时地监测试验流域沟道重力侵蚀指标，根据试验流域沟道重力侵蚀发生的特点，分别在罗玉沟流域的上、中、下游共布设重力侵蚀观测点21处，其中滑坡观测点13处，崩塌观测点6处，泻溜观测点2处。

4.2 沟道重力侵蚀方法

4.2.1 监测内容

主要监测沟谷侵蚀所导致的地形变化，包括通过对沟谷侧蚀、溯源侵蚀和沟床下切侵蚀和滑坡、崩塌及泻溜重力侵蚀进行监测，计算沟谷的侵蚀量和堆积量。

4.2.2 观测途径及计算方法

1）观测途径及方法

沟道土壤侵蚀观测的主要内容是沟道的产沙方式及侵蚀部位，从泥沙来源上说明沟道土壤侵蚀的基本特性。在观测中，采用以常规方法观测为主，结合先进的观测设施及方法，

以加快观测速度和提高观测精度。观测汛前、汛后各进行一次。

断面观测采用统一水准高程，按照国家四等水准测量方法及精度要求，采用高精度 GPS 配合全站仪进行观测。

2）计算方法

将 GPS 测量结果在 GIS 软件支持下，生成多期 TIN 模型（三角网数据模型），进一步生成 3 m 规则间距的 DEM（数字高程模型），在 GIS 下计算侵蚀量进而推求各断面组及各沟道的侵蚀量及淤积量，并通过可视化手段制作电子地图，显示侵蚀和堆积的空间分布，GIS 软件选择国际通用的 ESRI ® ARGIS9 或 ArcView3.1。

4.2.3 结果与分析

采用 GPS 对流域沟道重力侵蚀进行调查，能够准确掌握重力侵蚀发生的具体位置，同时对侵蚀面积、侵蚀量等参数的观测也比较可靠。罗玉沟流域重力侵蚀主要发生在中游区域，侵蚀类型多，分布比较普遍，详见表 4。

表 4 重力侵蚀观测结果汇总表

编号	所在流域	观测点位置	侵蚀类型	抽样面积（m^2）	高差（m）	侵蚀量（m^3）
1	罗玉沟流域上游	主河道上游	崩塌	84.71	0.025	2.118
2	罗玉沟流域中游	赵家河沟道	泻溜	89.24	0.032	2.856
3		刘家河河道	崩塌	76.52	0.028	2.143
4		赵家河沟中上游	滑坡	12.53	0.031	0.388
5		赵家河沟上游	滑坡	76.16	0.019	1.447
6		赵家河下游	滑坡	96.47	0.024	2.315
7		赵家河沟中游	滑坡	34.68	0.033	1.210
8	罗玉沟流域下游	干校河道	崩塌	62.47	0.026	1.624
9		芦家湾中上游	崩塌	23.47	0.022	0.516
10		芦家湾中游	滑坡	52.26	0.025	1.307
11		芦家湾下游	滑坡	42.73	0.034	1.453
12		芦家湾上游	滑坡	47.81	0.027	1.291
13		烟铺河道	崩塌	81.85	0.031	2.537
14	桥子西沟流域	桥子西沟上游	滑坡	83.43	0.027	2.253
15		桥子西沟下游	滑坡	24.69	0.031	0.765
16		桥子西沟下游	崩塌	21.64	0.032	0.692
17	李家园子河道	罗玉沟中下游	泻溜	60.25	0.032	1.928

5 应用遥感技术监测流域水土流失

5.1 土地利用类型调查

5.1.1 调查方法

土地利用调查可借助 IKONOS 全色影像以及 SPOT 多光谱影像资料，采用主成分变换方法对 IKONOS 1 m * 1 m 全色数据和 SPOT 多光谱数据进行融合。通过几何校正—资料分析、野外勘查、建立解译标志库—室内人机交互判读—建立拓扑、接边、裁切、整饰—分类信息提取、数据统计,形成流域土地利用调查结果。根据 2008 年遥感数据,对流域土地利用情况进行了调查。

5.1.2　调查结果与分析

全流域主要以旱地为主,占 60.76%;其次是草地、果园和林地,分别占 10.44%、10.43% 和 10.02%;空闲地最小,主要是陡坎等难利用地。详见表 5、图 1。

表 5　罗玉沟流域各土地利用类型面积统计表

代码	土地利用类型	面积(m^2)	所占比例(%)
013	旱地	43 339 979.54	60.76
021	果园	7 440 088.27	10.43
031	有林地	3 742 090.12	5.25
032	灌木林地	3 401 074.13	4.77
043	草地	7 448 549.46	10.44
062	采矿用地	94 188.15	0.13
071	城镇住宅用地	120 977.75	0.17
072	农村宅基地	2 584 759.96	3.62
102	公路用地	382 072.06	0.54
104	农村道路	853 224.40	1.20
116	内陆滩涂	730 645.24	1.02
121	空闲地	60 902.43	0.09
122	设施农用地	428 227.55	0.60
127	裸地	701 372.19	0.98
总面积		71 328 151.25	

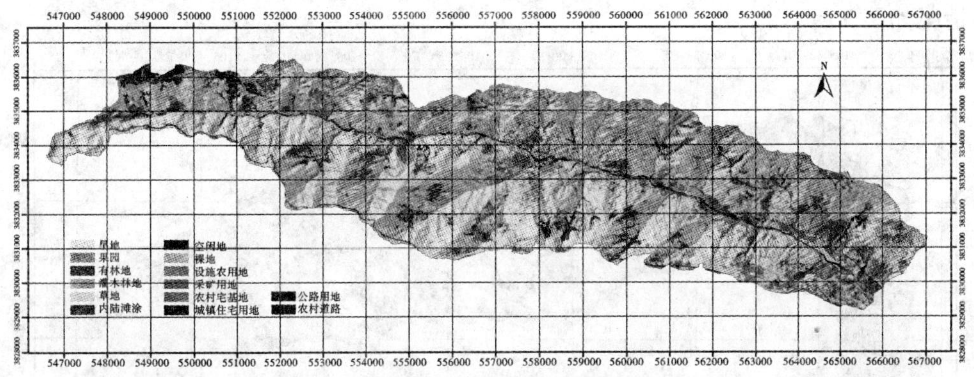

图 1　罗玉沟流域土地利用类型图

植被覆盖度调查项目的植被覆盖度主要基于 IKONOS 影像数据进行提取,利用遥感技术获取大面积区域的植被覆盖度,针对罗玉沟流域特性以及遥感数据源,使用植被指数法,把植被指数代入像元二分模型来计算植被覆盖度，即流域植被指数（NDVI）的计算——提取果园、有林地、灌木林地以及其他草地的 NDVI——计算 NDVIveg 和 NDVIsoil——计算植被覆盖度以及生成植被覆盖图，如图 2 所示。罗玉沟流域植被覆盖度统计表见表 6。

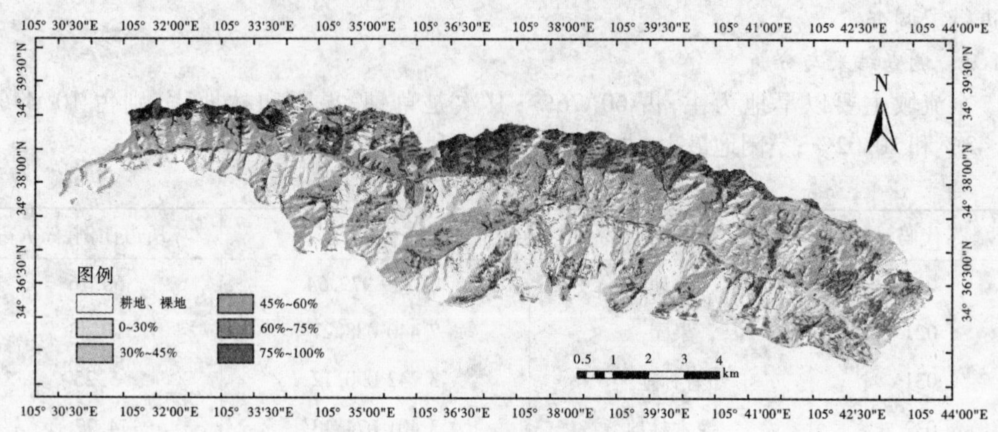

图 2 罗玉沟流域植被覆盖度图

表 6 罗玉沟流域植被覆盖度统计表

植被类型	覆盖度最小值（%）	覆盖度最大值（%）	覆盖度均值（%）
草地	14.21	89.53	66.04
有林地、灌木林地	27.76	92.33	74.53
果园	9.12	86.60	61.10
全流域	0	92.33	20.7

5.2 水土保持措施调查

5.2.1 调查方法

为了更加准确及时掌握及评价研究区各项水土保持措施，通过高精度 GPS、探查访问等对罗玉沟流域各项水土保持措施的位置、海拔以及利用状况进行了详细深入的调查核实。分布情况如图 3 所示。

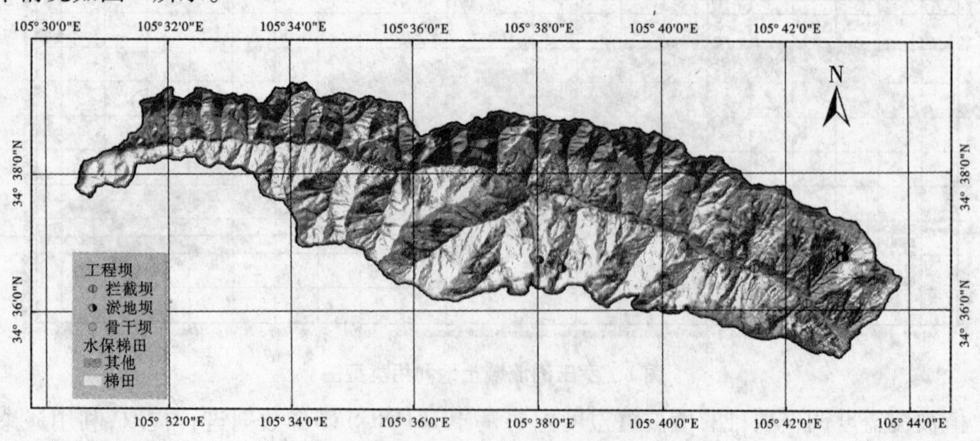

图 3 罗玉沟流域水土保持工程措施图

5.2.2 结果与分析

水土保持工程措施主要分布在流域中下游区域,以骨干坝、拦截坝和淤地坝为主。其中骨干坝3座、拦截坝2座、小型淤地坝21座。罗玉沟流域水利水保工程措施情况调查见表7。

表7 罗玉沟流域水利水保工程措施情况调查表

类型	经度	纬度	海拔(m)
骨干坝	105.534 2	34.642 1	1 607
	105.536 9	34.641 6	1 601
	105.535 6	34.641 4	1 605
拦截坝	105.705 5	34.601 6	1 171
	105.692 6	34.610 3	1 234
淤地坝	105.643 9	34.612 4	1 610
	105.713 4	34.613 2	1 352
	105.715 9	34.615 2	1 145
	105.715 4	34.613	1 387
	105.711 1	34.611 8	1 363
	105.711 9	34.612 7	1 368
	105.713 2	34.613 3	1 371
	105.713 9	34.614 1	1 347
	105.714 4	34.613 9	1 343
	105.712 7	34.615 2	1 441
	105.713 8	34.614 3	1 446
	105.640 2	34.600 8	1 513
	105.716 3	34.612 9	1 383
	105.714 3	34.613 8	1 352
	105.715 1	34.612 9	1 345
	105.714 5	34.615 3	1 146
	105.715 7	34.613 3	1 388
	105.715 2	34.616 3	1 432
	105.713 7	34.615 4	1 350
	105.714 7	34.615 2	1 141
	105.713 4	34.612 5	1 379

5.3 土壤侵蚀强度调查

5.3.1 调查方法

依据中华人民共和国水利部2008年批准公布的《土壤侵蚀分类分级标准》(SL 190—2007),利用已算出的DEM数据计算获得该流域的地形坡度,并结合利用植被指数法计算得到的植被覆盖度,经过综合处理分析最终获取该流域的土壤侵蚀强度。土壤侵蚀技术路线图见图4。

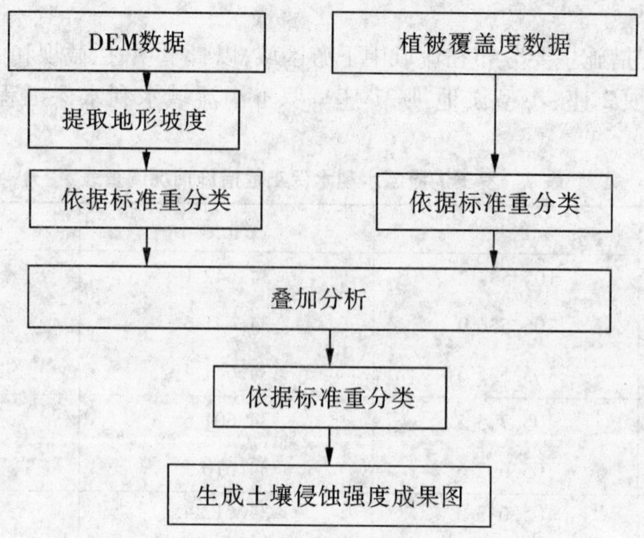

图4 土壤侵蚀强度技术路线图

5.3.2 结果与分析

全流域25°以上坡地侵蚀为极强烈和剧烈，占14.22%，主要分布在流域上游区域；5°~15°主要有中度侵蚀和强烈侵蚀，占71.80%，主要分布在流域中下游两岸的坡地上；轻度侵蚀分布在河谷地带，占13.98%，见表8~表10。罗玉沟流域土壤侵蚀强度成果见图5。

表8 非耕地土壤侵蚀强度面蚀(片蚀)分级指标

坡度(°)	0~5	5~8	8~15	15~25	25~35	>35
>75	轻度	轻度	轻度	轻度	中度	中度
60~75	轻度	轻度	轻度	轻度	中度	中度
45~60	轻度	轻度	轻度	中度	中度	强烈
30~45	轻度	轻度	中度	中度	强烈	极强烈
<30	轻度	中度	中度	强烈	极强烈	剧烈

表9 耕地土壤侵蚀强度面蚀(片蚀)分级指标

地面坡度(°)	0~5	5~8	8~15	15~25	25~35	>35
强度分级	轻度	轻度	中度	强烈	极强烈	剧烈

表10 罗玉沟流域土壤侵蚀强度等级面积统计

等级	罗玉沟	
	面积(km^2)	所占比例(%)
轻度	10.178 62	13.98
中度	31.796 23	43.68
强烈	20.467 56	28.12
极强烈	8.192 645	11.26
剧烈	2.154 944	2.96

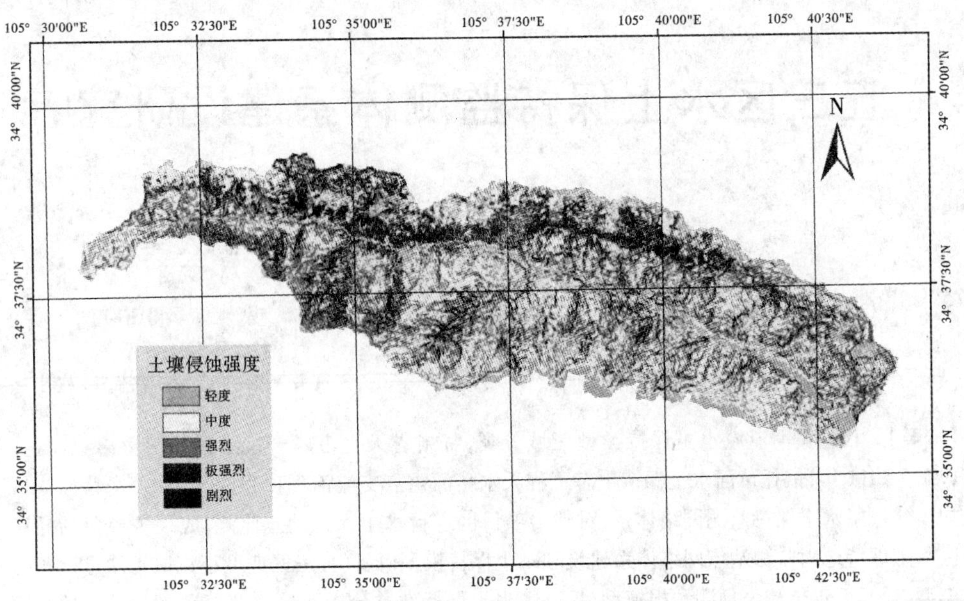

图5 罗玉沟流域土壤侵蚀强度成果图

6 结论

(1)小流域水土流失监测系统建设。建立水土流失监测网络系统,全面、准确地监测水土流失因子是今后水土流失监测的发展方向。通过对小流域监测技术、方法的有机整合,在罗玉沟试验流域建立了包括气象、土地利用、植被、沟道重力侵蚀、对比沟和卡口站监测为一体的小流域水土流失监测系统。

(2)罗玉沟试验流域在水土流失监测方面对新技术、新方法、新设备的引进使用,提高了水土流失监测的科技水平,增加了监测参数的精度,使水土流失监测由人工观测向自动化观测迈进。

(3)遥感技术的引用促使小流域水土流失因子监测把微观和宏观技术有机地结合在一起,使得观测指标更加全面,完全能够满足水土流失规律研究和小流域水土流失预测预报模型建立的需要。

(4)通过黄土丘陵沟壑第三副区罗玉沟试验流域水土流失监测方法技术的系统整合,监测技术、方法比较完善,监测内容全面,已经形成了适合土壤侵蚀和生态恢复治理综合研究的科研平台。

参考文献

[1] 张满良,等.黄土丘陵沟壑第Ⅲ副区罗玉沟流域水土流失监测体系建设[J].水土保持通报,2009,29(2):65-69.

[2] 伍飞舟,等.黄土高原丘陵沟壑区土地利用与水土保持措施的变化特征[J].水土保持研究,2010,17(2):10-14.

[3] 原翠萍,等.黄土丘陵沟壑区小流域治理对侵蚀产沙特征的影响[J].农业机械学报,2011,42(3):36-43.

【作者简介】 张琳玲,女,1974年10月生,硕士,2008年毕业于北京林业大学,现任黄河水利委员会天水水土保持科学试验站业务科副科长,工程师,主要从事水土保持科研、管理及水土流失监测工作。

丘三区水土保持监测体系建设研究

秦瑞杰　雷启祥　安乐平　杨丽萍　曹全意

（黄河水利委员会天水水土保持科学试验站　甘肃　天水　741000）

【摘要】 根据黄河流域水土保持类型区域分类，甘肃省天水市属于黄土丘陵沟壑第三副区。1998～2011年，依托黄河水土保持生态工程天水藉河重点支流治理项目的实施，天水藉河流域初步建成了水土保持监测网络体系，对整个项目区进行水土保持动态监测，成果主要包括丘三区水土保持地面监测站点网络体系建设和水土保持模型研发（丘三区典型小流域水沙过程预报模型、丘三区水土流失预测预报模型、丘三区水土保持效益分析评价模型）等。但是，对丘三区的水土保持监测体系建设今后还需作进一步的研究。

【关键词】 黄土丘陵沟壑第三副区　水土保持监测体系　水土保持模型体系

在黄河流域水土保持类型区域划分中，甘肃省天水市属于黄土丘陵沟壑第三副区（简称“丘三区”），该地区范围内的各支沟和支毛沟均正处于侵蚀发育阶段，沟道多为“V”字形，比降较大，建设大中型淤地坝的适宜条件有限，导致沟道治理相对薄弱。

目前，依托黄河水土保持生态工程天水藉河重点支流治理一、二期项目的实施，在项目区范围内初步建成了水土保持监测网络体系，主要包括水土保持地面监测站点网络体系建设、丘三区水土保持模型体系研发、水土保持监测综合数据库建设、系统运行环境建设、“数字藉河”三维可视化管理平台建设、三个业务系统开发（水土保持立体监测网络可视化管理信息系统、水土保持生态工程可视化系统、水土保持生态效益评价系统）等，对整个项目区进行水土保持动态监测。

1　国内外水土保持监测体系研究现状

1.1　国外水土保持监测研究现状

现代水土保持监测起始于18世纪末到19世纪初，1877～1895年德国土壤学家沃伦（Ewald Wollny）建立了第一个坡面径流小区，研究地形、土壤和植被对土壤侵蚀的影响，此后几十年该方法在世界侵蚀严重的国家和地区得到迅速推广和应用。20世纪40年代以后，随着许多国家对自然资源合理利用和水土保持工作的日益重视，水土保持监测和评价预报等研究得到了迅速发展。

美国等发达国家，已经形成了一套比较有效的水土流失监测机制，并将遥感、GIS等技术广泛应用于水土流失动态监测与预报，开发出了基于物理过程的土壤侵蚀预报模型。比如在水土流失预报方面，美国于1959年提出的通用土壤流失方程式（USLE），被USDA以农业手册的形式颁布执行。1985年以后，提出了修正土壤流失方程式（RUSLE），并于1994年被SCS确定为官方土壤保持预报和规划工具。USLE和RUSLE为田间水土保持规划提供了一个简单实用的工具,在世界范围内得到了广泛的应用。后来,为了克服USLE的不足和限制,美国1985年开始研发物理过程模型——水蚀预报模型(Water Erosion Prediction Project,WEPP),于1995年8月发布了WEPP 95,又相继于1998年、2000年和2001年分别颁布了不同版本,对早期版本进行修缮,WEPP的研究大大促进了人们对土壤侵蚀机制和过程的理解。国外开发的其他土壤侵蚀预报物理模型还包括EUROSEM、LISEM、GUEST、WEPS等。

1.2 国内水土保持监测研究现状

我国的水土保持监测工作起步较晚,但发展很快。自20世纪40年代初开始,我国相继在福建的长汀和黄河流域的天水等地建立了水土保持观测实验站,开展了水土流失规律研究。时至今日,中国水土保持监测工作在监测技术、网络和制度建设、技术标准制定、科技推广、水土流失动态监测与公告、预测模型及人才培养和队伍建设等方面取得了长足的进步和令人振奋的成绩,已发展成为面向国家生态文明战略需求、紧跟先进科技前沿、能够担当国家水土流失动态监测与预报、为水土保持生态建设提供决策依据的水土保持重要基础性工作。

1.2.1 监测技术的发展

20世纪20年代,我国在山西沁源等地建立首批径流小区,40年代开始在天水等地建立水土保持试验区,50年代陆续建立了一批实验站,开始坡面水土流失规律研究和小流域径流、泥沙观测。后来在长江、松辽、海河等流域先后设立水土保持研究所,中国科学院地理研究所和水土保持研究所还在黄土高原建立了试验基地。“七五”、“八五”期间,国家组织大批力量开展“黄土高原区域综合治理”攻关项目、综合考察项目,发展了以小流域为单元等各具特色的生态农业体系,水土保持监测全面发展。

20世纪80年代以来,我国将遥感、地理信息系统、数据库等技术应用到水土保持工作中,水土保持监测工作得以更进一步迅速发展。新技术(自动监测技术、遥测技术、数据传输技术、室内计算机管理系统等)在我国水土保持中应用日益普及,以水土保持监测点的自动化采集为基础,以快速典型抽样调查为补充,以“3S”技术、无线通信系统以及计算机网络技术为支撑,形成各级水土保持站点相互连接、高度集成的水土保持监测站网体系。

1.2.2 法制建设及监测网络的发展

1991年,《中华人民共和国水土保持法》首次颁布,经过20余年的实施运行,我国水土保持监测制度体系初步建立。在此基础上,为适应新形势,国家于2010年对原《中华人民共和国水土保持法》进行重新修订,并于2011年3月起正式实施新的水土保持法。

新水土保持法强化了水土保持监测的作用和地位,指出国务院水行政主管部门应当完善全国水土保持监测网络,对全国水土流失进行动态监测。截至2010年,全国已建成水利部水土保持监测中心、7个流域机构水土保持监测中心站、31个省(区、市)水土保持监测总站和186个水土流失重点防治区监测分站,初步建立起了水土流失观测与监测站网。

1.2.3 水土流失预报模型研究

水土流失预报模型是定量化开展监测、评价水土流失危害和水土保持设施防治效益的核心。近十年来,我国土壤侵蚀和水土保持工作者在黄土高原、长江中上游等地区开展的大量水土流失预测预报模型研究和应用试验都积累了丰富的经验。国内研究者早期建立的水土流失模型大都是经验性的，其中20世纪50～60年代初建立的经验模型仅限于进行坡耕地的侵蚀量计算。USLE模型被引进后，一些研究者陆续以它为原型，根据各自研究区实际情况进行相应的修正，建立了若干个地区性水土流失预报方程。如刘宝元等建立了中国土壤流失方程CSLE(Chinese Soil Loss Equation)，用于计算坡面上多年平均年土壤流失量。

2 丘三区水土保持监测体系建设成果

1998～2011年，依托黄河水土保持生态工程天水藉河重点支流治理一、二期项目的实施，天水藉河流域范围内初步建成了丘三区水土保持监测网络体系。监测体系主要包括丘三区水土保持地面监测站点网络体系和水土保持模型体系研发等，对整个项目区进行水土保持动态监测。

2.1 水土保持地面监测站点网络体系建设

在研究区域内选择典型坡面和流域，科学合理地规划建设了坡面水土流失规律观测系统、小流域水土流失梯级监测系统、中尺度流域水土流失监测系统、坡面及小流域综合治理效果评价监测系统、重力侵蚀监测点和植被监测点(见图1)，构建了一个科学、系统、规范的藉河流域水土保持地面监测站点网络体系，为丘三区水土流失规律研究奠定了良好的基础条件，并为水土流失模型的建立和验证以及水土保持综合治理效益评价提供了精确的数据。

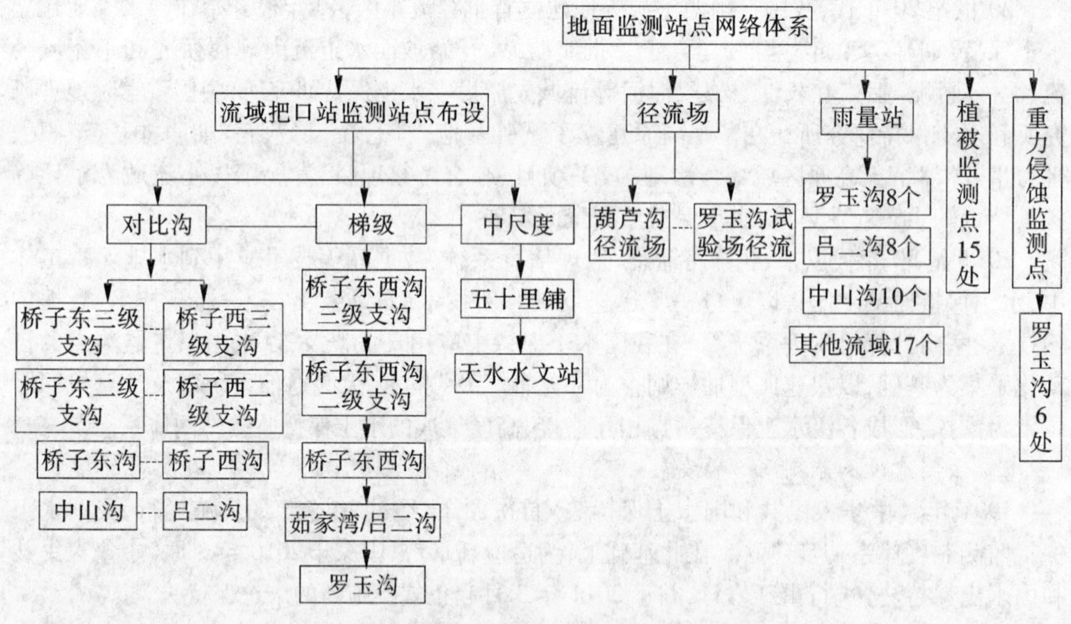

图1 藉河流域水土保持地面监测站点网络体系

2.2 丘三区水土保持模型体系研发

2.2.1 丘三区典型小流域水沙过程预报模型研究

以藉河项目范围内桥子东、西沟流域同一次降雨均出现产流记录为标准，筛选1987～2006年流域径流历史观测资料共19场次，结合Matlab软件计算模块，进行回归分析，计算得出流域产流随时间动态变化过程模型。计算结果按照时间变量t阶数变化规律，分别得出时间变量t从一阶至三阶计算模型，见式(1)～式(4)。

一阶计算公式
$$q = \frac{1}{a + b\mathrm{Exp}(t) + \frac{c}{t}} \tag{1}$$

二阶计算公式
$$q = \frac{t}{a\ (t + b)^2 + c} \tag{2}$$

三阶计算公式 1
$$q = \frac{1}{a + bt^3 + \frac{c}{t}} \tag{3}$$

三阶计算公式 2
$$q = \frac{1}{a + bt^3 + \frac{c}{t^{\frac{1}{2}}}} \tag{4}$$

式中:t 为降雨历时,h;q 为流域出口径流量,m^3/s; a、b、c 为拟合参数。

通过比较分析各模型模拟和计算结果,发现二阶模型能够较好地模拟径流动态变化过程。

2.2.2 丘三区水土流失预测预报模型研究

区域水土流失模型的主要功能在于分析区域尺度水土流失的时空趋势,为水土流失治理的宏观决策提供支持。因此,在区域水土流失计算中,为有效反映水土流失过程在空间和时间上的变化,同时方便径流和泥沙物质的汇流和输移计算,模拟计算以月为基本时间单元,将每个月总的降雨作为一场降雨来处理。将月降水量对应的总降水历时划分成若干时段(时间间隔)作为计算迭代的时间单元,构建一个月降水过程。在空间上,以 DEM 的栅格为基础,将研究区域离散化为一系列规则的网格,并以此为基本空间计算单元。将每月的降水划分为若干降水时段分别加以描述,计算其径流和侵蚀产沙过程的各个环节,包括降雨过程、植被截留、入渗、微洼地存储、地表径流量、径流流速及其相应的挟沙能力、径流的剥蚀、泥沙沉积等。然后根据水沙平衡原理,计算出任一单元格任一时段末的径流泥沙量,并在 GIS 支持下,通过对单元之间径流泥沙汇集进行计算,完成区域或者是流域每月土壤侵蚀的计算。利用该模型,输入主要计算参数(DEM、土地利用图、降水量、土壤抗冲和渗透系数等),通过计算,即可获得示范区土壤侵蚀模数图。具体计算包括空间单元划分与时段确定单元地表径流产生、侵蚀产沙和沉积过程计算、单元间径流泥沙汇集和运移计算。区域水土流失模型总体框架见图 2。

根据天水站的观测资料,输入各月模型相关数据,分别模拟 5 ~ 10 月每月的水土流失过程,得到的模拟结果和实际水土流失情况较为符合,可以模拟和预测区域水土流失状况。部分模拟值偏大,主要原因可能是没有考虑水土保持治理措施的减水减沙作用。

2.2.3 丘三区水土保持效益评价模型研究

根据丘三区水土流失与治理的特征,提出了适用于丘三区的水土保持效益评价指标体系,并主要参照《水土保持综合治理效益计算方法(GB/T 15774—2008)》规定的指标体系和计算方法,以水土保持工作对水土流失和径流的影响(基础效益)为主,兼顾对生态效益、社会效益和经济效益的评价,时间尺度上以二期工程(2007 ~ 2010 年)为主,展开了水土保持效益评价方法研究。水土保持效益评价指标见表 1。图 3 和图 4 为水土保持生态效益评价系统界面及水土保持生态效益评价参数输入界面。

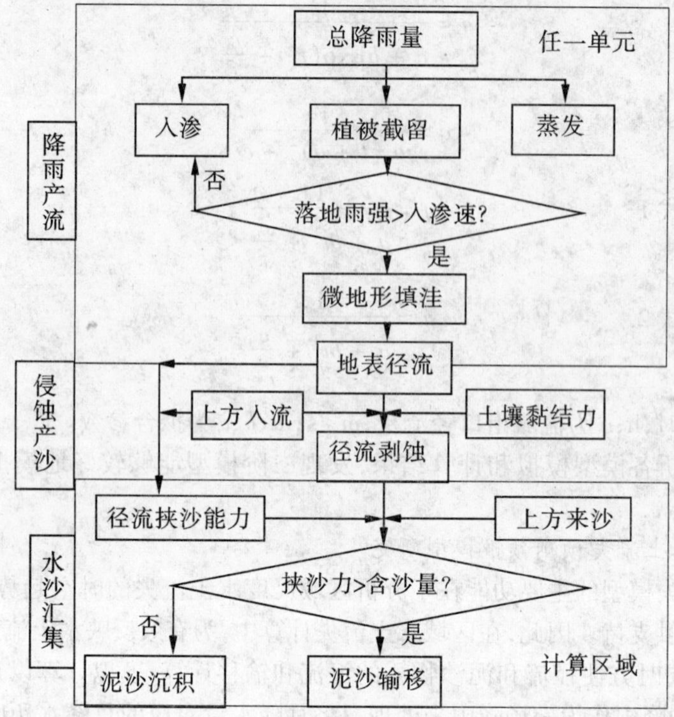

图2 区域水土流失模型总体框架图

表1 水土保持效益评价指标体系

效益类型	计算内容
调水保土效益	调水:增加土壤入渗,拦蓄地表径流,坡面排水,调节小流域径流
	保土:减轻土壤面蚀,减轻土壤沟蚀,拦蓄坡沟泥沙
经济效益	直接经济效益
	间接经济效益
社会效益	减轻自然灾害,促进社会进步
生态效益	调蓄地表径流,改良土壤理化性质,改善近地层气候,促进生物繁育

3 丘三区水土保持监测体系建设研究展望

依托天水藉河重点支流治理项目(一、二期),在藉河流域初步建成了水土保持监测网络体系,研发了三个丘三区水土保持模型。但从项目区土壤侵蚀类型区看,现有站点主要集中在中部黄土丘陵中度侵蚀区(Ⅲ区)。这种布局,虽然基本满足了三个模型的建模需要,但在项目区内表现为观测站点相对集中,分布极不均匀,其监测结果不能全面反映整个项目区的实际情况。同时,在区域土壤侵蚀评价、预报模型开发方面,缺少系列化观测与统计数据,已有的坡面土壤侵蚀模型等在通用性、适用性方面还存在不少问题。

随着监测技术的发展,遥感与 GIS 集成监测方法将成为快速监测的核心手段,该方法能够发挥遥感和 GIS 各自的优势,对地面监测站点监测数据和江河水文泥沙等数据与各种监测结果进行耦合,实现地面监测与遥感监测一体化。这将从根本上改变用人工方法进行监测的模式,使水土保持监测工作实时化、信息化,提高监测效率并能保证监测数据的精度,为水土流失模型的建立和验证以及水土保持综合治理效益评价提供技术支撑。

图3 水土保持生态效益评价系统界面

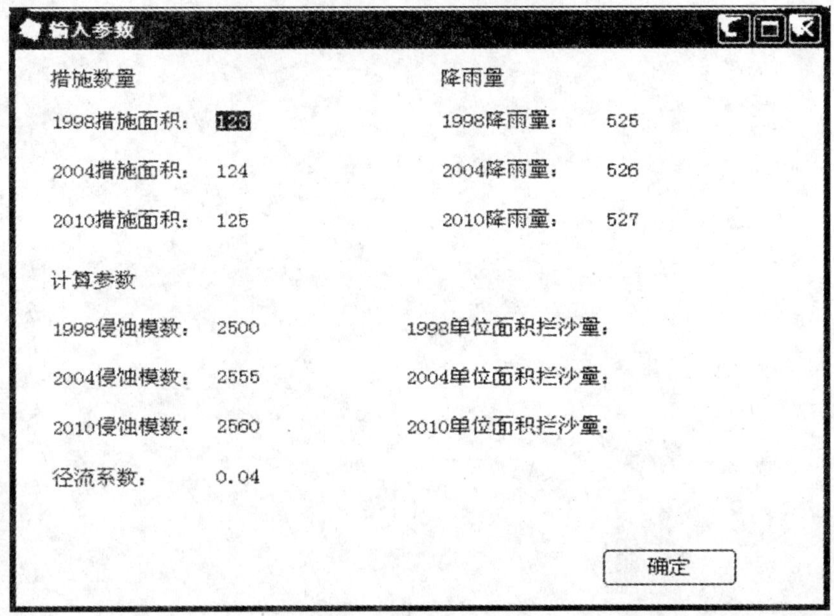

图4 水土保持生态效益评价参数输入界面

参考文献

[1] 黄河上中游管理局. 黄土高原水土保持实践与研究(1997～2000)[M]. 郑州:黄河水利出版社,2005.

[2] 郭索彦. 水土保持监测理论与方法[M]. 北京:中国水利水电出版社,2010.

[3] 孟庆枚. 黄土高原水土保持[M]. 郑州:黄河水利出版社,1997.

[4] 李智广,曹炜,刘秉正,等. 我国水土流失状况与发展趋势研究[J]. 中国水土保持科学. 2008,6(1):57-62.

[5] 许峰,郭索彦,张增祥. 20 世纪末中国土壤侵蚀的空间分布特征[J]. 地理学报,2003,58(1):139-146.

【作者简介】 秦瑞杰,女,1985 年生,硕士,2011 年 6 月毕业于西北农林科技大学,现工作在黄河水土保持天水治理监督局,助理工程师。

坝上风电场建设区水土流失特征及其防治措施体系

贾志军 甄宝艳

（河北省水利技术试验推广中心 石家庄 050061）

【摘要】 本文以坝上地区风电场为研究对象，开展了水土流失特征、水土保持设计原则与措施体系的研究。结果表明：风电场处于风蚀水蚀两相侵蚀相，其水土流失特征表现为风蚀、水蚀共有，点状与线状水土流失并存，植被破坏点多面广、恢复难度大，土石方弃渣相对少，水土流失重点在施工建设期的特点。风电场区的水土保持措施设计应遵循合理分区、综合防治，保护优先、措施可行，水土保持与经济发展相结合，生态景观与资源循环利用相结合等新理念，并由此形成一个以生物措施为主，工程措施为辅，临时措施为补充的水土流失防治措施综合体系。

【关键词】 坝上 风电场 水土流失 水土保持措施

能源是国家经济发展的基础，电气化是国家现代化的标志。我国是一个电力资源紧缺的国家，电力紧缺成为制约某些区域经济发展的瓶颈。风电作为一种清洁的可再生能源越来越受到国家的重视而得以迅猛发展。河北坝上地区位于内蒙古高原、燕山山地、华北平原之间，是河北省风能资源丰富区，也是国家授权开发的大型风电示范基地和河北省千万千瓦级风电基地建设的核心区域。近几年，风力发电在河北省坝上地区发展迅速，但开发建设项目不可避免地会造成植被退化、水土流失等一系列生态退化问题，特别是坝上地区处于农牧交错带内的典型脆弱生态环境区，如何科学合理地设计水土保持措施体系至关重要，是防治项目区水土流失的重要环节。笔者在前期研究的基础上，提出了风电场建设区水土流失的特点、水土保持措施设计原则与理念及其防治体系，以期为加快项目区的水土保持工作提供依据。

1 风电场建设区水土流失特征

坝上地区属于典型的农牧交错带生态脆弱区，区内风蚀与水蚀并存，为典型的风蚀水蚀二相侵蚀区。在该区建设风电场，因破坏土壤、植被而造成新的水土流失。其水土流失的特点与风电场建设项目的特性有关。

1.1 点状与线状水土流失并存

风电场的水土流失以点状与线状分布为主，点状侵蚀集中于风电机组区，包括风机的基础部分、箱式变压器的基础部分和临时吊装场地，基础的开挖造成原土壤与植被的破坏，而临时吊装场地则为吊装风机时碾压占地和对植被的破坏，这些点状的水土流失区域极易形成风蚀与水蚀的源地。线状侵蚀则发生于运输道路区，包括进场道路、场内道路及临时施工道路，道路系统在建设初期的水土流失量大，后期主要为临时施工道路，因设备运输需要而在道路上铺设了红色玄武岩砂砾或泥结碎石，因缺乏土壤基质而使得植被很难恢复，因而成为水土流失的主要对象。此外，点状水土流失区域还表现在升压站、施工生产生活区和弃渣场，特别是弃渣场，因土石方的裸露堆放而加大了水土流失的强度。

1.2　风力侵蚀与水力侵蚀共存

风电场建设项目一般历时1年以上，在施工建设期内，恰好经历冬春的大风季节和夏季多雨季节。坝上地处北方季风主风道，受蒙古高压控制，冬春季节大风频繁而强劲，土壤风蚀强烈。而在夏季雨水集中，常以暴雨形式降落，对风电场地表，特别是弃渣场及地表清理的土壤极易造成冲刷而形成水蚀。因雨水冲刷而造成的地表沟壑又成为新的风蚀源地。因而，风电场项目区受风力侵蚀与水力侵蚀的共同作用。

1.3　植被破坏点多面广，景观破碎，增加了植被恢复难度

对于风电场项目而言，尽管每个风机区破坏植被的面积不大，但每个风电场同时具有数十台风机，且风机间均有道路系统连接，道路沿线也会造成植被的破坏。在风电机组吊装、箱式变压器安装过程中，由于采用500 t履带吊和100 t汽车吊辅吊等大型机械作业和重型车辆运输，对风机区作业场地和道路频繁地扰动，造成地表的光板化，植被缺失，渗透减少，水土流失加剧。由于基础开挖、建筑材料准备和机械设备运输，清除了地表植被、表土和施工障碍物，破坏了原有的微域生态系统，产生了新地貌、新地物和新建筑，形成新的微域生态系统，这些地表的变化使得景观破碎化，形成以风机区为主体的“嵌块体”，并由道路景观相互连接。因此，风电场的植被破坏呈现点多面广的特征，地表光板化，增大了植被恢复难度。

1.4　土石方量较小，弃渣相对少

风电场的土石方工程主要发生于风机基础、箱式变压器基础、升压站和生活生产区的建设，产生的土石方量较小，且多数土石方能就地回填，如道路区就地挖高填低，达到填挖平衡，风机基础和箱式变压器基础的土石方有50%左右又重新回填，因而使得项目区土石方相对较小，弃渣相对少。如赤城风电的土石方总量为9.46万m^3，其中，挖方5.08万m^3、填方4.38万m^3，弃方仅0.70万m^3，占总土石方量的7.4%。

1.5　水土流失重点在施工建设期

风电场建设项目水土流失的重点在施工建设期，建设期内风蚀与水蚀共存，点状侵蚀与线状侵蚀并有。根据风电场工程特点及工程建设条件、工程施工工序等，工程建设对水土流失的影响主要集中在建设期，在此期间工程占地、基础开挖与回填、弃土等工程活动都会扰动或再塑地表，并使地表植被受到不同程度的破坏，地表抗蚀能力减弱，产生新的水土流失。

2　风电场建设区水土保持措施设计原则

2.1　合理划分点线工程区与综合治理原则

风电场建设项目区点状与线状水土流失并存，且呈现点散、线长的特点，各个区域存在自身的水土流失特点，其侵蚀强度、发生部位均有不同，应根据水土流失区内相似性最大和区间相异性最大的原则，合理区分点线工程，并按分区的水土流失差异实施不同的综合防治措施，形成点、线、面结合的水土保持措施防治体系，实现良好的水土保持防治效果。

2.2　保护原有设施与资源循环利用原则

坝上地区原始地貌、土壤及植被是自然界长期演化的结果，具有很强的稳定性和多种生

态功能，但受风电场占压、开挖等破坏过程的影响，植被与土壤遭到损坏，极易诱发水土流失。建设过程中，在风电机组基础开挖、道路三通一平、升压站建设时应尽可能保护原始地貌与植被，少占压、少开挖、少破坏，保持水土保持设施的完整性与多功能性，控制水土流失。在水土流失防治时，应考虑充分利用原有资源与自然资源，运用资源循环利用的理念，结合水土保持工程，加强水土资源的高效利用。对于风电场场区的表层土，主要采取表土剥离存放措施保存表土资源，待工程结束后进行表土回填覆盖，利用表土中的土壤种子库进行植被自然恢复，或利用表土资源进行植物重建，防止因采挖回填土而造成新的破坏。对于雨洪资源，则可通过硬化面或排水设施作为雨水收集设施，集中汇集在蓄水池或通过导流设施直接汇集至绿化地段，利用雨洪资源增加灌溉能力，降低绿化成本。

2.3 自然景观与人工景观统筹兼顾原则

风电场建设区进行植被恢复时，应注意与周边景观环境的协调性，利用原有植被，实现人工景观与自然景观的和谐。植被恢复时，应依托原有植被，进行生态景观设计、物种的配置、结构与功能的时空配置等。在物种选择上，以乡土树种为主体，采用多种乔灌草物种，提高项目建设区的物种多样性；在景观结构与功能搭配上，既要考虑防护功能，又要考虑景观要素与功能，力争做到“四季常青可防护、三季有花可欣赏”的生态景观特征与功能；在结构上，力争体现乔灌草一体、片带区结合的立体多层次结构，达到一种全新的景观。

2.4 水土保持与地方经济发展相结合

坝上风电场的施工除考虑保护风电场的安全和保持水土外，可以考虑将风电场作为景观资源来开发，做到施工建设与当地经济和谐发展。从大的区域看，采用景观生态的理念进行水土保持，可将风电场水土流失防治与生态旅游结合起来，形成一个新的生态增长点与经济增长点，实现区域生态经济的持续发展。目前，整个坝上地区，已将风电场的水土保持与该区的原生态草原、林场等景观有机结合起来，形成了一道“云的故乡、风电的广场、林草的海洋、动物的天堂、旅游者的梦想”生态景观。

3 风电场水土流失防治措施体系

根据各个水土流失分区的水土流失特点及各区发挥的功能，结合水土保持设计原则，设计风电场水土流失防治措施体系如图1所示，该体系包括工程措施、植物措施、临时措施。

3.1 工程措施

在升压站区、弃渣场区、风电机组区、施工生产生活区、运输道路区、集电线路区等各个水土流失区，首先采取表土剥离堆放措施收集表土，待工程结束后，清理现场，实施土地平整和覆土等土地整治措施，以便进行植被恢复或恢复原有植被。其次，根据各分区施工工艺及特点不同采取不同的工程措施设计。升压站区主要设防洪排水截水沟，以防止外部的雨水流入站区内；站内雨水采取下凹式绿地、蓄水池与沉沙池等雨洪资源利用工程，收集雨洪资源用于绿地、乔灌木树种的灌溉。道路区路基两侧设土质截排水沟，在沿道路走向的低洼处和填、挖方量较大的路段设急流槽，急流槽下方设消力池，消力池接排水沟，以便有利于雨水沿自然坡度排到道路两侧的沟道中，避免雨水对路基及边坡产生冲刷。弃渣场周边设置土质排水沟，以减少对周边的影响，沿弃渣场坡脚及边缘修建铅丝网笼坝，防止弃渣流失。

3.2 植物措施

风电场的水土保持植物措施主要包括植被防护工程、植物恢复工程及绿化美化工程等。

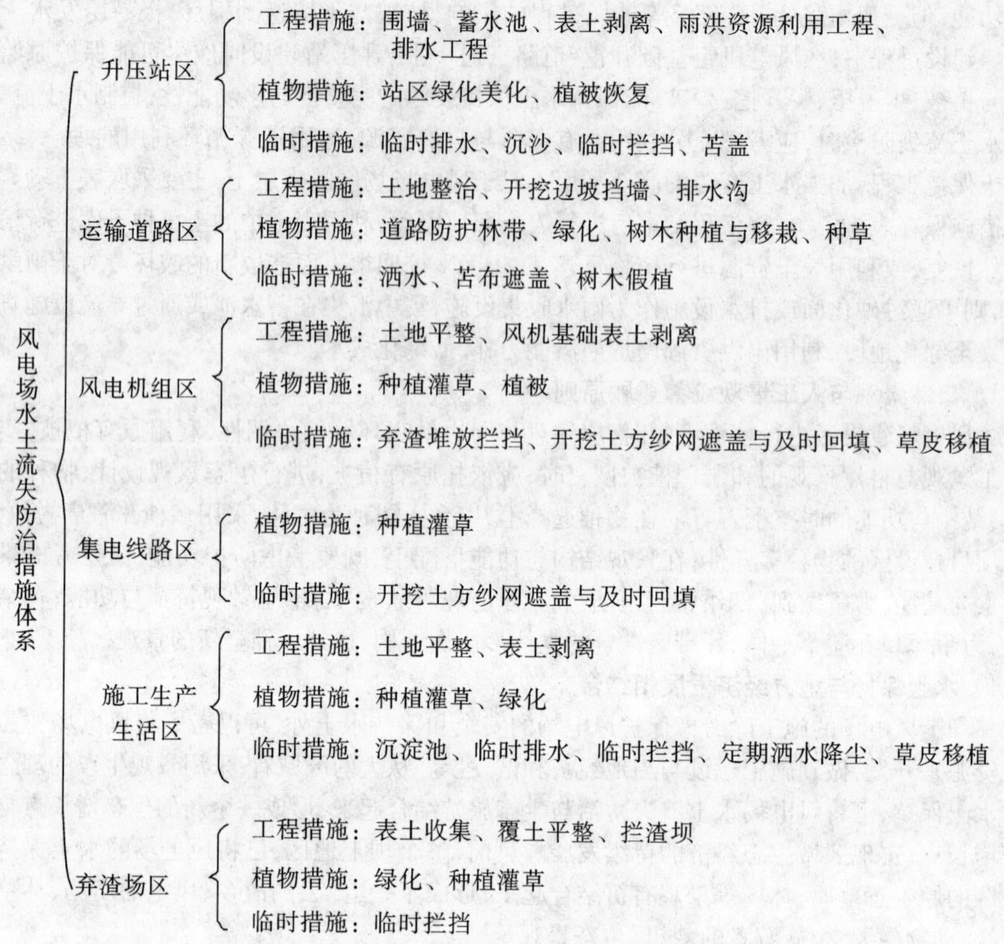

图1　风电场水土流失防治措施体系

对地形较缓或稳定边坡的运输道路,可采取封育管理恢复自然植被,并在路基两侧营造防护绿化林带,林带采用乔灌木结合的方式栽植。在地形起伏较大且开挖量大的路堑、路基边坡,可采用灌草结合的方式进行防护。对于升压站区,除硬化区域外的裸露地表外,均应绿化与美化,以便创建一个环境优美的管理区。绿化时,采用以灌为主,灌、草结合,适当配置乔木的原则,形成灌、草、乔一体,花、形、色兼备的防护体系,以营造出生动活泼的观赏空间及视觉效果。风电机组区、集电线路区、施工生产生活区、弃渣场区则以植草为主,适当增加灌木。结合区域立地特点,可供栽植的乔木树种有油松、白杆、落叶松、杨树、榆树、旱柳、白蜡等,灌木有沙棘、柠条、黄刺玫、榆叶梅、珍珠梅、沙地柏、丁香、玫瑰、枸杞、紫穗槐、胡枝子等,草本有披碱草、无芒雀麦、冰草、羊草、紫花苜蓿、沙打旺、草木樨等经济价值较高的牧草。

3.3　临时措施

考虑到风电场各分区工程特点及临时工程的短时效性,一般选择有效、简单易行、易于拆除且投资少的工程措施。在施工生产生活区修建排水沟、沉淀池。弃渣场、施工生产生活区弃土弃渣采用草袋装土筑坎临时拦挡。风电机组区的临时弃渣采取纱网遮盖和草袋装土拦挡措施;运输道路和施工生产生活区的水土流失防护一般采取排水沟和干砌石石坎或草袋装土拦挡措施。

综上所述,风电场的水土流失防治形成了一个以植物措施为主,工程措施为辅,临时措施为补充的综合体系,在该体系中,生物措施是最根本、效果最持久的一项措施,也是提高项目建设区植被覆盖率,提高景观价值的重要措施,更是恢复风电场建设区域退化生态系统的重要措施。

4 结语

(1)坝上地区风电场建设区水土流失表现出风蚀与水蚀兼有;点状与线状水土流失并存;植被破坏点多面广,植被恢复难度大;土石方量较小,弃渣相对少;水土流失重点在施工建设期等特点。这些水土流失特点与风电场建设项目的特征相吻合。

(2)风电场区的水土保持措施设计应遵循新原则与新理念,合理分区、综合防治,保护优先、措施可行,水土保持与经济发展相结合,生态景观与资源循环利用相结合。

(3)结合风电场的水土流失特征与水土保持设计原则,形成了生物措施为主,工程措施为辅,临时措施为补充的风电场的水土流失防治措施综合体系。

参考文献

[1] 江哲生. 我国电力发展的未来[J]. 发电设备,2006(1):1-5.
[2] 张祥军. 我国电力能源现状及其发展趋势[J]. 辽宁科技学院学报,2006,8(3):1-2.
[3] 海春兴. 河北坝上土地利用与土壤风蚀的动力学过程研究[D]. 北京:北京师范大学,2003.
[4] 胡立峰. 河北坝上土壤风蚀监测与防治对策研究[D]. 保定:河北农业大学,2003.
[5] 董智,贾志军,李红丽,等. 河北省坝上风电场建设区水土流失特点与植被恢复途径[J]. 中国水土保持科学,2009,7(5):82-86.

【作者简介】 贾志军,1974 年 1 月生,硕士,河北省水利技术试验推广中心科长,高级工程师。

沙漠明渠风积沙工程性质——湿陷性问题研究

李 江 王 旭

（新疆水利水电规划设计管理局 新疆 830000）

【摘要】 古尔班通古特沙漠风积沙是以细沙为主，含较多中沙和粉粒，厚度巨大的地质体。本文结合室内外试验研究成果，分析了沙漠风积沙的工程性质，认为沙漠风积沙具有典型的湿陷特征，是一种湿陷性土。在沙漠中修建水利工程应结合工程特点研究风积沙湿陷性所造成的问题。

【关键词】 沙漠风积沙 湿陷性 现场试验 水利工程

2007 年第 4 期《水利规划与设计》发表了闫宇、宋嶽两位学者撰写的“一种特殊性土——沙漠明渠风积沙工程性质研究”。文章从风积沙物质组成、接触性孔隙结构、压实等角度详细阐述了风积沙的工程性质，并得出了“沙漠风积沙是以细沙为主，含有较多中沙和粉粒，具有接触式孔隙结构，在静荷载条件下压缩变形量小，不易被压密，在振动荷载条件下，易于压密的特殊性松散土”的观点。

沙漠明渠全长 166.5 km，设计挖方 6 000 万 m^3，填方 4 043 万 m^3。明渠设计流量 55～49 m^3/s，渠深 4.1～3.8 m，纵坡 1/8 000～1/7 000。渠道内边坡 1∶2.5，开挖、填筑外边坡 1∶3，渠道断面自下而上依次布置 40 cm 厚砂砾石基层、3 cm 厚当地砂砂浆找平层、0.6 mm 厚一布一膜、砌板砂浆、6～7 cm 厚护面混凝土预制板。

风积沙的特殊性质决定了对其工程性质的划分及判断，也决定了针对风积沙的工程处理措施。研究表明，古尔班通古特沙漠风积沙属于湿陷性风积沙土。经室内外试验及计算分析，沙漠明渠及渠系建筑的湿陷变形较大，且渠道与渠系建筑物的湿陷变形相差不大，明渠北部与中、南部相比较小。所以，渠道全线均需采取一定的工程措施，在加强防渗的基础上，同时对挖方砂基进行加密、加固（表层压实及砂砾料换填），最终达到消除或减小湿陷变形的目的。

1 沙漠风积沙的基本特点

古尔班通古特沙漠区，除沙漠进口 2.2 km 的渠线属第四系更新统（Q_2^{al+pl}）冲洪积砂层外，其他均为上更新统-全新统风成砂层，室内外试验研究表明沙漠沙基本特点如下：

（1）沙漠风积沙，颜色为淡黄色。矿物成分以石英砂为主，颗粒比较均一，沙漠砂颗粒粒径组成北部粒径大于中部、中部略大于南部，粒径组成由北向南逐渐变细。地层垂直深度在 40 m 内，含水率及干密度变化不明显，含水率在 1% 左右，干密度在 1.58～1.62 g/cm^3，相对密度0.49～0.60,属中密状态。天然状态下不同地层砂的各项物理、力学性指标变化不大,基本相近。控制粒径 D_{60} 变化范围 0.25～0.12 mm,按《土工试验规程》(SL 237－1999),北部定义为级配不良砂(SP),中部、南部砂为含细粒土砂(SF)。沙漠风积沙不同位置颗粒粒径组成见表1,沙漠风积沙颗粒粒径曲线见图1。

表1　沙漠风积沙不同位置颗粒粒径组成

位置	桩号	含量(%)						有效粒径	不均匀系数	曲率系数
		>2.0	2～0.5	0.5～0.25	0.25～0.1	0.1～0.075	<0.075			
		砾	粗砂	中砂	细砂		粉粒	D_{10}	C_u	C_c
北部	30+568	0.5	12	22.5	48.5	15	1.5	0.084	2.56	0.85
	37+400		5	26	49	18.5	1.5	0.08	2.59	0.87
	42+600	0.5	21.5	23	29	24	2	0.076	3.82	0.55
中部	94+500			17.5	64	13.5	5	0.082	2.32	1
	94+600		3	15	66	9	7	0.084	2.02	1.18
	96+175			1.5	73	14	11.5	0.071	1.63	1.34
南部	151+625			3	70	9	18	0.052	2.15	1.54
	151+900			2.5	71	14.5	12	0.07	1.83	1.18
	152+100			1	58.5	26	14.5	0.067	1.79	1.1

注:位置划分:北部段0～49 km,中部段49～130 km,南部段130～166.5 km。

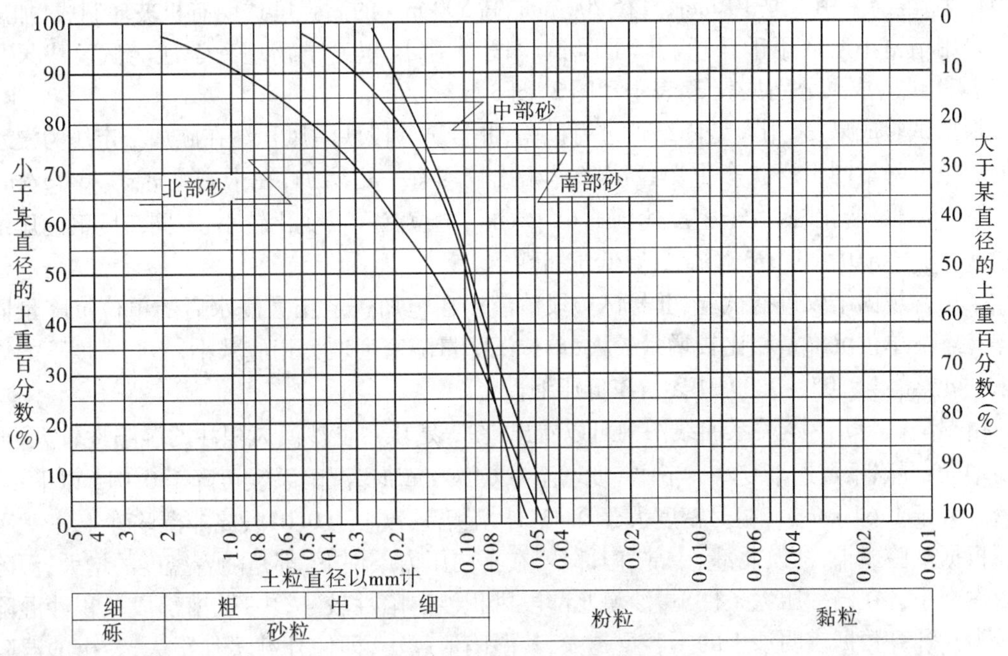

图1　沙漠沙颗粒粒径曲线图

(2)沙漠风积沙的含盐量低,易溶盐含量平均为0.55%。

(3)沙漠风积沙最优含水量一般为11.9%~13.6%,孔隙比为0.57~0.75,湿陷系数为0.012~0.048,抗剪强度$\varphi=32°\sim35°$,$C=9$ kPa,渗透系数$K=(4\sim9.0)\times10^{-3}$ cm/s。饱和状态下压缩系数$a_{v1-3}=0.05\sim0.09$ MPa^{-1},属低压缩性土。

(4)依据《土工试验规程》(SL 237—1999),相对密度试验适用于透水性良好的无黏性土,对细粒土含量较多的沙土,按美国水道试验站规定,应采用相对密度控制,按美国垦务局规定,应采用压实度控制。从实际施工试验碾压情况分析,以Ⅰ号试验段为代表,采用按相对密度的控制指标为1.79 g/cm^{3}大于采用按压实度控制的1.77 g/cm^3,偏安全,且沙土主要呈干燥状态,采用干压实是符合实际情况的,既能满足工程施工要求,又可以满足设计要求,故采用相对密度进行质量控制是合适的。

风积沙土具有湿陷性,在20世纪90年代出版的参考资料、设计手册、规程规范中已有相关规定,对风积沙土产生湿陷的原因、湿陷程度的判定、工程处理措施已有大量的研究成果可供借鉴。

风积沙土类同黄土的湿陷性,而沙土的湿陷性容易被忽视,沙漠中采用长距离明渠输水,塑膜防渗,渗漏在所难免,渗漏后沙漠沙的湿陷性对输水渠道的安全有无影响是需要研究的。作为工程地质勘察应结合设计进行风积沙土湿陷程度的研究及对工程的影响研究。

2 试验研究成果

2.1 室内试验成果

根据渠道引水方向由南至北分别针对北部(Ⅰ号试验段)、中部(Ⅱ号试验段)、南部(Ⅲ号试验段)的沙土进行了室内试验,试验仪器采用大型固结仪与常规固结仪进行了对比。大型固结仪主要由钢架、透明有机玻璃容器、加压设备、仪器、仪表等组成。钢架主要起支承反力作用,容器采用壁厚10 mm、直径200 mm、高500 mm的高强度透明有机玻璃制成,加压设备采用精密小型液压千斤顶。以Ⅱ号试验段为例,风积沙土物理力学性能试验成果见表2,湿陷试验成果分别见表3、表4和图2、图3。

根据试验成果,采用大型固结仪进行试验,Ⅱ号试验段沙漠风积沙在施加一定压力浸水后的单位沉降量同干燥状态下单位沉降量相比较均有一定增加,在干密度$r_d<1.64$ g/cm^{3}时,单位沉降量变化较大;干密度达到1.68 g/cm^3,单位沉降量变化较小,不同密度下的压缩系数$a_{v0.1-0.3}=0.01\sim0.075$ MPa^{-1},属低压缩性土。

采用常规固结仪进行试验,Ⅱ号试验段沙漠沙在施加一定压力浸水后的单位沉降量同干燥状态下单位沉降量相比较增量较大,单位沉降量随密度增大相应减小,不同密度下的压缩系数$a_{v0.1-0.3}=0.05\sim0.09$ MPa^{-1},属低压缩性土。

由湿陷试验$\rho_d\sim\delta_s$关系曲线图中可以看出:在一定压力下沙漠风积沙的湿陷性较大,随密度增大,湿陷系数逐渐变小,当干密度增至1.64 g/cm^{3}时,湿陷系数可降至0.015以下;当干密度大于1.64 g/cm^3,相对密度大于0.41时,湿陷系数小于0.015,属非湿陷性土。

根据试验成果,沙漠北部、中部、南部沙,在不同压力条件下,都是随着密度的增大,湿陷性依次减小。在同一压力条件下,沙漠北部段风积沙湿陷性较小,中部、南部段风积沙湿陷性较大。计算数据表明沙土的干容重越大,其湿陷值越小,即使外载增加很大,沙土的湿陷值增加也不大。

表2　沙漠渠线Ⅱ号试验段沙土物理力学性能试验成果

试样编号	取样深度	比重	天然干密度	抗剪强度 干燥状态 摩擦角	抗剪强度 干燥状态 凝聚力	抗剪强度 饱和状态 摩擦角	抗剪强度 饱和状态 凝聚力	压缩试验 饱和状态 压缩模量	压缩试验 饱和状态 压缩系数	湿陷系数	渗透系数	临界比降
		Δs	r_d	φ	C	φ	C	$E_{s_{0.1-0.3}}$	$a_{v_{0.1-0.3}}$	δ_s	K_{20}	i
	(m)	—	(g/cm³)	(°)	(kPa)	(°)	(kPa)	(MPa)	(MPa^{-1})		(cm/s)	
K94+600	2.0～3.0	2.69	1.61	33	5	31.5	5	25.28	0.065	0.012	8.0×10^{-3}	1.34
K94+600	5.0～7.0	2.69	1.6	33	10	31	5	23.67	0.07	0.015	1.47×10^{-2}	1.27
K95+225	1.0～2.0	2.69	1.59	31	15	29	5	19.7	0.085	0.012	1.07×10^{-2}	0.95
K95+225	7.0～9.0	2.69	1.61	35	3	31.5	3	22.28	0.075	0.016	8.62×10^{-3}	1.34
K96+175	1.0～2.0	2.69	1.59	32	5	29.5	15	21.15	0.08	0.014	9.7×10^{-3}	0.95
K96+175	5.0～7.0	2.69	1.57	30	10	28	8	19.04	0.09	0.015	1.17×10^{2}	0.89
K96+175	9.0～12.0	2.69	1.59	33	10	27.5	20	21.15	0.08	0.015	1.11×10^{-2}	0.95
K96+175	15.0～18.0	2.69	1.57	30	5	28	5	20.16	0.085	0.016	1.19×10^{-2}	0.89
K96+175	21.0～24.0	2.69	1.61	32.5	20	30	10	21.4	0.075	0.014	7.78×10^{-3}	1.34
K96+375	1.0～2.0	2.69	1.59	31	5	29	5	25.16	0.07	0.015	9.6×10^{-3}	0.95
K96+375	3.0～5.0	2.69	1.6	33	10	32	10	25.86	0.065	0.014	9.2×10^{-3}	1.27

表3　不同部位试验段压缩试验成果(大型固结仪)

试验段编号	干密度	不同压力下湿陷系数(δs) 0	100 kPa	200 kPa	300 kPa	400 kPa	不同压力下单位沉降量(mm/m) 状态	50 kPa	100 kPa	200 kPa	300 kPa	400 kPa	压缩试验 试验状态	压缩模量 $E_{s_{0.1-0.3}}$	压缩系数 $a_{v_{0.1-0.3}}$
	(g/cm³)													(MPa)	(MPa^{-1})
Ⅱ号试验段	1.56	0.016	0.027	0.027	0.027	0.026	干燥		8.21	12.51	15	16.72	干燥	29.46	0.06
							饱和		31.88	36.49	36.61	39.88	饱和	29.62	0.06
	1.60	0.006	0.018	0.018	0.018	0.018	干燥		2.88	5.53	7.63	9.28	干燥	42.1	0.04
							饱和		17.56	20.95	23.02	24.46	饱和	36.63	0.045
	1.64	0.004	0.011	0.011	0.012	0.012	干燥		2.09	4.7	6.58	7.93	干燥	44.5	0.035
							饱和		12.77	16.1	18.14	19.71	饱和	37.25	0.045
	1.68	0.003	0.002	0.002	0.004	0.005	干燥		0.79	1.81	2.93	3.64	干燥	93.46	0.01
							饱和		2.28	4.23	6.67	8.31	饱和	45.52	0.035

表4 不同部位试验段压缩试验成果(常规固结仪)

试验段编号	干密度 (g/cm³)	不同压力下湿陷系数(δ_s)					不同压力下单位沉降量(mm/m)						压缩试验		
		0 kPa	100 kPa	200 kPa	300 kPa	400 kPa	状态	50 kPa	100 kPa	200 kPa	300 kPa	400 kPa	试验状态	压缩模量 $E_{s0.1-0.3}$ (MPa)	压缩系数 $a_{v0.1-0.3}$ (MPa^{-1})
Ⅱ号试验段	1.56	0.027	0.022	0.023	–	–	干燥	7.8	13.2	17.3	20.4	23	干燥	28.73	0.06
							饱和	29.6	33.23	38.8	41.34	43.88	饱和	24.63	0.07
	1.58	0.023	0.019	0.02	–	–	干燥	5.65	10	14.9	19.75	19.75	干燥	26.18	0.065
							饱和	25.37	30.65	36.81	38.95	41.39	饱和	24.31	0.07
	1.6	0.02	0.016	0.017	–	–	干燥	5	9.75	14.25	16.75	19	干燥	28.02	0.06
							饱和	23.88	28.36	33.48	36.81	38.31	饱和	24.01	0.07
	1.62	0.017	0.015	0.015	–	–	干燥	4.8	9.3	14.05	16.25	18.3	干燥	30.18	0.055
							饱和	22.98	27	32.65	35.57	37.26	饱和	23.71	0.07
	1.64	0.014	0.014	0.013	–	–	干燥	4.6	9	14.03	15.67	17.31	干燥	29.82	0.055
							饱和	20.75	25.4	31.65	35	36.3	饱和	21.87	0.075
	1.68	0.012	0.012	0.01	–	–	干燥	4.4	8.56	12.69	15.02	17.16	干燥	32.03	0.05
							饱和	20.15	24.1	30	33.35	35.5	饱和	22.87	0.05

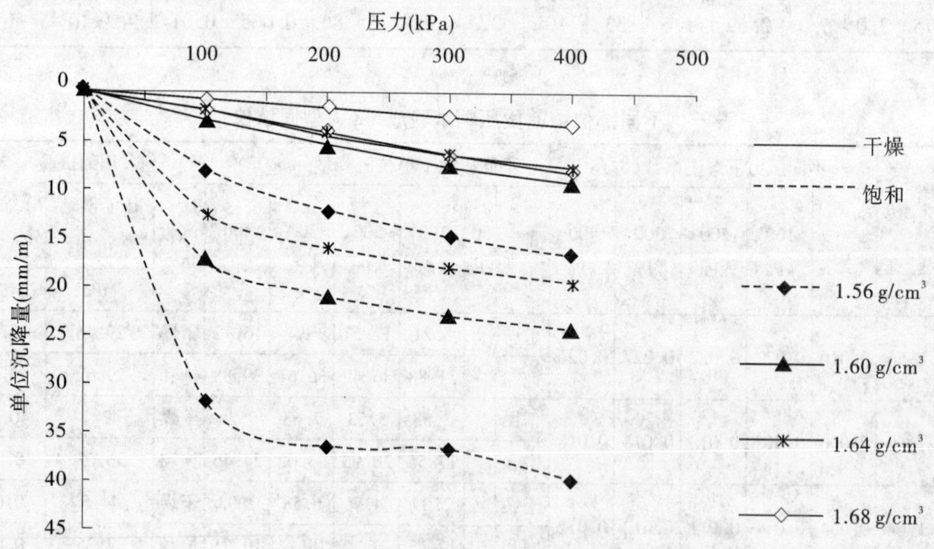

图2 Ⅱ号试验段不同密度下 P-S_i 关系曲线(大型固结仪)

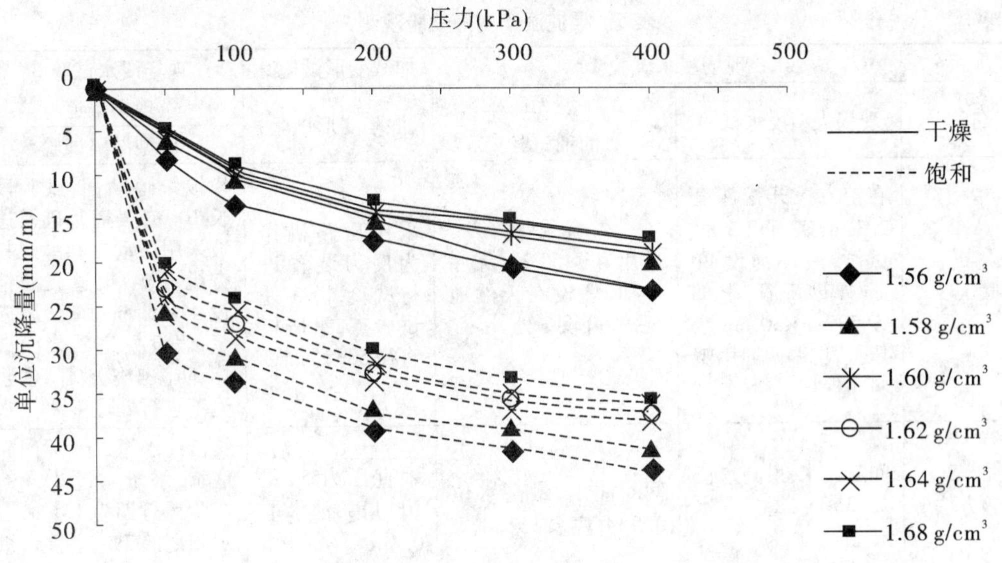

图3 Ⅱ号试验段不同密度下 $P \sim S_i$ 关系曲线(常规固结仪)

2.2 室外试验成果

天然状态沙土浸水后沉降变化值仅是室内试验,天然沙层分布广、厚度大,因受浸湿深度影响,应在现场进行沉降试验及载荷试验,在保持一定水头并有充足浸泡时间下进行,研究深层地基的沉降变化情况、天然地基压缩性及湿陷性,以对室内固结、湿陷试验的沉降值进一步验证,确定渠道的渠堤裂缝、边坡塌陷等建筑物的稳定性。

为获取渠道在天然状态下及运行状态下的湿陷变形情况,采用三种手段进行试验:①试坑浸水试验,采用大型试坑及无底油桶进行,其目的是确定实测自重湿陷量;②借助压路机作为反力装置进行天然砂层的现场压缩性试验;③模拟渠道设计断面,进行不同水位情况下的渠底、边坡湿陷变形观测。各试验方式及结论见表5。

上述试验中室外采用设计渠道断面模拟受水浸泡后渠道的湿陷变形,在挖方段渠道进行试验。试验时假设防渗层基本损坏,模拟水头维持3 m,经1个月注水试验表明,渠道底板最大湿陷位移4.22 mm,该变形对整个渠道底板安全运行不存在威胁;边坡最大湿陷位移10.78 mm,比底板稍大,基本不构成对渠道的危害。但渠坡的湿陷变形规律与渠底相同,也即湿陷变形与沙层的含水量大小有关,渠内运行充水初期,湿陷变形速率较大,因此要控制好渠道边坡碾压密实度。

由此看来,沙漠沙土在天然状态下按相对紧密度划分属于中密状,而通过碾压后,干密度值有一定提高,碾压后的沙土处于中密或致密状态,所以碾压后的沙土压缩性及湿陷性均较小。

3 消除风积沙土湿限性措施研究

沙漠沙按风积成因,应为次固结性沙土,由于地层固结时规律相当复杂,它不仅取决于沙的类别和状态,也随着边界条件、沙的骨架结构、密实度、饱和度、颗粒间凝聚力的大小、沙粒形状大小、排水条件和受荷方式等多因素影响,与固结均有一定关系。因该沙土形成是在高寒干旱地区,又长期处于非饱和及欠压密状态下,受荷载及浸水作用产生湿陷,应考虑湿陷性对工程的影响。

表5　渠道湿陷变形试验观测

项目	现场浸水试验1		加载变形试验2	现场浸水试验3
名称	现场大型浸水试验	小型无底油桶试验	湿陷变形试验	现场渠道原型湿陷试验
试验手段	在挖方段开挖深度2.3 m,底面积3 m^2试坑,底部铺设20 cm厚砂砾石,上部设50 cm×50 cm承压板,坑壁四周用塑料布保护	采用直径50 cm无底油桶,埋设在天然地层内,分设3处	采用压路机作为反力装置进行试验,设3处试验点,承压板直径50.5 cm	模拟渠道断面,试验段长10 m,渠道基础及边坡采用振动碾和蛙式打夯机夯实,渠道表面铺设4～5 cm厚水泥砂浆,假设渠道砂浆面破坏、防渗塑膜破坏,砂浆缝不进行处理
试验方法	间断注水,最高水位1.3 m,总注水51 m^3	间断注水,最高水位0.5 m,注水1.2 m^3	垂直压力50～100 kPa,注水量0.8 m^3	缓慢注水至3 m,后间断注水,水量超过2 m的试验运行期为131 h,占观测期的7%,总注水量1 506 m^3
观测仪器	水准仪、百分表	水准仪、百分表	百分表	水管式沉降仪,底板3支、边坡2支
时间	7 d	7 d	7 d	1个月
试验分析	注水1 d后,沉降量基本稳定,5 d后终止试验,最终沉降量为8 mm。实际渗透系数在8×10^{-3}～2×10^{-2},沙层不易达到饱和状态,试验结束后3 d取样观察,影响深度约6 m,并呈45°扇形分布	总沉降量为0.68～1.2 mm,浸湿影响深度2.4 m	垂直压力50 kPa,3组沉降量为0.71～0.73 mm,压缩沉降量1.7～2.4 mm;垂直压力100 kPa时,2组沉降量为0.9～0.95 mm,压缩沉降量2.1～3.25 mm	渠道渗水量与渠内水位呈正相关关系,水位达到一定范围后,下渗速率上升值明显降低,渠道将存在一个经济运行水位;湿陷变形随水位的升高而增加,但水位相对稳定时,湿陷变形将保持相对稳定
结论	天然沙层在无荷载时浸水后沉降量变化不大,沉降量的变化是随试验面积的大小和注入水量浸湿深度而变化的,渗水量大,浸湿深度深,沉降量即增大。总的沉降量较小,因湿陷问题对渠道稳定性不会产生较大影响		天然沙层在一定压力下,压缩量及湿陷变形量变化不大;沙土属低压缩性、弱湿陷性土,渠道漏水因湿陷问题不会对渠道产生很大影响	渠道基础湿陷主要受沙层含水率的影响大,水位影响较小,实测最大变形量4.22 mm;边坡最大湿陷位移10.78 mm,运行初期,湿陷变形速率相对较大,故边坡应控制好干密度
	注:该试验是在无侧限限制条件下所作			注:受试验季节及水量限制,未能进行长期高水头运行试验观测

湿陷性土地基受水浸湿至下沉稳定的总湿陷量Δs,应按下式计算:

$$\Delta s = \sum_{i=1}^{n} \beta \Delta F_{si} h_i$$

式中：ΔF_{si}为第 i 层土浸水载荷试验的附加湿陷量，cm；h_i 为第 i 层土的厚度，cm，从基础底面（初步勘察时自地面下 1.5 m）算起，$\Delta F_{si}/b<0.023$ 的不计入；β 为修正系数，cm^{-1}，当承压板面积为 0.50 m^2 时，$\beta=0.014$，当承压板面积为 0.25 m^2 时，$\beta=0.014$。

计算分析表明，12 m 高填方渠道在填料达到相对密度 0.7 时，总湿陷量为 8～12.8 cm，若干密度增加、相对密度提高，则湿陷量减小，即使在外荷载作用下，湿陷值增加也不大。

根据试验研究及结构分析，沙漠明渠结构设计是安全可行的，即使在渠道防渗层被破坏后，沙层湿陷变形对渠道影响也不是很大，但仍应注意处理好渠道建基面松散沙基，尽可能提高其干密度，以减少湿陷变形量。

针对填方渠道，采用相对密度 0.75 作为控制指标，重型振动碾分层碾压密实。经处理后，填方控制指标较高，湿陷量较小，基本不会对渠道构成威胁。

针对挖方渠道及渠道内边坡松散渠床，挖方段渠堤干密度较低，也不可能对深层沙土进行处理。采用在渠道内边坡垂直渠床厚度换填 40 cm 天然砂砾石垫层，并用水平土沙平起方式同步碾压，加固松散渠床。预制板下铺设 0.6 mm 厚一布一膜加强防渗。填方渠道内边坡处理措施采用与挖方段相同的工艺。

4 结语

沙漠风积沙在浸水后的单位沉降量明显大于天然状态下的单位沉降量，随着密度的增大，沉降量逐渐减少。天然状态的沙漠风积沙属低压缩性土。沙漠风积沙在不同密度下的自重湿陷系数均较小，但在一定压力下的湿陷系数较大，湿陷系数随密度增大而变小，当干密度达到 1.64 g/cm^3时，湿陷系数可降至 0.015 以下，即可消除湿陷性。

针对沙土湿陷，设计采取的提高渠床压实干密度，加强防渗的综合处理措施是合适的，这也是保证输水工程设施安全，正常运行，避免工程地质问题发生的关键。

参考文献

[1] 闫宇，宋嶽．一种特殊性土——沙漠明渠风积沙工程性质研究[J]．水利规划与设计，2007(4)：36-39.

【作者简介】 李江，男，1971 年 1 月生，1994 年毕业于新疆农业大学，新疆水利水电规划设计管理局副局长，提高待遇高级工程师。

水平旋流消能泄洪洞的研究与应用

周 恒 王卫国

（中国水电顾问集团西北勘测设计研究院 西安 710065）

【摘要】 本文是对公伯峡导流洞改建为水平旋流消能泄洪洞研究与应用成果的总结，重点介绍了水平旋流消能泄洪洞的布置、体型设计、水力条件、泄洪运行试验等主要研究论证成果。公伯峡水平旋流消能泄洪洞较好地适应了工程复杂的地形地质条件，具有布置灵活、结构简单、掺气减蚀条件好、消能率高、施工方便等特点。泄洪运行试验表明，公伯峡水平旋流消能泄洪洞的研究与应用是成功的，具有较好的推广价值。

【关键词】 导流洞改建 水平旋流消能泄洪洞

随着我国水电开发的不断发展以及西部高山峡谷地区一大批高坝大库工程的即将建设，泄洪消能建筑物设计已经成为许多工程的关键技术问题，泄洪消能建筑物的布置及型式选择论证受枢纽布置及地形地质条件、消能区抗冲刷能力及泄洪雾化对高边坡的影响、泄洪运行条件等诸多因素的制约，有的工程采用常规的泄洪消能建筑物布置及型式已不能满足工程技术性、安全性及经济性的要求，因此需要研究采用新型的泄洪消能建筑物。

另外，高土石坝施工期间一般采用隧洞导流，而大断面的导流隧洞造价往往比较高，如何合理利用导流洞后期改建为永久泄水建筑物，达到一洞多用的目的，是解决枢纽布置困难及降低工程造价的一个发展方向。大部分工程导流洞改建成“龙抬头”型式泄洪洞，但高水头泄洪洞洞内流速尤其是“龙抬头”段反弧末端流速很大，出口下游挑流消能冲刷和雾化严重，增加了洞内破坏、高边坡失稳等风险。

黄河公伯峡右岸泄洪洞原设计是利用导流洞改建为“龙抬头”泄洪洞，鉴于地质条件复杂，导流洞施工过程中发生多次塌方，为适应地质条件、减小施工风险、充分采用新技术，进行了导流洞改建为“龙抬头”泄洪洞、竖井旋流泄洪洞、水平旋流泄洪洞的方案比选；在选择采用水平旋流泄洪洞方案的基础上，又进行了水力学模型试验、掺气减蚀模型试验、结构数值计算分析及泄洪运行试验等大量科研论证工作。本文总结了公伯峡水平旋流消能泄洪洞的布置、体型设计、水力条件等重点设计问题，以期对这种新型消能泄洪建筑物的设计有所裨益。

1 泄洪洞布置

公伯峡水平旋流消能泄洪洞由溢流堰、竖井、起旋室、通气井、旋流洞段、水垫塘及退水洞等组成（泄洪洞布置见图1）。

泄洪洞进水口采圆弧型实用堰，溢流堰轴线与旋流洞轴线夹角45°。堰顶布置12.0 m×

16 m 的平板检修闸门和平板工作闸门各 1 扇。溢流堰面下游与直径 9 m 的圆形竖井连接，竖井中心线与旋流洞中心线错距 3.5 m。在竖井中上部设置坎高 0.8 m 的环形掺气坎，坎下设 5 个直径 0.63 m 的通气管通气，以防止竖井下部位压力较低的空蚀敏感部位发生气蚀破坏。竖井下部断面从直径 9.0 m 圆形断面渐变为矩形收缩断面，并通过椭圆曲线与起旋室圆弧曲线偏心相切，强迫产生旋转水流。为保证旋流洞内形成稳定的旋转水流并充分掺气，在起旋室上游端设直径 3.3 m 的通气井。旋流洞段长 45.5 m，由原导流洞城门形断面改建为直径 10.5 m 的圆形断面。水垫塘由原导流洞衬砌改建为 11.0 m×14.0 m 城门形断面，长 65.5 m，水垫塘末端通过三向收缩尾坎与退水洞连接。退水洞利用原导流洞，出口采用挑流鼻坎使水流归槽。

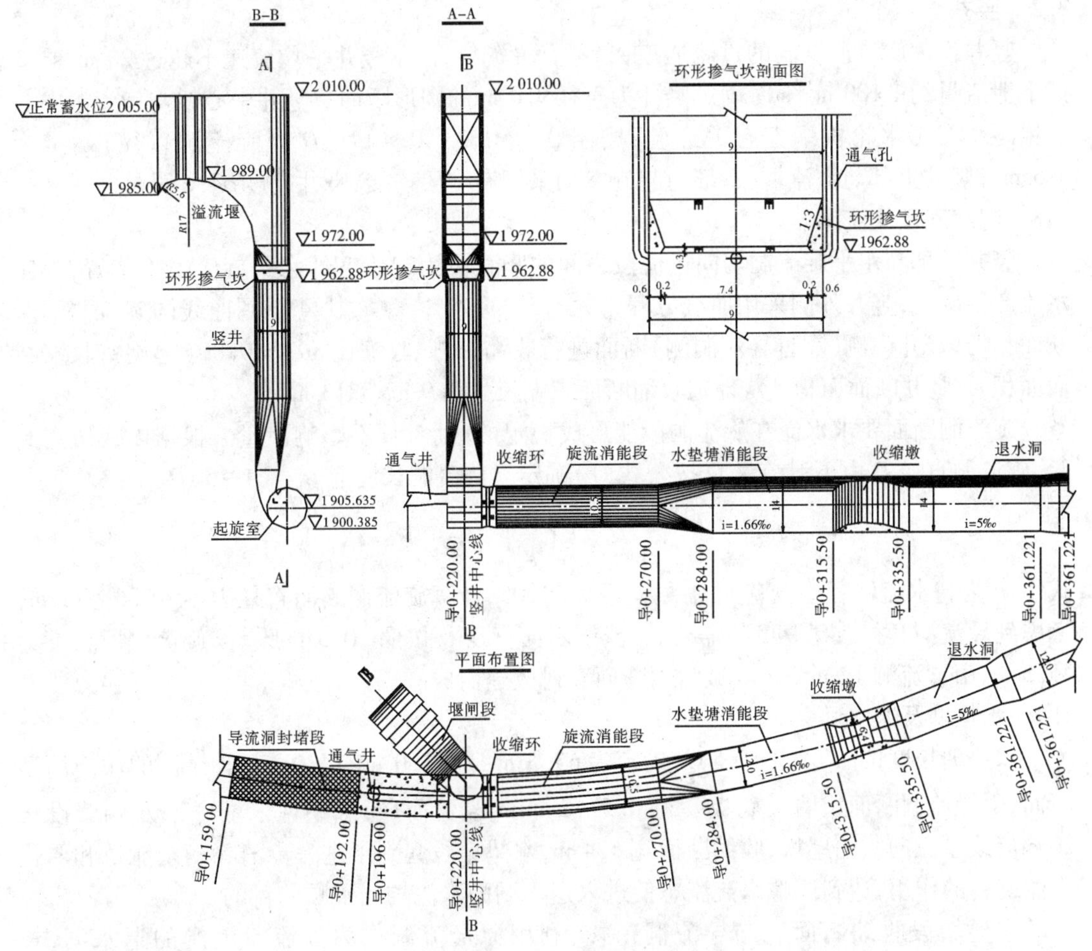

图 1　公伯峡水平旋流消能泄洪洞布置图

2　体型设计

2.1　溢流堰

溢流堰体型设计进行了堰后自由出流和淹没出流的方案比较。自由堰流孔口尺寸 9 m ×15 m，堰后水流以自由流形式进入竖井，水面跌落大，水舌前缘的冲击水跃顶冲竖井井壁，水舌两侧立轴涡带旋滚强烈，竖井水面波动大、井壁振动强烈。由于溢流堰基础为古全风化及强风化岩体,地质条件较差,自由堰流竖井上部水流流态较差,竖井振动强烈,结构安全风险大。

为改善堰后水流流态,提高结构安全性,加大溢流堰过流断面,采用堰后淹没出流。溢流堰闸孔面积ω_1与竖井断面面积的比值A按照3.0控制,进口宽度B与竖井直径D_1的比值按照1.3~1.4控制,即取$B=(1.3\sim1.4)D_1$,$\omega_1/A=3.0$。溢流堰进口宽度从9 m扩大到12 m,孔口尺寸调整为12 m×16 m,门槽下游两侧墙按照1:12的收缩角与竖井直径9 m的圆形过流断面相切连接。旋流洞底板以上竖井总水头108.25 m。淹没堰流竖井上部水面跌落小,水位基本与库水位齐平,水面平稳、无波动和涌浪,井壁振动很小,很好地适应了较差的地质条件。

2.2 竖井

竖井作为旋流泄洪洞的过流通道,参考有关资料,控制竖井平均流速不超过20 m/s,按照下泄流量约1 100 m^3/s时确定竖井为直径9.0 m的圆形断面。竖井中心线与旋流洞中心线错距Δ参考水轮机蜗壳有关经验,按照$\Delta=(0.2\sim0.4)D$(D为旋流洞直径)确定为3.5 m。竖井上部与溢流堰、下部与起旋室均采用渐变段连接,竖井高度64.6 m。

2.3 起旋室及旋流洞

竖井下部与水平旋流洞切向进流段之间的竖直段为起旋室即水平旋流发生装置,为使水流产生旋转,竖井外侧采用曲线方程为$x^2/15.2^2+y^2/8^2=1$的1/4椭圆曲线同旋流洞圆形断面相切,竖井内侧(靠近导流洞侧)断面进行收缩,按照收缩比$\omega/A=0.80$(起旋室收缩断面面积ω/竖井断面面积A),确定收缩断面末端尺寸为9 m×5.13 m。

旋流洞断面要求水流在旋流洞段能形成稳定的空腔,且不影响泄量。根据以往研究经验,旋流洞直径D由下泄流量Q按公式(1)确定。旋流洞长度取$L_1=(4\sim5)D$。

$$D=k\left(\frac{Q^2}{g}\right)^{0.2} \tag{1}$$

式中:k为与水流佛劳德数有关的系数,取$k=1.05$,计算旋流洞断面直径D为10.6~11 m,考虑到导流洞套衬改建厚度的需要,旋流洞断面直径D按照10.5 m设计,旋流洞段长度为45.5 m,由导流洞12 m×15 m城门洞断面套衬改建而成。

2.4 水垫塘及退水洞

水垫塘长度按照$L_2=(4\sim5)D$确定为65.5 m,为11.0 m×14.0 m(宽×高)的城门洞型断面,由导流洞断面边墙及底板套衬0.5~1.5 m改建而成,水垫塘末端底部及侧墙设长20 m流线型三向收缩尾坎,收缩断面宽6.4 m,水垫塘段起到了进一步消除剩余能量和弱化水流旋转的作用,达到调整水流流态使进入退水洞的水流均匀平顺的目的。

水垫塘段收缩断面后与导流洞衔接,利用原导流洞作为旋流泄洪洞的退水洞,长794.5 m,末端接挑流鼻坎。

3 水力条件

水平旋流消能泄洪洞是一种新型的泄洪消能建筑物,各部位水流流态变化大、水力条件较复杂。

进口水流以淹没堰流进入竖井,堰后及竖井上部水面跌落小,水面平稳,竖井内形成稳定的压力管流;竖井水流通过起旋室的导向作用,使水流在水平洞内作高速旋转运动,离心力的作用提高了边壁压力,水流与边壁的摩阻增加,流程增长,以达到增大水流的沿程损失,同时通过通气井向旋流空腔大量掺气,以维持稳定的空腔旋转流态、达到削减水流能量的目的;旋转水流在进入水垫塘后,由于旋转及突扩作用使水流和气流发生强烈的紊动和掺混作用,形成气水两相流流态;气水两相流出水垫塘后经过再次紊动扩散,掺气均匀。在退水洞内由于自由水面的存在,空气逐渐逸出,形成稳定的明流流态。

水平旋流消能泄洪洞的水力条件论证工作,重点研究论证了泄流能力、掺气减蚀、脉动压力以及消能率等水力学问题。

3.1 泄流能力

泄洪洞设计洪水位 2 005 m 和校核洪水位 2 008 m 工况泄流量分别为 1 032 m^3/s 和 1 060 m^3/s,泄洪运行最大水头差 105 m。

泄洪洞泄流能力问题的研究是要解决堰式进水口、竖井环形掺气坎断面、起旋室末端收缩断面以及旋流洞等部位泄流能力的协调匹配问题。虽然各个部位的泄流能力计算都有据可依,但其相互协调匹配问题需要通过水工模型试验验证。

通过水工模型试验论证,死水位 2 002 m 为溢流堰后自由流与淹没流的分界水位:当库水位低于 2 002 m 自由溢流情况时,泄洪洞泄流量由溢流堰控制;在 2 002 m 水位以上淹没流时泄流量由环形掺气坎断面和起旋室末端收缩断面控制。在环形掺气坎收缩断面直径 7.4 m、收缩坡度为 1∶3 的条件下,孔流流量系数为 0.854。起旋室收缩断面面积大于或等于环形收缩坎过流面积(即 $\omega/A_1 \geqslant 1$)时,环形掺气坎下周边才能形成掺气空腔和满足泄流量的要求 。

3.2 掺气减蚀

泄洪洞的掺气减蚀研究工作重点进行了旋流洞内旋转空腔掺气、起旋室与旋流洞连接段低压空蚀敏感区的掺气减蚀问题。

3.2.1 *旋流洞内旋转空腔掺气*

旋流洞段水流流速大(起旋室底板最大流速 33 m/s),起旋室体型复杂,为确保旋流洞形成稳定的空腔旋转流态、达到消能及减蚀的目的,最有效的办法是向起旋室及旋流洞内水流旋转空腔充分通气。为此,在旋流洞的上游设置通气井,使得空气进入起旋室和旋流洞,以便在旋流洞中心形成稳定的空腔。由于旋流洞内旋转水流受重力的作用,通气井应设在旋流洞中心偏上的位置。当通气井位置偏低或直径较小时,旋流洞内空腔减小甚至消失,旋转流可能遭到破坏,所以通气孔的合理设置是非常重要的。试验表明,通气井通气量达到 200 m^3/s 时就能维持旋流洞内形成稳定的空腔,在设计和校核工况通气井通气量分别为 258 m^3/s 和 287 m^3/s,通气量与泄流量的比值 $\beta=0.25\sim0.27$。通气井的中心位置布置在旋流洞中心上部 2.1 m 处,通气井采用圆形断面,直径采用 3.3 m,相应通气井内最大风速 34 m/s,通气井出口直径 2.5 m,断面处最大风速 58.5 m/s。

3.2.2 *旋流洞内低压空蚀敏感区体型优化及掺气*

起旋室与旋流洞的连接段处由于水流的强烈剪切作用,出现负压区,该部位水流空化数 $\sigma=0.25$,而导流升坎处的初生空化数约为 2,该区域是空蚀的敏感区。为防止该部位发生空蚀破坏,研究工作按照两种思路进行:首先以优化连接段体型提高敏感区的压力为主要手段,其次论证向低压区掺气进行减蚀的可能性。

起旋室与旋流洞的连接段体型优化主要进行了上游侧设盲洞、加设折流垫、设置导流升坎、导流升坎局部开口和削坡、升坎后加设收缩段等措施论证。最终采用体型见图2,即在连接段设置导流升坎,升坎采用开口和削坡的体型,并在升坎后旋流段进口处设一段长3.8 m、高0.3 m的收缩环,收缩环上游坡比1∶4,下游坡比1∶6。通过起旋室与旋流洞的连接段体型优化提高了起旋室及升坎部位的压力,升坎后边壁负压值仅为-0.6 m。

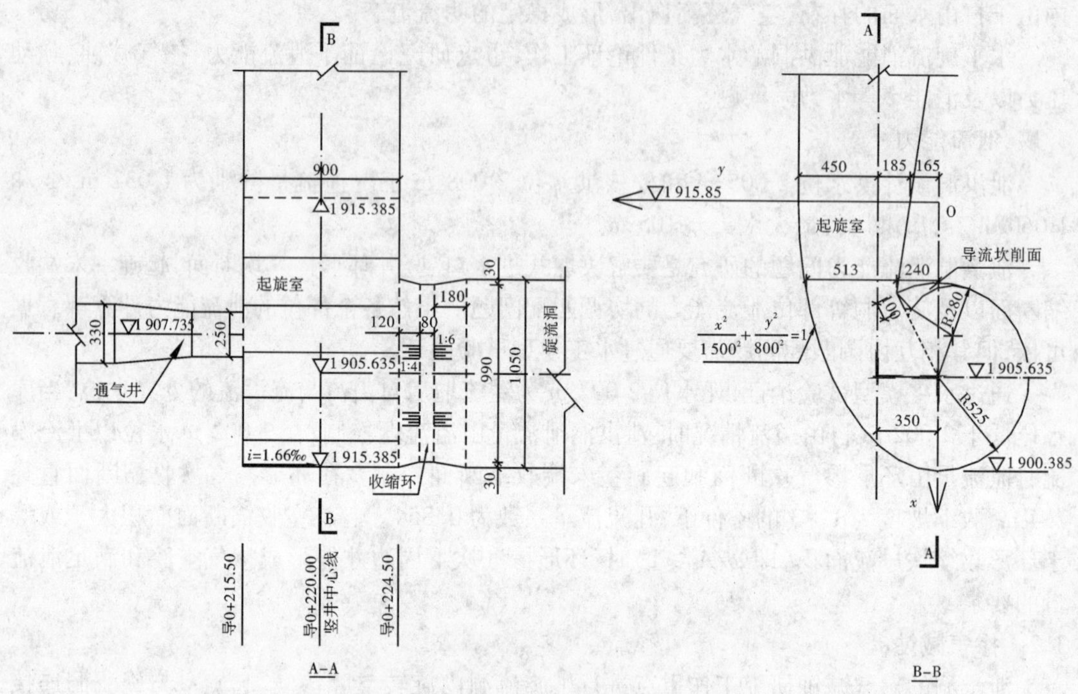

图2　起旋室与旋流洞的连接段体型图

对起旋室与旋流洞连接段处低压空蚀敏感区掺气问题经过试验反复论证,在过流竖井中上部设环形通气坎(见图1)可起到向竖井下部水流掺气,增加起旋室低压区掺气浓度的作用。设计中在竖井1 962.9 m高程设置坎高0.8 m、坡比1∶3的环形通气坎,环形掺气坎收缩断面直径7.4 m,并通过5个直径为0.63 m的通至坝顶的通气管进行通气。通气坎下单孔通气量在设计和校核工况分别为20.9 m^3/s和19.5 m^3/s,空腔长度分别为为5.0 m和5.8 m。

3.2.3　沿程掺气浓度分布及空化特性

起旋室连接段壁面低压区模型掺气浓度在设计和校核工况分别为3.1%和3%,考虑模型比尺及空气密度差异,换算到原型掺气浓度约为4%。水垫塘进出口边墙底部掺气浓度设计工况为2.1%、4.8%,校核工况2.4%、5.1%,换算到原型约3%和6%。从泄洪洞沿程掺气浓度的分布情况看,基本满足可以掺气减蚀的要求。

减压水工模型试验采用突体试验判断各部位的空化特性。试验表明:全洞各部位在突体高度为12 mm时,基本无空化现象产生;仅水垫塘出口段侧墙中上部在突体高度为12 mm时出现弱的空化云,在突体高度为7 mm时无空化现象产生。根据突体试验成果及掺气情况,表面不平整度要求控制在6~12 mm以内即满足安全运行要求。严格按照此要求来控制混凝土施工的不平制度。

3.3 脉动压力及结构振动

水工模型试验表明,竖井段最大脉动压力均方根值为53.1 kPa,起旋室段脉动压力均方根值小于25.0 kPa。泄洪运行试验监测表明,竖井段各测点最大脉动压力均方根值为45.3 kPa,起旋室段各测点脉动压力均方根值不超过26 kPa。泄洪洞洞壁各测点脉动压力优势频率不超过1 Hz,泄洪洞内水流为大尺度水流脉动。原型与模型脉动压力均方根的大小结果基本一致,泄洪洞过水前后,岩石开合度、缝展度和混凝土的应变没有明显的变化,即泄洪洞过水运行脉动压力对混凝土结构振动影响小,建筑物的结构设计是安全的。

3.4 消能率

公伯峡水平旋流消能泄洪洞进水口到各计算断面的分段消能率见表1。泄洪洞总的消能率达86%以上(从起旋室到水垫塘出口约占总消能率的77%),退水洞断面平均流速在15 m/s以下,出口鼻坎归槽流速在旋流泄洪洞单独泄洪运行时为10.1 m/s。

表1　水平旋流消能泄洪洞各段消能率汇总表

部位	消能率(%)	
	校核工况	设计工况
进口段(环形通气坎以上)	5.1	4.7
竖井段(环形通气坎至起旋室)	14.4	15.5
起旋室、旋流洞段和水垫塘(含100 m退水洞)	67.2	65.9
进口至水垫塘后100 m	86.7	86.1

泄洪运行试验成果表明,泄洪洞体型设计合理、进出口流态平稳、旋流空腔稳定、消能率高。洞内空化水流噪声较弱,未有明显的不良负压和空蚀迹象出现。通气设施通气良好、水流掺气充分,水流掺气对过流面起到有效保护作用。试验中无不良结构振动及动力响应,泄洪洞运行是安全的,泄洪洞的设计和研究取得了成功。

4 结语

公伯峡水电站水平旋流消能泄洪洞通过大量的设计论证和研究探索,在渐缩式竖井淹没进水口、起旋室、水平旋流洞、水垫塘末端流线型收缩墩、旋流洞掺气及竖井压力流段内布置环形掺气坎等体型设计和水力学研究方面取得了一定的创新性成果,发展了泄洪消能建筑物型式。

水平旋流消能泄洪洞作为一种内消能工,具有良好的水力特性和消能效果,对枢纽布置及地形地质条件适应性强,泄洪建筑物出口水流雾化、下游河道冲淤变形和岸坡稳定问题容易解决,是一种环境友好型泄洪消能建筑物;为提高工程枢纽布置灵活性、增加高水头泄洪建筑物可靠性、减小施工风险和降低工程造价提供了一个成功范例,也是利用导流洞改建为泄洪洞的一种较好方式。公伯峡水平旋流消能技术研究应用和建成投运,为推动新型泄洪消能建筑物的发展提供了工程实例,积累了实践经验,具有显著的经济效益和较大的推广应用价值。

由于水平旋流消能泄洪洞为新型建筑物,几乎没有可借鉴的实际工程经验,在设计与研究方面都处在尝试阶段,研究成果又是结合公伯峡具体条件而进行的,因此在消能机理、理论依据等方面还有待进一步完善和提高。

参考文献

[1] 周恒,王卫国,等.黄河公伯峡水电站利用导流洞改建为旋流消能泄洪洞专题研究报告[R].西安:中国水电顾问集团西北勘测设计研究院,2002.

[2] 周恒,王卫国,等.黄河公伯峡水电站右岸旋流泄洪洞选型专题报告[R].西安:中国水电顾问集团西北勘测设计研究院,2002.

[3] 董兴林,等.公伯峡水电站导流洞改建竖井——水平旋流洞综合试验研究报告[R].北京:中国水利水电科学研究院,2003.

[4] 卫勇,等.公伯峡水电站导流洞改建为水平旋流洞试验研究报告[R].西安:中国水电顾问集团西北勘测设计研究院,2003.

[5] 刘韩生,等.公伯峡水电站水平旋流泄洪洞减压水工模型试验研究报告[R].咸阳:西北水利科学研究所,2004.

【作者简介】 周恒,男,1970年1月生,工程硕士,1997年7月毕业于西安理工大学,教授级高级工程师,现任中国水电顾问集团西北勘测设计研究院副总工程师、公伯峡水电站项目副设总。

黄土高原土地利用/森林植被变化下流域水沙运移的尺度分异规律

张晓明[1] 曹文洪[1] 武思宏[2]

(1. 中国水利水电科学研究院 北京 100048；
2. 国家科技部科技评估中心 北京 100038)

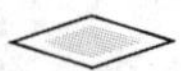

【摘要】 通过分析4个小流域土地利用/森林植被变化前后产流产沙响应，探讨土地利用结构及森林植被理水减沙的尺度变异规律。结果表明，在面积小于100 km^{2}范围的小尺度流域，其面积与产流、产沙均存在较好的正相关性；剔除降水影响，各流域由LUCC引起的径流量和产沙量分别减少20%～100%和10%～100%；不同降水水平的径流减少率不等，表现出尺度性；较大流域因沟道输沙能力强，坡面措施对流域减沙作用有限，森林植被的减沙效应只通过减水来体现。当森林植被覆盖率小于45%时，其减水减沙效应相对植被增加的响应才显著。

【关键词】 尺度效应 水沙输移 土地利用/森林植被变化 黄土高原

侵蚀产沙过程对空间尺度有着强烈的依赖性，侵蚀产沙模拟、预测研究的一个重要内容就是探析侵蚀产沙过程对尺度改变的响应。侵蚀产沙空间尺度效应产生的原因主要是影响侵蚀产沙过程的主要因子的空间异质性和不均匀性，以及新变量和新过程的出现。如小流域的侵蚀产沙计算，一般都假定流域内植被、地形、土质等下垫面因素相对均匀。事实上，尺度问题不仅是一个科学挑战，而且是一个流域管理和侵蚀产沙模型中的实际问题。产沙模数与流域面积之间的相关关系近年来引起广泛关注，认为产沙模数与流域面积之间存在较好的相关性。另外，国际上也有部分研究涉及侵蚀产沙规律及主导侵蚀产沙因子随时空尺度上的变异。但目前国内有关侵蚀产沙尺度问题的研究，多集中在对产沙模数与流域面积的关系探讨上，而侵蚀产沙因子对侵蚀产沙过程的作用机理研究还较少。由于不同尺度流域，坡沟系统产沙机制不同，影响侵蚀产沙的主控因子不同，以及次暴雨存在空间异质性，流域侵蚀产沙过程的尺度变异性有待于进一步研究。本研究通过对3个面积尺度上的4个小流域减水减沙对土地利用/森林植被变化响应分析，探讨不同尺度小流域产流产沙规律的差异性。

1 研究区概况

本研究以位于天水市的吕二沟(Leg)、罗玉沟(Lyg)、桥子东沟(Qzd)及桥子西沟(Qzx)4个不同尺度流域为研究对象，研究区位置及4个流域分布见图1。

天水市位于甘肃省东南部，地理坐标为东经104°35′～106°44′、北纬34°05′～35°10′，平均海拔1 100 m，多年平均(1951～2004年)降水量492 mm，土壤以山地灰褐土和黄绵土为

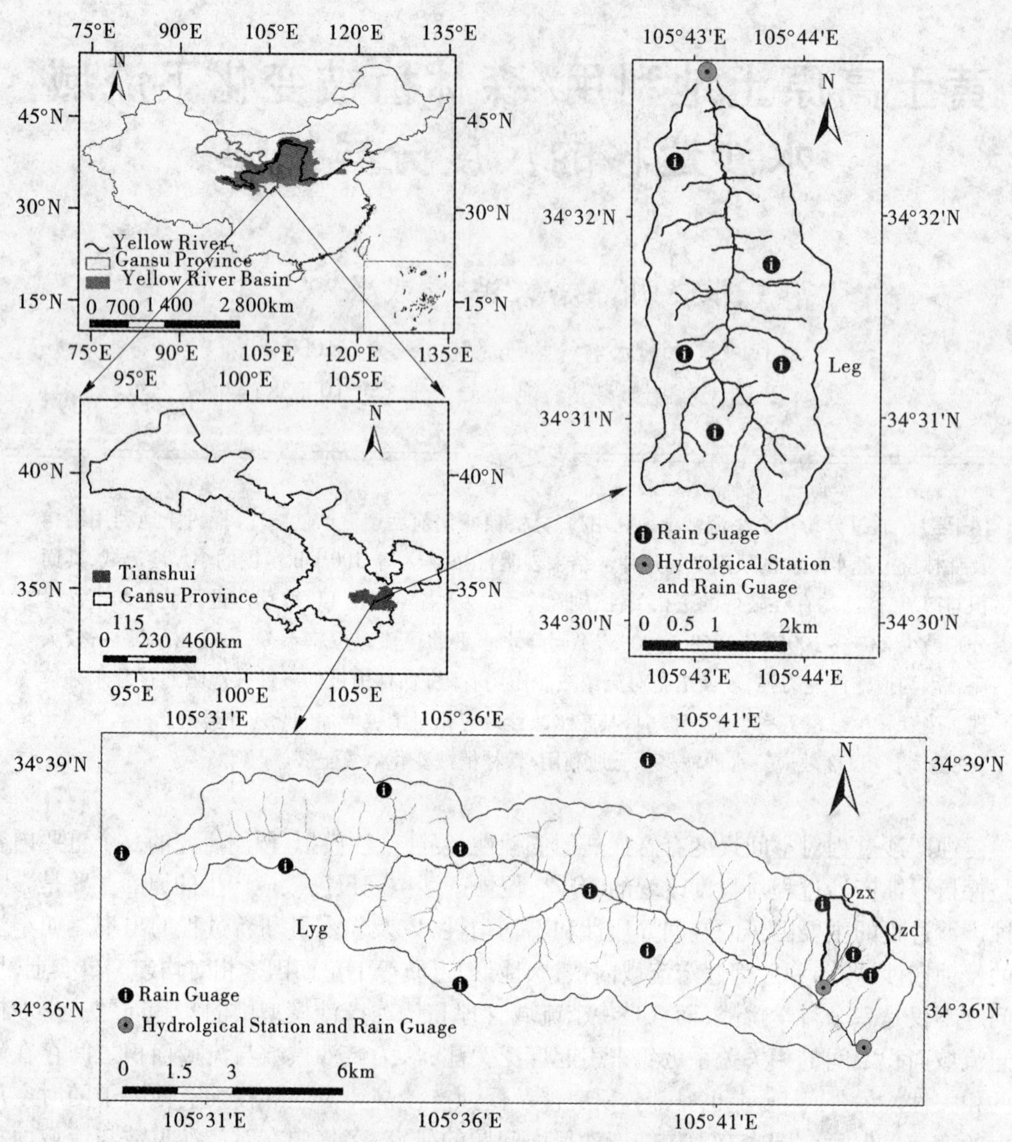

图1 研究区位置及4个试验流域分布图

主要类型。流域森林覆被率为29.4%，乔木均为人工植被。吕二沟是渭河支流耤河右岸的一条支沟，位于天水市南郊，多年(1982～2004年)平均降水量579 mm，6～9月降水量占年降水量的61.3%，年蒸发量1 293 mm。罗玉沟是渭河支流耤河左岸的一条支沟，位于天水市北郊，多年(1985～2004年)平均降水量549 mm，6～9月降水量占年降水量的60%以上，年蒸发量1 293 mm。桥子东沟、桥子西沟流域(合称桥子沟流域)是罗玉沟流域内的两个嵌套对比小流域。表1和表2分别为4个试验流域的基本地理特征。

2 研究材料和方法

2.1 图像处理

根据罗玉沟1986年、1995年的Landsat TM和2004年的Landsat ETM$^+$影像以及吕二沟

1982 年、1993 年 Landsat TM 影像和 2004 年 Landsat ETM$^+$影像数据，以 1∶10 000 地形图为依据，应用 Erdas 和 Arcgis 图像处理软件，采用人机交互的监督分类方法解译，并通过野外验证对其精度进行评价，最后生成罗玉沟 1986 年、1995 年和 2004 年及吕二沟 1982 年、1994 年和 2004 年的土地利用/覆被数据及图件。

表 1　流域主要地理特征值

流域名称	面积(km^2)	流域形状	形状系数	流域长度(km)	平均宽度(km)	沟道比降(%)	相对高差(m)	沟壑密度(km/km^2)	侵蚀模数($t/(km^2 \cdot a)$)
罗玉沟	72.79	羽状	0.16	21.63	3.37	2.3	275	3.54	4935
吕二沟	12.01	羽状	0.25	6.94	1.73	4.0	531	3.82	2476
桥子东沟	1.36	扇形	0.34	2.00	0.68	8.0	377	5.13	2310
桥子西沟	1.09	羽状	0.23	2.18	0.50	8.0	377	5.09	4270

表 2　流域地面坡度组成表　(%)

流域名称	0°～5°	5°～10°	10°～15°	15°～20°	20°～25°	25°以上
吕二沟	14.57	14.82	11.16	22.23	8.74	28.48
罗玉沟	7.20	13.10	28.10	15.80	12.20	12.20
桥子东沟	5.38	38.17	44.27	1.16	0.42	10.60
桥子西沟	6.46	63.83	14.60	2.22	3.14	9.75

2.2　土地利用时段划分

根据获得的试验流域 3 期土地利用/覆被统计数据和图件，采用统一土地利用类型分类标准进行对比分析，根据土地利用变化大小，流域土地利用类型占流域面积比例见表 3。

2.3　降水水平年划分

对试验流域研究时段的年降水进行频率统计，得到降水频率分别为 10%、50% 和 90% 的降水量，并认定其分别为丰、平、枯水年的降水量界值。

表 3　试验流域前、后期土地利用类型占流域面积比例　(%)

流域	土地利用时期	有林地	草地	梯田	坡耕地	灌木林地	疏林地	果园	裸地	居民用地
罗玉沟	前期	5.84	9.71	4.02	68.93	3.17	0.67	2.58	1.74	3.32
	后期	22.98	6.55	46.08	9.87	3.25	7.47	6.20	1.74	3.25
吕二沟	前期	11.27	24.32	11.55	25.21	1.66	15.42	1.45	8.29	0.83
	后期	24.26	28.35	15.19	21.88	1.05	5.66	1.84	0	1.77
桥子东沟	前期	13.96	2.42	16.08	54.20	1.48	9.16	1.63	0.84	1.23
	后期	18.59	0.63	26.97	18.04	1.34	17.31	15.94	1.17	0.02
桥子西沟	整个时期	0.88	0.48	2.83	77.46	1.10	0.00	0.57	14.54	2.14

2.4 数据处理

2.4.1 通径系数及其获取

通径系数是变量标准化后的偏回归系数，它能够有效地表示相关变量间原因对结果的直接效应（直接通径系数）和间接效应（间接通径系数）。通径系数可通过 DPS 6.55 软件对指标进行相关性分析来直接获取。

2.4.2 土地利用前、后期径流或侵蚀产沙量比较中降水影响的剔除

以罗玉沟流域为例，借助 SPSS 16.0 软件建立土地利用前、后期的径流量和侵蚀产沙量预测方程（见表4）。表4中的预测方程，径流量由降雨因子预测，侵蚀产沙量由降雨和径流两个因子共同预测，且相关系数显示预测方程具有较高的可信度。现将前期1986~1994年的逐年实测降雨量代入后期的径流预测方程，计算获得1986~1994年降雨条件下土地利用后期的预测径流量，将后期1995~2004年逐年实测降雨量带入前期的预测方程，得到1995~2004年降雨条件下土地利用前期的预测径流量。同样，分别将1986~1994年实测降雨量和径流量代入土地利用后期预测方程，将1995~2004年实测降雨量和径流量代入土地利用前期预测方程，最后得到相同降雨条件下不同土地利用状态的径流量，以及相同降雨和径流条件下不同土地利用状态的侵蚀产沙量。

表4 流域土地利用前、后期产流和产沙预测方程

流域	因子	土地利用前期			土地利用后期		
		回归方程	R^2	n	回归方程	R^2	n
罗玉沟	降雨—径流	$Q=180.4P-58\,298$	0.588	9	$Q=2\times10^{-7}P^{3.92}$	0.614	9
	降雨—输沙	$S=-42.5P+0.98Q-783.9$	0.832	9	$S=-26.2P+0.4Q+11\,447.8$	0.969	9
吕二沟	降雨—径流	$Q=787.4\ e^{0.006P}$	0.861	11	$Q=122.03P-54\,453$	0.807	12
	降雨—输沙	$S=-10.1P+0.13Q+3\,855.8$	0.700	11	$S=-3.1P+0.12Q+1\,407.7$	0.898	12
桥子东沟	降雨—径流	$Q=7\cdot10^{-22}\cdot P^{8.88}$	0.657	9	$Q=45.32P-19\,484$	0.940	10
	降雨-输沙	$S=-2.5P+0.3Q+1\,299.4$	0.991	9	$S=-10P+0.44Q+4\,652.18$	0.962	10

注：表中 Q 为径流量（$m^3/(km^2\cdot a)$），S 为侵蚀产沙量（$t/(km^2\cdot a)$），P 为降雨量（mm），下同。

3 结果与分析

3.1 流域径流、侵蚀产沙对面积变化的响应

本研究根据实测的年降雨量、径流量与侵蚀产沙量，以降雨量和径流量为预测因子，分析三者的相关性，见表5，径流量比降雨量对侵蚀产沙量影响显著，且降雨直接或通过影响径流来间接作用于侵蚀产沙。另外，表5显示，罗玉沟、吕二沟和桥子东、西沟的剩余通径系数分别为0.678 3、0.513 9、0.128 5、0.188 7，表明除降雨和径流外存在影响侵蚀产沙的潜在因素，且随流域面积增大其贡献率逐渐增强。图2显示了流域面积是影响侵蚀产沙的一个潜在因素，虽样本数据少，但流域面积与产流量和产沙量分别存在明显的正指数与线性关系。

3.2 土地利用/森林植被理水减沙的尺度效应

如表3所示，各试验流域土地利用均以草地、林地、坡耕地和梯田为主，各期占流域面积约80%。罗玉沟流域后期林地和梯田面积占流域面积比例比前期分别增加约17%、42%，坡耕地减少59%。吕二沟流域后期林地、草地和梯田占流域面积比例比前期分别增加约13%、4%和4%，坡耕地和裸地分别减少约3%和8%。桥子东沟后期林地、梯田、疏林地和

果园占流域面积比例比前期增加约 5%、11%、8% 和 14%，而坡耕地减少约 36%。表中具有较好水土保持效益的灌木林地和草地后期较前期有所减少，裸地略有所增加，但认为三者均占流域面积比例甚小且前后两期变化率小，因此产生的负面影响较小。总体来看，罗玉沟、吕二沟和桥子东沟土地利用后期林草面积增加，坡耕地和裸地减少，植被条件较前期有较大改善。而桥子西沟作为桥子东沟的对照流域，土地利用类型以坡耕地和裸地为主，其结构相对桥子东沟前期或后期均较差。

表 5　流域降雨、径流和侵蚀产沙量相关关系

流域	变量	相关系数分析			通径系数分析			
		P	Q	S	直接作用	间接作用		剩余通径系数
						通过 P	通过 Q	
罗玉沟	P	1	0.000 8*	0.091 0*	−0.207 9		0.606 5	0.678 3
	Q	0.700 8	1	0.000 5*	0.865 4	−0.145 7		
	S	0.398 6	0.719 7	1				
吕二沟	P	1	0.000 1*	0.001*	−0.129 9		0.770 9	0.516 4
	Q	0.805 1	1	0.000 1*	0.957 5	−0.104 6		
	S	0.641	0.852 9	1				
桥子东沟	P	1	0.008 1*	0.022 6*	−0.118 5		0.720 9	0.128 5
	Q	0.675 1	1	0.000 1*	1.067 8	−0.08		
	S	0.602 5	0.987 9	1				
桥子西沟	P	1	0.002 3*	0.000 5*	0.179 2		0.610 7	0.188 7
	Q	0.723	1	0.000 1*	0.844 6	0.129 6		
	S	0.789 9	0.974 2	1				

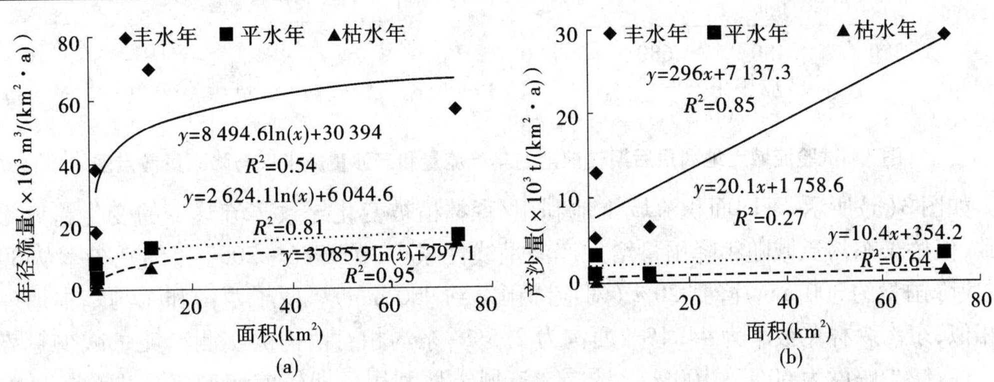

图 2　流域面积与年径流量和侵蚀产沙量的关系

表 6 显示了各流域在不同降水水平年土地利用后期较前期降雨量、径流量和侵蚀产沙量的变化率。从表中的整体趋势看，后期较前期的径流和侵蚀产沙显著降低，径流量和侵蚀产沙量的减少率分别为 40% ~90% 和 29% ~95%，不同尺度流域的径流量和侵蚀产沙量减少率的范围无明显界限。考虑到土地利用前、后期的降雨量不同，需剔除降水的影响来分析

由于流域土地利用结构优化和森林植被增加引起的产流产沙变化。根据表4的预测方程计算同一降水条件下的流域土地利用前、后期年产流量和同一降水—径流条件下流域年侵蚀产沙量，然后求算各流域土地利用后期的径流量和侵蚀产沙量相对前期的减少率，绘制各流域随降雨量变化的减少率散点图和趋势线，见图3。

表6　各流域土地利用后期较前期降雨、径流和侵蚀产沙减少率　(%)

流域		降雨量	径流量	径流系数	产沙量	平均输沙率
罗玉沟	丰水年	-4	-51	-47	-83	-29
	平水年	-16	-85	-83	-68	-68
	枯水年	-3	-84	-84	-37	-37
吕二沟	丰水年	-7	-64	-65	-51	-85
	平水年	-16	-86	-83	-91	-91
	枯水年	-4	-73	-73	-51	-51
桥子东沟	丰水年	-3	-56	-58	-63	-63
	平水年	-23	-90	-87	-95	-95
	枯水年	-7	-40	-36	-29	-20

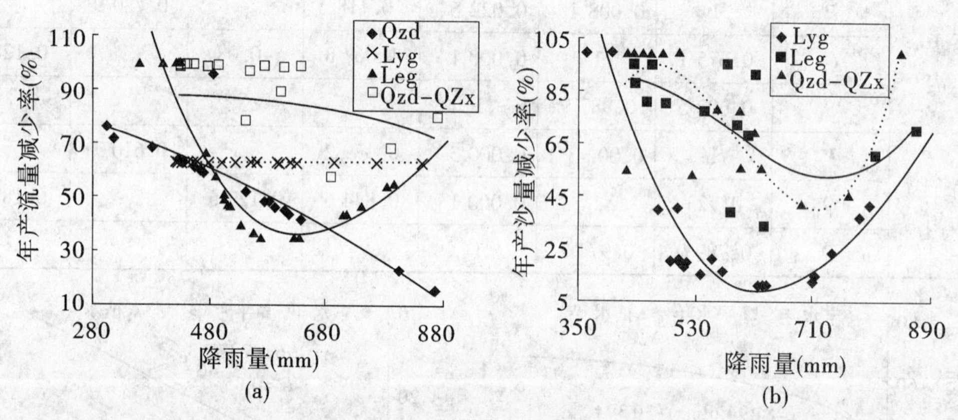

图3　试验流域土地利用后期较前期的年产流量和产沙量减少率与降雨量散点图

如图3(a)所示，不同面积流域土地利用/森林植被变化导致产流减少的变化趋势并不一致，仅桥子东沟流域随年降雨量增加产流呈线性减小，在1986～2004年间减少率从80%（对应降雨量约300 mm）降到20%（对应降雨量约900 mm）。桥子西沟和东沟基本地理特征相似，东沟森林覆被率为44.4%，西沟为2.5%，东沟相对西沟的实测产流量减少趋势亦呈线性，减少幅度为60%～100%。吕二沟流域土地利用后期较前期产流减少趋势呈单峰凹型抛物线，即减少率随降雨量的增加呈先减小后增大的趋势，减少率的最低值约为35%，此时降雨量值约为630 mm。即吕二沟流域在森林植被增加的情况下，平水年时的径流量减少率较小，为35%～45%，而枯水年和丰水年时较大，在50%～100%。罗玉沟流域无论年降雨量多寡，土地利用后期较前期的径流减少率均保持在63%左右，趋势线为一条近似平行于横轴的直线。以上分析说明，在土地利用结构优化和森林植被增加后，流域径流减少趋势具有尺度性，随年降雨量的变化，不同面积流域的径流量减少率或呈线性减小、或呈单峰凹型抛物线变化、或者保持在一定水平。

显然，流域土地利用后期较前期侵蚀产沙量的减少率变化趋势不同于径流量减少率的变化趋势(见图3(b))，各流域的年降雨量与产沙量减少率的散点虽分布较散，但整体都呈单峰凹型抛物线变化。不同流域面积谷底值和对应的降雨量值不一致，罗玉沟流域谷底值最小，吕二沟和桥子东沟—西沟(桥子东沟较桥子西沟年侵蚀产沙量的减少率)则依次增大，三流域减少率最小值分别约为10%、35%和45%；且从散点的分布来看，各流域产沙量减少率的最小值大都集中在平水年期。

3.3 流域森林植被变化对理水减沙的影响

根据以上分析知，流域森林植被覆盖对产流产沙影响较为显著。为此，根据天水水土保持试验站已记录的和通过图像处理获得的不同土地利用时期流域森林覆被率与实测产流、产沙数据，分别建立径流和侵蚀产沙与降雨量和森林覆被率的关系式，见表7。

表7 试验流域面积与年径流和输沙的关系

流域	径流—降雨、森林覆盖率		侵蚀产沙—降雨、森林覆盖率	
	关系式	R^2	关系式	R^2
罗玉沟	$Q=2708.1e^{0.006P-0.1L}$	0.903	$S=252.9e^{0.005P-0.053L}$	0.874
吕二沟	$Q=19110.6e^{0.007P-0.14L}$	0.929	$S=53.7e^{0.007P-0.045L}$	0.872
桥子东沟	$Q=13.9e^{0.01P-0.05L}$	0.946	$S=15.1e^{0.01P-0.048L}$	0.895

注：表中L为林地面积占流域面积比例(%)。

从表7可看出，流域径流量和侵蚀产沙量与降雨量和森林覆被率均呈指数关系，即

$$M=ae^{bP-cL} \tag{1}$$

式中：M为径流量，$m^3/(km^2\cdot a)$，或为侵蚀产沙量，$t/(km^2\cdot a)$；P为年降水量，mm；L为林地面积占流域面积比例(%)；a、b、c均为系数。

流域产流量或侵蚀产沙量随降雨量的增大而增加、随森林覆被率的增加而减少。假设流域年降水量保持在平水年水平，根据式(1)分别计算不同森林植被覆被率下的流域侵蚀产沙量，绘制如图4所示曲线。显然，当森林覆被率在45%以下时，其减沙效应相对森林植被率增加的响应才变得显著，即增加森林植被率并不可能无限制地减少流域产流产沙。

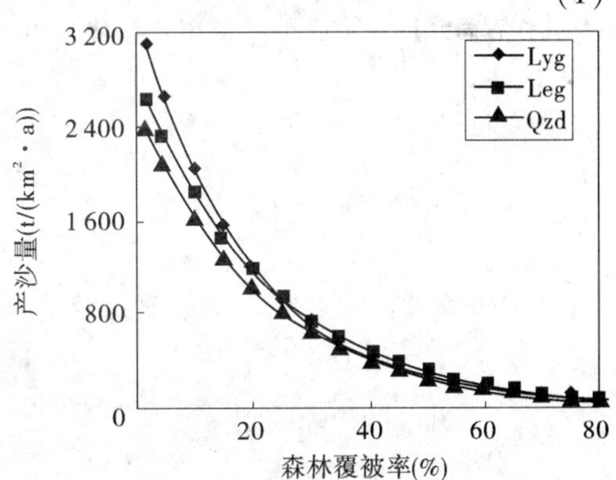

图4 流域森林覆被率与侵蚀产沙量关系

4 结论与讨论

流域降雨侵蚀产沙不仅受降雨、地形地貌、土地利用/土地覆被等的影响，流域面积也是潜在影响因素，通过分析知，流域面积与径流量和侵蚀产沙量分别存在较好的正指数关系和线性关系。

通过剔除降雨量的影响来探讨流域水沙输移对土地利用/森林植被变化响应的尺度效应，经分析知，试验流域森林植被增加及土地利用结构优化引起的径流量和侵蚀产沙量减少分别为20% ~100%和10% ~100%，不同降水水平的减少率存在差异，且在本研究所考察的流域尺度范围内表现出尺度性。小面积尺度流域（约1 km^2）的减少率随降雨量增加线性减小；中等面积尺度流域（12 km^2）则随降雨量增加先减少后增大，即在平水期径流减少率最低；而较大尺度流域（约73 km^2）的减少率则保持在一定水平不变。下垫面改变对侵蚀产沙的影响并不同步于对径流的影响，各试验流域土地利用后期的产沙量相对前期的减少率均随降雨量的增加表现出先减少后增大的趋势，即试验流域枯水期和丰水期时的森林植被减蚀作用强于平水期。同时，较大尺度流域土地利用/森林植被的减沙效应是通过减少产流来实现的，那么黄土高原的水沙治理，沟壑整治工程的实施是必不可少的。

森林植被的减水减沙功效不能无限扩大，由流域侵蚀产沙与森林植被覆被率的指数关系知，当森林植被覆盖率在45%以下时，其减水减沙效应相对森林植被增加响应才显著。

参考文献

[1] De Boer D H, Campbell I A. Spatial scale dependence of sediment dynamics in a semi-arid badland drainage basin[J]. Catena, 1989, 16(3): 277-290.

[2] 傅国斌，李丽娟，刘昌明. 遥感水文应用中的尺度问题[J]. 地球科学进展，2001，16（6）：755-760.

[3] Krishnaswamy J, Richter D D, Halpin P N, Hofmockel M S. Spatial patterns of suspended sediment yields in a humid tropical watershed inCosta Rica [J]. Hydrological Processes, 2001, 15: 2237-2257.

[4] De Vente J, Poesn J. Precditing soil erosion and sediment yield at the basin scale: scale issues and semi-quantitative models [J]. Earth-Science Reviews, 2005, 71: 95-125.

[5] Renschler C S, Harbor J. Soil erosion assessment tools from point to regional scales-the role of geomorphologists in land management research and implementation[J]. Geomorphology, 2002, 47: 189-209

【作者简介】 张晓明，1979年10月生，博士，2007年毕业于北京林业大学，现任中国水利水电科学研究院高级工程师。

波浪与波生流相互作用下的底部边界层数值模拟

张　弛　郑金海

（河海大学　港口海岸与近海工程学院　南京　210098）

【摘要】 推导出波生流引起的时均水平压力梯度项，建立了波浪与波生流相互作用下的底部边界层数学模型。基于实测数据的验证，成功地复演了浅化波浪边界层内的向岸时均流动、破碎波浪边界层内的离岸时均流动以及剪切应力在垂向上的线性分布特征。研究表明，所推出的时均水平压力梯度项对准确描述波浪与波生流相互作用下的近底水动力结构和泥沙运动有重要意义。

【关键词】 波浪　波生流　底部边界层　数学模型

在海岸泥沙从静止、起动再到推移和悬浮的过程中，直接施加于泥沙颗粒的作用力来自于床面上很薄的一层振荡水体，称为底部边界层。底部边界层的厚度虽然通常只有厘米级，但层内流动为黏性有旋且伴随着较大的紊动，是上方水体和下方海床相互作用的动力交换区域。在河口海岸地区，底部边界层通常受到波浪和水流的共同影响。其中，水流可以是近岸短波驱动的波生流或长波驱动的潮流，也可以是水面坡降引起的河口径流。研究波流相互作用下的底部边界层对于海岸防护、岸滩演变、港口建设与航道维护等都有重要意义。

在控制方程中，水流的影响主要体现为一个时均水平压力梯度项。已有的试验研究和数学模型大多针对平底上波流相互作用下的底部边界层，同时假定时均水平压力梯度项与波浪要素无关：有的忽略这一项，有的将其与某参考点流速联系起来（例如水深平均的流速）。这些模型在模拟平底上波流相互作用的物理试验时都取得了较好的结果，原因是在这些试验中，水流是独立于波浪由试验装置生成的，其平均流速为已知且流速剖面接近于对数剖面，所以时均水平压力梯度项可认为与波浪无关并根据已知的平均流速独立求解。这种情况适用于河口地区波浪和径流的相互作用。

然而，在波浪与波生流共存的实际海岸中，上述方法的应用存在困难。原因主要有4个方面：①波生流由波浪驱动，时均水平压力梯度项也应与波浪参数有关；②波生流的流速事先未知，因此不能作为已知条件输入到模型中；③流速和流向在垂向上剧烈变化并与对数曲线相去甚远，因此很难选取合适的参考点流速；④向、离岸方向上水深平均的流速为0以保持质量守恒，例如水体表层的向岸流与底层的离岸流相互平衡，因此水深平均流速不再具有代表性。尽管已有一些物理试验着眼于斜坡上波浪传播变形条件下的底部边界层，并观测到了一些重要的物理现象，但这方面的理论研究和数学模型还不完善。

本文通过推导波生流引起的时均水平压力梯度项表达式，建立了波浪与波生流相互作用下的底部边界层数学模型，通过与实测数据的对比，讨论波浪与波生流相互作用对底部边界层的影响。

1　数学模型

1.1　控制方程

底部边界层一阶动量守恒方程可写为

$$\frac{\partial u}{\partial t} = -\frac{1}{\rho}\frac{\partial p}{\partial x} + \frac{1}{\rho}\frac{\partial \tau}{\partial z} \tag{1}$$

式中：u 为水平速度分量；p 为压力；τ 为紊动剪切应力。在波流共存场中，可将这三个物理量分解为波浪分量和水流分量，即

$$u = \tilde{u} + \bar{u}, p = \tilde{p} + \bar{p}, \tau = \tilde{\tau} + \bar{\tau} \tag{2}$$

式中：波浪线上标代表波浪分量；直线上标代表水流分量。

根据边界层假定，式(1)右端第一项水平压力梯度在边界层内为一定值，其值等于边界层上边界处(自由流动层)的值，于是得到下列关系式

$$-\frac{1}{\rho}\frac{\partial \tilde{p}}{\partial x} = -\frac{1}{\rho}\frac{\partial \tilde{p}_\infty}{\partial x} = \frac{\partial \tilde{u}_\infty}{\partial t} \tag{3}$$

$$-\frac{1}{\rho}\frac{\partial \bar{p}}{\partial x} = -\frac{1}{\rho}\frac{\partial \bar{p}_\infty}{\partial x} = -\frac{1}{\rho}\frac{\partial \bar{\tau}_\infty}{\partial z} \tag{4}$$

式中：下标∞代表边界层上边界处的物理量；u_∞ 则是近底水质点自由流速。式(3)、式(4)分别表示波浪和水流引起的水平压力梯度，前者与近底波浪振荡速度有关，后者与时均剪切应力的垂向分布有关。又有边界层外的时均动量守恒方程为

$$\frac{\partial \bar{\tau}_\infty}{\partial z} = \frac{\partial}{\partial x}\rho\overline{(\tilde{u}_\infty^2 - \tilde{w}_\infty^2)} + \frac{\partial}{\partial x}\rho g\bar{\eta} \tag{5}$$

式中：$\tilde{w}_\infty$ 是垂向速度的波浪分量。式(5)右端第一项代表波浪变形引起的剩余动量流梯度，第二项代表时均水位变化引起的静水压力梯度。这两项的不平衡将造成时均剪切应力和水流的垂向变化。

郑金海推出了任意波向角下波浪剩余动量流沿水深分布表达式。在浅水中线性波正向入射的条件下，波谷下方的剩余动量流表达式为

$$\rho\overline{(\tilde{u}_\infty{}^2 - \tilde{w}_\infty^2)} = \frac{E_w}{h} = \frac{\rho g H^2}{8h} \tag{6}$$

联立式(4)、式(5)和式(6)，边界层内的时均水平压力梯度可表示为

$$-\frac{1}{\rho}\frac{\partial \bar{p}}{\partial x} = -\frac{gH}{4h}\frac{\partial H}{\partial x} - g\frac{\partial \bar{\eta}}{\partial x} \tag{7}$$

将式(3)和式(7)代入式(1)，并引入 Boussinesq 假定($\tau = \rho(v_t + v)\partial u/\partial z$)，波浪与波生流相互作用下的底部边界层控制方程可表示为

$$\frac{\partial u}{\partial t} = \frac{\partial \tilde{u}_\infty}{\partial t} - \left(\frac{gH}{4h}\frac{\partial H}{\partial x} + g\frac{\partial \bar{\eta}}{\partial x}\right) + \frac{\partial}{\partial z}\left[(v_t + v)\frac{\partial u}{\partial z}\right] \tag{8}$$

式中：v 为运动黏滞系数。

式(8)考虑波浪场和时均水面的沿程变化，时均水平压力梯度取决于波浪剩余动量流梯度和静水压力梯度之间的相互作用，在破波带内外都将得到非零值，反映了波生流对边界层的影响。

1.2 紊动闭合

通过一个修正的 k-ε 紊流模型来计算式(8)中的涡黏系数，该模型的优点是在低雷诺数黏性层中也可适用。紊动动能 k 和紊动耗散率 ε 的传播方程表示为

$$\frac{\partial k}{\partial t}=\frac{\partial}{\partial z}\left[\left(\frac{v_t}{\sigma_k}+v\right)\frac{\partial k}{\partial z}\right]+v_t\left(\frac{\partial u}{\partial z}\right)^2-\varepsilon \tag{9}$$

$$\frac{\partial \varepsilon}{\partial t}=\frac{\partial}{\partial z}\left[\left(\frac{v_t}{\sigma_\varepsilon}+v\right)\frac{\partial \varepsilon}{\partial z}\right]+f_1C_{\varepsilon 1}\frac{\varepsilon}{k}v_t\left(\frac{\partial u}{\partial z}\right)^2-f_2C_{\varepsilon 2}\frac{\varepsilon^2}{k} \tag{10}$$

紊动涡黏系数可用下式计算

$$v_t=f_uC_u\frac{k^2}{\varepsilon} \tag{11}$$

模型中的标准参数为：$C_u=0.09$；$C_{\varepsilon 1}=1.4$；$C_{\varepsilon 2}=1.8$；$\sigma_k=1.4$；$\sigma_\varepsilon=1.3$；$f_1=1.0$。另外两个函数为

$$f_u=(1+3.45/\sqrt{R_t})\times[1-\exp(-y^*/42.42)] \tag{12}$$

$$f_2=[1-2\exp(-R_t^{\ 2}/36)/9]\times[1-\exp(-y^*/3.03)] \tag{13}$$

式中：$R_t=k^2/\varepsilon v$ 是紊动雷诺数；$y^*=(v\varepsilon)^{1/4}z/v$ 是 Kolmogorov 长度尺度。

1.3 边界条件

在床面上，采用流速无滑移条件

$$u(z_0,t)=0\quad(z_0=k_s/30) \tag{14}$$

紊流模型的下边界条件为

$$k(z_0,t)=0,\quad \varepsilon(z_0,t)=2v\left(\frac{\partial\sqrt{k}}{\partial z}\right)^2 \tag{15}$$

在边界层的上边界处 $z=z_{max}$，流速等于近底水质点自由流速

$$u(z_{max},t)=u_\infty \tag{16}$$

紊流模型的上边界定义为零通量条件

$$\frac{\partial}{\partial z}k(z_{max},t)=0\quad \frac{\partial}{\partial z}\varepsilon(z_{max},t)=0 \tag{17}$$

2 计算结果与分析

使用两组物理试验数据对模型进行验证，包括波浪在斜坡上浅化和破碎条件下的底部边界层。验证数据包括流速剖面、紊动能量和床面剪切应力等，重点讨论波生流引起的时均水平压力梯度项对计算结果的影响及其物理机制。

Lin 和 Hwung 在试验水槽中测量了波浪在斜坡上浅水变形情况下的底部边界层。试验在国立成功大学台南水工试验所中进行，水槽长 9.5 m、宽 0.3 m、高 0.7 m。波高 0.053 m 以及周期 1.41 s 的规则波正向入射于 1∶15 的光滑底坡上。在破波带外布置 10 条测量断面，分别使用波高计和 LDV 测量水面高程和边界层内流速的时间序列。试验布置示意图见

图 1。

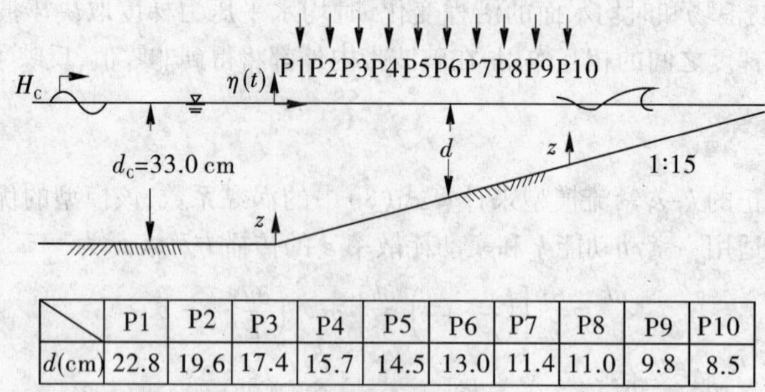

	P1	P2	P3	P4	P5	P6	P7	P8	P9	P10
d(cm)	22.8	19.6	17.4	15.7	14.5	13.0	11.4	11.0	9.8	8.5

图 1　Lin 和 Hwung 试验地形及测点位置示意图

本文选用 P4、P8 和 P10 三个断面的实测数据对模型进行验证。使用距底部 0.005 m 高程处的实测自由流速作为模型的流速上边界条件，床面粗糙高度为 0.002 m。使用式(7)计算波生流引起的时均水平压力梯度，计算结果在 P4、P8 和 P10 处分别为 0.002 m/s^2、0.005 m/s^2、0.01 m/s^2。可以看出，波浪浅水变形条件下底部边界层内的时均水平压力梯度总为正值且向岸逐渐变大，说明波浪减水引起的向岸静水压力梯度大于波高增大引起的离岸动量流梯度，这一趋势从外海向破波点逐渐增强。

图 2 给出了 P4 断面处一个波周期内 8 个相位的流速剖面验证。图中的实线和虚线分别表示考虑和不考虑时均水平压力梯度项的计算结果。如图所示，计算结果总体上与实测数据吻合较好。实测的流速剖面上端有较明显的倾斜趋势，同时速度不对称现象显著，而考虑了水平压力梯度的计算结果较好地模拟出这些现象。总体而言，引入的水平压力梯度改进了计算结果的精度，特别是在正向加速阶段(相位 A-C)。

图 3 给出了时均流速剖面的对比，可以看出时均水平压力梯度对计算结果产生了重要影响。当考虑时均水平压力梯度时，模型成功地模拟出靠近底床处的向岸时均流速；而若不考虑该压力梯度时，受上边界的影响，计算的时均流速剖面全部为离岸方向。近底局部向岸流速从 P4 点向 P10 点逐渐增大，这与同样向岸增大的时均水平压力梯度有关。引入的时均水平压力梯度对时均床面剪切应力也具有重要影响。考虑这个压力梯度时，时均床面剪切应力为向岸方向同时沿程递增，如果不考虑这个压力梯度，计算得到的时均床面剪切应力则为离岸方向，说明考虑波浪与波生流的相互作用对准确模拟泥沙输运起到重要作用。

Cox 和 Kobayashi 在 Delaware 大学的实验室中测量了破波带内的底部边界层。波浪水槽长 33 m，宽 0.6 m，高 1.5 m，周期 2.2 s 的规则崩破波入射于 1∶35 的斜坡上，床面铺上中值粒径 d_{50} = 1.0 mm 的天然泥沙，在 6 个断面处测量自由水面高程和流速剖面分布(从海向岸为 L1 ~ L6)。本文采用 L4 和 L5 断面的实测数据对模型进行验证。模型上边界取为实测的自由流速，分别位于距底部 0.121 m 处(L4)和 0.071 m 处(L5)。床面粗糙高度确定为 $k_s = 2d_{50}$。在 L4 和 L5 断面，计算的时均水平压力梯度都为负值(离岸方向)，分别为 -0.03 m/s^2 和 -0.018 m/s^2，意味着波浪增水引起的离岸静水压力梯度大于波高衰减引起的向岸动量流梯度，而其所引起的离岸时均水平压力梯度也正是破波带内底部离岸流的驱动力。

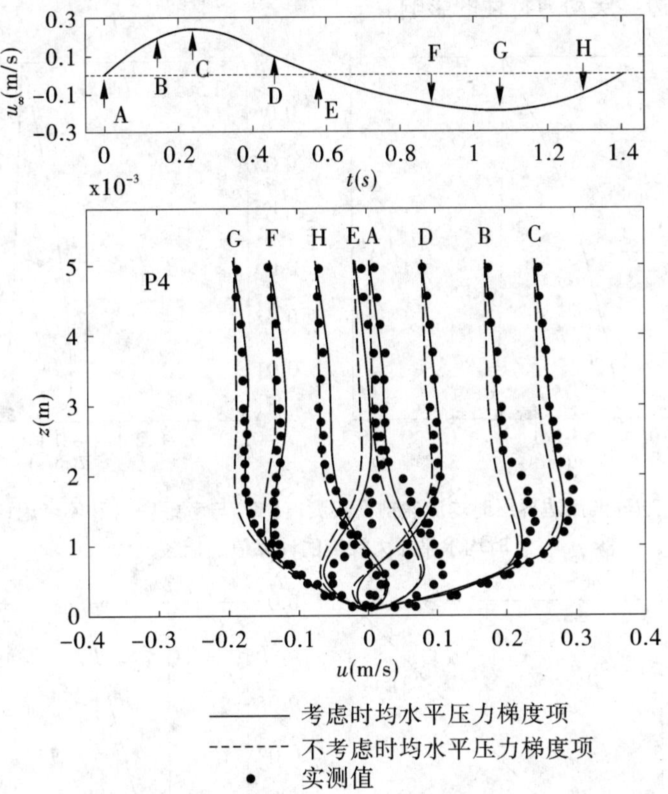

图 2　浅化波浪边界层内瞬时流速剖面计算值与实测值的对比

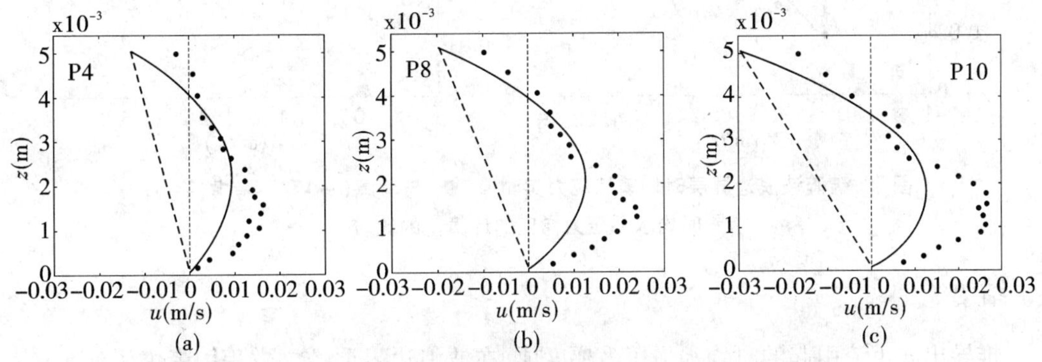

图 3　浅化波浪边界层时均流速剖面实测值(●)与考虑(—)和不考虑(---)时均水平压力梯度的计算值的比较

图 4 和图 5 分别给出了时均流速剖面和时均剪切应力的比较结果。从图 4 中可以看出,考虑时均水平压力梯度项之后计算结果有所改进,虽然计算的流速偏大于实测值,但模型仍较好地反映出流速剖面的垂向变化形式,离岸流速的最大值出现在临近底床处。如果不考虑这个水平压力梯度项,计算流速则偏小,且流速最大值出现在上边界处。从图 5 中可以看出,时均剪切应力的计算值与实测数据吻合很好。时均剪切应力在上边界为正值,由于存在离岸方向的时均水平压力梯度,向下逐渐减小至底床处为负值。若忽略该时均水平压力梯度,时均剪切应力垂向上则变为均匀分布。本文所推出的时均水平压力梯度项能较好

地描述波生流对边界层动力特性的影响。

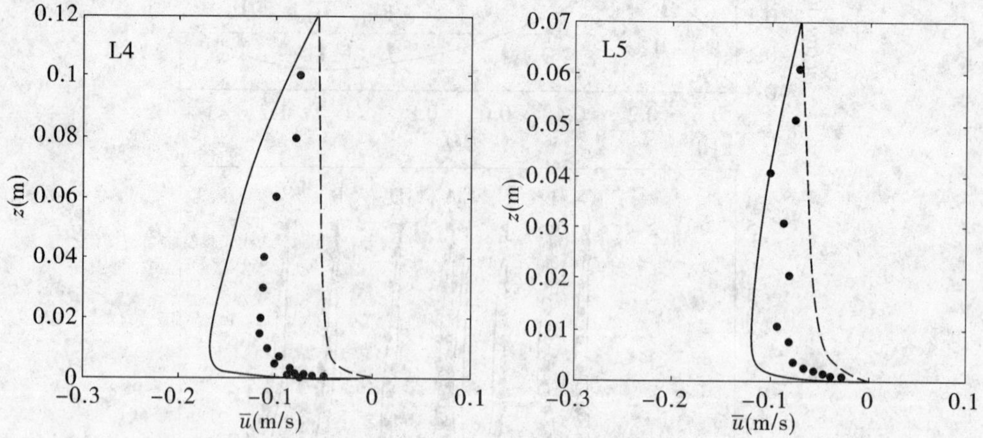

图 4　破碎波浪边界层时均流速剖面实测值(●)与考虑(—)和不考虑(---)时均水平压力梯度的计算值的比较

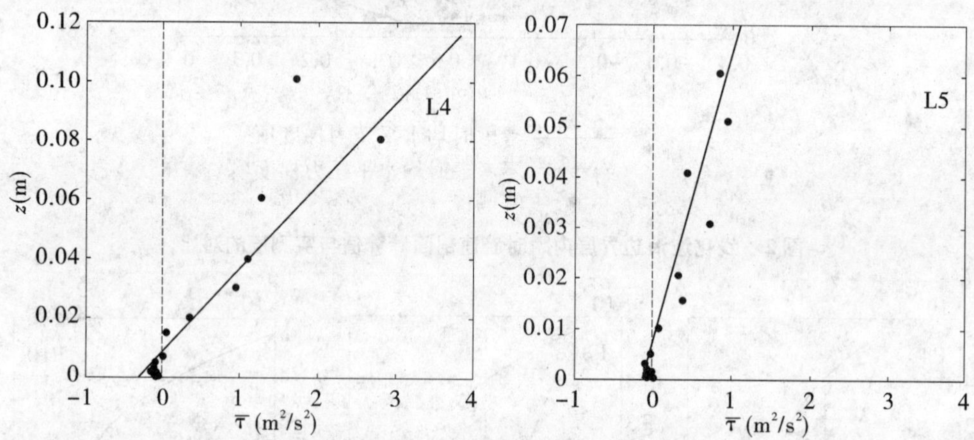

图 5　破碎波浪边界层时均剪切应力实测值(●)与考虑(—)和不考虑(---)时均水平压力梯度的计算值的比较

3　结语

推导出波生流引起的时均水平压力梯度项,在波浪边界层控制方程中考虑了波生流的影响,建立了波浪与波生流相互作用下的底部边界层数学模型。该水平压力梯度项与两个物理过程有关,一是波浪变形产生的剩余动量流梯度,二是波浪增减水引起的水位压力梯度。模型成功地复演了浅化波浪边界层内的向岸时均流动、破碎波浪边界层内的离岸时均流动以及剪切应力在垂向上的线性分布特征。所推出的时均水平压力梯度项反映了波浪与波生流相互作用的物理机制,对准确描述波浪变形条件下的近底水动力结构和泥沙运动有重要意义。

本文的英文版已在国际期刊《Geo-Marine Letters》2011 年第 3 期上发表。研究工作得到国家重点基础研究发展计划(973 计划)资助课题(2010CB429002)和河海大学水文水资源与水利工程科学国家重点实验室专项经费项目(2009585812)的资助。

参考文献

[1] Zhang C, Zheng J H, Wang Y G, Demirbilek Z. Modeling wave-current bottom boundary layers beneath shoaling and breaking waves[J]. Geo-Marine Letters, 2011, 31(3): 189-201.

[2] Zheng J H. Depth-dependent expression of obliquely incident wave induced radiation stress[J]. Progress in Natural Science, 2007, 17(9): 1067-1073.

[3] Sana A, Ghumman A R, Tanaka H. Modification of the damping function in the k-ε model to analyze oscillatory boundary layers[J]. Ocean Engineering, 2007, 34 (2): 320-326.

[4] Lin C, Hwung H H. Observation and measurement of the bottom boundary layer flow in the prebreaking zone of shoaling waves [J]. Ocean Engineering, 2002, 29 (12): 1479-1502.

[5] Cox D T, Kobayashi N. Undertow profiles in the bottom boundary layer under breaking waves. Proceedings of the 25th Conference on Coastal Engineering, ASCE, 1996: 3194-3206.

【作者简介】 张弛,男,1985 年 3 月生,2010 年毕业于河海大学港口、海岸及近海工程专业,获博士学位,现任河海大学港口海岸与近海工程学院讲师。

波斯湾北部某天然气液化项目温排放试验研究

钟伟强　倪培桐　苗　青　黄健东

(广东省水利水电科学研究院,广东省水动力应用研究重点实验室　广州　510610)

【摘要】　通过建立波斯湾北部天然气液化项目的物理模型,对该项目的温排水在潮汐过程中的运动情况及其对海水水温的影响进行研究、分析,经多方案对比试验,提出了较优的电厂取排水口布置方案。试验研究成果,对布置在近海区的电厂温排水工程研究具有参考价值。

【关键词】　液化项目　温排水　潮流　取水温升　试验研究

1　工程概况

某天然气液化项目位于波斯湾北部沿岸 Barikan(伊朗)附近海域,厂址坐标为北纬 28°20′57″,东经 51°11′02″(见图 1)。该项目的液化厂(LNG)工程、电厂循环冷却水工程均采用直流供水系统,冷却水取自波斯湾海水。工程规划容量的总冷却水量为 367 000 m^3/h,其中电厂项目冷却水流量为 147 000 m^3/h,液化厂为 220 000 m^3/h,取排水温升约为 9 ℃。工程排出的废热由潮流带至附近海域扩散冷却,其热效应对工程冷却水本身及波斯湾海区环境的影响如何,在很大程度上决定厂址的可行性,温排放的热影响问题自然成为各方关注的重点。因而,为验证取排水工程方案的可行性,获得最佳布置方案,需通过温排放试验寻求电厂取水温升低、冷却效果好、对环境影响小及工程投资小的取排水工程布置方案,为工程设计和环境评价提供依据。

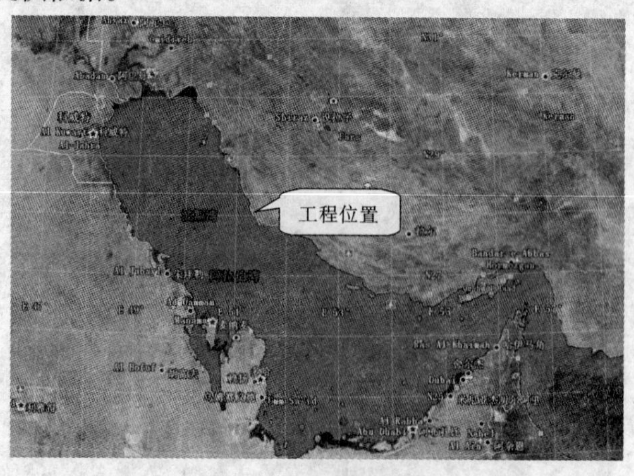

图 1　工程地理位置图

2 工程海域基本情况

2.1 海域自然条件特征

波斯湾海域水深一般不超过 90 m,伊朗沿岸水深较深处约 80 m,阿拉伯半岛一侧则一般浅于 35 m,湾口处最深达 110 m。工程地处波斯湾东北部,其附近海域的水下地形等深线基本上与岸线走向一致,呈西北—东南走向。

根据现场勘测资料,工程附近海域每日出现两次高潮和两次低潮,潮涨和潮落存在着明显的不等现象,潮汐性质属不规则半日潮。从流速分布情况来看,离岸水域流速较大,流向基本呈东北—西南方向。近岸水域流速较小,流向基本与岸线平行,呈西北—东南方向,落潮时流向东南,涨潮时流向西北,涨、落潮最大流速分别为 0.51 m/s 和 0.78 m/s。涨、落潮水流沿垂线分布,总体上呈上层大、下层小分布,最大流速基本出现在表层或 0.2h 层。潮位资料显示,从北向南,沿程高潮位逐渐抬高,沿程低潮位逐渐降低,沿程潮差则呈递增变化。

2.2 影响取排水口方案布置的主要因素

温排水注入工程海区后形成两种不同温度的水体,流场同时受水力和热力两种动力因素的作用,运动机制较为复杂。电厂及液化厂温排水的运动特性主要与工程后潮流流态、取排水口的布置等因素有关。由于液化项目对水温、水质要求较高,外方设计单位明确要求"液化厂取水口须布置在港池防波堤外南侧、取水温升满足最大不超过 1 ℃",而北区排水存在施工、造价等问题难以实施,因而不能按分列式原则布置取、排水口。同时,伊方要求湾内新建项目必须严格执行"排水口温升 200 m 范围内小于 3 ℃"的有关环境保护规定。因此,满足水域环保要求及工程安全运行要求是工程取排水口布置要考虑的主要因素。

3 模型设计要点

根据北帕斯海域地形和潮汐潮流特征,模型设计以水流运动相似和热力相似为主,因此本模型既是温差异重流模型又具有河工模型的特点,设计要兼顾黎氏定律和佛氏定律,才能满足模型试验要求。另外,为更真实反映温排水热量累积效应,要求温排放热水不溢出或少溢出模型开边界,综合多种因素考虑,本模型选取平面比尺 $L_r = 500$,垂向比尺 Z_r 亦须满足以下相似条件:

水流运动相似

$$V_r = \sqrt{\Delta\rho_r Z_r}$$

$$Q_r = V_r L_r Z_r$$

热平衡相似

$$\theta_r = K_r L_r^2$$

阻力平方区水流运动相似

$$R_{em} > R_{ecr}$$

式中:V_r、Q_r、Δ_{pr}、Z_r 分别为流速、流量、水的密度及垂直比尺;θ_r、K_r 为热量及散热系数比尺;R_{em}、R_{ecr} 分别为模型雷诺数及临界雷诺数。

经分析,垂向比尺 $Z_r = 100$ 满足各项要求。模型模拟的水域面积约 280 km^2,截取范围内含 7 个潮流观测站点且站点分布合适,能够反映模拟水域的流场状况,满足试验要求。

4 设计方案试验

4.1 方案布置

电厂及液化厂排、取水口采用分取合排方案。电厂取水口布置在码头港池内,采用一座12孔钢筋混凝土箱,单孔尺寸为3.5 m×3.5 m,设计流速为0.37 m/s。液化厂工程取水点放置在南防波堤外侧-10 m等深线处,为一座15孔箱式取水口,单孔几何尺寸也是3.5 m×3.5 m。排水口布置在南防波堤外东护岸侧。电厂工程采用四孔方涵,液化厂工程采用六孔方涵,排水孔单孔几何尺寸为宽×高=3.0 m×3.0 m,出口流速约1.51 m/s。排水口后接排水明渠和排水喇叭口,排水喇叭口扩散角为10°。

4.2 设计方案温排水流态

由于电厂及液化厂围填及码头防波堤的修建,切断了排水口位置潮流的流路,码头南防波堤与原岸线形成了一个半封闭的水域,且潮流流速较低,潮流的动力较弱。温排水南防波堤外东护岸侧排出主要靠潮流动力向外海输运和扩散,并通过大气散热。涨潮时受潮流顶托,大部分热水在排水口附近这一弱潮流区窝积,部分热水沿南防波堤向北推移,有少量热水进入港池。落潮时,从排水口直接排出的热水在落潮流的作用下,缓慢向东南方向输运扩散,随涨潮流北上的部分热水经主流通道折向东南扩散,此时亦有部分回归热水进入港区。

4.3 存在的问题

由于原设计方案取、排水口布置的思路是远(深)取近排,受排水口附近半封闭水域地形的影响,温排水扩散效果较差,同时由于温排水主要沿防波堤外围西北—东南向运动,回归热水也较接近取水口位置,在一个潮周期内涨潮温升要大于落潮温升。设计方案全潮平均取水温升在0.7~1.0 ℃,液化厂取水温升最大1.4 ℃,不能满足工程取水温升要求。从温升等值线形态看(见图2),由于排水口附近为半封闭水域,相应形成了扇形的温升场,全潮3 ℃以上温升面积3.61~4.7 km^2,排水口到3 ℃温升线的距离大大超过了200 m限值。为改善温排水扩散效果,必须使排水口布置在水体交换较快的水域,避免温排水窝积现象。

5 修改方案试验

5.1 方案布置

结合潮流及地形特点,修改方案取、排水口布置考虑远排近取方案。通过远排将热水掺入潮水中,随潮流向远处输运,工程取水口则布置在近岸取水。修改方案一电厂工程仍在码头港池内取水,综合考虑取水温升、水质等因素,液化厂工程取水口则布置在南防波堤外侧-7.5 m等深线附近,见图3。排水口布置在南防波堤外侧-9.5 m等深线附近,距离液化厂工程取水口约2.3 km,距离岸边约3.5 km,其结构型式与原设计一致。

5.2 试验成果

温排水运动特征表现为:电厂及液化厂排出的温排水在出流动量的推动下自西南方向向外海方向输送,出口水流呈射流流态,射流距离40~60 m,之后在水中强烈掺混并逐渐上浮并扩散,受潮流作用影响,排放热水随潮转向潮流方向并向外海扩散。在涨潮时段,排放热水在潮动力作用下随潮向北输移,落潮时段则相反随潮向南,呈椭圆带状温水带(见图3)。一次落潮时热水前锋最远能到北帕斯湾顶前沿海域,在涨潮流的作用下,回归热水大部分随潮流折回向北,部分热水随流向近岸移动、累积。

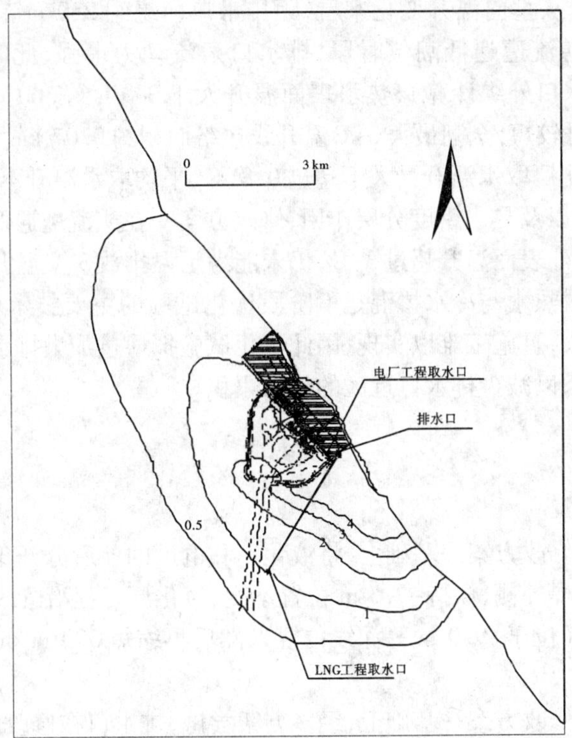

图2　原设计方案温升分布图

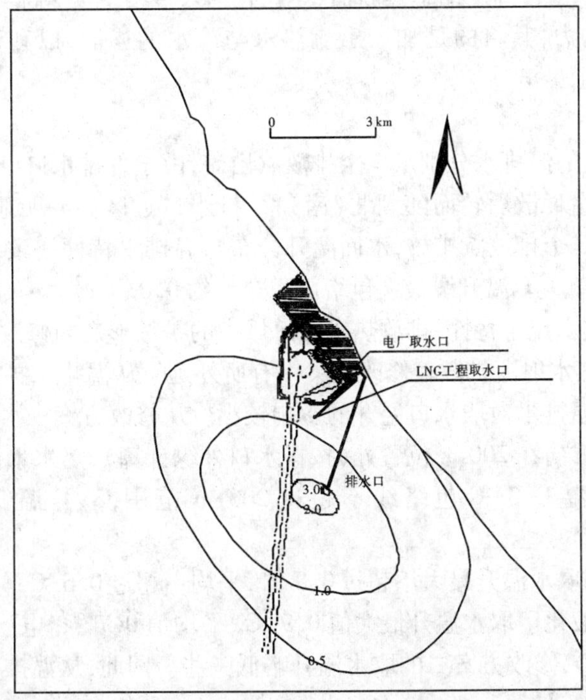

图3　修改方案一温升分布图

从温排放试验来看,从排水口排出的热水散热较快,由于出口的射流以及热水密度分层作用,热水上浮较快,从垂向温升变化来看,距排水口前缘约 250 m 范围内,温差值梯度较大,后逐渐减弱。在憩流遭遇低潮位时段,排水口射流动力最强,此时沿排水轴线方向在 200 m 位置处(按排水口外缘计算),热水表面温升大于 3 ℃,约 300 m 位置表面温升降至 3 ℃以内,且憩流历时较短,全潮最大 3 ℃温升线包络面积约 0.07 km^2。在电厂规划装机容量条件下,电厂和液化厂取水温升最大不超过 0.8 ℃,平均取水温升不超过 0.6 ℃。

由于出口的射流以及热水密度分层作用,修改方案一在憩流遭遇低潮位时段,沿排水轴线方向在 200 m 位置处表面温升超过 3 ℃,仍未能满足温排放 3 ℃温升线不超排水口200 m 的限制,继续加强排放热水与冷水掺混是下阶段优化试验的主要技术思路。再增加排水水深,不仅工程造价巨大,且施工难以实现,因此优化试验拟通过加大排水口出口的初始动量、多点分散排放等措施,以减少排水口近区的热水累积。

6　优化方案试验

6.1　方案布置

修改方案二是在修改方案一基础上,将液化厂和电厂的前后分开 3 点排放,电厂四孔排水口方涵放在南防波堤外侧,保持-9.5 m 位置不变,液化厂工程沿管线方向向内缩短200 m 布置三孔排水口方涵(位于-9.0 m 等深线位置),然后再缩短 300 m 布置三孔排水口(位于-8.0 m 等深线位置)。

修改方案三是在修改方案一基础上,经多方案尝试,排水口在修改方案一的基础上缩短 350 m,布置在南防波堤外-9.0 m 等深线附近,距离 LNG 工程取水口约 1.96 km,距离岸边约 3.2 km。压缩排水出口高度缩窄其断面,单孔几何尺寸变为宽×高=3.0 m×2.5 m,修改后增大了管排出流动力,其目的是加大出流热水与冷水的掺混,以利于热水向外海输移、散热。

6.2　试验成果分析

修改方案二排放的温排水分别从 3 个排放点出流,由于各排水口出流动量较小,其射流强度也较弱,待其动量耗散后转而随潮北(南)移,排水口近区温升明显较其他方案小。修改方案四采用 10 孔 3 个排放点排放,水面温升分布与温排水特性一致,各分排口前缘 3 ℃温升面积都较小,全潮 3 ℃温升线最大包络面积合计约 0.03 km^2。

修改方案三温排水流态特性:排放热水在海域中的扩散形态与修改方案一类同,因排水口头部压低 0.5 m,热水射流扩散较修改方案一有所外延。从温排水扩散流态看,该方案排水口位置基本合适,温排水对取水口温升的影响较小。从修改方案二水面温升变化来看,憩流时段,沿排水轴线方向在 200 m 位置处(按排水口外缘计算),热水水面温升接近 3 ℃,全潮排水口近区表面温升 3 ℃包络线不超过 200 m 范围,3 ℃温升线最大包络面积约0.035 km^2。

修改方案二全潮取水温升最大不超过 0.8 ℃,平均不超过 0.6 ℃;修改方案三取水温升小潮比大潮大,小潮液化厂取水温升最大值 0.9 ℃、平均值 0.7 ℃,电厂取水温升最大值为 0.8 ℃、平均值 0.7 ℃,修改方案二的取水温升略低一些。因此,从温排水效果来讲,两修改方案成立。

6.3　试验小结

(1)本水域的温排水运动特征及工程设计限制决定了取、排水口布置。“液化厂取水口位于港池南侧”、“排水管道不能穿过航道”、“排水口温升200 m范围内小于3 ℃”等设计要求,都是制约电厂排水口平面布置的主要因素。基于充分利用厂址前可利用水域对温排水进行稀释、散热的思路,优化阶段试验主要采取加大出口流速方案和分散排放两个修改方案进行了试验比较。

(2)修改方案二采用多点分散排放,扩大排水口附近水流掺混量,排水口附近温升影响大大减小。修改方案三采用缩窄出口断面的形式,增大了排放流速,加强了排放热水与底层冷水的掺混,使温排水尽快掺混冷却并向外输移;修改方案二和修改方案三,都能满足电厂安全运行要求,对排水口近区高温区的影响也较小,试验综合需考虑综合施工、造价及环境影响等多方面因素,最终推荐采用修改方案三。

7 结语

(1)本试验成功之处在于准确把握工程海域的环境流态,且利用温排水的浮射流特性,将排、取水口分别设在热水和冷水流道,尽量减少了两者之间的干扰,变相加大了排、取水口之间的距离,极大地降低了取水温升和表面温升。

(2)试验结果是在实验室条件下测得的,主要考虑了潮流作用及表面散热的影响,对于原体海域影响取水温升的其他不利因素主要有:不利风(浪)的影响,太阳辐射的影响,防波堤透水性的影响。从定性来看,上述因素对工程取水方案影响较小。

参考文献

[1] 陈惠泉.冷却水运动相似性原理和实践[J].水利学报,1964(4).

[2] 陈惠泉.火/核电厂冷却水试验研究50年的进展和体验[J].中国水利水电科学研究院学报,2008(4).

[3] 杨海燕.具有潮流影响的电厂温排放试验研究[J].贵州水力发电,2005(1):75-78.

【作者简介】 钟伟强,1975年9月生,毕业于河海大学水利水电工程学院,硕士学位,主要从事水力学及河流动力学应用研究,温排水热污染研究,水文及水资源、波浪研究等方面的工作,现为广东省水利水电科学研究院高级工程师。

淹没型旋流竖井泄洪洞流态过渡的数值模拟研究

张建民　胡小禹

（四川大学水力学及山区河流开发与保护国家重点试验室　成都　610065）

【摘要】　本文针对一种新型淹没出流型旋流竖井泄洪洞开展三维紊流模拟研究，获得了各个流段不同的水流流态、压力沿程分布及其特性，计算结果与试验结果符合良好。淹没型旋流竖井泄洪洞内水平洞段水流过渡过程流态可分为掺混消能区、气泡逸出区、气体排出区和平稳流动区，通过对数值计算所得流态、流速以及底板压强的分析，表明该新型旋流竖井泄洪洞在淹没条件下，气体能够顺利排出，流态转换平顺，压强分布平稳，并分析了强紊动掺气水流过渡到有压流动的转换机制及对结构的影响。为导流洞改建永久泄洪洞提供了一种新途径。

【关键词】　淹没型旋流竖井　流态过渡　紊流模型　水力特性

1　引言

导流洞改建为永久泄洪建筑物时常采用的内流消能工主要有三种方式：一是洞塞（孔板）技术。其进流方式为有压流，同时泄洪洞出口具有一定的淹没度。多级孔板已在黄河小浪底工程中应用。洞塞技术也在猴子岩和泸定设计中采用。二是水平旋流技术，其进流方式为明流，中间为有压流消能段，泄洪洞出口也为明流，该技术已在公伯峡工程中应用。三是旋流竖井技术。通常情况下，其流道中均为明流。该技术已在沙牌、仁宗海和狮子坪等工程中得到应用。第三种技术结构简单、布置灵活、消能率高，又能适应复杂的地形地质条件。但是，对于进口采用无压明流，出口为淹没出流的导流洞改建大流量、高水头大型泄洪洞，上述三种导流洞改建泄洪洞的技术无法直接应用。原因在于无压明流进口不宜采用洞塞（孔板）技术。采用旋流技术又面临两大难题：其一，该泄洪洞泄量大、水头高。因此，增加洞内消能效果，改善泄洪洞的流态，降低洞内流速并避免发生空蚀就显得尤为重要，也是技术难点之一。其二，下游水位较高，水流从无压内流泄水设施（竖井）过渡到有压内流设施（泄洪洞）时，水流流态转换时容易发生水跃等不利水力学现象，或者形成的强烈掺气及紊动水流是否影响泄洪洞的运行安全，有待深入细致的研究。

为解决下游高水位条件下导流洞改建为旋流竖井式泄洪洞的难题，孙双科等提出采用竖向压板技术优化洞内流态，并以多孔平板技术解决导流洞出口的排气问题。关于排气效果以及原模型水流之间可能存在的差异，不同学者尚存在不同看法。张建民等通过在泄洪洞（原导流洞）底部合适位置上设置挡坎、顶部设置排气孔和压坡等技术手段，先使水流在有限长度内剧烈混掺消能，其后通过挡坎和压坡的共同作用促使气体加速逸出，并将洞顶多余气体通过排气孔顺利排出，实现了在下游高水位情况下，竖井高掺气水流通过泄洪洞时由无压向有压的平稳过渡衔接。但是，该种泄洪洞内水流流态转换和过渡的详细过程尚待研究，为此，根据前人累积的经验，本文采用 VOF 法和 $k-\varepsilon$ 双方程紊流模型对淹没型旋流竖井泄洪洞水流流态和过渡过程进行了数值模拟。

2 数学模型

2.1 基本方程

连续方程

$$\frac{\partial\rho}{\partial t}+\frac{\partial\rho u_i}{\partial x_i}=0$$

动量方程

$$\frac{\partial\rho u_i}{\partial t}+\frac{\partial}{\partial x_j}(\rho u_i u_j)=-\frac{\partial p}{\partial x_i}+\frac{\partial}{\partial x_j}\left[(\mu+\mu_t)\left(\frac{\partial u_i}{\partial x_j}+\frac{\partial u_j}{\partial x_i}\right)\right]+\rho g_i$$

k 方程

$$\frac{\partial(\rho k)}{\partial t}+\frac{\partial(\rho u_i k)}{\partial x_i}=\frac{\partial}{\partial x_i}\left[\left(\mu+\frac{\mu_t}{\sigma_k}\right)\frac{\partial k}{\partial x_i}\right]+G-\rho\varepsilon$$

ε 方程

$$\frac{\partial(\rho\varepsilon)}{\partial t}+\frac{\partial(\rho u_i\varepsilon)}{\partial x_i}=\frac{\partial}{\partial x_i}\left[\left(k\mu+\frac{\mu_t}{\sigma_\varepsilon}\right)\frac{\partial\varepsilon}{\partial x_i}\right]+C_{1\varepsilon}\frac{\varepsilon}{k}G-C_{2\varepsilon}\rho\frac{\varepsilon^2}{k}$$

式中：ρ、μ 分别为体积分数加权平均的密度和分子黏性系数。

μ_t 为紊流黏性系数，它可由紊动能 G 和紊动耗散率 ε 求出。

$$\mu_t=\rho C_\mu\frac{k^2}{\varepsilon}$$

C_μ 为经验系数，取 $C_\mu=0.09$。

σ_k 和 σ_τ 分别为 k 和 ε 的紊流普朗特数

$$\sigma_k=1.0,\ \sigma_\varepsilon=1.3$$

$C_{1\tau}$ 和 $C_{2\tau}$ 为 ε 方程中的常数，

$$C_{1\varepsilon}=1.44,\ C_{2\varepsilon}=1.92$$

G 为平均速度梯度引起的紊动能产生项，定义为

$$G=\mu_t\left(\frac{\partial u_i}{\partial x_j}+\frac{\partial u_j}{\partial x_i}\right)\frac{\partial u_i}{\partial x_j}$$

基本方程采用控制体积法进行离散，压力和速度的耦合计算采用 PISO 算法，对固壁边界采用壁函数法处理。

2.2 基本体型及参数

本文采用文献[3]中模型试验采用的竖井泄洪洞体型及参数(见图 1)。其最大泄量 1 200 m^3/s，上下游水位差 125.23 m。竖井为双涡室掺气型旋流竖井。涡室直径均为 D=18 m，竖井直径 d=12 m，竖井高度 H=161 m，导流洞结合段长度 L=511.20 m，为城门洞型(16 m×15.38 m，顶高 4.62 m)。竖井与泄洪洞下水平段采用直接连接型式，竖井为无压垂直出口，泄洪洞进流为垂直淹没射流，取消传统的消力井；在 X=131 m 处设置梯形坎，高

3 m,顶宽 3 m,底面宽 6 m;$X=161$ m 处设置排气洞,内径为 3 m;$X=183$ m 处设置有压坡,高 5 m,孔口高度为 15 m;泄洪洞出口处也设置有压坡,高 8 m,孔口高度为 12 m。数值模型原点取竖井中心轴与相汇的泄洪洞中轴交点,高于泄洪洞底板 7.75 m,竖直向上设定为 Z 轴,沿泄洪洞指向下游方向设定为 X 轴,垂直于 XY 平面指向泄洪洞内水流左侧设定为 Y 轴方向。其具体参数如图 2 所示。

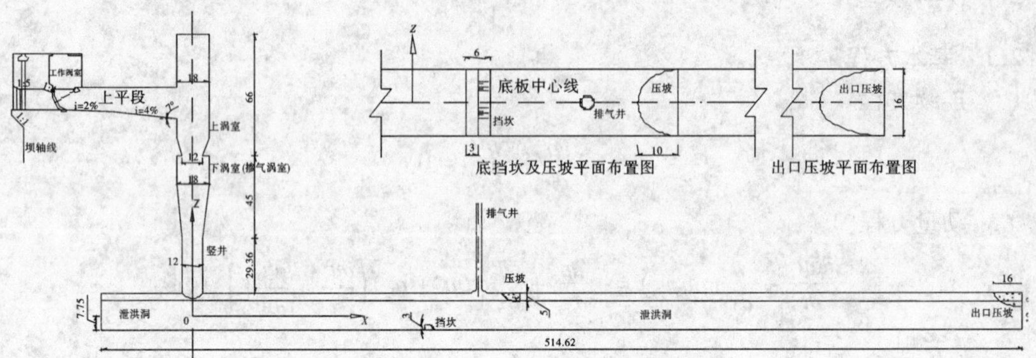

图 1　计算基本体型及相关参数示意图

2.3　网格划分及边界条件

数值模拟范围包括引水道、双涡室和泄洪洞以及下游引渠,利用结构化与非结构化的有机结合对旋流竖井泄洪洞进行合理的网格划分,网格尺寸范围为 0.05~2.00 m。对于几何结构相对规则的上库、引水道、泄洪洞和下游,采用容易实现区域边界拟合,易于计算且结果较准确的结构化网格划分,同时根据流态情况进行相应的指数加密;对于几何结构复杂的引水道与涡室连接段及竖井与泄洪洞连接段,采用更具灵活性的非结构化网格剖分,同时合理加密以适应流态的急剧变化,提高计算精度;而对于涡室和竖井段,则采用同心圆形结构化网格划分水流旋转的贴壁部分,其内部气核部分则采用非结构化网格,以期能较好地模拟内部螺旋流的水力学特性。

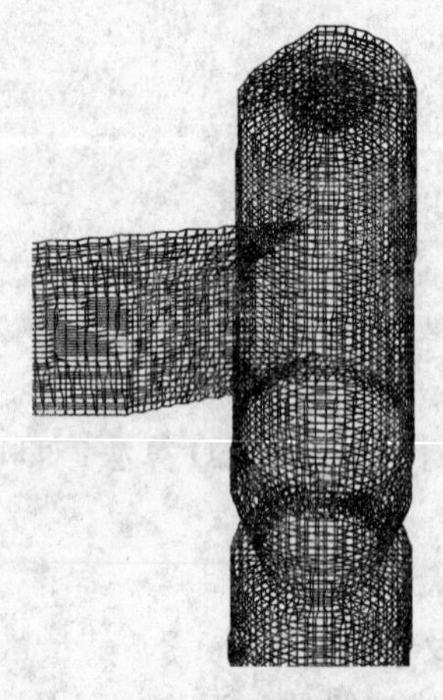

(a)涡室段网格

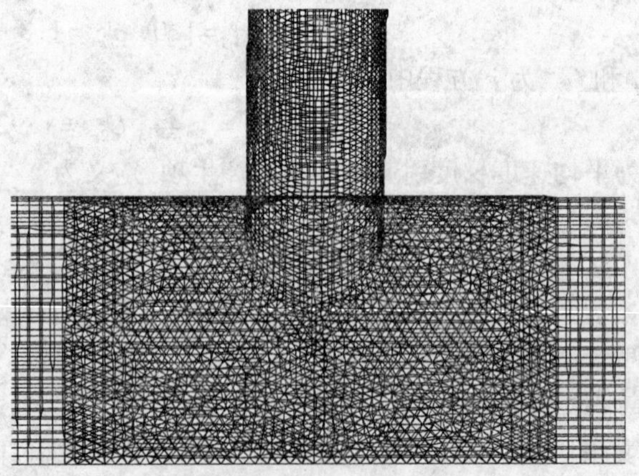

(b)竖井与泄洪洞连接段网格

图 2　主要计算区域的网格划分图

2.4 边界条件

(1)入口边界,给定库水位为设计水位,设计流量为1 200 m^3/s,有

$$v = Q/A$$

式中:v 为进口水流速度;Q 为进口流量;A 为进口面积。

(2)泄洪洞出口为淹没出流,在其出口按照下游水位设定自由出流边界。

(3)固壁采用无滑移边界,采用壁函数法。

3 计算结果与分析

3.1 压强特性

对于这种淹没出流式旋流竖井,泄洪洞底板压强特别值得关注。图3比较了底板及侧壁压强的模型试验结果与数值计算结果,其中底板中心压强指 $Y=0$ m,$Z=-7.75$ m处压强沿程分布,底板边缘压强指 $Y=5$ m,$Z=-7.75$ m处压强沿程分布,侧壁压强指 $Y=8$ m,$Z=5.5$ m处压强沿程分布。

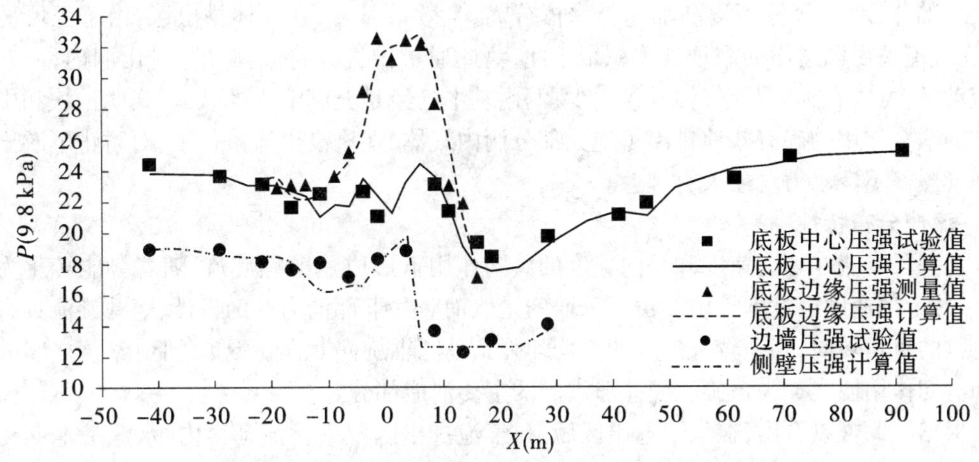

图3 泄洪洞掺混消能区底板及侧壁压强计算结果与试验结果对比图

由图3可知,总体看来,淹没型旋流竖井泄洪洞底板压强分布较为合理,数值计算的结果和试验结果的底板压强分布规律吻合良好,计算结果可信。

可以看出,泄洪洞底板没有特别突出的压强,其大小普遍在30×9.81 kPa以下,在不考虑底板分缝的条件下,竖井和泄洪洞采用直接衔接的体型是可行的;掺混消能区的底板上出现的极大值,基本分布在竖井边缘下部,分析是由于竖井内高速贴壁螺旋流冲击所致;掺混消能区竖井前方,由于出现水流的向上翻滚,产生一定的离心作用,同时气体的大量逸出导致水层较薄,此处压强出现了约为17×9.81 kPa的极小值。其后,水流的在洞内流态得到有效调整,由于下游为淹没出流条件,在洞内排气设施作用下,水流逐步由高掺气水流转化为满流,压强小幅度升高后保持相对稳定。

3.2 水流流态

图4是泄洪洞下平段流态转换过程数值计算结果。在下游淹没出流条件下,如何有效分离水中气体,保证泄洪洞内流态是极其重要的。竖井段水流旋转卷吸大量空气,形成强紊动掺气水流进入泄洪洞,随后气泡沿程不断溢出,泄洪洞一段范围内形成掺混消能区、气泡

逸出区、气体排出区及水流平稳区。

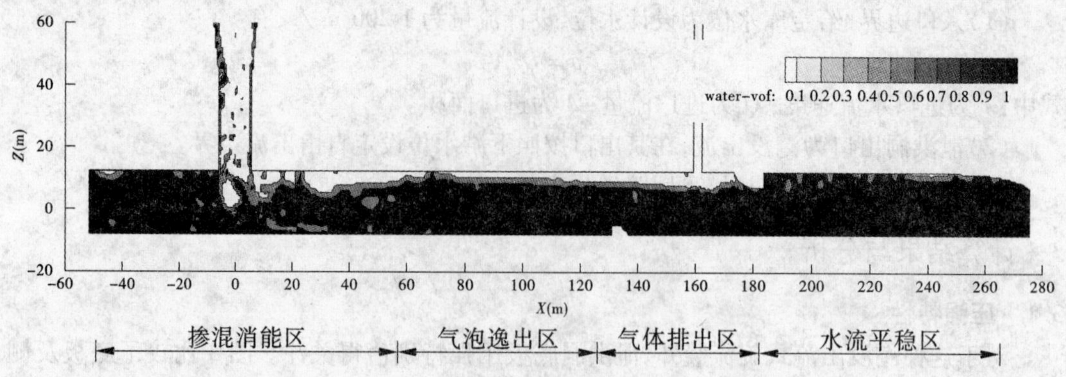

图4　洞内流态(水汽百分数)数值计算结果

在掺混消能区,水气之间形态上犬牙交错,说明在该区域水气之间相互旋滚剧烈,有利于能量的消刹;而在气泡逸出区,气体已经大部分逸出水面,水气初步分层,说明在泄洪洞底部挡坎的作用下,气体正逐渐加速与水体分离,且效果显著;在气体排出区,由于压坡的存在,在气泡逸出区逸出的洞顶气体积聚于此,藉此通过排气井将气体充分排出洞内;而在平稳流动区,可以看到,洞顶气体已经很少,洞内基本已经是充满清水的状态,说明挡坎和中部压坡的联合作用的确有效地排出了绝大部分洞内气体,在淹没出流条件下,对防止下游出口出现气爆等不良水力现象大有裨益。

3.3　流速分布特性

竖井段水流在切向速度和垂向速度的共同作用下形成贴壁螺旋流,随着势能转化为动能,竖井下部速度已达到极值。由于采取的是取消竖井下部消力井的设计,竖井旋流具有更加急剧的直角转向,存在较大尺度的不均匀强紊动三轴旋涡。在掺混消能区,紊流之间的旋转和剪切作用使之成为该旋流竖井泄洪洞的主要消能部分。

根据图5可以看出,在气泡逸出区后部、气泡排出区及平稳流动区内,水流基本恢复呈层流状态,实现了竖井与泄洪洞之间的平稳过渡衔接,尤其是在中部压坡的调整后,水流流态趋于均匀和稳定,且计算发现该区中水流流速大部分已经小于4.5 m/s,同试验结果吻合,该区洞顶残余的少量气囊在低流速下不足以对洞顶结构和洞内水力特性产生不良影响。

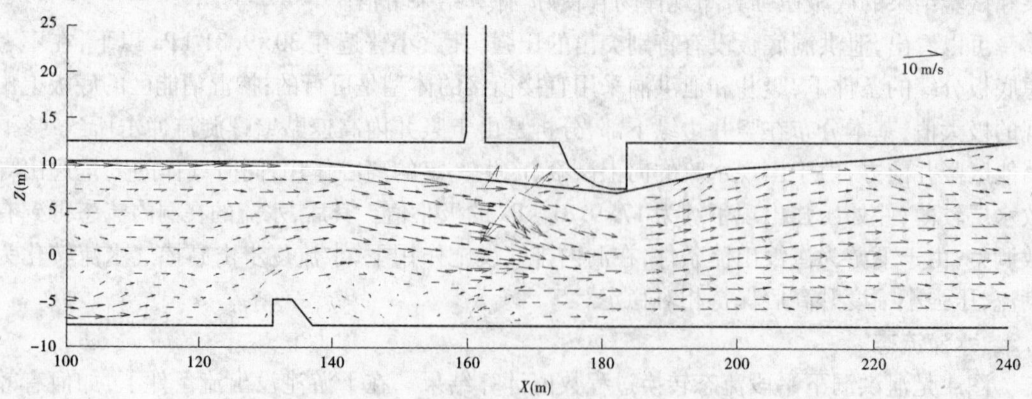

图5　气泡排出区及平稳流动区速度矢量示意图

4 结论

(1)利用数值模拟方法对淹没型旋流竖井泄洪的水流流态和过渡过程进行计算分析和研究,并将计算结果与试验结果进行对比。研究表明采用 VOF 法和 $k-\varepsilon$ 双方程模型模拟淹没型旋流竖井所得结果能够较为真实可靠地反映竖井及洞内水流流态转变及速度和压强分布与模型试验相比,体现了数值模拟技术的优越性。

(2)对淹没型旋流竖井泄洪洞的洞底压强进行了计算分析,在底板不分纵缝的情况下,稳定能够得到保证。

(3)对竖井与水平泄洪洞衔接段以及淹没条件下泄洪洞内流态进行了分析。研究表明,在淹没出流条件下,强掺气强紊动水流能够较为平顺地转换为有压水流,同时平稳地排出绝大部分气体,洞内及洞口不会产生不利水力现象。

(4)数值计算和模型试验的结果均表明,这种新型的淹没型旋流竖井泄洪洞流态过渡过程衔接平稳,没有出现不利的水力学条件,值得推广应用于类似工程。

参考文献

[1] 董兴林,高季章.超临界流旋涡竖井式溢洪道设计研究[J].水力发电,1996(1):44-48.

[2] 孙双科,等.下游高水位条件下导流洞改建为旋流竖井式泄洪洞的水力学问题研究[J].水利学报.2000(10):22-27.

[3] 雷刚,张建民,谢金元,等.一种新型掺气型旋流竖井的试验研究[J].水力发电学报,2011(5):86-91.

[4] Hua Yong Chen, Wei lin Xu, Jun Deng, et al. Theoretical and Experimental Studies of Hydraulic Characteristics of Discharge Tunnel with Vortex Drop[J]. Journal of Hydrodynamics, Ser. B, 2010, 22(4):582-589

【作者简介】 张建民,四川大学教授,博士生导师,研究方向为水工水力学。

六、岩土工程与结构材料

60 cm 厚混凝土碾压技术研究与应用

汪永剑

（广东水电二局股份有限公司　增城　511340）

【摘要】 为突破当前碾压混凝土压实厚度只有30 cm左右的限制，推动碾压混凝土施工技术的发展，引进国外新型垂直碾压设备和深层压实度检测仪器，通过两次生产性试验，对检测仪器进行改进，确定了60 cm厚层混凝土碾压工艺，并成功应用于108 m高碾压混凝土双曲拱坝局部坝段。60 cm厚层混凝土碾压技术可有效缩短碾压混凝土大坝施工工期，具有显著的经济效益和推广应用价值。

【关键词】 碾压混凝土坝　碾压层厚60 cm　垂直振动碾　ROCKY型改良密度仪

1 引言

因碾压混凝土（简称RCC）性价比的优势，越来越多的水电站挡水建筑物采用RCC坝，但受碾压设备和压实度检测仪器的限制，RCC坝施工均采用混凝土铺料厚约35 cm、压实厚约30 cm的薄层碾压工艺，使得RCC坝施工工期较长，且相对于本体混凝土属薄弱环节的碾压层面较多。为突破RCC常规薄层碾压的限制，缩短RCC坝施工工期，提高坝体施工质量，推动RCC筑坝技术的持续发展，近几年在国内RCC坝科研、设计、施工专家的推动下，引进国外新型垂直碾压设备和深层压实度检测仪器，对混凝土厚层碾压进行试验研究，取得了丰富的试验数据，并成功应用于黄花寨水电站108 m高碾压混凝土双曲拱坝局部坝段，为混凝土厚层碾压技术的推广应用奠定了一定的基础。

2 混凝土厚层碾压试验研究

为检验碾压设备和检测仪器的适用性、获取合适的碾压工艺参数，进行RCC厚层碾压工艺试验。

2.1 落脚河厚层碾压试验

为验证混凝土厚层碾压设备性能及压实度检测仪器的适用性，2006年9月在贵州落脚河水电站工地进行第一次生产性试验。采用新型SD451型垂直振动碾（性能见表1）和常规的BM202-AD型振动碾进行混凝土厚层（50 cm、75 cm、100 cm）碾压的效能对比，采用国产核子水分密度仪和新型插入式RI密度仪进行压实度检测对比。通过现场试验，得到2种碾压设备在不同铺料和压实厚度、碾压遍数时的压实数据，经长期从事RCC设计、施工、教学和科研工作的11名专家评审，认为50～100 cm厚混凝土采用SD451型垂直振动碾碾压，能满足压实度达到97%的要求。但是国产核子水分密度仪检测压实度时测深仅为30 cm，超过 30 cm 则需挖坑下卧检测，会扰动已碾实的混凝土，对 RCC 质量有一定影响；而插入型 RI 密度仪检测时，需在混凝土铺料时将成孔钢管水平埋入不同深度的检测部位（钢管要接出坝立面模板外，以检测时插入探测杆），不适用于目前规范条件下的实际施工。因此，RCC 碾压层厚要突破 30 cm，还需对碾压工艺进一步论证和研究。

表 1　SD451 型垂直振动碾主要性能

性能	工作质量（t）	全长（m）	全宽（m）	轮径（mm）	碾压宽（m）	功率（kW）	最大激振力（kN/轮）	振动频率（转/min）	最大振幅（mm）
指标	11	4.02	2.27	1 000	2.10	124	226	2600	1.4

2.2　马堵山厚层碾压试验

为进一步论证和研究混凝土厚层碾压工艺（特别是压实度检测工艺），获取合适的碾压工艺参数，2008 年 11 月在云南马堵山水电站工地进行第二次生产性试验。采用 SD451 型垂直振动碾碾压，采用 MC-4 核子湿度密度仪（30 cm 薄层检测）、MC-S-24 核子湿度密度仪（最大测深 60 cm）、大型表面透过型 RI 密度仪（最大测深 50 cm）、轻便 1 孔式 RI 密度计（最大测深 100 cm）4 种仪器进行对比检测。通过现场试验，取得了不同摊铺厚度（27 cm、41 cm、54 cm）、压实层厚（50 cm、75 cm、100 cm）和碾压遍数（2—8—2、2—10—2、2—12—2）的混凝土压实度数据 408 个，经来自中国水利学会、水力发电学会 RCC 筑坝技术分会及科研院校等单位的 13 名专家评审，认为 50 cm、75 cm 层厚混凝土采用 SD451 型垂直振动碾振压 10～12 遍条件下，相对压实度可以达到规范要求；三种新型的密度检测仪均能检测厚度超过 30 cm（最大测深 100 cm）RCC 的密度，但 M-S-24 型核子湿度密度仪为双杆式，铁钎打孔拔出后，两孔间的混凝土回弹使孔距缩小，两根测杆很难同时插入甚至无法插入孔内检测；单杆式 RI 密度计虽有改进（成孔钢管由水平预埋改为垂直钻孔后埋），但钻孔后埋钢管速度慢，且仍不能直接读取密度值，两类型的检测仪器应改进以适应工程检测需要。

2.3　压实度检测仪器改进

通过 2 次试验，压实度检测仪的生产厂家对大型表面透过型 RI 密度仪进行改进，改进后的仪器称为 ROCKY 型改良密度仪，其测深最大为 60 cm，可直接读取密度和压实度数据，单孔打孔式成孔速度快，其主要性能见表 2。

表 2　ROCKY 型改良密度仪性能

测定方法	检测范围（g/cm^3）	线源深度（cm）	测定深度（cm）	测定时间后台+计测	线源	检出器	显示	记录	机器尺寸 $W\times L\times H$	使用温度（℃）
伽马线透过型	1.6～2.5	35、45、55	40、50、60	1 min+1 min	60Co 3.7 MBq 以下	闪烁检出器	4 位数液晶显示×2 频道	数字打印	240×350×270	0～50

注：1. ROCKY 型改良密度仪电源为内置式充电电池，1 次充电可连续使用 10 h；

2. 仪器精度：测 60 cm 深为 0.02 g/cm^3、测 50 cm 深为 0.01 g/cm^3、测 40 cm 深为 0.004 g/cm^3。

3　混凝土厚层碾压应用

3.1　工程简介

黄花寨水电站位于贵州省蒙江流域干流格凸河上，装机容量 2×27 MW，水库正常蓄水位795.5 m,相应库容1.6亿m^3。电站枢纽大坝为RCC双曲拱坝,最大坝高108 m,坝顶高程800 m,坝顶宽6 m,坝顶弧长243.6 m,坝底最大宽度25.3 m。坝体设4条横向诱导缝,上游面采用二级配富胶材RCC($C_{90}20$)及二级配变态混凝土($C_{90}20$)防渗,下游面采用三级配RCC($C_{90}20$)及三级配变态混凝土($C_{90}20$),RCC工程量28.5万m^3。

3.2 应用部位

混凝土厚层碾压在右岸非溢流坝段789.5~800.0 m进行,分789.5~792.5 m、792.5~795.5 m、795.5~798.5 m、798.5~800.0 m四个仓块,仓面面积为640~778 m^2。因受混凝土综合运输条件限制,不能满足通仓厚层碾压施工要求,各仓仅能在长40~50 m约300 m^2范围进行厚层碾压施工(其余区域30 cm斜层碾压)。为减小施工干扰,加快作业速度,将坝体的4种混凝土均改为三级配RCC,上下游面50 cm范围采用三级配变态混凝土。

3.3 RCC厚层碾压施工

3.3.1 施工配合比

大坝工程所用混凝土配合比见表3。

表3 RCC配合比

强度等级	级配	水胶比	1 m^3混凝土材料用量(kg)								
			水	水泥	粉煤灰/掺量	减水剂ADD-3	氧化镁	砂/砂率	碎石(mm)		
									5~20	20~40	40~80
$C_{90}20W6F100$	三	0.53	87	82	82/50%	1.31	6.56	738/33%	449	599	449

采用畅达瑞安水泥厂P·O42.5水泥、安顺电厂Ⅱ级粉煤灰、北京科宁外加剂厂ADD-3型缓凝高效减水剂、辽宁海城东方滑镁公司轻烧氧化镁,采用人工砂石料(石灰岩母材),混凝土凝结时间为9 h 10 min,容重为2 480 kg/m^3。

3.3.2 压实度检测仪器的率定

30 cm深混凝土压实度采用HS-2002核子密度仪检测,40~60 cm深采用ROCKY型改良密度仪检测。按使用说明书要求,仪器在使用前率定。

率定块制作4块,尺寸为1 m×1 m×1 m。1#、2#率定块分别采用三级配、二级配RCC制作(分三层铺料,用插入式振捣棒分层振实),3#率定块用三级配RCC制作(分三层铺料但不予振捣);4#率定块采用人工砂填筑,每填20 cm高,用铁铲耙平再装上一层。率定块制作后,按使用说明书要求对两种仪器进行率定。

3.3.3 RCC生产和运输

混凝土采用DW240型双卧轴连续推进式搅拌站拌和,理论生产能力240 m^3/h,实际最大生产能力180 m^3/h。

仓面外水平运输:4台20 t自卸车将拌好的RCC运至大坝左坝头(运距2.6 km),沿坝顶及其交通桥倒运至1#闸墩顶(运距190 m),卸入抗分离溜管上部集料斗,运输能力约100 m^3/h。

垂直运输:采用管径500 mm的抗分离溜管,运输能力最小为248 m^3/h。

仓面水平运输:采用1台20 t自卸汽车,运距约80 m,因仓面行车宽度仅5.1 m,自卸车不能掉头,严重影响其运输速度,运输能力约50 m^3/h。

3.3.4 混凝土碾压工艺

根据落脚河和马堵山厚层碾压试验结果,确定 RCC 厚层碾压参数:铺料厚 66 cm,两次铺料,每次铺 33 cm;采用 SB-11 平仓机平仓、SD451 型垂直振动碾碾压,碾压速度控制在 2 km/h内,碾压遍数为 2—10—2(静压 2 遍+振压 10 遍+静压 2 遍),压实厚 60 cm。碾压条带搭接 10~20 cm,端头部位搭接约 100 cm。

3.4 施工检测

为验证厚层 RCC 的施工质量,对压实度、试块及实体抗压和抗渗等性能进行检测。

3.4.1 压实度检测

混凝土按规定遍数碾压后,先用 ROCKY 型改良密度仪进行 40 cm、50 cm、60 cm 深处混凝土压实度检测,再用 HS-2002 核子密度仪进行 30 cm 深处压实度检测,两种仪器检测为相近点检测。每个碾压层检测 5 点,789.5~792.5 m 仓块检测点布置见图 1,其余各仓块检测点布置参照图 1。检测结果汇总见表 4。

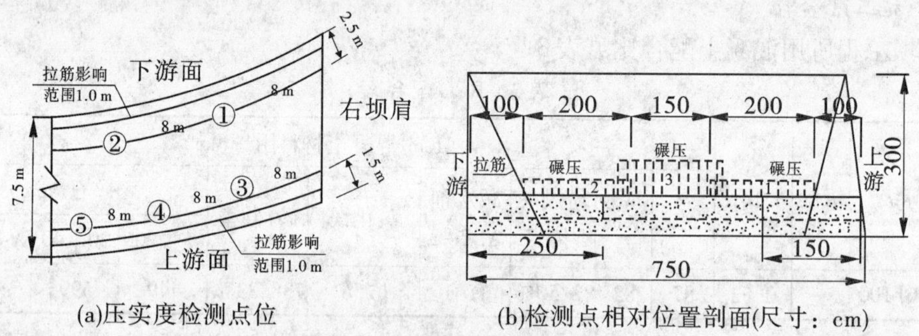

(a)压实度检测点位 (b)检测点相对位置剖面(尺寸:cm)

图 1 坝块压实度检测点位

表 4 厚层 RCC 压实度汇总

仓块高程(m)	检测深度(cm)	压实度(%)			检测数(点)	合格数(点)	每层合格率(%)	每仓合格率(%)
		最大	最小	平均				
789.5~792.5	30	103.6	93.5	99.7	25	24	96	99
	40	103.6	97.3	99.8	24	24	100	
	50	102.6	97.0	100.1	24	24	100	
	60	104.8	97.2	100.6	23	23	100	
792.5~795.5	30	99.3	97.1	98.2	25	25	100	97
	40	101.7	95.3	99.4	25	24	96	
	50	102.2	95.1	99.7	25	24	96	
	60	103.9	95.6	100.4	25	24	96	
795.5~798.5	30	101.2	94.3	97.8	25	24	96	99
	40	100.5	95.9	98.7	25	25	100	
	50	100.2	97.1	98.8	24	24	100	
	60	103.0	97.1	99.7	23	23	100	
798.5~800.0	30	101.4	97.3	98.3	10	10	100	100
	40	101.5	97.0	98.4	10	10	100	
	50	101.5	97.7	99.2	10	10	100	
	60	100.9	97.3	98.6	10	10	100	
总计					333	328	98.5	

从表 4 可知,一次压实度合格率达 98.5%。

3.4.2 混凝土试块及实体芯样强度检测

2011 年 2 ~4 月混凝土厚层碾压施工时，每个仓块 28 d、90 d 抗压和抗渗试块各取 1 组，经养护后检测；2011 年 5 ~6 月在坝体钻孔取芯，芯样直径 190 mm，钻 1#、2#、3#共 3 孔，进尺分别为 9.8 m、9.8 m、8.86 m，取得芯样长 9.65 m、9.65 m、8.73 m，芯样获得率分别为 98.4%、98.4%、98.5%。芯样胶结较好、骨料分布基本均匀，层面少量骨料架空，芯样层缝面折断率为 47.1%，单层碾压层内混凝土未断裂，混凝土芯样见图 2。混凝土试块及芯样强度均超过设计要求的 C_{90}20W6，检测结果见表 5。

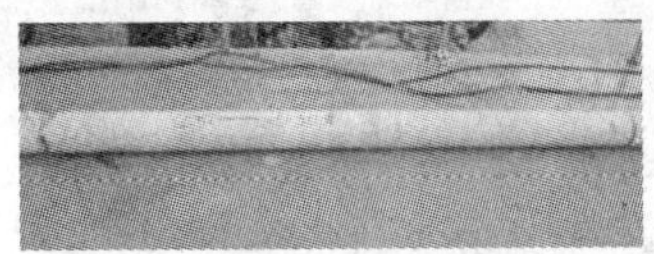
(a)1#孔 3 m 长芯样

(b)2#孔全孔芯样

(c)3#孔全孔芯样

图 2 混凝土芯样

表 5 混凝土芯样及试块抗压、抗渗强度

部位(m)	坝体芯样			混凝土试块			
	龄期(d)	抗压强度(MPa)	抗渗等级	28 d 抗压强度(MPa)	90 d 抗压强度(MPa)	28 d 抗渗等级	90 d 抗渗等级
789.5 ~792.5	113	23.4	W6	20.4	28.9	W6	W6
792.5 ~795.5	100	23.8	W6	17.7	26.0	W4	W6
795.5 ~798.5	86	24.2	W6	19.3	27.1	W6	W6
798.5 ~800.0	76	23.8	W6	19.6	27.5	W6	W6

4 结语

（1）新型垂直振动碾进行 60 cm 厚层混凝土碾压（铺料厚 66 cm，分两次铺，每次铺 33 cm），碾压遍数 2—10—2（静压 2 遍+振压 10 遍+静压 2 遍），混凝土压实度检测 333 点，超过 97.0% 的有 328 点，合格率为 98.5%。坝体芯样混凝土抗压及抗渗强度分别取试样 12 块和 4 组，均超过设计要求的 C_{90}20W6；芯样完整、胶结好、骨料分布基本均匀，层面少量骨料架空，芯样断裂在层缝面，单层碾压层内混凝土未断裂；检测结果表明，采用 60 cm 厚层碾压工艺施工的混凝土坝体各项质量指标满足标准要求，新型垂直振动碾碾压 60 cm 厚混凝土是可行的。

（2）改良后的 ROCKY 型密度仪检测深度可达 60 cm，可直接读取实测密度值及压实度数据，成孔方式采用单孔打孔，速度较快（从打孔到检测完 1 孔 40 cm、50 cm、60 cm 深各点压实度，平均时间 13 min），能满足 30 cm 以上厚层碾压混凝土在施工过程的检测要求。但改良型 ROCKY 密度仪只能检测 40 cm、50 cm、60 cm 深压实度，尚应继续改进，使其能检测 30 cm 深混凝土的压实度，以便与现有检测 30 cm 深密度检测仪检测的压实度进行对比，使其检测成果更具说服力。

（3）厚层混凝土碾压速度比常规薄层碾压快，对混凝土入仓能力要求较高。本次 60 cm 厚层碾压，由于仓面狭窄、作业设备多（包括振动碾、推土机、吊机、自卸车等）、施工工序多

(铺料、平仓、碾压、检测、铺预制块等),对混凝土施工速度有一定的影响,混凝土实际浇筑速度约35 m^3/h,使混凝土层间铺料间隔时间局部超过初凝时间而导致混凝土芯样层间断裂较多。因此,采用混凝土厚层碾压施工工艺时应根据不同的仓面面积及通行条件,选用合适的混凝土运输方法及碾压层厚,以保证混凝土入仓能力满足混凝土铺料层间间隔时间要求。

(4)混凝土60 cm厚层碾压技术突破了我国碾压混凝土常规30 cm层厚的碾压筑坝方法,可有效缩短碾压混凝土大坝施工工期,具有显著的经济效益和推广应用价值

参考文献

[1] 汪永剑,丁仕辉.混凝土厚层碾压试验研究[J].混凝土与水泥制品,2010(2):16-20.
[2] 李春敏,刘树坤.黄花寨水电站RCC坝大层厚现场试验成果及启示[J].水利水电技术,2007,38(4):39-43.
[3] 月本行则,岩限秀树,李春敏,等.垂直振动压路机的主要特性[J].中国水利,2007(21):67.
[4] 汪永剑,丁仕辉,等.高陡坡运输RCC抗分离溜管的研制与应用[J].水利水电技术,2011,42(10):91-93.

【作者简介】 汪永剑,1971年10月生,2004年毕业于武汉大学,本科,2010年获工程硕士学位,广东水电二局股份有限公司,主任工程师/副所长,高级工程师,广东省青年科学家协会会员。

水库大坝补强修复方案优选模型研究*

苏怀智[1,2] 胡 江[2]

(1. 河海大学水文水资源与水利工程科学国家重点实验室 南京 210098;
2. 河海大学水利水电学院 南京 210098)

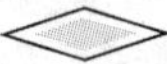

【摘要】 21世纪是病险水库补强修复的高峰期,补强修复方案优选方法成为亟待解决的热点问题。近年来,基于风险的补强修复方案优选方法逐步取代了传统确定性方法。本文介绍了结构风险率的计算方法,研究了包括生命损失在内的补强修复效益评价方法,提出了安全条件约束下以溃坝期望损失、平均失效风险差和补强修复投资为变量的综合优选方法。将方法用于某病险水库补强修复方案优选中,得到了最优方案,结果表明,对失效损失巨大的民生工程,最优补强修复方案是在可靠度将低于规定值时立即进行,文中理论和方法能为当前水利工作提供有力的技术支持。

【关键词】 病险水库 补强修复 风险率 效益评价 生命质量指数

基于实测结构信息,综合考虑结构长期性能和效益,融合解析、数值方法和风险理论寻求劣化在役结构补强修复最优方案,符合现代坝工理念,逐步取代了传统确定性方法。进入21世纪以后,我国大坝老化问题日益突出,病险水库补强修复方案优选方法成为亟待解决的热点问题。同时,《水库大坝安全鉴定办法》规定对鉴定为三类坝、二类坝的水库,应对潜在溃坝方式和期望损失进行评估,采取补强修复、降等或报废等措施予以处理。

针对上述问题,本文系统研究了水库大坝补强修复方案优选方法。探讨了补强修复后全寿命周期内风险率平均变化量的计算方法;以有无对比原则为基础,将全寿命周期效益引入大坝补强修复工程中,研究了在役大坝全寿命周期的包括生命损失在内的综合效益评价方法;以溃坝期望损失、平均失效风险率差和补强修复成本等为变量,提出了安全条件约束下的在役坝的全寿命补强修复方案优选模型;将上述方法应用到某病险土坝的补强修复方案优选过程中,验证了方法的可行性。

1 结构风险率

裂缝、渗漏溶蚀、冻融、疲劳荷载及材料性能退化等多因素耦合作用导致在役坝性能劣化,严重影响其安全水平缩短其使用寿命。不同补强修复措施下,在役坝的长期性能会存在较大差别。

* 基金项目:国家自然科学基金(51179066,51139001),新世纪优秀人才支持计划(NCET-10-0359),国家重点实验室专项经费资助项目(2009586912)。

假定结构失效模式的功能函数为

$$Z = g(x_1, x_2, \cdots, x_n) \tag{1}$$

式中：$x_1, x_2, \cdots, x_n$ 为与结构抗力、荷载效应等有关的基本随机变量。

失效概率可表示为

$$P_f = P(Z < 0) = \int_\Omega f(x_1, x_2, \cdots, x_n)\,dx_1 dx_2 \cdots dx_n \tag{2}$$

式中：Ω 为失效域；$f(x_1, x_2, \cdots, x_n)$ 为 $x_1, x_2, \cdots, x_n$ 的联合概率密度函数。

Ditlevsen 提出的广义可靠指标为

$$\beta = \Phi^{-1}(P_s) = -\Phi^{-1}(1 - P_s) = \Phi^{-1}(P_f) \tag{3}$$

式中：P_s、P_f分别为可靠、失效概率；$\Phi(\cdot)$为标准正态累积概率分布函数；β 为广义可靠指标。

若考虑结构性能和外荷载的动态变化，可得时变失效概率 $P_f(t)$。设第 j 次补强修复运行期间，结构的可靠概率 P_s在补强修复下和未补强修复条件下 P_{s0}的平均差值为

$$\Delta P_s^j(T) = \frac{\int_0^T P_s(t)\,dt}{\int_0^T dt} - \frac{\int_{(j-1)T}^{jT} P_{0s}(t)\,dt}{\int_{(j-1)T}^{jT} dt} \tag{4}$$

式中：T 为补强修复间隔。

从而全寿命可靠概率平均差值为

$$\Delta \overline{P}_s(T) = \frac{1}{\Gamma}\sum_{j=1}^{\Gamma} \Delta P_s^j \tag{5}$$

式中：Γ 为补强修复次数，$\Gamma = T_{sl}/T$；T_{sl}为服役全寿命。

2 病险水库补强修复效益评价

随着使用年限的增长，服役坝表现出显著时变性，失效风险率渐趋增加。补强修复的目的是降低安全隐患，效益则体现在减少溃坝概率降低期望损失。采用有无对比的效益计算原则，计算补强修复效益。

2.1 溃坝损失的计算

分别计算有、无补强修复措施水库发生溃坝的期望损失 $C_{failure}$ 及补强修复费用 C_r。期望损失由当年溃坝损失乘失效概率得到。溃坝损失包括生命损失、经济损失和非经济损失，即

$$C_{failure} = C_l + C_e + C_s \tag{6}$$

式中：$C_{failure}$为溃坝损失；C_l为生命损失；C_s为非经济损失；C_e为经济损失。

$$C_e = C_{de} + C_{ide} \tag{7}$$

式中：C_{de}、C_{ide}分别为直接、间接经济损失。

直接经济损失通过经济统计调查获得，财产损失率则由失效模式、受灾区地形特征、抢险措施等共同决定。间接经济损失 C_{ide}采用折算系数法

$$C_{ide} = \alpha C_{de} \tag{8}$$

式中：α 为溃坝间接经济损失系数。

经济分析一般可折算成货币成本予以考虑。

2.2 生命损失的计算

生命质量指数(LQI)是平衡有限社会资源和提高生命安全减少人身伤害的有效工具,表示如下[4]

$$L(g,l,q)=\frac{g^q}{q}l \tag{9}$$

式中:L 为 LQI;g 为人均 GDP;l 为人的预期寿命;q 为寿命中用于工作的比例,$q=w/(1-w)$,$w=0.1\sim0.12$。

实施减灾措施时,将对 LQI 产生影响,即

$$\mathrm{d}L(g,l,q)=\frac{\partial L(g,l,q)}{\partial g}\mathrm{d}g+\frac{\partial L(g,l,q)}{\partial l}\mathrm{d}l\geqslant 0 \tag{10}$$

式中:$\mathrm{d}L(g,l,q)$表示 LQI 指数的变化。

只有 LQI 变化为正时,减灾措施才是合理的,可得 LQI 准则

$$\frac{\mathrm{d}g}{g}+\frac{1}{q}\frac{\mathrm{d}l}{l}\geqslant 0 \tag{11}$$

式(11)应用到服役工程表示为挽救生命损失而愿意支付和负担得起的安全成本

$$\mathrm{d}C_1=\mathrm{d}g\geqslant-\frac{g}{q}K_{\Delta\bar{l}}N_{\mathrm{PE}}\mathrm{d}\mu=-\frac{g}{q}K_{\Delta\bar{l}}N_{\mathrm{PE}}k\mathrm{d}m \tag{12}$$

式中:$\mathrm{d}C_1$为年投资;$K_{\Delta\bar{l}}$ 为参数,由死亡降低率 Δ 和平均寿命人口 $\bar{l}$ 决定,$K_{\Delta\bar{l}}\approx 18$;$N_{\mathrm{PE}}$为溃坝风险人口数;$\mathrm{d}\mu$ 为补强修复后死亡率的变化量;k 为风险人口死亡率,取决于溃坝的失事模式;$\mathrm{d}m$ 为无条件失效概率。

计算生命损失风险的降低值(Δl)和安全成本的增加值(Δg),可得降低溃坝生命损失的价值(减轻生命风险的成本)。

2.3 考虑折现率的期望损失

结构补强修复投资的现值可以表示为

$$C_{\mathrm{r}}(t)=\sum_{i=1}^{u(t)}\frac{c(a_i)}{(1+r)^{t_i}} \tag{13}$$

式中:$u(t)$为至 t 时刻的补强修复次数;$c(a_i)$为第 i 次补强修复投资;t_i为第 i 次补强修复到计算时刻 t 的时间;r 为年折现率,取 $r=2.5\%$。

实施不同补强修复措施将得到不同的结构失效概率,失效期望损失是失效概率同失效损失的乘积,时刻 t 累积期望损失的现值可表示为

$$C_{\mathrm{failure}}(t)=\sum_{i=1}^{u(t)}\frac{c_{\mathrm{f}}P_{\mathrm{f}}(t_i)}{(1+r)^{t_i}} \tag{14}$$

式中:c_{f}为结构失效的经济损失;r 为年折现率;P_{f}为结构在决策时刻 t 下的失效概率。

2.4 常用补强修复方案及费用

水库病害主要为防洪不安全、抗震不安全、稳定性差及坝体坝基渗漏严重、输放水及泄洪建筑物老化等。补强修复方案则包括提高防洪能力、渗漏处理、抗震加固和附属设施加固。假设补强修复费用同大坝的风险率变化值呈线性关系,即

$$c=\eta\Delta P_{\mathrm{f}}C_1 \tag{15}$$

式中:η 为补强修复成本系数。

3 病险水库补强修复最优化模型

3.1 补强修复方案比较方法

补强修复方案关系到结构长期性能和成本,过频或可靠度过高将造成不必要的安全冗余。补强修复预期效果、间隔和次数是全寿命方案优选的主要内容。基于结构性能的补强修复方案优选方法反映了现代结构设计思想的一个重要转变,即从只注重结构安全,向全面注重结构性能、安全及效益等诸多方面发展。以大坝长期性能和效益最优平衡为原则,建立多目标的优化模型。

3.2 补强修复方案优选模型

补强修复方案可表示为补强修复的总实施次数 $\Gamma=\{t_i,i=1,2,\cdots,u(t)\}$ 和补强修复措施 $\Omega=\{a_i,i=1,2,\cdots,u(t)\}$。可接受最低可靠度或最大失效概率是优选中的重要约束,据《水利水电工程结构可靠度设计统一标准》(GB 50199—94)规定,Ⅰ、Ⅱ、Ⅲ级坝设计可靠度指标分别为3.7、3.2和2.7,对应的失效概率分别为 1.0×10^{-4}、6.8×10^{-4} 和 3.4×10^{-3}。将优选模型表示为

$$\left.\begin{aligned}&\text{Find}\quad \Gamma\&\Omega\\&\text{to}\quad \text{Minimize}C_{\text{r}}(t)\\&\text{and}\quad \text{Minimize}C_{\text{failure}}(t)\\&\text{s.t.}\quad \beta(t)>\beta_0\quad \forall t\in[t_1,T_{\text{f}}]\end{aligned}\right\}\tag{16}$$

式中:T_{f} 为结构服役寿命的终点;β_0 为允许的最小可靠度,β_t 为 t 时刻的可靠度。

病险水库补强修复方案优选方法流程如下:

(1)确定大坝失效标准和评判准则;

(2)基于实测资料或经验值确定结构时变风险率;

(3)调查收集数据,确定大坝的失效损失;

(4)研究大坝补强修复方案(包括可靠度阈值和补强修复间隔)及相应的费用;

(5)计算各方案补强后大坝结构的时变风险率;

(6)根据方案优选模型确定最优方案。

4 算例分析

以华东某水库均质土坝为例,水库已运行30年。防洪保护人口约20万,是一座综合性多功能大(2)型水库,灌溉面积约2 800 hm^2,下游多条高速公路、铁路穿过。防洪标准按100年一遇洪水设计,设计洪水位14.77 m,库容8 893万 m^3;2 000年一遇洪水校核,校核洪水位16.37 m,库容1.15亿 m^3。坝顶宽6.50 m,坝顶高程17.2 m;坝底高程0.0 m。

4.1 渗透破坏风险率的计算

基于渗流有限元和Monte Carlo法,对给定洪水位下失事概率进行抽样模拟,结合洪水位区间概率,计算得大坝破坏风险率,考虑影响渗透破坏的重要因素的时变效应,推求并拟合得到时变风险率。

忽略土体容重、孔隙率、最低水位值的变化,渗透破坏失事的功能函数为

$$Z=J_{\text{c}}-J=J_{\text{c}}(c,\varphi)-J(H_1,k)\tag{17}$$

式中：J 为渗透坡降；J_c 为临界渗透坡降；c 为土体的黏聚力；φ 为土体间的摩擦力；H_1 为计算失稳风险率最高洪水位；k 为渗透系数。

渗透破坏风险率可表示为

$$P = F_J(\overline{H}_i)\,f(H)\,\mathrm{d}H = \sum_{i=1}^{N} F_J(\overline{H}_i)\cdot \Delta F_H(\overline{H}_i) \tag{18}$$

式中：$F_J(\overline{H}_i)$ 为给定 $\overline{H}_i$ 时，J 大于 J_c 的概率；$f(H)$ 为洪水频率曲线概率密度函数；$\overline{H}_i$ 为洪水位频率曲线第 i 段区间水位平均值；$\Delta F_H(\overline{H}_i)$ 为 $\overline{H}_i$ 的区间概率；N 为计算段数。

图 1 为大坝有限元网格，表 1 列出了各主要随机变量的实测统计特征值，计算水位取值范围为 7.6～16.37 m。采用 Monte Carlo 法模拟不同洪水位下 $F_J(\overline{H}_i)$，由式(18)计算得到渗透破坏风险率。

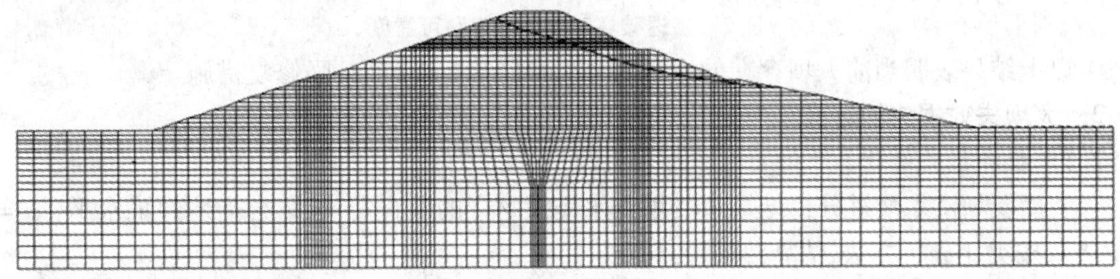

图 1　大坝典型断面有限元网格

表 1　随机变量统计特征值

项目	黏聚力 c	摩擦系数 $\tan\varphi$	上游库水位 H	坝体渗透系数 k
平均值	19 kg/m^2	0.363 9	12.0 m	4.675×10^{-5} cm/s
标准差	9.5kg/m^2	0.073	1.46 m	1.216×10^{-5} cm/s
变异系数	12.0m	0.2	0.12	0.26
分布型式	极值 I 型	对数正态	正态	正态

经检验，库水位服从正态分布，均值为 12.0 m。据上游洪水位频率曲线求取 $\Delta F_H(\overline{H}_i)$；采用 Monte Carlo 法求取相应 $F_J(\overline{H}_i)$。由式(18)计算渗透破坏风险率为 1.83×10^{-3}，可靠度指标为 2.9。

长时间服役后，大坝渗流边界条件、物理力学性质指标都将缓慢变化。抗剪强度指标变异性对坝体性能影响最为明显，J_c 的衰变特性用 c 和 $\tan\varphi$ 的时变性表示，即

$$c = c_0 e^{-0.005t} \qquad \tan\varphi = e^{-0.005}\tan\varphi_0 \tag{19}$$

式中：c_0 为初始黏聚力；φ_0 为初始内摩擦角；t 为服役时间，年。

坝体土体渗透率 k 反映了渗透特性，k 越大，越易发生渗透破坏，k 的时变特性表示为

$$k_t = k_0\left(1 + \frac{2}{\pi}\arctan t\right) \tag{20}$$

式中：k_0 为坝体建成时的土体渗透率。

计算得到的渗透破坏风险率如图 2 所示，拟合得到的渗透破坏时变曲线为

$$P_{\mathrm{fseep}}(t) = \frac{1}{1 + 1\,206.787 \times 0.966^t} \tag{21}$$

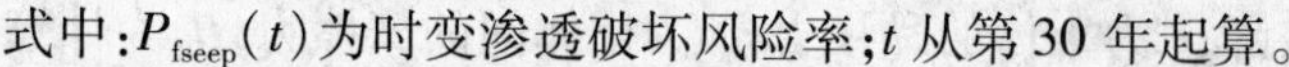

式中：$P_{\mathrm{fseep}}(t)$为时变渗透破坏风险率；t从第30年起算。

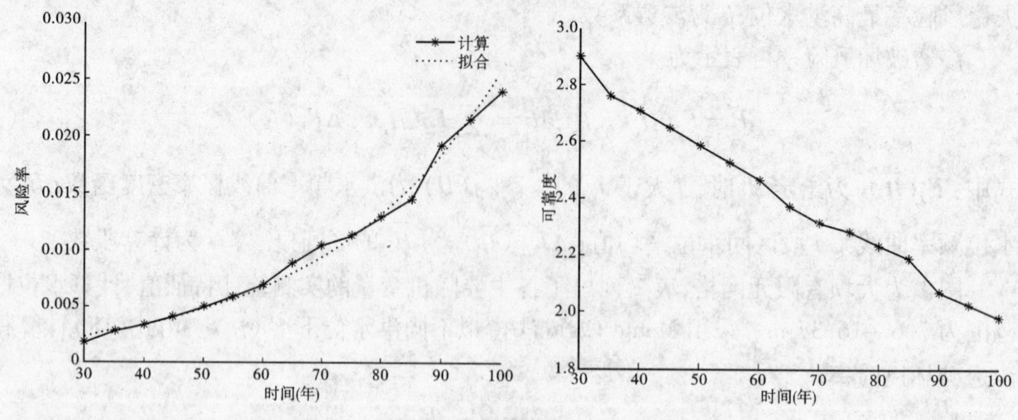

图2　渗透破坏时变风险率和可靠度

以上结果表明当前大坝存在的渗透破坏失事隐患，应采取补强修复措施。

4.2　大坝失效损失的计算

4.2.1　经济损失的计算

利用溃坝洪水演进分析结果和溃坝洪水淹没财产的调查资料，结合当地经济水平（α = 0.18），按类比分析方法，估算溃坝经济损失

$$C_{\mathrm{e}} = C_{\mathrm{de}} + C_{\mathrm{ide}} = (1+\alpha)C_{\mathrm{de}} = 1.18 \times 22.3 \times 10^{8} = 26.3 \times 10^{8}\ (\text{元}) \tag{22}$$

4.2.2　生命损失的计算

该区域l=75.5岁；g=44 700元，w=0.167。由式(12)计算为避免单个人员死亡的年安全成本为

$$C_{\mathrm{lca}} = \frac{\mathrm{d}l}{l}\frac{1-w}{w}g = \frac{1}{2} \times \frac{1-\left(\dfrac{1}{6}\right)}{\dfrac{1}{6}} \times 44\ 700 = 223\ 500\ (\text{元}) \tag{23}$$

遭遇100年一遇洪水溃坝时，淹没面积达76.11 km^2；风险人口为128 419人，风险人口死亡率与溃坝洪水严重性、警报时间等有关，取保守水平，死亡人数为1 147人，死亡率为8.9×10^{-3}。

4.3　补强修复费用

水库存在许多险工隐患，如坝体上部存在大量风化岩石体，压实性较差渗透系数大。补强修复措施包括坝趾压戗、坝体材料替换等，总投资为554.1×10^{4}元。补强修复后，可靠度提高到3.4。

4.4　补强修复方案优选

考虑3种定期补强修复方案：Ⅰ为"有坏就修"（小于目标可靠指标$\beta_{\min}$即进行补强修复），Ⅱ为"快坏才修"（大坝已不能安全工作时才补强修复），Ⅲ为介于Ⅰ和Ⅱ间的方案。据《水利水电工程结构可靠度设计统一标准》(GB 50199—94)，将已不能安全工作即$\beta_0 \leqslant 0.85\beta_{\min}$=2.72定义为"快坏才修"的准则；$\beta_{\min}$=3.2，对应的系统失效概率为$P_{\mathrm{fsys}}=6.8\times10^{-4}$；将$\beta \leqslant \beta_{\min}$定义为"有坏就修"的判断准则。

不同方案下大坝长期性能不同，假定补强修复后的可靠度略低于设计时的水平。补强

修复的目标为 $\beta=3.4$,相应 $P_f=3.4\times10^{-4}$,通过现值估算及 4.3 部分的分析,式(15)中系数 $\eta C_1=36.9\times10^8$ 元。据补强修复效果(结构风险率变化值),结合式(21),本文确定了以下 3 种补强修复方案:

方案Ⅰ,当 $\beta=3.2$ 时进行补强修复,间隔为 20 年;

方案Ⅱ,当 $\beta=2.72$ 时进行补强修复,间隔为 40 年;

方案Ⅲ,当 $\beta=3.06$ 时进行补强修复,间隔为 30 年。

据式(16)模型进行方案优选,计算得到不同方案下大坝后续服役期内的风险率如图 3 所示。

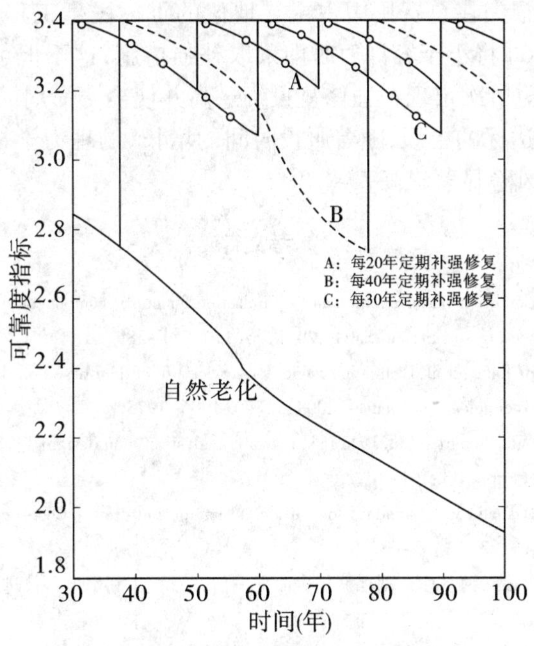

图 3 不同补强修复方案对大坝可靠度的影响

据式(4)和式(5)计算 3 种方案下大坝在后续服役期间可靠概率(失效风险率)的平均差值,据式(13)和式(15)结合 4.2 部分分析成果计算 3 种方案下全寿命周期内的成本现值;据式(14)、式(22)和式(23)计算 3 种方案下累积期望损失的现值,结果列于表 2。

由表 2 可看出,3 个补强修复方案中,方案Ⅰ中大坝始终处于较好的安全水平,且补强修复费用、失效损失最低,方案Ⅰ为最优方案。

表 2 补强修复费用现值和期望损失现值

方案	ΔP_f	C_r(万元)	$C_{failure}$(万元)
Ⅰ	1.39×10^{-4}	763.830	5 428.2
Ⅱ	16.96×10^{-4}	1 213.693	60 717.2
Ⅲ	5.54×10^{-4}	975.636	14 526.0

5 结论

本文提出了基于大坝时变风险率和包括生命损失评价在内的效益评价的病险水库补强修复方案优选方法,运用该方法对某病险土坝进行了计算分析。

(1)计算结果表明,大坝全寿命周期补强修复方案优选可将大坝工程安全和下游经济发展水平相联系,把大坝风险降低到下游能够承受的水平上。

(2)借鉴国外研究成果,将生命损失风险看做一个复合的社会指数,较合理地将大坝失效经济损失、生命损失和非经济损失综合考虑。

(3)对大坝等失效损失在总费用中占主导地位的重大民生工程,最优补强修复方案一般是在结构可靠度指标即将小于容许值时即采取补强措施,这不但可使大坝始终保持在较好的安全水平,还能避免单次出现补强修复投资过高并使全寿命周期内投资最小。

(4)对土粒级配不达标的土坝,随着服役时间的增长,物理力学指标和渗透系数的变异性逐渐增大,渗透破坏风险显著提高。

参考文献

[1] Tian Zhigang, Liao Haitao. Condition based maintenance optimization for multi-component systems using proportional hazards model[J]. Reliability Engineering and System Safety, 2011, 96(11): 581-589.

[2] Su Huaizhi, Wen Zhiping, Hu Jiang, et al. Evaluation model for service life of dam based on time-varying risk probability[J]. Science in China Series E-Technological Sciences, 2009, 52(7): 1966-1973.

[3] He Xianfeng, Gu Chongshi, Wu Zhongru, et al. Dam risk assistant analysis system design[J]. Science in China Series E-Technological Sciences, 2008, 51(Supp. II): 101-109.

[4] Pandey M D, Nathwani J S. Canada wide standard for particulate matter and ozone: Cost-benefit analysis using a life quality index[J]. Risk Analysis, 2003, 23(1): 55-67.

[5] 李雷,彭雪辉,周克发. 溧阳市沙河水库东副坝溃坝生命损失估算[J]. 水利水电科技进展,2008,28(1):46-49.

【作者简介】 苏怀智,1973 年生,博士,河海大学教授,博士生导师,河海大学水工结构研究所副所长。

黄河龙口水利枢纽坝基深层抗滑稳定问题研究

余伦创　苏红瑞

（中水北方勘测设计研究有限责任公司　天津　300222）

【摘要】 龙口水利枢纽工程是黄河中游的一座Ⅱ等大(2)型水利水电工程,坝基岩体质量较好,但存在数层近水平的软弱夹层,影响坝基的抗滑稳定性。本文简单介绍了坝基工程地质条件和岩体分类,重点阐述了坝基深层抗滑稳定问题与处理方法。

【关键词】 坝基岩体　软弱夹层　抗滑稳定　处理方法　龙口水利枢纽

龙口水利枢纽工程位于山西省河曲县与内蒙古自治区准格尔旗交界处,是一个综合利用的水利水电枢纽工程,其主要任务对万家寨水电站调峰流量进行反调节,使黄河万家寨—天桥区间不断流,并参与晋蒙电网调峰发电。大坝为混凝土重力坝,设计最大坝高 51 m,坝顶全长 408 m,坝顶高程 900 m,正常蓄水位 898 m,总库容 1.96 亿 m^3,采用河床式电站厂房,总装机容量为 420 MW。

1　坝址工程地质条件

1.1　地形地貌

坝区处于黄河托克托—龙口峡谷段的出口处。黄河由东向西流经坝区,河谷为“箱”型,宽 360 ~ 400 m。河床大部分为岩质,两岸为岩石裸露的陡壁,高度 50 ~ 70 m。

1.2　地层岩性

坝基地层为奥陶系中统上马家沟组(O_2m_2)地层,根据岩性和工程地质特征,可分为三段共十一个小层,岩性见表 1。

1.3　地质构造

坝区构造变动微弱,地层呈平缓的单斜,总体走向 NW315° ~ 350°,倾向 SW,倾角 2° ~ 6°。中小型断层(Ⅲ、Ⅳ级断裂)及裂隙是坝址区主要的断裂构造。坝区构造裂隙主要有四组,以 NE20° ~ 40°和 NW275° ~ 295°两组相对较发育,NE70° ~ 80°和 NW300° ~ 355°两组次之。

1.4　风化与卸荷

坝区岩体风化作用以物理风化为主,化学风化相对较微弱。两岸和河床岩体均有一定程度的卸荷,卸荷带厚度大体与弱风化带相当。左岸弱风化带平均厚度为 3.8 m;右岸局部存在强风化,平均厚度为 3.2 m,弱风化带平均厚度为 4.5 m;河床部位弱风化带平均厚度为 3.2 m。

表 1　上马家沟组地层主要岩性

段	地层代号	厚度(m)	主要特征
第三段	$O_2m_2{}^3$	40.00	中厚、厚层灰岩、豹皮灰岩夹薄层灰岩和白云岩、泥质白云岩。发育 NJ401 泥化夹层
第二段	$O_2m_2{}^{2-5}$	26.65	中厚层、厚层棕灰色灰岩、豹皮灰岩。发育有八条泥化夹层
	$O_2m_2{}^{2-4}$	2.36	上部为薄层灰岩,下部为薄层白云岩。发育有两条泥化夹层
	$O_2m_2{}^{2-3}$	14.87	中厚、厚层灰岩、豹皮灰岩。发育有四条泥化夹层
	$O_2m_2{}^{2-2}$	1.12	薄层灰岩,灰色,底部普遍含有燧石结核。发育有两条泥化夹层
	$O_2m_2{}^{2-1}$	41.90	中厚层、厚层状棕灰色灰岩、豹皮灰岩。发育有六条泥化夹层
第一段	$O_2m_2{}^{1-5}$	9.18	薄层、中厚层灰黄色白云岩、中厚层棕灰色及灰黄色泥灰岩
	$O_2m_2{}^{1-4}$	5.65	薄层白云岩、泥质白云岩
	$O_2m_2{}^{1-3}$	19.81	黄绿色角砾状泥灰岩夹中厚层棕灰色灰岩。泥灰岩遇水可塑;灰岩中蜂窝状溶孔发育
	$O_2m_2{}^{1-2}$	13.08	薄层灰白、灰黄色白云岩、泥质白云岩
	$O_2m_2{}^{1-1}$	9.42	中厚层、厚层白云岩夹泥质白云岩,浅灰、灰黄色

1.5　含水层类型

根据埋藏条件和含水介质特征坝址区地下水划分为 $O_2m_2{}^{2-3}$ 岩溶裂隙潜水、$O_2m_2{}^{2-1}$ 岩溶裂隙承压水和 $O_2m_2{}^{1-3}$ 岩溶承压水三类。

$O_2m_2{}^{2-3}$ 岩溶裂隙潜水:含水层厚度为 2.60 ~ 15.58 m,在坝线附近残留小部分,以上被剥蚀掉,以溶隙、裂隙含水为主,属于弱 ~ 中等透水性岩体,水位基本与河水位持平。

$O_2m_2{}^{2-1}$ 岩溶裂隙承压水:含水层厚 40.00 ~ 43.80 m,在河床坝线处顶板埋深 2 ~ 16 m,属于弱 ~ 中等透水性岩体,其相对隔水顶板为 $O_2m_2{}^{2-2}$ 薄层灰岩夹泥化夹层,厚度为 0.80 ~ 1.45 m。水位为 864.4 ~ 864.6 m。

$O_2m_2{}^{1-3}$ 岩溶承压水:含水层总厚度为 19.22 ~ 20.44 m,在坝线河床处埋深 59 ~ 70 m。以溶孔、孔洞含水为主,裂隙和溶隙不发育。本层岩体透水性不均一。承压水位在 864.5 m 左右。$O_2m_2{}^{1-4}$、$O_2m_2{}^{1-5}$ 层为 $O_2m_2{}^{1-3}$ 承压含水层相对隔水顶板,总厚 13 ~ 15 m。

1.6　岩体(石)物理力学性质

1.6.1　岩石力学特征

$O_2m_2{}^{2-1}$、$O_2m_2{}^{2-3}$、$O_2m_2{}^{2-5}$ 层岩性相同,均为致密坚硬岩石类,其饱和抗压强度平均值为 87 ~ 120 MPa。

$O_2m_2{}^{2-2}$、$O_2m_2{}^{2-4}$ 层岩性分别为薄层灰岩和薄层白云岩,饱和抗压强度平均值分别为 104 MPa、125 MPa,岩块弹性模量平均值分别为 85.67 GPa、44.09 GPa,属于致密坚硬岩石。

$O_2m_2{}^{1-3}$ 层岩石干密度平均值为 2.48 g/cm^3,干抗压强度平均值为 29 MPa,饱和抗压强度平均值为 17 MPa,表明该层岩石属于软岩类,并为易软化岩石。

1.6.2　岩体力学特征

$O_2m_2{}^{2-1}$、$O_2m_2{}^{2-3}$、$O_2m_2{}^{2-5}$:新鲜岩体具有较高的强度,且各向异性不明显。作为河床坝

基主要持力层的$O_2m_2^{2-1}$层，其垂直静弹性模量平均值为18.07 GPa，水平方向为20.20 GPa；垂直方向变形模量平均值为10.24 GPa，水平方向为10.94 GPa。

$O_2m_2^{2-2}$、$O_2m_2^{2-4}$层薄层灰岩、白云岩岩体力学强度相对稍低，且各向异性较明显。以点荷载强度为例，平行层面方向点荷载强度仅为垂直层面方向的60%～70%。

$O_2m_2^{1-3}$层岩质软弱，表现出强度低，抗变形能力差的特点，弹性模量、变形模量均较低。该层在坝基下埋深较大。

1.6.3 岩体地球物理特性

弱风化与卸荷带岩体：$O_2m_2^{2-1}$、$O_2m_2^{2-3}$、$O_2m_2^{2-5}$层平硐地震波纵波速度平均值为2 120 m/s，$O_2m_2^{2-2}$层平硐地震波纵波速度平均值为2 290 m/s，$O_2m_2^{2-4}$层平硐地震波纵波速度平均值为2 010 m/s。

微风化—新鲜岩体：$O_2m_2^{2-1}$、$O_2m_2^{2-3}$、$O_2m_2^{2-5}$层，平硐地震波纵波速度平均值为4 522 m/s；竖井地震波纵波速度平均值为3 894～4 171 m/s；钻孔声波纵波速度平均值为5 840 m/s。

2 坝基岩体分类

依据岩性、岩体风化与卸荷程度、岩体结构类型、岩体纵波速度（地震波）、岩体完整性系数和结构面的发育特征等因素和指标，并重点考虑了软弱夹层分布与性状特征及其对工程的影响，坝基岩体可以分为三大类六小类（见表2）。

表2 坝基岩体工程地质分类

岩体工程地质类别		Ⅲ		Ⅳ			Ⅴ
		$A_{Ⅲ-1}$	$A_{Ⅲ-2}$	$A_{Ⅳ-1}$	$A_{Ⅳ-2}$	$C_{Ⅳ}$	
主要岩性		厚层、中厚层灰岩	厚层、中厚层灰岩	薄层灰岩、白云岩、泥质白云岩	灰岩、白云岩	角砾状泥灰岩为主	裂隙密集带、断层破碎带
饱和抗压强度（MPa）		87～157		80～169		17	
风化、卸荷程度		微风化—新鲜	弱风化中下部	弱风化—新鲜状	强风化—弱风化上部	新鲜	
岩体结构类型		厚层结构	厚层、次块状结构	薄层结构	碎裂镶嵌结构	碎裂镶嵌结构	散体结构
结构面间距（cm）	裂隙	2/30～100	3/30～50	3/30～50	<30		一般<5
	层面	30～120	30～120	<2			
岩溶发育特征		以溶隙为主	以溶隙为主	$O_2m_2^{2-4}$层中发育溶洞、溶孔，其他地层不发育	河床部位较发育	蜂窝状溶孔发育	

续表 2

岩体工程地质类别	Ⅲ		Ⅳ			Ⅴ
	$A_{Ⅲ-1}$	$A_{Ⅲ-2}$	$A_{Ⅳ-1}$	$A_{Ⅳ-2}$	$C_{Ⅳ}$	
透水性	弱—中等	中等	中等	强	中等	
地震波均值(m/s)	4 522	2 120	3 214	1 500 ~ 2 500		
声波均值(m/s)	5 840				3 330	
RQD(%)	60 ~ 85	20 ~ 40	<50	<20	<25	0
主要岩层及分布位置	$O_2m_2^{2-1.3.5}$和$O_2m_2^3$层，大范围分布于河床及两岸坝基	$O_2m_2^{2-1.3.5}$层，位于浅表层	$O_2m_2^{2-2}$和$O_2m_2^{2-4}$层，在河床和两岸少量分布；$O_2m_2^{1-4}$和$O_2m_2^{1-5}$层分布于河床深度 43 ~ 50 m 以下。	分布于河床及两岸表部	$O_2m_2^{1-3}$层，分布在河床深度 59 ~ 70 m 以下	仅在局部分布
岩体工程地质评价	岩体较完整，具有较高承载力和抗变形能力，为良好坝基持力层，其间发育有产状平缓的软弱夹层，控制坝基抗滑稳定，必须采取针对性工程处理措施	岩体完整性较差，但仍具有较高承载力，但变形指标较低，应采取灌浆加固等措施，以提高其抗变形能力	岩体完整性差，并发育缓倾角、贯通性的泥化夹层。其中$O_2m_2^{2-2}$层分布于河床浅部，其间发育的泥化夹层明显影响坝基抗滑稳定，故不宜直接作为坝基，建议挖除或采取其他工程措施	张开裂隙发育且多夹泥，透水性较强，岩体破碎，不宜直接作为大坝地基。因集中在表层，建议挖除	岩体强度和变形指标均较低，但两层在坝线处埋藏深，工程影响有限	岩体破碎，分布于坝基的局部地段，建议挖除

3 坝基深层抗滑稳定问题

3.1 软弱夹层发育特征

根据勘察期间及施工揭露情况，坝址区发育有多层软弱夹层，根据物质组成与成因，将坝区发育的软弱夹层划分为三类，即岩屑岩块状夹层、钙质充填状夹层（原称糜棱岩状夹层)和泥化夹层,其中泥化夹层又细分为泥质类、泥夹岩屑类和钙质胶结物与泥质混合类,各类夹层的一般特征和典型夹层编号见表3。

表3　各类夹层的一般特征和典型夹层编号

<table>
<tr><th colspan="2">类型</th><th>一般特征</th><th>典型夹层编号</th></tr>
<tr><td colspan="2">岩屑岩块状夹层</td><td>由层间剪切破碎形成的岩块、岩屑构成,延伸短,强度高</td><td></td></tr>
<tr><td colspan="2">钙质充填状夹层</td><td>沿层面发育,主要由钙质胶结物构成,一般厚度为1~5 mm,延伸性差,主要分布于厚层灰岩中</td><td></td></tr>
<tr><td rowspan="3">泥化夹层</td><td>泥质类</td><td>沿层面发育,主要由泥质构成,一般厚度为5~20 mm,延伸性好</td><td>NJ_{305}、NJ_{305-1}、NJ_{306-2}、NJ_{307}、NJ_{307-1}、NJ_{401}</td></tr>
<tr><td>泥夹岩屑类</td><td>沿层面发育,主要由泥质和灰岩碎屑构成,一般厚度为5~20 mm,延伸性好,发育于厚层灰岩中</td><td>NJ_{301}、NJ_{302}、NJ_{303}、NJ_{304}、NJ_{304-1}、NJ_{306}、NJ_{308}、NJ_{308-1}~NJ_{308-7}</td></tr>
<tr><td>钙质胶结物与泥质混合类</td><td>沿层面发育,主要由泥质和钙质胶结物构成,厚度为5~20 mm,延伸性好,发育于厚层灰岩中</td><td>NJ_{304-2}</td></tr>
</table>

各夹层基本沿层面发育,其中泥化夹层的延伸性较好,且厚度较大,一般厚度为5~10 cm,大于围岩界面起伏度。泥化夹层(泥化带)常与两侧的劈理带、节理带构成三元结构。泥质类泥化夹层泥化程度较高,泥质含量一般在90%以上,但不均一,局部含有较多岩屑。泥化夹层中岩屑的矿物成分主要为方解石和白云石,与围岩相同;泥质物的矿物成分主要为伊利石,局部含有微量或少量的高岭土和蒙脱石。致使泥化夹层的抗剪强度较低。

3.2　软弱夹层抗剪强度指标建议值

勘察期进行了大量的室内外试验工作,采用的试验方法主要有:原位大型抗剪试验(自然固结快剪、饱和固结快剪)、室内中型剪试验(饱和固结快剪)、原状饱和固结快剪试验与重塑土饱和固结快剪试验。结合软弱夹层性状特征和试验成果,坝基软弱夹层抗剪强度指标建议值见表4。

表4　软弱夹层抗剪强度指标建议值

<table>
<tr><th colspan="2">软弱夹层</th><th>摩擦系数</th><th>凝聚力(kPa)</th><th>说明</th></tr>
<tr><td colspan="2" rowspan="2">岩屑岩块状夹层钙质充填夹层</td><td>f=0.5~0.55</td><td></td><td rowspan="8">1. f、c 为纯摩指标,f'、c' 为剪摩指标。
2. 泥质类包括 NJ_{305}、NJ_{305-1}、NJ_{307}、NJ_{307-1} 等。泥夹岩屑类包括:NJ_{301}、NJ_{302}、NJ_{303}、NJ_{304}、NJ_{304-1}、NJ_{306}、NJ_{306-1}、NJ_{306-2}、NJ_{308}~NJ_{308-7}、NJ_{401} 等。钙质充填物与泥质混合类包括 NJ_{304-2}</td></tr>
<tr><td>f'=0.6~0.65</td><td>c'=40~100</td></tr>
<tr><td rowspan="6">泥化夹层</td><td rowspan="2">泥质类</td><td></td><td>f=0.25</td></tr>
<tr><td>c'=10~20</td><td>f'=0.25</td></tr>
<tr><td rowspan="2">泥夹岩屑类</td><td></td><td>f=0.25~0.3</td></tr>
<tr><td>c'=15~50</td><td>f'=0.25~0.32</td></tr>
<tr><td rowspan="2">钙质充填物与泥质混合类</td><td></td><td>f=0.35</td></tr>
<tr><td>c'=35~60</td><td>f'=0.35~0.4</td></tr>
</table>

3.3 河床坝基深层滑移边界条件

根据河床坝基各种结构面组合和强度特征分析,河床坝基深层滑移模型为坝后有抗力体的软弱夹层控制形式。

上游拉裂面:滑移体上游方向的拉裂面主要由 NE20° ~40°裂隙构成。

两侧切割面:滑移体两侧切割面主要由 NW275° ~295°裂隙构成。

滑动面:由于软弱夹层分布的多层性和性状的差异性,河床坝基控制性滑动面的确定较为复杂。在河床部位,由上而下依次发育有 NJ_{306-1}、NJ_{306}、NJ_{306-2}、NJ_{305}、NJ_{305-1}、NJ_{304-2}、NJ_{304-1}、NJ_{304}、NJ_{303} 等泥化夹层,其连续性好,强度低;$O_2m_2^{2-1}$层上部钙质充填夹层较发育,竖井揭露到其平均分布间距仅 0.8 ~1.7 m,其抗剪强度明显低于大坝混凝土/基岩胶结面指标;此外,尚有个别岩屑岩块状夹层和三级泥化夹层发育。这些夹层分布深度不同,抗剪强度存在明显差异,对坝基抗滑稳定的影响程度也不相同。

尾岩抗力体:在河床坝基及坝后抗力体范围未发现倾向上游的缓倾角断层发育。坝区倾向上游的缓倾角裂隙不甚发育。河床坝基抗滑稳定将主要依赖于坝后抗力体。

4 针对深层抗滑稳定的处理措施

由于坝基岩层倾向左岸偏下游,在设计过程中最终选择左岸布置电站厂房,右岸布置泄洪建筑物的方案,致使开挖深度和坝基软弱夹层埋藏深度相适应,减少为挖除软弱夹层额外增加工程量。

国内外已建和在建的大、中型水利水电工程中,在坝基存在软弱夹层的复杂地基处理方面,取得了丰富的实践经验。根据这些经验,对软弱夹层的处理措施,应按其产状、埋深、夹层性状及其对坝体的影响程度,结合工程规模进行研究,按照施工条件和工程进度综合分析确定。已建工程坝基深层软弱夹层已采用而又行之有效的处理措施有明挖、洞挖、大口径混凝土桩,深齿槽(坝踵处,坝趾处)和预应力锚索等,其作用归纳起来为以下三大类:

(1)提高软弱夹层抗剪指标;

(2)增加尾岩抗力;

(3)同时提高软弱夹层抗剪指标与增加尾岩抗力。

龙口水利枢纽坝基及坝肩地层主要由奥陶系马家沟组(O_2m)及第四系(Q_4)地层组成。坝区地层中发育有连续性较好的软弱夹层共 23 层。在河床部位,地层中的 NJ_{305-1}、NJ_{305}、NJ_{306-1}、NJ_{306-2} 泥化夹层,因其埋藏浅,已经挖除,因而不控制坝的抗滑稳定。而 NJ_{304-1}、NJ_{304}、NJ_{303} 则是河床坝基中的控制滑动面,其摩擦系数 $f=0.25 \sim 0.35$,凝聚力 $C=10 \sim 35$ kPa。

由于坝基泥化夹层多且连续性好,抗剪断强度低,故大坝深层抗滑稳定问题是本工程的重大技术问题。为确保大坝安全,本工程设计采取了如下工程措施:

(1)挖除:对浅层连续泥化夹层(如 NJ_{305})因其埋藏浅,尾岩不能形成对坝体的有效支撑,采取挖除的方法。

(2)坝踵设置齿槽。

在坝踵设置齿槽。齿槽底宽 12 m,齿槽深入 NJ_{304} 下 1 m 左右,利用齿槽的抗剪断作用提高 NJ_{304} 以上各夹泥层的抗滑稳定性,并利用齿槽控制坝体沿 NJ_{304} 以下夹泥层滑动。电站及安装间坝段结合厂房开挖,在坝基中部设置齿槽,齿槽底宽 22 m,齿槽深入 NJ_{303} 夹层 1 m

以下。

(3)枢纽泄水建筑物采用二级底流消能,确保尾岩的厚度和稳定。

因龙口坝基存在多层泥化夹层,必须依靠坝趾下游尾岩抗力作用才能维持坝的稳定,因此在一定长度内保护尾岩免遭破坏是十分重要的。为此表孔、底孔采用二级底流消能。为保证抗力体厚度,选用尾坎式消力池,一、二级池池底高程分别为858 m和857 m。底孔、表孔一级消力池长分别为75 m和80.34 m,二级消力池底、表孔共用,池长均为64 m。消力池纵横缝设有止水,消力池水重有利于增大尾岩抗力。

(4)加强坝的整体稳定性。

龙口坝基分布有多层连续的泥化夹层及不连续的钙质充填物与泥质混合类夹层,夹层倾向下游及左岸。由于夹层埋藏深度及力学性能的不均一及钙质充填物与泥质混合类夹层的不连续性,所以各坝段稳定安全度不同。为使各坝段相互帮助,提高坝的整体稳定性,将部分坝段间横缝均做成铰接缝,横缝基础部分共有3段灌浆(1#~3#、10#~16#、17#~19#共3段),这样使各坝段整体作用加强。

(5)安装间坝段及岸坡坝段下游利用废渣压重,增大尾岩抗力。

左、右岸下游利用废渣分别回填至高程872.9 m和866.0 m,除满足布置要求外,还可增大尾岩抗力,提高两岸边坡坝段及安装间坝段的抗滑稳定性。

(6)坝基防渗帷幕及抽排系统。

坝基设上游帷幕和坝基排水。上游帷幕设主副两排,主帷幕深入相对隔水层5 m,上游帷幕向两岸坝肩延伸约50 m。主排水孔深入控制滑动面以下,确保坝基扬压力符合设计要求。抗滑稳定计算时不考虑抽排作用。

(7)坝基及尾岩进行固结灌浆,并在尾岩布置预应力锚索,加固基岩和尾岩。

坝基、尾岩根据岩石条件有针对性地进行固结灌浆处理,以提高坝基、尾岩的承载能力、整体性和均一性。在底孔、表坝段坝趾下游一级消力池前半部布置预应力锚索:即在坝体下游地表20 m范围之内,通过预应力锚索施加20 000 kN的垂直压力。

5 结语

龙口水利枢纽的坝基工程地质条件较复杂,其中软弱夹层大量发育,致使深层抗滑稳定问题成为该工程的主要工程地质问题之一。在工程设计过程中,对该问题已引起足够的重视,进行了针对性的设计。在施工过程中,根据开挖所揭露的软弱夹层实际情况已对前期设计进行了优化调整。在以后的运行过程中,仍需对大坝进行必要的变形监测,发现问题及时分析研究。

【**作者简介**】 余伦创,1972年1月生,大学本科学历,1994年7月毕业于成都科技大学,学士学位,现任中水北方勘测设计研究有限责任公司水工设计处副处长,担任黄河龙口水利枢纽工程设计总工程师,高级工程师。

黄河龙口水利枢纽工程主要技术难题及措施

任　杰　任智锋

（中水北方勘测设计研究有限责任公司　天津　300222）

【摘要】 龙口水利枢纽为万家寨水利枢纽的配套工程，主要任务是发电和对万家寨电站调峰流量进行反调节。本文对龙口工程防沙排沙、大坝深层抗滑稳定、泄水建筑物消能防冲和厂房下部结构设计等主要技术问题进行了研究。针对其特殊的地形地质条件和泥沙问题，采用以底孔为主的泄洪排沙型式，在电站坝段布置排沙洞保证电站进口“门前清”。根据大坝深层抗滑稳定需要，泄水建筑物消能采用二级消能方式，并在坝基设置齿槽、预应力锚索加固尾岩和横缝部分灌浆等工程措施，有效地解决了大坝深层抗滑滑动稳定问题。

【关键词】 反调节　排沙设施　深层抗滑稳定　消能防冲　厂房下部结构

龙口水利枢纽位于黄河北干流托克托至龙口段的末端，坝址距上游万家寨水利枢纽 25.6 km，下游距天桥水电站约 70 km。枢纽工程的主要任务是发电和对万家寨电站调峰流量进行反调节。水库总库容 1.96 亿 m^3，电站装机容量 420 MW，多年平均发电量 13.02 亿 kW · h，为Ⅱ等大(2)型工程。枢纽基本坝型为混凝土重力坝，主要建筑物从左到右依次为左岸挡水坝、河床式厂房、隔墩坝、泄洪底孔、表孔及右岸挡水坝，其中河床式电站安装 4 台单机容量为 100 MW 和 1 台单机容量为 20 MW 的水轮发电机组，泄洪建筑物由 10 个 4.5 m ×6.5 m 的底孔和 2 个 12 m×12 m 的表孔组成，大坝、电站厂房、泄洪建筑物按 2 级建筑物设计。

枢纽主体工程于 2006 年 6 月开工，2007 年 4 月二期截流，2009 年 7 月下闸蓄水，2009 年 9 月第一台机组发电，2010 年 6 月 5 台机组全部并网发电。

1　对万家寨下泄流量进行反调节

万家寨水利枢纽总库容 8.96 亿 m^3，电站装机容量 1 080 MW，单机额定流量 301 m^3/s。万家寨电站在晋蒙电网中担负调峰任务，每天不同时段下泄流量很不均匀，电站满出力发电时最大下泄流量 1 806 m^3/s，而不发电时则没有下泄流量，致使万家寨—天桥区间黄河干流有 17 ~ 18 h 的断流，不能满足区间生态基流的要求；同时，万家寨电站调峰运行时的不稳定流到达天桥电站时尚不能完全坦化，而天桥电站因受调节库容的限制，不能对万家寨电站调峰发电流量进行完全调节。

龙口水利枢纽位于万家寨—天桥之间，正常蓄水位与万家寨电站尾水相衔接，具有 3 400 万m^3左右的日调节库容，能承上启下，对万家寨电站调峰下泄流量进行反调节。龙口电站设计满出力发电流量 1 520 m^3/s，小于万家寨电站最大下泄流量，龙口电站发电时间可比万家寨电站长一些，通过合理的联合调度，其反调节的作用更加明显，有利于坦化流量过程，减小流量波动幅度，改善下游河道生态条件和天桥电站运行条件。

为确保非调峰时段河道不断流，满足枯水年份瞬时流量不小于 50 m^3/s，日平均流量不小于 100 m^3/s 的要求，设计采用了大小机组方案，即 4 台大机组（单机容量 100 MW，单机额定流量 365 m^3/s）与万家寨电站联合调度调峰，1 台小机组（单机容量 20 MW，单机额定流量 60 m^3/s）发电下泄基流，以提高梯级电站的综合经济效益。

2 排沙设施和运行方式

龙口水利枢纽位于万家寨水利枢纽下游 25.6 km，库区有支流偏关河在左岸距坝址约 13.5 km 处汇入。龙口坝址控制流域面积 39.7 万 km^2，其中支流偏关河控制流域面积 2 089 km^2。万家寨库区泥沙达到冲淤平衡后，多年平均下泄悬移质泥沙量 1.32 亿 t，万家寨坝址到龙口坝址区间悬移质多年平均入库 0.188 亿 t，推移质多年平均入库 1 万 t。万家寨水库出库泥沙颗粒较细，多年平均 d_{50} 为 0.023 mm。万家寨—龙口区间入库悬移质泥沙颗粒较粗，多年平均 d_{50} 为 0.039 mm。多泥沙河流上修建水利枢纽必须重视和妥善处理泥沙问题，采取必要的防沙排沙措施。

2.1 泄洪排沙底孔

本工程库容较小，库沙比约为 1.6，水库淤积平衡年限较短，枢纽必须具有较强的排沙能力，结合枢纽泄洪建筑物的要求，采用以底孔为主的泄洪排沙布置型式，在主河床布置了 10 个 4.5 m×6.5 m 的底孔，进口底高程为 863 m，略高于原河床底高程，汛期降低水位排沙。

同时，10 个泄洪排沙底孔在施工期还兼作二期导流底孔使用，有利于节省工程投资和加快施工进度。

2.2 电站排沙洞

为保证电站进水口“门前清”，根据已建工程的实践经验，在电站坝段设置了排沙洞。排水洞位置和高程的选定应使排沙漏斗足以控制进水口，经比较和模型试验验证，采取了分散排沙布置方式，即在每个大机组段布置 2 个排沙洞，副安装间坝段设 1 个排沙洞，共 9 个排沙洞。

排沙洞进口位于电站进水口下方，按照尽量压低排沙洞进口底高程的原则，确定底高程为 860.0 m，进口设有检修闸门，孔口尺寸 5.9 m×3.0 m。排沙洞出口高程与下游运行水位有关，按满足完全淹没出流确定，出口底高程为 860.0 m，排沙洞出口设有工作闸门和检修闸门，出口断面尺寸为 1.9 m×1.9 m。单孔设计流量为 71 m^3/s。

排沙洞进口流速选择考虑以下两个方面因素：

（1）为了使电站进水口保留一定的过水断面，要求排沙洞进口处有足够大的流速，能将厂房前的推移质泥沙带进排沙洞，排至电站下游。

（2）为了避免影响水轮机出力，排沙洞进口与机组进口两者的流速不宜相差过大，应保持适当比例，工程经验表明，两者流速比为 1.8 时，机组运行较为稳定；当两者流速比达到 3.9 时，会严重影响水轮机出力。本工程经比选排沙洞进口与机组进口两者的流速比为 3.1。

2.3 拦沙坎

在电站进水口前设置拦沙坎，以加大进水口与主河槽的高差，保证水流能将厂房前的推移质泥沙带进底孔排走。本工程拦沙坎高度不低于3.0 m,利用上游围堰改建而成。

2.4 **“蓄清排浑”**

从控制水库泥沙淤积末端和水库经济指标两方面考虑,确定龙口水库的运行方式为“蓄清排浑”,考虑到上、下游梯级排沙及发电运行的同步性,龙口水库排沙期设定为8月、9月。

3 大坝深层抗滑稳定

龙口水利枢纽坝址区地层主要由奥陶系马家沟组(O_2m)、石炭系(C)及第四系(Q_3+Q_4)地层组成。河床坝段建基面岩性主要为中厚层、厚层灰岩,少量为薄层灰岩,岩体完整坚硬,适合修建混凝土坝。但是,在坝基岩体中发育多层软弱夹层,控制着河床坝基深层抗滑稳定性,是本枢纽工程的主要工程地质问题。

根据物质组成与成因,软弱夹层可划分为三类,即岩屑岩块状夹层、钙质充填夹层和泥化夹层,其中泥化夹层又细分为泥质类、泥夹岩屑类和钙质充填物与泥质混合类。这些不同类型的软弱夹层,连续性好,抗剪强度低,特别是NJ_{304-1}、NJ_{304}、NJ_{303}是河床坝基中的控制滑动面,故大坝深层抗滑稳定问题是本工程的重大技术问题。为确保大坝安全,本工程设计采取了如下综合工程措施。

(1)挖除。

对埋藏较浅的软弱夹层(如NJ_{305}),采取挖除的方法处理。

(2)坝基设置齿槽。

底孔、表孔坝段在坝踵设置齿槽。齿槽底宽12 m,齿槽深入NJ_{304}下1.0 m;电站及安装间坝段结合厂房开挖,在坝基中部设置齿槽,齿槽上口宽22 m,齿槽深入NJ_{303}以下1.0 m。

(3)充分利用尾岩抗力。

河床坝基内的NJ_{304}是底孔、表孔坝段稳定的控制滑动面,为典型的双面滑动。底孔、表孔坝段需依靠坝趾下游尾岩支撑才能维持稳定,为了保护尾岩免遭破坏,泄水建筑物采用二级底流消能的方式,以充分发挥尾岩的抗力作用。

(4)横缝部分灌浆。

由于夹层埋藏深度及力学性能的不均一及钙质充填夹层的不连续性,所以各坝段稳定安全度不同。为了使各坝段相互帮助,提高坝的整体稳定性,将左岸河床坝段横缝做成铰接缝,对部分横缝进行灌浆。

(5)利用开挖弃渣压重。

在安装间坝段及左、右岸坡坝段下游,利用开挖弃渣分别回填至高程872.9 m和866.0 m,除满足布置要求外,还可增大尾岩抗力,提高两岸边坡坝段及安装间坝段的抗滑稳定性。

(6)坝基设上、下游帷幕及排水。

本工程除按常规设置上游防渗帷幕和排水外,在坝基下游侧及岸边均布设帷幕与上游帷幕形成封闭系统。上游帷幕深入相对隔水层$O_2m_2^{1-5}$层3 m,下游帷幕及左右岸边帷幕深入$O_2m_2^{1-5}$层1 m。主排水孔深入控制滑动面以下。抗滑稳定计算时不考虑抽排作用,作为安全储备。

(7)坝基及尾岩固结灌浆。

对全坝基和部分尾岩进行固结灌浆处理，以提高坝基、尾岩的承载能力、整体性和均一性。

(8)预应力锚索加固尾岩。

在底孔、表孔消力池下面设置预应力锚索，提高尾岩抗力。

4 消能防冲

龙口水利枢纽为大(2)型工程，拦河坝、泄水建筑物和电站厂房均为2级建筑物，其洪水标准按100年一遇洪水设计，1 000年一遇洪水校核。水库运用方式是蓄清排浑，汛期排沙。泄水建筑物除应满足泄洪要求外，还应满足冲沙、排污、排冰凌等要求。规划对泄水建筑物泄流能力的要求是：

(1)校核洪水位时，下泄流量大于8 276 m^3/s；

(2)设计洪水位时，下泄流量大于7 561 m^3/s；

(3)汛期排沙水位为888.0 m时，下泄流量大于5 000 m^3/s。

由于本工程冲沙流量大，汛期要求泄洪兼排沙，故泄水建筑物以底孔为主。经反复比较，选用10个孔口尺寸4.5 m×6.5 m(宽×高)的泄洪底孔，孔口处最大单宽流量为137.8 $m^3/(s \cdot m)$；2个12 m×12 m的泄洪表孔，兼作排污排冰之用，最大单宽流量为76.7 $m^3/(s \cdot m)$；为保持电站坝段“门前清”，还设有9个1.9 m×1.9 m排沙洞。

因坝基存在多层泥化夹层，从大坝稳定性考虑不宜采用挑流消能方式，故表孔和底孔均采用底流消能。同时，为了保护尾岩免遭破坏，维持大坝稳定需要的抗力，消力池不宜深挖，为此采用二级消能方式。消能建筑物主要由一级消力池、一级消力坎、二级消力池、差动尾坎、海漫、消力池左边墙、消力池右边墙及消力池中隔墙等建筑物组成。

一级消力池池长75.0 m，表孔和底孔消力池池宽分别为29.0 m、95.0 m，之间用中隔墙分开，在一级消力池末端设置梯形断面消力坎，坎高7.0 m。

二级消力池为底孔、表孔共用，池长53.0 m。为了减少消力池末端的单宽流量，平面上二级消力池的左、右边墙分别向两侧扩散(左侧1∶6.67坡度、右侧1∶8坡度向外扩散)，池宽由131.0 m扩散为147.0 m。二级消力池末端设差动尾坎，高坎体型为直角梯形断面，坎高5.0 m，低坎体型为等腰三角形断面，坎高3.0 m，两种形式每隔3.5 m交叉布置。为了防止出池水流淘刷，差动尾坎末端设置一齿槽，齿槽深5.5 m，齿槽后接20.0 m海漫。

为了防止下泄水流对消力池底板表面混凝土的冲刷，在消力池底板顶面铺设0.4 m厚的抗冲磨混凝土。

经整体水工模型试验验证，在宣泄各级洪水情况下，消能充分，水流平顺，流态较好，出池水流最大流速控制在9.0 m/s以内。

5 厂房下部结构

龙口电站为河床式厂房，总装机容量420 MW，年发电量13亿kW·h，共装4台单机容量为100 MW轴流转桨式机组和1台单机容量为20 MW的混流式水轮发电机组，额定水头31.0 m，最大、最小净水头分别为36.1 m和23.6 m。厂房长187.0 m，下部宽81.0 m，上部宽30.0 m，高67.2 m，采用一机一缝，大机组段宽30.0 m，小机组段宽15.0 m。

大机组转轮直径7.1 m，单机引用流量359 m^3/s，采用特殊的、上下伸式的钢筋混凝土蜗壳,蜗壳包角216°,上下伸角各为15°,流道体型极其复杂。同时,因排沙需要,在每个机组段布置有两个排沙洞,排沙洞进口位于电站进水口下方,呈重叠布置,排沙洞从蜗壳两侧下方穿过,在尾水管上方进入尾水渠,致使厂房下部结构孔洞多,上下左右交叉,体型复杂。

厂房下部结构采用平面框架法和三维有限元法进行分析计算。平面框架方法简便、实用,但忽略了空间作用,只能确定顶板和侧墙径向内力和配筋,无法计算环向应力,致使蜗壳顶板径向钢筋和侧墙竖向钢筋偏多,而环向钢筋不足。三维有限元方法可以弥补结构力学方法的不足,能精确反映各个部位的应力状况。两种方法相互验证,实际配筋综合了两种方法的计算结果,对转角、孔口交叉或应力突变的部位配置了加强筋。蜗壳进口段净跨22.0 m,净高11.5 m,因受水轮机层布置和蜗壳上伸角的限制,该段顶板混凝土仅2.5 m厚,为保证结构强度在蜗壳进口段混凝土顶板内沿顺水流方向和横水流方向各布置2道暗梁。

本工程蜗壳最大内水压力(计入水击压力后)达46.0 m,为提高防渗性能,在蜗壳顶板内壁,正X轴方向13.65 m,负X轴方向7.85 m的范围内,设置薄钢板衬砌,其他部位采用10 mm厚环氧砂浆抹面。

6 结语

本工程坝虽不高,电站装机容量不大,但其特殊的地形地质条件和泥沙问题,仍然存在一些重大的技术问题需解决,针对这些问题采取的相应工程措施值得总结。

(1)采取大小机组方案不仅能满足下游河道生态流量的要求,而且有利于坦化流量过程,增强反调节的作用。

(2)本工程枢纽布置紧凑、合理,泄洪排沙建筑物采用以底孔为主,表孔为辅的方式,可有效地控制库区淤积形态。同时,在电站坝段分散布置排沙洞可以保证电站进口"门前清"。

(3)针对坝基存在的软弱夹层,采用多种工程措施进行处理,有效地解决了大坝深层抗滑滑动稳定问题。

(4)根据大坝深层抗滑稳定需要,为保护尾岩,避免消力池深挖,泄水建筑物消能采用二级消能方式,经整体水工模型试验验证,消能充分,水流平顺,流态较好。

(5)针对河床式厂房下部结构孔洞多,体型复杂的特点,采用平面框架法和三维有限元法两种方法进行分析计算,相互验证。钢筋混凝土蜗壳承受的内水压力较高,为提高防渗性能,在蜗壳顶板内壁局部设置薄钢板衬护。

【作者简介】 任杰,女,1970年4月生,大学本科,1992年6月毕业于天津大学,获学士学位,中水北方勘测设计研究有限责任公司,高级工程师。

随机波浪作用下土吸力对钢悬链线立管动力响应的影响*

刘庆海　郭海燕　张　莉　李　朋

（中国海洋大学工程学院　青岛　266100）

【摘要】 采用细长柔性杆模型，用二次弹簧模拟立管触地区管道压缩土体的过程，考虑土体吸力影响，在随机波浪作用下，计算钢悬链线立管的动力响应，用 MATLAB 编制相应的计算程序，获取立管关键点处位移和有效张力，探讨土吸力对立管动力响应的影响；结果表明：由于土吸力的存在，立管位移发生变化，触地点位置发生改变，立管张力亦有增加。

【关键词】 钢悬链线立管　土体吸力　有效张力　动力响应分析

1　引言

钢悬链线立管（SCR）是深水开发的新型立管系统，在海洋环境荷载的作用下受力极为复杂，特别是钢悬链线立管与海床的接触处，触地点（TDP）一直是海洋工程界研究的热点。

对钢悬链线立管触地区的研究，主要有试验和数值模拟两种方法。Bridge C 在和墨西哥湾海床相似的一个潮汐港湾进行试验，提出了土体吸力模型。P Aubeny 等研究钢悬链线立管与海床的相互作用，建议使用基于 P-y 曲线和等价梁-弹簧模型模拟管道和土体的相互作用。杜金新等利用等效弹簧单元模拟土体并建立立管与土体有限元接触模型，讨论了触地点位置和触地区与土体受力变化之间的关系。郭海燕等利用 ANSYS 中非线性弹簧单元和接触单元模拟海床土体，考虑海床土体刚度退化和土体吸力对管道的作用进行了有限元分析。

对于钢悬链线立管动力响应的数值模拟通常用细长柔性杆力学模型，该模型的控制方程直接在全局坐标系中建立，可以考虑大变形，能较好地模拟海洋工程中各种柔性杆件结构。陈海飞推导了考虑内流作用的细长柔性杆模型，并且将海床的作用用弹簧阻尼力和摩擦力来模拟，计算分析了钢悬链线立管在海洋环境中的力学性能，给出了钢悬链线立管的有效张力。

本文采用二次非线性弹簧模拟海床土体，并同时考虑土体吸力对管道作用，建立三阶段线性土吸力模型，利用柔性杆理论，对钢悬链线立管整体在波浪力作用下的动力响应进行数值模拟，分析土体吸力对管道动力响应的影响，其中对有效张力的大小、触地点位置和管道触地区位移的影响显著。

---

* 基金项目：国家高技术研究发展计划项目（2010AA09Z303），国家自然科学重点项目（50739004）。

2 细长柔性杆理论

柔性杆模型最早由 Garrett 提出，Paulling 和 Webster 允许小伸缩变形对其进行了进一步改进。陈海飞将海床土体的作用用二次非线性弹簧模拟，在柔性杆坐标系中建立了细长杆动力学模型，即

$$\rho\ddot{r} + C_a\ddot{r}^n + (EIr'')'' - (\lambda r')' = \overline{w} + \overline{F}^d \tag{1}$$

式中：$r(s,t)$为三维笛卡尔坐标系统中一条空间曲线，用来描述杆轴线变形后的位置，该位置向量是弧长 s 和时间 t 的函数。杆件的有效张力 λ 和单位长度水动力荷载 $\overline{F}^d$ 分别为

$$\lambda = (T + p_o A_o - p_i A_i) - EI\kappa^2 \tag{2}$$

$$\overline{F}^d = \rho_w \frac{\pi D^2}{4}(\dot{V} + C_a \dot{V}^n) + \frac{1}{2}C_d \rho_w D \mid V^n - \dot{r}^n \mid (V^n - \dot{r}^n) \tag{3}$$

式中：ρ 为杆件单位长度质量；ρ_w 为海水密度；C_a为单位长度附加质量；C_d为法向曳力系数；EI 为抗弯刚度；κ 为曲率 $\kappa = r''$；p_o为外部静水压力；p_i 为内部静水压力，$\dot{V}$、$\dot{V}^n$、V^n分别为水质点的加速度、加速度法向分量和速度法向分量；$\overline{w}$ 为有效重力；T 为杆的张力。

3 海床土体模型

文献[5]用弹性地基模型模拟立管对土体的压缩作用。立管压缩土体过程中，土体对管道的作用力可以表示成

$$f_1 = f_2 = 0$$

$$f_3 = \begin{cases} c\left[R - (r_3 - Z_s)\right]^2 & R - (r_3 - Z_s) > 0 \\ 0 & R - (r_3 - Z_s) \leqslant 0 \end{cases} \tag{4}$$

式中：f_1、f_2、f_3 分别为土体对管道 x、y、z 向作用力；R 为立管半径；r_3为立管位置向量的 z 向分量；Z_s为海底 z 向坐标；c 为系数，$c=w_0/R^2$；w_0为立管单位长度的湿重。

本文根据 Watchet 港口的模型试验得出的土体吸力模型，对上述模型进行修正，建立土体对管道的作用力模型。

当立管向上运动时，土体会产生一个对管道的吸力作用，抵抗立管向上运动。Watchet 港口的模型试验得出的土体吸力模型（见图 1）可以分为三个线性阶段：

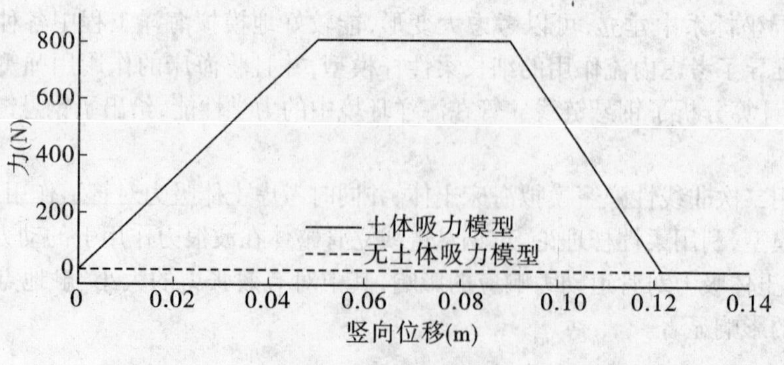

图 1 土体吸力模型

第一阶段：吸力产生扩大期。当立管向上运动时，土体抵抗力逐渐消失，吸力从 0 增加

到最大值;

第二阶段:吸力平稳期。立管继续向上运动,在一定范围内,吸力保持不变;

第三阶段:吸力消退期。立管继续向上运动,吸力由最大逐渐减小,在破坏位置处变为0。

土体吸力变为0后,立管继续向上运动时,土体不再对立管产生作用。

由此,本文采取的海床土体模型如下:管道压缩土体时,土体对管道的作用力采用二次弹簧模拟,即式(4);管道向上拔起运动时,土体对管道产生一个吸力,按照图1分三个阶段对土体产生的吸力取值。

4 钢悬链线立管动力响应分析

根据 Rayleigh - Ritz 方法对控制方程和海床土体作用模型进行有限元离散,利用 Galerkin 方法得到矩阵形式的立管运动微分方程,选择 P-M 谱作为立管动力作用的波浪谱,采用迭代法求解上述非线性方程,编制相应的 MATLAB 计算程序。

4.1 钢悬链线立管参数

钢悬链线立管参数取自 Chai,海流表面流速 1.2 m/s,海底流速 0.2 m/s,法向曳力系数 1.2,附加质量系数取为 1.0。随机波浪显著波高 15.7 m,周期是 13.5 s。计算钢悬链线立管的有关参数见表1。

表1 钢悬链线立管有关参数

立管总长(m)	3 750	海水密度(kg/m^3)	1 025
外径(m)	0.355	水平投影(m)	3 150
内径(m)	0.305	垂直投影(m)	1 300
干重(无内流)(kg/m)	204.209	水深(m)	1 300
轴向刚度(N)	$5.442\ 8\times10^{-9}$	抗弯刚度($N\cdot m^2$)	$7.451\ 5\times10^{-7}$

4.2 钢悬链线立管有效张力计算验证

同时考虑随机波浪、海流的作用,不考虑土吸力,用本文编制的计算程序求解钢悬链线立管的动力响应,并用式(2)计算立管的有效张力,和文献[5]计算的立管有效张力作了比较,见图2。由图2可以看出,两者吻合较好。

4.3 土吸力对钢悬链线立管动力响应的影响

4.3.1 土体吸力对有效张力的影响

用本文编制的计算程序计算有土体吸力作用时钢悬链线立管两端节点和触地点处有效张力均方根值,同不考虑土体吸力作用的钢悬链线立管两端节点和触地点处的有效张力均方根值一起绘成图3。从图3可以看出,土吸力使立管两端结点有效张力均方根值增加了400 kN,触地点有效张力均方根值的增量比两端结点的略小。可见,由于土体吸力的作用,使得钢悬链线立管的有效张力加大,所以在工程设计和安全评估时应该考虑土吸力的影响。

4.3.2 土体吸力对触地点位置和管道位移的影响

按照上述柔性杆模型及本文建立的海床土体模型对本算例进行求解,将立管触地点处运动的平均位置绘成图4。从图4可以看出,土体吸力的存在,使触地点的位置向靠近顶部端结点的方向移动,管道浸入土体的深度因为土体吸力的缘故,较无土体吸力时要大。考虑土体吸力时,管道触地区运动的平均位置示意图和文献[2]中给出的结果相近。

综上所述,土体吸力的存在增加了立管的有效张力,使触地点位置发生改变,管道浸入土体的深度增加,立管动力响应比无土体吸力时要大。

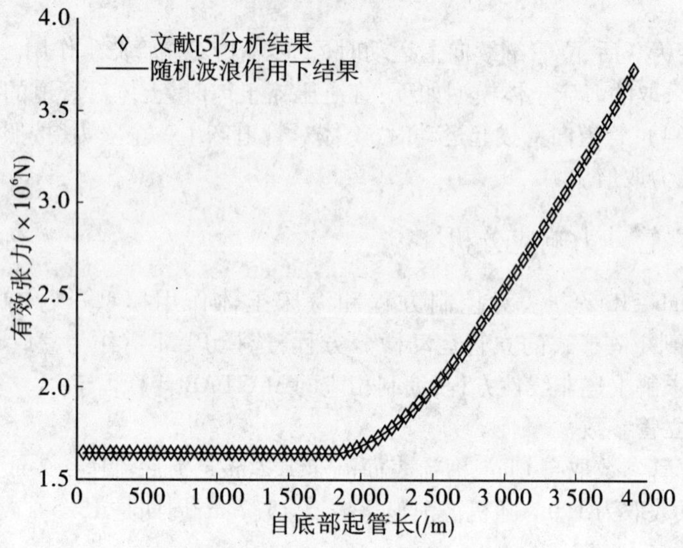

图2 沿管长有效张力分布

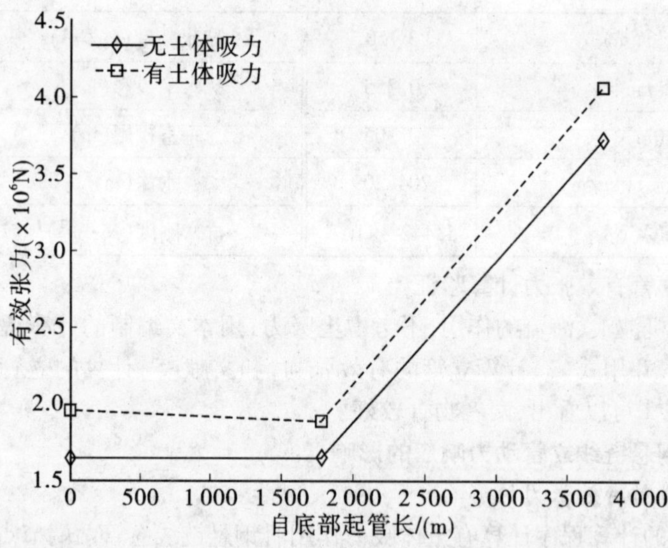

图3 立管有效张力均方根值

5 结语

采用细长柔性杆模型,同时考虑触地端弹性地基模型的压缩作用和三阶段土体吸力的作用,对钢悬链线立管整体在波浪力作用下的动力响应进行数值模拟,通过 MATLAB 编程和算例计算分析,研究土体吸力对立管动力响应的影响,结果表明:土体吸力的作用,使得钢悬链线立管的有效张力加大;土吸力对立管位移影响显著,考虑土吸力时,触地点位置向靠近顶端结点的方向移动,同时加深了管道的入土深度。所以在工程设计和安全评估时应该考虑土吸力的影响。

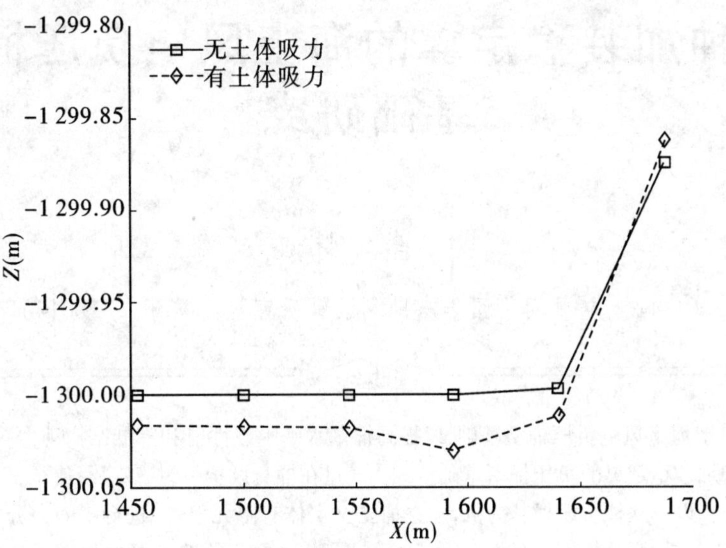

图4 触地点附近管道平均位置示意图

参考文献

[1] Bridge C, Willis N. Steel Catenary Risers results and conclusions from large scale simulations of seabed interaction[C]. 14th Annual Conferen ce Deep Offshore Technology, 2002.

[2] Charles P Aubeny, Giovanna Biscontin, Jun Zhang. Seafloor Interaction with Steel Catenary Risers[D]. Texas: Texas A&M University, 2006.

[3] 杜金新, Low Y M. 海洋立管-海床土体接触作用数值分析[J]. 工程地质计算机应用, 2008, 52(4): 1-6.

[4] 郭海燕, 高秦岭, 王小东. 钢悬链线立管与海床土体接触问题的 ANSYS 有限元分析[J]. 中国海洋大学学报, 2009, 39(3): 521-525.

[5] 陈海飞. 柔性深水立管非线性有限元分析[D]. 青岛: 中国海洋大学, 2011.

【作者简介】 刘庆海, 1987 年 9 月生, 中国海洋大学研究生, 2012 年 6 月毕业, 硕士。

一种加装稳定翼的海上风电负压桶型基础研究*

李 炜

（中国水电顾问集团华东勘测设计研究院 杭州 310014）

【摘要】 针对海上风电负压桶型基础，以提高桶基水平承载性能和降低结构动力响应为目的，提出了一种加装稳定翼的负压桶型基础型式。通过在桶身设置一组稳定翼，使得桶周土抗力得以充分利用。以单立柱负压桶基为例，建立桶土全实体有限元模型，进行了静力分析和包括模态分析、瞬态分析、谱分析在内的动力分析。结果表明：稳定翼的设置增强了桶基水平承载性能；水平位移和动力响应显著减小；结构低阶固有频率略有提高。

【关键词】 海上风电 负压桶 稳定翼 动力响应

伴随对于开发可再生能源的全球性推进，海上风能的开发成为时下的热点。我国的海上风电开发也已全面展开。在海上风电场的建设中，基础结构成本占总造价的比例较高，加之所处环境复杂，成为研究的重点和难点。

负压桶型基础（Suction Bucket）又称吸力筒基础，分为单桶和多桶等结构型式。浅海、深海皆可运用，其中浅海中的负压桶实际上是传统桩基和重力式基础的结合；在深海海域则是作为张力腿浮体支撑的锚固系统。

它根据渗流理论和桶基负压沉贯原理，将传统导管架平台的桩基础改变成一个短粗的刚性开口薄壁圆筒壳构造的桶型基础，利用桶基自重和上部结构重量，将桶基压入海底一定深度，形成密封条件，然后用泵抽吸桶基内部，造成桶内外的负压，并通过该压力差将桶体压入海底预定深度达到固定的作用。

负压桶基础设计需要考虑的因素也较多，设计难度较大，DNV 规范推荐适用水深为 0 ~ 25 m。负压桶基础起步较晚，发展时间也不长，应用还不成熟，一些风险分析不全面，因此暂不推荐此种方案，可以说，负压桶基础尚处于研究阶段，但随着研究的深入，必将推进其在海上风电基础设计中的应用。

与其他类型的海上风电基础结构一样，负压桶型基础也需要满足基础刚度和转角控制两个方面的要求。

众所周知，桩侧土抗力大部分由近地表的浅层土发挥，因此如何充分利用浅层土的土抗力成为提高基桩水平承载性能的关键。本文给出了一种对海上风电负压桶型基础加装翼板

* 基金项目：中国水电顾问集团华东勘测设计研究院院立项课题：海上风电基础结构关键技术研究（KY2010-02-19），海上风电基础结构动力响应及疲劳寿命分析（KY2011-02-07-02）。

的构思(以下称为加翼负压桶型基础),通过在常规负压桶型基础外周设置稳定翼,使浅层土的抗力得到充分发挥,进而提高基础刚度,降低结构的动力响应。并借助数值模拟,对其进行了静、动力学分析。

1 加翼负压桶型基础

1.1 基本构造

加翼负压桶型基础结构布置图见图1,其中图1(a)、图1(b)为大直径加翼负压桶型基础立面图;图1(c)为小直径加翼负压桶型基础立面图。

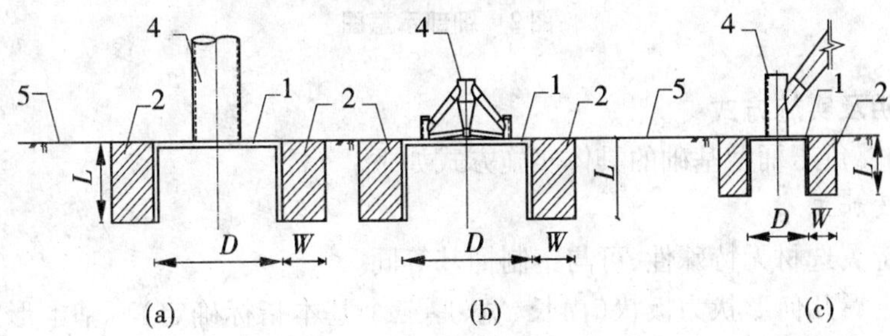

图1 基本构造示意图

加翼负压桶基础的负压桶部分与传统钢制负压桶型基础无异。根据负压桶桶径,区分为大直径(D=10~25 m)、小直径(D=4~10 m)两种情况:对于前者,可用钢筋混凝土制作,上部单立柱(见图1(a))或导管架整体(见图1(b))固定于桶顶,此种情况下,一个基础(安放一台风机)用一个大型加翼负压桶;对于后者,可用钢材质制作,此种情况下,一个基础用多个加翼负压桶分别布置在基础结构支撑位置,共同承担上部结构(见图1(c))。

对于钢制负压桶,稳定翼采用与桶基同材质的钢板(特殊要求时另行对待);对于钢筋混凝土制负压桶,稳定翼可采用钢板,且在主体桶基浇筑时嵌入固定,也可采用钢筋混凝土构造,与主体桶基同步浇筑而成。布设方式为:以桶基为中心轴线呈放射状布置,长度(L)方向与桶基轴向平行,宽度(W)方向与桶基轴向垂直。

对于大直径负压桶(D=10~25 m),根据桶径及构造需要设置4~8个翼板(见图1),宽度(W)的推荐取值为:$D<15$ m时,取$D/3 \leqslant W \leqslant 2D/3$;$D>15$ m时,取$5 \leqslant W \leqslant D/3$;长度($L$)取$H/2 \leqslant L \leqslant H$($H$为桶基高度),具体参数可通过仿真计算优化确定,翼板从泥面开始设置,并注意考虑冲刷影响。

图2为细部构造或处理的示意图,图2(a)为大直径加翼负压桶型基础加8个稳定翼的剖面示意图;图2(b)为小直径加翼负压桶型基础加4个稳定翼的剖面示意图;图2(c)为对稳定翼进行优化的示意图。

在图1、图2中:1为负压桶;2为稳定翼;3为加强支撑构件;4为上部结构(单桩时为塔筒,多桩时为导管架);5为泥面;L为稳定翼长度;W为稳定翼宽度;θ_1为减阻角(降低沉桩阻力);θ_2为减缓应力集中坡角(降低应力集中)。

考虑到沉桩过程中稳定翼入土时所受阻力,在稳定翼与桶基连接处增设加强支撑构件3;并对稳定翼先入泥一侧翼缘进行坡度化处理(设置减阻角θ_1);考虑到稳定翼上翼缘与桶基连接处会出现应力集中,可在此增设减缓应力集中坡脚(θ_2)。

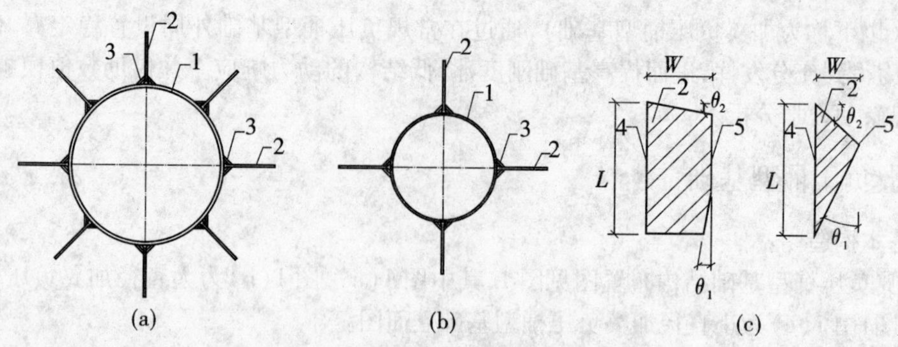

图 2　细部示意图

1.2　结构说明及实施方式

本文中加翼负压桶型基础的具体实施方式如下。

1.2.1　预制及加工

选材:稳定翼选材无特殊性,可与钢制桶基等同。

选型:稳定翼几何形状为板状(由长、宽、厚三个基本指标确定)。基本形状为矩形,考虑打桩阻力及应力集中,可将上、下翼缘进行坡度化处理:考虑稳定翼会增加沉贯阻力,可将稳定翼下翼缘坡度化处理(加工成带减阻角 θ_1);考虑稳定翼上翼缘与基桩连接处易发生应力集中,进行坡度化处理(加工成带减缓应力集中坡角 θ_2);沉贯阻力的增加需通过相应计算确定,进而对 θ_1 和 θ_2 进行调整。

加工:稳定翼除上述对于上、下翼缘必要的坡度化处理之外,无特殊加工要求。

装配:稳定翼与桶基连接采用焊接,并辅助设置必要的支撑构件。

1.2.2　沉桩、调平施工

沉桩施工同传统无稳定翼负压桶基础的负压沉贯、调平过程。

1.2.3　上部构件施工

由于沉桩施工完成后稳定翼完全位于泥面以下,因此不会对上部结构的施工造成任何影响。

2　加翼负压桶型基础效果检验

2.1　有限元模型

负压桶直径 $D=20$ m(壁厚 $t=70$ mm),入土(高度)$H=8$ m;稳定翼设置 8 片,以桶基轴线为中心沿外桶呈放射状等间距布置,尺寸:$L=H$、$W=D/4$、$T=t$;上部采用图 1(a)所示的单桩塔筒型式,直径 4.5 m。桶周土体黏聚力 $c=20$ kPa,摩擦角 $\varphi=32°$,容重 $\gamma=20$ kN/m^3。下文图表中分别以“Wing-Bucket”和“Bucket”表示“加翼”和“无翼”负压桶型基础。

本例负压桶直径较大,实际设计时通常会有桶内的“分舱”及考虑桶体用钢筋混凝土制作,本例旨在对负压桶加翼的效果进行数值验证,故暂用钢材质,且忽略桶内的分舱设计。另外,本例仅对一种加翼设计与无翼情况进行对比,暂未涉及翼板形状参数的敏感性分析。

有限元模型采用大型有限元分析软件 ANSYS 建立,如图 3 所示,其中图 3(a)为负压桶型基础及桶周土体有限元模型及网格,用于下述静力分析;图 3(b)为在图 3(a)模型基础上补充上部塔筒、机舱、轮毂、叶片等之后的整体结构模型及网格,用于下述动力分析;桶基、土体均采用实体单元(SOLID)模拟,并在交界面上设置接触单元;土体本构模型遵循 D-P 准则。

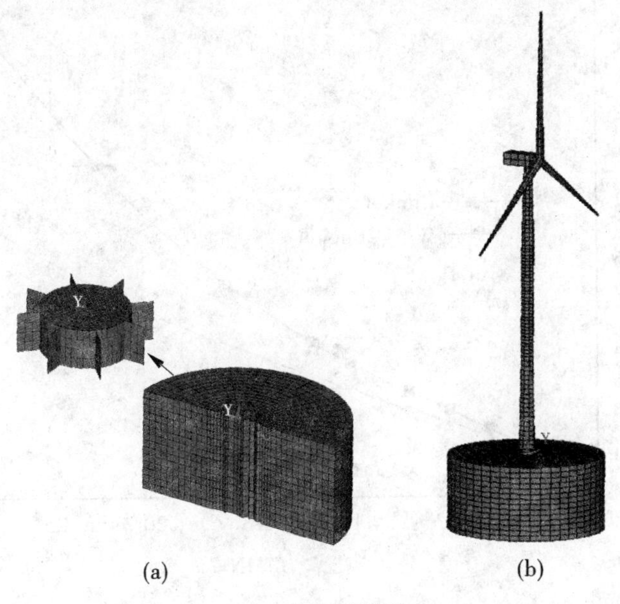

图3 有限元模型

2.2 静力分析

静力分析重点针对桶基进行,建立如图 3(a)所示的有限元模型并划分网格。其中,采用0.5~30 MN 范围内的九级静荷载,并施加在桶基顶(与泥面同高程)以上 2 m 位置。荷载—位移及荷载—转角曲线如图 4、图 5 所示。可见,加翼后的桶基水平位移得到有效控制,降低的幅度随水平载荷的增加而增大。

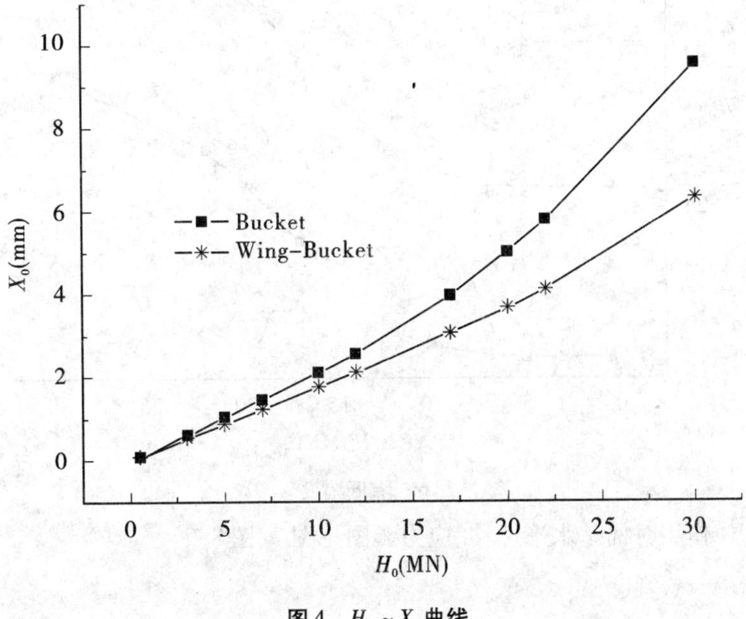

图4 $H_0 \sim X_0$ 曲线

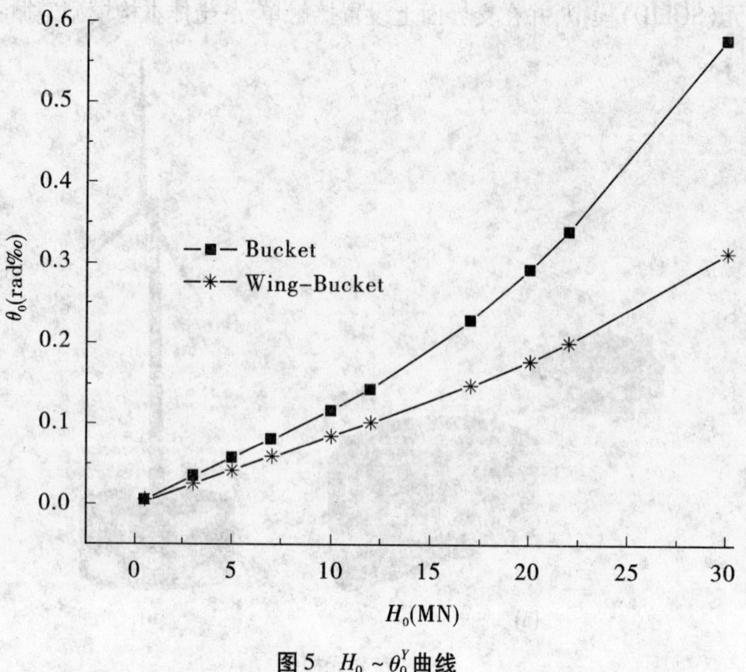

图5　$H_0 \sim \theta_0^Y$ 曲线

各荷载作用下，位移、转角降低幅度如图6所示。可见，伴随荷载的增加，稳定翼对于降低单桩的水平位移、转角的作用愈加明显。

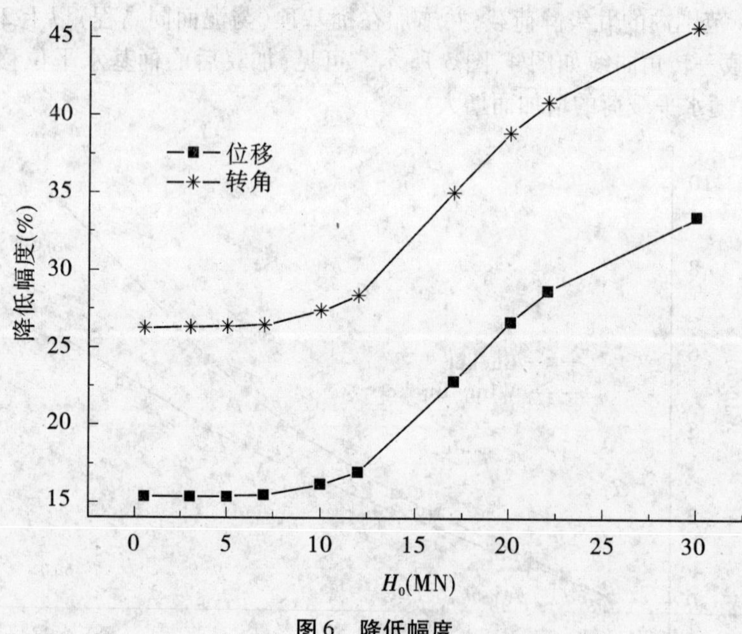

图6　降低幅度

截取30 MN作用时桶基水平位移、应力云图见图7、图8。

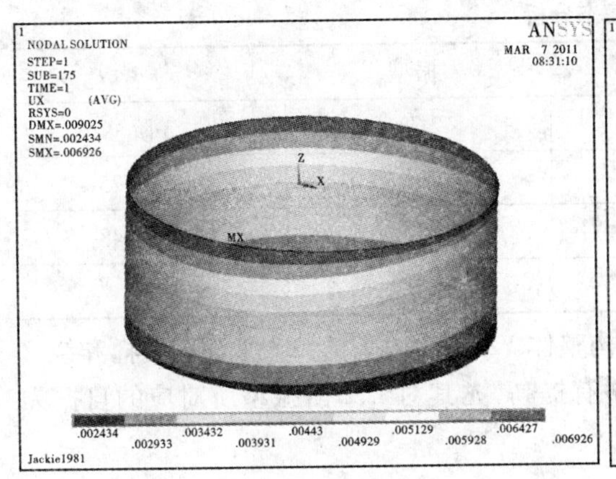

(a)Bucket

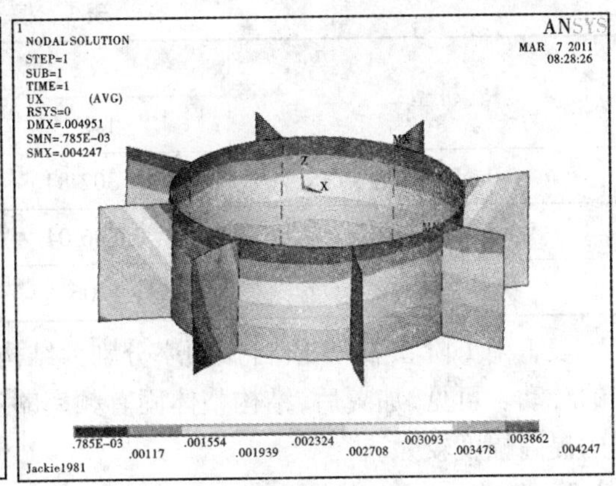

(b)Wing-Bucket

图 7 桶基 X 方向水平位移

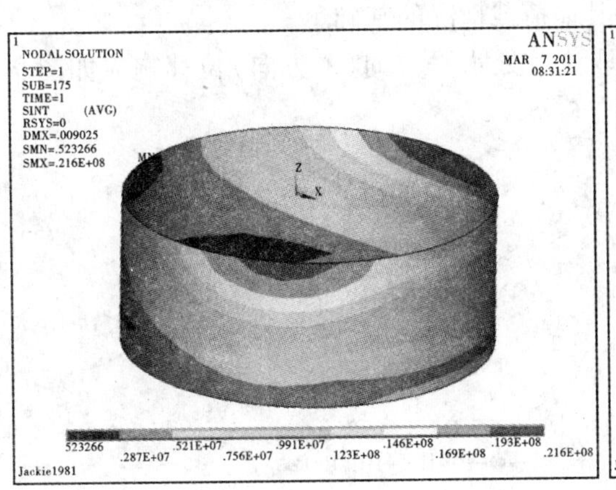

(a)Bucket

(b)Wing-Bucket

图 8 桶基应力

由图 7 可见,加翼后桶基水平位移显著降低;由图 8 可见,加翼后桶基最大应力提高,在翼板与主桶体连接处应力较为集中,最大值出现在荷载作用方向正前方一片稳定翼与桶基连接处的上缘(49.7 MPa),但该应力水平在桶基可承受的应力范围内,通过翼板的形状优化可以降低该应力。

2.3 动力分析

动力分析采用图 3(b)的整体结构模型。本例假定平均海平面为 0 高程,泥面高程为 -24.9 m,法兰高程 5 m,塔筒顶部高程 87 m。以模态分析为基础,包括谐响应分析、冲击荷载作用下的瞬态分析、波浪谱分析。其中,动力荷载均假定作用在 3 m 高程位置。

2.3.1 模态分析

提取结构前 3 阶自振频率如表 1 所示。

表 1　模态分析

桩型	振型		
	1 阶	2 阶	3 阶
Bucket(Hz)	0.302 84	0.304 73	1.264 325
Wing-Bucket(Hz)	0.306 04	0.307 99	1.264 348
提高幅度(%)	1.06	1.07	0.002

表 1 中 1 阶、2 阶、3 阶自振频率分别为结构整体 YZ 平面内平摇、XZ 平面内平摇和绕 Z 轴的转动。可见,加翼后,结构整体固有频率略有提高,尤其对 1、2 阶振型所对应的自振频率的提高幅度较大。

2.3.2　谐响应分析

复杂荷载作用时的情况可通过 Fourier 变换近似成简谐荷载的求和形式,在系统线性的假定条件下,单自由度体系承受简谐荷载时的特性可以方便地推广到任意荷载条件下单自由度系统的情况,研究简谐荷载的情况是开展其他荷载情况的基础。

施加水平向(X 方向)简谐荷载作用。提取法兰处水平方向(X 方向)位移响应如图 9 所示。

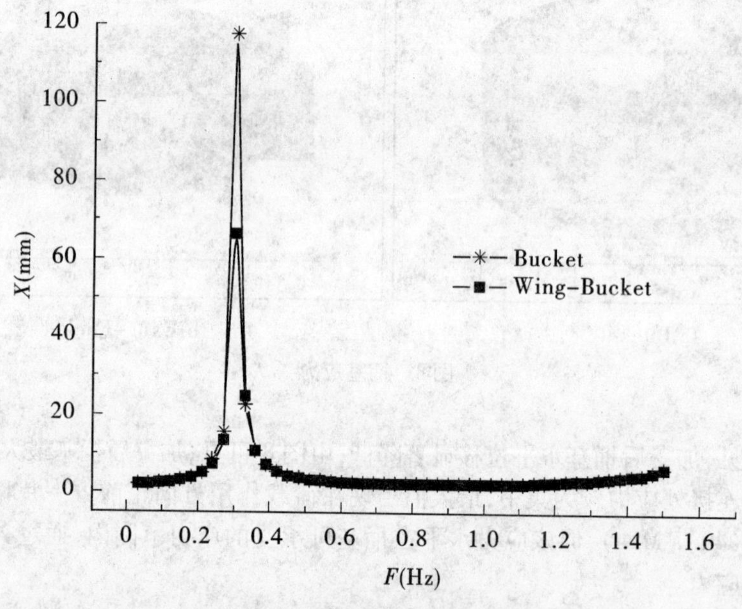

图 9　位移响应(简谐荷载)

可见,X 方向位移响应峰值与结构 1 阶自振频率对应;加翼后位移响应峰值显著降低。

2.3.3　瞬态分析——冲击荷载作用

假定结构受如图 10 所示的 X 方向冲击荷载作用,提取法兰及塔筒顶部的 X 方向位移响应如图 11 所示,最大位移响应列于表 2。

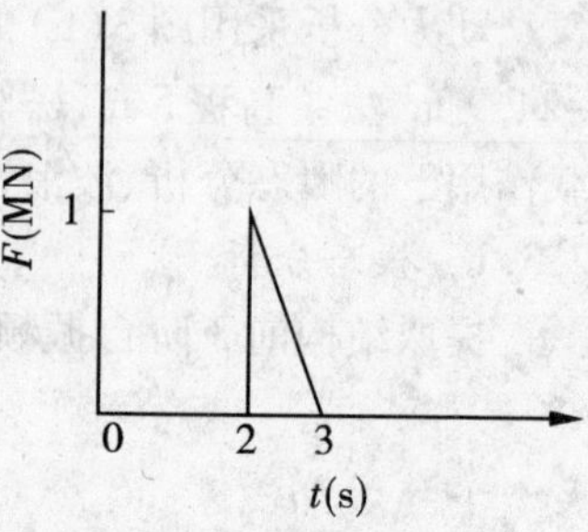

图 10　冲击荷载时程曲线

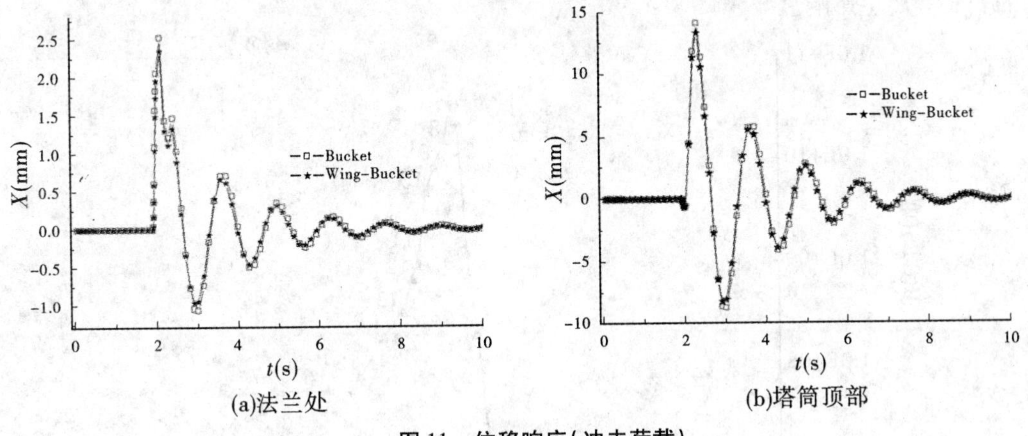

(a)法兰处 (b)塔筒顶部

图11 位移响应(冲击荷载)

表2 计算结果(冲击荷载)

桩型	位移响应峰值(mm)	
	法兰处	塔筒顶
Bucket	2.522 45	14.119
Wing-Bucket	2.346 21	13.334 3
降低幅度(%)	6.987	5.558

从响应曲线趋势角度观察可见:对于距离冲击位置较近的法兰处而言,紧随冲击荷载开始作用的 $t=2$ 时刻开始出现峰值($t=2.1$),而对于距离冲击位置较远的塔筒顶部而言,其位移响应峰值发生时刻有所滞后($t=2.3$),这是符合实际的;从响应曲线峰值大小角度观察,加翼后单桩的位移响应显著降低。

2.3.4 波浪谱分析

假定结构受到如图12所示波浪谱作用,作用方向为 X 方向,法兰及塔筒顶处的 X 方向位移响应如图13所示。

可见,0~1 Hz 范围内,响应曲线出现两个较为明显的峰值,且分别与波浪谱频率及结构1阶自振频率对应,两者之间的差异决定了结构发生共振的可能性;从响应曲线峰值观察,加翼后桶基的位移响应得到显著降低。

3 结语

以充分利用浅层土的土抗力为设计思路,对一种加装稳定翼的海上风电负压桶型基础进行了研究,对其构造、效果及实施方法进行了详细阐述。通过有限元仿真,从静力学、动力学角度对稳定翼的设置效果进行了验证。

加翼负压桶型基础有益效果可以总结为:

(1)充分发挥近地表浅层桩前土的抗力,增强桶基水平承载性能;加翼对结构整体固有频率影响较小(略有提高)。

(2)达到与无稳定翼普通负压桶基础同等的承载性能时,加翼负压桶基的负压桶尺寸得以减小,相应的材料、施工成本得以降低。

(3)不会影响负压沉贯施工,且由于沉桩后稳定翼位于海床泥面以下,因此不会对上部结构的安装等产生影响。

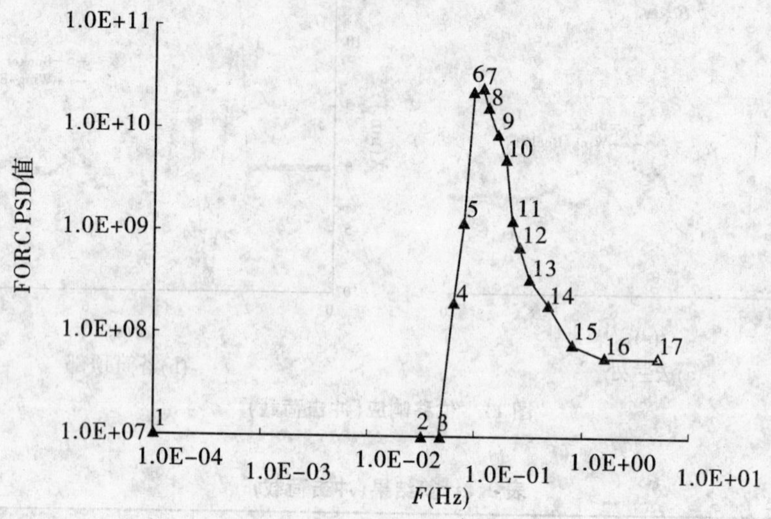

图12 波浪力谱

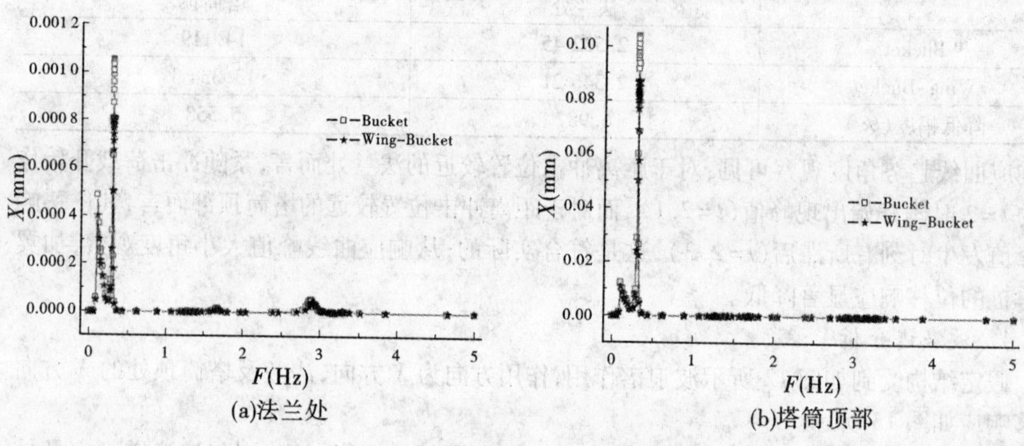

(a)法兰处　(b)塔筒顶部

图13 位移响应(波浪荷载)

参考文献

[1] Kuo Y-S, Achmus M, Kao C-S. Practical design considerations of monopile foundations with respect to scour [C]. Global Wind Power 2008, Peking.

[2] Li Wei. Study on the Fuzziness in Fatigue Life Estimation of the Foundation of Offshore Wind Turbine[J]. Advanced Materials Research, 2011, 243-249: 4741-4745.

[3] 李炜,李华军,郑永明. 海上风电基础结构大直径钢管桩水平静载荷试验数值仿真[J]. 水利水电科技进展,2011,31(4): 69-72.

[4] 王靖,许涛. 海上桶形基础采油平台综合分析[J]. 海洋技术,2003(3):27-28.

[5] 张枢文. 海洋平台结构损伤识别与健康监测技术研究[D]. 镇江:江苏科技大学,2007.

[6] 李炜,李华军,郑永明,等. 海上风电基础结构疲劳寿命分析[J]. 水利水运工程学报,2011(3):70-76.

【作者简介】 李炜,1981年生,2010年毕业于大连理工大学,获工学博士学位,中国水电顾问集团华东勘测设计研究院,高级工程师。

水泥土平面应变试验研究

徐海波[1] 宋新江[1] 崔 飞[2]

(1. 水利部淮河水利委员会水利科学研究院 蚌埠 233000；
2. 中水淮河规划设计研究有限公司 蚌埠 233001)

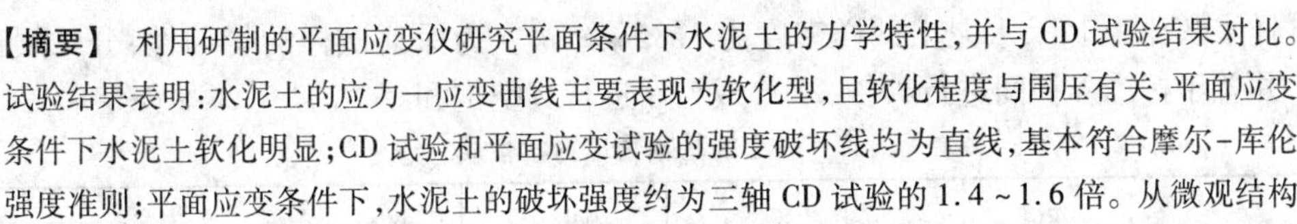

【摘要】 利用研制的平面应变仪研究平面条件下水泥土的力学特性，并与CD试验结果对比。试验结果表明：水泥土的应力—应变曲线主要表现为软化型，且软化程度与围压有关，平面应变条件下水泥土软化明显；CD试验和平面应变试验的强度破坏线均为直线，基本符合摩尔-库伦强度准则；平面应变条件下，水泥土的破坏强度约为三轴CD试验的1.4～1.6倍。从微观结构和能量守恒原理解释了CD试验和平面应变试验水泥土软化成因，认为水泥水化物等凝胶材料是水泥土软化的主要因素，中主应力的存在使水泥土软化更加明显。

【关键词】 水泥土 平面应变试验 三轴固结排水试验 强度

平面应变是岩土工程中常见的一种受力状态，是土力学中古老的课题之一，如边坡、堤坝截渗墙、基坑、路基、挡土墙等工程。由于试验仪器等方面的限制，工程中常利用常规三轴试验成果解决平面应变问题。常规三轴试验忽略了中主应力的影响，即试验过程中保持中主应力与小主应力相等，主要用于解决轴对称问题；平面应变试验始终保持中主应变 $\varepsilon_2 = 0$，而不限制中主应力 σ_2，即 σ_2 不等于 σ_3，主要解决工程中的平面应变问题。CD试验和平面应变试验条件不同，对于相同材料，若采用常规三轴试验结果求解平面应变问题，其计算结果与工程实际状态必然存在偏差。

目前，平面应变试验研究成果主要集中于黏性土和砂等材料方面，平面应变条件下沙土峰值强度的内摩擦角比三轴试验高2°～8°，破坏时的中主应力参数 $b = (\sigma_2 - \sigma_3)/(\sigma_1 - \sigma_3) = 0.25 \sim 0.35$，并建立了平面应变条件下沙土和黏性土的本构模型。水泥土具有低渗透性和高强度特点，是近年来堤坝截渗墙工程常用的新型截渗材料之一。当前水泥土研究成果基本来自于无侧限抗压强度试验及三轴试验等常规试验成果，而水泥土平面应变试验方面的研究成果甚少。本文利用研制的水泥土平面应变仪，研究平面应变条件下水泥土的力学强度特性；并与相同条件下三轴固结排水剪(CD)试验成果对比，分析了两种试验方式下水泥土的强度特点。

1 试验仪器

试验仪器采用水利部淮河水利委员会水利科学研究院研制的水泥土平面应变仪。该仪器由压力室、上部应力控制系统、压力室内试样反力系统、孔隙水压力量测系统和数据采集系统等五部分组成，可根据需要进行应变或应力控制式试验。压力室由合金钢铸成，最大承受压力 4.0 MPa,压力室通过阀门与气缸相连,利用液态氮气为试验提供周围压力。压力室内反力架两端各设置一块侧压板,其中一侧压板主要约束被测试样中主应力方向的变形,钢板通过螺杆与压力室外部手柄相接,试验前调整手柄位置使试样与侧压板接触,确保试验过程中中主应变始终为 0。试验过程中,通过反力架钢板后布置的应力传感器量测试样的中主应力。试验数据由计算机自动采集。

平面应变试验试样为 10 cm×5 cm×10 cm 的长方体,试验过程中始终保持 σ_2 方向应变为 0,即中主应变 $\varepsilon_2=0$。

2 水泥土试样制备及试验

水泥掺入量根据质量比原则计算,即 α_w =(掺和的水泥质量/土体烘干质量)×100%。本试验水泥掺入量为 12%,水泥为 P·C 32.5,土料为低液限黏土,土料物理力学性质见表 1。试样水灰比为 1.0,水泥土试样密度为 1.85 g/cm^3。为保证试样的均匀性,分 4 层进行击实,并置于标准养护间进行养护,28 d 成型后进行真空饱和。

表 1 土的基本物理力学性质

土粒比重 *Gs*	液限 (%)	塑限 (%)	塑性指数	砂粒粒径 >0.05 mm (%)	粉粒粒径为 0.05~0.005 mm (%)	黏粒粒径 <0.005 mm (%)
2.67	28.2	18.4	9.8	16.0	75.0	9.0

常规 CD 试验试样直径为 3.91 cm,高 8.0 cm,分 3 次击实,水泥土试样密度与平面应变试验要求相同。

2.1 试样安装

(1)将饱和后的试样安装在下部试样架上,套上橡皮膜,加盖上部试样架,拧紧螺栓使橡皮膜与试样架紧密接触。

(2)将套有橡皮膜的试样置于压力室底部,安装侧向应力量测设备,接通上、下排水管路,打开上、下排水管,并从下排水管将水压入试样中,将橡皮膜与试样之间的空气从上排水管排除,直至上排水管内无气泡排出,关闭下排水阀。

(3)降低排水管位置,使水面至试样中心高度以下 20~40 cm,吸出试样与橡皮膜之间多余的水分,然后关闭排水阀;安装侧向位移传感器。

(4)加盖压力室上盖并用螺栓拧紧,通过压力室外部活动旋钮调整压力室内部钢板位置,使钢板与试样接触,通过侧向压力计判断钢板与试样是否接触。

2.2 试验方案

为便于平面应变试验与常规 CD 试验成果比较,两试验固结压力均选取为 100 kPa、200 kPa、400 kPa、600 kPa,试验类型为固结排水剪,剪切过程中保持围压不变,增加轴向压力,直至试样出现破坏,试验剪切速率为 0.006 4 mm/min。

3 试验成果分析

水泥土是水泥和土的混合物,两者在水环境中发生一系列复杂的物理化学反应,水泥水化物改变了土体原有结构,形成具有较高强度和低渗透性的混合材料。水泥水化物主要包括氢氧化钙($Ca(OH)_2$)、含水硅酸钙($3CaO \cdot 2SiO_2 \cdot 3H_2O$)、含水铝酸钙($3CaO \cdot Al_2O_3 \cdot 6H_2O$)、含水铁酸钙($3CaO \cdot Fe_2O_3 \cdot 6H_2O$)和水化硫铝酸钙($3CaO \cdot Al_2O_3 \cdot 3CaSO_4 \cdot 32H_2O$)等细颗粒胶体。

3.1 试验结果

图1为不同围压水泥土平面应变试验和常规CD试验应力—应变曲线对比图。图中可知,平面应变条件下水泥土应力—应变曲线为软化型驼峰曲线,并随着围压的增大曲线软化程度减小,破坏应变为3%~5%,曲线存在明显拐点,属脆性破坏。试样破坏后,强度迅速降低,最终趋于一稳定值。

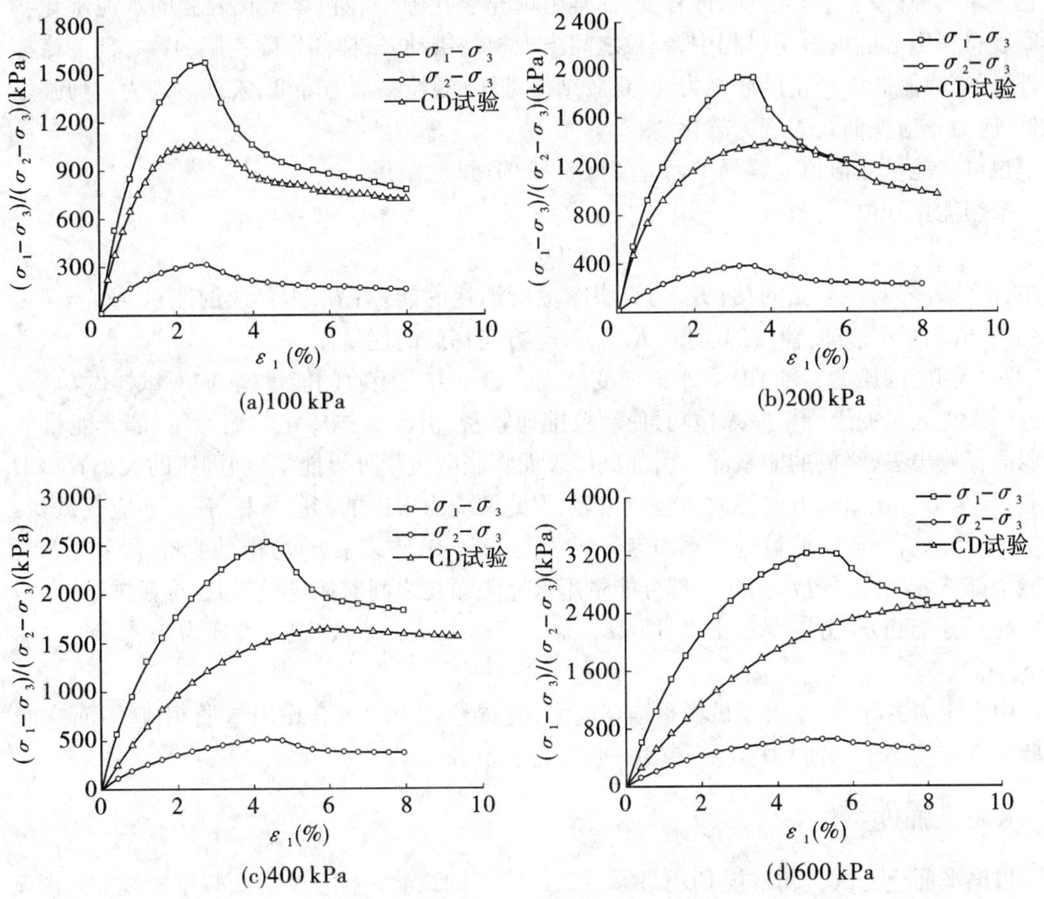

图1 平面应变试验和三轴CD试验应力—应变曲线对比

水泥土CD试验应力—应变曲线有软化和硬化两种型态,围压≤400 kPa时,曲线为软化型,围压>400 kPa时,曲线为硬化型,破坏应变为3%~8%。

相同围压,由于中主应力的影响,平面应变试验比三轴CD试验应力—应变曲线软化明显,初始弹性模量增加,峰值强度增大,平面应变条件下水泥土破坏强度约为三轴CD试验的1.4~1.6倍。围压较小时,两种试验方式破坏应变基本相同;围压较大时,CD试验破坏应变略大于平面应变试验。

3.2 试验成果分析

水泥水化物充填土颗粒之间的孔隙,使颗粒之间黏结力增加,同时胶体材料粘在土体颗粒表面,使颗粒变粗,接触面积增大,凝聚分量增大,使水泥土内摩擦角和黏聚力相对于土体材料增大。

研究表明,沙土的软化与颗粒的松紧程度有关,沙土颗粒排列紧密,其应力—应变曲线为软化型;颗粒排列松散,应力—应变曲线则为硬化型。正常固结黏土的应力—应变曲线多认为是硬化型。从图 1 可以看出,低围压作用下,水泥土平面应变试验和常规 CD 试验的应力—应变曲线均为软化型。水泥土与沙土出现软化现象的原因不同,本文认为引起水泥土软化现象的原因不是颗粒的松散程度,而是由填充于孔隙中和包裹土粒表面的水泥水化物等凝胶材料引起的。剪切过程中,颗粒之间出现相互错动,试样变形除克服颗粒之间的摩擦力外,还需克服胶体产生的胶结力,一旦胶结体破坏,颗粒黏结力降低,宏观表现为抗剪强度减低,应力—应变曲线表现为软化。

也可从能量守恒角度解释平面应变试验和 CD 试验结果。

根据能量守恒方程,有

$$W = E_{克} + E_{释} \tag{1}$$

式中:W 为外界对试样做的功; $E_{克}$ 为克服摩擦所消耗的能量; $E_{释}$ 为释放的能量, $E_{释} = E_{释2} + E_{释3}$, $E_{释2}$ 为 σ_2 方向释放的能量, $E_{释3}$ 为 σ_3 方向释放的能量。

对于相同围压的三轴 CD 和平面应变试验,从 σ_1 方向吸收的能量(轴向对试样做功)一部分用来克服水泥土颗粒摩擦做功,使颗粒排列紧密,引起试样体积收缩;另一部分能量从薄弱面释放出去,释放的能量能否引起试样表面膨胀取决于外界能量(即围压的大小)。CD 试验:$\sigma_2 = \sigma_3$, σ_2、σ_3 方向释放的能量相等,因此两应力方向的膨胀量相等;平面应变试验:$\sigma_1 > \sigma_2 > \sigma_3$,释放的能量进行不均匀分配,由于 $\varepsilon_2 = 0$,约束了 σ_2 方向的变形,限制了 $E_{释2}$ 不能全部从 σ_2 方向释放,而是一部分能量用来克服颗粒之间摩擦,进行颗粒重新排序,另一部分能量分配到 σ_3 方向释放出去,即 $2E_{释3CD} > E_{释3} > E_{释3CD}$,宏观上表现为 $2\varepsilon_{3CD} > \varepsilon_{3平面} > \varepsilon_{3CD}$ 。

中主应力的存在,增大了试样平均球应力,使颗粒约束和咬合作用增强,引起平面应变试验试样体积小于三轴 CD 试验,应力—应变曲线软化显著。

4　水泥土强度特性

根据平面应变试验和常规 CD 试验结果,研究不同试验条件下水泥土强度特性等。图 2 和图 3 分别为 CD 试验和平面应变试验强度包线,两种试验条件下水泥土破坏线均为直线,符合摩尔-库伦强度准则。CD 试验:凝聚力 $c = 250$ kPa ,摩擦角 $\varphi = 30°$;平面应变试验抗剪能力略高于 CD 试验结果,黏聚力 $c = 300$ kPa ,摩擦角 $\varphi = 37°$ 。

5　结语

(1)水泥土 CD 试验应力—应变曲线呈现软化和硬化两种形式,低围压下为软化型,高围压趋于硬化型。水泥水化物等凝胶材料是引起水泥土软化现象的最主要原因。

(2)当 $\sigma_3 \leqslant 600$ kPa 时,水泥土平面应变试验应力—应变曲线为软化型,随围压的增大软化程度减弱,中主应力的影响使水泥土软化现象更加明显。根据能量守恒原理,阐述了水泥土 CD 试验和平面应变试验结果。

(3)4 种不同围压下,平面应变试验水泥土破坏强度约为常规 CD 试验的 1.4~1.6 倍,水泥土 CD 试验和平面应变试验破坏线均为直线,符合摩尔-库仑强度准则。CD 试验: c = 250 kPa, φ = 30°,平面应变试验: c =300 kPa, φ = 37°。

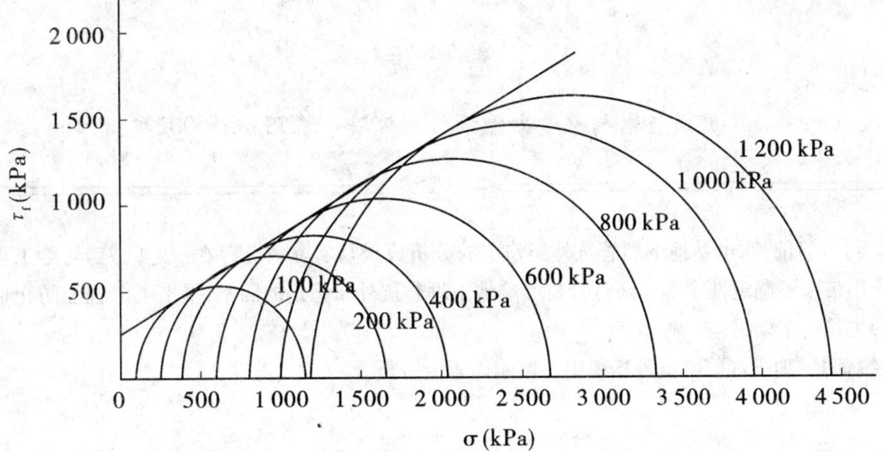

图2 三轴 CD 试验强度包线

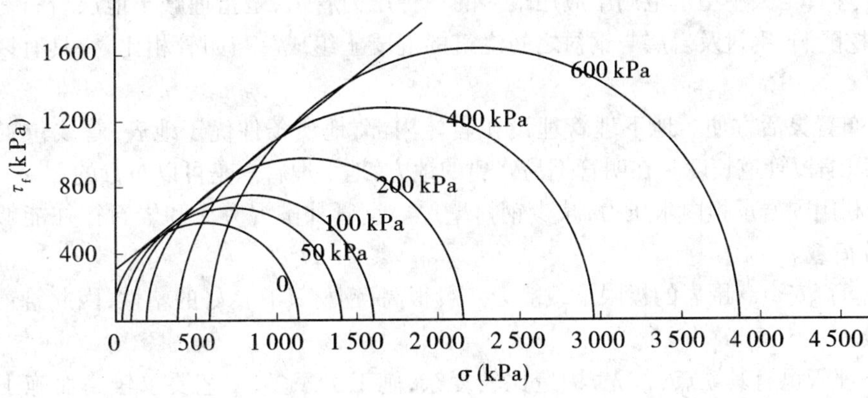

图3 平面应变试验强度破坏线

参考文献

[1] Cornforth D H. Some experiments on the influence of strain condition on the strength of sand [J]. Geotechnique, 1964, 14 (2): 143-167.

[2] Bishop A W. The strength of soil as engineering material [J]. Geotechnique, 1966, 16(2):89-130.

[3] 郑楚健,郑颖人,朱建凯. 平面应变条件下 M-C 材料屈服时的中主应力公式[J]. 岩土力学, 2008, 29(2):310-314.

[4] 李广信,黄永男,张其光. 土体平面应变方向上的主应力[J]. 岩土工程学报, 2001, 23(3):358-361.

[5] 马险峰,望月秋利,温玉君. 基于改良型平面应变仪的沙土特性研究[J]. 岩石力学与工程学报, 2006, 25(9): 1745-1754.

【作者简介】 徐海波,1981 年 5 月生,工学硕士,2009 年 6 月毕业于河海大学岩土工程专业,现任职于水利部淮河水利委员会水利科学研究院,助理工程师。

水电站地下埋管抗外压稳定影响因素分析

曹　骏

（贵州省水利水电勘测设计研究院　贵阳　550002）

【摘要】 目前，水电站地下埋管抗外压稳定的分析成果与实际情况存在一定的差异，给工程上的应用带来不确定性。本文通过分析影响地下埋管抗外压稳定的因素，提出了工程上防止地下埋管失稳的主要措施。

【关键词】 地下埋管　抗外压稳定　影响因素　分析

1　地下埋管工作特点

地下埋管是大中型水电站中应用最多的一种压力管道，是指埋藏于地层岩石之中的钢管，由开挖隧洞、钢衬及岩层与钢衬之间浇筑的混凝土组成。与明管相比，它具有以下突出的优点：

(1)布置灵活方便。地下埋管埋设在岩体内部，地质条件优于地表，管线位置选择自由，且可以缩短管道长度。在明管不易修建的地方，地下埋管总是可以布置的。

(2)利用围岩承担内水压力，减少钢衬壁厚。钢管外围混凝土和岩石往往能够分担很大一部分荷载。

(3)运行安全。围岩的极限承载能力一般很高，钢材又有良好的塑性，因此管道内压超载能力较大。

地下埋管也有其缺点：首先，构造比较复杂，施工工序多，工艺要求较高而施工条件较差，施工质量不易保证。其次，由于钢衬管壁薄，在放空检修或灌浆施工时，钢衬有可能承受较大的外水压力或灌浆压力而失去稳定，即通常所说的管壁发生鼓包或压瘪的问题。在某些条件下，外压稳定是决定管壁厚度的主要因素。从国内外有关资料来看，埋藏式压力钢管外压失稳的事故不少，有些还造成了相当严重的后果。

2　地下埋管抗外压稳定分析及存在的问题

国内外学者对压力钢管的抗外压稳定问题进行了长期的研究和探索，并取得了一系列研究成果。比较有影响的计算理论和方法有伏汉、包罗特、阿姆斯图兹、孟泰尔、米赛斯、雅可比森、斯沃依斯基和我国压力钢管设计相关规范推荐的公式。

但是由于采用的一些基本假定有所不同，因此所得到的计算结果与模型试验和实际情况存在一定的差异。据《水利水电技术》(1980 年第 2、3 期)，上述公式计算结果与模型试验成果均存在误差，误差在±70%之间，且规律性不强。因此，给工程上的运用带来一定的不确定性,或偏于冒险,给工程留下隐患,或偏于保守采用较厚的钢板,甚至给施工带来困难,使工程量加大;因此,为减少风险,压力钢管设计相关规范要求的安全系数较大,达1.8~2.0,而一般的钢筋混凝土结构安全系数仅1.4~1.6。

即便如此,国内外地下埋管失稳事故仍有发生,如绿水河电站、响水电站。究其原因,影响地下埋管抗外压稳定的因素比较多且较复杂,从理论上难以完全模拟分析这些因素。上述解析法和经验公式都只能考虑部分因素,因此存在一定的误差。即使是在结构计算中运用比较好的数值分析方法,也未得到钢管设计规范的认可。

概括上述计算模型,其主要原则为钢衬与混凝土之间的缝隙为均匀的缝隙δ。在内水作用下,钢衬变形充填缝隙,使混凝土衬圈开裂,并与围岩按变位相容条件联合受力;在外压作用下,钢衬圆周上出现初始波,但当圆周变形即径向变位达到δ时,受混凝土的约束,变形不再发展。如果此时钢衬最大应力未达到钢材屈服强度,钢衬在外压作用下还可保持稳定状态。如果外压继续增大,钢衬应力值也相应增加,如超过钢材屈服强度,钢衬将发生塑性变形而导致发生多波屈曲(伏汉-包罗特公式)或在某个薄弱部位单波屈曲(阿姆斯图兹公式)。

无论是现行设计规范还是国内外学者提出的地下埋管结构分析计算理论,都假定钢衬与混凝土之间的缝隙作为均匀的缝隙,而实际上,钢衬与混凝土之间的缝隙并非均匀的缝隙δ(见图1)。受重力的作用或由灌浆施工等引起的差异,部分钢衬实际上是紧贴混凝土,而部分钢衬圆周与混凝土之间的缝隙值可达$(1\sim2)\delta$,因而导致理论模型中的缝隙值δ比实际情况小,从而临界外压P_{cr}计算与实际情况存在着差异。

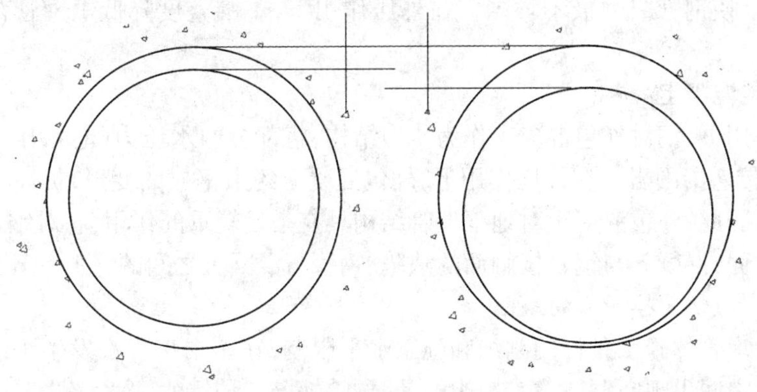

图1　地下埋管初始缝隙

3　影响地下埋管抗外压稳定因素分析

由于理论计算与实际存在的差异,因此应从方法论上重新寻找新的思路。与其在结构分析方法上去追求解析解的计算精度,不如在已有的分析计算方法基础上去分析影响地下埋管抗外压稳定的因素,使这些影响因素的不确定性,通过一定的工程措施,控制为可基本确定的因素,从而保证钢管抗外压的可靠性,减少其失稳的风险,即在单一的理论计算基础上,运用综合方法分析影响地下埋管抗外压稳定性能的因素,以合理判断其抗外压稳定的安全性。

从目前已知的因素分析,影响压力埋管抗外压稳定的因素主要有四方面,即钢管结构尺寸、钢管圆度偏差、外包混凝土、钢管材料和外压力。钢管结构尺寸上述公式已有论述，在此不再赘述。现对其他影响因素进行分析。

3.1 钢管圆度偏差

圆度偏差对地下埋管的抗外压稳定性能有较大的影响：一是圆度偏差降低了钢管对外压的抵抗能力。对地下埋管而言，按《水电站压力钢管设计规范》(DL/T 5141—2001)公式计算 $P_{cr}=2E_s(t/D)^3$，圆度偏差控制在5D/1 000以内，局部曲率半径按钢管半径的5/1 000计，则 P_{cr} 也要下降1.5%。按《水电站压力钢管设计规范(试行)》(SL 281—2003)推荐的经验公式计算，则 P_{cr} 也要下降0.85%。

绿水河水电站1#斜井钢管失稳的一个主要原因即为钢管安装后的圆度偏差过大，实测1#斜井未发生失稳的上半段的圆度偏差均不合格，最大达到了58D/1 000。按孟氏公式计算，不计圆度偏差的临界外压为2.5 MPa，而计入圆度偏差的临界外压仅1.64 MPa，可见圆度偏差对抗外压稳定性能的影响是比较大的。

二是在内压作用下，圆度偏差使钢管受力不均匀，局部应力过大，甚至超过了钢管的屈服强度，使管材进入塑性变形区。在内水放空后，产生的残余变形加大了钢管的圆度偏差，使其抗外压能力进一步降低。绿水河水电站实测1#斜井未发生失稳的上半段的圆度偏差最大达到了58D/1 000，其实就包含了钢管内压作用下的残余变形，降低了抗外压能力，最终在灌浆时发生了的失稳。

以礼河三级套管试验报告指出："钢管不圆整度(即椭圆度)是钢管在内压和外压下影响钢管工作性能的主要因素之一。在内压作用下，钢管不圆整度与局部缝隙一样，会导致弯曲应力的出现，使钢管应力极不均匀。在外压作用下，不圆整度降低了钢管对外压的抵抗能力。"

3.2 钢管外包混凝土

在内压作用下，钢衬外包混凝土作为传力结构，将部分内水压力(总水压力减去钢衬承担的水压力)传递给围岩。在外压作用下，外包混凝土约束钢衬的变形，提高了钢衬的抗外压稳定性能。因此，外包混凝土对地下埋管结构具有举足轻重的作用，它对钢管抗外压的影响主要有两方面：混凝土与钢管接触面的缺陷、钢管与混凝土之间的缝隙。

3.2.1 混凝土与钢管接触面的缺陷

由于地下埋管的施工条件的制约和施工水平限制，在多个工程都发生了地下埋管外包混凝土与钢衬之间出现空洞、蜂窝带、沟槽、凹坑等缺陷。如绿水河水电站3#平硐失稳段钢管底部1.3 m的混凝土和响水电站失稳段的大面积混凝土缺陷。

混凝土缺陷对地下埋管结构产生了较为不利的影响：一方面，减少了对钢衬的约束作用，因此降低了钢管的抗外压稳定能力；另一方面，在内压作用下，在混凝土缺陷边缘就会使钢衬产生较大的局部应力。在绿水河失稳前的观测资料就说明3#平硐失稳段钢管底部1.3 m的范围空洞边缘钢衬应力达3 588 kg/cm²，而按明管计算应力仅为1 740 kg/cm²。较大的局部应力会产生严重的后果。如果局部应力超过钢材的屈服强度后使钢衬进入塑性变形，在钢管放空后产生残余变形，加大了圆度偏差，降低了抗外压能力。因此，钢管极易在有混凝土缺陷的地方发生失稳破坏。

3.2.2 钢管与混凝土之间的缝隙

缝隙值取值是否合理，关系到钢衬与围岩之间承担内水压的分配比例。钢衬与混凝土之间的缝隙并非均匀的缝隙δ(见图1),因而导致理论计算模型中的缝隙值δ比实际情况小,就会使钢衬实际比计算值承担过多的内水压力,甚至超过了屈服强度,使管材进入塑性变形区,产生残余变形,加大了圆度偏差,降低了抗外压能力。

3.3 钢管材料

钢管的失稳是其受力后产生了较大的应力。当应力超过了钢管材料的屈服强度后,钢衬将发生塑性变形而导致发生大变形(屈曲)而失稳。因此,钢管材料屈服强度愈高,其抗外压的能力就愈强。

从地下埋管失稳的工程实例可提供的经验教训,内水外渗后对钢管的抗外压稳定也将造成十分不利的工况,因此钢管材料应首先保证在内水压工况下钢管不被拉裂。这就要求钢材本身的性能满足一定的要求,如强度、冲击韧性、化学成分、延伸率等。

3.4 地下水压力

地下埋管的外压力主要包括大气压、混凝土浇筑时流态混凝土的压力、地下水压力及钢管施工时的灌浆压力,下面主要论述地下水压力。

地下埋管在开挖过程中,隧洞成为地下水的主要排泄区,地下水位将随着开挖的过程而逐渐下降。因此,不能仅凭此时的地下水位而确定地下埋管的外水压力。地下埋管施工完成后,又将堵塞隧洞开挖形成的排泄通道,地下水位将逐渐恢复,地下埋管的外水压力将逐渐增大。

工程运行期间,水库和引水隧洞的渗漏作为地下水的补给区的一部分,对地下埋管的工作环境也会有一定的影响。影响的大小和工程的地质条件有着密切的关系。因此,应结合隧道开挖前的天然地下水位(至少应有一个水文年的观测资料)、隧洞施工中的出水情况、并结合运行后水库和引水隧洞对地下水位的抬升等因素综合确定合理的外水压力。

此外,引水隧洞内水或地下埋管内水外渗后,内水压力将传递到山体中的地下水中,造成地下水位抬升后,较高的内水压力储存于岩层的裂隙中,一旦钢管快速放空,钢衬上的裂缝在水压下迅速合拢,此时储存于山体中地下水压力也就成为了地下埋管的外水压力。对高压埋管而言,内水压常常高于天然地下水压,因此内水外渗所形成的外压工况常常造成地下埋管的破坏。例如巴斯康蒂抽水蓄能电站在首次充水时就观测到引水隧洞发生严重渗漏,渗透压力压屈折裂了13.5 m长的一段压力钢管。响水电站在钢管内水外渗后,又经多次充放水造成大范围的失稳破坏。

4 防止地下埋管失稳的主要措施

首先,地下埋管相关设计规范和施工规范对影响钢管抗外压稳定的一些因素已作了一定要求,如对抗外压稳定的安全系数的规定、对地下埋管灌浆的要求、圆度偏差的要求以及光面管隔一定距离必须设置加劲环等。这些措施和要求对防止地下埋管的失稳都起到了很好的效果。

其次,地下埋管还应在降低外水压力上采取措施。

地下埋管钢衬的失稳事故,多发生在地下水作用的情况下。采取有效措施,降低地下水压力,是防止钢衬失稳的有效方法,特别是在水电站运行后地下水位有可能增加时,采取这种措施更有重要意义。在工程实践中,广泛地运用排水廊道结合排水孔的方法降低地下水位。排水廊道大都是水平隧洞,可利用施工时开挖的交通洞及勘探平硐。

另外,在钢衬壁周围设置排水系统,更是直接降低钢管外水压力的有效方法。排水系统大多采用排水管,排水管可采用钢管、PVC 管、无砂混凝土管甚至草绳。采用此方法虽直接有效,但应防止运行时排水系统的堵塞,运行时应经常用高压空气对排水管进行冲洗。

此外,运行单位应健全运行管理制度,以保证钢管的安全运行。对地下埋管的充、放水应缓慢,不得快速充、放水。钢管不要长时间处于空管状态,应及时充水保压。运行时应加强排水系统排水效果(流量变化)的观测,充、放水时和放空后应加强对钢管应力和外水压力的观测。

参考文献

[1] 中华人民共和国水利部. SL 281—2003 水电站压力钢管设计规范(试行)[S]. 北京:中国水利水电出版社,2003.

[2] 中华人民共和国国家经济贸易委员会. DL/T 5141—2001 水电站压力钢管设计规范[S]. 北京:中国电力出版社,2002.

[3] 汪易森,庞进武,刘世煌. 水利水电工程若干问题的调研与探讨[M]. 北京:中国水利水电出版社,2006.

[4] 曹骏,杨卫中,将锁红. 云贵响水电站地下埋管失稳原因初步分析[J]. 贵州水力发电,2003(4):34-41.

[5] 中华人民共和国国家发展和改革委员会. DL/T 5017—2007 水利水电工程压力钢管制造安装及验收规范[S],北京:中国电力出版社,2007.

【作者简介】 曹骏,1968 年 12 月生,2005 年毕业于武汉大学,获硕士学位,贵州省水利水电勘测设计研究院,高级工程师。

北疆某碾压混凝土坝施工配合比中掺合料的试验研究

孙根民

（新疆额尔齐斯河流域开发工程建设管理局　乌鲁木齐　830000）

【摘要】 针对北疆某碾压混凝土重力坝施工，通过对混凝土配合比设计中掺加粉煤灰和石粉后力学性能、热学性能及变形性能的试验研究，提出在不同条件下碾压混凝土配合比设计中掺合料的选择和运用，以提高碾压混凝土自身的温控防裂性能，取得良好的施工效果。

【关键词】 碾压混凝土坝　大坝施工　混凝土配合比　掺合料　北疆

1　引言

碾压混凝土筑坝技术是在20世纪70年代开始研究应用的。碾压混凝土与常态混凝土的主要施工条件、拌和设备及施工程序基本相同，但改变了混凝土的配合比和施工工艺，它采用土石坝常用的施工设备运输、铺筑及碾压，是在强力振动和碾压的共同作用下成形的一种干硬性混凝土。与土石坝相比，碾压混凝土坝具有体积小、强度高、防渗性能好的优点。与常态混凝土相比，碾压混凝土具有施工工序简单，碾压机械可进行全断面通仓碾压、连续上升、快速施工的特点。因此，这种筑坝技术得到了广泛采用。

为了满足碾压混凝土力学性能、热学性能、变形、耐久性及施工可碾性等各类指标要求，在碾压混凝土中都加入了一定数量的掺合料。掺合料主要有粉煤灰、矿粉、石粉等。北疆某碾压混凝土重力坝地处我国西北地区，属大陆性寒温带气候，其特征是气候干燥，冬季长且严寒，四季温差悬殊。为满足碾压混凝土的抗冻、抗渗、防裂、耐久性及经济安全等技术指标，对碾压混凝土的掺合料及配合比进行了深入研究，本文着重对配合比设计中掺合料的优选进行论述。

2　粉煤灰品质比选及掺量选择

碾压混凝土中粉煤灰与水泥之和一般控制在150 kg/m^3左右，其中粉煤灰的掺量可占总胶凝材料的30%～70%。由于优质粉煤灰中含有大量的圆珠形玻璃晶体，对混凝土有改性作用，混凝土的施工性能可得到较大的改善，还可提高混凝土的抗渗性、抗裂性等耐久性指标，所以碾压混凝土中应掺入优质粉煤灰，粉煤灰品质应在Ⅱ级以上。由于不同品质的粉煤灰活性及需水量都不尽相同，其在混凝土中的掺量需试验后确定。

2.1　粉煤灰的选择

北疆某工程碾压混凝土坝选用了两家电厂的粉煤灰，其中玛纳斯电厂的粉煤灰品质指标达到了《水工混凝土掺用粉煤灰技术规范》(DL/T 5055—2007)中Ⅰ级粉煤灰标准,独山子电厂的粉煤灰需水量比为96%,烧失量为6.38%,未达到Ⅰ级粉煤灰指标要求,属于品质较好的Ⅱ级粉煤灰。因此,这两家电厂的粉煤灰均可用于本工程碾压混凝土中。玛纳斯电厂的粉煤灰可优先使用在抗冻和抗渗要求高的部位。

2.2 粉煤灰掺量对混凝土发热量的影响

不同粉煤灰掺量下的胶凝材料水化热试验成果表明,掺入粉煤灰后胶凝材料的水化热降低,且掺量越高水化热降低越多,但水化热的降低率低于粉煤灰的取代率,水化热试验结果见表1。

表1 水化热试验结果

粉煤灰掺量(%)	水泥水化热(kJ/kg)/水化热降低率(%)					
	3 d			7 d		
	普通水泥	中热水泥	高抗硫水泥	普通水泥	中热水泥	高抗硫水泥
—	239	199	213	288	247	260
30	191/20%	—	182/15%	222/23%	—	224/14%
40	177/26%	157/21%	172/19%	204/29%	181/27%	201/23%
50	161/33%	146/27%	—	190/34%	173/30%	—
60	138/42%	130/35%	—	167/42%	153/38%	—

2.3 粉煤灰掺量对混凝土强度的影响

通过对不同粉煤灰掺量、不同水胶比、固定石粉掺量16%、掺入缓凝高效减水剂及引气剂的配比试验研究,可得出以下结论:在碾压混凝土中掺入粉煤灰,早期强度较低,且发展较慢,但后期强度增长率较大;在相同的水胶比下,碾压混凝土抗压强度随粉煤灰掺量的增加而降低,在相同的粉煤灰掺量下,碾压混凝土抗压强度随水胶比的增大而减小,试验结果见表2。

表2 粉煤灰强度试验结果

编号	水胶比	粉煤灰掺量(%)	抗压强度(MPa)			劈拉强度(MPa)			抗拉强度(MPa)		极限拉伸值($\times10^{-6}$)		弹性模量(GPa)
			7 d	28 d	90 d	7 d	28 d	90 d	28 d	90 d	28 d	90 d	28 d
XK25	0.55	30	20.2	26.9	37.2	1.76	2.29	2.79	2.21	2.96	92	99	25.0
XK26		45	18.7	27.9	36.9	1.40	2.10	2.67	2.18	2.86	84	86	25.4
XK27		60	17.0	25.8	35.0	1.36	2.10	2.60	2.08	2.65	84	85	25.9
XK28	0.50	30	24.2	32.9	40.9	2.00	2.54	3.00	2.30	3.27	97	99	26.9
XK29		45	22.0	32.1	39.6	1.68	2.38	2.89	2.23	3.01	93	94	24.1
XK30		60	19.9	31.6	38.9	1.48	2.26	2.74	2.08	2.90	92	93	24.2
XK31	0.45	30	25.6	34.3	44.5	2.05	2.64	3.20	2.39	3.46	102	110	25.4
XK32		45	23.0	33.4	42.9	1.83	2.46	2.97	2.27	3.27	91	108	25.4
XK33		60	21.5	32.8	41.0	1.52	2.28	2.87	2.24	3.13	84	105	24.7
XK34	0.40	30	28.7	36.1	48.6	2.24	2.85	3.33	2.55	3.70	106	125	28.1
XK35		45	25.9	34.9	45.2	1.90	2.60	3.27	2.40	3.56	99	120	26.8
XK36		60	22.3	34.1	43.8	1.81	2.77	3.30	2.38	3.43	93	120	25.9

碾压混凝土28 d的抗渗性能均能满足W10的等级要求,试验渗水高度随粉煤灰掺量降低而降低,抗渗性能随着龄期的增长将有进一步提高,试验结果见表3。

表3 抗渗性能试验结果

编号	水胶比	粉煤灰掺量（%）	级配	试验压力（MPa）	渗水高度（cm）	抗渗等级
XK1	0.55	60	三	1.1	5.5	≥W10
XK2		50			4.6	≥W10
XK3		40			4.1	≥W10
XK4	0.50	60	三	1.1	4.5	≥W10
XK5		50			2.8	≥W10
XK6		40			2.5	≥W10
XK7	0.45	60	三	1.1	2.8	≥W10
XK8		50			2.5	≥W10
XK9		40			2.4	≥W10

在不同粉煤灰掺量下，含气量大于3%，碾压混凝土均可满足抗冻等级F50的设计要求，含气量大于4%，水胶比小于0.50的碾压混凝土可满足抗冻等级F300的设计要求。试验结果见表4。

表4 抗冻性能试验结果

编号	水胶比	粉煤灰（%）	级配	含气量（%）	各冻融次数质量损失（%）				各冻融次数相对动弹模量（%）				抗冻等级
					50	100	200	300	50	100	200	300	
XK11	0.55	60	三	3.2	0.3	1.2	—	—	90.6	85.0	—	—	>F100
XK12		50		3.2	0.2	0.9	—	—	93.8	92.9	—	—	>F100
XK13		40		3.8	0.1	0.5	—	—	94.5	93.4	—	—	>F100
XK14	0.45	60	三	4.5	0	0	0.3	—	97.6	95.9	91.6	—	>F200
XK15		50		4.3	0	0	1.0	—	97.9	96.0	92.2	—	>F200
XK16		40		5.2	0	0	0.3	—	97.3	96.0	93.9	—	>F200
XK17	0.50	60	二	4.3	0	0	0.3	1.3	97.3	95.4	93.9	81.9	>F300
XK18		50		4.2	0	0	1.0	2.5	95.7	94.3	91.5	78.9	>F300
XK19		40		5.6	0	0.2	0.8	2.4	96.7	96.6	96.5	90.0	>F300

3 对石粉掺量的研究

石粉是人工砂中含有的微细颗粒，不同行业对石粉的定义也有差别。我国水利水电行业对石粉的规定为人工砂中颗粒小于0.16 mm的细颗粒。石粉含量以石粉占粗骨料或人工砂质量的百分数来表示。

石粉作为砂中的微细颗粒，对碾压混凝土的性能有重要作用。由于碾压混凝土的用水量和胶凝材料用量较少，而石粉的细度与水泥、粉煤灰相当，可共同发挥包裹和充填作用。经研究及工程实践表明，适当增加石粉含量，相当于增加胶凝材料浆体，能在一定程度上改善碾压混凝土拌和物的和易性，增进碾压混凝土的均质性、密实性及抗渗性，提高碾压混凝土的强度和断裂韧性，改善碾压混凝土施工层面的胶结性能。

我国目前在建的几个百米级的高坝，人工砂中的石粉含量均控制在22%以下。北疆某工程对掺石粉的碾压混凝土性能试验结果表明：在保持混凝土拌和物的V_c值和含气量在设计范围的前提下，石粉含量每增加4%，需减少0.5%～1.0%的砂率。随着砂中石粉含量的增加，碾压混凝土的抗压强度先增加后降低，并随石粉含量的增加，碾压混凝土中灰浆/砂浆比增大，其抗拉强度明显提高，但石粉含量过大时，灰粉浆体强度下降，碾压混凝土强度降低。该工程碾压混凝土石粉的掺量不宜超过16%（见表5）。

表5　不同石粉掺量试验结果

编号	石粉掺量（%）	级配	抗压强度（MPa）			抗拉强度（MPa）		
			7 d	28 d	90 d	7 d	28 d	90 d
XSO-1	0	二	100	100	100	100	100	100
XS1	8		109	104	108	110	113	109
XS2	12		102	96	104	109	116	115
XS3	16		110	106	101	106	116	104
XS4	20		97	95	95	96	103	101

4　配合比基本参数的优选

该碾压混凝土重力坝因其建坝位置的特殊性，受冷、热、风、干等不利环境影响较大，如果采取的温控防裂措施不当，极易造成碾压混凝土表面裂缝，甚至向深层发展。为此，建设单位与国内专业研究机构合作，除采用常规的差分法分析温度场及应力变化外，还采用三维有限元法进行坝体温度控制分析，对碾压混凝土的力学性能、热学性能及变形性能进行试验研究，提出并采取了一系列坝体温度控制及防裂具体措施，并取得了良好的效果。其中，对碾压混凝土主要原材料（包括掺合料）的性质、配合比的分析研究是温控防裂措施一项重要内容，配合比试验基本参数见表6。

表6　配合比试验基本参数

使用部位	混凝土等级	级配	配合比参数						
			水胶比	砂率（%）	减水剂（%）	引气剂（%）	水泥（%）	粉煤灰（%）	石粉（%）
大坝底部	R_{90}20F100W8	三	0.45	32	0.85	0.08	70	30	6
迎水面上部	R_{180}20F300W10	二	0.45	35	0.85	0.12	60	40	4
650以上内部	R_{180}15F50W4	三	0.56	32	0.80	0.04	38	62	8
650以下内部	R_{180}20F50W4	三	0.53	32	0.80	0.04	38	62	8
背水面上部	R_{180}20F200W6	三	0.45	32	0.80	0.10	50	50	6

由于碾压混凝土抗裂性能的主要指标为极限拉伸值和抗拉强度，而碾压混凝土的主要原材料性质、配合比及施工质量是影响碾压混凝土极限拉伸值各种因素中的主要因素。因此，对碾压混凝土配合比设计中主要原材料及掺合料进行优选，是保证和提高碾压混凝土极限拉伸值和抗拉强度，降低其弹性模量，保证碾压混凝土坝质量和安全的重要保证。

【作者简介】 孙根民，男，1966年4月生，本科，新疆额尔齐斯河流域开发工程建设管理局，副处长，高级工程师。

全级配、高性能常态混凝土在严寒高蒸发地区高拱坝建设中的试验研究和应用经济评价

丁照祥　李新江　徐元禄

（新疆额尔齐斯河流域开发工程建设管理局　乌鲁木齐　830000）

【摘要】 在新疆布尔津山口水电站高拱坝建设时，为解决严寒、高蒸发地区水工混凝土的耐久性、抗裂性、抗渗性等问题，确定了混凝土双掺减水剂和引气剂、高掺粉煤灰、低水胶比、较低用水量、高含气、较低坍落度控制的研究路线，展开了全集配混凝土配比的试验研究工作，从原材料开始，对粉煤灰不同掺量与水泥胶砂强度关系、不同配合比混凝土的耐久性、力学性能等指标展开试验研究，取得了较好的进展，该混凝土的试验成果达到了设计要求，满足了工程建设需要。

【关键词】 全级配混凝土　高性能　配合比　试验研究　评价

1　概述

1.1　工程概况

布尔津山口水电站总库容为2.22亿 m^3，属大(2)型Ⅱ等工程。大坝为常态混凝土双曲拱坝，最大坝高94 m，厚高比0.266，混凝土总方量为40万 m^3。混凝土拱坝每隔15 m设置一条横缝，共分为21个坝段，不设纵缝通仓浇筑，每块的浇筑量从270～1 300 m^3不等。坝址地区极端最高气温39.4 ℃，极端最低气温-41.2 ℃，年最大温差80.6 ℃；多年平均降水量153.4 mm；多年平均蒸发量1 619.5 mm；极端最大风速32.1 m/s；最大冻土深127 cm。主要气候特征：夏季炎热，高蒸发，冬季严寒，日温差大，年气温差较悬殊，寒潮出现频繁。

1.2　大坝混凝土需要解决的问题

在严寒、高蒸发、年气温差较大的环境恶劣地区建一座高拱坝，且有大体积混凝土浇筑施工，需要解决拱坝混凝土的抗冻耐久性、高抗渗性和抗裂性等问题，结合当地情况，采取措施提高混凝土的工作性能，实现该坝的高品质性，能够有较长的使用寿命和长期运行安全，对拱坝混凝土进行试验研究和应用分析，以保证拱坝混凝土建设的顺利进行。

1.3　混凝土研究思路及指标

考虑当地的自然气候条件，该拱坝采用高性能混凝土，主要以耐久性作为设计的主要指标，重点保证混凝土的耐久性、工作性、适用性、强度、体积稳定性和经济性。综合已建成类似大坝的经验，大坝混凝土配合比设计试验的技术路线为：双掺减水剂和引气剂、高掺粉煤灰、低水胶比、较低用水量、高含气量、较低坍落度控制，达到适宜强度、较好施工性、高性能化和较好抗裂的目的。在大坝的建设中拟采用高性能全级配混凝土，即四级配常态混凝土，

主要设计指标确定为：C_{90}25W10F400、C_{90}30W10F400，强度保证率为 80%，抗拉强度分为 1.7 MPa、2.0 MPa，极限拉伸大于 0.85×10^{-4}。

2 原材料

2.1 水泥

采用专家会确定的布尔津水泥厂生产的 P · Ⅰ42.5 型硅酸盐水泥，水泥物理力学和化学指标均满足《通用硅酸盐水泥》（GB 175—2007）中 P · Ⅰ42.5 水泥技术要求。

2.2 粉煤灰

试验采用玛纳斯电厂生产的翔天和牌粉煤灰，粉煤灰品质检测结果满足《水工混凝土掺用粉煤灰技术规范》（DL/T 5055—2007）中Ⅰ级粉煤灰要求。

为了便于经济有效地利用粉煤灰，进行了粉煤灰掺量与水泥胶砂强度关系试验，粉煤灰掺量分别为 0、20%、30%、40%、50%、60%，其试验成果见表 1。由试验成果表可知，水泥胶砂强度随粉煤灰掺量增加而降低。粉煤灰掺量 30% 以内胶砂强度降低幅度不大，掺量 50% 以上胶砂强度降低幅度明显。

表 1 粉煤灰掺量与水泥胶砂强度关系试验成果

编号	水泥用量（g）	粉煤灰		抗压强度（MPa）					抗折强度（MPa）				
		掺量（%）	用量（g）	3 d	7 d	28 d	90 d	180 d	3 d	7 d	28 d	90 d	180 d
F09-31	450	0	—	20.7	31.0	47.8	58.1	58.2	5.0	6.5	7.7	8.9	9.0
F09-31-1	360	20	90	17.3	25.0	40.6	56.4	66.2	4.1	5.4	7.3	9.0	10.1
F09-31-2	315	30	135	14.6	22.5	36.9	55.8	64.0	3.6	4.9	7.2	8.7	9.9
F09-31-3	270	40	180	12.5	18.3	30.1	48.9	60.0	2.9	4.2	6.3	8.3	9.5
F09-32-2	247.5	45	202.5	11.2	16.7	24.0	44.8	53.4	2.9	4.1	5.4	7.5	9.2
F09-31-4	225	50	225	7.5	13.2	23.2	39.9	49.9	2.3	3.4	5.1	7.3	8.7
F09-31-5	180	60	270	4.8	9.2	16.4	28.5	39.8	1.8	2.7	3.8	6.1	8.1

胶凝材料水化热试验结果见表 2，粉煤灰掺量分别为 0、20%、30%、40%、50%。

表 2 粉煤灰不同掺量的胶凝材料水化热试验成果 （%）

材料用量（%）		水泥水化热（kJ/kg）/ 水化热降低率（%）	
布尔津 P · Ⅰ42.5 水泥	玛纳斯粉煤灰	3 d	7d
100	0	232	264
80	20	210	244
70	30	182	215
60	40	177	215
50	50	140	177

分析试验成果,水化热随粉煤灰掺量增加而降低;粉煤灰掺量为30% ~50%时水化热降幅较大,对降低混凝土发热量有利。

根据上述成果,设计龄期28 d的混凝土配合比中粉煤灰掺量可按30%左右考虑,设计龄期90 d的混凝土配合比中粉煤灰掺量宜控制在40% ~50%。

2.3 砂石骨料

试验骨料由C1料场生产提供,细骨料按天然砂40%、人工砂60%的比例配制成混合砂,细度模数满足2.6±0.1的试验控制范围,各关键粒径累计筛余量均在中砂区域内。粗骨料小石、中石、大石、特大石的各项检测结果符合《水工混凝土施工规范》(DL/T 5144—2001)的控制标准。粗骨料级配组合比例选择原则:根据最大密度和最小空隙率优选粗骨料的级配比例,尽量少用小石、多用大石,通过对不同级配的粗骨料进行组合试验,选择了振实密度最大2 046 kg/m^3、振实孔隙率最小25.1%的骨料级配组合,四级配骨料比例为小石:中石:大石:特大石=20:20:30:30。

2.4 外加剂

试验采用新疆五杰外加剂厂生产的NF-2型缓凝高效减水剂和PMS-NEA3型引气剂。通过外加剂匀质性检验、布尔津P·I 42.5水泥与萘系减水剂的适应性试验、掺加外加剂混凝土适应性试验以及外加剂掺量与混凝土性能关系试验,结果表明,缓凝高效减水剂和引气剂品质均满足《混凝土外加剂》(GB 8076—2008)中该类产品要求,NF-2缓凝高效减水剂与P·I 42.5水泥适应性较好,效果较佳掺量为0.07% ~1.2%。

3 混凝土配合比设计及试验

3.1 混凝土配制强度的确定

根据《水工混凝土施工规范》(DL/T 5144—2001),山口大坝全级配混凝土配制强度及设计要求如下:

设计标号分为C_{90}25W10 F400、C_{90}30W10 F400;强度保证率为80%;概率度系数t均为0.84;标准差σ为4.0、4.5;配制强度分别为28.4 MPa、33.8 MPa;抗拉强度为1.7 MPa、2.0 MPa;抗压弹30 GPa;含气量为5%;密度≥2 400 kg/m^3;极限拉伸值>0.85×10^{-4}。

3.2 配合比设计试验原则

配合比设计试验按照《水工混凝土试验规程》(SL 352—2006)中绝对体积法计算。配合比试验采用的主要参数见表3。

表3 混凝土配合比调试试验参数

项目	水胶比	粉煤灰掺量(%)	用水量(kg/m^3)	砂率(%)	小石:中石:大石:特大石	理论计算含气量(%)
三级配	0.38、0.43、0.48	35、45、55	92	28~29	20:30:50	3~4
四级配	0.38、0.43	45	82	25~26	20:20:30:30	2.5~3

注:骨料均以饱和面干为计算基准。

根据大坝混凝土耐久性要求和混凝土高性能化目标,配合比试验严格控制混凝土拌和物的含气量。按混凝土拌和物出机30 min、含气量测值在控制范围内、坍落度基本符合要求时成型各类试件。

3.3 三级配混凝土配合比

三级配混凝土配合比试验结果见表4，三个配合比28 d试件的抗冻等级达到F400，力学指标及相关性能均满足设计要求。

表4 大坝三级配混凝土配合比试验参数

强度等级	配合比参数							
	水胶比	粉煤灰(%)	砂率(%)	减水剂(%)	引气剂(%)	坍落度(cm)	含气量(%)	
C_{28}25W10 F300	0.43	35	29	0.70	0.018	3~6	4~5	
C_{90}25W6 F300	0.43	45	29	0.70	0.020	3~6	4~5	
C_{180}25W6 F300	0.48	45	29	0.70	0.020	3~6	4~5	
强度等级	材料用量(kg/m^3)							
	用水量	总胶材	水泥	粉煤灰	砂	小石	中石	大石
C_{28}25W10 F300	92	214	139	75	620	310	465	775
C_{90}25W6 F300	92	214	118	96	618	309	464	773
C_{180}25W6 F300	92	192	105	86	625	312	468	781

注：含气量按混凝土拌和物出机后30 min、坍落度30~60 mm时，以4.5%~5.5%控制。

3.4 四级配混凝土配合比试验

参照三级配的试验情况，结合掺粉煤灰胶砂试验结果和其他工程配合比参数，四级配混凝土按掺45%粉煤灰设计，具体试验参数见表5，试验成果见表6~表8。试验结果表明，混凝土拌和物出机后30 min、坍落度在30~60 mm、含气量在4.5%~5.5%范围时，成型试件的各项混凝土性能指标均满足设计要求。

表 5　大坝四级配混凝土配合比试验参数表

试件编号	混凝土等级	级配	配合比参数							材料用量(kg/m^3)												
			水胶比	粉煤灰(%)	砂率(%)	减水剂(%)	引气剂(%)	坍落度(cm)	含气量(%)	用水量	总胶材	水泥	粉煤灰	砂	天然砂	人工砂	粗骨料				减水剂	引气剂
															40%	60%	小石	中石	大石	特大石		
SQ4-3	C_{90}25W10F300	四	0.43	45	26	0.70	0.020	3~6	5~5.5	82	191	105	86	570	228	342	332	332	498	498	1.335	0.038
SQ6-1	C_{90}30W10F300	四	0.38	45	25	0.70	0.020	3~6	5~5.5	82	216	119	97	542	217	325	332	332	498	498	1.511	0.043

表 6　大坝四级配混凝土配合比拌和物性能与抗压强度试验成果

试件编号	混凝土等级	级配	坍落度(cm)				含气量(%)				推算容重(kg/m^3)				凝结时间(时:分)		抗压强度(MPa)				
			出机	15 min	30 min	1 h	出机	15 min	30 min	1 h	出机	15 min	30 min	1 h	初凝	终凝	3 d	7 d	28 d	90 d	180 d
SQ4-3	C_{90}25W10F300	四	10.1	6.1	3.7	—	7.9	6.3	4.5	—	2441	2470	2493	—	15:15	21:48	10.8	15.2	26.8	37.5	42.6
SQ6-1	C_{90}30W10F300	四	11.7	9.6	7.4	—	8.4	7.0	5.4	—	2447	2478	2495	—	15:12	20:06	13.0	18.8	28.8	40.7	45.9

表 7　大坝四级配混凝土配合比力学性能与变形性能试验成果

试件编号	混凝土等级	级配	劈拉强度(MPa)				轴心抗压强度(MPa)			抗压弹模(GPa)			轴心抗拉强度(MPa)			轴心抗拉弹模(GPa)			极拉(10^{-6})		
			7 d	28 d	90 d	180 d	28 d	90 d	180 d	28 d	90 d	180 d	28 d	90 d	180 d	28 d	90 d	180 d	28 d	90 d	180 d
SQ4-3	C_{90}25W10F300	四	1.24	1.89	2.47	2.66	25.6	32.2	38.1	22.8	29.5	31.1	2.10	2.49	3.30	25.7	31.1	34.5	91.4	95.2	99.1
SQ6-1	C_{90}30W10F300	四	1.26	2.21	2.60	2.92	27.3	38.4	49.6	25.6	29.8	35.8	2.21	3.26	3.41	31.1	30.7	32.3	79.2	115.6	117.7

表 8　大坝四级配混凝土配合比耐久性能试验成果

试件编号	混凝土等级	级配	龄期(d)	抗渗		抗冻试验																						
				等级	渗水高度(cm)	质量损失(%)											动弹模量损失(%)											等级
						50	100	150	200	250	300	350	400	500	600	750	50	100	150	200	250	300	350	400	500	600	750	
SQ4-3	C_{90}25 W10F300	四	90	>W10	0.5	0.0	0.2	0.2	0.2	0.2	0.3	0.4	0.5				96.9	96.5	96.3	96.0	95.7	95.3	94.8	93.3				>F400
			180	>W10	0.2	0.1	0.2	0.2	0.3	0.3	0.7	0.7	0.8	0.9	1.0	1.8	98.6	97.6	96.8	95.8	94.9	93.7	97.2	97.0	96.5	96.1	95.2	>F750
SQ6-1	C_{90}30 W10F300	四	28	>W10	2.5	0.0	0.0	0.5	0.6	0.7	0.8	0.9	1.1	—	—	—	99.3	98.5	97.7	97.1	96.4	95.7	95.1	94.3	—	—	—	>F400
			90	>W10	0.5	0.2	0.4	0.5	0.7	0.8	0.9	1.0	1.0	1.1	1.3	1.6	99.0	98.1	97.0	95.9	94.8	93.8	97.7	97.4	96.2	94.9	92.9	>F750

4 试验结果配合比

根据上述试验成果,布尔津山口电站大坝混凝土试验成果配合比见表9。

表9 布尔津山口电站大坝全级配混凝土试验成果配合比

混凝土等级	配合比参数							材料用量(kg/m³)							
	水胶比	粉煤灰(%)	砂率(%)	减水剂(%)	引气剂(%)	坍落度(cm)	含气量(%)	水	水泥	粉煤灰	砂	粗骨料			
												小石	中石	大石	特大石
$C_{90}25$ W10F400	0.43	45	26	0.7	0.02	3~6	4.5~5.5	82	105	86	570	332	332	498	498
$C_{90}30$ W10F400	0.38	45	25	0.7	0.02	3~6	4.5~5.5	82	119	97	542	332	332	498	498

5 全级配混凝土的经济技术评价

5.1 全级配混凝土配合比技术评价

通过试验验证,$C_{90}25$W10 F400 、$C_{90}30$W10 F400 配合比 90 d 龄期的抗冻等级均达到F400以上,抗渗达到W10以上,耐久性满足设计要求;90 d 龄期抗压强度分别为37.5 MPa、40.7 MPa,均大于配置强度为28.4 MPa、33.8 MPa;抗拉强度分别为2.47 MPa、2.60 MPa,均大于设计值1.7 MPa、2.0 MPa;极限拉伸值为0.952×10^{-4}、1.15×10^{-4},大于设计值0.85×10^{-4}。该配合比设计合理,混凝土性能满足耐久性、抗裂、抗渗设计要求,能够安全用于拱坝的施工建设。

5.2 确定了粉煤灰的最优掺量

通过对粉煤灰不同掺量的胶砂强度及水化热试验可知,水泥胶砂强度随粉煤灰掺量增加而降低,粉煤灰掺量30%以内胶砂强度降低幅度不大,掺量50%以上胶砂强度降低幅度明显;水化热随粉煤灰掺量增加而降低,粉煤灰掺量为30%~50%水化热降幅较大,对降低混凝土发热量有利。分析试验成果,确定粉煤灰的最优掺量为45%。

5.3 选择了骨料级配最优比例

C1料场天然骨料级配较差,小石很少,中石不多,特大石很多,需要破碎大石补充小、中石,考虑在配合比中少用小、中石,将大幅降低骨料生产成本,通过各级配骨料组合与振实试验,选择四级配骨料比例为小石∶中石∶大石∶特大石=20∶20∶30∶30。

5.4 混凝土含气量的控制要求

含气量是混凝土高性能化的关键控制参数,试验成果也反映了该问题,成型含气量4.5%~5.5%时可满足F400以上抗冻要求。含气量在施工过程中的控制,以混凝土现场入仓施工时(出机约30 min)按5%~5.5%控制,引气剂掺量根据含气量测值进行调整。

5.5 全级配的经济评价

混凝土配合比中采用全级配,把一般作为弃料的超大石作为骨料用到混凝土中,减少了小、中石的用量,减少了补充破碎小、中石的成本,减少了弃料,每立方混凝土用特大石0.3 m³,骨料均价为60元/m³,每立方混凝土降低费用18元,其中的35万m³全级配混凝土节约费用525万;对用量较大的C30全级配混凝土来说,采用高掺粉煤灰达到45%的配比,

1 m^3混凝土总胶材 216 kg,粉煤灰代替水泥 97 kg,现场水泥供应单价每吨 650 元,粉煤灰每吨 150 元,每立方混凝土降低费用 48.5 元,35 万 m^3混凝土节约费用 1 697.5 万元,合计节约建设费用 2 222 万元,经济效益非常可观,同时混凝土掺加粉煤灰从一方面解决了电厂产生粉煤灰污染环境的问题,有较大的生态环保效益。

参考文献

[1] 李文伟,郑丹.溪洛渡大坝混凝土特性及防裂措施[J].水利水电技术,2010(2):48-51.

【作者简介】 丁照祥,男,1969 年 9 月生,本科,西安科技大学土木工程,高级工程师。

Hardfill 坝地震破坏模式与抗震安全性研究*

熊　堃[1,2]　翁永红[1,2]　何蕴龙[3]

(1. 长江勘测规划设计研究院　湖北　武汉　430010；
2. 国家大坝安全工程技术研究中心　湖北　武汉　430010；
3. 武汉大学水资源与水电工程科学国家重点实验室　湖北　武汉　430072)

【摘要】 基于细观损伤理论和有限元法，以 Weibull 分布表征 Hardfill 材料力学性能的随机分布，对典型 Hardfill 坝遭遇 8 度地震情况及强震条件下进行动力反应分析，研究其在地震过程中的破坏模式与破坏机制。结果表明，在 8 度地震荷载作用下，Hardfill 坝应力水平较低，处于未损伤或轻微损伤状态；强震作用下靠近坝踵与坝趾的坝体因抗拉强度不足而发生从坝面萌生且不断向坝体内部发展的拉裂缝。与传统重力坝相比，梯形的体形使 Hardfill 坝具有优良的抗震性能，大坝抗震安全度高于传统重力坝。

【关键词】 Hardfill 坝　地震破坏模式　抗震安全性　细观损伤

1　引言

Hardfill 坝是一种新坝型，其基本剖面是上下游坝坡呈基本对称的梯形，上游坝面采用面板或其他设施防渗，筑坝材料为价格低廉的低强度胶凝砂砾石料，可采用坝址附近易于得到的河床砂砾石或开挖弃渣并加入水和少量水泥，经简单拌和而成。Hardfill 坝的倡导者认为这种坝型具有高安全性、高抗震性和对地基条件要求低的优点，而且施工简便、快速，造价低廉，对环境的负面影响小。

在 Hardfill 材料的制作过程中，其粗放的施工工艺决定了这种材料具有较强的离散性与非均匀性，材料中含有微裂纹甚至明显的宏观缺陷。本文借鉴唐春安等应用 Weibull 模型反映细观不均匀性影响研究岩石破裂过程的思想，在材料宏观均质假定基础上引入随机分布函数考虑 Hardfill 材料细观不均匀性的影响，建立了能够体现材料非均匀特征的损伤本构模型，对典型 Hardfill 坝在 8 度地震工况下及强震条件下进行动力反应分析，并与相同高度的传统重力坝进行了对比计算，研究两种坝型在地震过程中的破坏特点及破坏模式，探讨 Hardfill 坝的地震破坏机制。

* 基金项目：国家大坝安全工程技术研究中心研发课题(2011NDS021)，湖北省博士后创新岗位资助。

2 考虑材料非均匀性的细观损伤模型

2.1 材料非均匀性表征

将 Hardfill 材料的弹性模量与强度参数作为随机变量，认为其服从 Weibull 分布，该分布的概率密度函数为

$$f(x)=\frac{m}{x_0}\left(\frac{x}{x_0}\right)^{m-1}\exp\left(-\left(\frac{x}{x_0}\right)^{m}\right)\ (x>0) \tag{1}$$

式中：x 为满足 Weibull 分布的材料参数值；x_0 为与材料参数均值相关的参数，m 为 Weibull 分布密度函数曲线的形状参数，反映参数的离散程度，当由小到大变化时，材料细观单元的参数分布密度函数曲线由矮而宽到高而窄变化。因此，形状参数 m 反映了数值模型中材料的均质性，本文将它称为均匀系数，m 越大，组成试件的材料性质越均匀。

2.2 细观单元损伤本构关系

每一个细观单元采用连续介质的损伤本构模型，细观单元在单轴应力状态下

$$\sigma=E_0(1-D)\varepsilon \tag{2}$$

式中：D 为损伤变量；E_0 为初始弹性模量，即未损伤状态下的弹性模量。

图 1 给出了所采用的 Hardfill 材料细观拉伸与剪切（受压）损伤本构关系，损伤变量表达式如式(4)与式(6)，其中应力应变以受拉为正。主要考虑两种损伤准则，即最大拉应变准则与摩尔库仑准则，在计算过程中对每个细观单元首先按最大拉应变准则进行判断。

摩尔库仑准则可变换为

$$F=\frac{1+\sin\varphi}{1-\sin\varphi}\sigma_1-\sigma_3\geqslant f_c \tag{3}$$

式中：φ 为细观单元的内摩擦角；f_c 为单轴抗压强度。

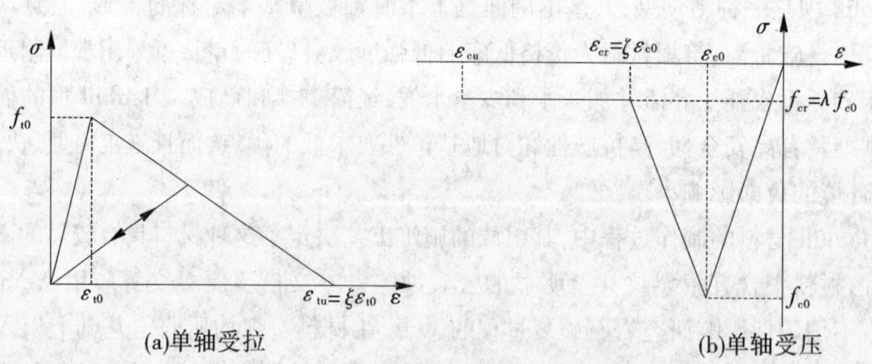

图 1 Hardfill 材料细观损伤本构关系

$$D_t=\begin{cases}0 & (0\leqslant\varepsilon_t\leqslant\varepsilon_{t0})\\ 1-\left(\frac{\xi}{\xi-1}\frac{\varepsilon_{t0}}{\varepsilon_t}-\frac{1}{\xi-1}\right) & (\varepsilon_{t0}<\varepsilon_t\leqslant\varepsilon_{tu})\\ 1 & (\varepsilon_t>\varepsilon_{tu})\end{cases} \tag{4}$$

图 1(a)中 f_{t0} 为材料单轴抗拉强度，ε_{t0} 为抗拉强度所对应的拉应变，单元拉应变达到 ε_{t0} 时进入损伤阶段；ε_{tu} 为极限拉应变，$\varepsilon_{tu}=\xi\varepsilon_{t0}$，$\xi$ 为极限拉应变系数，当单元拉应变达到 ε_{tu} 时发生完全损伤，此时损伤变量 $D_t=1$，单元完全破坏。

$$D_c=\begin{cases}0 & (\varepsilon_{c0}\leqslant\varepsilon_c\leqslant 0)\\ 1-(\dfrac{\xi-\lambda}{\xi-1}\dfrac{\varepsilon_{c0}}{\varepsilon_c}-\dfrac{\lambda-1}{\xi-1}) & (\varepsilon_{cr}\leqslant\varepsilon_c<\varepsilon_{c0})\\ 1-\dfrac{\lambda\varepsilon_{c0}}{\varepsilon_c} & (\varepsilon_{cu}\leqslant\varepsilon_c<\varepsilon_{cr})\\ 1 & (\varepsilon_c<\varepsilon_{cu})\end{cases} \tag{5}$$

图 1(b)中f_{c0}为材料单轴抗压强度，ε_{c0}为抗压强度对应的应变，单元压应变达到ε_{c0}时进入损伤阶段；f_{cr}为材料的残余强度，ε_{cr}为残余强度对应的应变，其中$f_{cr}=\lambda f_{c0}$，$\varepsilon_{cr}=r\varepsilon_{c0}$，$\lambda$称为残余强度系数，$r$称为残余应变系数；$\varepsilon_{cu}$为极限压应变，$\varepsilon_{cu}=\zeta\varepsilon_{c0}$，$\zeta$为极限压应变系数，当压应变达到$\varepsilon_{cu}$时发生完全损伤，此时$D_c=1$。

建立一个单轴受拉平面应力数值试件，按照同一组宏观材料参数和均匀系数m生成6组细观单元材料参数样本，计算得到材料细观结构的随机性使试件的宏观断裂形态亦具有了随机性，但破坏模式是相同的，而且随机性对宏观弹性模量与宏观强度等材料的宏观特性影响较小。为探讨材料细观非均匀程度对其宏观性质的影响，再将均匀系数m分别取为1、1.5、2、3、5和10进行计算，结果表明不同均匀程度的试件表现出的宏观裂缝并没有大的区别。图2(a)给出了应力—应变关系曲线，比较明显的变化是随着均匀系数的增大，在其他细观参数相同的情况下，材料的宏观抗拉强度逐渐增大，弹性模量也逐渐增大，即材料越均匀，宏细观之间的性质差异越小。图2(b)给出了由单轴拉伸数值计算所得出的材料细观与宏观弹性模量、强度比值与均匀系数之间的关系曲线，并且其与均匀系数的关系可以用对数曲线较好地拟合。对单轴压缩数值计算也可得出一致的结论。

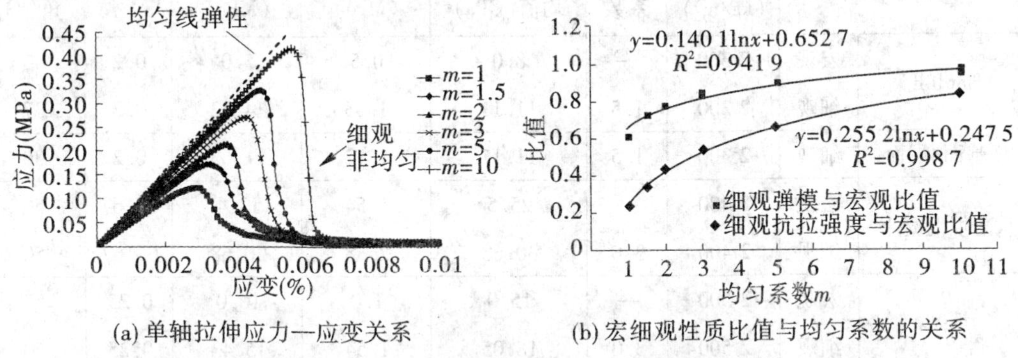

(a) 单轴拉伸应力—应变关系　　(b) 宏细观性质比值与均匀系数的关系

图2　不同均匀系数时 Hardfill 试件单轴拉伸

3　基于细观损伤模型的 Hardfill 坝地震动力分析

3.1　计算模型

所分析的典型 Hardfill 坝坝高 70 m，坝坡坡比取为1：0.7，有限元模型中模拟了坝基面及坝体内部概化的填筑层面见图3(a)，每层高7 m。坝体网格基本尺寸为0.5 m×0.5 m，坝基上下游及深度方向均取1.5倍坝高范围，坝体有限元网格见图3(b)。

静力分析主要考虑了坝体自重、静水压力和扬压力等荷载，大坝承受齐顶水压力，下游无水。其中假定扬压力在坝踵折减为水头的1/2，在坝趾处为零，扬压力在坝底呈线性分

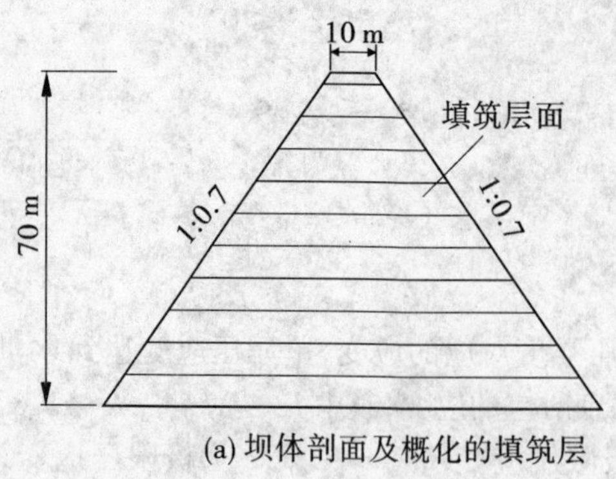

(a) 坝体剖面及概化的填筑层

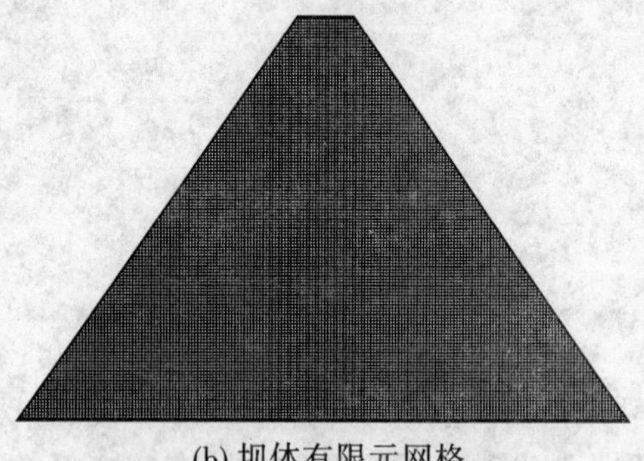

(b) 坝体有限元网格

图3 坝体模型

布。地震动力分析采用时程法,选取 Taft 地震波,主震周期调整为 0.2 s,计算取地震历时为 12 s,时间步 $\Delta t=0.02$ s,地震最大加速度按照 OBE 标准取为 0.24g,采取顺河向与竖直向地震波同时输入,竖直向峰值加速度为水平向的 2/3。为消除地基的放大作用,将其按无质量处理。动力计算中考虑坝体自重、水荷载和地震的共同作用,库水动水压力的影响采用附加质量矩阵考虑,附加质量按 Westergaard 公式计算。宏细观材料参数见表 1,其中材料细观弹性模量与强度的均值按照图 2(b)所示的宏细观力学参数比值与均匀系数的关系曲线换算而来。在动力计算中,动弹性模量与动强度在静态参数的基础上提高 30%,泊松比不变,阻尼比均取为 0.05,采取瑞利阻尼。

表1 各种材料宏细观参数

材料		密度(kg/m³)	均匀系数	弹性模量均值(GPa)	抗拉强度(MPa)	抗压强度(MPa)	泊松比	内摩擦角(°)
Hardfill	宏观	2 300	—	8.0	0.5	5.0	0.2	40
	细观	2 300	1.5	11.11	1.45	14.53	0.2	37
坝基面/层面	细观	2 300	1.5	11.11	1.12	11.63	0.2	30
重力坝混凝土	宏观	2 400	—	25.5	1.54	13.4	0.167	50
	细观	2 400	3.0	30.68	2.38	20.68	0.167	50
坝基	宏观	2 500	—	15.0	1.0	10.0	0.25	50
	细观	2 500	3.0	18.05	1.54	15.43	0.25	50

3.2 遭遇 8 度地震情况下 Hardfill 坝动力响应

图 4 为遭遇 8 度地震情况下坝体动力反应状况,其中应力正值表示拉应力,负值表示压应力。在 8 度地震荷载作用下,基于细观损伤模型的坝体动力反应规律与线弹性模型的结果基本相同,特别是加速度与动位移等值线的分布规律与数值大小相当一致,均在坝顶有加速度和位移的最大值,说明本文细观损伤本构模型可以正确地应用于大坝的静动力分析中。只是当使用细观损伤本构模型时,由于材料的非均匀导致了变形等值线不是很光滑。

坝体内的应力分布受材料非均匀性的影响更为明显,图 4(c)、(d)中可见,坝体应力分布呈现出显著的不均匀性,细观单元与单元之间应力差异较大,总体上在地震过程中坝踵以及下游坝面附近出现了拉应力区，但拉应力值较小，而坝踵坝趾以及上下游坝面附近主压应力较大，这与线弹性计算所得的坝体总体应力规律是一致的。在遭遇8度地震时，Hardfill坝由于坝体应力水平较低而基本处于未损伤或轻微损伤状态。

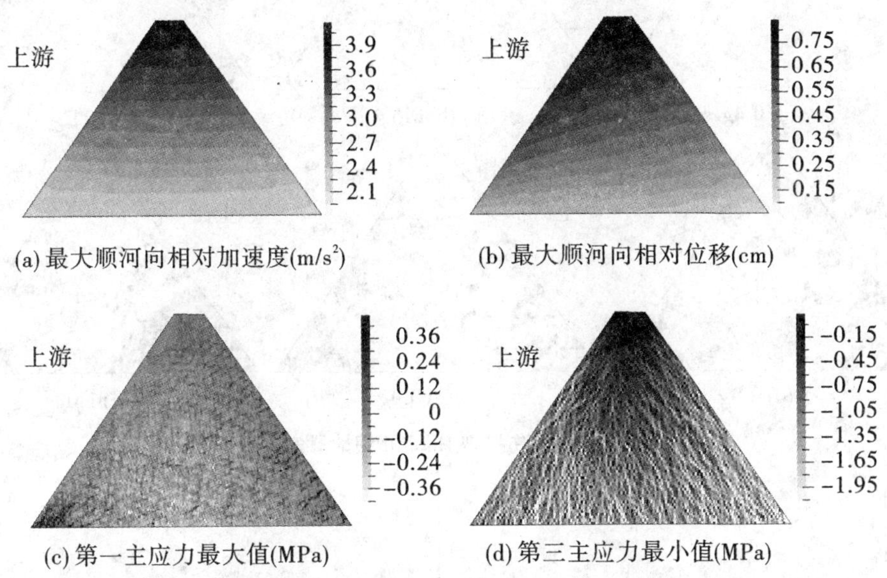

(a) 最大顺河向相对加速度(m/s^2) (b) 最大顺河向相对位移(cm)

(c) 第一主应力最大值(MPa) (d) 第三主应力最小值(MPa)

图4 8度地震情况下大坝动力反应

3.3 地震荷载超载法大坝破坏模式

采用地震加速度超载法分析大坝的动力破坏模式，并将Hardfill坝与传统重力坝进行了对比分析。重力坝坝高与Hardfill坝相同，下游坡比也是1∶0.7，并位于相同的地基条件上，混凝土的力学参数见表1，有限元网格尺寸与Hardfill坝模型基本相同。

在地震过程中，重力坝的坝踵、上游面及下游面均出现了一定程度的主拉应力，坝踵位置应力集中，随着地震强度的增大，拉应力迅速增大，表明这些部位极易发生开裂破坏。在强震作用下，Hardfill坝主要在大坝上下游表面拉应力值较大，坝踵和坝趾是拉应力集中的位置，在强震中可能发生破坏。图5、图6分别给出了采用细观损伤本构模型，不同地震峰值加速度时典型Hardfill坝与传统重力坝的破坏情况，其中黑色区域表示宏观裂缝。

由图5可知，8度地震荷载作用下，Hardfill坝的坝体基本没有破坏的区域；9度地震，即地震峰值加速度为0.4g时，仅坝踵沿坝基面有一定区域的破坏，下游坝面附近的层面出现了裂缝；随着地震加速度的增大，坝踵沿坝基面的裂缝向下游扩展，坝趾出现的垂直于下游坝面的裂缝到达坝基面，上、下游坝面也产生了垂直于坝面向坝内发展的局部裂缝，这些裂缝多从层面的位置萌生，但没有沿层面发展，上游坝面发育的裂缝要多于下游；地震强度极度增大后，从坝面开始的裂缝将层面的局部裂缝连接并逐渐扩展到坝基面，形成Hardfill坝最终的地震破坏形态，此时坝顶部有较明显的水平向地震永久变形。

典型重力坝结构在8度地震荷载作用下在坝踵处出现了较少的开裂；随着地震加速度的增大，坝踵裂缝沿坝基面向下游扩展，但达到一定深度后基本稳定，在坝体上下游坝面靠近坝颈的部位出现开裂，并且裂缝基本垂直于坝面向坝体内部延伸；地震加速度峰值为0.6g时，坝颈部的损伤区域连通，而上下游坝面有两条宏观裂缝也几乎相连，同时上游坝面也发

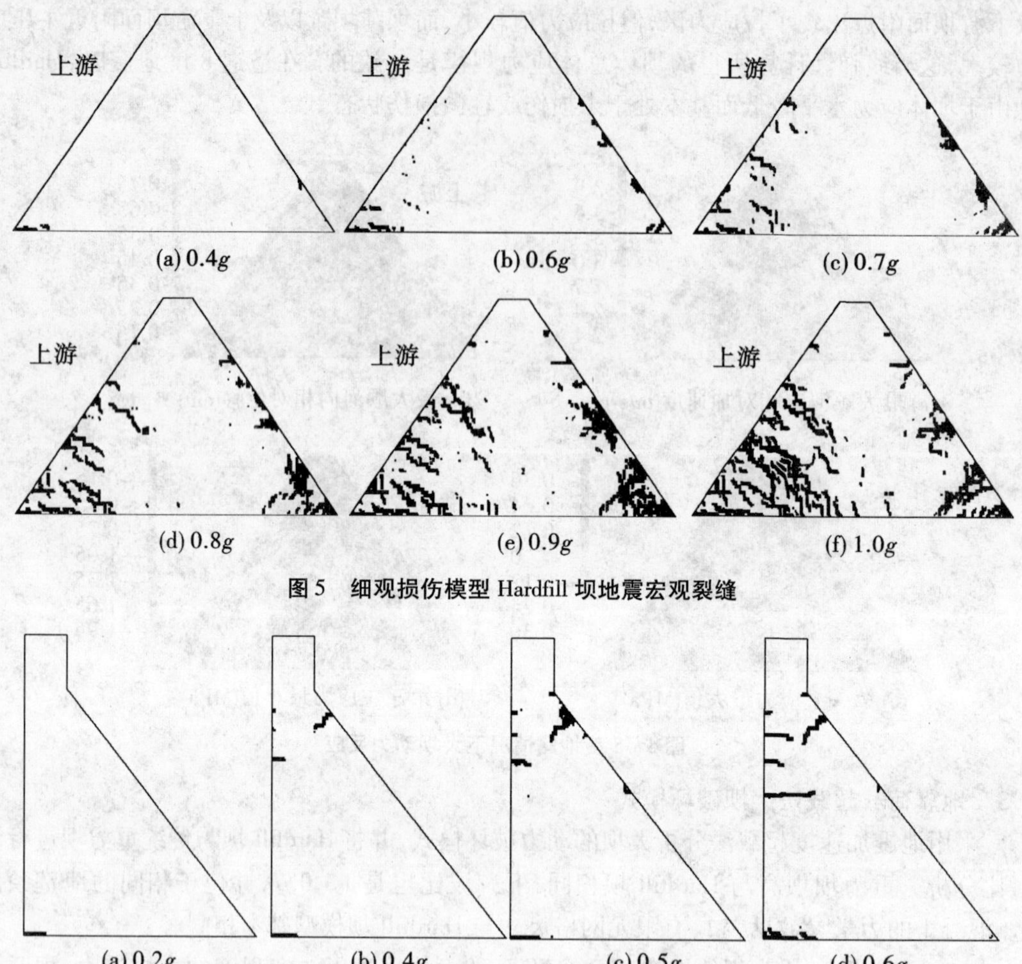

图5 细观损伤模型 Hardfill 坝地震宏观裂缝

图6 细观损伤模型重力坝地震宏观裂缝

展了数条具有一定深度的水平裂缝，形成重力坝典型的地震破坏形态。地震后重力坝因坝颈部断裂而顶部有明显的永久变形，坝颈以下变形不显著。

与重力坝相比，Hardfill 筑坝材料强度低，但由于大坝剖面较肥大使得其抗震性能较好。从两种坝型细观损伤模型的地震动力破坏形态可以看出，Hardfill 坝的坝体裂缝较多：一方面是由于在计算中模拟了强度较低的填筑层面，造成"薄弱缺口"较多；另一方面是坝体材料比重力坝不均匀性要显著的缘故。

4 结语

(1)8 度地震荷载作用下，基于细观损伤模型的 Hardfill 坝体动力反应规律与线弹性模型的结果基本相同，说明本文细观损伤本构模型可以正确地应用于大坝的静动力分析中。

(2)在遭遇 8 度地震时，典型 Hardfill 坝由于坝体应力水平较低而基本处于未损伤或轻微损伤状态，坝体基本没有破坏的区域，但坝体细观单元间的应力值差异较大，受材料非均匀性的影响较为明显。

(3)Hardfill 坝在往复变化的地震惯性力作用下因剖面对称而在上下游坝体均出现破坏

区域,上游坝体破坏程度大于下游;不断增大的地震荷载将使靠近坝踵与坝趾的坝体因抗拉强度不足且发生从坝面萌生而不断向坝体内部发展的拉裂缝。因此,在地震中 Hardfill 坝的破坏机制为坝踵与坝趾附近区域的拉伸断裂。

(4)梯形体型使 Hardfill 坝具有优良的抗震性能,大坝的抗震安全度高于传统重力坝。

参考文献

[1] P. Londe and M. Lino. The faced symmetrical Hardfill dam: a new concept for RCC[C]. International Water Power & Dam Construction,1992,44(2):19-24.

[2] 唐春安,朱万成. 混凝土损伤与断裂:数值试验[M]. 北京:科学出版社,2003.

[3] Zhu W C,Tang Ch A. Numerical simulation on shear fracture process of concrete using mesoscopic mechanical model[C]. Construction and Building Materials,2002,16(8):453-463.

【作者简介】 熊堃,1984 年 6 月出生,博士,2011 年毕业于武汉大学,工学博士,长江勘测规划设计研究院博士后科研工作站,在站博士后,工程师。

特高拱坝施工期数字监控方法、系统与工程应用

刘 毅[1] 张国新[1] 王继敏[2] 周 钟[3]

(1. 中国水利水电科学研究院 流域水循环模拟与调控国家重点实验室 北京 100038;
2. 二滩水电开发有限公司 四川 610051;
3. 中国水电工程顾问集团成都勘测设计研究院 成都 610072)

【摘要】 近10年来,我国有一批300 m级特高拱坝建成、在建或即将开工。由于特高拱坝与一般拱坝存在较大差异,因而其分期蓄水、分期施工、施工程序及蓄水过程对坝体应力的影响需要专门研究。本文依据混凝土坝数字监控的理论,从主要目的、基本内容和工作模式三个方面论述了特高拱坝施工期数字监控方法,介绍了特高拱坝施工期数字监控系统,提出了利用该系统开展特高拱坝施工期数字监控的工程应用模式。特高拱坝施工期数字监控系统的工程应用表明,运用该系统可以实时开展大坝工作性态评估,降低事故风险,同时可以为施工期动态设计提供决策支持。

【关键词】 特高拱坝 施工期 数字监控

1 引言

在近10年来的我国水电开发高峰期中,特高拱坝的建设占据着显著而重要的位置。小湾拱坝(坝高294.5 m)、拉西瓦拱坝(250 m)已经投产发电,并即将接受正常蓄水位的考验;锦屏一级拱坝(坝高305 m)、溪洛渡拱坝(坝高285.5 m)浇筑高程已经过半,将于2013年投产发电;白鹤滩拱坝(坝高289 m)、乌东德拱坝(坝高265 m)已经进入可行性研究阶段,将陆续开工建设。在上述拱坝建设之前,我国已经竣工投产的最高拱坝是1998年竣工的二滩拱坝(坝高240 m),这10多年的特高拱坝建设将我国的拱坝建设实践从240 m提升至300 m级。

在巨大的成绩面前,我们也应该清醒地认识到300 m级特高拱坝与普通拱坝存在的显著差异,以及由于这些差异而可能导致的工程风险。首先,300 m级特高拱坝应力水平高,按照规范方法计算得到的压应力接近或达到10 MPa,与普通拱坝主要受拉应力控制不同,特高拱坝的拉压应力均达到规范允许应力的临界值,整体安全裕度小于普通拱坝;其次,出于提前发电的需要,分期浇筑、封拱与蓄水以及在施工过程中的诸多调整使得施工期的工作性态与初始设计状态出现较大差异,施工期拱坝工作性态十分复杂;最后,由于特高拱坝底宽较大,混凝土后期温度回升较大,温控防裂的难度加大,它对通水冷却时空间方面的梯度要求高,导致拱坝施工期悬臂高度增加,可能会导致施工期拱坝工作状态的恶化。为此,规范特别指出:“应研究分期蓄水、分期施工、施工程序及蓄水过程对坝体应力的影响”,“对于

200 m 以上的高坝,其拉应力控制标准应做专门研究”。

2008 年初,朱伯芳院士提出了混凝土坝数字监控的理论,其基本思想是:考虑到目前大坝仪器监控测点少且不能给出大坝应力场和安全系数等缺点,在仪器监控的基础上增加数字监控,基于仪器观测资料进行反分析,利用全坝全过程仿真分析,在施工期即可给出当时的温度场和应力场,并可预报运行期的温度场、应力场及安全系数,如发现问题可及时采取对策;在运行期可以充分反映施工中各种因素的影响,对大坝作出比较符合实际的安全评估。

本文结合在建特高拱坝工程,从主要目的、基本内容和工作模式三个方面论述了特高拱坝施工期数字监控方法,介绍了特高拱坝施工期数字监控系统,提出了利用该系统开展特高拱坝施工期数字监控的工程应用模式,即利用数字监控系统跟踪分析大坝施工期的工作性态,评估它与设计工作性态的差异与风险,在施工条件变更时,分析变更对大坝工作性态的影响,提出相应的应对措施。

2 特高拱坝施工期数字监控方法

本节从特高拱坝施工期数字监控的基本目的、主要内容和工作模式三个方面来论述特高拱坝施工期的数字监控方法。

2.1 基本目的

结合在建特高拱坝的实际情况,特高拱坝施工期数字监控的目的包括以下两个方面:

(1)评估施工期大坝工作性态与设计工作状态的差异及其风险。在大坝施工前,设计会针对特高拱坝可能存在的各种工况进行分析。尽管如此,由于水电工程的复杂性,施工后的各种条件会与设计状况存在一定差异,施工状态与设计状态会存在一定差别。此外,由于受限于我们对大坝工作性态认知水平的限制,在大坝施工前我们对大坝工作性态的认知与实际状况肯定会有一定偏差,对于中低坝而言这一偏差可能涵盖在较大的安全系数内,对于特高拱坝而言这一偏差可能导致工程风险。因此,运用数字监控在仪器监控的基础上进行反分析,进而真正把握特高拱坝施工期的工作性态,评估其与设计工作状态的差异,降低风险。

(2)评估施工方案调整的可行性与相应对策。在大坝浇筑过程中,由于施工条件发生变化,往往需要调整施工进度和施工方式,比如浇筑层厚度、浇筑层间歇期等,这些施工措施调整会带来哪些影响,需要进行符合实际的评估。在大坝施工过程中,由于施工进度落后于设计进度,度汛形象面貌达不到设计要求,为此需要对悬臂挡水等各种情况进行分析研究,提出度汛安全的相应措施。

2.2 主要内容

特高拱坝施工期面临的主要荷载是自重和温度,为避免拱坝在自重和温度荷载作用下的开裂风险,数字监控的主要内容包括温度状态监控和应力状态监控。

(1)大坝温度状态监控。根据其他高拱坝工程的经验,目前各工程对于高拱坝施工期温度控制的设计要求都非常详细,对每个浇筑仓从出机口温度到封拱温度的全过程以及各浇筑仓在空间上的温度梯度均作出了规定。大坝温度状态监控的第一项内容就是按照设计要求全面评估出机口温度、浇筑温度、最高温度、各期冷却目标温度是否符合设计要求以及符合的比例,在高度方向是否按照要求设置了同冷区、过渡区和盖重区,对超出设计要求项进行报警。大坝温度状态监控的第二项内容是对大坝的温度过程作出预测,对可能超标现象提出预警,并提出相应的应对措施。

(2)大坝应力状态监控。按照大坝实际浇筑过程和实测气温资料,对大坝浇筑全过程进行跟踪仿真,在仪器监控资料的支持下,开展关键热力学参数的反演分析,进而得到符合实际情况的大坝应力状态。考虑设计拟定的进度,通过仿真分析对大坝未来一段时间的应力状态进行预测,如应力超标,则提出应对措施。

2.3 工作模式

混凝土坝的数字监控是一项新的工作模式,要最大限度地发挥数字监控的作用,必须要将数字监控融入现有的由业主、设计、施工和监理各方组成的工作模式。本文提出了一套混凝土坝数字监控的工作模式与流程,如图1所示。

工作模式流程如下:

(1)在业主的统一领导下,建立由业主、设计、监理、施工和科研参建各方组成的工作小组。

(2)按照设计文件,进行现场施工,并埋设监测仪器。

(3)利用专门开发的施工与监测信息集成系统将施工信息、监测信息统一纳入数据库。

(4)在现场实测数据的基础上,利用有限元正反分析软件系统进行跟踪反演仿真分析。

(5)基于仿真成果,进行现场预警报警和科研服务,提供评估意见和应对措施,为参建各方提供决策支持。

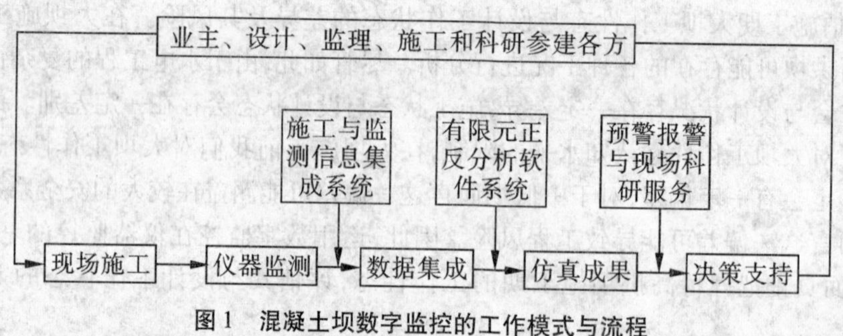

图1 混凝土坝数字监控的工作模式与流程

3 特高拱坝施工期数字监控系统

在二滩水电开发有限公司和成都勘测设计研究院的支持下,中国水利水电科学研究院和天津大学合作开发了锦屏一级水电站高拱坝施工实时控制系统,其中中国水利水电科学研究院承担了温控信息集成与预警系统。本节介绍这一系统的主要功能和模块。

3.1 系统主要功能

该系统的功能包括以下几个方面:

(1)施工与监测信息集成。采取尽可能方便的方式,将施工和监测信息录入数据库,并进行高效管理。主要信息包括环境信息(气温、水温、上下游水位、气象信息等)、浇筑仓信息(浇筑高程、开仓时间、收仓时间、混凝土方量、浇筑方式、浇筑温度等)、通水冷却信息(冷却水管布置、通水流量、进水口水温、出水口水温等)、封拱信息(封拱高程、封拱时间、封拱温度等)、浇筑仓温控信息等。

(2)施工与监测信息管理。海量信息进入数据库后,需用图表等直观的形式对信息进行高效的管理。图形包括实测温度过程线图、浇筑仓综合信息曲线图、上游面综合信息立视图,沿高程温度分布图等。表格包括出机口温度统计表、入仓温度统计表、浇筑温度统计表、通水冷却统计表等。

(3)有限元实时跟踪反演仿真分析。利用数据库中的信息,按需要形成仿真计算文件,提交后台服务器进行有限元仿真分析,并依据实测温度数据,反演有关参数,确保计算符合实际情况。

(4)混凝土开裂风险预测预报。依据设计要求,对超出设计要求的部分进行报警。包括最高温度报警,间歇期超标报警,超冷报警、冷却速率超标报警等。依据有限元计算成果对一段时间以后的大坝混凝土温度和应力状况进行预测,对开裂风险进行预测,提出应对措施。

3.2 系统的主要模块

本系统包括以下四个主要模块,典型模块的截面如图2所示。

(1)温控信息管理模块。该模块实现施工与监测信息数据的集成与查询功能。

(2)图表信息模块。该模块实现数据的统计表格与图形表示功能。

(3)预警报警模块。该模块实现根据数据进行预警和报警的功能。

(4)仿真分析模块。该模块实现有限元仿真分析和反演分析的功能。

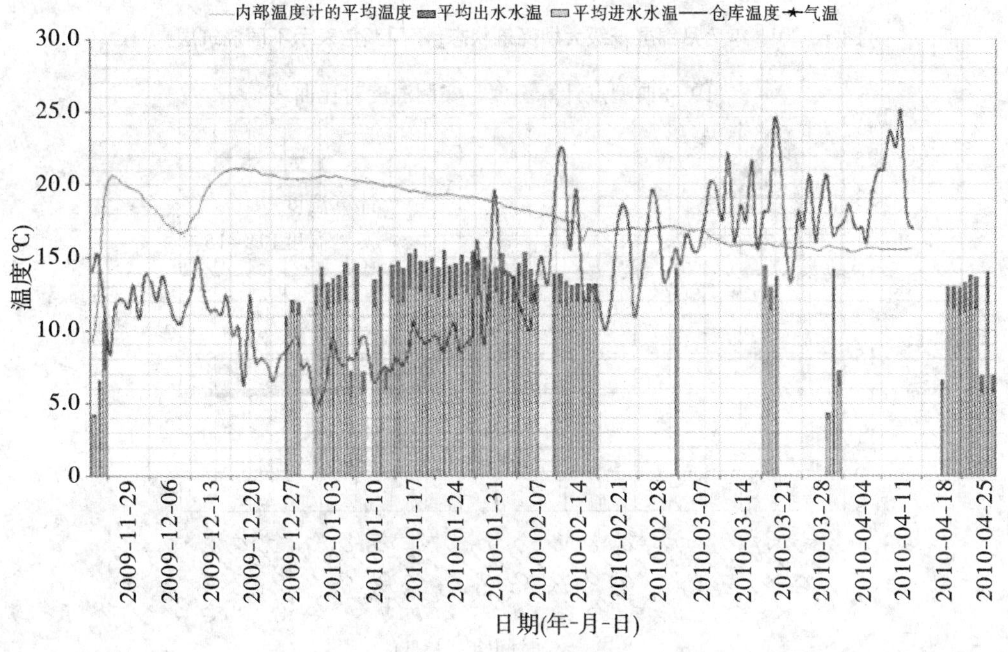

图2 系统典型界面图

4 特高拱坝施工期数字监控工程应用

自2009年3月起,中国水利水电科学研究院的科研团队在锦屏一级拱坝现场开展施工期数字监控科研工作。工作的主要内容包括两个方面:一是评估大坝实际工作性态及其开裂风险,每月提交一次评估报告;二是条件变化时进行跟踪仿真分析,并提供专题报告,为动态反馈设计提供技术支持。本节举例说明数字监控在特高拱坝施工中发挥的作用。

4.1　实际工作性态评估

按照大坝实际浇筑过程对大坝截至 2011 年 7 月 15 日的大坝工作性态进行仿真模拟，如图 3 所示。通过实测温度过程反馈混凝土热学参数，确保计算温度过程与实测吻合，如图 4所示为计算温度与实测温度过程对比。从坝体应力、位移和横缝开度三个方面评估大坝实际工作性态，基本结论如下：

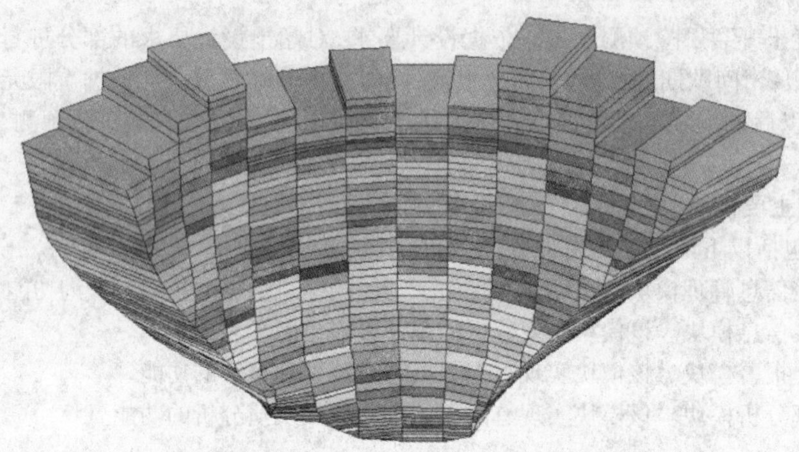

图 3　2011 年 7 月锦屏一级大坝浇筑状态（不同颜色表示不同浇筑层）

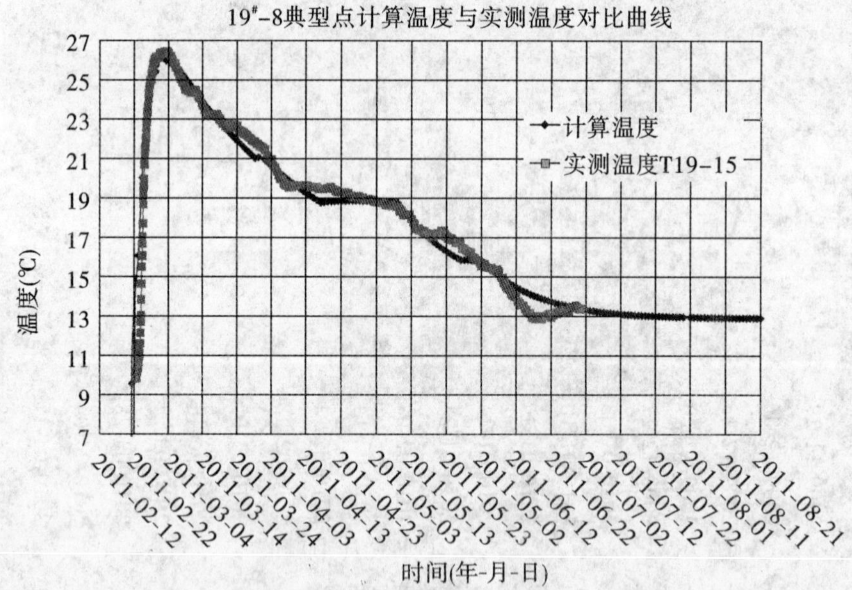

图 4　典型测点实测温度过程与计算温度过程对比

（1）坝体最上部由于刚浇筑，位移为 0，下部受自重的影响，向下变形，坝踵向下竖直向位移大于坝趾；顺河向，底部高程受降温影响上游面向下游变形，下游面向上游变形，中部高程由于自重的影响，大坝整体倾向上游。

（2）竖直向应力主要由自重决定，主要为压应力。由于拱坝体型倒悬的影响，河床坝段上游面坝踵部位有较大压应力，坝踵部位最大压应力为 5.5 MPa。

(3)顺河向应力主要受施工期温度荷载影响,约束区温度应力较大,且陡坡坝段约束区的温度应力大于河床坝段,约束区最大应力为2.0 MPa,脱离约束区后温度应力小于1.3 MPa,内部温度应力均小于允许拉应力。

(4)横河向应力主要受施工期温度荷载影响,约束区温度应力较大,且陡坡坝段约束区的温度应力大于河床坝段,约束区最大应力为1.9 MPa,脱离约束区后温度应力小于1.0 MPa,内部温度应力均小于允许拉应力。

(5)底部强约束区横缝最大开度较小,沿高程方向随着地基约束力的减弱横缝最大开度逐渐增大,二冷结束后脱离强约束区的横缝最大开度一般可达到1 mm以上。截至2011年7月15日横缝最大开度监测值中约78%的数值≥0.5 mm,最大缝开度平均值约为1.51 mm,计算所得缝开度平均值为1.42 mm,计算结果与监测结果基本吻合。

4.2 辅助反馈设计

冬季浇筑薄层混凝土长间歇存在较大的开裂风险,2009年过冬前,中国水利水电科学研究院对固结灌浆长间歇对混凝土开裂风险的影响进行了研究,分别考虑了“三进三出”和“一进一出”两种固结灌浆方式和是否保温两种情况,典型计算结果如图5所示。研究表明:在不保温情况下,固结灌浆面上超过20 d的长间歇时,长短周期温度荷载叠加都会使得最大应力远远超过容许拉应力,安全系数为1.05,开裂风险较大;设计保温条件下,20 d间歇时,长短周期温度荷载叠加后的温度应力小于容许拉应力,满足要求;设计保温条件下,50 d间歇时,超过30 d间歇期后,长短周期温度荷载叠加,应力有所超标,安全系数为1.35,有一定的开裂风险。

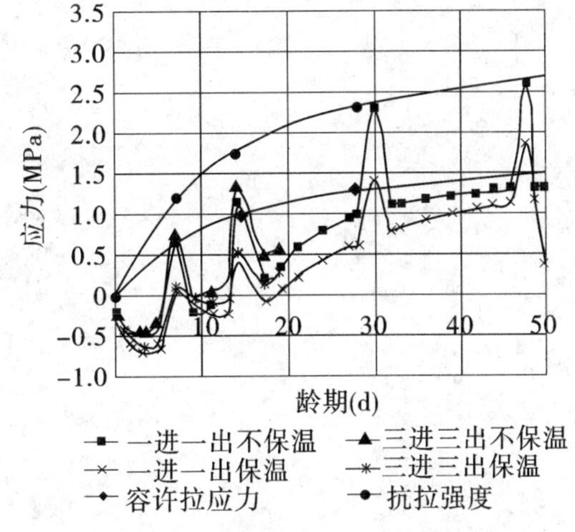

图5 各工况层中表面点长短周期温度荷载叠加温度应力值

根据这一研究成果,设计院将原设计中的盖重固结灌浆方式改为下部盖重灌浆加表层引管灌浆的方式,避开了冬季长间歇问题。2010年冬季,采用无长间歇的固结灌浆方式后,浇筑仓面几乎未发生裂缝。

5 结语

混凝土坝数字监控是近几年来提出的新理论,其基本思想是基于仪器观测资料进行反分析,利用全坝全过程仿真分析,预测预报未来一段时间混凝土坝的应力状态,评估混凝土坝施工期和运行期的工作性态与开裂风险,在条件变化时辅助反馈设计,确保工程安全。

本文结合锦屏一级拱坝的工程应用实践从基本目的、主要内容和工作模式三个方面阐述了特高拱坝施工期数字监控方法,特高拱坝施工期数字监控的主要目的在于评估施工期大坝工作性态与设计工作状态的差异及其风险、评估施工方案调整的可行性与相应对策,其主要内容包括大坝温度状态监控和大坝应力状态监控,并提出了现场科研团队跟踪反馈仿真、业主与设计进行技术决策、参建各方共同参与的数字监控工作模式。

笔者所在的科研团队在锦屏一级拱坝开展的施工期数字监控工程应用表明,运用特高拱坝施工期数字监控系统，可以实时开展大坝工作性态评估，降低事故风险，同时可以为施工期动态设计提供决策支持。

参考文献

[1] 中华人民共和国发展和改革委员会. DL/T 5346—2006 混凝土拱坝设计规范[S]. 北京：中国电力出版社，2007.

[2] 中华人民共和国水利部. SL 282—2003 混凝土拱坝设计规范[S]. 北京：中国水利水电出版社，2003.

[3] 朱伯芳. 混凝土坝的数字监控[J]. 水利水电技术，2008，39(2)：15-18.

[4] 朱伯芳. 大坝数字监控的作用和设想[J]. 大坝与安全，2009(6)：8-11.

[5] 朱伯芳，张国新，贾金生，等. 混凝土坝的数字监控——提高大坝监控水平的新途径[J]. 水力发电学报，2009，28(1)：130-136.

【作者简介】 刘毅，1979年9月生，博士，2005年7月毕业于清华大学水利水电工程系，获工学博士学位，现任中国水利水电科学研究院结构材料所副所长兼可视化仿真研究室主任，获中国水利水电科学研究院科技英才称号。

碾压混凝土坝浇筑进度三维仿真

李　江[1]　李秀琳[2]　夏世法[2]

(1. 新疆水利水电规划设计管理局　乌鲁木齐　830000;
2. 中国水利水电科学研究院　北京　100044)

【摘要】 随着虚拟现实技术的发展,三维仿真作为一种全新角度的管理方法和技术手段在工程进度管理中起到越来越重要的作用。本文讨论了基于Qt应用程序开发框架和OSG图形引擎应用程序设计接口的浇筑仿真系统的开发方法。本系统采用的方法动态、灵活,可根据施工采集的高程数据来生成对应的浇筑情况。基于上述方法设计实现了一个混凝土坝浇筑过程三维仿真系统。结果验证了提出的方法可以满足工程需要。

【关键词】 混凝土坝　仿真　虚拟现实　Qt　OSG

1　引言

水利水电工程是一项复杂的系统工程,针对工程项目建立三维的、动态的、可视的虚拟仿真环境,将大坝的浇筑面貌和仿真结果用图形反映出来,就可以将用户的视野带入三维主体工程空间。使参与项目建设各方都能在此环境下直观地了解工程实际施工状况,分析它与施工计划的差异,了解差异产生的原因,预测未来施工发展趋势,调整未来的施工计划,又可控制实施过程中各种因素对施工进程的影响。这将会对大坝混凝土浇筑的施工设计和施工过程起指导性作用。

随着计算机计算能力的提高,很多图形学及图像处理的算法得以在消费级计算机硬件中实现,其绘制速度大大加快,会使得其应用面更加广泛。目前,混凝土坝的实时跟踪均以数字量或二维图形的方式表示,缺乏整体性和直观感,提供的决策支持力度不高、效果不佳。实现实时三维仿真后,由于其直观性、整体性、时间连续性,容易梳理思路,触发灵感,加快决策结果的产生,特别是类似抢险救灾这样时效性强的决策有非常高的价值。另外,施工过程中不可控的因素很多,不可能完全按照预期计划执行。这样以往的针对不同时期建造不同模型的仿真方式以及其他的动态模型生成的方式,不能快速、准确地反映出当前的浇筑情况。

1973年第11届国际大坝会议上,D. H. Bassgen首先结合混凝土重力坝施工提出了混凝土浇筑过程模拟,并在奥地利修建施立格坝时,采用了确定性数字仿真技术对缆机浇筑混凝土方案进行优选。其后,计算机仿真技术逐步在水利水电工程和建筑工程施工中进行应用。我国水利水电行业应用计算机仿真技术始于20世纪80年代初,而应用多目标、多任务的虚拟仿真系统目前才刚刚起步。近期国内的研究成果主要有:1999年,天津大学建筑工程学院考虑了随机因素对龙滩碾压混凝土双曲拱坝施工过程的影响，建立了随机模拟模型，通过理论分析模拟了降雨对施工过程的影响。1999 年，三峡开发总公司和成都勘测设计研究院为三峡工程二期大坝施工开发了二期混凝土大坝浇筑过程计算机模拟系统，该系统能够对技术措施进行质量分析，快速比较多种浇筑方案，可对混凝土施工进行各种因素敏感性分析，可及时对工程进行实时控制，此外可对浇筑过程进行三维动态显示等。2002 年以来，武汉大学、天津大学等高校和设计、科研、建设单位等合作，在各大型水利水电工程建设中，应用计算机仿真技术对水利水电工程施工全过程进行动态模拟，并推动了新技术，如随机循环网。

总体来讲，国内对混凝土坝浇筑进度模拟方面的研究刚刚起步不久，而且还都在起步阶段，并且大多着重于浇筑过程中温度、应力的计算与控制，在三维真实感绘制方面做的工作很少。

2 系统结构

本系统是一个跨平台的基于 C/S 体系结构的大坝浇筑进度可视化系统，其服务器端和客户端均可运行于任何主流的 Linux/Unix/Windows 操作系统之上。系统整体结构如图 1 所示。

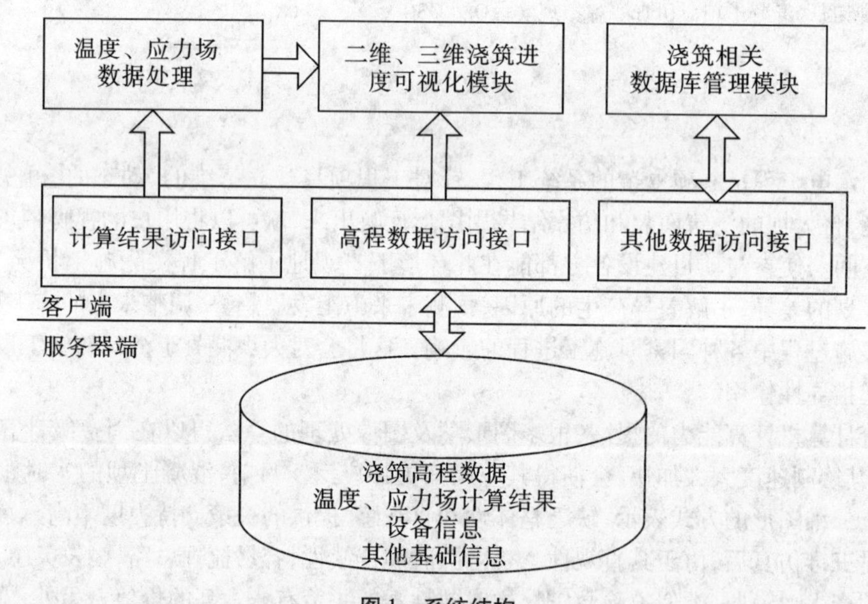

图 1 系统结构

服务器端使用数据库管理系统存储浇筑高程数据、温度(应力)计算结果、设备信息以及其他基础信息。客户端包括三维浇筑进度显示、二维浇筑进度显示、温度(应力)场显示以及数据库管理维护等模块。温度、应力场计算数据经过处理与二维浇筑进度可视化模块融合在一起显示。

由于三维仿真系统对系统实时性的要求，选用了 C++作为本系统的编程语言；数据库选用 PostgreSQL 对象-关系型数据库系统；基于 Qt 应用程序开发框架来搭建系统的整体框架和界面的实现；基于 OSG 图形绘制引擎来进行相关的三维渲染。其中，PostgreSQL 是最富特色的自由数据库管理系统，有最丰富的数据类型的支持，支持海量数据存储和查询；Qt

是一个多平台的 C++图形用户界面应用程序框架,是完全面向对象的很容易扩展,并且允许真正地组件编程;OSG 是一款开源的高效的 3D 图形开发包,它对 OpenGL 进行了完全的类封装,而且实现了目前速度最快的场景图。另外,由于 OSG 对目前绝大多数的三维模型存储格式提供了支持,对所选用的建模软件没有限制,本文选用了 3D Max 8 对大坝进行建模。

上述软件平台都是开源的项目,无需支付使用费用,降低了平台搭建的成本。更重要的是,基于上面软件平台的选择,本文的仿真系统具有良好的跨平台性、开放性以及可扩展性。通过这种 C/S 架构的设计,用户可以通过修改服务器端的数据,客户端的绘制效果会相应的发生变化。

3 三维浇筑动态仿真

本系统的核心部分为浇筑过程的三维仿真,这一节详细介绍本系统的实现方法。

3.1 建模

在预处理阶段,通过每个坝段的 AutoCAD 图纸,在建模软件中建立坝段的三维模型,并给模型加上相应的材质和纹理。图 2 为某典型坝段的几何模型。图 3 为大坝的整体三维模型以线框模式绘制出来的效果,其中每个坝段包含的几何体以组为单位组织,便于以后的计算。

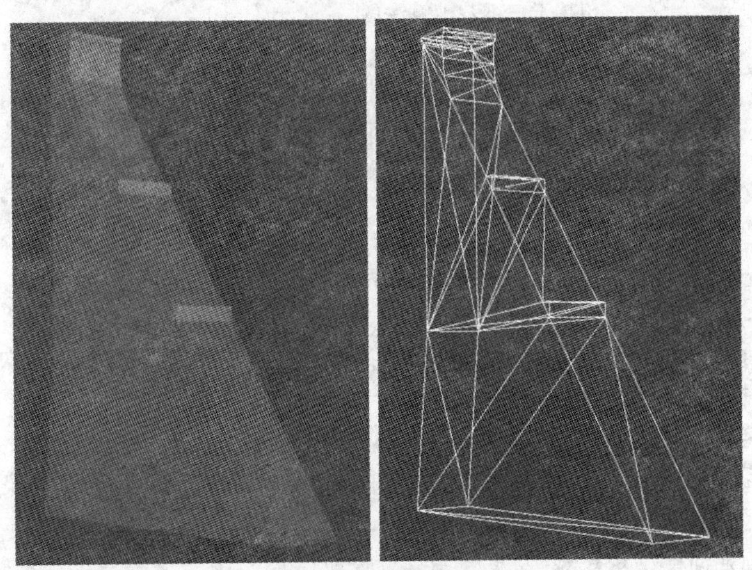

图 2 典型坝段模型

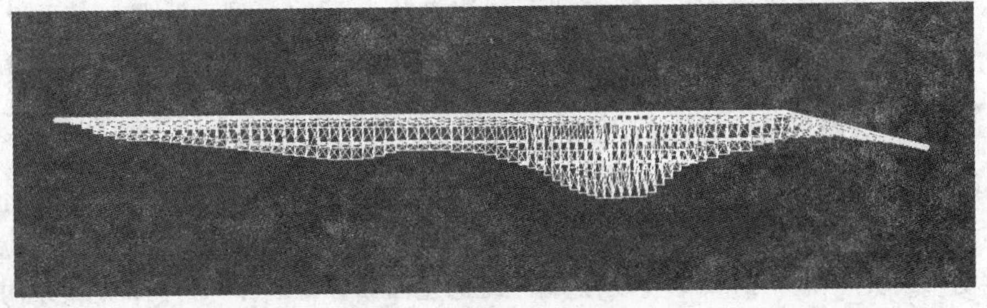

图 3 大坝整体几何模型

3.2 坝基开挖效果实现

大坝浇筑开始时，首先要开挖坝段的坝基。本文通过地形网格形状的改变来实现这一效果。地形网格通过高度图通过程序进行 Delaunay 三角剖分得到，本文利用 OSG 工具包中提供的 Delaunay 剖分工具来实现。根据坝基开挖的位置、大体形状，定义相应的剖分约束；根据相应的约束，对地形重新剖分，即可以实现坝基开挖的效果。这种方式需要一定的计算量，但是动态灵活，开挖的效果可以通过采集的参数控制。

3.3 三维浇筑仿真流程

整个仿真以仿真时间做为输入，最终渲染出输入的指定时间的大坝浇筑状况。三维浇筑仿真流程如图 4 所示。

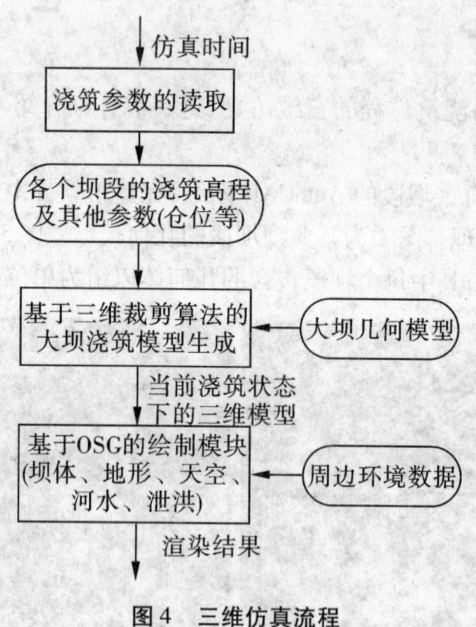

图4 三维仿真流程

系统运行时，首先，加载大坝的几何模型以及周边环境的相关模型。其次，输入制定的浇筑时间，根据浇筑时间从数据库中读取各个坝段的浇筑高程以及仓位数据，数据可以是实地采集的浇筑数据，也可以是计划或者仿真的数据。再次，基于三维裁剪算法计算出大坝浇筑模型。最后，通过 OSG 绘制模块对整个仿真的三维场景进行渲染，得到渲染结果。

3.4 浇筑模型的动态生成算法

由于大坝几何模型相对规则，本文引入计算机图形学中经典的三维裁剪算法通过对大坝整体几何模型进行裁剪计算出大坝在指定浇筑状态下的几何模型。这样做只需一次建模，通过输入高程、仓位数据，计算得到不同浇筑状态下的模型。对经典的 Sutherland－Hodgman（编码）线段裁剪算法进行扩展，将其扩展到三维，将“逐边裁剪”扩展为“逐面裁剪”。二维裁剪中要计算线段与直线的交点，而现在要计算线段与多边形的交点。空间直线段与任一平面的求交运算可以通过将裁剪面方程与直线段方程联立求出。

整个大坝的三维模型是以坝段为单位分成组的，每个组对应一个坝段，对模型的裁剪也是以组为单位进行的。大坝整体几何模型作为被裁剪体，高程和仓位数据生成相应的裁剪面。遍历整个大坝模型，对每个坝段，进行三维的逐面裁剪。即可得到在特定浇筑时间的大坝三维模型，直接交由后面的绘制模块进行绘制即可。

图5、图6为不同时期大坝浇筑状态的直观显示。图5为第n天浇筑尚未完成的情况，其中河水被截流，故而未显示水流效果；图6为浇筑完成以后的浇筑情况。可以看出，对浇筑进度的仿真非常直观。在实际应用中，通过对计划安排的高程数据和实际采集的高程数据进行对比仿真，对施工进度以及计划安排起到辅助决策和分析的作用。

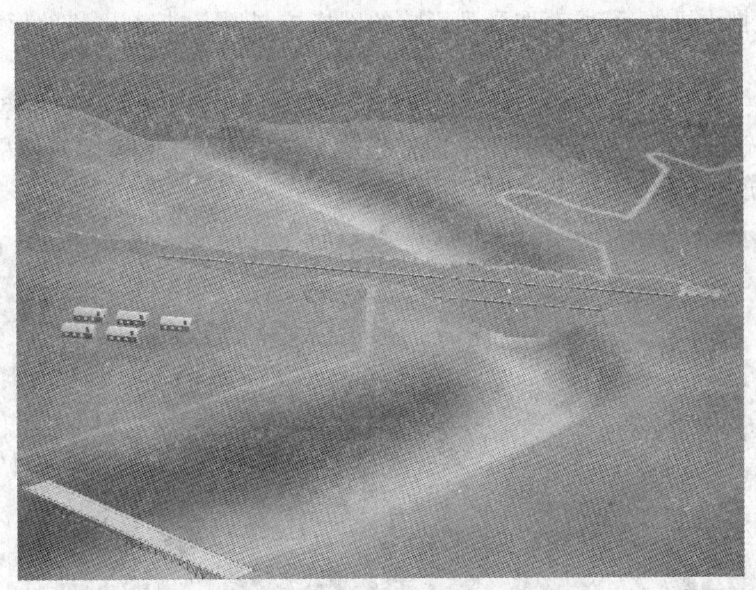

图5　第n天浇筑情况

图6　浇筑完成情况

3.5　碾压过程的仿真

通过本文浇筑模型的生成算法可以生成混凝土坝整体模型浇筑情况，但是正在浇筑的微观细节以及动态浇筑情况还无法体现出来。鉴于此，通过仓位碾压过程的仿真，在混凝土坝整体模型基础上，实现仓位碾压过程的动画，为整个施工模拟增加真实感。按以下步骤进行仿真：

(1)针对正在碾压的仓位,根据仓位信息(仓位的位置、形状)计算相应的参数,根据参数控制仓位上生成的网格形状来实现混凝土堆的模拟。另外,根据仓位信息计算出推土机运动的关键路径点,通过对关键路径点进行插值,用插值结果驱动推土机模型进行坐标变换,实现推土机的运动。

(2)通过推土机的运动,获取受到影响的网格控制顶点,控制网格的形变,模拟碾压过程。

绘制模块包括混凝土坝体绘制、地形绘制等许多功能。混凝土坝浇筑模型由 CPU 计算生成,考虑到碾压过程网格形变的计算复杂度,本系统采用基于 CPU 实现网格形变来模拟碾压过程的方法,仓位碾压过程如图 7 所示。

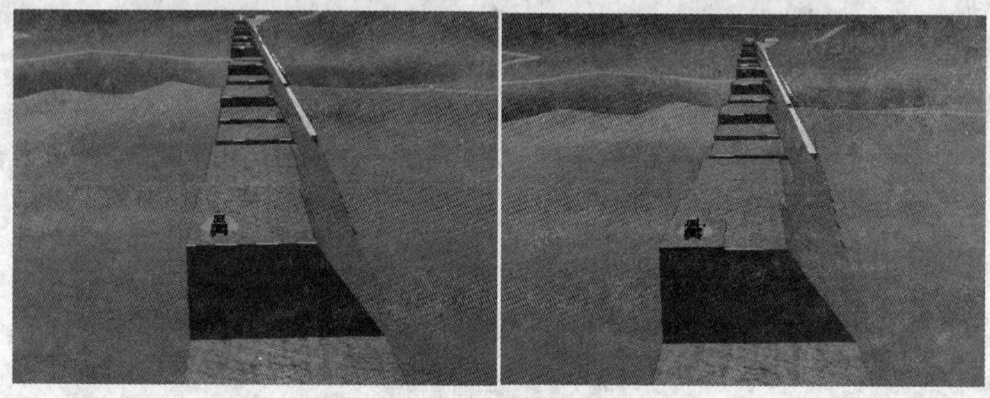

图 7　仓位碾压过程

4　基于粒子的大坝泄洪仿真

除坝体本身的绘制外,大坝泄洪仿真是很重要的部分,便于用户直观地观察大坝泄洪情况以及上下游水位变化情况。坝孔水流由许多微小的颗粒组成,其外观是由大量的颗粒泄洪时所产生的不同速度表现出来的。水流特点就是没有规则的外观、随机性和动态性强。这些特性使得它很难用传统的表面建模方式进行描述,而粒子系统的特点决定了它是描述坝孔的最佳模型。

在泄洪的仿真中,上、下游的水位会根据数据库中的相关数据发生相应的变化,水位的变化会影响坝孔水流的速度。粒子分类为两种特性,一类为本身的特性,比如粒子的大小、材质、纹理、寿命等;另一类为受外界影响可以变化而表现的特性比如重力、风速、方向、本身衰减速度等。

如图 8 所示,水流的抛射曲线公式,上、下抛射曲线方程为:

$$y_{-\mathrm{up}} = a_1 z^2 + b_1 z + c_1$$

$$y_{-\mathrm{down}} = a_2 z^2 + b_2 z + c_2$$

本系统根据上下抛射曲线方程设计粒子系统的运动规律。

根据水位的变化来控制上下抛射曲线的参数,使水流的形状发生变化,以体现出水位的变化。图 9 为坝孔闸门开放时泄洪的效果,可以看出,达到了一定的视觉真实感。

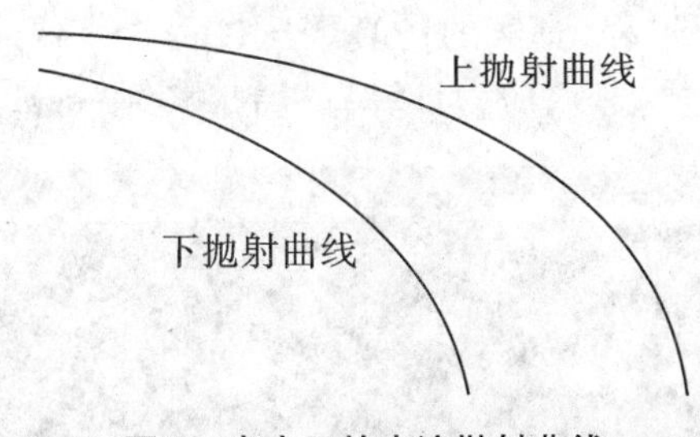

图 8　出水口的水流抛射曲线

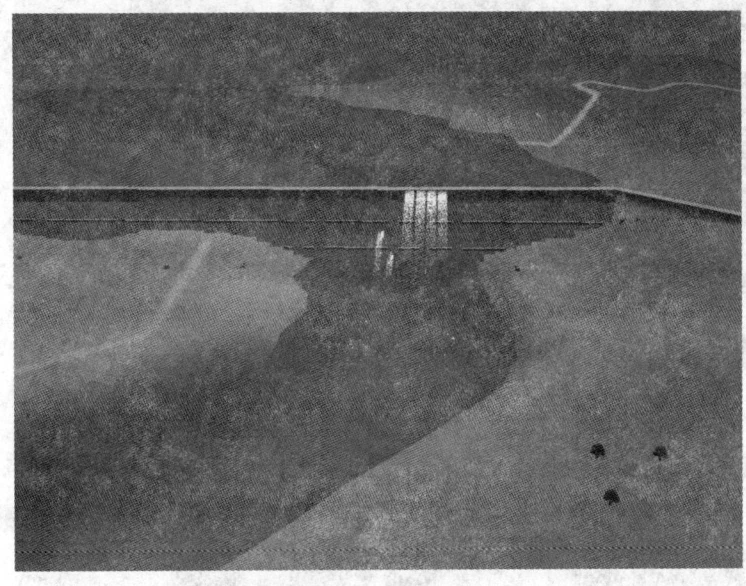

图 9　大坝泄洪

5　二维动态仿真

除了上文介绍的三维显示方式,本系统还具有传统的二维显示模块,温度、应力场显示模块以及数据库管理模块。在大坝的浇筑过程中,温控是最重要的工作。通常使用 Ansys 等有限元分析软件来计算坝体温度数据。本文的可视化系统中提供了接口,读取计算分析的温度数据,将得到的离散数据进行处理,并将散乱点重建为三角网格,最终和二维浇筑仿真结合在一起,以温度云图的形式呈现给用户,二维整体浇筑情况见图 10,典型坝段浇筑情况见图 11,温度场温度分布的云图显示见图 12。

6　结语

本文提出了一种大坝浇筑进度仿真方法,可以根据施工进度信息动态灵活地模拟大坝的浇筑进度。另外,本文仿真系统是针对国内某混凝土坝而设计实现的,系统已用于对实际的工程项目进行辅助决策和相关的数据管理。为用户提供了直观的绘制效果,使决策、管理人员不必往返于施工现场,即可身临其境的感受到施工的现状。并且系统很好的与传统的施工进度管理、温度(应力)计算等系统结合在了一起,避免了数据的重复采集。

系统对碾压过程仿真的真实感不够,仅能做碾压的过程的示意,下一步拟采用基于

GPU 的粒子系统来模拟碾压过程，真实的仿真碾压过程。另外，本文的浇筑过程建模与绘制应该和传统混凝土坝仿真中温度、应力计算有机的融合在一起，提供更专业的施工指导。

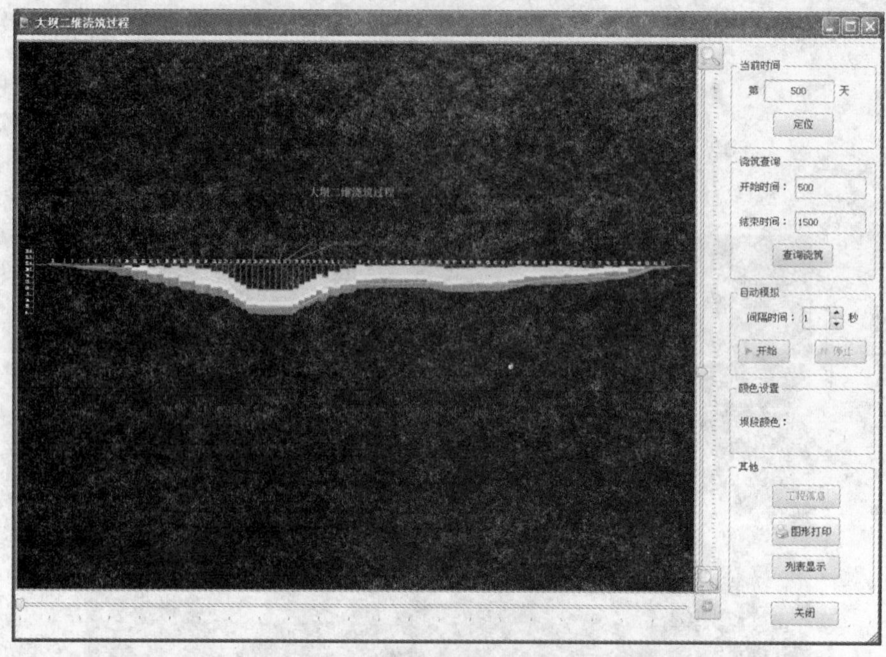

图 10　二维整体浇筑情况

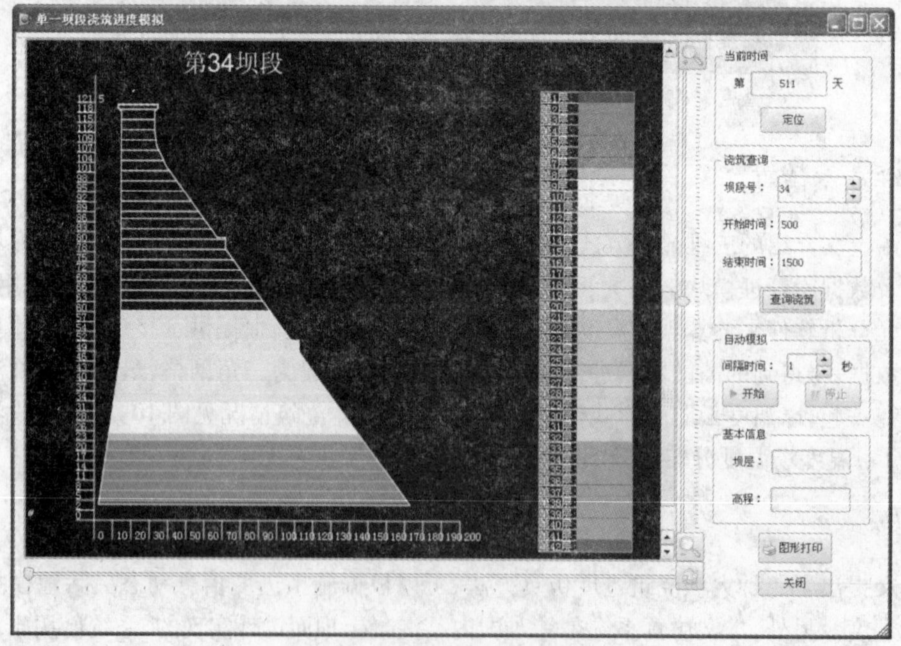

图 11　典型坝段浇筑情况

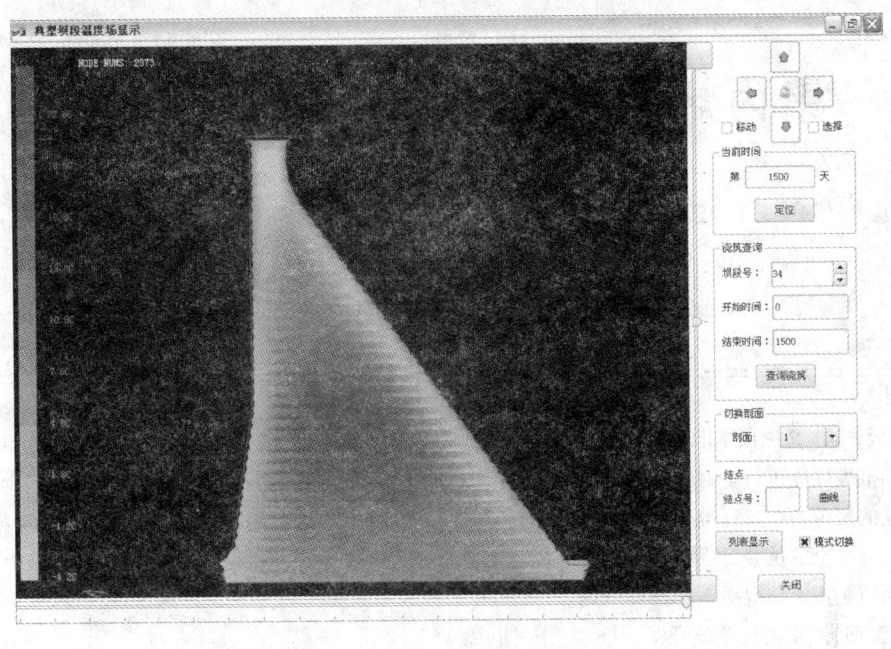

图 12 温度场温度分布的云图显示

参考文献

[1] 钟登华.可视化仿真技术及其应用[M].北京：中国水利水电出版社,2002.

[2] 钟登华,李景茹.复杂地下洞室群施工交通运输系统仿真与优化研究[J].系统仿真学报,2002,14(2)：140-145.

[3] 陈金水,颜伟琼.基于 OpenGL 的三维建模在水利行业中的应用[J].南京：计算机技术与发展,2006(3)：197-199.

[4] 孙锡衡,齐东海.水利水电工程施工计算机模拟与程序设计[M].北京：中国水利水电出版社,1997.

[5] 李勇刚,胡志根,燕乔.混凝土拱坝浇筑仿真的可视化技术研究[J].湖北：武汉水利电力大学学报,2000(1)：33-36.

[6]《三峡水利枢纽混凝土工程温度控制研究》编辑委员会.三峡水利枢纽混凝土工程温度控制研究[M].北京：中国水利水电出版社,2001.

[7] 倪明田,吴良芝.计算机图形学[M].北京：北京大学出版社,2002.

【作者简介】 李江,男,1971 年 1 月生,1994 年毕业于新疆农业大学,新疆水利水电规划设计管理局副局长/提高待遇高级工程师。

聚丙烯单丝纤维混凝土试验研究

蒋巧玲　朱　琦

（广西壮族自治区水利科学研究院　南宁　530023）

【摘要】 本文选用了四种聚丙烯单丝纤维，对混凝土拌和物性能、硬化后的力学性能、抗渗性能和抗裂性能进行了对比试验，分析了其作用机制。研究结果表明：在普通混凝土中掺入少量改性的聚丙烯单丝纤维，能改善混凝土的和易性和变形性能，并能有效减少混凝土的微细裂缝，提高混凝土的抗裂、抗渗等性能，纤维混凝土所具有的这些特点，非常适合在抗渗抗裂要求较高的渠道防渗工程混凝土中推广应用。

【关键词】 聚丙烯单丝纤维　抗渗　抗裂　裂缝

1　引言

20世纪80年代，国外学者尝试将合成纤维掺入混凝土中，研究纤维对混凝土性能的影响。发现聚合物纤维特别是聚丙烯单丝纤维材料对减少混凝土微细裂缝，提高抗裂抗冲击性能效果较好。于是，将这种材料逐渐推广到飞机跑道、高速公路、桥面、屋面板、军事工程等有高耐磨性、高抗冲击性和有抗裂要求的部位和构件中，并取得了良好的技术效果。

未经改性的聚丙烯纤维表面是憎水的，掺入混凝土后，不利于形成良好的纤维混凝土界面，采用硅氧烷、烷基磷酸盐等表面处理剂对传统的聚丙烯纤维进行表面处理后，可以改善纤维在混凝土基体中的分散性，提高纤维基体的黏结强度。改性后的聚丙烯纤维具有较高的细度和单位体积分布率，在混凝土搅拌过程中，受到水泥、砂、石的冲击而散开，几百万根纤维丝乱向均匀分布在混凝土基体中，当混凝土出现裂缝后，高度分散的纤维在混凝土基体中发挥搭接和牵制作用，形成一个乱向支撑体系，起到次级加强筋的作用，可以防止细骨料离析，减少早期泌水，抑制混凝土的早期收缩变形，有效阻碍微细裂缝的形成和扩展，从而提高混凝土的抗渗性能。在水利工程渠道防渗混凝土中，其设计强度等级一般在C20以下，抗渗等级W4以上，普通混凝土虽然可以达到上述技术指标，但从近年来我区已完工的混凝土防渗衬砌渠道情况看，由于衬砌面板厚度一般较薄，养护条件较差，渠基填筑往往达不到设计要求，衬砌混凝土有不同程度的开裂渗漏现象。裂缝导致渠道的渗漏损失加大，工程效益降低。有着良好抗裂性能的纤维混凝土正好可以弥补普通混凝土的这种不足，起到改善混凝土的抗裂性能，达到提高抗渗性的目的。为了取得聚丙烯纤维在改善混凝土性能方面的数据资料及推广应用中需要解决的施工技术问题，本文选了四种改性聚丙烯单丝纤维掺入混凝土中，与普通混凝土进行对比试验，摸索聚丙烯纤维混凝土的施工工艺情况，对混凝土力学性能，抗渗抗裂性能的影响。试验的目的是在渠道混凝土中推广应用聚丙烯纤维寻找技术依据。

2 试验原材料

(1)胶凝材料:采用广西扶绥海螺水泥有限责任公司生产的海螺牌 P·C32.5 水泥。

(2)骨料:细骨料采用细度模数为 3.13 的人工砂,粗骨料采用二级配的石灰岩碎石。

(3)聚丙烯单丝纤维:试验选用四种聚丙烯单丝纤维,分别是江苏省丹阳合成纤维厂生产的丹强丝纤维(以下简称丹强丝纤维)、深圳市恒悦达建筑材料有限公司的抗裂王系列聚丙烯单丝纤维(以下简称抗裂王纤维)、上海山容实业有限公司的聚丙烯单丝纤维(以下简称上海纤维)及湖南长沙市博赛特建筑工程材料有限公司的博赛特聚丙烯单丝纤维(以下简称博赛特纤维)。四种纤维的主要性能见表 1。

表 1 聚丙烯单丝纤维性能

性能指标	纤维品种			
	丹强丝纤维	抗裂王纤维	上海纤维	博赛特纤维
抗拉强度(MPa)	≥300	633	—	494
拉伸极限(%)	≤30	—	—	—
弹性模量(MPa)	≥3 795	4 427	—	4 241
纤维直径(μm)	10 ~ 100	34.25	18	34.41
断裂伸长率(%)	—	25.6	15 ~ 20	40.4
拉伸强度(Cn/dtex)	—	—	7.67	—

3 试验结果与分析

3.1 纤维混凝土拌和工艺试验

纤维在混凝土中分布的均匀性对混凝土的质量有着决定性的影响,而混凝土的拌和工艺是影响纤维在混凝土中分布均匀性的重要因素,以丹强丝纤维为例,掺量为 0.9 kg/m^3,用不同的投料方式、不同的拌和时间,不同的拌和方法拌和混凝土,测试其坍落度值,观察混凝土黏聚性、保水性、振捣密实性,1 h 后收面情况及纤维在混凝土中的分布均匀性情况,找出最佳的拌和方式是强制式搅拌机拌和,搅拌时间大于等于 3 min。再用其他三种纤维进行复核,所得结论与掺丹强丝纤维一致。试验结果详见表 2。

从表 2 中可以看出,与不掺纤维的混凝土相比,掺聚丙烯单丝纤维后的混凝土,坍落度从 80 mm 下降至 10 ~ 30 mm,坍落度明显下降。黏聚性和保水性均比不掺纤维时要好。这主要是因为聚丙烯单丝纤维在混凝土中经过搅拌后,在混凝土拌和物中呈三维乱向分布,对骨料起到承托作用,乱向分布的纤维限制了水分从混凝土拌和物内部移向表面的速度和规模,因此聚丙烯单丝纤维的掺入降低了混凝土表面的析水和集料的离析,改善了拌和物的黏聚性和保水性。另外,掺入的聚丙烯单丝纤维比表面积较大,使得包裹骨料和纤维的水泥浆层变薄,降低了水泥浆的润滑作用,从而使得拌和物的流动性大幅降低。

表 2　纤维混凝土和易性、施工性和分布均匀性试验结果

拌和方式	拌和工艺	和易性			振捣密实性	1 h 后收面情况	纤维在混凝土中分布均匀性
		坍落度(mm)	黏聚性	保水性			
强制式搅拌机拌和	不掺丹强丝纤维的普通混凝土拌和 3 min	80	一般	少量泌水	易振	较好	—
	掺丹强丝纤维湿拌 2 min	20	较好	较好	易振	较好	基本均匀
	掺丹强丝纤维湿拌 3 min	25	好	好	易振	较好	均匀
	掺丹强丝纤维干拌 0.5 min,再湿拌 2.5 min	25	好	好	易振	较好	均匀
人工拌和	先将砂、石、水泥干拌均匀,加丹强丝纤维干拌,再加水湿拌	10	较好	较好	易振	较好	不均匀,少量结团
强制机拌和	掺抗裂王纤维湿拌 3min	15	好	好	易振	较好	均匀
	掺上海纤维湿拌 3min	15	好	好	易振	较好	均匀
	掺博赛特纤维湿拌 3min	30	好	好	易振	较好	均匀

从试验中观察到:坍落度明显降低的纤维混凝土,均很容易被振捣密实,拌和物在静置 1 h 后也很容易收面抹平。这说明混凝土掺纤维后,坍落度虽然降低了,但并不影响纤维混凝土振捣的密实性,也不影响其浇筑和收面等施工作业。

从拌和方式看:使用强制式搅拌机拌和,拌和时间 3 min 以上,就能保证纤维在混凝土中分布均匀。人工拌和方式则很难将纤维拌和均匀。因为纤维遇水后容易粘在一起,难以分散,加上人工拌和混凝土,劳动强度非常大,增加了施工作业的难度,所以人工拌和难以保证纤维在混凝土中分布均匀。

3.2　聚丙烯单丝纤维混凝土力学性能、抗渗性能试验

四种聚丙烯单丝纤维混凝土与不掺纤维的混凝土,采用相同的配合比,相同的拌和时间,用强制式搅拌机拌和(四种纤维掺量均为 0.9 kg/m^3),成型抗压强度、劈裂抗拉强度及抗渗试件,进行 7 d、28 d 的抗压,劈裂抗拉及 28 d 抗渗试验,试验结果见表 3、图 1 和图 2。

表 3　聚丙烯纤维混凝土力学性能及耐久性能试验结果

序号	纤维		水灰比	抗压强度(MPa)		劈裂抗拉强度(MPa)		渗透高度比(%)	抗压强度比(%)		劈裂抗拉强度比(%)	
	品种	掺量(%)		7 d	28 d	7 d	28 d		7 d	28 d	7 d	28 d
1	不掺纤维	0	0.62	15.0	22.2	1.38	1.68	100	100	100	100	100
2	丹强丝纤维	0.9	0.62	15.9	23.8	1.42	1.92	7.5	106.0	107.2	102.9	114.3
3	抗裂王纤维	0.9	0.62	14.9	23.0	1.48	1.96	7.5	99.3	103.6	107.2	116.7
4	上海纤维	0.9	0.62	15.7	23.3	1.41	1.84	15.0	104.7	105.0	102.2	109.5
5	博赛特纤维	0.9	0.62	15.2	23.1	1.40	1.81	15.0	101.3	104.1	101.4	107.7

注:表中强度增长率是以不掺纤维的混凝土强度为基准,四种不同纤维混凝土的强度与之作对比。

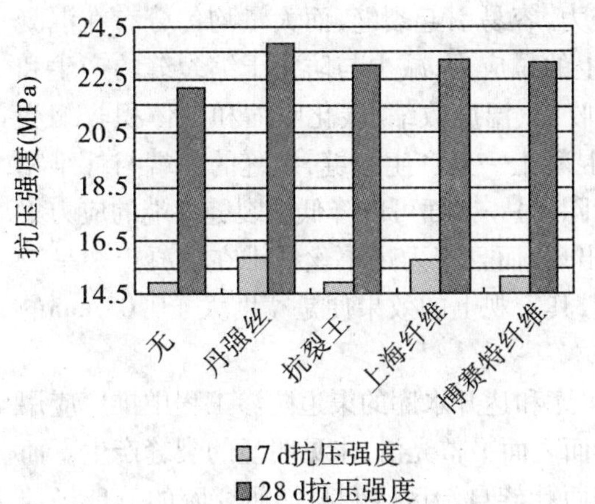

图1 聚丙烯纤维对抗压强度的影响

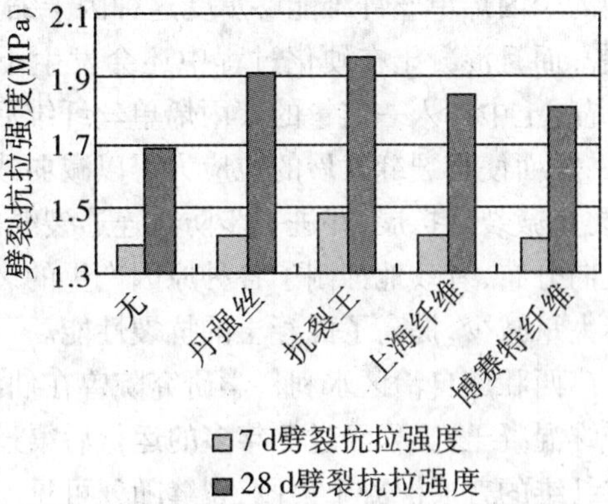

图2 聚丙烯纤维对劈裂抗拉强度的影响

从表3、图1和图2中可以看出：

(1)四种掺聚丙烯单丝纤维的混凝土7 d、28 d劈裂抗拉强度比分别为101.4%～107.2%、107.7%～116.7%纤维的掺入提高了混凝土的劈裂抗拉强度，原因在于纤维的阻裂机制。混凝土的破坏是由于在外力作用下，裂缝形成、扩展而导致的，混凝土掺入纤维后，在荷载作用下，纤维产生拉拔阻力，阻止了裂纹的进一步扩大和发展，在混凝土破坏前其内部会存在大范围的缓慢的稳定裂纹，而且由于纤维在混凝土中是三维乱向分布，这种阻裂效应也是乱向的，这就增加了裂纹扩展的曲折性，使纤维混凝土在荷载作用下表现为裂纹缓慢增长，进而延缓了整个破坏过程，宏观上提高了混凝土的劈裂强度和韧性。

(2)四种掺聚丙烯单丝纤维的混凝土7 d、28 d抗压强度比分别为99.3%～106.0%、103.6%～107.2%，与劈裂抗拉强度相比，聚丙烯单丝纤维对混凝土抗压强度影响不大。这是由于混凝土中裂纹的扩展速度对劈裂强度的影响要远远大于对抗压强度的影响，总体上说，纤维的掺入提高了混凝土的韧性，减缓了混凝土的脆性破坏过程。

(3)四种掺聚丙烯单丝纤维的混凝土28 d渗透高度比是不掺纤维混凝土的7.5%～15.0%，抗渗性能显著提高。聚丙烯单丝纤维掺入到混凝土中，一方面，减少了混凝土表面的析水和集料沉降，大大减少了混凝土中的孔隙；另一方面，混凝土微裂缝的发展受阻，只能在混凝土内部形成类似于无害孔洞的封闭空腔或者内径非常细小的孔，大量聚丙烯单丝纤维的存在减少了裂缝的数量、宽度和长度，降低了生成相互贯通孔洞和裂隙的可能性，阻断了渗水的通道，从而有效地提高了混凝土的抗渗性能。

3.3 聚丙烯单丝纤维混凝土早期抗裂性试验

聚丙烯单丝纤维混凝土早期抗裂性试验，在对比样配合比中加纤维，用600 mm×600 mm平板式混凝土早期塑性收缩和开裂模具成型，试件在浇筑、振实、抹平后即开始暴露，用100 W电风扇平行吹试件表面。吹24 h后观测裂缝的数量、宽度和长度。不掺纤维的混凝土有2条明显的细长裂缝。掺纤维的四种混凝土表面均没有裂纹，说明掺纤维的混凝土抗裂效果明显好于不掺纤维的混凝土。

从前面的试验中可以看出，均匀分散的聚丙烯单丝纤维在混凝土中呈现三维网状结构，拌和时能阻止粗、细集料的沉降，防止混凝土产生离析，提高了混凝土的匀质性，而且纤维的掺入减少了混凝土的表面析水，也就抑制了混凝土因表面失水引发的沉降裂缝；混凝土在塑性状态时强度极低，会因水分蒸发产生收缩拉应力，容易引起裂缝，而大量均匀分布在混凝土中的聚丙烯单丝纤维能够承受这种因干缩产生的拉应力，减少与防止干缩裂缝的产生和发展。而且混凝土在硬化过程中还会发生干燥收缩、温度收缩、碳化收缩和自体积收缩变形，混凝土中掺入一定量的聚丙烯单丝纤维后，混凝土一旦产生裂缝，裂缝的前端与纤维相交，一方面使得裂缝发展的拉应力得以减弱或消除；另一方面可以降低微裂缝尖端的应力集中，防止微裂缝扩展，并进一步防止连通裂缝的出现，而且聚丙烯单丝纤维在混凝土中呈三维乱向分布，有效地抑制了各种原因产生的裂缝，其宏观上的效果便是宽度大于0.05 mm的裂缝大量减少，提高了混凝土的抗裂性能。

广西壮族自治区水利科学研究院曾在仙湖水库和达开水库的渠道防渗工程中推广应用过纤维混凝土，工程经过 7 年多的运行后渠道断面表面平整光滑，未见明显的裂缝产生。而不掺纤维的普通混凝土渠段，裂缝随处可见。广西壮族自治区水利科学研究院的篮球场地面，原来是普通混凝土地面，使用不久后四处开裂，改用纤维混凝土地面后，已使用数年，地面依然平整，少有裂缝。抗裂效果非常好。

4 试验结论

(1)聚丙烯单丝纤维混凝土，采用强制式搅拌机，拌和 3 min 以上，即可将纤维拌和均匀，人工方式很难将纤维拌和均匀。

(2)掺入纤维后引起的坍落度下降不影响混凝土的振实和收面作业。

(3)与普通混凝土的性能相比，掺聚丙烯单丝纤维混凝土的 28 d 劈裂抗拉强度提高了 7.7% ~16.7%。渗透高度比是不掺纤维时的 7.5% ~15.0%，混凝土的抗渗性能显著提高。

(4)在混凝土中掺入聚丙烯单丝纤维能有效地抑制裂缝的产生和扩展，进一步改善混凝土早期收缩，显著提高混凝土的抗裂性能。

5 聚丙烯纤维混凝土在工程应用中需要注意的问题

可以预见聚丙烯纤维在渠道防渗混凝土工程中必将有广阔的推广应用前景。纤维混凝土要达到最佳的抗渗抗裂效果，在应用中必须注意以下问题：

(1)合理选择原材，对水泥、砂、石、外加剂、纤维等材料，特别是纤维，要选择聚丙烯纤维和聚丙烯腈纤维品种，这两种纤维对改善混凝土的抗渗抗裂性能效果较好。所用原材料应按规范要求取样和检验，选择符合规范技术要求的原材料，纤维应按《水泥混凝土和砂浆用合成纤维》(GB/T 21120—2007)标准检验断裂强度和断裂伸长率，检验结果应符合有关技术要求。

(2)应通过试验，确定合理的纤维混凝土拌和工艺和施工配合比，并严格按配合比施工。纤维混凝土的坍落度以不掺纤维时混凝土的坍落度为 60 ~80 mm 标准进行控制。

(3)纤维的掺入会使混凝土坍落度降低，流动度变小，但不影响混凝土的振实和收面，不需要采取措施增加流动度。

(4)纤维混凝土不能起到减少水泥用量的作用，施工中不应降低单位水泥用量，也不宜减小渠道混凝土的设计厚度。

(5)纤维混凝土应尽量采用强制式搅拌机拌和。投料顺序：石子→水泥→纤维→砂→

水拌和3 min,如采用自落式搅拌机拌和,按上面的方法一次性加入材料后,需在强制式搅拌机所用拌和时间的基础上延长0.5 ~1 min。人工拌和很难将纤维拌和均匀,不宜采用。

(6)纤维混凝土浇筑、振捣和养护可按普通混凝土施工的方法进行。

(7)浇筑纤维混凝土前,应对渠道地基特别是填方渠段进行夯实,预防基础的不均匀沉陷导致混凝土的拉裂破坏。

参考文献

[1] 吕军政,周民祥,宋晗,等.聚丙烯纤维混凝土在工程中的应用[J].陕西建筑,2007,142(4):22-24.

[2] 张伟.改性聚丙烯纤维增强混凝土抗疲劳性能的研究[D].大连:大连交通大学,2008.

[3] 陈卫东.聚丙烯纤维对混凝土防渗性能影响的研究[J].山西水利科技,2008,167(1):17-21.

[4] 刘永胜.纤维混凝土增强机理的界面力学分析[J].混凝土,2008,222(4):34-35.

[5] 孙玉龙.聚丙烯纤维对混凝土干缩性影响的试验研究[J].黄河水利职业技术学院学报,2009,21(2):40-42.

[6] 宗荣.聚丙烯纤维混凝土使用性能研究[D].西安:长安大学,2004.

【作者简介】 蒋巧玲,1966年生,女,广西全州人,工程师,主要从事水利工程试验研究工作。

基于层次分析法的鹅公带古滑坡体高速滑坡危险度评价*

黄锦林[1,2] 赵吉国[1,2] 张 婷[1,2] 李嘉琳[1,2]

(1. 广东省水利水电科学研究院 广州 510630;
2. 广东省岩土工程技术研究中心 广州 510630)

【摘要】 高速滑坡的判定是滑坡治理工作中的一项重要内容,但由于目前尚没有统一规范的滑速计算方法,给高速滑坡的预测带来一定困难。层次分析法是一种定性与定量相结合的多目标决策分析方法,本文针对库岸高速滑坡发生的各影响因子,采用层次分析法对鹅公带古滑坡体高速滑坡危险度进行了分析评价,为库岸高速滑坡的预测提供了一条新思路。

【关键词】 层次分析法 高速滑坡 影响因子 评价体系 危险度评价

滑坡是指斜坡上的土体或者岩体,受河流冲刷、地下水活动、地震及人工切坡等因素影响,在重力作用下,沿着一定的软弱面或软弱带,整体地或者分散地顺坡向下滑动的自然现象,俗称“走山”、“垮山”、“地滑”、“土溜”等。根据滑坡启动后的运动速度大小,可将滑坡分为高速滑坡、快速滑坡、中速滑坡和慢速滑坡,其中运动速度大于5 m/s的为高速滑坡。对于水库而言,危害性最大的滑坡就是高速滑坡,国内外已发生多起重大库岸滑坡灾害的实例。由于目前尚没有统一规范的滑坡速度计算方法,给高速滑坡的预测带来一定困难。滑坡危险度评价是将产生滑坡的内部因子、外部因子等影响因素进行数学模型分析,定量或半定量地评价发生滑坡的危险性等级。文献[3]至文献[5]采用层次分析法对滑坡的危险度进行了评价,但对于高速滑坡问题,尚未见采用层次分析法进行危险度评价的相关文献。

鹅公带古滑坡体位于某新建水利枢纽坝址上游1.3 km处,分布在右岸。根据地质测绘和钻孔勘探,认为鹅公带属古滑坡体,钻孔和地表测绘揭露存在古滑动带,地貌上存在“双沟同源”现象,地下水活跃,有多处泉水出露点,地表水下渗,前、后缘滑坡体特征明显。地面产生的裂缝为受强降雨影响产生的浅层开裂和局部塌陷,多沿基岩与坡积、风化土界面分段开裂,影响范围较小,未见整体变形迹象,目前处于稳定状态。为了研究鹅公带古滑坡体在水库蓄水运行后发生高速滑坡的可能性大小,本文采用层次分析法,对鹅公带古滑坡体高速滑坡危险度进行了评价。

1 层次分析法基本原理及分析步骤

层次分析法(The Analytic Hierarchy Process,简称AHP)是20世纪70年代由美国运筹

* **基金项目**:广东省水利科技创新项目(项目编号:2009-36)。

学家 Saaty 提出的,经过多年的发展已成为一种较为成熟的决策方法。

1.1 层次分析法基本原理

层次分析法将要决策或判断的问题看做受多种因素影响的系统,各因素间依据隶属关系可分解成目标、准则、方案等层次,在此基础上进行定性和定量分析,确定不同层次中各因素的排序权重,来辅助进行决策或判断。层次分析法的主要特点是将人的主观判断用数量形式表达出来并进行科学处理,同时,这一方法虽然有深刻的理论基础,但表现形式却很简单,容易被人理解和接受。

1.2 层次分析法分析步骤

运用 AHP 解决问题,大体可以分为 4 个步骤。

1.2.1 建立递阶层次结构模型

针对需要决策或判断的问题,将其分解为不同的影响因素,并根据各因素间的隶属关系,将因素进行组合,形成一个多层次的分析结构模型,由高层次到低层次分别为目标层、准则层、方案层等。

1.2.2 构造两两比较判断矩阵

对同一层次的一系列成对因素两两进行比较,不同因素间的相互比较结果采用 1 ~9 的标度方法进行打分,不同重要程度分别赋予不同的分值(见表 1),得到两两比较判断矩阵 $\boldsymbol{A}$,$\boldsymbol{A}=(a_{ij})_{n\times n}$。

表 1 重要性标度含义

重要性标度	含义
1	表示 a_i 与 a_j 两个元素相比,具有同等重要性
3	表示 a_i 与 a_j 两个元素相比,前者比后者稍重要
5	表示 a_i 与 a_j 两个元素相比,前者比后者明显重要
7	表示 a_i 与 a_j 两个元素相比,前者比后者强烈重要
9	表示 a_i 与 a_j 两个元素相比,前者比后者极端重要
2,4,6,8	2,4,6,8 分别表示相邻判断 1 ~3,3 ~5,5 ~7,7 ~9 的中间值
倒数	若元素 a_i 与元素 a_j 的重要性之比为 a_{ij},则元素 a_j 与元素 a_i 的重要性之比为 $a_{ji}=1/a_{ij}$

1.2.3 计算单一准则下元素的相对权重

这一步要解决在某一准则下的 n 个元素 $A1,A2,\cdots,An$ 排序权重的计算问题,并进行判断矩阵一致性检验。计算判断矩阵各因素针对其准则的相对权重的方法有和法、根法、幂法等,本文采用根法。判断矩阵一致性指标用 CI 表示,$CI=\dfrac{\lambda_{\max}-n}{n-1}$。当 $n<3$ 时,矩阵永远具有完全一致性,$CI=0$;当 $n\geqslant 3$ 时,需通过一致性比例 CR 来进行一致性检验,当 $CR<0.1$ 时,认为判断矩阵的一致性是可以接受的。$CR=\dfrac{CI}{RI}$,RI 为平均随机一致性指标,可查表 2 确定。

表 2　平均随机一致性指标 *RI*

阶数	1	2	3	4	5	6	7	8	9	10	11	12	13	14	15
RI	0	0	0.52	0.89	1.12	1.26	1.36	1.41	1.46	1.49	1.52	1.54	1.56	1.58	1.59

1.2.4　计算各层元素的组合权重

为了得到递阶层次结构中每一层次中所有元素相对于总目标的相对权重,需将第三步的计算结果进行适当的组合,计算得出最低层次元素(即决策方案)相对于总目标的相对权重,并进行整个递阶层次模型的一致性检验。

2　鹅公带古滑坡体高速滑坡危险度评价指标体系

鹅公带古滑坡体高速滑坡危险度评价的递阶层次结构可分为 3 个层次:目标层次(A)、类指标层次(B)和基础指标层次(C),其中:目标层次指鹅公带古滑坡体高速滑坡危险度;类指标层次指鹅公带古滑坡体高速滑坡危险度评价中的一级评价指标;基础指标层次指鹅公带古滑坡体高速滑坡危险度评价中的二级评价指标。

对于高速滑坡而言,文献[7]将其产生的因子划分为内部条件(因子)、外部条件(因子)和斜坡变形现状因子三部分,并给出了进一步的判别因子和指标。结合文献[7]的研究成果及鹅公带古滑坡体现状和今后的蓄水运行条件,确定的鹅公带古滑坡体高速滑坡危险度评价指标见表 3。

表 3　鹅公带古滑坡体高速滑坡危险度评价指标

目标(A)	一级评价指标(B)		二级评价指标(C)	
鹅公带古滑坡体高速滑坡危险度评价	B1	内部因子	C1	相对坡高
			C2	平均坡度
			C3	纵向坡形
			C4	横向坡形
			C5	基本岩性
			C6	风化程度
			C7	斜坡岩性组合
			C8	斜坡结构及构造
	B2	外部因子	C9	地下水作用
			C10	库水位升降作用
			C11	降雨作用
	B3	斜坡变形现状因子	C12	斜坡变形现状

需要说明的是,在外部因子中,地震是个不确定因子,很难在一个具体的地方确定其作用程度和指标,是个动态因子,且发生的概率很小,虽然它是高速滑坡形成、发生的一个重要外部条件,但本文与文献[7]一样,不将其列为鹅公带古滑坡体高速滑坡的判别因子。此

外，由于鹅公带古滑坡体在水库蓄水后不存在水流削弱坡脚作用，且斜坡体也不存在自然加载和人为加载作用，因此与文献[7]不同，在外部因子中取消了这两个因子。考虑到鹅公带古滑坡体受水库蓄水的影响，以及该水库所在区域降雨量强度较大这一事实，在外部因子中增加了库水位升降作用和降雨作用两个因子，这样的处理符合鹅公带古滑坡体在水库建成蓄水后的实际情况，也更能满足其高速滑坡危险度评价的需要。表4～表6给出了各评价指标的判断矩阵和特征向量。

表4　一级评价指标判断矩阵

A	$B1$	$B2$	$B3$	ω_B
$B1$	1	3	5	0.648
$B2$	1/3	1	2	0.230
$B3$	1/5	1/2	1	0.122

$\lambda_{max} = 3.004, CI_B = 0.001\,85$

$RI_B = 0.52, CR_B = 0.003\,6 < 0.1$

表5　二级评价指标判断矩阵(1)

$B1$	$C1$	$C2$	$C3$	$C4$	$C5$	$C6$	$C7$	$C8$	ω_{C1}
$C1$	1	3	3	5	2	4	6	1	0.263
$C2$	1/3	1	1	2	1	2	3	1/3	0.104
$C3$	1/3	1	1	2	1	2	3	1/3	0.104
$C4$	1/5	1/2	1/2	1	1/2	1	1	1/5	0.052
$C5$	1/2	1	1	2	1	2	3	1/2	0.115
$C6$	1/4	1/2	1/2	1	1/2	1	2	1/4	0.060
$C7$	1/6	1/3	1/3	1	1/3	1/2	1	1/6	0.039
$C8$	1	3	3	5	2	4	6	1	0.263

$\lambda_{max} = 8.668, CI_{C1} = 0.095\,5$

$RI_{C1} = 1.41, CR_{C1} = 0.068 < 0.1$

表6　二级评价指标判断矩阵(2)

$B2$	$C9$	$C10$	$C11$	ω_{C2}
$C9$	1	1	1	0.333
$C10$	1	1	1	0.333
$C11$	1	1	1	0.333

$\lambda_{max} = 3, CI_{C2} = 0$

$RI_{C2} = 0.52, CR_{C2} = 0 < 0.1$

此外，由于一级评价指标 $B3$ 下的二级评价指标仅有 $C12$ 一项，$n<3$，矩阵具有完全一致性，$\omega_{C3} = 1.0, CI_{C3} = 0, RI_{C3} = 0$。

由以上计算结果可推得鹅公带古滑坡体二级评价指标相对于总目标的权重向量为

$\omega_C = (0.170, 0.068, 0.068, 0.034, 0.075, 0.039, 0.025, 0.170, 0.077, 0.077, 0.077,$

0.122)

整个递阶层次模型的一致性检验如下

$CI_C=(CI_{C1},CI_{C2},CI_{C3})\omega_B=0.0619$

$RI_C=(RI_{C1},RI_{C2},RI_{C3})\omega_B=1.033$

$$CR_C = CR_B + \frac{CI_C}{RI_C} = 0.0036 + \frac{0.0619}{1.033} = 0.0635 < 0.1$$

说明整个递阶层次模型具有较好的一致性,判断合理。

3 鹅公带古滑坡体高速滑坡危险度评价

鹅公带古滑坡体高速滑坡危险度评价模型为:

$$R = \sum_{i=1}^{n}(C_i\omega_{ci})$$

式中:R 为高速滑坡危险度值;C_i 为各二级评价指标的定量打分;ω_{ci}为各二级评价指标的权重。

为了避免二级评价指标评分的随意性,本文采用"黄金分割"的原理来确定各指标中不同等级的危险分值,即危险性大的分值取1,危险性中等的分值取0.618,危险性小的分值取0.382。各二级评价指标分类及定量评分见表7。

表7 二级评价指标分类及定量评分

评价项目		指标分类及评分		
C1	相对坡高	>300 m	100~300 m	<100 m
	定量评分	1	0.618	0.382
C2	平均坡度	>45°	25°~45°	<25°
	定量评分	1	0.618	0.382
C3	纵向坡形	凸线型	缓坡陡坡型(含阶状陡坡)	直线状陡坡型
	定量评分	1	0.618	0.382
C4	横向坡形	"凸"形岸(坡)	平直岸(坡)	"凹"形岸(坡)
	定量评分	1	0.618	0.382
C5	基本岩性	具脆性坚硬岩	具脆性软岩	一般软岩土地层
	定量评分	1	0.618	0.382
C6	风化程度	全、强风化岩体	弱风化岩体	微风化岩体
	定量评分	1	0.618	0.382
C7	斜坡岩性组合	软硬相间顺向坡	软硬相间逆向坡	同一岩性斜坡
	定量评分	1	0.618	0.382
C8	斜坡结构及构造	完善结构面	较完善结构面	缺少结构面
	定量评分	1	0.618	0.382

续表 7

评价项目		指标分类及评分		
C9	地下水作用	在基岩面以上或坡脚处地下水呈带状溢出	在基岩面以上或坡脚处地下水有少量溢出	无明显地下水溢出
	定量评分	1	0.618	0.382
C10	库水位升降作用	升降速率>5 m/d	升降速率 2~5 m/d	升降速率<2 m/d
	定量评分	1	0.618	0.382
C11	降水作用	年降水量>800 mm	年降水量 400~800 mm	年降水量<400 mm
	定量评分	1	0.618	0.382
C12	斜坡变形现状	强变形斜坡	中强变形斜坡	弱变形斜坡
	定量评分	1	0.618	0.382

结合鹅公带古滑坡体的现状，表 8 给出了该古滑坡体高速滑坡危险度二级评价指标定量评分值 Ci 及考虑权重后的综合评分值 $Ci\omega_{ci}$，由此可计算出该古滑坡体高速滑坡危险度值为

$$R = \sum_{i=1}^{n} (Ci\omega_{ci}) = 0.788$$

表 8　鹅公带古滑坡体高速滑坡危险度值计算

i	1	2	3	4	5	6	7	8	9	10	11	12	合计
Ci	0.618	0.618	0.618	0.382	0.618	1	1	1	1	1	1	0.618	
ω_{ci}	0.170	0.068	0.068	0.034	0.075	0.039	0.025	0.170	0.077	0.077	0.077	0.122	1.0
$Ci\omega_{ci}$	0.105	0.042	0.042	0.013	0.046	0.039	0.025	0.170	0.077	0.077	0.077	0.075	0.788

据调查和资料统计分析，可将高速滑坡发生的斜坡危险度划分为以下三级：

极危险斜坡 $R>0.7$；危险斜坡 $0.4\leqslant R\leqslant 0.7$；相对稳定斜坡 $R<0.4$。

鹅公带古滑坡体高速滑坡危险度值 $R=0.788$，属于极危险斜坡，说明发生高速滑坡的概率极高，必须考虑相应的滑坡防治方案和措施，避免由于鹅公带这一近坝古滑坡体在水库蓄水运行后发生高速滑坡，给大坝和水库带来不利影响。鹅公带古滑坡体稳定分析表明，在库水位骤降工况下，其安全系数小于 1.0（为 0.95），将产生失稳滑动。采用常规条分法的最大滑速计算结果约 8 m/s，属于高速滑坡，这与层次分析法的结论基本一致，说明采用层次分析法进行库岸高速滑坡预测是可行的。

4　结语

层次分析法是一种定性与定量相结合的多目标决策分析方法。针对库岸高速滑坡发生问题，采用层次分析法可分析确定影响库岸高速滑坡发生各因子的影响权重。对于具体的库岸滑坡体而言，根据其实际状况能够对各影响因子进行定量评分，通过高速滑坡危险度评

价模型,可实现对库岸高速滑坡发生的斜坡危险度的预测和评价。采用该方法能够在不开展大范围地质勘探工作的前提下实现对库岸高速滑坡的初步判定,为进一步的研究和处理工作指明方向。高速滑坡的判定是滑坡治理工作中的一个难点,本文提出的方法为库岸高速滑坡预测提供了一条新思路。

参考文献

[1] 张倬元,王士天,王兰生.工程地质分析原理[M].北京:地质出版社,1997.

[2] 胡杰,王道熊,胡斌.库岸滑坡灾害及其涌浪分析[J].华东交通大学学报,2003,20(5):26-29.

[3] 樊晓一,乔建平,陈永波.层次分析法在典型滑坡危险度评价中的应用[J].自然灾害学报,2004,13(1):72-76.

[4] 樊晓一,乔建平.滑坡危险度评价的地形判别法[J].山地学报,2004,22(6):730-734.

[5] 武立新,王建国,杨计准.层次分析法在山体滑坡危险度评价中的应用[J].有色金属(矿山部分),2009,61(4):66-69.

[6] 许树柏.实用决策方法——层次分析法原理[M].天津:天津大学出版社,1988.

[7] 王成华,孔纪名.高速滑坡发生的危险斜坡判别[J].工程地质学报,2001,9(2):127-132.

【作者简介】 黄锦林,1971年生,男,江西赣县人,博士,教授级高工,从事水利水电工程及防洪减灾研究工作。

新拌混凝土水灰比测定方法试验研究

张　振　王建波　郝志香　袁春波

（天津市水利科学研究院）

【摘要】 在混凝土搅拌过程中，胶凝材料、石、外加剂等材料均可被准确地计量，偏差很小，这种偏差对混凝土质量影响较小，但砂的含水率受货源供应及存储方式的影响，在生产过程中，对混凝土单方用水量影响较大，因此水灰比成为了决定混凝土质量的关键因素。试验表明，新拌混凝土水灰比测定仪设计原理合理、准确度高、操作简单、检测时间短，测试结果能很好的反映混凝土拌合物组分与设计配合比之间的偏差，是混凝土在生产过程控制环节中一种稳定准确的检测方法。

【关键词】 新拌混凝土　水灰比　含气量　测定仪

混凝土是以胶结材料、石子、砂等建筑材料按一定比例搅拌均匀，经过水化硬化形成的人工石材。由于混凝土具有强度较高和耐久性好等特点，混凝土结构被广泛应用于国民经济建设的各个领域，混凝土也成为最重要的建筑材料之一。

目前，在施工现场检测混凝土时，只做坍落度和含气量试验。混凝土质量的传统评价方法是以混凝土标养 28 d 抗压、抗渗、抗碳化、抗冻等试验结果作为评定混凝土质量的依据。实质上，检测结果与工程主体进度相比具有严重的滞后性，一方面不能及时发现生产中存在的不足，对于可能出现的风险或事故，没有采取积极有效的措施进行预防和防范；另一方面无法及时对生产过程进行合理的调整，即使发现问题，对整个施工也是于事无补。可见，混凝土质量的早期测试与评价技术成为混凝土技术发展和质量检测的重要组成部分。

1　新拌混凝土水灰比测定仪测定原理

众所周知，混凝土拌合物是由水、胶凝材料、骨料、外加剂等按照一定的配合比组成并伴有一定空气含量的混合多相人工符合材料。按《混凝土配合比设计规程》(JGJ 55—2000)中体积法设计配合比，容重和含气量的关系见下列公式：

$$\gamma = \frac{\gamma_0}{V_0} = \frac{M_B + M_W + M_S + M_G}{\frac{M_B}{\rho_B} + \frac{M_W}{\rho_W} + \frac{M_S}{\rho_S} + \frac{M_G}{\rho_G} + 0.01\alpha}$$

式中：γ_0 为混凝土质量，kg；V_0 为混凝土体积，m^3；γ 为混凝土体积密度即容重，kg/m^3；M_B、ρ_B 为胶凝材料质量和密度，kg、kg/m^3；M_W、ρ_W 为水质量和密度，kg、kg/m^3；M_S、ρ_S 为砂子质量和密度，kg、kg/m^3；M_G、ρ_G 为石子重量和密度，kg、kg/m^3；α 为混凝土内的含气量(%)。

由此得出图示关系见图 1。

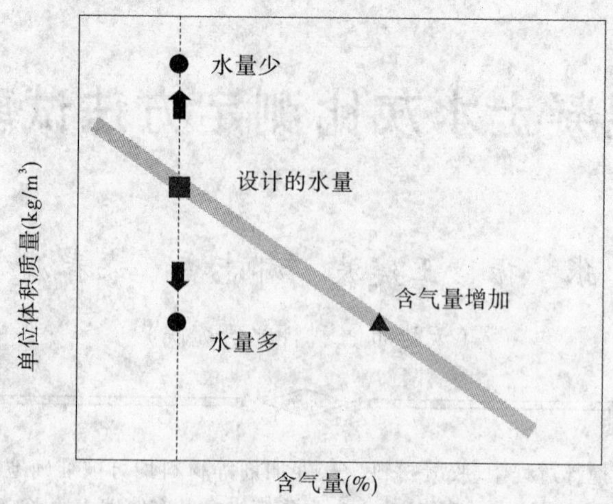

图 1 混凝土拌合物密度、含气量与用水量的关系

通过图 1 可分析到,混凝土拌合物中的总用水量与容重的大小和含气量的多少存在相关关系。而在容重、含气量、水量三个参数中,可以很容易检测出容重和含气量的数值,再通过由大量实验数据回归编制好的软件,计算被检测拌合物中单位体积内的总水量是否在控制范围之内及水灰比与强度之间的关系,从而达到控制混凝土拌合物中关键组分——水的目的。

2 新拌混凝土水灰比测定仪在国内使用中所遇问题分析

新拌混凝土水灰比测定仪 2006 年由北京科源公司引进中国并在公路和铁道工程中得以应用,但在使用中,出现了一些实际检测数据与原始设计数据不相符的结果,其具体原因有以下两种:

2.1 “假定容重法”对当前混凝土的配制已不再有普适性

过去使用的水泥、砂、石的表观密度变化不大,所配制混凝土的容重变化也不大,但是如今普遍使用较大掺量的矿物掺合料,粉煤灰表观密度为 1.90 ~ 2.40 kg/m^3,磨细矿渣表观密度为 2.60 ~ 2.80 kg/m^3,水泥表观密度约为 3.0 kg/m^3,按假定的混凝土表观密度计算,则体积可能会大于 1 m^3,以这样的配合比输入为设计配合比,测试结果必然会产生偏差。因此,在使用仪器时,应输入使用“绝对体积法”设计的 1 m^3 混凝土配合比。

2.2 骨料的含水率和表观密度的取值存在差异

在日本购买的砂子都要求含水饱和以上,进场后储存于封闭仓中,使用前检测骨料饱和面干状态下的表观密度,用以按体积法计算混凝土配合比。因此,使用仪器测试时,砂、石的表观密度应采用饱和面干状态下的测试值,砂、石的含水率是指饱和面干以外的含水率,即应减掉饱和面干的含水率。

3 仪器性能验证

3.1 仪器原理验证试验

为了验证仪器设计原理的正确性和检测的准确性,我们对引进的仪器进行实验室验证,数据分析结果如表 1 所示。

表1 试验结果分析

序号	项目	水灰比	容重(kg/m³)	含气量(%)	用水量(kg/m³)	胶凝材料(kg/m³)
1	偏差值	0.005	-4	0.05	1.5	-1
	偏差率(%)	1.19	-0.17	1.25	0.94	-0.26
2	偏差值	-0.008	6	-0.08	-2.7	0.3
	偏差率(%)	-1.90	0.26	-2.00	-1.69	0.08
3	偏差值	-0.009	9	-0.15	-3.2	1
	偏差率(%)	-2.14	0.39	-3.75	-2.00	0.26
4	偏差值	-0.007	5	-0.05	-2.7	0.2
	偏差率(%)	-1.67	0.22	-1.25	-1.69	0.05
5	偏差值	-0.013	7	-0.04	-4	0.8
	偏差率(%)	-2.89	0.30	-1.00	-2.50	0.22
6	偏差值	-0.009	7	-0.11	-2.8	0.4
	偏差率(%)	-2	0.30	-2.75	-1.75	0.11
7	偏差值	0.006	-6	0.12	1.6	-1.3
	偏差率(%)	1.32	-0.26	2.91	0.99	-0.37
8	偏差值	0.007	-8	0.17	1.9	-1.4
	偏差率(%)	1.56	-0.34	4.25	1.19	-0.39
9	偏差值	0.003	-4	0.09	0.5	-0.8
	偏差率(%)	0.63	-0.17	2.25	0.31	-0.24
10	偏差值	-0.009	8	-0.13	-3	0.4
	偏差率(%)	-1.88	0.34	-3.25	-1.88	0.12

由试验结果分析可见：

(1)用水量是影响混凝土性能的关键因素，此次检测结果总用水量最大偏差-4 kg/m³，偏差率-2.5%，综合分析抗压强度等因素，偏差率不到±3%的水对强度没有实质性的影响，从对比值分析该仪器准确性满足使用要求。

(2)混凝土强度是建筑构件承载力的基本保证。此次检测结果水灰比最大偏差-0.013，偏差率-2.89%，而引进的水灰比测定仪计量偏差为±5%，说明该仪器的检测精度满足要求。

(3)检测容重是分析混凝土拌和物中各种材料在单位体积下砂浆与骨料之间的位置与分布是否合理，现检测结果每种材料占总量中的比值容重偏差最小-0.34%，说明与实际输入值基本吻合，证明仪器的可靠性。

(4)合理的空气含量不仅能改善混凝土拌和物的施工性，还能改善结构件的外观质量，从而提高混凝土表面抗风化的能力，而且有利于抵抗不同原因引起的膨胀型的腐蚀。此次检测最大偏差率+4.25%。

3.2　现场试验

该仪器主要是检测混凝土在正常生产供应中,所以现场试验选择了天津市南水北调混凝土工程,其结果见表2。

表2　双掺混凝土实际检测值(P·O42.5)

水灰比	容重 (kg/m^3)	含气量 (%)	用水量 (kg/m^3)	胶凝材料 (kg/m^3)	28 d 强度 (MPa)	坍落度 (mm)
0.430	2 314	4.21	162.9	379.1	35.2	192
0.422	2 320	4.15	160.2	379.9	37.5	184
0.406	2 336	3.87	154.9	382.0	43.4	180
0.414	2 330	3.92	157.8	381.2	38.8	182
0.409	2 337	3.75	156.3	382.1	40.1	180
0.429	2 321	3.95	162.9	380.0	35.5	190
0.424	2 319	4.14	161.0	379.8	37.7	188
0.435	2 315	4.05	165.0	379.3	35.0	195
0.430	2 310	4.37	162.7	378.6	35.4	190
0.414	2 332	3.85	157.7	381.5	38.2	187
0.407	2 340	3.68	155.6	382.5	39.5	183
0.413	2 327	4.06	157.3	380.8	38.6	185
0.428	2 318	4.09	162.5	379.7	35.8	193
0.428	2 322	3.93	162.6	380.2	36.0	194
0.424	2 317	4.21	161.1	379.5	37.0	187
0.407	2 328	4.17	154.9	381.0	41.6	182
0.408	2 336	3.82	155.8	382.0	42.7	181
0.428	2 314	4.26	162.1	379.1	36.3	193
0.412	2 335	3.77	157.2	381.9	38.4	187
0.420	2 322	4.11	159.6	380.2	37.1	190

分析主要参数

(1)坍落度分析,被检拌和物均能满足施工泵送工艺要求(基本要求)。

(2)实测28 d抗压强度结果均超出委托设计检测值,平均值为37.99 MPa,最低值为35.0 MPa(达到设计值的117%以上),最高值为43.4 MPa,标准差为2.58 MPa,整体性稳定。

(3)既然坍落度都能满足要求,单从抗压强度指标分析,实际投入的水最好小于设计值,这样强度保证率会高。分析总用水量数据,平均用水量为159.5 kg/m^3,平均值小于设计值0.5 kg/m^3,说明作为强度较高的混凝土配合比主要指标控制在满意状态下。

3.3 综合分析

(1)所检拌和物坍落度在 180 ~200 mm,和易性基本满足要求,符合施工技术要求。

(2)检测容重最大值波动在-0.7% ~ +0.7%,总体分析混凝土拌和物浆骨分布基本均匀,计量准确,总用水量在控制范围之内。

(3)含气量对于普通泵送混凝土的作用主要是增加润滑、减少摩擦、提高可泵性。从实际检测坍落度结果均在 180 mm 以上,说明没有影响拌和物的可泵性,施工可振性均能满足工艺要求,而 *R*28 的平均强度值最小也达到设计值的 127%,所以含气量偏大 10% 以内不会降低抗压强度。

(4)分析总用水量数据,平均用水量 160 kg/m^3 与设计值相同,最大用水量超出设计值 5.4 kg/m^3,但都能超出设计强度等级,所以在线控制以 6 kg/m^3 为偏差强度等级能够满足设计要求。

(5)在以上被检拌和物现状基础上成型的试块,标准养护 28 d 后,全部满足委托设计强度指标,最低达到设计值的 117%,平均达到设计值 127% 以上。

4 结语

(1)对新拌混凝土主要组分进行分析,以新拌混凝土的水灰比判断其浇筑成型后一系列性能是否能够满足设计要求,从而可避免很多工程质量问题甚至是质量事故的发生和纠纷。

(2)新拌混凝土水灰比测定仪设计原理合理、准确度高、操作简单、检测时间短,测试结果能很好地反映混凝土拌合物组分与设计配比之间的偏差,既为混凝土生产企业在生产过程控制环节中提供了一种稳定准确的检测设备;也为建筑工程混凝土结构施工提供了便捷可靠的验收检测手段,从源头上保证了混凝土结构的施工质量。

(3)用新拌混凝土容重与含气量的检测值推定水灰比,相对于以 28 d 强度为基础的评判混凝土质量的方法具有可控性,同时对于混凝土拌和物的质量保证性更具有实际性。

参考文献

[1] 徐福春,倪宇光,沈淑红. 浅谈混凝土强度与水灰比的关系[J]. 水利天地,2002(6):47.

[2] 王忠德,等. 实用建筑材料试验手册[M]. 2 版. 北京:中国建筑工业出版社,2003.

[3] 普通混凝土配合比设计规程[J]. 北京:北京建筑工业出版社,2001.

[4] 韩小华,李玉琳. 新拌混凝土单位用水量快速测定方法的试验研究[J]. 科技导航,2008,8(4),42-47.

[5] 廉慧珍,李玉琳. 当前混凝土配合比"设计"存在的问题[J]. 混凝土,2009,3:1-5.

【作者简介】 张振,1976 年 10 月生,学历:硕士研究生,于 2002 年 3 月毕业于河北工业大学,学位:硕士,工作单位:天津市水利科学研究院、职务:水工程研究所所长、职称:高级工程师。

北方寒冷地区碾压混凝土重力坝病害缺陷及修补对策

夏世法[1] 阿孜古丽[2]

(1.中国水利水电科学研究院 北京 100038;
2.新疆水利水电勘测设计研究院 乌鲁木齐 830000)

【摘要】 由于北方寒冷地区独特的气候条件,在运行一段时间以后,碾压混凝土重力坝往往存在诸如裂缝、渗漏、冻融剥蚀等老化病害现象,这将严重影响大坝的安全性和耐久性,必须加以处理。本文总结了北方寒冷地区碾压混凝土重力坝的老化病害及其成因,并提出了相应的修补对策,以供类似工程参考。

【关键词】 寒冷地区 RCC重力坝 病害 修补

1 北方寒冷地区碾压混凝土重力坝的建坝情况

辽宁观音阁碾压混凝土坝是我国在北方严寒地区修建的第一座碾压混凝土高坝,是我国首次引入日本的RCD技术而修建的大型水利工程,也是当时世界上规模最大、碾压混凝土方量最多的工程。迄今为止,我国在寒冷地区已建的碾压混凝土重力坝共有11座:观音阁、桃林口、松月、阎王鼻子、玉石、满台城、和龙、白石、特克斯山口、喀腊塑克、冲乎儿,其中7座位于东北地区,3座位于新疆地区,1座位于河北。目前,丰满大坝重建也已选定碾压混凝土重力坝坝型。

2 主要的病害缺陷

2.1 裂缝

2.1.1 温度裂缝

我国北方寒冷地区的气候往往具有冬季严寒,夏季炎热的特点,气温年较差较大,日较差明显,这些气候特点加上碾压混凝土重力坝通仓浇筑、分层摊铺碾压、越冬间歇等施工特点,使寒冷地区碾压混凝土重力坝温度场和应力场的分布具有独特的规律。大坝防裂对坝体材料和温度控制的要求很高,坝体极易出现裂缝。从国内已建工程出现的裂缝来看,北方寒冷地区碾压混凝土重力坝出现的温度裂缝可归结为以下5类:①上、下游坝面的劈头裂缝;②强约束区长间歇顶面(包括越冬面)的纵向裂缝;③永久底孔、导流底孔四周的环形裂缝;④溢流坝反弧段的纵向裂缝;⑤越冬层面附近上、下游侧水平施工缝的开裂。在上述5种裂缝中,上、下游坝面的劈头裂缝属于横缝,这种裂缝的危害在于它既破坏了坝体的整体性,又破坏了坝体的防渗性.第②、③、④三种裂缝统属纵缝,其中底孔环形裂缝和反弧段裂缝位于高速水流过水断面上,很可能发展成深层或贯穿性裂缝,危害严重。第⑤种裂缝即越冬面水平缝,越冬面水平缝是北方寒冷地区大坝因为冬季停浇越冬而导致的独特的温度裂缝。大坝冬季长间歇后,来年开春又浇筑新的混凝土,由于新、老混凝土的上、下层温差及老混凝土对新混凝土强约束,在越冬水平面附近会产生较大的温度应力,从而导致越冬水平面张开。越冬面水平缝上、下游侧可分别沿着坝轴线方向贯穿于整个越冬层面,破坏坝体的防渗结构,影响坝体的整体稳定性,对坝体的危害较大。

2.1.2 水平施工缝

由于碾压混凝土筑坝采用分层摊铺碾压的施工工艺,以往研究表明:水平施工缝抗拉强度只有整浇混凝土的41% ~86%,如果处理不当,水平施工缝就会变成水平裂缝,进而形成水平缝渗水。

2.1.3 其他裂缝

如结构缝、灌浆抬动裂缝等。

2.2 渗漏

坝体渗漏的通道主要包括以下4个方面。

2.2.1 裂缝形成的渗水通道

比较典型的为大坝的劈头裂缝、越冬面水平缝贯穿上、下游而引起的渗漏,西北某RCC重力坝下游面劈头裂缝渗漏及挂冰如图1所示。

图1 西北某RCC重力坝下游面劈头裂缝渗漏及挂冰

2.2.2 水平施工缝形成的渗水通道

碾压混凝土分层浇筑,如碾压层面处理不当,则会形成水平施工缝。水平施工缝渗漏,虽然对库内存储的水量损失影响不大,可是沿渗水的层面对坝体进行侵蚀,削弱坝体在该层面的抗剪强度,影响大坝的稳定性和安全性。因此,层间渗水危害最大。

2.2.3 坝体施工薄弱部位形成的渗水通道

坝体施工的薄弱部位主要是异种混凝土结合处,如廊道周围的碾压混凝土与变态混凝土结合部位(图2)。在观音阁水库施工期间,上游灌浆廊道沿层缝有几处渗水,经过分析,这些水是养生水从常态混凝土与碾压混凝土结合不良的部位渗入,又从廊道周围常态混凝土的水平缝渗到廊道内的。

图 2　廊道周围碾压混凝土及变态混凝土施工

2.2.4　伸缩缝止水失效

北方寒冷地区修建的碾压混凝土重力坝伸缩缝止水容易出现问题(见图 3),主要原因是严寒地区年内气温变化剧烈,根据新疆某 RCC 重力坝现场实测资料,伸缩缝开度年内变化幅度在 1 ~2 cm,在夏季时开度较小,在冬季时开度较大。如此大的开度变化将导致伸缩缝止水部分破坏,从而出现渗水现象,而伸缩缝一旦出现渗水,在冬季时结冰膨胀又会进一步加剧止水破坏,最后导致伸缩缝漏水。

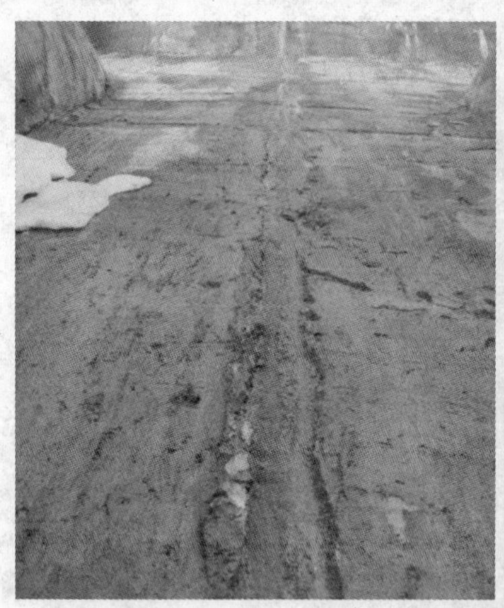

图 3　某大坝伸缩缝渗水照片

2.3　冻融剥蚀破坏

北方寒冷地区的碾压混凝土重力坝,冻融剥蚀主要出现在上游坝面水位变化区、大坝溢流面及下游坝面。由于冬季气温较低,坝体一旦出现渗漏现象,则会导致冻融剥蚀见图 4。

2.4　溢流面空蚀破坏

一些实际工程调查结果表明,凡泄水建筑物表面有明显的凹凸不平,当过流流速大于 15 ~20 m/s 时,一般都有可能在其下游发生空蚀破坏。流速愈大,发生破坏的可能性也愈大。某些试验表明,空蚀破坏强度与水流速度的 5 ~7 次方成正比。北方寒冷地区碾压混凝

图4　东北满台城大坝上游面水位变化区冻融剥蚀照片

土溢流面错台、蜂窝、麻面、冻胀隆起等会导致表面不平整，高速泄水时出现空蚀现象，并进一步导致冲蚀破坏。

3　修补对策及工程实例

3.1　裂缝的修补

裂缝修补目的有两方面：一是恢复结构的整体性，二是堵漏防渗。大坝裂缝的修补方法见表1。

表1　不同大坝裂缝的处理方法

修补加固方法	适宜修补的裂缝
内部修补	削弱结构整体性、强度、抗渗能力和导致钢筋产生锈蚀的裂缝
表面修补	浅层或表现裂缝，有防风化、防渗漏、抗冲磨要求的裂缝
预应力锚固	整体受力的裂缝、危及建筑物安全运用和正常功能发挥的裂缝
内外联合修补法	建筑物迎水面裂缝

上静水坝位于美国犹他州杜克森（DuChesne）北部72.42 km处，该坝采用碾压混凝土技术施工，于1987年8月竣工。1988年6月4日，水库开始蓄水时，便在25+20坝段的基础廊道发现了一条贯穿缝。当水库继续蓄水时，裂缝变宽且产生了不容许的渗漏现象，水渗入基础廊道并从下游坝面渗出，水库蓄水至最高位时，经测量，裂缝宽度约有6.6 mm，渗入基础廊道的水量约4.9 m^3/min，从下游坝面裂缝中渗出的水量约6.6 m^3/min。最终裂缝从基础扩展至坝顶，从上游坝面扩展到下游坝面。坝基变形和混凝土冷却后产生的温度应力是形成大坝裂缝的原因。最后采用裂缝灌浆的方式对裂缝进行处理。灌注裂缝的树脂选用软性亲水聚氨酯树脂，所选用的方法是：分三个基本步骤用聚氨酯树脂灌注裂缝，第一、第二步的工作都在廊道和水库水位高程以下的下游面进行，第三步即最后一个阶段的工作是用浮船或从坝顶悬吊星形平台对上游坝面进行施工，对位于水库水位以上的裂缝带进行灌脂。经过上述处理后，裂缝的渗流量减少到3 m^3/min以下。

观音阁水库大坝采用碾压混凝土RCD筑坝技术，拦河坝段为碾压混凝土重力坝。坝顶长1 040 m，共分65个坝段，坝体结构采用“金包银”形式，即上游3.0 m防渗层，下游2.5 m

保护层和2.0 m基础垫层为常态混凝土，廊道周围采用1.0 m厚钢筋混凝土，其余为含粉煤灰30%～35%的干贫碾压混凝土。大坝于1990年5月正式开始兴建、1995年9月大坝主体工程结束。

据1996年4月的调查结果，坝体裂缝共326条。上游坝面共有裂缝53条，除2条裂缝外，其余均为水平施工缝，且主要集中在1991～1994年3个越冬结合面(约209 m、218 m、233 m高程及其上下两层范围内，每层0.75 m。缝长大小不等，缝宽0.2～3 mm，最大缝深达6.0 m)，裂缝贯穿多个坝段，如233 m高程越冬面水平裂缝贯穿42个坝段，长度约640 m。观音阁大坝不同裂缝的处理方案见表2。

表2 观音阁大坝不同裂缝的处理方案

序号	部位及形态	宽度(mm)	长度(mm)	处理方法
1	侧面	>0.25		方案1:缝上贴3 mm厚橡胶板，埋入灌浆管并引至指定高程后灌浆
2	平面	≤0.25	>2 500	方案3:重要部位扣管，铺单层浇筑常态混凝土 方案6:一般部位扣管浇筑常态混凝土
		>0.25	≤2 500	方案4:骑缝凿槽浇筑常态混凝土
		>0.25	>2 500	方案2:重要部位扣200 mm半圆管，铺双层钢筋，浇常态混凝土，后灌水泥浆 方案7:一般部位扣管，铺双层钢筋浇筑常态混凝土 方案5:孔洞底板或顶部平面裂缝凿槽，扣200 mm半圆管，铺三层钢筋后灌水泥浆
3	平面深层裂缝或上游劈头缝的平面部分			方案8:凿槽，回填砂浆，钻缝头孔，铺双层钢筋，浇常态混凝土灌浆
4	底孔环缝底孔缝			方案9:凿槽埋管、回填塑性嵌缝材料，PCC砂浆止缝，磨细水泥灌浆或聚氨脂化灌浆
5	上游水平缝和上游劈头立面部分			方案10:沥青混凝土防渗面板

其中，越冬面水平缝采用了粘贴T1密封带、三元乙丙卷材并辅以锚固措施，处理后渗漏量降低了70%，效果明显。

3.2 渗漏的处理

对大坝渗漏的处理，除了针对裂缝表面封闭、灌浆等，如大坝水平施工层面渗漏较严重，目前采取的方法有两种。

3.2.1 从坝顶靠近上游面部位打深孔灌水泥浆进行防渗

福建的水东大坝，大坝最大坝高63 m，坝顶宽8 m，于1991年4月开工，自1993年11月下闸蓄水后，大坝廊道便开始出现漏水，并随着水位的升高漏水量不断增大，整个下游面呈湿润状态，部分区域可见明显的漏水点。后经研究决定在坝顶距上游面2.5 m处钻灌一排防渗帷幕，灌浆孔布置：左岸挡水坝段灌浆孔孔距为500 mm，右岸挡水坝段及泄洪闸坝段灌浆孔孔距400 mm。左、右岸挡水坝段补强灌浆孔，采用两排布置。溢流坝段补强灌浆孔采用单排布置。灌浆孔从坝顶一直深入到基岩面以下 3 m,灌浆孔孔径为 75 mm 。灌浆方法采用自下而上分段孔口封闭、孔内循环式灌浆,灌段长 2.5 ~ 3.0 m。灌浆材料选用优质的普通硅酸盐水泥,要求水泥标号不低于 42.5。水东大坝补强灌浆于 2002 年 10 月开始施工,2004 年 7 月全部完成。根据现场运行观测,大坝渗漏量已达到低温期 $Q_{漏} \leqslant 1$ L/s 控制标准,表明水东大坝补强灌浆后已满足正常坝的要求。

东北延吉满台城水电站,1997 年开始发电。因坝体碾压混凝土水平层面尤其是越冬面出现渗漏,在 2000 ~ 2001 年也采用从坝顶打深孔灌水泥浆的方法进行了处理,处理后效果较好,越冬水平面基本不再渗水,并消除了越冬面附近下游坝面挂冰的现象。

3.2.2 高分子卷材或防渗涂层对上游坝面进行封闭防渗

近 20 年来,随着碾压混凝土筑坝技术、老坝除险加固技术的发展,混凝土坝表面柔性防渗技术取得了较大进步。根据 A. M. Scuero 等的统计,世界范围内 27 座 RCC 坝采用 PVC 薄膜防渗,其中 47% 在设计阶段采用,53% 是在后期修补过程中采用 PVC 薄膜进行修补。27 座采用薄膜防渗的 RCC 坝中,有 10 座在美国。目前,采用坝面薄膜防渗最高的 RCC 混凝土坝是哥伦比亚 188 m 高的 Miel I 坝,该坝坝高 188 m,2002 年投入运行。选用 Carpi 的 PVC 土工膜,薄膜采用 Carpi 的专利技术进行连接和安装固定。蒙古 52 m 高的 taishir RCC 大坝,位于距首都乌兰巴托 1 000 km 远的偏远地区,那里气温范围是-50 ~ +40 ℃,年平均气温 0 ℃,气候条件严酷。为了解决大坝出现温度裂缝后的渗水问题,设计者选择了一个不透水的上游面 PVC 土工膜护面,它可以降低极低温度对大坝的损伤并加强大坝在极端气候条件下的水密性。泰西尔 RCC 坝防渗膜的安装比较理想地解决了坝体的防渗问题,截至 2007 年冬季,库水位上升稳定至 1 677.10 m(达到最高蓄水位的 45%),坝后没有发现任何渗漏。

我国大坝工程建设中也开发应用了多种防水技术。例如,温泉堡为了防止渗漏在上游坝面铺设了 4 000 m^2 PVC 复合土工膜;130 m 高的沙牌碾压混凝土拱坝在死水位以下部位使用了二布六涂弹性防水涂层等,但我国在薄膜连接、排水细部构造上需要继续完善。聚脲防渗技术自 2000 年左右引入国内水利界以来,近年来发展迅速,已在多个水利工程中进行了应用。小湾高拱坝坝基以上 30 m 基础强约束区采用喷涂双组分聚脲的方式进行防渗及防止高压水对裂缝的劈裂作用,2008 年及 2009 年分两年施工,共计喷涂约 2 万 m^2,目前监测资料表明其防渗效果良好。

3.3 冻融剥蚀的处理

在北方寒冷地区,碾压混凝土大坝冻融剥蚀表现为混凝土表面剥落、深层冻胀破坏及冰冻裂缝等三种形式。

混凝土冻融剥蚀的修补应根据剥蚀部位、剥蚀程度、剥蚀深度等选择合适的修补材料,通常采用聚合物砂浆及聚合物混凝土,其配合比应经试验确定,并考虑掺入纤维、外加剂等增加其抗冻性能。施工时,宜采用合适的界面剂,必要时应在混凝土基面设置插筋。

另外,可考虑在混凝土基面上覆盖涂层(如聚脲或其他封闭材料)来提高其抗冻性,其主要机制是利用材料的高抗渗性,杜绝外水进入混凝土内部,从而提高其抗冻性。

3.4 溢流面空蚀破坏处理

目前,对溢流面空蚀破坏的处理主要分为无机材料及有机材料。

无机抗磨蚀材料包括普通高强混凝土及砂浆、高强硅灰混凝土、高强硅灰铁矿石混凝土及砂浆、高强硅灰铸石混凝土及砂浆、高强硅灰钢纤维混凝土等。

有机材料有环氧树脂砂浆及混凝土、不饱和聚酯树脂砂浆及混凝土、丙烯酸环氧树脂砂浆及混凝土、聚氨酯砂浆(混凝土)等 。

纤维增强混凝土是目前溢流面抗冲蚀材料的发展趋势,目前常用的是钢纤维,随着纤维技术的发展,高强高弹模(初始模量大于5 000 MPa)的聚丙烯纤维在水工混凝土中开始应用(在巴家嘴水库溢洪道工程中的运用)。目前,纤维增强混凝土在应用中还存在着一些问题,一是生产过程中纤维不易在混凝土中均匀分散而易缠绕成团,影响了混凝土的性能及和易性;二是具有较好增强效果的纤维价格较高,增加了混凝土的成本。

4 结论及建议

(1)在严寒地区修建碾压混凝土重力坝,由于其独特的气候及坝体施工工艺特点,极易出现裂缝、渗漏、冻融剥蚀、溢流面空蚀等缺陷,这些缺陷将严重影响大坝的安全性和耐久性,必须引起重视。

(2)裂缝、渗漏、冻融剥蚀、溢流面冲蚀等缺陷之间具有一定的联系,如裂缝可导致坝体的渗漏,渗漏会导致坝体混凝土的冻融剥蚀,而溢流面的冻胀又会引起严重的冲蚀。因此,在选定大坝的修补措施时,必须统筹兼顾,综合考虑。

(3)无论是对何种缺陷的处理,在选择修补材料和修补技术时,应考虑其施工条件和周围的环境。施工条件包括:①基层的湿度情况;②施工时的温度;③通风情况;④施工操作空间;⑤修补位置(垂直,水平);⑥施工周期;⑦修补区域厚度(不同厚度要求不同的修补材料系列);⑧边缘接合形式等。修补处理时还需要考虑建筑物周围的一般环境(例如有无侵蚀性液体、气体等),需要考虑材料表面的微观环境和材料内部的微观环境以及材料对基层环境条件的要求等。

(4)碾压混凝土重力坝水平碾压层面可能会出现裂缝、渗漏等问题是其区别于常态混凝土坝的一个重要特征。在大坝的上游面专门设置防渗结构(如土工膜、聚脲涂层等)可有效解决这个问题。但如在大坝运行期进行防护,需要放空水库,如能在设计阶段即充分考虑坝体混凝土和防渗结构各负其责、体现坝体功能分开的原则,可大大降低大坝运行期病害的发生,此问题应引起各方关注并需继续深入研究。

(5)由于剧烈的气温变化,寒冷地区碾压混凝土重力坝伸缩缝年内开度变化大,容易引起止水失效从而导致渗漏,应进一步研究改进横缝止水形式。

参考文献

[1] 王永存. 寒冷地区碾压混凝土坝几个问题的探讨[J]. 东北水利水电,1997(04).

[2] 王成山. 严寒地区碾压混凝土重力坝温度应力研究与温控防裂技术[D]. 大连:大连理工大学,2003.

[3] 宋玉普,魏春明. 混凝土施工缝接缝面劈拉强度试验研究[J]. 混凝土,2006(6).

[4] W GLENN SMOAK. 上静水坝的裂缝修补方法[J]. 水电建设,1992(02).

[5] 黄柏洪,陈俊喜. 观音阁水库大坝裂缝处理[J]. 吉林水利,2005(08).

[6] 王国秉,鲁一晖,等. 福建水东水电站大坝补强加固的工程经验[J]. 山西水利科技,2007(02).

[7] 李新宇. 高拱坝坝踵防渗措施及高重力坝水力劈裂研究[R]. 北京:中国水利水电科学研究院.

【作者简介】 夏世法,男,1976年8月生,山东高密人,现为中国水科院高级工程师,主要从事碾压混凝土坝温控仿真分析,水工混凝土建筑物的无损检测、评估及修补加固工作。

南水北调中线工程总干渠渠道衬砌嵌缝材料选择浅谈

李宁博　吴剑疆

（水利部　水利水电规划设计总院　北京　100120）

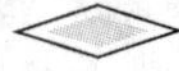

【摘要】 南水北调中线一期工程是解决我国北方地区严重缺水问题的特大型跨流域调水工程，属于特大型供水渠道，渠道衬砌嵌缝材料的选择对渠道运行安全和工程投资影响较大，而已有的规范并不能完全适应其特点。为探讨衬砌嵌缝材料选取的合理性，本文根据南水北调工程特点，参考国内外其他工程实例，结合有关计算、分析和工程类比，对南水北调中线工程渠道衬砌嵌缝材料的选取进行了分析研究。

【关键词】 南水北调　渠道　嵌缝材料

南水北调中线一期工程是解决我国北方地区严重缺水问题的特大型跨流域调水工程。工程由水源工程、输水工程和汉江中下游治理工程三大部分组成，其中总干渠主要采用明渠输水型式，渠线全长约1 276 km，梯形明渠长约1 105 km。为满足总干渠输水需要，中线一期工程总干渠全线采用全断面混凝土衬砌。土质渠道边坡混凝土衬砌厚10 cm，底板混凝土衬砌厚8 cm，均为现浇混凝土，下设复合土工膜防渗，现浇混凝土板衬砌一般间隔4 m设一道纵、横伸缩缝，缝宽为1～2 cm，缝上部2 cm为聚硫密封胶或聚氨酯密封胶嵌缝，下部为闭孔泡沫板填缝。石质渠道段过水断面部分一般采用模筑减糙混凝土衬砌，厚度为20 cm或25 cm，分缝型式同土质渠道。由于南水北调中线工程总干渠为特大型供水渠道工程，水深大，衬砌板承受荷载较大，沿线渠基地质条件变化也大，衬砌嵌缝材料选取是否合适对总干渠渠道运行安全和工程投资影响较大。

1　嵌缝材料的作用及要求

伸缩缝是混凝土防渗渠道可能漏水的通道之一，是渠道防渗工程产生病害的第一发生点，从伸缩缝中渗入渠基的水分不仅会造成水量损失，而且使渠基土可能处于饱和状态，易引起渠体塌陷、破坏。在北方寒冷地区，还会发生由于渠基土冻胀而引起的渠床冻胀破坏。嵌缝材料的作用一是为了杜绝或减少由渠道渗入渠床而造成水量流失；二是为了在混凝土衬砌板下地基出现不均匀变形或由于温度变化造成混凝土变形时，使衬砌板能够适应变形。

以往，渠道嵌缝材料多采用沥青砂浆、油毡涂沥青等材料，这类材料造价较低，但因其适应温度的变形性能差，与混凝土黏结不好，气温高时，缝宽缩窄，填料被挤出，气温低时，填料不能复位，因而形成渗水缝。对嵌缝材料的要求，除具有一定的防渗性能外，还必须具有较好的耐热性和抗冻性，以保证其在高温时不流淌或被挤出，低温时不开裂，抗老化性能好；另外还必须与混凝土基面有良好的黏结，回弹性优良，以保证在变形时黏结面不被拉开。由于渠道伸缩缝在承受不同的水压、不同的温度条件下会产生张合作用，尤其是渡槽的伸缩缝其变形位移相对较大，对止水材料的要求非常高，再加上渠道中的水流都有一定的流速，因此对于伸缩缝的嵌缝材料要求具有一定的抗冲刷性能。老的渠道常采用沥青或沥青砂浆作为止水材料，因为其造价相对较低。但这种材料在高温时易流淌或被挤出，故目前的渠道衬砌嵌缝材料中一般不采用这种材料。

对南水北调工程嵌缝密封材料来说，除满足上述嵌缝材料特性外，还应不含焦油、无毒无挥发性溶剂，不溶于水和不渗水，并具有优良的耐冻融性及耐水、耐化学介质性等特点，施工技术简便且接近混凝土本色，美观、经济的填缝材料，适合渠道输水、饮用水等系统的防渗材料是本次选材的重要依据。

2 规程规范的规定及其适应性分析

对渠道衬砌分缝设计，目前国内可参考的规范有《灌溉与排水工程设计规范》（GB 50288—99）、《渠道防渗工程技术规范》（SL 18—2004）、《水工建筑物抗冰冻设计规范》（SL 211—98）、《聚硫、聚氨酯密封胶给水排水工程应用技术规程》（CECS 217:2006）和《水工建筑物塑性嵌缝密封材料技术标准》（DL/T 949—2005）。各规程规范中关于缝间距和缝宽的规定汇总见表 1。

表 1 规程规范中关于嵌缝材料选取和适用范围规定汇总

规范名称	嵌缝材料选取规定	适用范围
《灌溉与排水工程设计规范》（GB 50288—99）	焦油塑料胶泥	灌溉、排水工程
《渠道防渗工程技术规范》（SL 18—2004）	焦油塑料胶泥、聚硫密封胶、聚氨酯密封胶	灌溉、排水、供水工程
《水工建筑物抗冰冻设计规范》（SL 211—98）	黏结力强、变形性能大，在当地最高气温下不流淌、最低气温下仍具有柔性	抗冰冻水工建筑物
《聚硫、聚氨酯密封胶给水排水工程应用技术规程》（CECS 217:2006）和《水工建筑物塑性嵌缝密封材料技术标准》（DL/T 949—2005）	聚硫、聚氨酯密封材料	市政工程

2.1 《灌溉与排水工程设计规范》（GB 50288—99）

该规范主要适用于灌溉和排水渠道。南水北调中线工程为特大型供水渠道，供水保证率高，对水质要求也高，且需长期不间断运行，和灌排及排水渠道相比有较大区别。衬砌混凝土板的设计应做到安全可靠，在长期运行下衬砌混凝土板不开裂；嵌缝材料需要有良好的力学性能、耐久性及使用寿命。因而，该规范不能完全适用于南水北调中线工程总干渠的特点，只可参考。该规范推荐的嵌缝材料采用焦油塑料胶泥。南水北调中线工程目前嵌缝材料采用聚硫密封胶，和该规范相比，标准有一定程度提高。考虑到南水北调中线工程的重要性，这是合理的。

2.2 《渠道防渗工程技术规范》（SL 18—2004）

该规范和《渠道防渗工程技术规范》（SL 18—91）相比，在条文说明中增加了聚硫密封胶和聚氨酯密封胶的选择，但考虑到聚硫密封胶和聚氨酯密封胶价格较高，规范中还是推荐采用焦油塑料胶泥等材料。该规范和《灌溉与排水工程设计规范》(GB 50288—99)相比,适用范围增加了供水渠道,嵌缝材料选择的规定更为详细,标准也更高,对南水北调中线工程总干渠的适应性更好。

2.3 《水工建筑物抗冰冻设计规范》(SL 211—98)

该规范对衬砌分缝间距规定和《渠道防渗工程技术规范》(SL 18—2004)基本一致,嵌缝材料要求其“黏结力强、变形性能大,在当地最高气温下不流淌、最低气温下仍具有柔性”,但未明确推荐采用哪种材料。该规范主要针对水工建筑物的抗冰冻来对分缝设计和嵌缝材料进行规定,对黄河以北渠道分缝设计和嵌缝材料设计有一定参考价值。

2.4 《聚硫、聚氨酯密封胶给水排水工程应用技术规程》(CECS 217:2006)和《水工建筑物塑性嵌缝密封材料技术标准》(DL/T 949—2005)

这两个规范主要针对在衬砌材料已选定密封胶的情况下从材料配置、施工工艺及试验测试等方面提出的技术要求。其中,《聚硫、聚氨酯密封胶给水排水工程应用技术规程》(CECS 217:2006)规定国内可以见到的聚硫、聚氨酯密封材料用于给排水工程的唯一标准,一般适用于市政工程。

从上述分析可知,目前已有规范可作为南水北调中线工程参考,但还不能完全适应南水北调中线工程总干渠的特点和要求。为论证嵌缝材料选择的合理性,需结合南水北调工程特点,对嵌缝材料做进一步分析。嵌缝材料的选择需做全面的技术经济比较。

3 工程实例及分析

南水北调中线一期工程总干渠规模大,在国内外很难找到与其规模大小相近且任务以供水为主的工程实例。国内规模比较大的渠道中,除南水北调东线工程外,主要为过去已建的一些以灌溉为主的大型灌区的干渠,建设年代都比较久远,标准都比较低。近年来,国家投资对这些渠道进行了改造,对防渗衬砌结构进行了整修,标准有所提高,但和南水北调中线工程总干渠相比,规模和标准还是有一些差距。国内部分规模较大的渠道衬砌分缝设计实例见表2。

表2 部分大型供水渠道衬砌嵌缝材料统计

序号	工程名称	建筑物级别	设计流量(m^3/s)	渠道水深(m)	衬砌结构型式	嵌缝材料	备注
1	南水北调中线一期引江济汉工程	1	350	6	现浇混凝土	对需防渗渠段上部采用聚氨酯密封胶,下部采用闭孔塑料泡沫板	在建
2	南水北调东线一期工程济平干渠工程	1	50		现浇混凝土	缝口2 cm采用高分子黏合剂,下部采用闭孔泡沫塑料板	已建

续表 2

序号	工程名称	建筑物级别	设计流量(m^3/s)	渠道水深(m)	衬砌结构型式	嵌缝材料	备注
3	南水北调东线一期济南—引黄济青段输水明渠段工程	1	50	3.0~3.2	现浇混凝土	上部采用 2 cm 聚硫密封胶,下部采用闭孔泡沫板	初设
4	南水北调东线一期工程鲁北段小运河输水渠	1	50	3.4~5.4	现浇混凝土	上部采用 2 cm 塑料胶泥,下部采用闭孔泡沫板	初设
5	北疆供水工程某干渠	2			预制混凝土板	顶部 3 cm 采用聚氨酯砂浆嵌缝,下部采用水泥砂浆勾缝	已建
6	新疆伊犁河北岸干渠	3	60	5	预制混凝土板	聚氨酯砂浆	可研
7	新疆恰甫其海二期南岸干渠	2	74	6	预制混凝土板	上部采用 3 cm 水泥砂浆,下部采用 3 cm 聚氨酯砂浆	在建
8	吉林省松原灌区引水总干渠工程	2	175	3.66~5	现浇混凝土	聚乙烯泡沫塑料板	在建
9	内蒙古河套灌区渠道	大型灌区			预制混凝土板	焦油塑料胶泥	已建
10	陕西宝鸡峡灌区渠道	大型灌区	渠首 95		现浇混凝土	焦油塑料胶泥	已建
11	陕西泾惠渠渠道	大型灌区			预制或现浇混凝土	焦油塑料胶泥	已建
12	东深供水改造工程输水渡槽	1	90	4.41	现浇混凝土	临水侧 2.5 cm 采用聚硫密封胶,下部采用聚乙烯板	已建

从上述实例可看出,对嵌缝材料的选择,可以分为两大类,一类是已有大型灌区改造中的渠道,这类渠道在进行防渗衬砌改造时,嵌缝材料一般采用《渠道防渗工程技术规范》(SL 18—2004)等规范推荐的材料焦油塑料胶泥,该材料可基本满足灌区运行要求;另一类是近年来新建的一些大型供水渠道,这类渠道供水保证率高,标准高,一般选用造价较高,但黏结力强、变形性能较好、耐老化的聚硫密封胶或聚氨酯等新型材料。

4 嵌缝材料技术经济比较

南水北调工程为特大型跨流域调水工程,同时也是向北京、天津等主要城市提供生活、工业用水,解决北京、天津等城市生产、生活的大型供水工程,嵌缝材料应性能良好、经济合理,应满足以下要求:一是材料应环保,即嵌缝材料应安全无毒、满足饮用水有关标准。二是材料应具有较好的性能指标,即具有良好的伸缩性,在缝口张大时,填料不裂缝;缝口缩小时,填料不被挤出。与混凝土面具有良好的黏接力,在负温下不脱开;同时还应具有一定的耐热性,当气温最高时,填料不发生流淌现象;具有良好的抗冻性,当气温较低时,填料不冻裂或剥落;耐久性好。三是施工简单方便。四是工程投资不要太高,在满足上述性能要求的前提下,力求经济。

因篇幅有限,仅罗列目前工程较为常用的三种嵌缝材料性能指标及施工方法,见表3。

表3 嵌缝材料性能测试比较

测试项目	指标	聚硫密封胶	聚氨酯密封胶	焦油塑料胶泥
密度（g/cm^3）	规定值±0.1	1.7	1.3	1.2
适用期(h)	≥3	4.8	3	无
表干时间(h)	≤24	≤24	6	无
下垂度(mm)	≤3	1	1	<42.5
低温柔性	无裂纹	无裂纹	无裂纹	无裂纹
拉伸黏结强度(MPa)	≥0.2	0.5	0.6	无
断裂伸长率(%)	≥200	350	300	无
室温定伸黏结性(%)	160	160	160	无
弹性恢复率(%)	90	93	96	88.3
有无毒性	无毒性	无毒性	无毒性	有毒性
造价(万元/t)		1.8	1.8	0.9

4.1 聚硫密封胶

(1)施工前先清除基层杂质。

(2)在缝中预填背衬材料,确保密封材料自由伸缩变形。

(3)在基面上预涂双组分胶粘剂。

(4)按比例配制密封膏,充分搅拌均匀。

(5)将配制好的聚硫密封胶与缝壁黏结牢固,防止形成气泡。

4.2 聚氨酯密封胶

(1)伸缩缝的表面清理。

(2)装衬垫材料。

(3)粘贴不粘纸。

(4)嵌缝施工。

4.3 聚氯乙烯胶泥(焦油塑料胶泥)

(1)清除缝内杂质。

(2)将聚氯乙烯胶泥缓慢升温并控制温度,保持填缝不起泡。

(3)停工后,清除剩余填缝料。

从施工工艺难易程度排序为:聚氯乙烯胶泥(焦油塑料胶泥)最复杂,聚硫密封胶和聚氨酯密封胶相对较容易，遇水膨胀橡胶居中。

从性能指标比较看，遇水膨胀橡胶止水条、聚硫密封胶、聚氨酯密封胶具有一定的优势。

从嵌缝材料造价看，遇水膨胀橡胶止水条最贵，聚氯乙烯胶泥最便宜，聚硫密封胶和聚氨酯密封胶居中且价格相当。聚氯乙烯胶泥虽然价格低，但其热施工工艺复杂，并且耐老化及耐低温性能不佳，一般在经历一两个冬夏循环后就会产生渗漏情况，这对于要求较高的工程及水资源严重缺乏的地区不是一种好的选择。遇水膨胀橡胶止水条施工时要在缝两侧预留孔隙，下部垫托，施工要求高，价格也太高，不适应目前我国渠道工程经济性要求，因此也不是渠道防渗工程用的理想填缝材料。

5　结语

目前，聚硫密封胶和聚氨酯密封胶的品牌、性能和质量存在一定差异，在价格比选时一定要注意性能和质量要求。另外，由于密封胶的施工质量是保证其功能发挥的重要环节，其嵌缝施工工艺要求较高，具有较强的专业性，建议施工时加强质量控制，以保证密封胶的性能能够正常发挥。

南水北调中线工程总干渠为特大型供水渠道，水深大，流量大，目前已有的规范主要适用于灌溉与排水工程或从施工角度方面提出技术要求，和南水北调中线工程总干渠的规模和以供水为主的特点并不太相适应。目前，国内已建的工程中，其规模一般都比中线一期工程的规模要小，标准也比南水北调中线工程要低。综合考虑性能指标、造价、施工等因素，对于南水北调中线工程，聚硫密封胶和聚氨酯密封胶与其他密封材料相比具有较强的竞争优势。

参考文献

[1] 江河水利水电咨询中心. 南水北调中线工程总干渠衬砌分缝及嵌缝材料选择研究报告[R]. 2010.

[2] 长江勘测规划设计研究院. 南水北调中线一期工程可行性研究总报告[R]. 2007.

【作者简介】 李宁博，男，1981 年生，陕西延安人，水利水电建筑工程专业，工程师。

高拱坝坝肩稳定破坏试验研究*

郑钦月　张　林　陈　媛　胡成秋

（四川大学水力学与山区河流开发保护国家重点实验室　四川　成都　610065）

【摘要】 立洲水电站是木里河干流水电规划的重要梯级电站，该工程的大坝为碾压混凝土双曲拱坝，最大坝高132 m，坝址区地质条件较为复杂，需要开展拱坝与地基的整体稳定问题研究。本文采用三维地质力学模型超载法试验，对立洲拱坝与地基的整体稳定问题进行研究，确定拱坝与地基的超载安全系数，获得了坝肩破坏形态。将光纤光栅传感器布置在坝顶及上游坝面，监测坝体超载过程中的应变，从而分析坝体的开裂破坏过程。

【关键词】 地质力学模型　超载法试验　光纤传感技术　立洲拱坝

1　引言

随着大型水电工程建设的发展，越来越多的高坝修建在地质构造复杂的地基上，其坝址区内往往存在诸多不良地质构造，如锦屏一级拱坝坝肩发育有X煌斑岩脉、断层、层间挤压带、深部卸荷裂隙等；小湾拱坝坝址区断裂构造较发育，存在不同规模的断层、挤压带、蚀变岩带及节理裂隙等地质缺陷；大岗山拱坝坝址区存在小断层、沿岩脉发育的挤压破碎带和节理裂隙等地质构造，为了保证工程的顺利建成和安全运行，必须解决好高坝与地基的整体稳定性和安全性问题。地质力学模型试验是解决上述问题的有效途径之一。地质力学模型出现在20世纪60年代，当时世界高坝建设迅速发展，成功地进行了多项地质力学模型试验，如伊泰普重力坝，坝高196 m，在意大利贝加莫结构模型试验所（ISMES）进行了地质力学模型试验，以确定混凝土坝和基岩联合系统中的最软弱部位，为工程加固处理提供依据。自20世纪70年代以来，我国开展了大坝地质力学模型试验研究，如清华大学完成了溪洛渡拱坝的地质力学模型试验，倡导的λ_1、λ_2、λ_3安全度评价体系被广泛应用；四川大学也开展了对锦屏一级高拱坝的地质力学模型试验研究，得到了拱坝稳定安全度，为工程的加固处理提供了依据。

四川木里河立洲拱坝坝肩地质条件较为复杂，存在断层、层间剪切带、裂隙等不良地质，本文针对立洲拱坝坝基、坝肩稳定问题开展三维地质力学模型试验研究，通过模型试验分析拱坝与坝肩的变形破坏特征，坝肩失稳破坏过程，确定坝肩稳定超载安全系数，为工程的设计和施工提供科学依据。

* 基金项目：国家自然科学基金资助项目（50879050，51109152）；国家重点基础研究发展计划（973）项目（2010CB226802）；博士学科点专项科研基金（20100181110077）

2 试验方法

地质力学模型试验属破坏试验，其试验方法主要有 3 种：超载法、强度储备法、超载法与强度储备法相结合的综合法。超载法是考虑到工程上可能遇到的洪水对坝基承载能力的影响，在假定除水荷载外的所有因素在整个破坏过程中不变的前提下，逐步增加上游荷载直至坝基破坏失稳，由此得到的超载倍数称为超载安全系数；强度储备法考虑坝肩坝基岩体本身具有一定的强度储备能力，要求得它的强度储备能力有多大，可以逐步降低岩体的力学参数直到基础破坏失稳，由此求得的安全系数叫强度储备系数；综合法是超载法和强度储备法的结合，也就是在一个模型上既采用超载法又采用强度储备法试验，由此得到的安全系数叫做综合安全系数。其中，超载法是当前地质力学模型试验中最常用的一种试验方法，在多年的工程实践中得以广泛应用。

2.1 模型试验相似理论

基本原理：结构模型试验主要是确定在外荷载作用下，施工建筑物表面和内部的应力及位移分布状态。首先按模型规律选择材料及适当的比尺，做成相似的模型；其次把作用在原型水工建筑物的各项荷载按一定的相似比换算成相当的荷载加在结构模型上；再次通过各种量测仪器，测量出模型的应变、位移等数据；最后根据模型相似律，将所测数据换算为原型相应的数据，得到原型水工建筑物在外荷载作用下的应力和位移。

相似定律：为使模型上产生的物理现象与原型相似，模型材料、模型形状和荷载必须遵循一定的规律，这个规律就是相似定律。在原型和模型两个系统中，存在有几何相似、物理相似、力学相似等，即 $C_\gamma = 1$，$C_\mu = 1$，$C_f = 1$，$C_c = C_L$，$C_E = C_L$ 等。

2.2 超载法安全系数关系式

在超载法试验中，模型破坏时的超载倍数 K_p，即超载破坏时的荷载 P'_m 与设计荷载 P_m 的比值，其表达式为

$$K_p = \frac{P'_m}{P_m} = \frac{\gamma'_m}{\gamma_m}$$

式中：γ_m 为模型加压液体的设计容重；γ'_m 为模型破坏时加压液体的容重。

由地质力学模型试验的相似条件 $C_\tau = C_\sigma = C_E = C_\gamma C_L$ 可得

$$C_\tau = \frac{\tau_p}{\tau_m} = \frac{\gamma_p}{\gamma_m} C_L$$

则
$$\frac{1}{\gamma_m} = \frac{\tau_p}{\tau_m \gamma_p C_L}$$

得到超载安全系数 K_p 的表达式为

$$K_p = \frac{\tau_p \gamma'_m}{\tau_m \gamma_p C_L}$$

式中：C_E、C_γ、C_L、C_σ 及 C_c 分别为变模比、容重比、几何比、应力比、黏结力比及集中力比等，C_μ、C_ε、C_f 分别为泊桑比、应变比及摩擦系数比。

3 坝肩地质概况及岩体力学参数

立洲水电站位于四川省凉山彝族自治州，是木里河干流水电规划“一库六级”的第六个

梯级，开发任务以发电为主，电站装机容量 355 MW。拦河大坝为抛物线双曲拱坝，最大坝高 132 m。立洲拱坝地处高山峡谷，两岸坝肩地质条件复杂，断层、层间剪切带、长大裂隙与裂隙带等多种地质构造纵横交错。断层 f4、f5 横贯左右两岸，层间剪切带 fj1 ~ fj4 平行于层面发育，裂隙密集带 L1、L2 与长大裂隙 Lp285 发育于左坝肩中部。两岸地质条件具有较明显的不对称性：左坝肩集中发育有 f5、L1、L2、Lp285，削弱了左岸抗力体的完整性，右坝肩抗力体相对较完整。

3.1 模型几何比及模拟范围

立洲拱坝三维地质力学模型超载法试验综合考虑了坝址区河谷的地形特点、坝基及坝肩主要地质构造特性、拱坝枢纽布置特点及试验任务要求等多种因素，确定模型几何比 $C_L = 150$，模型尺寸为 2.6 m×2.8 m×2 m（纵向×横向×高度），相当于原型工程 390 m×420 m×300 m 范围。

3.2 模型材料及力学特性

根据原型坝体材料及坝肩坝岩体材料的力学参数，通过相似关系 $C_\gamma = 1$，$C_E = C_L = 150$ 计算，可得模型坝体材料的容重为 $\gamma_m = 2.4\ \mathrm{g/cm^3}$，$E_m = 160$ MPa。模型各类岩体及软弱结构面模型力学参数见表 1、表 2。

表 1 模型岩体材料主要力学参数表

地层代号	地层岩性	密度 ($\mathrm{g/cm^3}$)	风化程度	μ	E_0 (MPa)	岩/岩			岩/混凝土	
						f	f'	c' (10^{-3} MPa)	f'	c' (10^{-3} MPa)
Pk	厚层状灰岩、大理岩化灰岩	2.6	卸荷岩体		20	—	—	—	—	—
		2.65	弱风化下部	0.25	53.33	0.55	0.8	4	0.80	4
		2.7	微新	0.23	80	0.65	1.2	6.667	1.05	6
D_1yj	极薄、薄层炭硅质板岩	2.67	微新	0.30	27	—	0.8	4.667	—	—
F_{10} 断层及影响带	左岸	2.5	微新		20	—	0.8	4.667	—	—
			弱风化			—	0.5	0.333	—	—
	右岸		弱至微新			—	0.5	0.333	—	—

表 2 模型结构面主要力学参数

结构面类型	结构面性状	抗剪断强度		变形模量 (MPa)
		f'	c' (10^{-3} MPa)	
fj1 及 fj2 层间剪切带	岩屑充填型	0.65	0.533	—
fj3、fj4 层间剪切带	岩屑夹泥型	0.45	0.2	—
陡倾裂隙 Lp285	泥质充填	0.20	0.0333	—
卸荷裂隙	微张	0.35 ~ 0.45	0	—
L1、L2 裂隙带	裂隙多紧密或少量方解石薄片或泥膜充填	0.65	0.4	20 ~ 26.7
f4、f5 断层带	岩屑、方解石夹泥充填型	0.45	0.333	20 ~ 26.7

4 模型试验及成果分析

4.1 模型加载与量测系统

在拱坝上游坝面采用小型油压千斤顶进行分层、分块加载，对坝体施加水压力、淤沙压力、温升荷载，其中自重通过模型材料与原型材料容重相等来实现，温升荷载通过当量水荷载进行近似模拟。根据立洲拱坝荷载分布特点，全坝共分为 13 块，分别由 13 支不同吨位的油压千斤顶加载。

地质力学模型试验主要有三大量测系统，即拱坝与坝肩表面变位 δ 量测、结构面内部相对变位 $\Delta\delta$ 量测、坝体下游坝面应变 ε 量测系统。在本次模型试验中，结合立洲拱坝的地质条件及空间分布关系，由于坝体上游侧布置有加压和传压系统，受空间限制，上游基岩面和坝体迎水面没有布置表面测点，在上游坝面沿建基面及坝顶拱圈处布置了三条分布式传感光纤；同时在下游坝面 4 个典型高程的拱冠及拱端处，共布置了 12 个应变测点，每个测点在水平向、竖向及 45°方向各布置一张电阻应变片，共布置了 36 张电阻应变片，用于监测坝体下游坝面应变随超载倍数增加的相应变化情况。立洲拱坝坝顶拱圈处光纤光栅布点见图 1。

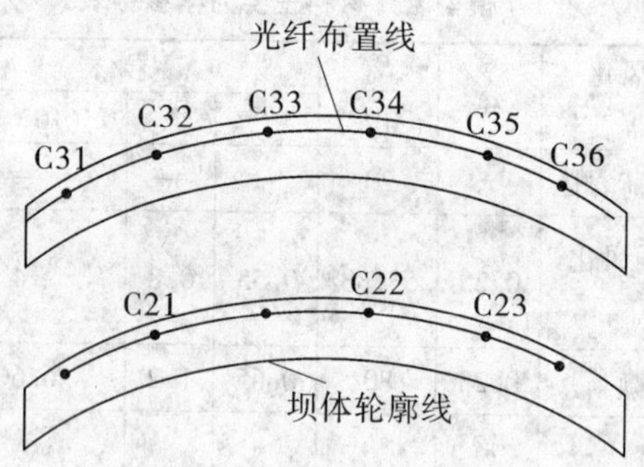

图 1 拱坝坝顶光纤光栅布点

4.2 试验成果分析

试验过程中，首先将荷载逐步加载至 1 倍正常荷载，测试在正常工况下坝与地基的工作性态，然后对上游水荷载按 $(0.2 \sim 0.3)P_0$（P_0为正常工况下的水荷载）进行分级超载，直至坝与地基发生大变形、出现整体失稳趋势，则停止加载、终止试验，试验中记录各级荷载下的测试数据，观测坝基坝肩的变形特征、破坏过程和破坏形态。

4.2.1 坝肩位移及破坏分析

在正常工况下，两坝肩抗力体变位均较小，呈现出顺河向变位向下游、横河向变位向河谷、左岸变位大于右岸变位的规律；在超载阶段，变位逐步增大，位移值由拱端附近最大往下游逐步递减；两坝肩表面变位沿高程方向的分布规律为：左坝肩以 fj2、fj3 及坝肩中部高程的岩体表面变位较大；右坝肩以坝肩上部高程 fj3、fj4 附近岩体的表面变位值较大。由此可见，层间剪切带 fj2、fj3、fj4 及发育在左坝肩中部的断层、裂隙对拱坝与坝肩的变形和稳定有较大影响。

由于地质条件的不对称，左右坝肩的破坏形态及特征也呈现出明显的不对称：左坝肩破坏程度比右坝肩稍大，左坝肩的破坏范围自坝顶拱端向下游延伸约 81 m，主要是坝肩中上部的结构面及附近岩体发生破坏：断层 f5 沿结构面开裂，从河床到坝顶完全贯通；层间剪切带 fj3、fj4 沿结构面开裂、扩展；坝肩中部 fj2—fj4 之间的岩体表面有多条裂缝产生。右坝肩的破坏范围自坝顶拱端向下游延伸约 57 m，主要是坝肩上部岩体发生破坏，其中层间剪切带 fj3、fj4 沿结构面发生局部开裂。左岸上游侧主要是 L1、L2、Lp285 沿结构面发生开裂，右岸上游侧主要为岩体拉裂破坏，这些裂缝沿坝踵相互交汇、贯通左右两岸，其破坏形态见图 2。

4.2.2 坝体应变及开裂分析

光纤光栅传感器成功采集了坝体应变数据，如图 3 所示，对坝体应变情况进行分析：在正常工况下，即当 K_p = 1.0 时，上下游坝面应变总体较小，在超载阶段坝体应变随超载系数的增加而逐渐增大；当 K_p = 1.4 ~ 2.2 时，上下游坝面的监测数据同时出现波动，上游部分测点率先出现转折，说明拱坝上游侧坝踵附近发生初裂，但裂缝尚未贯穿至下游坝面；当 K_p = 3.4 ~ 4.3 时，两套监测数据同时出现较大波动，均形成了较大的拐点，尤其是下游坝面监测数据发生了明显的波动和转折，表明大坝下游侧出现压剪破坏；当 K_p = 5.0 ~ 6.3 时，上下游监测数据陆续发生了转折，裂缝逐步发展，表明大坝出现整体失稳的趋势。现场观察，左半拱裂缝继续向上扩展，开裂至拱顶约 1/2 左弧长附近，右半拱在建基面附近出现一条裂缝，裂缝位于 f5 与坝体交汇的坝址处，并逐渐向上扩展；当 K_p = 6.3 ~ 6.6 时，坝体裂缝贯通至坝顶，坝肩岩体表面裂缝相互交汇、贯通，坝体、坝肩岩体及结构面出现变形不稳定状态，拱坝与地基逐渐呈现出整体失稳的趋势。

图 2 坝肩最终破坏形态

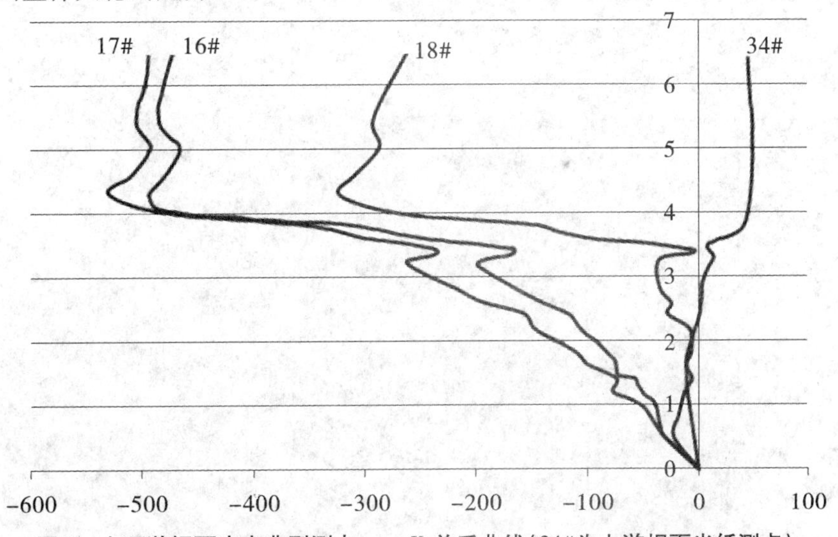

图 3 上下游坝面应变典型测点 μ_ε—K_P 关系曲线（34#为上游坝面光纤测点）

5 结语

(1)通过三维地质力学模型试验,获得了大坝各阶段的超载安全系数为:起裂超载安全系数 $K_1=1.4\sim2.2$,非线性变形超载安全系数 $K_2=3.4\sim4.3$,极限超载安全系数 $K_3=6.3\sim6.6$。

(2)模型试验过程中监测到坝肩的最终破坏形态,左坝肩中上部及右坝肩上部的岩体及结构面变形较大、开裂破坏较严重,为工程设计及施工提供科学依据。

(3)通过上游坝面及坝顶拱圈处布置的光纤光栅传感器及下游坝面布置的电阻应变片,成功地监测到了坝体的应变过程及破坏形态。

参考文献

[1] 王毓泰,周维垣,毛健全,等. 拱坝坝肩岩体稳定分析[M]. 贵阳:贵州人民出版社,1982.
[2] 潘家铮,何璟. 中国大坝50年[M]. 北京:中国水利水电出版社,2000.
[3] 张林,陈建康,张立勇,等. 溪洛渡高拱坝坝肩稳定三维地质力学模型试验研究[C]//中国岩石力学与工程学会. 第八次全国岩石力学与工程学术大会论文集[C]. 北京:科学出版社,2004:946-950.
[4] 杨庚鑫,吕文龙,张林,等.高拱坝坝肩稳定三维地质力学模型破坏试验研究[J].水力发电学报,2010(5):82-86.
[5] 张林, 陈建康, 张立勇, 等. 溪洛渡高拱坝坝肩稳定三维地质力学模型试验研究[C] //中国岩石力学与工程学会. 第八次全国岩石力学与工程学术大会论文集.北京: 科学出版社, 2004: 946-950.

【作者简介】 郑钦月,女,1990年2月生,2011年6月毕业于四川大学,四川大学在读硕士研究生。
通讯作者:张林,女,1955年生,博士生导师,主要从事水工结构模型试验方面的教学与研究工作。

七、管理、政策及其他

关于贵州省水利水电工程农村移民安置人口确定实例分析与探讨

刘永新　罗　政

（贵州省水利水电勘测设计研究院　贵阳　550002）

【摘要】 水利水电工程建设征地移民安置是我国水利水电工程建设的重要组成部分，移民安置的好坏直接关系到社会稳定、经济发展和工程成败。其中，安置人口的确定尤为关键。如何才能既严格执行政策、法规、规范，又充分尊重移民意愿，还能结合移民区和移民安置区的具体情况，较为准确地反映安置人口数量？笔者以省内几个水利水电工程为实例，力图在这方面作一点探讨与尝试，以期更好地开展相关工作。

【关键词】 规范　移民　生产安置　搬迁安置

1　引言

贵州省水资源总量814.6亿m^3，人均水资源量2 059.6 m^3，属水力资源较丰富的省份。随着西部大开发的实施，贵州省水利水电事业得以大力发展。与此同时，为确保库区移民群众“搬得出、稳得住、逐步能致富”，贵州省委省政府高度重视，有关部门做了大量扎实细致的工作。

2　政策规范

指导水利水电工程移民安置的根本政策法规是《大中型水利水电工程建设征地补偿和移民安置条例》（简称新条例）。同时，由于投资主体和工程建成后功能、性质不同，现行移民安置规划设计规范有两套：一套为2009年7月水利部发布的水利行业标准（简称水口），另一套为2007年7月国家发展和改革委员会发布的电力行业标准（简称电口）。

2.1　水口规范

2.1.1　生产安置人口

生产安置人口指因工程建设征收或影响主要生产资料需进行生产安置的人口。对以耕园地为主要生活来源者，按库区涉淹村、组受淹没耕园地，除以该村、组征地前人均占有的耕园地数量计算，必要时还应考虑库内外土地质量级差因素，计算公式如下。

2.1.1.1　质量系数法

考虑土地质量级差系数，用计算单元征收或影响耕园地比重计算，即

$$R = AM_{征地影响}/(M_{征地前}/R_{基准}) \tag{1}$$

式中：R为基准年生产安置人口；A为质量级差系数，为计算单元征收或影响耕园地质量与

该计算单元总耕园地质量比值；$M_{征地影响}$为基准年征收或影响耕园地面积；$M_{征地前}$为基准年征地前耕园地面积；$R_{基准}$为基准年农业人口。

2.1.1.2 标准地法

将计算单元征收或影响耕园地和该计算单元总耕园地用土地质量系数折算为标准地后计算生产安置人口，即

$$R = AM_{b征地影响}/(D_{b征地前}/R_{b基准}) \tag{2}$$

式中：$M_{b征地影响}$为基准年征收或影响标准地面积；$M_{b征地前}$为基准年征地前标准地总面积。

2.1.2 搬迁安置人口

搬迁安置人口指居住在水库淹没区的人口；坍岸、滑坡、孤岛、浸没等影响区中须迁移人口；移民迁移线上的零星住户，受水库淹没影响后，交通难以恢复或生产生活条件明显恶化，须搬迁安置人口；在淹地不淹房居民中，无法就近生产安置须易地安置的人口。

2.2 电口规范

2.2.1 生产安置人口

生产安置人口指工程土地征收线内因原有土地资源丧失，或其他原因造成线外原有土地资源不能使用，需要重新配置土地资源或解决生存出路的农村安置人口。对以耕地为主要农业收入来源者，按照被征收耕园地面积除以征地前被征地人均耕园地面积计算。应根据占地影响程度、土地质量、剩余资源情况和安置条件综合确定。必要时，还需考虑征地处理范围内外土地质量级差因素，计算公式为

$$R = \sum R_i(1 + k)^{(n_1 - n_2)} \tag{3}$$

$$R_i = \frac{S_{i,z} + S_q}{S_{i,zq}/R_{i,j}} \times N_{i,n} \tag{4}$$

式中：R为规划设计水平年生产安置总人口数；R_i为计算单元设计基准年生产安置人口数；$S_{i,z}$为计算单元设计基准年征收耕园地面积；S_q为其他原因造成原有土地资源不能使用耕园地面积；$S_{i,zq}$为计算单元设计基准年征收前耕园地总面积；$R_{i,j}$为计算单元设计基准年农业人口数；i为计算单元数量；k为人口自然增长率；n_1为移民安置规划水平年；n_2为移民安置规划基准年；$N_{i,n}$为该计算单元征地处理范围内耕园地质量与该计算单元耕园地质量级差系数，可采用亩产值差异进行分析计算。

根据上述方法计算出的规划水平年生产安置人口须满足下列条件

$$R \leqslant \sum R_j(1 + k)^{(n_1 - n_2)} \tag{5}$$

式中：R_j为设计基准年农业人口总数。

2.2.2 搬迁安置人口

搬迁安置人口指水电工程居民迁移线内因原有居住房屋拆迁，或居民迁移线外因生产安置或其他原因造成原有房屋不方便居住，需重新建房或解决居住条件的农村移民安置人口。

2.3 规范比较

对比可知，两套规范在农村移民安置人口的计算和确定上，除设计阶段、文字描述和公式形式略有差异外，其实质是相同的。生产安置人口均以其主要生产资料受征地影响程度为依据计算。搬迁安置人口均考虑了直迁人口和随迁人口，同时结合生产安置方案综合

考虑。

3 工程实例

移民安置工作艰巨、复杂、涉及面广、影响深远，牵涉到移民的切身利益和各级部门、各行各业。工程建设，将不可避免地征占土地，使原住地居民丧失全部（或部分）生活来源，改变生活模式、生存环境。因此，要想妥善安置移民，就必须因地制宜、实事求是、科学合理地确定安置人口。移民安置规划设计，一要严格依照规范；二来，由于水利水电工程的特殊性，导致移民除了非自愿性，还有依赖性、复杂性和长期性等特性。因此，确定安置人口，不可脱离实际、生搬硬套，要结合安置方式，具体情况具体分析。

3.1 五嘎冲水电站

五嘎冲水电站地处普安与盘县交界的隔界河上，主要任务是发电，并为煤化工循环经济产业园生产运行提供水源保障。水库总库容 1.151 亿 m^3，正常蓄水位 1 340 m，属大(2)型水库，坝后电站装机 18 MW。

项目总征地 4.680 km^2（淹没区 3.858 km^2、影响区 0.375 km^2、枢纽区 0.447 km^2），以淹没区为主，占 82.44%。其中淹没耕地 2 020.37 亩（1 亩 = 1/15 hm^2），占受淹土地面积的 28.79%。淹没使当地土地面积减少，人均拥有耕地指标下降，从而增大了安置区承载力。经统计，征占耕地中水田 211.4 亩（占 10.46%）、一般旱地 1 430.03 亩（占 70.78%）、25°以上旱地 378.94 亩（占 18.76%）。考虑到不同耕地年产值的差异，在计算生产安置人口时，贵州省水利水电勘测设计研究院首次引入了土地质量级差系数 A。通过测算，A 值为 0.85 ~ 1.19，经 A 值修正后的生产安置人口数更加切合实际，有效地处理了耕地数量与质量之间的关系，即良田好土受淹多的生产安置人口适当调增，反之则酌量削减。以下为土地质量级差系数计算公式

$$A = \frac{A_{征}}{A_{总}}$$

$$A_{征} = \frac{S_{征1}F_1 + S_{征2}F_2 + \cdots + S_{征n}F_n}{S_{征}}$$

$$A_{总} = \frac{S_{总1}F_1 + S_{总2}F_2 + \cdots + S_{总n}F_n}{S_{总}}$$

式中：$S_{征1}, S_{征2}, \cdots, S_{征n}$ 为各地类征地面积；$S_{总1}, S_{总2}, \cdots, S_{总n}$ 为各地类总面积；$F_1, F_2, \cdots, F_n$ 为各地类亩产值。

3.2 黔中水利枢纽一期工程

黔中水利枢纽一期工程是贵州省迄今为止最大的水利工程，总库容 10.89 亿 m^3，正常蓄水位 1 131 m，为大(1)型工程。建设任务以灌溉、城市供水为主，兼顾发电、改善生态环境等。坝址位于乌江上游三岔河平寨河段，建成后可解决 7 县 40 个乡镇 51.17 万亩农灌、5 个县城和 27 个乡镇供水、农村 34.99 万人和 31.52 万头牲畜饮水以及贵阳市供水，电站装机 136 MW。

在可行性研究阶段，水源区规划水平年共计生产安置人口 14 835 人，搬迁安置人口 15 785人（含扩迁 5 789 人）。通过环境容量分析，初拟纳雍县新猴场、织金县凤飞和广顺农场为移民集中安置点，其他移民分散安置。结合移民意愿调查情况，生产安置以种植业为

主,第二、三产业为辅。搬迁安置以外迁安置为主(广顺农场8 037 人),县内安置为辅(后靠1 917人、分散4 328 人,集中1 503 人)。

贵州省人均耕地面积偏低,加之近年来因"西电东送"等工程建设,征占了大量耕地,截至2006 年6 月底,省内水库移民已达43.69 万人。由于耕地资源有限,传统的"有土安置"方式实施难度加剧,为妥善解决工程建设对移民的影响,加快社会经济可持续发展步伐,贵州省一直在积极探索新的移民安置补偿方式。经过充分调研论证,在大量征求移民意愿的基础上提出了"长期补偿"方式,其核心是根据农户被征占耕地面积,按照政府每年公布的耕地年产值,逐年补偿给被征占耕地的农民。通过双河口等工程几年"试点",取得了地方政府、移民、业主三方满意的效果。

鉴于此,为削减黔中的搬迁压力,在初步设计阶段,水源区移民安置方式改为长期补偿,搬迁安置人口锐减5 875 人,其优点明显;第一,可以减少前期征地移民安置资金投入,有利于工程的顺利建设,发挥工程的社会效益;第二,可以减少移民安置任务,不用考虑随迁人口搬迁安置;第三,有利于移民长远生计保障,减少地方政府移民安置难度,有效解决淹没剩余土地处理问题、移民长途搬迁以及在安置区的适应性等问题。由此,还能解放生产力,使失地农民有条件从事第二、三产业,有利于提高家庭收入和库区社会稳定。同时,也为农村移民安置人口的确定带来了新的思路与途径。

3.3 马家水库工程

威宁是国家级喀斯特地区扶贫开发综合治理试点县,其经济水平落后。马家水库位于威宁县迤那镇,正常蓄水位2 198 m,相应库容107 万 m^3,为小(1)型水利工程,其主要任务是解决水库周边人畜供水及水库下游的烟田灌溉用水。

马家水库淹没区涉及威宁县1 个镇1 个村。根据实际调查,有6 户房屋处在20 年一遇回水线上,即房屋迁移线略高于屋基。根据新条令第12 条和电口规范《水电工程建设征地实物指标调查规范》(DL/T 5377—2007)5.3.4 第1 条,经走访附近移民和现场核实,水库蓄水将造成住户出行不便,同时这6 户的田土位置偏低,有超过60% ~95% 的耕地受淹,导致其主要生活来源中断,因此这6 户居民被列为随迁人口,纳入移民安置规划。这也符合水口规范《水利水电工程建设农村移民安置规划设计规范》(SL 440—2009)7.2.3:"主要生产资料部分征收户,可根据本组的生产安置人口,通过调整生产资料的方式,原则上将剩余的生产安置人口,落实到主要生产资料被征收,占本户全部生产资料比例大的农户作为生产安置对象,保证远迁户的完整性。"

3.4 芦家洞水电站

芦家洞水电站属于改扩建工程,位于铜仁市锦江中游,是一个以发电为主,具有改善水环境、防洪、发电、旅游、航运等综合效益的工程。原电站由于冲沙底孔淤堵、右岸船闸损坏、库容和发电量偏小等原因需改扩建。改扩建后,水库总库容3 433 万 m^3,最大坝高22.50 m,总装机21 MW,对提高铜仁城区防洪标准,促进地方经济发展有着十分积极的意义。

电站位于城集镇,当地经济发展水平较高,人均耕地低,农民对土地的依赖程度低,若仍按规范公式计算,则计算出的生产安置人口偏多,误差很大。为如实反映实情,在计算人均耕地时,应用土地依赖程度系数 m 进行修正,m 主要受两个因素影响:一是计算单元内农民从耕园地获得收入占总收入的比重,二是计算单元内移民迁建去向。经分析计算被征地农民收入构成情况，种植业占0～35%，养殖业占17%～29%，商业占21%～34%，外出务工占25%～40%，结合库区耕地资源情况分析，m 为0.05～0.40。修正后，规划水平年生产安置人口折减率达83.1%。生产安置人口数与实际情况吻合，移民对此并无异议。

3.5 金阳水电站

金阳水电站位于习水县习酒镇坪头村，地处赤水河一级支流桐梓河下游。水库总库容541.5万 m^3，正常蓄水位360 m，属小(1)型水库，电站装机20 MW。水库淹没土地0.118 km^2，其中耕地23.2亩，园地8.65亩；枢纽工程建设征占土地36.29亩，其中永久征地占耕地13.60亩。

电站征收耕地比重仅占涉及村的0.11%和0.09%，占涉及组的1.32%和0.89%。当地第二、三产业相对发达，比较突出的是酿酒业。另外，居民大多从事饮食、副食、建筑、家电维修、汽车运输和修理等行业，农业收入只占当地居民家庭收入的30%。经分析计算，到规划水平年水库淹没区生产安置人口14人，枢纽建设区6人，数量有限且无直迁人口。当地农民人均耕地较低，且开垦流转耕地困难，如仍实行农业安置十分困难，也与移民意愿相违背(移民均要求进行一次性补偿)。因此，根据本工程特点，按水口规范《水利水电工程建设农村移民安置规划设计规范》(SL 440—2009)7.6.4，在充分尊重移民意愿的前提下，主要考虑一次性补偿，并积极引导移民发展第二、三产业。获得补偿的农户，既可以用补偿补助资金改良线外剩余耕地，提高其剩余耕地质量，又可以作为扩大再生产投入，为第二、三产业发展注入资本金，加速地方经济发展。由于该安置方案符合当地实际情况，获得了当地干部群众的拥戴，得以顺利实施。

4 结语

移民安置人口的认定，必须紧扣移民安置方案，针对不同项目实际情况，除考虑农业安置外，还可酌情考虑非农业安置，农业安置与非农业安置相结合的安置形式(复合或兼业安置)以及其他安置方式。本文所列举的工程，其移民安置规划均已通过审批，有的正在实施，有的已竣工运行。从反馈情况看，计算的移民安置人口数量基本上与实际吻合，移民区和移民安置区群众思想稳定，对工程建设予以支持配合，为工程的顺利推进，打下了良好的基础。

参考文献

[1] 甘盟. 乐昌峡水利枢纽工程生产安置人口计算方法探讨[J]. 广东水利电力职业技术学院学报，2009，7(1)：72-74.

[2] 赵彪，郭琦. 水库移民生产安置人口计算问题思考[J]. 水利经济，2007，25(5)：72-74.

[3] 蒋建东，李文军. 浅谈水库淹没处理的人口问题[J]. 人民长江，2008，39(13)：22-24.

[4] 尹忠武. 水库农村移民生产安置人口计算方法初探[J]. 人民长江，1999，30(11)：3-4.

[5] 包广静，杨子生. 大型水电工程移民人口影响研究[J]. 水电能源科学，2008，26(2)：107-109.

【作者简介】 刘永新，1969年12月生，1991年毕业于贵州工学院，大学本科，学士学位，贵州省水利水电勘测设计研究院，高级工程师，注册监理工程师，注册咨询工程师。

完善水利工程建设国库集中支付方式的相关对策

刘建树

（淮河水利委员会治淮工程建设管理局　蚌埠　233001）

【摘要】 2002年，财政部、水利部确定淮河中游临淮岗洪水控制工程为水利国库资金支付试点第一个项目，随后在全国逐步推广。该项制度至今已有10年时间，对水利工程建设具有较好的促进作用，但在实践中也发现或存在着一些问题和不足。本文从治淮工程建设实际出发，针对发现的配套政策措施不够完善、与水利工程实际情况不相适应和财政直接支付周期长等问题，提出了完善水利工程建设国库集中支付的一些对策，以更好地促进水利工程建设。

【关键词】 水利工程　国库集中支付　对策

2001年，党的十五届六中全会明确提出"推行和完善国库集中收付制度"。建立国库集中支付制度，是将财政资金实行集中收缴和支付，对防范财政收入流失，财政支出监督弱化以及财政资金使用过程可能出现的被截留、挤占、挪用等问题具有重要意义。2002年，财政部、水利部确定淮河中游临淮岗洪水控制工程为水利国库资金支付试点项目，至今已有10年时间。随着水利工程建设国库集中支付制度逐步建立，水利资金已更快、更直接地用于水利工程建设，加快了水利工程建设速度。作者亲身参加和实践这一重大改革，回顾10年来的实践与体会，浅谈如下。

1　水利工程建设国库集中支付的主要特点

水利建设工程具有投资主体的多层性，涉及的利益相关方多，资金支付容易受到外界因素的干扰。水利建设工程国库集中支付有以下特点。

1.1　国库集中支付的多级性

淮河水利委员会（简称淮委）作为项目法人实施的水利工程建设项目，大多是中央同地方拼盘的项目，中央财政投资通过水利部下达到淮委，省级地方配套投资通过省级水利部门下达，市级地方配套资金通过市级水利部门下达，中央、省级和市级财政层级不同，相互交叉。

1.2　国库资金支付的不均衡性

水利工程建设国库资金支付不同于部门预算的基本支出（人员经费和公用经费）。基本支出的特点是每月支出具有均衡性特征，而水利工程在建设过程中，影响资金支付的因素较多。例如，移民拆迁、工程设计变更、工程度汛等因素，都会直接影响水利工程建设的进度，进而直接影响资金的均衡支付；在淮河流域水利工程施工的黄金季节在每年汛后到第二年汛前期间,施工任务较重,资金支付较多,施工年度与财政年度不一致,说明水利工程建设国库集中支付具有特殊性。

1.3 国库集中支付的复杂性

按照国务院《关于加强公益性水利工程建设若干意见的通知》,水利工程建设要建立健全项目法人责任制。开工建设前要组建项目法人,特别是一些大型流域控制性水利工程建设,涉及省与省之间、市与市之间的矛盾。为便于工程建设,在一个项目法人的基础上,根据不同地方工程建设管理需要,组建专项工程实施机构,涉及的层次和单位较多,在办理国库资金支付时,管理和协调难度大,决定了国库集中支付的复杂性。

2 水利工程建设国库集中支付的主要效果及存在问题

2.1 国库集中支付的主要效果

随着水利建设工程国库集中支付的实施,参加改革试点的项目范围逐步扩大,取得了一定的积极效果。

2.1.1 有利于资金监督管理

水利工程建设国库集中支付试点工程项目必须依照项目预算和投资计划申报财政直接支付用款计划,然后在批复的财政直接支付用款计划额度内申请直接支付。这种支付模式大大加强了财政部门对用款计划额度使用情况的监督,有利于保证水利工程建设资金安全。

2.1.2 有利于严格资金管理

以往水利工程建设资金支付的审核程序相对简单,主要按投资计划和预算安排逐级下拨到主管部门或建设单位。由于资金流经环节多,资金账户重复分散设置,导致财政监督乏力,预算约束软化,财政资金运行信息反馈不畅,资金使用效率低下、安全性不高,影响了水利工程建设。实行国库集中支付,特别是财政直接支付方式后,建设单位必须对项目合同协议每笔款项进行事前审核,并记录直接支付项目每个合同的合同额、收款人信息、本次支付金额、累计支付金额等支付信息。审核后,工程建设类进度款需根据监理签证的工程量清单进行支付;服务、物品采购类资金要根据工作量清单、阶段性成果及货物、物品的验收合格单支付,进一步加强了水利工程项目合同管理和资金管理。

2.1.3 有利于落实地方配套资金

对于拼盘建设的水利工程建设项目,在实行财政直接支付前,中央财政资金下拨后,地方配套资金到位率低,到位时间迟缓。由于地方资金不能按时到位或拖欠,导致有些项目不能按计划及时开工,或者已开工的项目进展缓慢,甚至停工,造成工程建设任务不能按时完成,不能及时竣工验收,投资效益不能充分发挥。实行财政直接支付后,财政资金的支付始终在上级管理部门与财政专员办的监控之下,有利于建设单位催促地方配套资金及时、足额到位。

2.2 存在的主要问题

经过 10 年的实践,水利工程建设国库集中支付改革已取得显著成效,但仍存在一些问题和不足,这既影响水利工程建设的顺利进行,也影响水利工程建设国库集中支付改革的推进。

2.2.1 配套政策措施尚不完善

根据现行水利工程建设管理办法,工程建设项目结余资金可按一定比例由建设单位留用,但在实际操作中,直接支付项目结余资金留在财政部门。依照财政国库管理规定,零余额账户资金不能够向自有资金账户划拨款项。另外,直接支付方式要求每一笔款项均需通过财政部门审核后才能使用。由于政策不完善的影响,致使单位留用资金无法实现,影响水利工程建设单位的积极性。

2.2.2 工程投资计划管理体制与预算管理体制不协调

原水利工程项目投资额审批权限在投资计划管理部门,财政部门根据投资计划管理部门拨付投资,投资额核定管理权限在投资计划管理部门,资金支付在财政部门,管理权限比较明确。目前,水利工程建设进展到一定阶段后,一些地方财政部门对水利工程建设项目进行投资评审,目的是重新确定工程投资,这一做法直接导致水利工程建设项目投资额的不确定性。特别是中央与地方拼盘项目,在中央投资全部到位并全部使用后,地方财政部门进行投资评审,核减投资,并直接抵减地方投资,致使资金不能按照批复概算全额到位,工程建设任务难以完成,而且变相加大了中央投资。由于管理体制不顺,增加了水利工程建设的困难。

2.2.3 与水利工程建设实际情况不相适应

根据规定,财政部门在审核施工单位资料时,要求施工单位必须提供施工发票。施工单位在先开据发票,又长时间不能收到工程款的情况下,必须先行垫付税款。当水利工程施工工期要求紧,资金支付集中时,往往会出现无资金垫付税款的情况。此外,直接支付手续办理的间断性与工程建设资金连续需求相矛盾,致使资金不能及时到位,造成施工单位不得不垫付资金,加大了施工单位的负担。有时由于无力垫付资金,造成推迟支付材料款和人员工资等问题。

2.2.4 财政直接支付周期长

水利工程建设项目资金需求具有连续性、及时性等特点,特别是防洪度汛项目。施工任务紧、施工时间短,与国库资金支付周期较长相矛盾。一是水利工程参建单位多,报送资料时间难统一,每次都有部分单位支付资料推延,影响报批;二是财政部驻地方财政监察专员办事处审核人员少,身兼数职,工作任务重,出差多、每月办理审核的次数和时间受到制约。此外,大多数水利工程处在边远地区,对外信息不畅,交通不便,工地现场、建设单位与各级管理部门不在同一地区,跨区域支付审核手续周期长,加大支付成本,降低支付效率。

3 完善水利工程建设国库集中支付的对策

3.1 加强宣传和培训,提高认识和业务水平

通过宣传、培训等方式,进一步提高水利建设单位各级领导和财务人员对水利工程建设国库集中支付改革重要性的认识。正确认识改革的重要性和必要性,使财务人员尽快熟悉和掌握国库集中支付的操作程序,弄清各支付环节要做的具体工作和时间要求,确保用款计划及时上报和支付申请渠道畅通。各单位要配备专职人员,具体负责和办理本单位或指导下级单位的国库集中支付工作,从实施财政国库管理制度提供组织和人员保障。与此同时,财务人员要及时掌握国库集中支付改革后的新制度、新业务、新方法,强化财会人员的操作技能;国库支付涉及预算编制、用款申请、计划下达、项目管理等多个环节,要求财会人员不仅要做好本职工作,还要熟悉工程建设管理工作。

3.2 各职能部门要充分发挥职能作用

为了避免多个部门对水利工程项目投资额审批权限的管理矛盾，各职能部门要各负其责，充分发挥职能作用。建议由计划管理部门继续负责水利工程投资额的审批，财政部门负责资金拨付和监管，水利部门负责工程质量监督，项目法人对工程投资、质量和进度负责。计划部门应改进对工程投资额的核定方式，在批复工程投资初步设计时，采用投资复核的科学方法，发挥专家的作用，合理确定工程施工方案，准确核定工程投资，也可采取“静态投资，动态管理”的方法，提高国库集中支付效率。

3.3 强化项目预算管理，合理确定支付方式

财政直接支付较财政授权支付更便于对资金的监督和管理，而财政部国库集中支付管理试点办法规定，财政直接支付项目中只有建设单位管理费能够实行财政授权支付方式，对直接支付范围的设备采购费、建筑安装工程费、工程监理和设计费等支出实行直接支付。但在具体办理过程中，零星支付业务较多，全部实行直接支付，支付成本较高、效率低，不符合水利工程建设管理的实际。因此，需要合理确定直接支付与授权支付项目范围，对直接支付项目合理确定直接支付的起点，便于办理支付手续，既能抓住重点，又方便基层。

3.4 减少直接支付审核环节，提高支付效率

从财政直接支付涉及的部门责任看，项目法人是资金直接使用者，对项目建设的全过程负责，对资金使用和管理承担最主要和最直接的责任。在财政直接支付过程中，不仅要遵循部门的管理要求，还应按照监管成本最小和资金效益最大化的原则，设计优化水利财政资金直接支付方案，减少直接支付审核环节，提高直接支付效率，使其更符合水利工程建设的要求。

3.5 加强财政专员办建设，提高工作效率

财政部驻地方专员办事处应设置财政直接支付审核固定岗位，并明确第一经办人和第二经办人。当第一经办人不在时，第二经办人可继续办理直接支付审核手续。财政专员办工作人员应经常深入水利工程建设一线，深入了解水利工程建设的实际情况，避免因对水利工程建设不了解，而影响直接支付审核工作。财政专员办也应严格按财政部文件要求，简化手续，取消一些不符合实际的规定，确保在 2 个工作日内审核完毕。例如，在办理国库支付审核中，取消对水利建设单位先出具设备采购发票和施工企业发票的要求，以提高工作效率。

【作者简介】 刘建树，男，1970 年 12 月生，2010 年 1 月毕业于北京工商大学函授，大学本科学历。现任淮河水利委员会治淮工程建设管理局财务处副处长，高级会计师。

中央一号文件对水管事业单位科技工作的启示

李良庚

（深圳市北部水源工程管理处　深圳　518110）

【摘要】 本文从科研管理、技术装备发展两个方面入手，试对中央一号文件关于水利科技工作的有关要求予以解读和探索，以期为水管事业单位科技工作出谋划策，积极推动水利事业的可持续发展。

【关键词】 中央一号文件　水管事业单位　水利工程　科研　技术装备

1 引言

随着社会经济的不断发展，供水局势日益趋紧，水利工程兴建规模越来越大；水污染形势日益严峻，人工湿地等水环境治理工程应势兴起；水安全态势日益紧张，水利工程建在城市建成区的比例已呈扩大趋势。显然，新时期水利发展对水利科技提出了新的、更高的要求和挑战，水利科技面临的任务更加艰巨。2011 年中央一号文件（简称一号文件）聚焦水利，并将水利科技提高到水利基础工作的高度，强调了其支撑作用，无疑为水利行业注入了一针强心剂，予以水利科技工作诸多启示。本文抛砖引玉，以纯公益性水利工程管理单位（简称水管事业单位）的角度，试从一号文件重点涉及的科研管理、技术装备发展两个方面，解读和探索水利科技工作的开展思路和相关要求。

2 科研管理

科研是科技工作的基本表现形式，不仅是提升科研管理单位自身能力的需要，也关系着科研管理单位技术装备的发展。大到国家级的基础研究计划、科技支撑计划、科技基础平台条件计划、863 计划，水利部的 948 计划，小到各省市级科技计划，都充分体现了各级科技主管部门对科研的重视。同时，有关体制机制、转化推广、宣传普及、队伍培养等保障措施的成功实践也全面反映了各级科技主管部门对科研的政策扶持和行动支持，无疑为水管事业单位开展科研管理工作提供了重要的借鉴与探索。

2.1 研究领域

（1）供水保障。水管事业单位所辖工程面广、线长、点多，周情（特别是城市水利工程）复杂，维修量大、难度大，运行调度复杂、乱占乱用等水事纠纷多，有必要开展水利工程安全预警与监测、用水量调配与原水价格调整联动机制、联合调度决策、水利工程保护与应急机制等研究工作，以提升水管事业单位原水供应的水平和应变处理的能力。

(2)水资源保护。城市发展对水利工程管理造成巨大冲击力,恐怖事件、面源污染无时无刻不在考验水管事业单位的"神经",有必要开展河流生态修复技术、水源保护区面源污染治理、水库排洪河治理、水源水质在线监测与毒性检测、水利工程沿线安全监测等研究工作,以实现水管事业单位原水水质的安全稳定与运行成本的合理低平。

(3)水资源开发利用。水利工程周边地区一方面城市发展迅猛,用水需求剧增,另一方面,工业、农业与生活排水存在乱排乱放现象,极大污染了当地天然水资源,从而造成水资源短缺和开发利用难的双输局面,有必要开展雨洪等非常规资源利用、水利工程沿线渗漏与减糙(糙率)监测、水利工程沿线水锤防护、人工湿地运用等研究工作,以缓解水管事业单位水资源供水与需水之间的矛盾和经济与社会利益之间的冲突。

(4)三防。水利工程特别是水库工程,一方面为供水受益区提供了水源保障;另一方面则为水利工程区域特别是下游区带来三防压力,而防大洪、抗大险为重中之重,有必要开展洪水保险机制、三防管理信息系统、雨量站技术改造与资料整编等研究工作,以确保水管事业单位水利工程的运行安全和应变处理能力的提升。

(5)水土保持。水土流失的产生无非是由于生产或建设的过程中出现了一定的水土扰动,为避免较大的水土扰动,确保水利工程所辖区域生态和谐,有必要开展工程治理、生物修复、预警监测等研究工作,以促成水管事业单位社会效益与生态效益的兼顾与建设和管理的同步。

2.2 体制机制

(1)水利科技管理体制。把水利科技工作纳入日常工作,明确水利科技分管领导、统筹部门和专职人员,制订规划计划、项目管理制度及考核办法,以规范水利科技工作行为,增强水利科技工作执行力。

(2)水利科技运用机制。重视统计总结、资料归档与监督检查,并将"四新"(新产品、新材料、新工艺、新技术)的运用纳入水利工程的规划、设计、施工、管理、运行、维护等各个环节,以提升水利工程技术含量和管理水平。

(3)水利科技竞争机制。通过公开招标、竞争性谈判等政府采购形式,切实引进技术水平高、资金投入优的科研队伍,充分利用水利科技基础条件平台(包括水利科研设施设备共享平台、水利科技数据共享平台、水利科技成果转化公共服务平台等),以实现科研目标,促进水管事业单位可持续发展。

2.3 转化推广

(1)水利科技论文。水利科技人员结合科研项目的学习心得和开展体会撰写论文,不仅能够及时转化工作成果,系统梳理知识架构,而且能够有效提高经验总结和解决问题等方面的能力。

(2)水利科技成果示范工程(园区)。通过建立示范工程(园区),不仅促成了水利科技成果的中试转化,而且健全了以行业为主体,市场为导向,"政、产、学、研、用"相结合的水利科技创新体系。

(3)水利科技运用机制。做好成果登记、专利申请及奖项申报,并根据水利行业技术标准、水利先进应用技术重点指导目录,建立合乎水管事业单位实际需求的运用机制,全面实现水利科技成果的价值。

2.4 宣传普及

(1)水利科技展览与培训。举办展览与培训宣传普及了水利科技知识和应用愿景,为宣传普及对象培养了科学观念。

(2)水利科技交流与推介。通过参与研讨、提交论文及推介“四新”,营造水利科技促进水利发展的良好社会氛围。

(3)水利科普宣传。灵活采用宣传周(日)、博览会(博物馆)、示范工程(园区)、亲水平台(廊道)以及媒体等形式,并重点对三防、用水安全、节水、水环境保护、水生态修复、水土保持、水行政执法、水利信息化等内容以及水利工程布局、功用等方面,精心策划、全面到位,增强宣传普及对象的亲水、护水、节水意识。

2.5 队伍培养

(1)政策环境。通过及时解读和运用水利科技扶持相关政策,并制订以人为本的人才培养和使用计划和管理办法,切实保证人尽其才,工作开展顺畅。

(2)考评与激励机制。重点从组织管理(体制机制、宣传普及、队伍培养等)、项目管理(立项、采购、实施、验收、后评价等)等两个方面入手,建立工作评价指标体系并相应制订工作考核办法,确保人才选拔公正、人才流动公平,人才结构合理。

(3)学术氛围。重视弘扬大胆探索、勇于实践的科学精神,注意集思广益、传帮带和互相成就,以营造和谐、公正、民主的学术氛围。

3 技术装备发展

技术装备主要是指用于工程建设和基础性工作的各种仪器、设备、机械和金属结构,涉及机械、电气、信息化、建筑等多个领域,既是科技研究的基础条件,也是科技研究的成果转化,还是各行业寻求国际交流合作与应对 WTO/TBT 的支撑条件,更是各行业自身能力和发展方向的集中体现。就水管事业单位而言,机械类的主要有泵、闸、阀、流量计等,电气类的主要有电机、高中低压开关等,信息化类的主要有网络通信系统、电子巡更与对讲系统、计算机监控系统、水情(工情)自动测报系统、质量检测(探测)工器具、水质检测(监测)设备(设施)等,建筑类的主要有三防抢险装备(含物料、设备、仓库等)、除险加固设备、土(岩)工试验仪器等。随着社会经济的发展,水利基础建设对技术装备发展提出了更高的要求,但现有市场还普遍存在结构(性能、品种、规格等)单一、标准不统一、准入不规范等现象,另外,水利技术装备尚未纳入水利规划,严重影响了政策扶持力度和资金投入程度。水管事业单位需要在摸清现有水利技术装备技术法规(标准)、合格评定的基础上,在规划和招标环节,把好相应技术要求和招标方案关,真正采购到标准高、质量好、品种多元化的水利技术装备,而在制造和安装环节,则把好相应监造和质监关,如实发挥出水利技术装备的功效。

3.1 运用领域

由于水利技术装备发展尚未正式纳入水利发展规划,因而水利技术装备的运用领域无论是深度上还是广度上还远未形成体系化,造成水利技术装备的引入、研制及运用缺乏系统性的政策支持和技术指导,从而不利于水利技术装备的可持续发展。基于此,水管事业单位需在摸清现有水利技术装备相关技术法规、技术标准、合格评定程序的基础上,借助科学研究平台,对水管事业单位运用需求予以系统分析、论证,以此确定出技术装备的运用领域,具体如下:

(1)水利工程施工。如原位监测仪器(监测施工条件)、建筑物质量监测仪器(监测施工质量)等。

(2)水土保持监测。如泥沙采样仪(监测生态效益)、土壤抗冲仪(监测适应性)、土壤剖面水分测定仪(监测生态条件)等。

(3)水环境监测应用领域。如生物操纵技术装备(食物链操纵)、生物滤池(生物膜吸附)、底泥疏浚装备(机械清淤)等。

(4)工情监测应用领域。如大坝安全监测系统(观测渗压、渗流等情况)、地质雷达探测仪(探测建筑物内部填筑材料情况)、土(岩)工试验测试仪(检测地质情况)等。

(5)水情监测应用领域。如水情测报系统(测报水文、雨情、气象等情况)、洪水预报系统(预报洪水情况)等。

(6)防汛抢险救灾应用领域。如防汛指挥车船(调度物资和发布信息)、防汛抢险成套装备(含物料、机具、仓库等)等。

3.2 技术规程

技术规程是为了让管理人员更好地掌握装备的结构、特性和运行特征,安全、科学地使用装备,确保装备寿命的最大化,编制和运用要点具体如下:

(1)格式要统一。包括版本和幅面、字体和段落、编号和符号、样式和格式等。

(2)内容要准确。包括目的、使用范围、定义和术语、性能和参数、规程内容、监督和检查、编制依据和参考文献、工作记录和表格式样、说明和附则等。

(3)程序要讲求可操作性。灵活采用 5W1H(指原因(WHY)、对象(WHAT)、地点(WHERE)、时间(WHEN)、人员(WHO)、方法(HOW)等六“何”)分析法、图表法、量化法等方式。

3.3 维护机制

要实现维护的既定需求目标,需要水管事业单位在摸清现有水利技术装备标准化实施情况的基础上,维护任务下达之前的计划和招标等环节,把好相应技术要求和招标条件关,真正采购到标准高、质量好、品种多元化的水利技术装备,确保维护过程中真正实现技术规程的执行到位,从而实现维护到位,充分发挥出技术装备的科技效用。水利技术装备现有标准化实施有关事项具体如下:

(1)技术法规。所涉及的领域包括机械、电气、信息化、建筑等,应当注意的是,这些领域与其他行业有业务上的交叉,如机械领域,与农业部门、机械制造行业有交叉;电气领域,与农业部门、电力行业有交叉;信息化领域,与气象海洋部门、环保部门、工控行业有交叉,建筑领域,则与建筑行业、交通部门有交叉。基于此,水管事业单位应深入了解交叉行业相关的技术法规实施情况,以争取到更大的政策支持和技术依据,进而把握好开展方向和运用重心。

(2)技术标准。水利技术装备涉及的机械、电气领域由于市场成熟,水利行业相关的技术条件、安装与验收规程、试验方法较齐备完善,基本能够满足应用需要;而涉及的信息化、建筑领域尽管引入技术相对成熟,但水利行业相关的技术条件、安装与验收规程、试验方法相对缺乏,因而存在产出效益差、维护难到位等问题,不利于应用、维护和普及推广。基于此,水管事业单位应在广泛调研交叉行业已出台的相关技术法规、技术标准的基础上,结合已实践的成功案例,并反复论证技术条件,确定出满足应用和维护需求的技术要求和工作考核指标，助力于水利技术装备的普及推广。

（3）合格评定程序。它包括产品合格性评定、企业合格性评定和中介技术机构合格性评定等三个方面。考虑到水利技术装备合格评定程序实施尚处于初级阶段，基于此，水管事业单位应在借鉴相关行业合格评定程序实施经验的基础上，重点在水利技术装备招标和出厂验收环节，认真核对生产许可、使用（推荐）许可、市场准入等证明文件，以考察投标文件技术标的符合性，比选出标准高的产品；同时，仔细查看投标方 ISO9000、ISO14000 等企业合格性评定证明文件，计量认证、审查认可（验收）、实验室认可等中介技术机构合格性评定证明文件，以考察投标文件商务标的适用性，比选出质量好、品种多样化的产品。

4 结语

（1）科研项目管理、水利技术装备发展是一号文件涉及的水利科技工作的要点，予以水管事业单位颇多工作开展与研究的启示。

（2）科研项目管理是科技工作的本质要求和技术装备发展的基本依据，水管事业单位要在创新研究领域的同时，灵活选用开展方法，重视工作措施的落实。

（3）技术装备发展是科研的基础条件和科研的成果转化，水管事业单位要在创新运用领域的同时，重视技术规程，完善维护机制。

（4）研究科研管理和技术装备发展是顺应我国社会经济蓬勃发展对水利现代化要求剧增的需要，也是提升水管事业单位自身能力建设的需要，意义重大，任重道远。

【作者简介】 李良庚，男，1974 年 7 月生，硕士学历，2002 年 6 月毕业于武汉大学动力与机械学院，流体机械及工程专业，工作单位为深圳市北部水源工程管理处，任工程技术部副部长，高级工程师。

深圳水务工程施工质量管理系统研究

张传雷[1] 宋 军[1] 陈军强[2]

(1.深圳市水务工程质量监督站 深圳 518036;
2.广东华南水电高新技术开发有限公司 广州 510611)

【摘要】 本文对深圳市水务工程施工质量管理系统从研究背景、内容、系统结构等方面进行阐述,并通过该系统在部分水务工程中的使用效果可以看出,该系统可实现水务工程施工质量的信息化管理,并可提高工作效率和监控力度。

【关键词】 水务工程 施工管理 系统

目前,深圳市在建水务工程点多、线长、面广,近3年来在深圳市水务工程质量监督站办理质量监督手续的项目数量剧增(2007年度177项、2008年度181项 、2009年度215项),工程质量、安全监管工作日益繁重,如果依靠传统的监管手段,工程管理难度将非常大,因此迫切需要用信息技术提高管理水平和效能,以适应新时期治水思路的总要求。本研究项目得到了广东省、深圳市主管部门的大力支持,并于2007年在“广东省水利科技成果推广计划”立项,由深圳市水务工程质量监督站负责项目的具体实施,该科研项目获2009年度广东省水利学会水利科学技术奖三等奖,目前研究成果是深圳水务工程质量监管的重要手段之一,大大提高了质量监督效率和工程质量控制效果。

1 研究背景

结合网络信息化建设和用户的实际,在网络平台上,利用软件工程技术、Web技术、GIS技术等,搭建一套深圳市水务工程施工质量管理系统,这既是对工程质量管理的要求,也是工程质量管理与国际接轨的重要手段,通过计算机网络技术辅助进行工程质量数据的动态管理,提高工作效率,通过网络发布工程质量管理信息,起到社会监督作用。通过项目的研究,实现了水务工程动态实时监管,为工程管理单位提供了一个信息化管理手段。

2 研究内容

通过对水务工程质量管理工作内容的分析,该系统一方面解决了质量管理工作内容及工作的规范化、程序化问题。针对某一个单位工程,通过系统软件以国家和水利部相关规程、规范、强制性条文,以法规文件为依据,完成工程质量项目划分、质量评定、项目验收、管理信息工作。

另外,系统研究解决了工程监管过程中所采用的技术手段问题。系统通过“水利数码通”提供强大的现场检查手段,质量管理人员可以在施工现场完成质量信息(照片、录像、文字）的实时采集后，将这些信息以及同步的卫星定位信息进行加密，传输到信息接收及 GIS 信息服务系统，为提高水务工程质量监管效率提供技术手段。

3　系统特性

系统开发过程中紧扣以下 5 个特性。

3.1　集成性

要求系统是一个高度集成的综合性系统，系统各功能模块、软硬件之间相互配合，数据相互交换，具有统一的人机交互界面设计风格。

3.2　先进性

先进性是在需求分析基础上，充分掌握水务工程施工管理的业务流程，对系统进行合理设计，并尽可能采用先进适用的技术对系统进行部署实施，以先进的 CRM 技术进行系统集成，数据库平台，系统集成平台系统，不仅在单项技术上超前、先进，系统在整体性能上表现出先进的特性，给系统用户提供方便的功能特性。

3.3　可扩展性

要求系统使用最新的信息技术，使用三层架构的框架保证了系统以后的扩展，能自定义系统表格，保证数据的可扩充性。

3.4　易用性

系统要具有非常友好、简洁的操作界面，输入/输出界面清晰，随时提供联机帮助，使用者在短时间内即能掌握系统的使用方法。

3.5　安全性

在系统管理上，考虑到作为主要用户的监督单位、施工单位、监理单位、设计单位、建设单位等 5 种角色，根据角色再对应到单位工程，为每种角色分配不同的使用权限，保障系统数据内容的高度安全。从操作系统到数据库访问，从软件功能到信息流等，软件系统应提供全面的安全机制，一方面防止系统出现异常丢失数据的现象，另一方面防止非法使用、修改或查看其中的内容，从而保障系统安全可靠。

4　系统结构

系统结构见图 1。

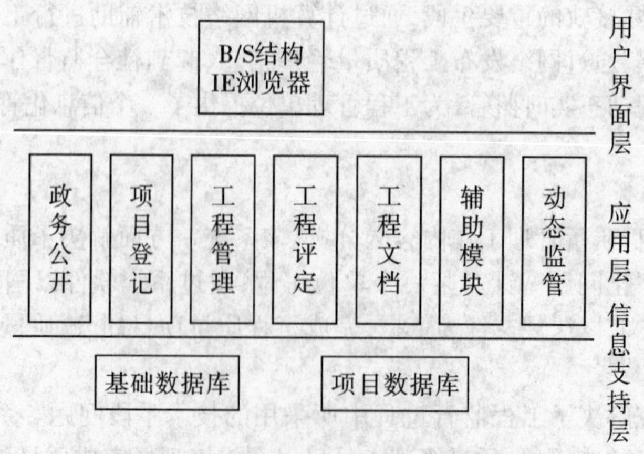

图 1　深圳市水务工程施工质量管理系统结构

5 功能模块

系统采用 GPS 技术、无线传输技术、多线程技术、动态 Web GIS 技术和数据库技术、集成信息服务、事务处理和业务管理，实现了工程质量监督登记、项目划分、项目结构图及评定表选择、工程评定以及评定表流程化管理、动态监管、用户角色和用户信息管理、站内信息发送、常用文件上传下载、操作日志管理等功能。

系统实现的主要功能见图 2。

6 主要成果

6.1 多用户流程化管理

多角色多用户参与使用，工程文档相互之间可以进行信息共享。各单位只负责各单位的任务和需求，做到职责分明，根据具体的评定及管理用表实行具体的流程传递。

6.2 直观性

通过系统评定、管理表表格统计分析，可以直观地判断项目是否达到或具备验收的条件。并可通过系统功能查看工程管理信息是否及时有效。

6.3 可扩展性

监督单位可以自定义表单流程，可以无限扩展表格库，适应将来工程评定、验收文档变化的需求。

6.4 实时动态管理

采用水利数码通对施工现场数据进行采集，对施工过程关键部位进行实时动态管理，为工程质量管理提供远程、实时监管的技术手段。

7 结语

该项目构建了一个水务工程施工质量动态监管系统平台，通过在部分水务工程中使用，该系统辅助工程参建单位人员实时在线掌握水务工程的施工情况，实现了对水务工程施工过程的远程、实时、动态管理，解决了施工管理数据采集难、时效性差的难题，提高了工作效率和监控力度。

下一步在推广应用系统的过程中，不断对系统进行升级维护的同时，拟在该项成果的基础上开展“深圳市水务工程施工质量动态监管系统”项目研究，通过空间数据库的建立、GIS 模块的开发和移动信息采集的集成，形成水务工程施工质量安全网络化管理模式和施工现场的对接和互动机制。

【作者简介】 张传雷，1980 年 3 月生，2005 年 4 月毕业于西安理工大学水利水电学院，硕士研究生，深圳市水务工程质量监督站，高级工程师。

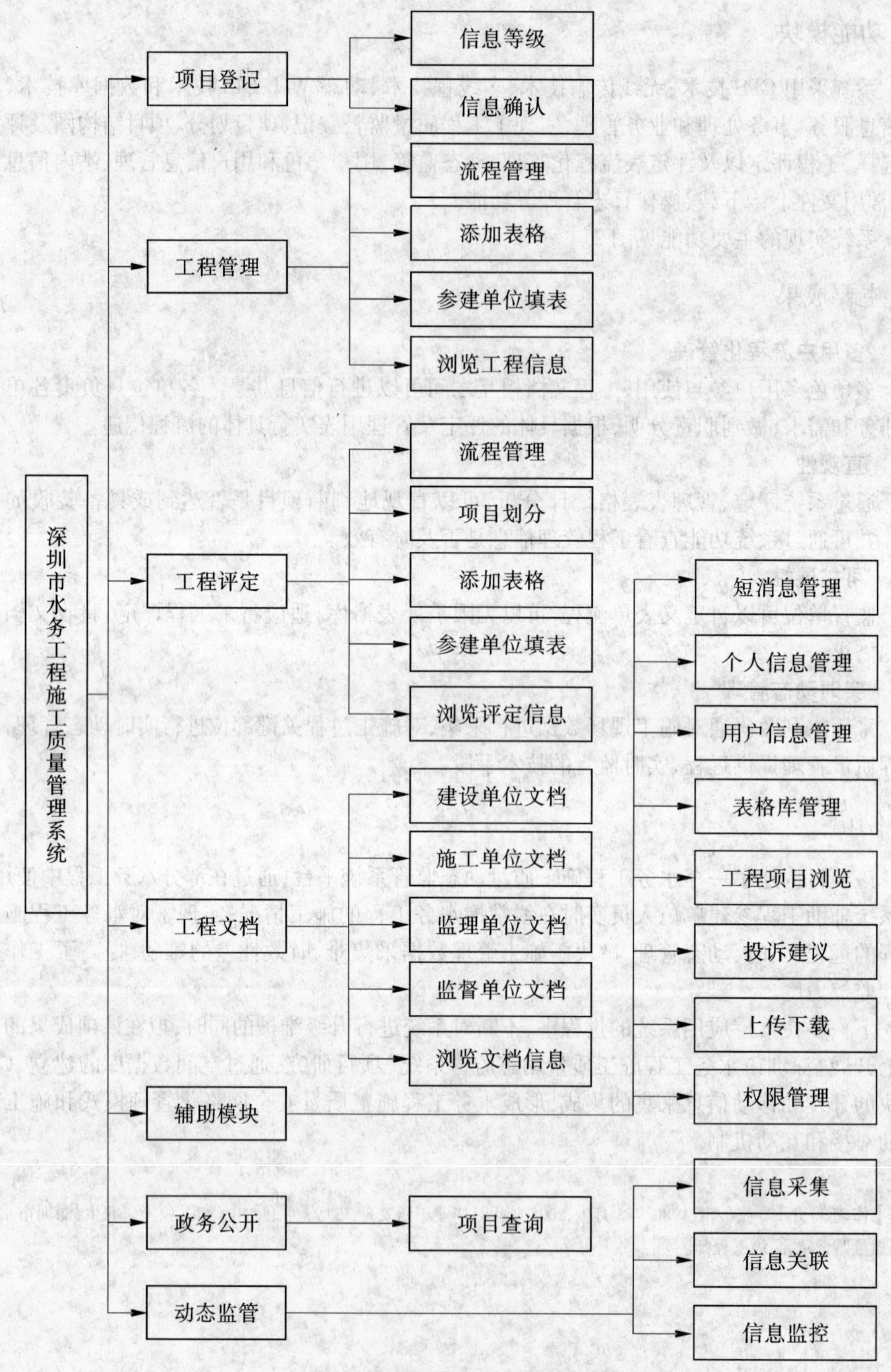

图2　深圳市水务工程施工质量管理系统模块

“635”水电站综合自动化技术改造和集控中心建设

刘振龙　刘小鹏　金　光

（新疆额尔齐斯河流域开发工程建设管理局六三五水电站　阿勒泰　836006）

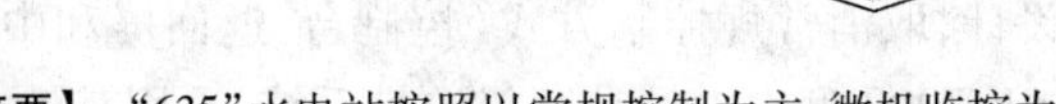

【摘要】“635”水电站按照以常规控制为主、微机监控为辅原则设计，受当时技术水平限制，监控系统功能很不完善，而且经过多年的运行，励磁、调速器、继电保护、自动化元件等设备因为老化也经常出现故障。对电站进行技术改造并建设集中控制中心，全面提高系统运行的稳定性和可靠性，目前已经实现了“无人值班（少人值守）”。

【关键词】 水电站　综合自动化　技术改造　集控中心建设

1　引言

“635”水电站位于新疆阿勒泰地区的额尔齐斯河中上游，装机容量 32 MW，是地区电网的主力电站。首台机组于 2000 年 9 月发电，电站控制原则按照以常规控制为主、微机监控为辅进行设计。经过近 10 年的运行，设备逐渐老化，已经不能满足电站及电网安全运行的需要，主要存在以下问题。

（1）微机监控系统因为受当时技术条件的限制功能很不完善，且运行不可靠，一直没有完全投入运行，仅能对功率、电压、电流、温度进行监视，没有参数的处理功能和通信控制功能，机组的所有运行工况和运行参数全部由人工监视并记录统计。

（2）机组和 10 kV 馈线保护均采用常规继电器控制和保护设备，动作准确性和可靠性差，检修工作量大，硬接线回路使局部改造困难。110 kV 线路保护虽已经更换为微机保护，但是存在保护信息不能上传至监控的问题。

（3）励磁和调速器系统没有数据通信和记录分析功能，存在元件老化现象。

（4）计量表计为普通电度表，计量精度低，而且二次电流、电压电缆截面不符合现行规程的要求，造成损耗电量较高，且无通信功能。

（5）还存在自动化元件动作可靠性差，常出现拒动和误动现象。远动和通信设备不能满足电网调度的要求。辅助设备不能实现自动控制，信号不能上传。防误操作设备不能满足行业要求。当前需要运行人数较多，设备操作、巡检等工作量大，工作效率低等问题。

（6）喀腊塑克水电站即将投入运行，山口水电站已经开工建设，如何顺利接管两个电站，实现集中控制以达到减员增效和发电效益最大化之目的，对提高流域电站群在市场中的竞争力至关重要。

2 综合自动化技术改造

2.1 改造方案

针对以上存在的问题,制订了改造方案,改造方案分为两个部分:第一部分针对电站设备进行改造,主要包括监控系统、继电保护装置、励磁装置、调速器电气控制柜、中压机和低压机、深井泵控制柜、远动和通信系统、直流充电装置屏、更换电锅炉房 PLC 并接入监控、自动化元件、计量表计和全厂二次电缆;第二部分结合后期运行需要增加了集控中心、工业电视、微机五防系统、机组状态监测装置。

2.2 监控系统

2.2.1 监控系统结构

监控系统按照全计算机控制方式设计,取消常规控制方式(控制台、返回屏和中央音响屏)。监控系统采用南瑞公司生产的 SSJ-3000 型水电站计算机监控系统,采用基于 Unix 操作系统跨平台的全分布开放系统结构、NC2000 监控系统软件。

电站计算机监控系统分为主站级(电站控制级)和现地控制单元级。其中电站控制级为多机功能分布型式。电站控制级和现地控制单元级经 100 Mbps 以太网相连。以太网采用双光纤冗余的星型结构。

其中,主站级监控系统通过网络双绞线与冗余交换机相连,现地控制单元级通过光缆与冗余交换机相连。网络通信协议为 TCP/IP,网络速度为 100 Mbps。监控系统主要由系统工作站、操作员工作站、工程师工作站、系统通信工作站等部分组成。现地 LCU 柜有 1# ~4#机组 LCU 柜,另外设公用开关站 LCU 柜,其结构见图 1。

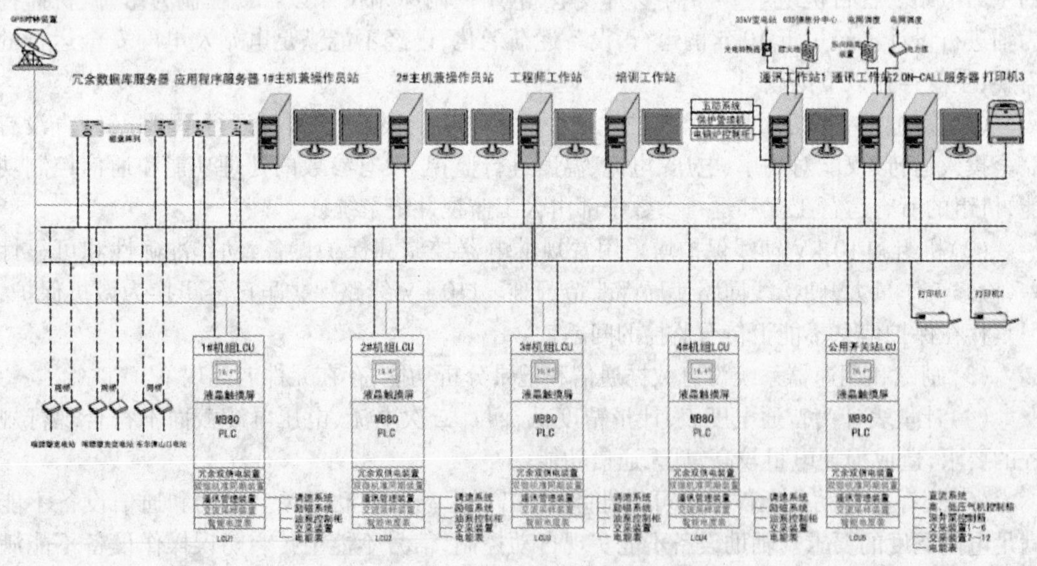

图 1 635 水电站计算机监控系统结构图

监控系统具有数据采集与处理、综合计算、安全监视、人机联系、指令操作与调节、自动控制、生产统计、运行日志与报表打印、通信、防误操作的功能;还具有系统数据库、系统时钟同步、系统自诊断与恢复、系统的授权管理等功能。

2.2.2 现地控制单元(LCU)

LCU 共 5 套,包括机组 LCU 4 套、开关站及公用 LCU 1 套,采用双 CPU、双网络的冗余结构。

LCU 主要完成对被监控设备就地数据的采集及监控功能,其设计能保证当它与主站级系统脱离后仍然能在当地实现对有关设备的监视和控制功能。当它与主站级恢复联系后又能自动地服从主站级系统的控制和管理。LCU 采用 PLC 直接联网的方式,提高了系统的可靠性。机组 LCU 包括主控制器(PLC)、电源、同期装置、测速装置、温度巡检保护装置、串口通信装置、液晶触摸屏、交流采集装置、多功能电度表等设备。开关站及公用 LCU 主要负责对厂用电设备、辅机控制设备、开关站的各断路器、隔离刀闸、接地刀闸等主要和辅助设备进行监视和控制。

2.2.3 计算机监控系统主要监控对象

4 台 4×8 MW 水轮发电机组及调速、励磁调节器等辅助设备;

2 台 110 kV 主变压器及 2 台厂用变压器;

110 kV 出线及其 110 kV 母线,及 10 kV 馈线;

全厂公用设备(直流系统,电锅炉,高、低压气机系统,供排水系统,厂用电系统); 35 kV 变电站。

2.3 保护系统

保护系统的改造主要包括电站和 35 kV 变电所的保护装置,原则是取消常规的发电机保护屏、主变保护屏、10 kV 线路保护屏、10 kV 厂变保护屏、110 kV 母差保护屏,除两条 110 kV线路微机保护未更换外,其余全部进行改造。

2.4 励磁系统

原励磁系统属于模拟调节控制,没有数据采集分析功能和通信上传功能。改造后的励磁系统在“自动” 运行方式下,能与监控配合完成发电机电压的升、降压等操作,运行稳定,并能满足发电机运行时,对电压质量的要求;并网后可接收监控的无功调节控制令,具备远方“自动”调节功能,具有“手动”、“自动”运行切换方式,可满足发电机在“手动”方式下运行 。

2.5 调速系统

调速器采用在原电气柜、机械柜升级时改造方式,布置位置和原来一致。电气柜内采用 PLC/PCC,CPU 等级不低于 486,位数为 32 位,主频达 100 MHz。机械柜要求取消杠杆及中间环节,采用德国 BOSCH 伺服比例阀作为电液转换器;同时,将油压装置的压力油罐的液位计及自动补气装置、回油箱的液位计进行改造并增加一个压力变送器。

改造后的调速器调节部分按照调速器和监控的需要配置。使调速器在“自动” 运行方式下,能与监控配合完成水轮机的开、停机操作,运行稳定,并能满足发电机运行时,对频率质量的要求;并网后可接收监控的有功调节控制令,具备远方调节功能,具有“手动”、“自动”运行切换方式,满足发电机在“手动”方式下运行 。

2.6 辅机系统

辅机系统主要包括调速器油泵、蝶阀油泵、深井泵、高低压气机、直流充电装置、电锅炉(对 PLC 进行了更换)等,该部分的改造均以各自的 PLC 为核心,实现它所负责的信息的处理,电机的启停及工作、备用状态的自动切换,并能通过通信方式将各种状态上传到监控系统。

2.7 自动化元件

随着监控系统及辅机控制系统的改造，对不能满足监控系统及辅机系统要求的所有自动化元件进行了更换，包括压力开关、液位计、传感器、示流器、电磁阀等。另外，原设计的电压和电流二次回路导线截面均为 1.5 mm^2 和 2.5 mm^2，不符合现行规程要求，对电站和 35 kV变电所的控制、计量和保护用的二次电缆进行了更换。

2.8 工业电视

为了实现“无人值班(少人值守)”的目标，此次改造增加了工业电视系统，在电站和 35 kV变电所的主要设备和重点区域共安装了 23 个室内、室外监控点，其中包括 18 个高速球机、5 个恒速球机。运行人员在集控中心的监视器上通过网络对所属厂站的设备进行远程实时视频监控。

2.9 机组状态监测装置

根据行业要求，为了做到由“事故检修”模式和“计划检修”转变为“状态检修”提供依据，此次改造安装了机组状态检测装置，通过该装置信号采集和分析处理功能，为机组状态分析提供定量的依据。

2.10 微机五防系统

另外，根据《国网公司十八项反措》的要求增加了电站和 35 kV 变电所的微机五防系统，实现了防止误操作事故的发生，提高了运行的安全性。

3 集控中心的建设

3.1 网络结构

考虑所属流域电站群地理位置以及流域内电站实现“无人值班(少人值守)”模式的需要，结合流域内各电站的地理位置，集控中心采用扩大厂站模式方案，即将集控中心设备和“635”水电站站控层设备合二为一，将集控中心设置在流域内人员较为集中且地理位置相对居中的“635”水电站，对喀腊塑克水电站、布尔津水电站进行远方监控，这样可减少运行维护人员及投入经费，可充分体现集中调度的优势。设备布置在“635”水电站，其网络结构见图 2。

3.2 连接方式

在集控中心层和站控层均设置网络交换机，集控中心层的所有计算机设备均接入集控中心侧的网络交换机中，而“635”水电站内的各现地控制单元 LCU 和作为简化站控层的一台操作员工作站均接入水电站侧的网络交换机中，集控中心侧的网络交换机和“635”水电站侧的网络交换机之间采用高速光纤连接。

3.3 机房设备

集控中心计算机房按照标准计算机房进行装修改造，机房内布置系统工作站(即主机)、工程师工作站、调度通信工作站、打印设备、电话语音报警工作站、网络交换机、各种横向/纵向网络隔离装置、UPS 不间断电源等主要设备。

4 结语

“635”水电站的综合自动化改造和集控中心建成投运后，系统运行稳定，集控中心各种

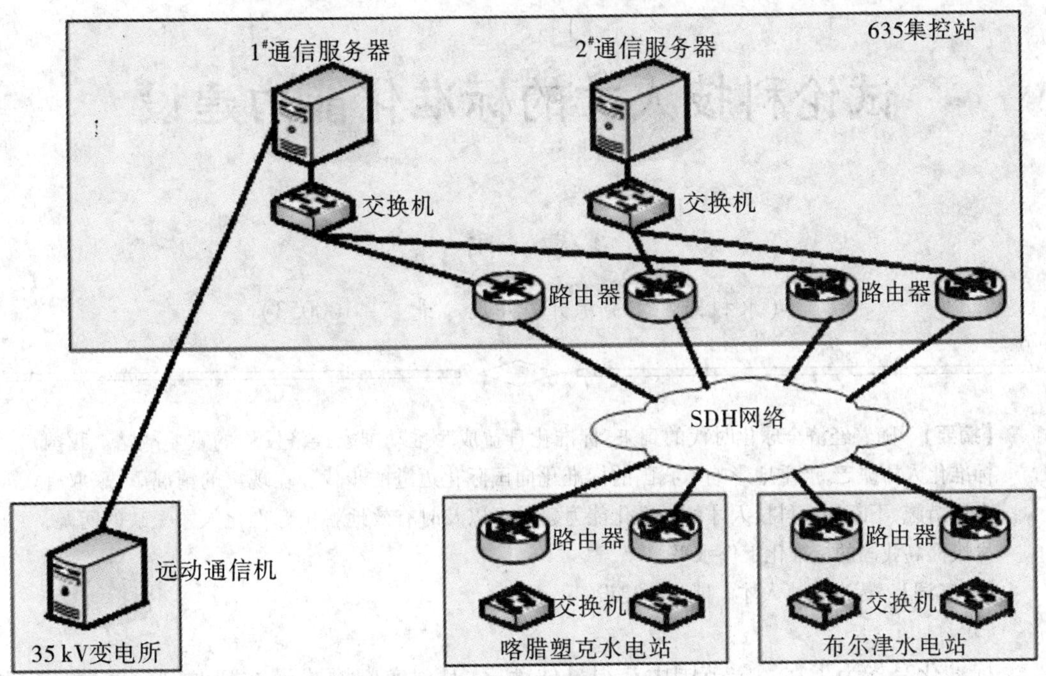

图2 集控中心网络拓扑结构

操作方便可靠、视频监控系统画面清晰，所有开停机操作及运行监视均由集控中心统一进行。改造后，电能质量得到了提高，设备故障率降低，自动化程度提高，实现了水资源的优化配置，人员劳动强度降低，工作效率提高，工作环境改善，对电网的运行稳定性和安全起到了保障作用。

参考文献

[1] 方辉钦. 现代水电厂计算机监控技术与试验[M]. 北京：中国电力出版社，2004.

[2] 中华人民共和国水利部. SL 193—97 小型水电站技术改造规程[S]. 北京：中国水利水电出版社，1997.

【作者简介】 刘振龙，1973 年 10 月生，大专，2003 年毕业于新疆高等工业专科学校，新疆额尔齐斯河流域开发工程建设管理局六三五水电站副厂长，工程师。

试论科技人才的标准化能力建设

曹 阳

（水利部人才资源开发中心 北京 100053）

【摘要】 随着经济全球化时代的到来，标准化日益成为推动国家经济发展的重要战略。我国标准化人才缺乏，严重地影响了我国的标准化向国际化迈进的步伐。在现实的情况下，采取有力地措施，不断加大科技人才的标准化能力建设可以及时有效地解决标准化人才短缺的问题，尽快提高我国的标准化工作水平。

【关键词】 标准化 人才 能力 建设

标准化是指为了在一定范围内获得最佳秩序，对现实问题或潜在问题制定共同使用和重复使用的条款的活动（①上述活动主要包括编制、发布和实施标准的过程。②标准化的主要作用在于为了其预期目的改进产品、过程或服务的适用性，防止贸易壁垒，并促进技术合作）。在经济全球化的大背景下，标准化已成为国家核心竞争力的重要体现和企业发展壮大的关键路径，也成为保证产品、过程或服务质量，沟通技术发展，促进国际贸易的重要手段。在我国，标准化人才的缺乏极大地限制了标准化前进的步伐。现实的情况下，加强科技人才的标准化能力建设可以在短时间内有效地解决标准化人才不足的问题。

1 引言

1.1 标准化成为国家核心竞争力的重要体现

自主创新能力是国家的核心竞争力，增强自主创新能力是我国应对未来挑战，实现建设创新型国家目标的根本途径。自主创新能力主要体现在专利和标准上。近年来，通过专利垄断技术，通过标准获得收益，通过标准国际化垄断市场，在各国经济竞争中已成为不争的事实。"技术专利化，专利标准化，标准国际化"成为各经济大国不断扩充经济实力，获得更大利益的有效战略。但在专利与标准两者中，前者影响的只是一个或若干个企业，后者影响的却是一个行业或产业。相较于前者，后者在一个国家竞争力的衡量指标中占据更重要的位置。

1.2 标准化成为企业发展壮大的关键路径

标准化是企业保证产品质量，提高产品市场竞争力的根本保证；是提高企业管理水平，增加经济效益的重要手段。加入世界贸易组织以来，国际市场森严的技术壁垒，让我国企业的标准化意识不断增强，开始认识到将标准化作为主要竞争手段来获得市场主导地位的重要性。三流企业做产品、二流企业做品牌、一流企业做标准，已成为许多企业成功的宝典，共为企业所认同。

1.3　人是决定标准化水平的关键因素

国外在技术标准研制方面的经验表明,国际标准的竞争实际上是各国产业实力的竞争。没有先进的产业技术、过硬的产业实力作为充分条件,在国际标准竞争中不可能胜出。而在标准的国际化过程中,标准化的行为主体——人,作为必要条件则起着关键作用。标准化工作与社会经济发展的不协调问题,需要有人来发现和研究解决方法;标准水平的提高需要有人来改进;标准的实施需要有了解标准化的人来执行;标准化工作向国际的迈进需要有对我国标准化和国际标准化工作模式和运行方式极为了解的人才。

在我国,能够按照标准化工作需要来完成标准化工作的人还远远不够。标准化人才的缺乏是我国标准化水平整体水平有待提高、国际标准化参与度不高的直接原因。目前,我国标准化人才的市场需求很大。据中国标准化研究院相关课题调研资料显示,如果按产值500万及以上的企业设置一名标准化人员计算,该类人才我国至少缺口32万。如果还考虑中小企业,则标准化人才将缺口100万以上。

2　科技人才标准化能力建设的必要性

根据"十五"期间国家的12项重大科技专项之一"国家技术标准体系建设研究"中的子课题"技术标准保障体系子课题"研究成果,将标准化人才分为高级复合型标准化人才、专业型标准化人才和管理型标准化人才三类。

标准化人才培养的重点是国际化的高级复合型标准化人才,因为复合型标准化人才可以提高国家在世界范围标准化工作中的整体地位,全面提升我国标准化的国际竞争能力,并最终实现在重要、核心领域关键技术标准的领先水平和领导地位。但标准化人才培养的基础和基本着力点是专业型标准化人才,因为该类人才会使整个标准化战略、技术法规、标准和相关体系得以在全国范围广泛、有效的实施。管理型标准化人才,在于能够发挥自身的管理优势,高效、有序地组织相关标准化人才在相关领域的技术研发、人力和财力资源管理,从而促进标准化系统自身以及相关经济、社会领域的良性、持续发展。同时,与复合型人才一起,能就标准化新出现领域和特点做出迅速、敏捷的判断和反应,组织研发力量进行科研攻关,并在尽可能短的周期内提出和制定相关的标准,以指导新领域的有关工作。

高级复合型标准化人才只能产生于专业型标准化人才或管理型标准化人才。而专业型标准化人才和管理型标准化人才在短期内无法通过高等教育解决,只能通过在职培养实现。因为标准化工作不是一个专业或者一个传统意义的领域,标准化工作涉及几乎所有的行业、领域或专业,覆盖社会、经济生活的每一个方面。标准化本身在一个具体的领域或行业中起一种"统领"的作用。因此,标准化人才必然首先是(该方面)专业技术人才或管理人才,也就是对某个领域的知识、技术具备了一定的基础后才有可能培养成为合格的标准化人才。换句话说,通过对科技人才的标准化能力建设,在科技人才中能很快地培训出一批通晓专业、掌握标准化知识及标准化运行规律的人才,有效地充实我国的标准化人才队伍,从而提升国家的标准化实力。

因此,科技人才的标准化能力建设是事关未来我国标准化水平的大事。

3　科技人才标准化能力的内涵

根据标准化的概念,结合实际情况,科技人才标准化能力应包括准确理解标准、精确执行标准、适时提出新标准、正确编制标准、有效推广标准等五个方面的能力。

3.1　准确理解标准的能力

我国已经发布的大量标准,就同一对象的不同方面或不同条件下的同一对象都有不同的标准规定,了解这些标准的内容、追踪标准的变化、熟悉标准的适用条件及范围,是科技人才准确理解标准的前提。在工作中,科技人才如果能对专业领域里的标准做到如数家珍,首先得了解标准对于工作的重要性,然后对现行专业标准的状况、内容及背景有一定的了解,进而才能准确理解标准在专业领域的作用与意义。

3.2　精确执行标准的能力

尽管我国有大量的标准,由于人们对标准理解上存在偏差,在执行中产生了各式各样的问题,以至于不能够保证标准执行的有效性。要做到精确的执行标准,就要对标准的条文有一定的了解,对条文的应用条件、注意事项、关键内容都有正确的理解,才能在实践工作中精准把握各种环境下标准的执行。因此,精确执行标准不仅需要准确理解标准条文,还需要熟悉标准条文的执行条件,从而才能让标准指导实际工作的作用得以发挥。

3.3　适时提出新标准的能力

我国的标准体系的构建是一个动态过程,国家、行业、地方、企业还需要大量的标准,但是这些标准何时出台、如何出台、规定什么,不了解技术发展、行业趋势,就不可能把新标准适时地提出。作为一名科技人才,光理解、执行标准还不够,还需要结合工作,以专业的眼光和技术的远见,对专业领域中所需标准化的内容提出来,把握制定标准的需求,从而为行业技术的发展和专业工作的规范化提供推动。

3.4　正确编制标准的能力

标准的需求提出来了,但如果没有足够了解标准制定要求的人员去制定这些标准,没有了解关键技术的人员将技术标准化,没有人将标准化技术作规范的条文化提炼,制定的标准就会存在各式各样的问题。因此,参与标准化工作的技术人员应有足够的标准化专业知识和标准编写的基本知识,对技术的程序化、数据化、实用化、推广化有深入的理解与把握能力,对文字有准确的驾驭能力,对标准的特点有充分的认识,对标准的构成有深入的认知,才能提高所编写标准的质量,增加标准对现实的指导力。

3.5　有效推广标准的能力

标准虽然发布了,也在一定程度上给予了实施,但对标准的实施情况如没有足够的人员给以监督,没有足够的机制保证对标准执行情况的信息反馈,那么标准化工作的效果也就大打折扣。科技人才应充分发挥熟知技术的优势,洞察标准的执行情况,广泛收集标准执行中的意见和建议,追踪标准实施过程中的重要事件,不断增加标准的适用性及指导性。

4　加强科技人才标准化能力建设的建议

4.1　在岗位条件中增加标准化的要求

目前,我国各行业都在大力推行岗位执业资格证制度,执业资格已成为许多岗位的上岗必备条件。如果将对标准的熟悉掌握程度作为执业资格的重要条件加以限定,并不断增加这一限定条件的权重,一方面可以提高上岗人员的质量,另一方面可以增加科技人才学习标准、使用标准、推广标准的自觉性,进而加强科技人才标准化能力建设。

4.2 单位需不断加强对标准化工作的重视

各单位应增强对标准化工作的重视程度，加大对标准化工作的管理，把贯标、采标及技术标准化、标准化活动的参与作为扩大行业影响力，增加企业市场竞争力的重要工作来抓，以此来引导科技人才参与标准化工作的积极性，激发他们投身标准化的热情。

4.3 用积极的奖励政策引导行业标准化工作

各行业标准化主管机构可采用积极的政策鼓励科技人才参与标准化工作，如设立专项奖励、强化参与标准化活动在评先选优体系中的指标权重等，吸引更多的科技人才重视标准化工作，投入、热爱标准化工作，最终促进标准化工作多出成果，出高质量的成果。

4.4 在国家标准化人才培养战略中体现

“十五”期间，国家启动的12项重大科技专项之一“国家技术标准体系建设研究”中的子课题“技术标准保障体系子课题”将标准化人才培养作为重要内容。但其中偏重通过高等教育和培训对标准化人才的培养，针对科技人才在工作中直接进行的标准化能力建设并未提及，而恰恰是这部分人可以在最短的时间内弥补标准化人才不足的问题。因此，在国家标准化战略中，应将科技人才的标准化能力建设作为重要工作列入其中，从而扩大标准化人才备选范围，将更多优秀的科技人才引入标准化工作领域。

参考文献

[1] 中华人民共和国国家质量监督检验检疫总局. GB/T 20000.1—2002 标准化工作指南　第1部分：标准化和相关活动的通用词汇[S]. 北京：中国标准出版社，2002.

[2] 王颖，顾佳隽. 标准化人才全国缺口32万 新专业就业前景广阔[N]. 浙江在线-今日早报，2009-02-03.

[3] 沈同，邢造宇. 标准化理论与实践[M]. 北京：中国计量出版社，2008.

【作者简介】 曹阳，1972年7月生，1995年6月毕业于北京师范大学，硕士，水利部人才资源开发中心，副处长，副编审。

从基层看取用水管理工作任重道远

李平雷鸣

（黄河水土保持天水治理监督局　天水　741000）

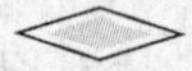

【摘要】 黄河水利委员会天水局作为基层单位，取用水管理辖区范围为青海、甘肃两省，依据权限及分工对辖区内9个限额以上工农业取水口和黄河水利委员会发证的黄河干流16座水电站进行日常监督管理。随着基层工作深入开展，掌握的情况更加详细，我们发现取用水管理逐步向好的方向发展，但是也遇到了一系列问题，意识到存在问题的严重性，并对问题进行研究探讨。

【关键词】 基层　取用水　取水许可　最严格

2011年中央发布一号文件，作出加快水利改革发展的决定。文件指出人多水少、水资源时空分布不均是我国的基本国情，阐明了水利改革发展的指导思想、目标任务和基本原则，即是坚持民生，统筹兼顾，依靠科学治水、凭借法律管水，实现改变当前水利建设滞后的局面，强调了实行最严格的水资源管理制度，不断创新水利发展机制，为我们继续开展取用水及其管理活动奠定了基调，指明了方向。

黄河水利委员会天水局管理辖区范围内青海、甘肃两省水资源时空分布不均匀非常明显：两省为黄河流域主要产水区，其中兰州以上水资源量占全流域的47.3%，由于地理因素，地表水集中于狭长河谷地带，空间分布不均匀；两省水资源分布不均匀还表现在时间上，由于区域水资源主要以降水形式形成，而每年降水集中于夏季短短数月，且多以暴雨出现，致使河流径流集中于汛期，年内分配很不均匀。两省属相对落后地区，取用水管理机制有待进一步优化。笔者了解到，两省取用水基本上实现收支两条线，这样能规范取用水管理，有效杜绝一些违规取用水情况的发生。但是两省取用水管理还存在不少问题，面临的普遍问题是水利用效率低下，尤其是农业取用水，渠系无衬砌，漏水严重，输水效率低下，农灌方式粗放，浪费水比较严重；一些单位取用水计量设施安装不到位，取用水日常管理维护不及时等问题亦困扰着取用水管理工作有效开展。这些问题亟须在加快水利改革发展的大环境下，通过各方不断工作逐步解决。

在从事取用水管理工作中，笔者亲身体验了民勤人如何利用有限水资源开展生产并与沙漠抗争，实现沙退人进的壮举的。民勤，位于内陆河石羊河流域下游，其西、北、东三面被巴丹吉林和腾格里两沙漠包围，总土地面积15 907 km^2，耕地面积96万亩，人口29万人。民勤地势平坦，土层深厚，集中连片，气候温和，光热条件好，有发展农、林、牧业的巨大潜力，但是年均降雨量只有70 mm，地表水量有限，缺水成为该地经济发展的一道紧箍咒。鉴于此，该县采取多种措施，加强取用水管理，大力发展生态农业，提倡节水。开源，民勤人能想敢想,将部分景电灌溉提水引入民勤,用黄河水滋润干旱的土地;节流,渠系改造工程、田间节水工程及种植业结构优化调整等手段提高水利用效率;管理,规划引导、政策扶持、技术服务促进农业高效化,采用水资源信息化管理手段,极大提高灌区管理效率。通过多种措施配合实施,灌区灌溉水利用率明显提高。

民勤是一个小系统,更大的系统,黄河流域上中游地区乃至全国的取用水管理工作,有待我们水利人在不断创新中努力完善管理机制。随着我国的宏观形势的变化,水利工作的许多观念和思路也在发生变化,水利行业进入了一个新的历史阶段,形势的发展使取用水管理工作面临着新的发展机遇,面对这些机遇,取用水管理工作必须坚定不移地贯彻一号文件的指导思想,从水资源宏观调配到取用水管理基础实践各方面均必须切实落实相关要求,这就要求取用水管理单位及其管理对象从不同角度思量自身行为是否符合相应要求,并考虑如何通过协调组织,合理利用现有的水资源满足社会经济发展和生态平衡需要。

对于取用水管理单位,从管理角度讲,主要以国家政策导向为依据,宏观控制、引导为主,对水资源分配利用进行监督管理。当前取用水管理工作框架结构是,水利部负责全国范围内水资源宏观管理,流域机构和地方水行政主管部门在各自职责范围内进行取用水管理实践活动。为落实最严格的水资源管理制度,取用水管理决策单位必须规范水资源论证,切实落实取水许可制度,抓好水资源优化配置和统一调度并抓好取用水管理的各项基础工作。

一是要规范和完善水资源论证管理制度。要进一步规范水资源论证资质申请、受理和审批行为,不断提高水资源论证工作质量。各水行政主管部门要依法严格审批,切实为水资源的开发利用把好关口。

二是要进一步贯彻落实取水许可制度。要加强取用水登记管理,完善取用水报表统计,做好水资源公报;抓好取水工程验收,全面开展计划用水工作,强化涉水事务的行政管理,依法治水。

三是要进一步抓好水资源的优化配置和统一调度,构建节水型社会。水资源的配置必须采用行政、经济、法律等多种调节手段来进行规范,各级水行政主管部门要高度重视水资源的优化配置和统一调度,特别是在用水矛盾突出的地方,要通过工程与非工程措施,对有限的水资源进行科学合理的分配,解决各部门、各行业之间的用水协调问题,满足经济社会发展对水资源的需求。要加大水资源配置和统一调度力度,通过制度建设引导水资源的节约和优化配置,通过节水规划的实施,促进经济和产业结构的优化升级,通过采用节水技术、设备、工艺和标准,提高水资源的利用效率。

四是要切实抓好取用水管理的各项基础工作。水资源调查评价、规划、监测、取水户档案资料等基础工作是取用水管理的重要依据,要抓好取用水管理计量设施、设备、监测手段、试点建设、规划编制、政策制定和技术改造等基础工作,提高取用水管理基础设施和技术装备水平,抓好信息系统建设,提高基础工作效能。

对于取用水管理单位,从被管理者角度讲,要按规定完成各取水工程项目水资源论证,配合取水许可监督管理,对不合理要求可以不予执行或者申请复议;从取用水角度讲,应当依托技术创新和完善管理方式,提高水资源利用效率,减少浪费。

一是按照各项法规要求,积极开展并完善水资源论证工作和取水许可申请工作,配合取水许可监督管理。水资源是一种有限的资源,国家对水资源利用实施宏观规划控制,统一调度。作为系统中的受调控对象,各个取用水单位必须服从调度安排,积极配合监督管理,以保障取用水有序,实现社会利益和生态效益的最大化。以民勤为例,为满足经济社会发展和生态用水需求,跨流域跨地区实现水资源统一调配亦不失为一种方法,但前提是要做好充分的论证工作,兼顾经济效益、社会效益及生态效益。

二是通过产业结构调整、技术创新等手段,多途径提高水资源利用效率,减少浪费。由于水资源有限,直接开辟新的水源已经不现实,最合理有效的途径就是提高水利用效率,一水多用,废污水循环处理重复利用等。同时通过产业结构调整,加强技术创新,提高用水效率,淘汰落后用水设备、工艺,促进节水观念深入贯彻。民勤地表水有限,必须从自身内部挖潜,提高水利用效率和单位水产值,采取的办法有多种:整治渠系,降低输水损失;调整农作物种植结构;推广节水技术。事实证明,上述多种措施配合实施,提高了灌区灌溉水利用系数,起到了良好的节水效果。

三是加强取用水单位自身管理,做到有序取用水。管理出效益,加强取用水管理,完善计量设施,构建取用水信息化系统,实现科学监控、计量、调度水量。以民勤为例,民勤取用水采取精细化管理,明晰水权、过程控制、强化监督、调整水价等,其核心观点就是将取用水效能直接与管理单位和用水户切身利益关联,实现节水观念深入人心,更有效地开展工作。

参考文献

[1] 取水许可和水资源费征收管理条例(国务院令第 460 号).

[2] 取水许可管理办法(水利部令第 34 号).

[3] 关于授权黄河水利委员会取水许可管理权限的通知(水利部水政资[1994]197 号).

[4] 黄河取水许可管理实施细则(黄河水利委员会黄水调[2009]12 号).

[5] 黄河上中游流域取水许可监督管理实施办法(试行)(黄河上中游管理局黄水政发[2009]16 号).

【作者简介】 李平,1985 年 3 月生,湖北麻城人,学士,2008 年毕业于复旦大学,现工作于黄河水土保持天水治理监督局,主要从事基层水政监察和水资源管理工作。

基于流域统一管理的黄河源区立法研究

王瑞芳　高小平　董光敏　韩玉峰　张　颖

（黄河水土保持天水治理监督局　甘肃　天水　741000）

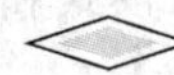

【摘要】 本文通过调查分析源区的流域特性、水资源保护立法和实践现状以及水资源开发利用现状及存在的问题，阐述了源区统一立法的必要性和可行性，并基于流域统一管理提出源区立法构想，建立适合源区特性的流域管理体制和事权明晰的水行政管理运行机制，实现由规范源区水资源开发利用，水生态环境的恢复为目的管理转变为可持续性的流域水资源-环境-生态的综合管理。

【关键词】 流域管理　黄河源区　立法　研究

2002年修订的《中华人民共和国水法》（以下简称《水法》）确立了流域管理体制框架，所规定的流域管理基本原则，适应了我国的水资源管理背景、各流域水问题的特殊性以及立法传统，成为加快流域管理法规体系建设的依据和有利契机。黄河源区对黄河流域生态安全具有无可替代的重要战略地位，但现有水生态环境保护和立法明显滞后于社会经济发展的步伐，制约了生态环境保护建设和水资源的可持续利用。针对黄河源区这一生态特区的特性进行专门立法，从流域整体出发对流域内一切涉及环境资源使用及保护的问题进行综合管理，是对源区水生态环境保护的一个重要途径，也是我国水资源流域管理的必然趋势。

1　源区水资源在我国可持续发展中的战略地位亟待统一管理

黄河源区被誉为“中华水塔”，源区生态系统对黄河水资源环境的变化将产生重要影响，对维持黄河中下游生态平衡起着积极的保护作用。近年来，受自然和人为因素双重作用的影响，源区生态环境急剧恶化，径流减少和湖泊湿地萎缩，草甸草场持续退化，植被覆盖率下降，土地荒漠化和沙化速度加快，人口增加和不合理开发利用造成的人为新增水土流失十分严重，直接或间接地加剧了黄河中下游地区水土流失、河水断流和河床的增高，特别是加大了洪涝灾害的威胁，也将威胁到西北和华北地区的生态安全，关系到黄河流域社会经济的可持续发展。因此，通过流域管理立法保障流域管理机构广泛的管理权力，制定效力及黄河流域和源区的法律法规体系，从全局利益出发打破行政区域条框分割管理，增强源区生态特区统一管理的针对性和可操作性，实现“维持黄河健康生命”的目标，以黄河源区水土资源的可持续利用保障流域社会经济的可持续发展。

2　源区水资源保护立法和实践现状要求统一立法

我国现行的水行政管理法律《中华人民共和国水法》、《中华人民共和国环境保护法》、

《中华人民共和国水污染防治法》、《中华人民共和国水土保持法》和配套法规在源区生态环境建设和水资源保护方面发挥了很大作用，但仍未能有效遏制源区生态环境恶化的趋势。流域统一管理法律依据的不足，水资源开发利用与保护的管理体制和机制不健全导致源区水资源开发利用缺乏有效的监督管理。

2.1 水行政管理的法规体系不健全要求专门立法

一是现有法律在立法上未明确法律效力等级，不利于对不同的行为形成规范体系。上述四部法律虽具有同等法律效力，但从理论与实践上看，水资源管理与水土保持、环境保护与水污染防治显然都不是同一层次的问题，在立法上也应有不同的法律效力等级，才有利于对不同的行为形成规范体系，目前这种立法模式和运行机制显然不能满足源区水生态环境保护的需要。

二是法律之间关系的不协调制约着流域统一管理。四部法律都对管理体制作出了规定，由于在立法时缺乏综合平衡，加之各个法律规定过于原则，可操作性不够，缺乏相应的配套法规和程序性规定，致使一些法律制度的适用范围不明，各自所确立的主管部门和协管部门在具体实施时困难重重，难以协调。

三是管理体制不完善，监督管理机制不健全在一定程度上限制了流域统一管理。《水法》对流域管理机构与地方行政区域及相关部门之间的事权划分零星且不明确，流域管理与区域管理相结合的管理体制尚不健全，缺乏协商沟通机制。行政区域管理注重区域利益，对区域之间、上下游、左右岸的利益关系考虑不够，综合管理网络尚未形成，没有从监督管理机制上形成对源区生态环境和水资源保护的全面综合管理。况且流域机构只是水利部的派出机构，是事业单位，不属于政府系列，受宏观管理职能限制，面对黄河源区水行政管理这项跨行政区，涉及水利、农牧、林业、环保、土地等部门的系统工程时，流域机构缺乏对多部门、多行业开展协调、合作及联合执法的法律法规支撑，从而使源区水生态环境保护的统一管理得不到充分保障。

四是流域管理机构的法律地位不明确，制约了黄河源区统一管理的实施。由于《水法》没有明确监督管理的执法主体，目前流域机构行政和执法主要依据于水利部的“三定”方案及有关文件，后来颁布的《中华人民共和国防洪法》、《河道管理条例》和《取水许可制度实施办法》虽然对流域机构的部分职责和权限作出了规定，但仅是单项授权。流域管理及法律地位不明、权力缺乏就有的强制力，致使流域机构对源区水资源保护的统一管理举步维艰。

2.2 源区水生态环境保护的流域特性需要实行统一管理

黄河源区资源既是一种环境资源又是一种经济资源，具有生态系统的完整性、跨行政区域性和使用的多元性特征，而无论是对流域资源的何种使用都涉及对资源的保护与管理问题。资源环境管理体制作为政府行政管理体制的组成部分，长期以来实行的是国土矿产、水利水电、草原畜牧、环境保护等各个资源管理部门的分散管理，这种监督管理体制虽然在一定程度上有利于发挥管理部门的专长，但从整体上来看不利于生态环境的保护。另外，从效率原则看，生态环境的统一监督管理也是必要的。对现有水资源进行有效保护、合理配置、科学利用，需要水利部门与森林、草原、水土保持、矿产、土地等部门间建立统一协调的管理秩序，黄河源区良好的水事秩序必须通过立法来保障。

2.3 源区资源开发利用现状及存在的问题要求加强统一管理

2.3.1 生态环境问题突出，水源涵养能力下降

近年来，受自然和人为因素双重作用的影响，源区生态环境急剧恶化，草场退化严重，生物多样性锐减，人类日益增长的物质需求，使天然草场的承载压力不断加大；湿地面积萎缩，水源涵养能力下降，河流补给量急剧减少，对全流域的供水和生态安全构成了严重威胁；草原鼠害猖獗，随着经济发展和人口增加，修建水电站、修路、开矿、采药、采砂等人类活动破坏植被、弃土弃渣，造成新的人为水土流失，随着国家西部大开发战略的实施，区域经济快速发展，资源过度开发利用，水电开发建设项目缺乏统一管理，生态移民逐年增加，加剧了生态环境的恶化。

2.3.2 综合规划滞后，流域管理缺乏依据

一是综合规划滞后，流域管理缺乏依据。源区在缺少黄河源区流域综合规划的情况下，地方部门修订了区域规划，水电部门编制了黄河源区干流水电梯级开发规划，导致了目前开展的水工程建设程序虽不符合《水法》要求，但因流域管理缺乏基本依据而相对滞后，以行政区域管理为主的局面。

二是基础工作薄弱，流域管理缺乏必要的技术支撑。黄河源区生态系统是一个整体，其水生态环境保护是一项科学性、技术性很强的系统工程。目前，还没有形成对源区生态环境和自然资源进行综合考察、调查、评价、监视、监测的网络和信息平台，各行业之间缺乏数据共享，限制了对源区水生态环境的有效保护。

3 流域通过立法管理是确保源区水生态环境可持续发展的必然选择

3.1 依法治国、依法治水为源区水生态环境保护创造了有利环境

国家依法治国战略的推进，新时期依法治水方针的贯彻落实，为黄河源区立法创造了良好的法制环境。黄河源区水资源情势的变化、生态环境的恶化和水电梯级开发的加快，已引起党和国家领导人的高度重视。国家科学技术部、水利部、黄委、长委等单位在源区开展的科考、调研和规划等工作，为黄河源区立法提供了技术依据。近年来，国家及地方各级政府相继实施了一系列生态保护与建设工程，取得了显著成效。源区群众切身体会到生态文明带来的社会进步，增强了保护环境的自觉性。

3.2 现有法律法规、管理机构为源区统一管理提供了支撑

目前，国家已颁布实施了多部涉水法律法规和部门规章，三省也出台了相关法律法规的实施办法和细则，形成了以法律为主体、法规和规章为配套、三省地方法规为辅的法律法规体系，为控制源区生态恶化、遏制人为水土流失发挥了重要的作用，为源区开展立法奠定了基础。

黄河水利委员会具有法律赋予的对流域实施统一管理的职能，具有完善的组织机构及各级水行政执法队伍、健全的规章制度，并开展了流域性的立法、河道建设项目管理、水资源保护与管理及水土保持预防监督等工作，取得了明显成效，积累了丰富的流域管理工作经验，为下一步在源区立法及实施奠定了坚实基础。

4　源区流域管理立法的基本构想

4.1　源区立法方案

黄河源区立法方案的不同，将直接影响立法实施的效果。立法的关键是提出具有针对性的管理方案，平衡和协调源区各行业、各部门利益，使现有管理弱化的局面通过上升到法律层面，使源区水行政管理工作得到强化，促进源区生态环境管理的重大改进和取得突破性进展。

4.1.1　第一方案——《黄河法》

由全国人大常委会制定统一协调、切实可行的《黄河法》。根据黄河源区特殊的地理位置和社会经济情况，针对黄河源区水事活动管理的实际和管理工作的紧迫性，把黄河源区管理纳入《黄河法》范畴统一考虑，制定黄河源区流域统一管理法规。

4.1.2　第二方案——《黄河源区管理条例》

由国务院制定《黄河源区管理条例》，以国家专门立法的形式理顺各部门的责、权、利的关系，确定源区重要地位、保护范围、保护政策、补偿机制、监管措施等，依法对黄河源区生态环境保护、水资源开发利用与保护等各个方面进行有效管理。

4.1.3　第三方案——《黄河源区管理办法》

由水利部颁布实施《黄河源区管理办法》，以完善流域与行政区域相结合的管理体制，进一步明晰流域与区域的事权划分。确保黄河源区管理机构法定化、规范化、投资多元化、实施主体市场化、评估科学化，并引入水资源管理成本制约机制和生态补偿机制，增强法律的超前性、针对性和可操作性。

4.1.4　第四方案——建立流域机构和源区各省联席会议制度

在坚持行政区域管理服从流域管理，国家利益和流域整体利益优先的前提下，形成流域管理机构与青、川、甘地方政府多方参与、民主协商、共同决策、分工负责的管理运行机制，确保源区保护的各项法律法规有效实施。

4.1.5　方案比选

以上四个方案中，第一方案的制定、颁布及实施尚待时日，需要解决黄河全流域的问题，针对黄河源区的管理不可能具体深入，操作性不足；第二方案的制定，需要水利、环保、土地、农牧、林业等多部门联合开展立法调研，工作量大、协调任务重、时效性差；第三方案既体现了流域管理与行政区域管理相结合的管理体制，也符合黄河源区实行水行政统一管理的实际需要，能够确立流域机构的主体地位，从而起到较好的水行政管理效果；第四方案以现有国家法律法规和地方法规为主，实行源区各省联席会议制度，在具体实施管理中，存在整体性、同步性和协调性局限等问题。

综上所述，第三方案《黄河源区管理办法》切实可行，建议在前期立法调查研究的基础上，加快立法进程，制定具有科学性、预见性、可操作性和权威性的《黄河源区管理办法》。为黄河源区生态环境的改善建立法律层面的管理依据。

4.2　《黄河源区管理办法》的主要内容分析与拟订

根据立法的一般要求和管理办法的需要，黄河源区管理办法应主要由总则、综合规划、环境保护与水土保持、水资源管理、河道建设项目管理、监督管理罚则和附则等八大部分构成。

在明确《黄河源区管理办法》制定的目的意义、立法依据、原则、管理范围、职责、内容等基础上，根据黄河流域综合规划制定源区规划，指导黄河源区综合治理与开发。按照法律法规的规定和授权，黄河水利委员会及其所属机构与青海、四川、甘肃三省人民政府各级水行政主管部门在各自辖区范围内负责黄河源区的水土保持生态环境建设、水资源的合理开发、优化配置、高效利用和有效保护以及河道建设项目的监督管理，建立健全水土保持预防监督、水资源监督和水资源开发利用保护等制度，建立适合源区特性的流域管理体制和事权明晰的水行政管理运行机制，加大源区流域统一管理和保护力度。

5 结论与建议

(1)黄河源区流域统一管理通过立法程序确立，既是必要的也是可行的。

(2)《黄河源区管理办法》的颁布实施将依法保障黄河源区面临的生态环境和水资源严重危机需要解决的主要问题和建立的法律制度，为黄河源区生态环境的改善建立法律层面的管理依据。

(3)目前开展的黄河源区管理立法调查研究为进一步开展源区管理立法奠定了基础，建议加大源区立法进程，实现由规范源区水资源开发利用，水生态环境的恢复为目的管理转变为可持续性的流域水资源-环境-生态的综合管理。

参考文献

[1] 王国永.流域管理的法规体系构建[J].人民黄河,2011,33(8):60-61.
[2] 雷玉桃,谢建春,王雅鹏.我国水资源流域管理的创新对策[J].水利经济,2003,21(6):12-13.
[3] 徐军.我国流域管理立法现状及反思[J].河海大学学报:哲学社会科学版,2004,6(4):20-24.
[4] 吕忠梅.长江流域水资源保护立法问题研究[J].中国法学,1999(02).

【作者简介】 王瑞芳,女,1968年生,1990年毕业于河海大学水资源水文系,获工学学士学位。现任黄河水土保持天水治理监督局水政水资源科副科长,高级工程师。

数学模型在县级土地利用总体规划中的应用

——以武山县土地利用总体规划中的人口预测为例

刘 晓 张满良 张琳玲

（黄河水利委员会天水水土保持科学试验站 甘肃 天水 741000）

【摘要】 人口预测是土地利用规划的首要工作，合理预测人口对土地利用规划和土地可持续发展有着十分重要的意义。本文运行马尔萨斯人口模型、GM(1,1)灰色模型、线性回归分析预测法等三种预测模型，采用1990～2009年《武山县统计年鉴》人口资料，对武山县2006～2020年人口发展规模进行了预测。预测结果显示，三种预测模型都取得了较好的模拟效果，但马尔萨斯人口模型和GM(1,1)灰色模型预测的结果比回归分析预测的结果更理想、误差更小。采用两种非线性回归模型预测结果的平均值作为预测结果，2010年武山县总人口规模达到459 473人，2020年武山县总人口规模达到510 963人。

【关键词】 人口预测 马尔萨斯人口模型 GM(1,1) 灰色模型 线性回归分析预测

人口预测是土地利用规划的首要工作，它既是规划的目标，又是确定土地利用规划用地规模与布局的前提和依据。合理预测人口对土地利用规划和土地可持续发展有着十分重要的意义。人口预测是指以规划区域或单位现有人口现状为基础，并对未来人口发展的趋势提出合理的控制要求和假定条件即参数条件，来获得对未来人口数据提出预报的技术或方法。一般需要充分收集资料、确定预测参数，通过建立预测模型来进行，包括人口数量、人口性别和年龄构成等。人口预测的方法有很多，常用的有人口自然增长法、Logistic增长模型、时间序列法、带眷系数法、马尔萨斯人口模型、GM(1,1)灰色模型、线性回归分析预测法、劳动平衡法、修正指数模型预测法等。

武山县属甘肃省天水市，是“关中-天水”经济区最西端的三级城市。本文运用线性回归分析预测的方法，对武山县人口规模在未来15年的发展趋势做出预测，旨在为武山县今后土地利用规划提供参考依据。

1 武山县历年总人口及其增长率变化趋势分析

武山县总人口由1991年的370 878人增加到2009年的451 408人，19年间增加了80 530人，平均年增长率10.40‰。总体呈现缓慢下降的发展趋势，但个别年份变化较大。

武山县总人口1991～2009年的变化趋势是：总人口数缓慢增长，各年增长的人口数相对平稳。1991～1996年总人口增长幅度较大，平均年增长率达13.97‰，其中1996年较1995年人口增加较多，年增长率25.23‰，是人口统计误差造成的；1996～2003年总人口增长速度较为平缓，年增长率9.20‰；2003年起人口开始缓慢下降，2004～2009年，年增长率稳定在7‰左右，武山县1991～2009年历年总人口见表1，1991年以来武山县人口比上年增

长率见图1。

表1 武山县1991~2009年历年总人口(人)

年份	总人口	年份	总人口	年份	总人口	年份	总人口
1991	370 878	1996	403 069	2001	426 300	2006	442 300
1992	377 676	1997	406 852	2002	429 400	2007	445 242
1993	382 664	1998	411 089	2003	433 700	2008	448 268
1994	388 697	1999	414 339	2004	436 400	2009	451 408
1995	393 151	2000	419 470	2005	439 500		

注:资料来源于武山县统计年鉴(1991~2005)。

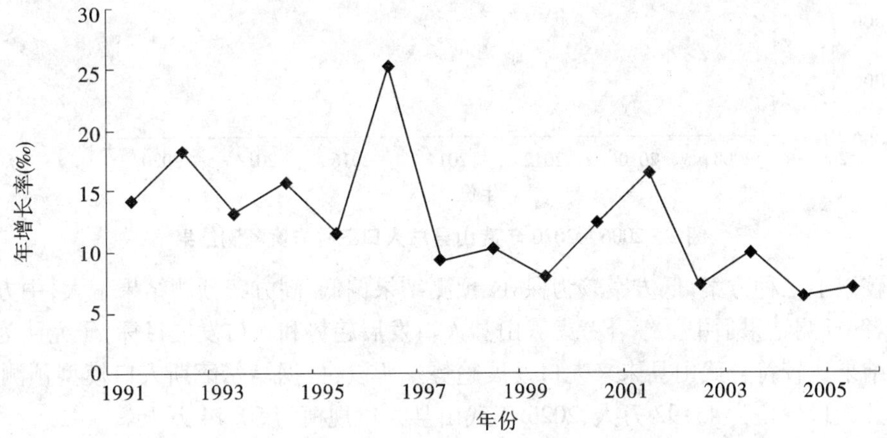

图1 1991年以来武山县人口比上年增长率

2 模型预测

武山县的人口统计资料较为完整,为人口预测提供了较为充分的依据。本文选择武山县1991~2005年的统计数据来建立模型,选用2006~2009年的统计数据来验证模型,分别采用马尔萨斯人口模型、GM(1,1)灰色模型、线性回归分析来预测武山县规划期内的总人口数。

2.1 马尔萨斯人口模型

英国人口学家马尔萨斯根据百余年的人口统计资料,于1798年提出了著名的马尔萨斯人口模型,其建模思路如下:在简单情况下,人口的(相对)增长率是常数,人口预测采用指数增长函数。该模型的预测公式为:

$$P(t) = P_0 (1 + r)^{(t-t_0)}$$

式中:t 为预测年份;$P(t)$ 为 t 年的总人口规模;r 为人口增长率;P_0 为初始年的总人口规模;t_0 为初始年份。

武山县1991~2005年的年人口平均增长率为11.38‰,综合考虑武山县人口发展现状,认为2005~2020年武山县人口增长率呈现逐渐降低的趋势,设定2005~2020年人口的增长率分别做高、中、低三个方案预测。高方案认为2005~2020年武山县的人口增长率保持较高增长,取1991~2005年的年人口平均增长率11.38‰;低方案取2001~2005年武山县的人口平均增长率6.12‰;中方案取其平均值8.82‰。

对应上述三种人口自然增长率方案,运用马尔萨斯人口模型,以2005年为基期年,可以得到2005~2020年武山县总人口高、中、低三个不同方案的预测值,见图2。

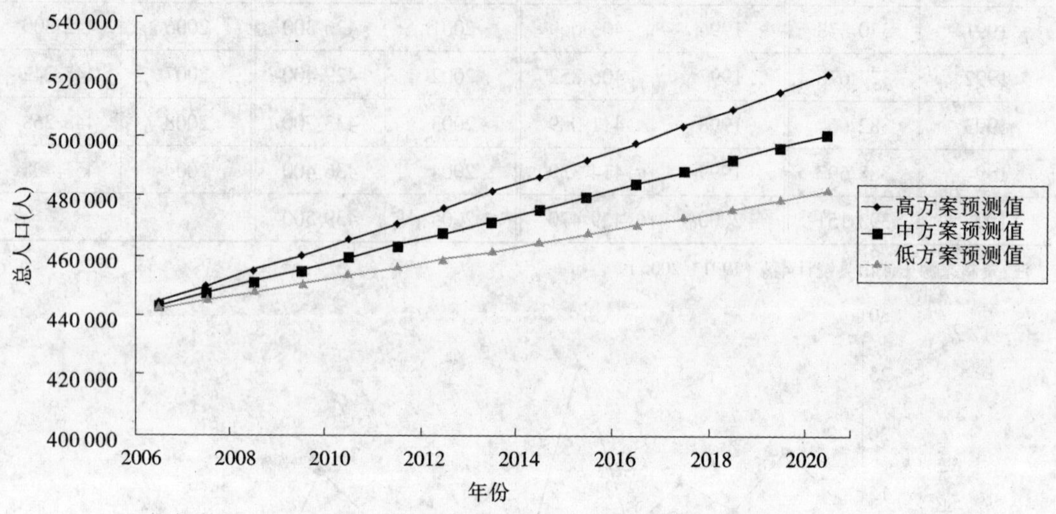

图2 2006~2020年武山县总人口三种方案预测结果

比较以上三种方案,低方案较为保守,预测结果偏低,高方案预测结果偏大,中方案预测介于两者,预测结果适中。综合考虑武山县人口发展趋势和人口发展目标,研究认为中方案的预测结果比较符合武山县未来人口发展趋势。所以,根据马尔萨斯人口模型预测到2010年武山县人口规模为45.92万人,2020年武山县人口规模为50.14万人。

2.2 灰色系统 GM(1,1)模型

灰色系统动态模型GM(1,1)是依据灰色系统理论原理,利用系统的离散采集数据建立其动态微分方程,以灰色模块为基础,微观拟合分析为核心的建模方法。主要特点是能够用较短的基础数据序列,将抽象的问题实体化、量化,将变化规律不明显的情况,找出规律,通过规律分析事情的发展变化,揭示发展过程的优劣,找出对策,以促进系统的满意、高效发展,为中长期规划编制、决策创造条件,用它可以通过对系统过去和现在采集的数据集对系统未来的发展趋势做出预测。GM(1,1)不直接利用原始数据,而是通过累加生成灰色模型,滤去原始数据中可能混入的随机量或者其他参杂数据,从上下波动的时间数列中寻找某种隐含规律。

根据武山县1996~2005年的总人口数据,利用统计软件,可以得出拟合方程:

$$x(t+1) = 438\ 860.471\ 4e^{0.010\ 623t} - 83\ 026.625\ 1$$

软件对当前模型的评价为优,模型可信度高。

利用灰色系统预测GM(1,1)模型计算出武山县2006~2020年的总人口数,见表2。

2.3 线性回归模型

回归分析实际上就是建立某种数学模型并做检验的一种数学预测方法,已经广泛的应用于人口、经济、土地利用等方面。它假定一列(或多列)数据的变化同另一列数据的变化呈某种函数关系,衡量数据联系强度的指标,并通过指标检验其符合的程度,就称为回归分析,线性回归分析是建立预测模型最常见、也是最简单、最常用的方法之一。

表 2　灰色系统预测 GM(1,1) 模型下武山县 2006 ~ 2020 年的人口数

年份	总人口	年份	总人口	年份	总人口
2006 年	437 140	2011 年	465 516	2016 年	495 439
2007 年	442 695	2012 年	471 374	2017 年	501 617
2008 年	448 310	2013 年	477 295	2018 年	507 861
2009 年	453 984	2014 年	483 279	2019 年	514 171
2010 年	459 719	2015 年	489 327	2020 年	520 549

一元线性回归模型建立如下：

$$P(t) = a + bt$$

式中：t 为年份序号（1991 年为 1，1992 年为 2，以此类推，2006 年为 16，2020 年为 30）；$P(t)$ 为 t 年的总人口规模；a、b 都是模型参数，可根据历史数据（样本），用最小二乘法求出。

将 1996 ~ 2005 年武山县总人口数据代入，进行线性拟合，求得：$P(t) = 4\ 965.8\ t + 369\ 153$，$R^2 = 0.989\ 2$，精度较高，模型建立较为理想。

将 2006 ~ 2020 年的序号 16 ~ 30 分别代入模型，预测出武山县 2006 ~ 2020 年的总人口数见表 3。

表 3　一元线性回归模型下武山县 2006 ~ 2020 年总人口预测

年份	总人口	年份	总人口	年份	总人口
2006	448 606	2011	473 435	2016	498 264
2007	453 572	2012	478 401	2017	503 230
2008	458 537	2013	483 366	2018	508 195
2009	463 503	2014	488 332	2019	513 161
2010	468 469	2015	493 298	2020	518 127

2.4　模型的验证情况

通过利用武山县 1991 ~ 2005 年人口统计数据建立预测模型，预测武山县 2006 ~ 2020 年的人口发展规模，通过比较 2006 ~ 2009 年人口预测数据与实际统计值，得出预测误差的大小（见表 4）。

表 4　三种模型的预测值及误差

年份	实际统计值	马尔萨斯人口模型		灰色系统 GM(1,1) 模型		线性回归模型	
		预测值	相对误差（%）	预测值	相对误差（%）	预测值	相对误差（%）
2006	442 300	443 346	0.24	443 556	0.28	448 606	1.43
2007	445 242	447 225	0.45	448 293	0.69	453 572	1.87
2008	448 268	451 138	0.64	453 080	1.07	458 537	2.29
2009	451 408	455 086	0.81	457 919	1.44	463 503	2.68

通过误差对比分析，马尔萨斯人口模型的平均相对误差为 0.53%，灰色系统 GM(1,1) 模型的平均相对误差为 0.87%，线性回归模型的平均相对误差为 2.07%。说明马尔萨斯人口模型与灰色系统 GM(1,1) 模型比线性回归模型的误差更小而且相近，模拟精度好，故采

用上述两种非线性回归模型预测结果的平均值作为预测结果。

3 未来人口规模预测结果

本研究将马尔萨斯人口模型与灰色系统 GM(1,1)模型预测结果的平均值作为预测结果,预测在 2010 年武山县总人口规模达到 459 473 人,2020 年武山县总人口规模达到 510 963人。说明武山县人口增长呈现较稳定的增长状态。

4 结语

总结上述三种模型的模拟及预测结果,可以发现:

(1)马尔萨斯人口模型、灰色系统 GM(1,1)模型和线性回归模型都可以满足人口预测的精度要求。

(2)线性回归模型精度较低,但直观易行、实际操作简便,适宜于对精度要求较低的预测研究。

(3)马尔萨斯人口模型、灰色系统 GM(1,1)模型预测精度较高,预测结果相近,适宜于对于精度要求较高的预测研究。

人口的增长受多方面因素的影响,任何一种模型都不能完整地预测其发展结果,具体采用何种模型进行预测,需要根据实际情况加以选择。

参考文献

[1] 石培基,祝璇.甘肃省人口预测与可持续发展研究[J].干旱区资源与环境,2007(9):1-5.
[2] 郭成利,董晓峰,潘竟虎,等.城市建设用地规模预测与分析——以兰州市为例[J].河北农业科学,2009(1):57-59.
[3] 邹自力,刘珊红.人口预测方法及可靠性探讨[J].华东地质学院学报,2002(2):142-146.
[4]《武山统计年鉴》(1990~2009 年),武山县统计局.
[5] 清远市清新县土地利用总体规划(2006~2020 年),清新县人民政府.

【作者简介】 刘晓,男,1979 年生,2002 年毕业于西南科技大学,现就职于黄河水利委员会天水水土保持科学试验站,助理工程师,长期从事水土保持、土地利用规划工作。

黄河拉西瓦水电站精密水准网关键技术研究

李祖锋[1,2] 巨天力[1] 张成增[1] 缪志选[1]

(1. 中国水电顾问集团西北勘测设计研究院工程勘察研究分院 西安 710065;
2. 甘肃农业大学工学院 兰州 730070)

【摘要】 为解决拉西瓦水电站特殊的工程环境给水准网建网带来的诸多难题,进行了多项技术攻关。为了提高水准点的稳定性,对水准点的建造方法进行了改进;为了解决精密水准网环线闭合差系统性偏大的问题,对隧洞内微气象条件下精密水准测量的影响特征进行了初步研究。

【关键词】 精密水准 基准点 隧洞 折光差

1 引言

拉西瓦水电站精密水准网是建立和维持该工程沉降变形监测系统的基准网。为解决拉西瓦水电站特殊的工程环境给水准网建网带来的诸多难题,进行了多项技术攻关。现将该精密水准网建网过程中所采用的关键技术进行阐述。

(1)沉降监测点的变形信息是相对于参考点或一定高程基准的,没有稳定的高程基准就无法获取高精度的高程数据,本工程通过建造双金属标对基准点的稳定性问题进行了有效解决。

(2)拉西瓦水电站冬季极端温度低,冰冻期长,据类似工程经验,其冻土膨胀对水准点所造成的影响严重。为了保证水准点的稳定性,并兼顾建网的经济性,制定了冻土基础上水准点的建造方法。

(3)为了解决施测过程中遇到的各次测量水准环线闭合差系统性超限的问题,开展了大量的试验研究工作,找到了系统误差源,并针对误差源特性制定了测量方案。

2 水准网点的建造

2.1 一等精密水准网的布置

拉西瓦水电站一等精密水准网,共计布设 35 个水准点,总体沿黄河两岸的高、中、低线进场公路布设,共由 3 个水准环线组成。水准网布置见图 1。

2.2 水准基准点的建造

高程基准点的稳定是保证沉降监测成果质量的前提。为了保证基准点不受外界因素的影响,基准点采用双金属标建造。

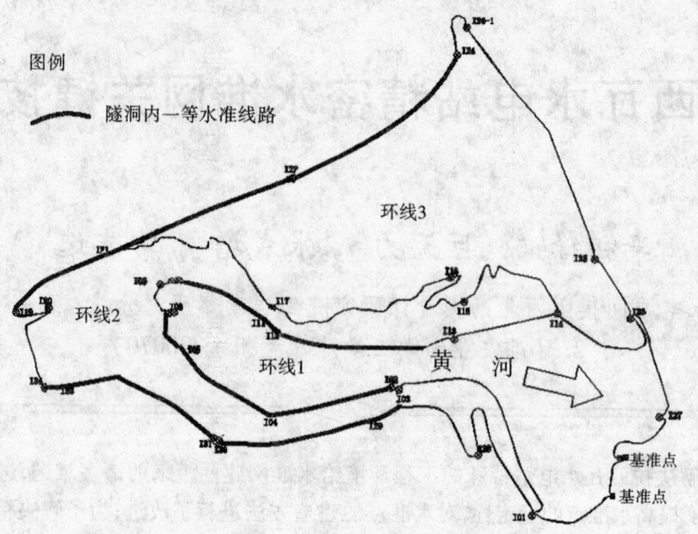

图 1　拉西瓦水电站一等精密水准网布置图

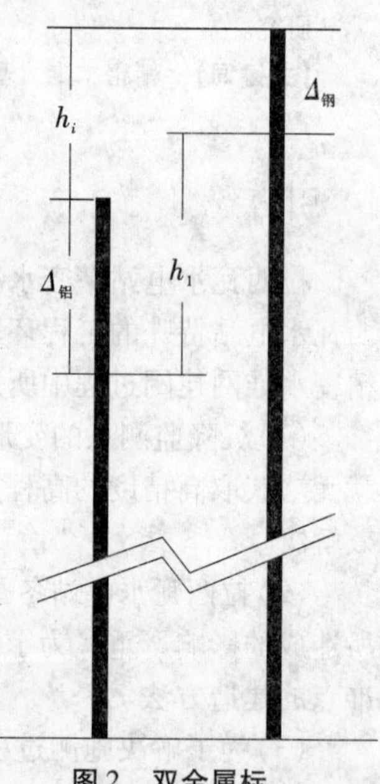

图 2　双金属标

双金属标是由膨胀系数不同的两根金属管组成的水准标志(见图 2),可根据两管长度变化的差数修正管顶的高程,它的优点是能对温度所产生的标点变形进行修正,从而提高垂直位移的观测精度。

2.3　工作水准点的建造

拉西瓦水电站位于冬季极端温度较低的高原地区,冰冻时间长,其冻土膨胀所造成的影响严重,土质及砂砾混合地基吸水后具有显著的湿陷或者膨胀(冰冻条件下)特性,可能会导致基础下沉或者上升。当冬季冻土厚度较大时,在埋设标石的土壤冰冻期间,冰晶和冰体在标石侧面冻结,会出现显著的膨胀特性,地基膨胀挤压会导致点位上升,但消融时一般不会回归到原始位置,这两种结果都会威胁到水准点的稳定。在我们已经完成的某类似工程中,5 次测量中发现了 3 个水准点发生了显著上升,平均上升达到了 8.5 mm,通过分析我们发现这 3 个点所在位置存在一个共性特征,就是其基础均为土质或者砂砾混合物,再通过数据比对,发现点位的上升一般发生在一个冻融周期之后,其属于冻土造成的点位逐年上升。为了解决这些问题,在建造过程中采取了以下两项措施:

(1)对于土质或砂砾混合基础,避开含水率较高的区域,水准点周边采取地表防水措施;水准点基础采用梯形结构建造,使冻土膨胀作用方向与基础面成锐角。

水准点埋设深度有着较具体的要求,但当出现冻土厚度较大的情况,冻土层的冻拔作用可能导致处于冻土层以下的基础不能平衡处于其上部的冻土作用,导致水准点失稳。

(2)为保护水准点而修建的围护要与水准点受力结构分离,以保证围护发生较小变形不破坏水准点的稳定性。

3　隧洞约束环境下大气折光对精密水准测量的影响

3.1　隧洞内精密水准测量存在的问题

在拉西瓦水电站精密水准网最初的3次测量过程中,均遇到了水准环线闭合差较大,其顺时针推算的环线闭合差均为正的问题,3次之间的闭合差偏差均较小,表现出系统特性。

针对闭合差超限的问题,一般所采取的传统措施是返工,但应充分尊重第一手资料,盲目重测会给人虚假的精度信息。

虽然水准高差值不同季节,甚至上下午测量结果不同,但其所构成的环线闭合差偏差变化较小,但是闭合差是重要的较全面反映水准测量误差的指标,下面我们的讨论将侧重于闭合差指标。

为了说明环线闭合差大且具有一定的系统性这个问题,我们对3次测量的部分数据进行了统计,统计的部分环线闭合差及高差绝对值见表1、表2。

表1　主要环线闭合差统计

项目	闭合差1(mm)	闭合差2(mm)	闭合差3(mm)	路线长度(km)	闭合差平均值(mm)
环线2	8.9	6.9	7.3	15.3	7.7

注:环线闭合差指顺时针推算的环线2闭合差。

表2　主要测段高差统计

项目	高差1(m)	高差2(m)	高差3(m)	路线长度(km)	高差平均值
测段1	34.003 4	34.001 0	34.001 5	1.2	34.001 9
测段2	101.033 9	101.033 4	101.033 7	1.7	101.033 7

相对高差指右岸高线隧洞内试验所测的部分水准高差绝对值。受坡度因素影响,其中测段1视距平均值约为20 m,各站视线高度平均在0.8 m以上,最高视线高度为2.4 m;测段2视距平均值约为12.8 m,各站视线高度主要集中在0.6~0.7 m,最高视线高度为2.6 m。

从表1、表2的统计数据可以看出,三次测量的环线闭合差均接近限差,且其符号相同,大小接近。环线闭合差最大相差2.0 mm,同一测段的三次测量结果最大偏差为2.4 mm。

3.2　水准环线闭合差系统性偏大成因分析

3.2.1　存在更大近地大气折光假设的提出

为了解决环线闭合差系统性偏大的问题,采取了调换仪器及观测人员等措施,但测量结果与前期一致。我们认为可能存在左、右岸分布上不对称,由测量路线环境等因素引起的系统性误差,从而造成水准环线闭合差偏大。

拉西瓦水电站一等精密水准15 km以上的水准测量工作在隧洞内进行,其中闭合差较大的环线近55%的测量工作在隧洞内进行,这是区别于其他项目最显著之处,所以我们重点对隧洞内水准路线可能存在的误差源进行了分析、排查。针对隧洞内的环境情况,我们分析造成上述问题的因素可能有以下两方面。

(1)遍布于隧洞内的输电线路、照明电路、大型载重汽车的运行等,均会产生一定强度的人工磁场。人工磁场是否会导致自动安平装置-光学机械重力摆作用视线的准绝对水平受到破坏,从而产生磁致误差?目前使用的水准仪重力摆一般均是无铁磁材料,同时人工磁场属于交流磁场,磁致影响很小,汽车等产生的干扰源本身具有偶然特性,而我们的测量结果显示该误差属于系统性误差。说明磁致误差不会是主要的误差源。

(2)隧洞内没有日光照射,不会具备和隧洞外相似的近地大气折光影响。这些气象特征,促使我们对隧洞内近地温度梯度变化的特性进行研究。是不是存在更大近地大气折光差的影响,从而导致隧洞内水准测量成果的系统性变化?

于是我们做出了在隧洞内存在更大大气折光影响的假设。为了验证折光对成果影响的显著性,在随后的测量工作中做了针对这一假设的试验。

3.2.2 隧洞内存在更大近地大气折光影响的成因分析

近地大气折光的大小主要是由近地气层温度场结构决定的,影响近地气层温度场结构的主要因素都是由地面热交换造成的,在露天状态下由于地面的热辐射差额导致热量的剩余和不足,从而形成地面与大气间的温度差异,为了达到热平衡,便发生热交换,如果近地面温度高于上方的温度,则发生地面向大气的热传递,这时温度梯度为负,感热通量向上(正值),即处于这一特性气层的折光的视线方向会向上弯曲,导致测量高差的绝对值变大(如图3所示)。

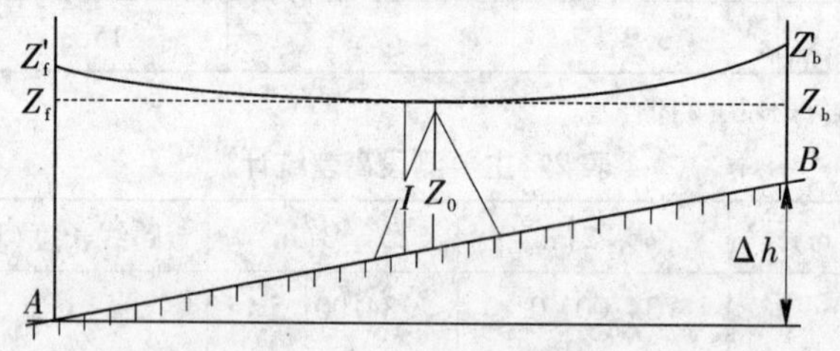

图3 感热通量向上时的水准视线弯曲示意图

由于隧洞内不存在日光照射,且空气运动受到隧洞洞壁的约束,其空气运动状态会和洞外存在非常大的差异。空气的物理特性决定了热空气会向上运动,在受到上层洞壁的约束后逐渐聚集在上部运动,其结果会导致感热通量向下(负值),温度梯度为正(如图4所示)。

如果近地面温度低于上方温度,则情况恰好相反,如图4所示。

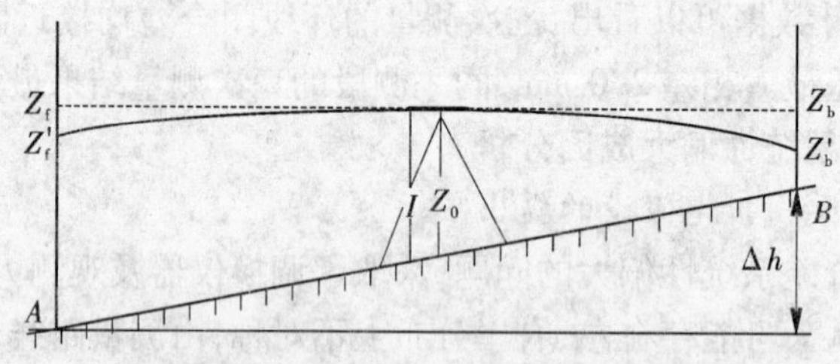

图4 感热通量向下时的水准视线弯曲示意图

图3中Z_f、Z_b系前后尺理论读数,Z'_f、Z'_b系前、后尺实测读数,故A、B间的理论高差绝对值为$|\Delta h| = |Z_b - Z_f|$,而测量的高差为$|\Delta h'| = |Z'_f - Z'_b|$。

按照芬兰大地测量研究所库卡梅奇的大气折光改正公式,在地面倾斜的情况下,只要近地气层存在梯度,折光差就会系统的积累。折光差影响的大小与视距的平方、温度梯度成正比,一般情况下,距离地面越近,温度梯度变化就越大,库卡梅奇这一结论得到了大量试验资料的证实。库氏单站折光改正公式如下:

$$R = 10^{-5} A \left(\frac{S}{50}\right)^2 \Delta h \Delta t$$

式中:S 为视距长度;Δh 为测站高差;Δt 为温差。

拉西瓦交通隧洞的条件就具备了地面倾斜的情况,按照库氏结论,折光差会系统的积累。隧洞内大气折光的分布特征如图 3 所示,且 $|\Delta h'| \geqslant |\Delta h|$。也就是说,隧洞内折光的影响总是导致水准高差绝对值变大,这个趋势与我们的测量结果一致。为了验证这个分析结果,我们对右岸水准进行了相关削弱近地大气折光的试验。

3.3 削弱隧洞内近地大气折光影响的试验

试验选取的隧洞内水准测段,主要为环线 2 高线隧洞的部分水准线路。

为了对上面我们提出的假设进行验证,我们共做了 2 次试验,按照分析结论,影响折光大小的主要因素为视线长度和视线高度,所以试验主要削弱这两方面的影响。方法如下:

将隧洞内水准测量最低视线高度从规范规定的 0.6 m 提高至 0.8 m 和 1.0 m 两个高度,将视距从之前的测量所采用的 20 米减少至 15 m 和 12 m,分别对隧洞内的水准进行测量。如果我们假设的隧洞内存在更显著的大气折光差影响成立,则其测量会出现如下结果:

(1)相应的右岸分布的测段上高差绝对值相对会变小。

(2)环线顺时针推算的闭合差会变小。

按照上面调整后的方法,我们做了 2 次测量试验,测量的结果统计如表 3 和表 4 所示。

表 3 限差调整后的环线闭合差统计

项目	闭合差 1(mm)	闭合差 2(mm)	路线长度(km)	平均值(mm)
环线 2	1.7	3.8	15.3	2.7

注:环线闭合差指顺时针推算的高线一等水准环线闭合差。

表 4 限差调整后的测段高差统计

项目	高差 1(m)	高差 2(m)	路线长度(km)	平均值(m)
测段 1	33.998 8	33.998 9	1.2	33.998 8
测段 2	101.032 9	101.030 8	1.7	101.031 8

受坡度因素影响,测段 1 视距平均值约为 12.2 m,各站最低视线高度主要集中在 1.1 ~ 1.5 m,最高视线高度为 1.9 m;测段 2 视距平均值约为 7.8 m,各站最低视线高度主要集中在 0.8 ~ 1.1 m,最高视线高度为 2.3 m。

通过表 3 和表 4 可以看出,调整后的环线闭合差平均值为 2.7 mm,比调整前减小了 5.0 mm,测段 1 减少 3.1 mm,测段 2 减少 2.0 mm。试验数据有力的支持了关于隧洞内存在更大大气折光影响的假设。

3.4 试验结果分析

试验结果显示,近地气层在受到如隧洞等外围约束的条件下,在一定高度范围内存在更大的温度梯度变化。当隧洞具有一定坡度时,其前后大气折光影响差异较大,当进行连续的上坡或下坡水准测量时大气折光所造成的误差将系统性的积累。

4 结语

通过对拉西瓦水电站精密水准网连续 4 年的测量成果进行分析显示,我们所采取的技术措施可靠、有效,对于提高精密水准网的整体可靠性及测量精度效果显著。基准点所采用的双金属标钢管及铝管底部深入基岩以下,提高了基准点的稳定性,其能对温度所产生的标点变形进行修正,提高了垂直位移的观测精度,4 年中对基准点点间高差进行了 7 次校测,校测最大偏差仅为 0.3 mm。本工程所采取的水准点建造方法有效地削弱了拉西瓦周期性冻融对水准点的沉降影响,水准点均未发生显著的局部沉降变形。近地气层在受到如隧洞等外围约束的条件下,存在更复杂且显著的近地大气折光影响,在长距离、大高差的隧洞内进行精密水准测量,需要采取措施削弱近地大气折光影响。

参考文献

[1] 梁振英,董鸿闻,姬恒炼,等.精密水准测量理论与实践[M].北京:测绘出版社,2004.

[2] 梁振英.特高精度水准测量方法的研究[J].测绘通报,1992(6):3-8..

[3] 梁振英.关于精密水准折光改正的若干问题[J].测绘学报,1991(1):66-75.

[4] 李祖锋,高建军,鹿思锋,等.隧洞约束环境下大气折光对精密水准测量的影响[J].测绘工程,2010(2).

【作者简介】 李祖锋,1981 年 11 月生,现工作于中国水电顾问集团西北勘测设计研究院,工程师,主要从事高等级控制测量及数据处理工作。

加强信息系统运行管理 支撑服务水利信息化

詹全忠

（水利部水利信息中心 北京 100053）

【摘要】 随着水利信息化的快速发展，信息系统运行管理的重要性突显。本文对水利信息系统运行管理面临的问题进行了总结，提出了水利信息系统运行管理体系架构，并对体系建设的关键因素进行了分析。

【关键词】 水利信息系统 运行管理

1 引言

随着信息技术的不断发展和广泛应用，信息化对全球经济社会发展的影响愈加深刻，信息化水平已成为衡量一个国家和地区现代化水平的重要标志之一。2011 年《中共中央 国务院关于加快水利改革发展的决定》提出了推进水利信息化建设，全面实施“金水工程”，以水利信息化带动水利现代化的明确要求。当前和今后一个时期，加快推进水利信息化，支撑和保障水利改革发展，促进并带动水利现代化，是一项事关水利发展全局的重大战略任务。

近年来，水利信息化快速发展，效益日益显著，已成为创新和提升水利工作的重要手段，促进了传统水利向现代水利、可持续发展水利的转变。目前，水利信息系统内容覆盖了水利工作的方方面面，是水利日常工作中不可或缺的重要组成部分。随着水利信息化的发展，水利信息系统规模越来越大，信息系统运行管理任务越来越重、难度越来越大、要求越来越高，因此有效地管理和维护现有的信息系统，保证其高效、稳定、安全运行，降低运营成本，提高工作效率及资源利用率，确保其发挥效益、支撑水利工作具有重要的现实意义。

2 水利信息系统运行管理面临的问题

一直以来，水利系统很多单位只重视信息化基础设施和业务应用系统建设，而忽视系统运行管理和保障，“重建设、轻维护”，“重技术、轻管理”的现象非常严重，已成为制约水利信息化进一步发展的瓶颈。水利信息系统运行管理存在的问题主要体现在思想认识、人员、经费、管理措施及技术手段等方面。

(1)思想认识方面。水利信息化建设和行业管理得到了一定的重视，但后期的运行维护和管理却远未达到相应的重视程度，信息系统运行管理经常被认为是简单“操作”。

(2)人员方面。强有力的运行管理机构、专业的运行管理人员是信息系统运行管理的前提。水利系统在运行管理机构及人员上还有很大不足，有近一半单位没有专门的信息系统运行管理机构，省级以上水利部门也有个别没有专门运行管理机构或运行管理机构不统一。平均每个单位不足 1 名专职运行管理人员，省级以上单位每单位平均也不到 4 人。机构的缺失、人员的匮乏，难以满足快速发展的水利信息系统运行管理的需要。

(3)经费方面。信息系统运行管理需要投入大量的人力、物力，足额的运行维护经费是保证运行管理工作顺利开展的必要条件。目前很多单位运行经费得不到保证，无法购买设备的保修服务及一些必要的备品备件，在设备发生故障时无法及时修复，有近 80% 的单位专项运行维护经费不足，甚至有近 40% 的单位根本没有专项运行维护经费，影响系统的运行。经费不足严重制约水利信息系统的运行管理工作。

(4)管理措施方面。完善的运行管理制度、规范是信息系统运行管理的基础。据统计，信息系统运行中只有 20% 的故障是由纯粹的技术问题造成的，而 80% 的故障来源于流程的缺乏和管理上的疏漏。水利系统在运行管理制度上非常不完善，特别是在规范化、流程化管理，应急处理等方面较为薄弱。维护工作存在很多不规范的地方，如日常维护工作缺乏规范，对人的依赖程度过高；资产管理混乱，配置不清；故障处理流程不规范，存在故障漏报、跟踪处理不及时的可能；变更存在随意性大、变更前的风险评估和应对措施不够，造成风险难以控制和防范；缺乏应急处置机制，突发事件处置不及时，造成较大影响；缺少知识积累手段，缺乏信息共享，能力越强工作越多；缺少运行管理数据资料，难以为系统优化、服务优化提供支持。

(5)技术措施方面。完备的技术手段能提高信息系统运行管理效率和质量。目前，水利系统很多单位建设了网管系统、工单系统等运行管理技术措施，对信息系统运行管理起到一定的作用，但是存在独立运行、无法集成、信息孤立、功能不全等问题，如监控监视不全面，存在盲区；预测预警机制不健全，难以发现隐患；维护操作缺乏监督、管理，工作疏漏、无操作难以避免，且不易发现；应急处置手段缺乏，不能及时处置；缺乏分析评估手段。技术手段不足，难以提高水利信息系统运行管理的能力和水平。

3　水利信息系统运行管理体系架构

信息系统运行管理应该说是一门组织科学，集 IT 技术、管理科学于一体。水利信息系统的运行管理是保障水利网络与业务系统正常、安全、有效运行而采取的生产组织管理活动，其核心在于提升运维能力，保障信息系统运行，支撑水利工作。水利信息系统运行管理工作涉及面广，既包括物理环境、计算机网络、主机系统、存储备份系统、数据库系统、中间件系统等基础设施的运行管理；又涉及防汛抗旱应用，电子政务应用，水土保持监测管理信息系统，城市水资源实时监控，重要水库、枢纽、灌区、供水、排水、调水等水利工程设施调度系统，水利政府网站等业务应用系统的运行管理。工作内容繁杂，包括监视监控、日常维护、故障处理、配置变更、应急处置和安全管理等。水利信息系统运行管理体系架构涵盖运行管理的各个方面，通过运行管理措施(人员、经费保障，技术手段保障，标准、规范、制度保障)，实现对运行管理对象(信息系统基础设施和业务应用系统)的管理，从而实现支撑保障水利信息系统，为水利专业人员、决策人员及社会公众提供服务。水利信息系统运行管理是一套完整的体系，其体系架构如图 1 所示。

运行管理措施包括人员队伍、运行经费、运行保障平台、运作模式、标准规范、管理制度。人员队伍指运行管理人员(包括技术人员、管理人员)构成、人员培训等；运行经费指运行管理工作所需的各种费用，包括人员经费、运行材料动力消耗费用、维护维修经费及其他有关

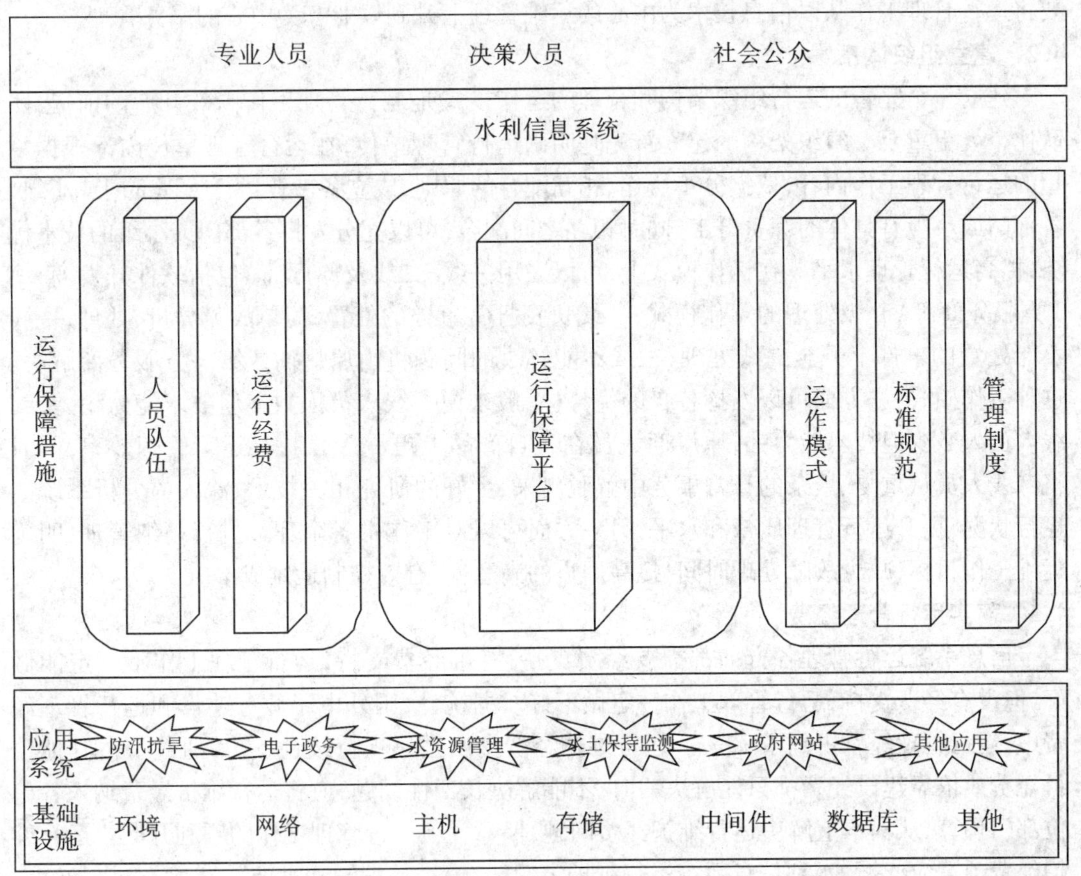

图1　水利信息系统运行管理体系架构

费用支出。人员队伍、运行经费是信息系统运行管理工作的前提;运行保障平台是信息系统运行管理工作的技术支撑平台,为运行管理工作涉及的监控监视、预测预警、维护操作、应急处置和安全管理等各项工作提供技术手段和管理工具,运行保障平台是信息系统运行管理的技术手段;运作模式指运行管理工作的组织体系、角色分工、工作流程等;标准规范主要包括各类运行管理标准,是行业性的运行管理规范;管理制度是运行管理工作的具体工作制度、守则等。运作模式、标准规范及管理制度是信息系统运行管理高效开展的基础。

4　加强水利信息系统运行管理的关键因素

4.1　转变观念

加强信息系统运行管理要转变三个观念。首先是转变“重建设、轻管理”的观念。从信息系统的生命周期来看,系统建设阶段只占20%,80%的时间是系统运行期,信息系统的建设只是完成功能,信息系统功效的发挥还需要后期的运行管理来实现。当前,全国水利信息化发展从基础设施和基本业务应用建设阶段逐步进入到深化应用、提高应用水平、提升服务能力的新阶段,运行管理的重要性一点不亚于系统建设的重要性。其次是转变“重技术、轻管理”的观念。信息系统运行管理不仅仅是技术问题,更多的是管理问题,仅仅靠购买监控管理类工具软件解决不了复杂的运行管理问题。最后是转变“重维护、轻服务”的观念。信息系统的运行管理不是单纯的软硬件系统的运行维护,而是为业务提供高质量的IT服务,要将运行管理工作从以信息技术为中心的运维管理上升到以业务为中心的服务管理。

4.2 建立组织体系

建立一个完善的运行组织架构和合理的运作模式是运行管理工作高效开展的前提，团队比个人更重要。组织架构的建立首先是明确运行管理单位，在运行管理单位内部根据承担的运行管理工作内容，依据分工协作、相互配合的原则，成立各专业运行管理部门或小组，各部门或小组有具体的职责分工，同时相互之间配合，可以充分发挥各部门或小组的技术优势，提高运行管理水平。在运作模式上，可以采用一线、二线及原厂商（包括软件开发商、第3方服务商等）分级处理的工作模式，一线负责日常事务的处置，简单故障分析、处理，一线不能处置的事件上升至二线处理，二线不能解决的问题将由原厂商或第3方服务商解决。这种工作模式，一方面可以发挥各自的优势（一般来说一线人员人数多但技术能力稍差，二线人员人数少但技术能力强，原厂商人员在各自产品上更深入），提高运行管理效率，同时将二线人员从重复、繁杂的日常事务中解脱出来，更好的研究相关技术、深入的分析系统的运行状况，提升运行管理质量和水平；另一方面可以对工作中各个环节进行精细管理，明确运维工作目标，减少故障处理时间，提高用户的满意度，有效控制运维成本。

4.3 落实运行管理经费

信息系统运行管理各方面都需要经费支持，有了经费支持，运行管理工作不一定能做好，但没有经费支持，运行管理工作一定做不好。据统计，国外电子政务建设和运行管理经费中，运行管理经费占到了75%。运行维护经费支出是一项持续性支出，通过挤占其他项目经费或依靠建设经费都只能解决一时，不能解决长期性问题，应将运行维护经费纳入各单位部门预算，从根本上解决运行维护经费问题，保障各项运行管理工作的顺利开展。运行维护经费预算应依据《水利信息系统运行维护定额标准》进行编制和批准。

4.4 完善规章制度

应对信息系统运行管理工作的各个方面建设和完善运行管理规章制度，形成包括标准规范、工作制度、行为规范、工作流程及应急处置预案等组成的运行管理制度体系，规范各项运行管理工作。标准规范主要面向行业或区域，从总体上规范行业或区域的运行管理工作；行为规范主要对运行管理人员的日常行为进行规范；运行管理工作制度主要规范各种运行管理工作（如值班、巡检等），保证运行管理工作的标准化；运行管理工作流程主要规范各类运行管理流程（如故障处理流程、变更流程等），实现运行管理工作流程化；应急预案主要是在《水利网络与信息安全事件应急预案》的基础上制定各信息系统的专项应急处置预案，规范各系统可能发生事件的应急处置措施，提高突发事件应急处置措施。通过制度体系建设，规范运行管理行为，理顺运行管理流程，完善应急处置机制，实现运行管理工作的规范化、制度化和流程化。

4.5 加强技术手段

优质、高效、高水平的运行管理还需要技术手段来提供支撑，科学、完善的管理制度还需要技术手段来保障落实，因此加强技术手段建设是运行管理体系建设的一项重要内容。通过建设和完善水利信息系统运行管理技术手段，对信息系统运行管理工作中涉及的监控监视、风险预警、维护操作、应急处置及分析评估等工作进行全面的管理，提供了全面的、细粒度的系统监管能力，以及规范的运行服务管理能力，实现运行管理工作的信息化管理。

5 结语

完善的水利信息系统运行管理体系通过集中监视、规范流程、事前预警、事后评估和智能化管理，实现最小故障率、最短反应期、最低运行成本、最好运行状态、最佳用户满意度，从而规范水利信息系统的运行管理工作，提高运行管理效率，提升运行管理水平，促进信息系统资源的合理利用，充分发挥已建信息系统的作用，促进水利信息化的健康发展，为水利发展提供支撑保障。

【作者简介】 詹全忠，男，1974 年 2 月生，研究生学历，1999 年毕业于北京理工大学、硕士学位，现工作于水利部水利信息中心，副处长/高级工程师。

新形势下水利工程建设监理工作的思考

李创团

（广西壮族自治区水利科学研究院　南宁　530023）

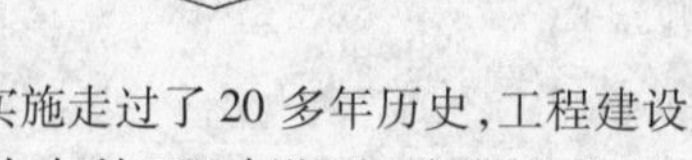

【摘要】 我国工程建设监理制度的实施走过了20多年历史，工程建设监理体系基本形成。工程建设监理制度实施以来，监理活动在水利工程建设过程中发挥了积极作用，取得了巨大成绩。随着经济社会的快速发展，水利基础设施在支撑经济社会可持续发展、促进社会和谐稳定、确保国家粮食生产安全的战略地位更加突出，国家对水利基础设施的投入和建设规模不断加大，对水利工程建设监理工作提出了新的更高的要求。分析了新形势下水利工程建设监理面临的主要问题和挑战，对新形势下水利工程建设监理工作进行了思考。

【关键词】 水利工程　建设监理　新形势　问题和挑战　思考

我国工程建设监理制度的推行始于20世纪80年代末，历经20多年的发展，基本形成了比较完整的工程建设监理体系。现行《中华人民共和国建筑法》明确规定国家推行建设监理，并对建设监理的内容、程序、依据、范围等作出了规定。2003年10月23日，中华人民共和国水利部正式颁布并从2004年1月1日起实施《水利工程建设项目施工监理规范》（SL 288—2003），进一步规范和加强了对水利工程建设监理活动的管理，标志着水利工程建设监理步入了规范化、制度化新阶段。2003年11月24日，国务院正式颁布并从2004年2月1日起实施《建设工程安全生产管理条例》（国务院令第393号），将建设工程安全生产监理责任明确纳入了建设监理的工作范围，赋于了工程建设监理新内涵。《水利工程建设项目施工监理规范》（SL 288—2003）和《建设工程安全生产管理条例》（国务院令第393号）实施以来，极大地推进了水利工程建设施工监理工作的开展，在促进工程施工质量、确保工期实现、提高资金使用效率、加强施工安全、建立工程施工参建各方和谐共赢关系等方面取得了巨大成绩，但与此同时，随着经济社会的快速发展，水利基础设施在支撑经济社会可持续发展、促进社会和谐稳定、确保国家粮食生产安全的战略地位更加突出，国家对水利基础设施的投入和建设规模不断加大，建设内容从工程实体逐渐向工程管理信息化建设推进。此外，工程建设领域新技术、新材料、新工艺发展迅速，对工程建设监理工作提出了新的更高的要求，水利工程建设监理工作面临着新挑战。新形势下如何进一步加强和完善对水利工程建设监理的监督管理、促进监理活动能力和水平的提高，适应市场经济发展的要求，适应水利工程建设发展的需要，更好地服务于水利工程建设值得思考。

1　新形势下水利工程建设监理面临的主要问题和挑战

1.1　监督管理机制不健全

《水利工程建设监理规定》（水利部第28号部长令）中明确水利部对全国水利工程建设

监理实施统一的监督管理,县级以上地方人民政府水行政主管部门和流域管理机构对其所管辖的水利工程建设监理实施监督管理,项目法人依据监理合同对监理活动进行检查,并对监理单位和监理人员违规行为的处罚作了明确规定。目前,对监理单位资质、从业人员资格的管理制度比较完善,监管比较到位,但对工程建设阶段监理活动的监督管理比较薄弱,没有建立起有效的动态监督管理机制,县级以上地方人民政府水行政主管部门、流域管理机构和项目法人对工程建设阶段监理活动的监督管理机制尚不健全,缺乏有效、有力的动态监管手段,监理单位超范围承揽监理业务、监理分包、总监理工程师或者监理工程师不到位、不按照监理合同承诺投入监理力量、监理工程师或者监理员监理活动超越授权范围、监理人员同时在两个以上的工程项目中从事监理活动、施工实施阶段的监理工作不力导致工程质量和安全事故发生等违反《水利工程建设项目施工监理规范》、《建设工程安全生产管理条例》和《水利工程建设监理规定》的问题时有发生,随着近年来水利工程建设投入的不断加大,水利工程项目建设规模和强度前所未有,监理活动过程中出现的上述问题越发突出,行之有效地对监理活动进行监督管理显得更加迫切和重要,健全对监理活动的监督管理机制势在必行。

1.2 监理活动的范围不能适应工程建设发展的需要

目前,业主委托监理单位的监理活动绝大多数仅限于工程建设施工阶段的监理工作,对工程项目前期工作开展监理的较少。工程建设前期工作的质量、进度和投资规模的控制与工程建设施工阶段同等重要,对于社会经济发展有重大影响、涉及公共利益重大的民生项目的前期工作甚至比工程建设施工阶段的工作显得更加重要。随着经济社会的快速发展,工程建设规模的持续增加,社会公众对工程项目前期工作的关注度越来越高。与此同时,工程建设领域的各种违规、违法现象依然存在,一些违规、违法现象发生在项目的前期工作中,具有更大的隐蔽性和危害性,在前期工作中夸大项目经济效益和社会效益、为谋取局部的经济利益扩大投资规模造成资金浪费、甚至套取国家财政资金等情况时有发生,而目前对项目前期工作的监督绝大部分仅由业主上级主管部门组织专家对前期工作报告进行评审,缺乏对过程的监督,更谈不上独立公正的第三方对中间过程的监督和管理,仅凭专家组的评审结论显然缺乏社会公信力,也无法确保项目前期工作质量、进度,更谈不上杜绝各种违规、违法现象的发生。近几年来,由于前期工作不到位,工作深度不够,造成项目建设计划受阻、资金浪费等问题越发突出,也导致施工阶段设计变更增多,给施工阶段的质量、进度及投资控制、合同管理造成更多困难和压力。因此,对项目前期工作进行监督和管理显得十分重要和迫切,处于第三方独立公正地位的社会监理单位应当承担起对项目前期规划、勘察和设计工作的监理活动,使监理活动的业务范围贯穿到工程建设项目的全过程,真正实现工程建设项目全过程的监督和管理,适应工程建设发展的需要,更好、更全面地促进工程项目建设的健康发展。

1.3 监理资源及监理活动能力总体不足

随着水利工程建设规模和强度的不断加大,水利工程建设对监理资源及监理活动能力的需求更加强烈。据《中国水利统计年鉴2009》统计,2008年水利工程施工项目总计7 529个,当年全国共有水利工程监理单位782家(分级的水利工程施工监理、水土保持工程施工监理、机电及金属结构设备制造监理共706家,不分级的水利工程建设环境保护监理单位76家),1个监理单位平均承担9.6个项目,当年监理从业人员25 237人(含监理员),1个项目平均3.4个监理从业人员。据《中国水利统计年鉴2010》统计,2009年水利工程施工项目总计10 715个,当年全国共有水利工程监理单位986家,(分级的水利工程施工监理、水土保持工程施工监理及机电和金属结构设备制造监理共847家,不分级的水利工程建设环境保护监理单位139家),1个监理单位平均承担10.7个项目,当年监理从业人员29 469人(含监理员),1个项目平均2.8个监理从业人员。2008年和2009年两个年度的监理单位和监理从业人员显然不能满足当年工程建设的需要,并且监理单位和监理从业人员的增长滞后于工程建设规模的增长。此外,截至2009年底,总计986家水利工程监理单位中,具有甲级资质的238家,占当年分级的24.1%;乙级资质的276家,占当年分级的28.0%;丙级资质的472家,占当年分级的47.9%,具有甲级资质的监理单位较少,70%以上为乙级和丙级,丙级监理单位几乎占据半数。由于乙级和丙级监理单位规模较小,高素质的监理人员较少,尤其是丙级监理单位更加缺少高素质的监理人员,甚至在一些乙级和丙级监理单位注册的监理人员中有不少是没有从事过监理工作或者监理经验不足的人员,过多的低资质监理单位和低素质监理人员削弱了监理活动能力。

根据中国水利工程协会最新统计,截至2010年底,水利工程建设监理从业人员总计50 570人,即使按2009年水利工程施工项目总数10 715个计算,1个项目平均也只有4.7个监理从业人员,监理从业人员还是略显不足。从监理从业人员结构看,2010年具有总监理工程师资格的7 141人,仅占14.1 %,即使按2009年水利工程施工项目总数10 715个计算,1个项目平均0.67个总监理工程师,还达不到1个施工项目1个总监理工程师的基本要求;具有监理工程师资格的40 199人,占79.5%;监理员3 230人,占6.4%。具有总监理工程师资格的人员较少,具有多专业技能、较高组织领导能力、协调能力及丰富的工程建设管理经验等综合能力的高素质总监理工程师更少,从业人员素质总体不高。因此,无论是监理单位数量、监理单位资质等级结构、监理从业人员数量和从业人员素质,都不能满足当前和今后水利工程建设发展的需要,监理资源及监理活动能力总体不足。

1.4 监理单位社会性、独立性和公正性未能得到充分体现

水利工程建设监理中具有真正意义上独立的社会监理单位很少,有不少监理单位依附于水行政管理部门、科研院所、勘察设计院等为母体的企事业单位中,缺乏市场竞争主体和法人实体地位,全国上规模的具有甲级资质监理单位有相当一部分为水行政管理部门、科研院所、勘察设计院所兼营,在实际的监理活动中,甚至存在设计单位兼营的监理单位承揽本单位设计项目的监理工作;此外,在工程建设活动中,项目法人对工程的建设管理与委托的工程建设监理共存,业主在工程的建设管理中往往高度关注工程进度,在工程进度控制方面给监理施加过多的压力,业主将自己的意愿强加于监理的现象不少,而由于监理处于业主委托的从属位置,给监理在处理工程质量和进度这对主要矛盾关系方面,以及在投资控制和合同管理上造成较大的障碍,从某种程度上削弱了监理单位在监理活动中的独立性和公正性,监理单位在监理活动中的社会性、独立性和公正性未能得到应有的尊重和充分的体现,也是在一定程度上造成参建各方之间关系不和谐的最主要因素。

1.5 制约监理活动良性发展的因素依然存在

促进监理活动的良性发展,除具有完善的监督管理和行为的规范约束外,有序竞争的市场环境、优胜劣汰的市场准入和清出机制不可或缺。我国工程建设监理制度的推行虽然走过了20多年,但有序竞争的市场环境还没有形成。2007年5月1日国家发展和改革委员

会、建设部联合颁布并实施了《建设工程监理与相关服务收费管理规定》([2007]670号),规定了实行政府指导价的建设工程施工监理服务收费基准价的计算方法,以及监理服务收费基准价费率浮动幅度为上下20%。在施工监理招标过程中,几乎所有的业主都采取监理服务收费基准价费率下浮的方式确定监理服务费用,有的监理服务收费基准价下浮费率甚至低于《建设工程监理与相关服务收费管理规定》中允许的下浮费率,没有严格执行该规定。业主在选择监理中标候选人时不是优先考虑监理单位的资信程度、监理方案的优劣等技术因素,而是过于关注投标人监理服务费用的高低,削弱了对监理单位综合实力和人员素质的要求,从实质上改变了招标的真正目的,无形中助推了监理单位无序竞争、恶性竞争、争相压低报价的行为,导致资质高、综合实力强、人员素质好的监理单位生存和发展困难,反而催生了一些靠压低报价承揽监理业务、资质低、综合实力弱、人员素质差的监理单位,造成监理单位无法吸引和留住高素质监理人才,降低了服务质量,影响了监理的信誉,从总体上导致监理单位出现生存和发展危机。

在水利建设市场主体诚信体系建设方面,水利部2009年10月15日发布了《水利建设市场主体信用信息管理暂行办法》(水建管[2009]496号),2009年12月1日颁布并实施《水利建设市场主体不良记录公告暂行办法》(水建管[2009]518号),2009年12月2日中国水利工程协会颁布和实施了行业自律管理办法《水利建设市场主体信用评价暂行办法》(中水协[2009]39号),初步建立了水利建设市场诚信体系,工程施工和监理市场主体信用监管工作在一定程度上得到了加强。2010年,中国水利工程协会组织开展了首批水利建设市场主体信用评价工作,但采取的是工程施工和监理市场主体自愿申请参加的方式,仅有137家水利建设市场主体自愿申报进行信用评级,其中监理市场主体只有67家自愿申报进行信用评级,信用评级是在市场主体申报的材料基础上,由全国性行业自律组织——中国水利工程协会进行初审、复审和组织评审委员会终审,信用评价缺乏强制性和约束性,信用评价结果缺乏公信力,实际的信用监管效果还不够理想,优胜劣汰的市场准入和清出机制没有形成,制约监理活动良性发展的因素依然存在。

2 新形势下水利工程建设监理工作的思考

2.1 创新监督管理机制,加强和提高对监理工作的监管力度和水平,促进监理单位管理水平和服务水平的全面提高

随着国家对水利基础设施投入的不断加大,对工程建设参建各方的监督管理提出了新的更高要求,各级水行政管理部门必须创新监督管理模式,以新机制、新方法、新举措加强对监理工作的监督管理。对监理活动的监督管理,要全部贯穿到准入前的资格考试、资质审查、准入后的从业管理等全部的监理活动中,重点加强工程建设阶段监理活动的监督管理,要从制度上和机制上形成高效有力的监督管理手段。当前对工程建设阶段的监理活动缺乏科学有效的跟踪监管,要加强工程建设阶段监理活动跟踪监管的政策措施研究,建立工程建设阶段监理单位派出的驻地监理部人员组成、人员身份核查和现场监理活动规范性及成效性等的日常监督检查制度,并建立检查结果信息发布平台,通过公开透明的实时信息管理平台对监理单位和监理人员在工程建设阶段监理活动的质量和信誉进行跟踪监管和评价,以创新的监督管理方式,全面加强监理工作的监管,规范监理行为,促进监理单位管理水平和服务水平的全面提高。

2.2 完善各个阶段监理活动的规范和制度，创新监理模式，建立适应工程建设发展需要的现代监理制度

现行的《水利工程建设项目施工监理规范》（SL 288—2003）仅针对施工阶段的监理活动，没有涵盖到项目前期阶段的规划及勘察设计阶段，缺乏全面、系统、完善的水利工程监理法规体系。由于对项目前期阶段的规划及勘察设计阶段监理没有强制性要求和规范的约束，很少有业主将项目前期阶段的工作进行委托监理，即使有些项目进行了前期阶段的委托监理，但由于没有相应的规范约束，委托方和监理方的合同缺乏足够的技术行为的可控性，往往在实施监理的实际过程中双方在监理工作程序、方法及控制手段上出现争议，同时也不利于政府部门对项目前期阶段监理活动的监督和管理，监理活动的质量和效果难以保证，这也是目前项目前期阶段监理工作很少进行，没能有效开展的主要原因。真正意义上的工程建设监理包括从项目立项到工程建成的全过程，随着经济社会的快速发展，工程建设规模的不断加大，制定项目前期阶段监理活动约束性规范和相关规定十分迫切和必要，强制性推行项目前期阶段监理工作的时机和条件已经具备，全面实现从项目立项到工程建成的全过程监理。此外，监理活动是一项高智能综合性的技术活动，除对从业人员有较高的业务能力要求外，采取必要的高科技设备进行监理活动过程中必要的检查和检测是必须的，可以更好地提高监理活动的权威性和公正性，为此要突破当前监理工作程序化、旁站式的模式制约，加强监理工作措施和手段的研究，创新监理模式，通过更加全面和更加科学有效的措施和手段加强和提高监理活动事前预防、事中控制的能力和水平。同时，积极推进工程建设管理和监理一位化模式，将业主管理和监理单位的监理统一委托监理单位进行，建立适应工程建设发展需要的现代监理制度。

2.3 创新对监理从业人员的管理模式，全面促进监理人员业务水平和职业素质的提高

新形势下，水利工程建设监理工作对监理从业人员业务水平和职业素质提出了新的更高要求。政府监管部门和监理单位有责任和义务共同促进和提高监理从业人员的业务水平和职业素质，推动建设监理工作能力和水平的全面提升。目前对监理从业人员的管理模式已经无法适应新形势下水利工程建设对监理工作的需要，随着水利工程建设规模的不断增大、工程建设复杂程度的不断增加、工程施工新技术、新材料的不断涌现，以及对工程质量、工程进度、投资控制、合同管理和组织协调能力等的更高要求，对监理从业人员的业务水平和职业素质要求更高，必须创新管理模式，建立以提高从业人员业务素质和能力水平为目的、通过加强对监理单位的管理能力指导、提高从业人员准入条件、完善从业人员后期继续教育机制，创新后期继续教育方式，以在建工程为基地，以监理工作经验交流为平台，着重提高监理人员现场解决关键问题、重点问题、难点问题和组织协调能力，全面促进监理人员业务水平和职业素质的提高。

2.4 建立政府引导、政策扶持的机制和体制，加快推进建立适应市场经济发展要求的社会监理单位，促进监理工作的社会化、公正性和独立性

监理单位社会化、咨询化是市场经济发展的必然要求，也是监理单位确立其市场主体地位，建立适应市场经济发展要求的社会化监理咨询单位的必由之路。我国推行建设监理制度虽然走过了20多年，但总体上监理行业的市场主体地位还比较脆弱，监理工作的社会化、公正性和独立性还没有得到充分的尊重，需要认真总结经验，深入分析监理单位在市场经济发展过程中面临的问题和挑战，建立政府引导、政策扶持的机制和体制，制订引导监理单位

确立、巩固和发展市场主导地位的政策措施，加快推进建立适应市场经济发展要求的社会监理单位的步伐，促进监理工作的社会化、公正性和独立性。

2.5 加强监理服务收费的管理和监督力度，加快推进信用体系建设，建立适应市场发展规律的市场环境和优胜劣汰机制，促进水利工程建设监理市场良性发展

加强监理服务收费的管理和监督力度，项目法人和监理单位严格执行《建设工程监理与相关服务收费管理规定》([2007]670号)，将扰乱建设监理市场秩序、违反规定以压低监理服务收费的恶性竞争手段承揽监理业务的行为纳入不良行为记录的诚信管理体系，并制订相应的处罚措施。建立完全由政府主导的、强制性的水利工程建设市场主体信用评价体系，加快编制统一的水利建设市场主体信用信息标准，积极推动信用信息平台互联互通和资源共享；进一步完善信用评价办法、程序和指标体系，强制性的全面开展水利建设市场主体信用评价工作；建立监理单位和从业人员行为信息共享系统，以动态透明的信息管理措施和手段建立和完善守信、失信信息公开制度，激励制度和失信惩戒制度。完善行业信用信息发布管理、良好行为和不良行为认定、信誉评价管理等相关规定，加快推进守信激励和失信惩戒制度建设。研究制定信用信息采集、审核、发布、更正和使用的相关管理规定，加强对评价结果的动态管理，促进信用信息在水利工程建设领域诚信体系建设中的共享应用；完善市场准入和退出机制，加强守信激励和失信惩戒，形成诚信监理社会氛围，建立适应市场发展规律的市场环境和优胜劣汰机制，创建有序、公平的市场环境，促进水利工程建设监理市场良性发展。

3 结语

新形势下，国家把水利放在更加突出的战略地位，水利工程建设又将迎来新一轮大规模、高强度建设高潮，各级水行政管理部门应抓住机遇，总结经验，创新监督管理机制，构建适应市场经济发展要求，满足水利工程建设发展需要的工程建设监督管理机制，加强和完善对水利工程建设监理活动的监督管理，从广度和深度上进一步规范监理活动，积极营造竞争有序、良性发展的工程建设监理市场，以更加科学有效的措施和手段，引导和推动监理单位加强自身能力建设和人才建设，促进监理单位经营管理水平和监理活动能力的提高，更好服务于水利工程建设。

【作者简介】 李创团，男，1965年生，广西横县人，高级工程师，主要从事建设监理及规划设计工作。

基于 ArcGIS 的栅格地图配准方法的研究

——以山西省汾河河道管理信息系统专题地图制作为例

刘向杰　张保生

（中国水利水电出版社　北京　100038）

【摘要】 本文针对汾河河道管理信息系统专题地图制作过程中栅格图像配准中存在的问题，探讨了基于 ArcGIS 的栅格图像配准的方法。通过各种试验表明：大中比例尺图像采用公里网配准，小比例尺图像采用经纬度配准。上述方法有效地提高了栅格图像数据配准的效率，满足了水利专题地图配准精度及数字化要求，实现了栅格图像数据的坐标匹配，为水利数字化制图提供了方便。

【关键词】 水利数字化　栅格图像数据　ArcGIS　配准

1　引言

目前，绝大多数的水利数字化地图都采用扫描数字化制作，即利用扫描仪对地图扫描得到栅格图像，进行矢量化和赋属性值得到水利数字化地图。然而，地图在扫描过程中不可避免地产生了图纸变形误差和扫描仪误差。人们往往不对栅格数据进行处理就直接使用，这样做显然不够准确。

ArcGIS 是美国环境系统研究所（ESRI）开发的新一代地理信息系统软件，它已经成为中国用户群体最大、功能最强、应用领域最广的 GIS 技术平台之一。ArcGIS 拥有强大的交互式栅格数据配准和矢量数据配准工具，分别是地理配准模块（Georeferencing）和空间校正模块（Spatial Adjustment）。水利工作者不须具有深厚的地理知识，通过其自带的配准功能，能够快速、准确地对各种栅格数据进行坐标匹配和几何校正。本文介绍了 ArcGIS 配准相关的地理知识及栅格图像的配准方法，以便为更多水利专题地图制作用户提供快速解决各种配准问题的可能。

2　ArcGIS 配准的地理基础知识

2.1　大地基准面

基准面是利用特定椭球体对特定地球表面地区的逼近，每个国家或地区都有各自的基准面。我国的两个大地基准面分别是参照苏联采用克拉索夫斯基（Krassovsky）椭球体建立的北京 54 坐标系和采用国际大地测量协会推荐的 1975 地球椭球体（IAG75）建立的西安 80 坐标系。此外，我国 GPS 测量数据多采用 WGS84 椭球体建立的地心坐标系，即 WGS1984 基准面。

2.2 地图投影

地球是一个球体,球面上的位置以经纬度坐标表示而地图上的位置以平面坐标表示。地图投影就是将地球上某一位置的经纬度坐标根据一定的数学法则转化为平面坐标。显而易见,投影必须基于一个椭球(基准面)且必须有一个投影算法。

我国大中比例尺地图均采用高斯—克吕格投影(Gauss Kruger),1 ∶25 000 ~1∶500 000比例尺地形图采用经差 6 度分带,1∶10 000 比例尺的地形图采用经差 3 度分带。我国大部分省区的 1 ∶1 000 000 地图多采用兰勃特投影(Lambert Conformal Conic)和属于同一投影系统的 Albert 投影(正等轴面积割圆锥投影) 。同时,我国位于中纬度地区,中央经线一般取 105°E,两条标准纬线分别选取北纬 25°和北纬 47°,应用时应根据各省区地理位置和轮廓形状的不同对各参数加以修定。例如,山西省中央经线为 111°E,两条标准纬线分别选取北纬 34.5°和北纬 41°。

3 各种比例尺地图的配准

配准的目的是通过控制点来校正栅格数据和给予栅格数据现实的地理意义,进而使处理得到的矢量数据具有地理空间坐标。配准的过程并没有改变栅格数据的投影数据,我们只是定义一个地图坐标系,然后把栅格数据找到合适的位置和角度放在这个坐标系内使之得到校正和坐标匹配。这就要求我们必须获取正确的投影和控制点。

3.1 地图投影的获取

一幅标准地图,一般都拥有投影方式、比例尺等地图参数,但现实中我们往往受多种条件限制,不能找到标准地图而选取其他图来代替,一般这些地图并未注明地图投影的种类和参数,为了实现配准,我们可通过如下方式获取所需信息。

(1)查阅地图相关资料获得投影类型。如查阅地图设计书、编图大纲或询问地图作者来获得地图投影信息。

(2)根据地图投影选择的一般常识和规律获得地图投影的类型(详见 2.2 节)。

(3)借助量算地图投影的变形获得地图投影的类型。即量算一些分布比较均匀的点上微小段经线和纬线的长度以及夹角,将之与地球仪上相应的长度和夹角相比,确定变形的性质和分布,判定其投影类型。

3.2 控制点的获取

控制点的获取是配准必不可少的环节之一。在一幅图中,公里网、经纬网的交点是配准首选控制点。我们总体把握均匀布置、局部控制、减小残差的原则,即控制点比较均匀的分布在图形各处,以避免图形校正不能满幅,如果某一区域需加强校正,可在这一区域多加控制,但一定注意对残差的控制。有时我们所能找到的图中不具备公里、经纬网格(如行政区图、各种专题图),我们可通过下面方法来获取控制点信息。

(1)找到其他相关地形图(可算得实际坐标),选取与配准图共同的特征点(典型建筑物、十字路口、桥梁等),然后借助相关图来获取典型地物的坐标。

(2)选取一些特征点(典型建筑物、十字路口、桥梁等),利用 GPS 或 GPRS 测定其坐标。由于测量数据投影方式与配准图一般不符,需要将数据通过 7 参数投影转换为配准图下投影方式的坐标。若对配准精度要求不太高(如宏观控制图),可将测量数据直接动态投影(系统自带 5 参数转换投影)到 ArcMap 中获取特征点坐标。

3.3　公里网配准

公里网是由平行于投影坐标轴的两组平行线所构成的方格网。我国 1∶10 000～1∶250 000 比例尺的地形图上,在内外图廓间绘有分划短线即“分度带”,有的图中将对应短线相连构成公里网,有的只以十字表示出各公里网的交点。

公里网配准的基本步骤如下:

(1)制定坐标系统。在 ArcCatalog 里制定栅格数据投影坐标系,如图 1(a)所示。在此定义的坐标系统并没有使数据本身发生投影变换,制定的仅仅是对栅格数据坐标系统信息的一个描述。

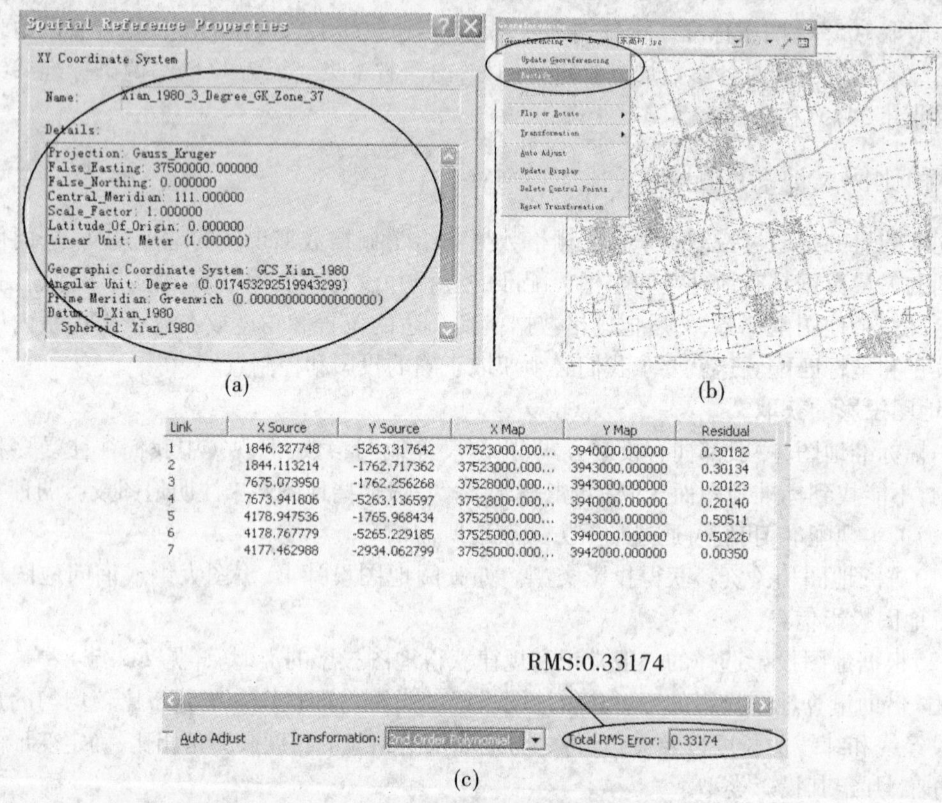

Link	X Source	Y Source	X Map	Y Map	Residual
1	1846.327748	-5263.317642	37523000.000...	3940000.000000	0.30182
2	1844.113214	-1762.717362	37523000.000...	3943000.000000	0.30134
3	7675.073950	-1762.256268	37528000.000...	3943000.000000	0.20123
4	7673.941806	-5263.136597	37528000.000...	3940000.000000	0.20140
5	4178.947536	-1765.968434	37525000.000...	3943000.000000	0.50511
6	4178.767779	-5265.229185	37525000.000...	3940000.000000	0.50226
7	4177.462988	-2934.062799	37525000.000...	3942000.000000	0.00350

(a)　(b)　(c)

图 1　公里网配准示意图

(2)添加控制点。添加 Georeferncing 工具条,将该工具条菜单下 Auto Adjust 处于不选择状态。在栅格数据图像上找到选定的控制点并输入该点实际坐标。例如,选定图上控制点为(37 523,3 940),那么应输入(37 523 000,3 940 000)。由于配准主要是对栅格图像变形的校正,因此我们一般应采用二次非线性多项式模型进行配准,这就要求控制点一般多于 7 个点。用相同的方法,在栅格数据图像上输入所有选定控制点的实际坐标,然后点击 Update Display 就可看到配准后的栅格图。

(3)局部控制、减小残差。理论上,控制点越多越均匀,配准效果越好,但控制点越多,残差值(RMS)也越大。残差值是配准精度的主要参考值,因此如果某一控制点残差值太大就应删除或修正。对局部引起的残差值偏大,可针对局部进行配准,若仍对残差值不满意,则继续选取控制点并配准,直到满意为止,如图 1(b)所示。

(4)保存,如图 1(c)所示。Georeferencing 工具提供两种保存配准信息的方法:一种是

Update Georeferencing 方法，它是将原图添加至 Word 文件或 aux 文件，将配准的结果保存在原图上；另一种是 Rectify 方法，它是在当前配准后栅格图显示的基础上重新采样，生成正确的具有空间参考信息的新栅格数据。

3.4 经纬度配准

1∶500 000～1∶1 000 000 地形图上绘有经纬线网，一般可采用经纬度配准。根据项目实践经验，经纬度配准可分为两种方式：一种是直接输入特征点的经纬度坐标来配准；一种是建立特征点与实际位置的关联来配准，本文简称为经纬度坐标配准和关联配准。两种配准方式各有其特点，应用时根据具体情况选择。

3.4.1 经纬度坐标配准

(1)制定坐标系统。在 ArcCatalog 里制定栅格数据的地理坐标系(只给定基准面)。此时配准可直接输入十进制的经纬度坐标。

(2)添加控制点。输入选定控制点(经纬网格交点或特征点)的经纬度坐标值。1∶500 000～1∶1 000 000 地图图幅较大，控制点均匀布置在 20 个左右为宜。

(3)局部控制、减小残差(同上)。

(4)保存(同上)。此时配准后的栅格数据有明显的拉伸变形，这是由于栅格数据只有基准面而没有投影算法的结果，如图 2(a)所示。

(5)设置视图投影。打开视图窗口属性，制定与原图相同的投影方式(具体参数见 2.2 节)，为配准的栅格数据进行投影，如图 2(b)所示。投影后的栅格数据具有真实的地理意义，可进行矢量化处理，如图 2(c)所示。

3.4.2 关联配准

(1)定制经纬网。一般情况下，只要栅格数据具有四角坐标值就可利用 ArcToolbox 中的 Create Fishnet 模块创建经纬网，如图 3(a)所示，如果栅格数据具有经纬网，首选制定与原数据相同的经纬网格，然后在 ArcCatalog 里为其制定一个地理坐标系后加载进 Arcmap 中显示。

(2)设置视图投影。打开视图窗口属性，制定与原图相同的投影方式，将自制经纬网进行投影，如图 3(b)所示。经纬网格有明显变形，这说明此时经纬网已不再是球面坐标而是平面坐标。

(3)建立特征点与实际位置的关联。以投影后的经纬网作为参考，将配准图所有控制点(经纬网格交点或特征点)与经纬网中实际坐标位置关联，即点击某一特征点后，直接移动至经纬网实际坐标位置处再次点击，建立两点之间的关联。

(4)显示保存(同上)，最终结果如图 3(c)所示。

4 结语

栅格图像数据的配准在水利数字化制图过程中具有重要的意义，它实现了栅格图像的精确定位，从而便于我们将采集到的水利相关坐标数据定位到相应的位置，同时也可以为水利专题地图中的每一点提供准确的地理坐标。通过实践证明了上述方法的正确性和可行性，其操作方便，易接受和掌握，为水利数字化制图提供了帮助。

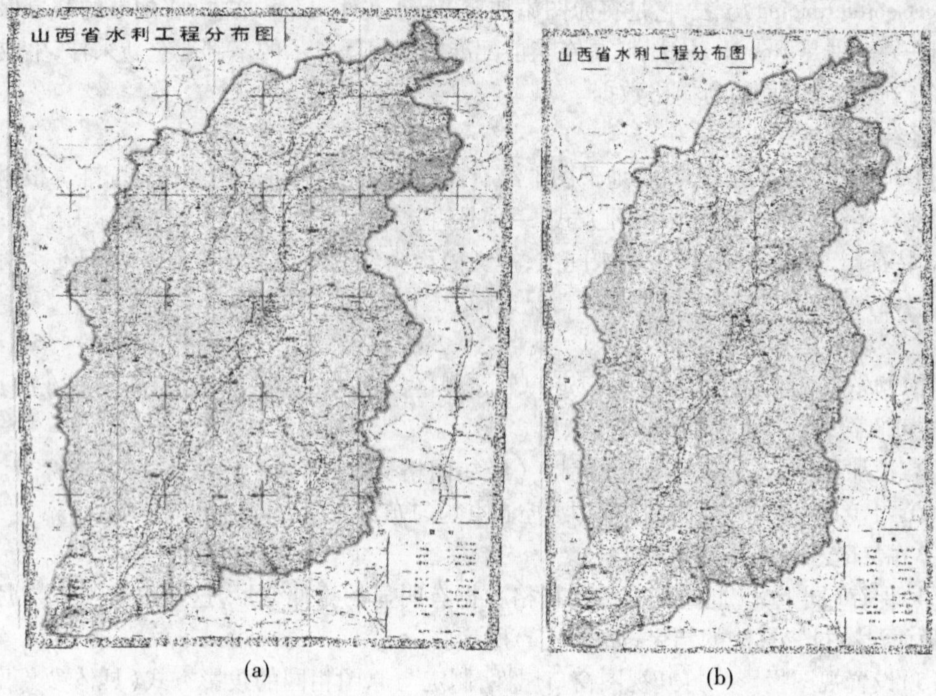

(a)　　　　　　　　　　　　(b)

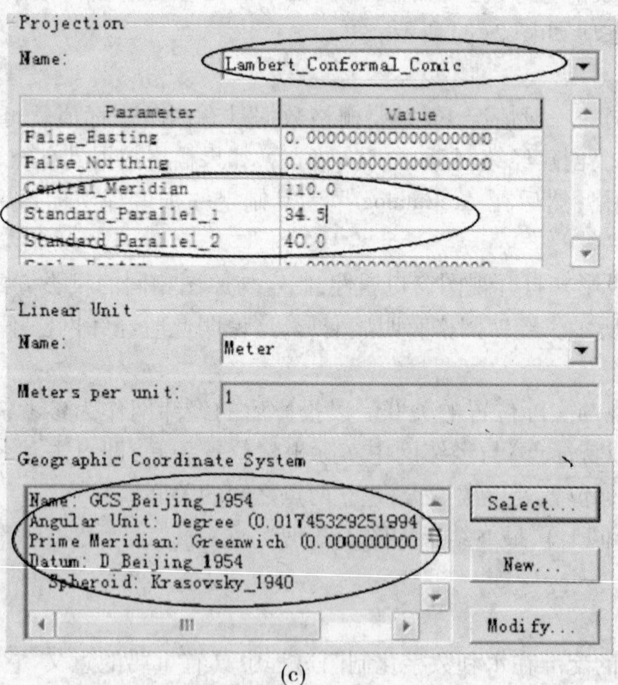

(c)

图 2　经纬度坐标配准示意图

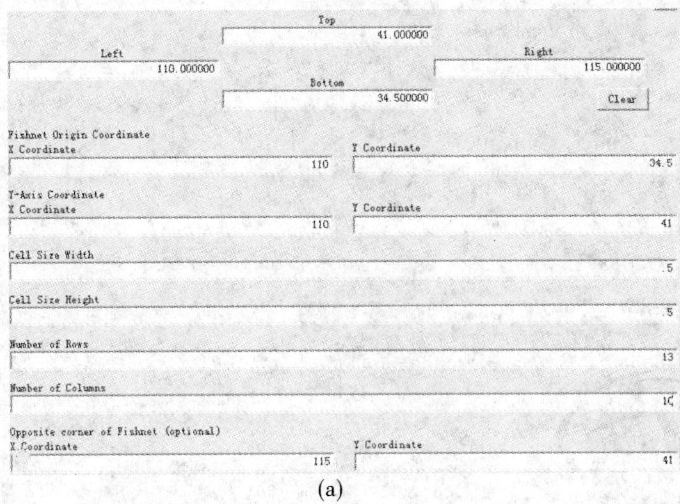

(a)

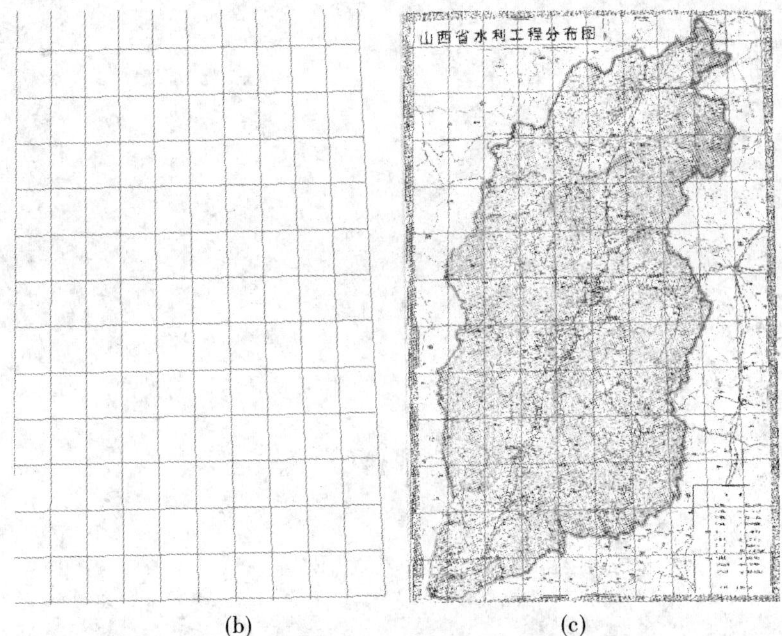

(b) (c)

图 3 关联配准示意图

参考文献

[1] 吴秀芹,张洪岩,李瑞改,等. ArcGIS9 地理信息系统应用与实践[M]. 北京:清华大学出版社,2007.

[2] 徐智. GIS 中地图投影的应用[J]. 水土保持研究,2004 (11):55-58.

【作者简介】 刘向杰,1984 年生,2010 年 7 月于太原理工大学取得水利水电专业硕士学位,现就职于中国水利水电出版社。